Photosynergetic Responses in Molecules and Molecular Aggregates

Hiroshi Miyasaka · Kenji Matsuda ·
Jiro Abe · Tsuyoshi Kawai
Editors

Photosynergetic Responses in Molecules and Molecular Aggregates

Springer

Editors
Hiroshi Miyasaka
Graduate School of Engineering Science
Osaka University
Toyonaka, Osaka, Japan

Kenji Matsuda
Graduate School of Engineering
Kyoto University
Kyoto, Japan

Jiro Abe
Department of Chemistry
Aoyama Gakuin University
Sagamihara, Kanagawa, Japan

Tsuyoshi Kawai
Graduate School of Materials Science
Nara Institute of Science and Technology
Ikoma, Nara, Japan

ISBN 978-981-15-5450-6 ISBN 978-981-15-5451-3 (eBook)
https://doi.org/10.1007/978-981-15-5451-3

This Springer imprint is published by the registered company Springer Nature Singapore Pte Ltd.
The registered company address is: 152 Beach Road, #21-01/04 Gateway East, Singapore 189721, Singapore

Preface

Molecules in the electronic excited state take important roles in various processes. For the excited molecules in condensed phase, however, there exist general restrictions in the efficient utilization of light energies. First, higher electronically excited states of rather large molecules very rapidly relax to lower electronic states (Kasha's rule), and some portion of the photon energy is diminished in this relaxation. Second, a large number of the molecules excited in assemblies undergo fast annihilation, and only a small number of excited state molecules can remain, leading to the loss of the number of photons absorbed in the system. In addition, the electronic state accessible by the photoabsorption is limited by the optical selection rule, and we cannot access various dark electronic excited states of molecules. These restrictions have confined most of the photochemical responses into the framework of "one-photon and one-molecule outcomes in the lowest excited state" and have been preventing us from efficiently and cooperatively utilizing a number of photons and excited molecules.

To explore principles to realize responses induced by a number of photons and excited molecules (photosynergetic responses) beyond the limit of "one-photon and one-molecule outcomes in the lowest excited state" leading to the new paradigm, we organized a Grant-in-Aid for Scientific Research (Kakenhi) Project on Innovative Areas "Photosynergetics: Application of Cooperative Excitation into Innovative Molecular Systems with High-Order Photofunctions" (Hiroshi Miyasaka, project leader and 34 research groups) from 2014 to 2019, supported by the Ministry of Education, Culture, Sports, Science and Technology, Japan.

The research project focused on the three subjects for the realization of the photosynergetic responses.

(1) Exploration and elucidation of reactions from higher electronic excited states, modulation of electronic states by the local field and effective utilization of multiple excitons.
(2) Development of high-order photofunctions based on the synergetic action of photoresponsive molecules.

(3) Advanced construction of meso- and macroscopic cooperative photoresponsive molecular assemblies and elucidation of their mechanisms.

This book compiles the accomplishments of this research project conducted with these subjects. In Part I, we introduce the research results of the first subject directly relating to the three general restrictions. By developing and applying multiple and multiphoton excitations, nano-environments, new materials and theoretical methods, we could find new phenomena beyond the three restrictions and elucidated their mechanisms, which can lead to the realization of photosynergetic responsive materials. In addition, the developments of new materials, experimental apparatuses and theoretical methods in the study have ensured further progress of the research field on the photosynergetics.

Part II bundles research outcomes on the second subject aiming to explore photoresponsive systems where the photoreaction in a molecular level is integrated and amplified. By designing and synthesizing molecules and assemblies, reaction chains and molecular networks overcoming the loss of the photon energy were achieved by the amplification. Moreover, many examples were realized for cooperative photoresponses of molecular aggregates arising from the integration of molecular photoreactions. From these results, we could obtain the principles for the realization of responses through the multiple photons and multiple excited molecules, which can guide future research.

Part III compiles the photoresponse of materials in rather large scale, where the individual reaction of molecules in the excited state is accumulated in the material, leading to the hierarchical photoresponse of meso- and macroscopic materials. Researchers in various fields of materials chemistry such as molecular synthesis, designing of molecular crystals, time- and space-resolved detection of reaction dynamics developed photoresponsive meso- and macroscopic systems and elucidated their mechanisms. These new phenomena, materials, mechanisms and experimental methods can lead to the further advancement of photosynergetics.

We hope that the achievement of the project is a significant landmark and principles in the stream of studies on the photochemistry of molecules and materials.

The editors express their sincere appreciation to outstanding contributors and to the Ministry of Education, Culture, Sports, Science and Technology, Japan, for their support of the project.

Toyonaka, Japan — Hiroshi Miyasaka
Kyoto, Japan — Kenji Matsuda
Sagamihara, Japan — Jiro Abe
Ikoma, Japan — Tsuyoshi Kawai

Contents

Part I
Exploration and Elucidation of Reactions from Higher Electronic Excited States, Modulation of Electronic States by the Local Field, and Effective Utilization of Multiple Excitons

Chapter 1
Advanced Control of Photochemical Reactions Leading to Synergetic Responses in Molecules and Mesoscopic Materials

Hiroshi Miyasaka, Hikaru Sotome, and Syoji Ito

Abstract In the present chapter, we first introduce the mechanism of the multiphoton-gated photochromic reactions in diarylethene and fulgide derivatives by pulsed laser excitation, where the multiple photons lead to the specific photoresponse beyond the Kasha's rule. In the second part, we discuss the one-color control of the activation, deactivation, and fluorescence excitation in the single-molecule detection. One-color of light takes various roles resulting in the effective single-molecule tracking via the photon-material synergetic response. In the third section, photochromic reaction is applied to the controlled translational motion of nano- to micron-sized materials under the laser trapping. Integrated photoresponses of materials, such as absorption, photochemical reaction, and the photon pressure, lead to the mesoscopic motion synchronized with the photochemical reaction.

Keywords Multiphoton excitation · Multiple excitation · Diarylethene · Fulgide · Ultrafast laser pulse · Single-molecule tracking · Photon pressure

1.1 Introduction

The property and reactivity of molecules are dependent on the electronic states and, accordingly, we can expect various photoresponses of molecules by changing the irradiation wavelength and the mode of the excitation. However, specific reaction depending on the high electronic state is very rare for polyatomic molecules due to the rapid internal conversion in the highly excited state (Kasha's rule). In the present chapter, we first introduce the multiphoton-gated cycloreversion reaction of diarylethene and fulgide derivatives as one of photosynergetic responses beyond the Kasha's rule. In these systems, marked enhancement of the cycloreversion reaction takes place in specific electronic states attained only by the two-photon absorption. We discuss the mechanism of the multiphoton-gated reaction on the basis of the

H. Miyasaka (✉) · H. Sotome · S. Ito
Division of Frontier Materials Science, Graduate School of Engineering Science, Osaka University, Toyonaka, Osaka 560-8531, Japan
e-mail: miyasaka@chem.es.oasaka-u.ac.jp

H. Miyasaka et al. (eds.), *Photosynergetic Responses in Molecules and Molecular Aggregates*, https://doi.org/10.1007/978-981-15-5451-3_1

experimental results by the two-color stepwise excitation and simultaneous two-photon excitation.

As the second topic, we show the one-color control of fluorescence activation, deactivation, and excitation of a fluorescent diarylethene derivative. In this system, a one-color laser light takes various roles through the synergetic interaction between the material and the light and enables the long-term fluorescence imaging with super-resolution.

In the third topic, we introduce the mesoscopic motion of nano- to micron-sized materials under the laser trapping, which is controlled by the photochromic reactions of molecules inside the materials. Integration of lasers with the photochemical reaction demonstrates the controlled mesoscopic motion which can amplify the chemical reaction in molecular levels into the mesoscopic translational movements.

1.2 Multiphoton-Gated Cycloreversion Reaction at Higher Excited States of 6π Electron Systems

Cyclization and cycloreversion between 1,3,5-hexatriene and 1,3-cyclohexadiene are typical reactions of 6π electron systems, where the relation between the structure of the product and the orbital symmetry of the reactant is regulated by the Woodward–Hoffmann rules [1]. Diarylethene and fulgide derivatives are typical photochromic molecules undergoing photoinduced cyclization and cycloreversion, which can be regarded as 6π electrocyclic reactions in principle [2–4]. From viewpoints of the elucidation of fundamental photochemical reactions as well as the application to photofunctional molecular materials, these photochromic systems have been long investigated.

We reported that the cycloreversion reaction of these derivatives in solution, polymer matrices, and in crystalline phase takes place very efficiently in higher excited state attained by the visible stepwise two-photon excitation under the pulsed laser excitation [5–12]. For derivatives with the low cycloreversion reaction yield of <few % in the lowest excited state, the yield in higher excited state was found to be more than 50% [5, 6, 12]. Interestingly, this efficient cycloreversion does not take place by the one-photon excitation whose photon energy is almost the same with that of the two-photon of the visible light, as shown in Fig. 1.1. This result strongly suggested that the excited state accessible only by the two-photon absorption takes an important role in the drastic enhancement of the cycloreversion reaction. This multiphoton-gated reaction is one of the rare responses in polyatomic molecules depending on the specific high electronic state that are not easily accessible by the one-photon absorption. The elucidation of the mechanism is quite important for the control of the photochemical reaction by the multiple laser excitations [13–17] and development of the photo-function of these molecular materials. In this project, we applied femtosecond two-color two-pulse excitation to precisely elucidate the mechanism of this specific reaction channel.

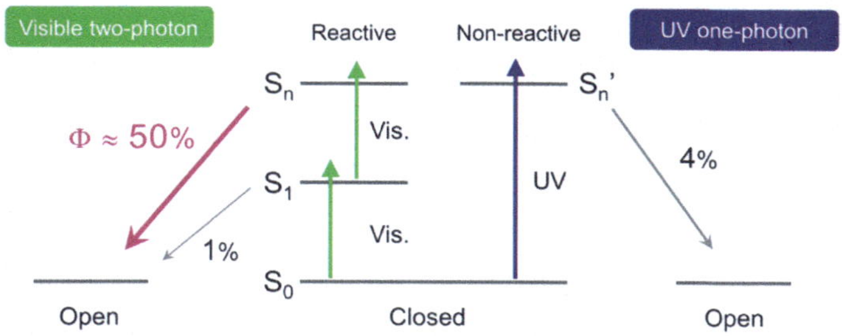

Fig. 1.1 Stepwise two-photon cycloreversion reaction of diarylethene and fulgide derivatives

1.2.1 Stepwise Two-Photon Cycloreversion Reactions in Diarylethene Derivatives

Figure 1.2a shows photochromic reactions of one of diarylethene derivatives, PT, used in the present investigation. PT undergoes the photoinduced cyclization and cycloreversion between the open-ring isomer, PT(o), and the closed-ring one, PT(c). As shown in Fig. 1.2b, PT(o) has absorption only in the UV region while PT(c) has absorption bands also in the visible region. PT(c) shows broad fluorescence spectrum with rather large Stokes shift (4200 cm^{-1}) even in nonpolar *n*-hexane solution, strongly suggesting large geometrical rearrangement in the excited state. The cyclization quantum yield of PT(o) by the 280-nm irradiation is 0.59, while the cycloreversion yield by the 492-nm excitation is 0.013 [3].

In the stepwise two-photon absorption process, the actual intermediate state produced by the first photon absorption is the initial state of the second photon

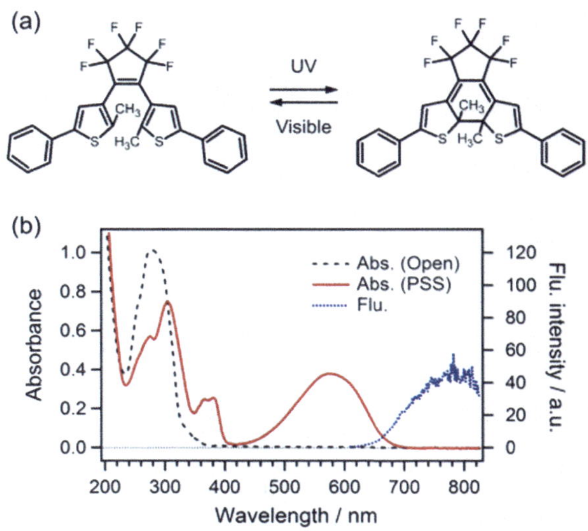

Fig. 1.2 **a** Molecular structure and photochromic reactions of PT. **b** Steady-state absorption spectra of PT(o) and PT(c) and fluorescence spectrum of PT(c) in n-hexane solution. Excitation wavelength for the fluorescence measurement was 580 nm. Adapted with permission from ref 18. Copyright (2017) American Chemical Society

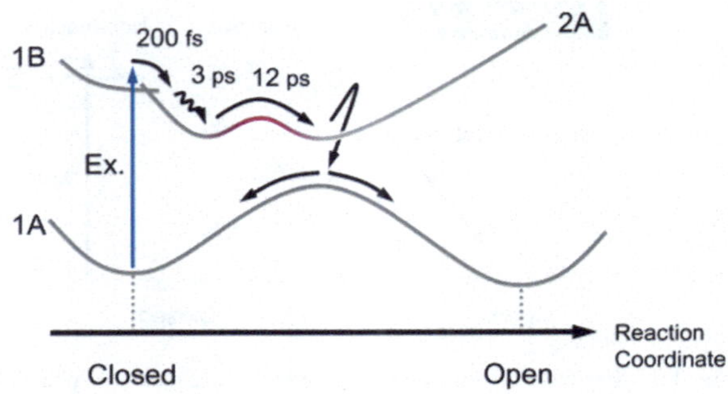

Fig. 1.3 Potential energy curves and cycloreversion reaction of PT(c). Adapted with permission from Ref. [18]. Copyright (2017) American Chemical Society

excitation. In the present case, the second photon is absorbed by the excited state pumped from the ground state by the visible light. Because the excited state undergoes various relaxation processes during its lifetime, it is indispensable to elucidate the dynamics of the excited state (initial state) responsible for the second photon absorption. Figure 1.3 shows potential energy curves and reaction profiles of PT(c) in *n*-hexane solution excited by the visible one-photon absorption [18]. The 1B state produced by the visible excitation of the ground state interconverts into the 2A state with the time constant of 200 fs. This 2A state undergoes vibrational relaxation accompanied with the geometrical rearrangement with the time constant of 3 ps, and the relaxed 2A state shows the transition into the conical intersection (CI) in competition with the direct deactivation to the ground state with the time constant 12 ps.

In order to elucidate the mechanism of the efficient cycloreversion reaction in higher excited states of PT(c), we employed a double-pulse pump and probe system, whose conceptual setup is shown in Fig. 1.4. The first pump pulse (Pump 1) excites the ground state molecule, and the second pump pulse (Pump 2) is introduced after the delay time of Δt from Pump 1. The amount of the reaction is monitored as negative transient absorbance of the closed-ring isomer at the delay time much longer than the lifetime of the excited state. In the present case, this delay time was set to be 400 ps. By subtracting the reaction amounts by the one-photon absorption of Pump 1 and Pump 2, we can obtain the reaction amount induced by the successive two-photon excitation. In the case where Pump 2 is introduced immediately after Pump 1, we can pump the 1B state to higher excited states. On the other hand, when Pump 2 is introduced with Δt ~ ca. 1 ps, geometrically unrelaxed 2A state can be excited.

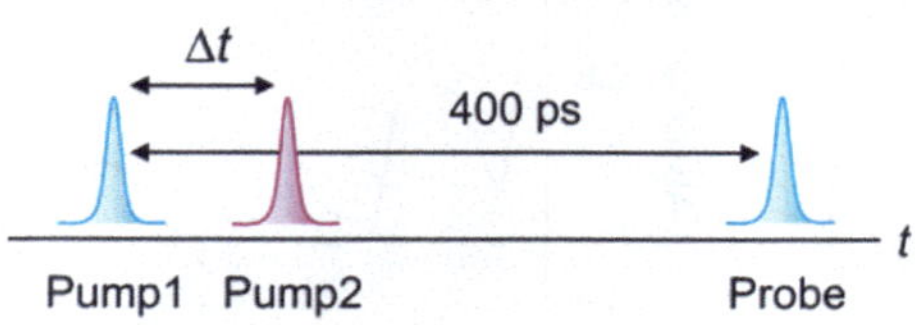

Fig. 1.4 Schematic illustration of the experimental setup for the stepwise two-photon cycloreversion reaction amount measurement

In the condition with much longer time interval, the relaxed 2A state is excited. Accordingly, we can choose the initial state of the second photon absorption by changing Δt. In addition to the selection of the initial state leading to the formation of the different final state, the wavelength of Pump 2 also determines the final state. Thus, the detailed information on the enhancement of the cycloreversion can be obtained as functions of the time interval, Δt, and the wavelength of Pump 2.

Figure 1.5 shows one of the experimental results, where Pump 1 at 600 nm and Pump 2 at 530 nm were employed. The abscissa in this figure is the time interval, Δt, between the Pump 1 and 2 pulses, and the ordinate is the reaction amount induced by the successive two-photon excitation detected as ΔΔAbsorbance. As shown in this figure, the amount of the two-photon reaction at $\Delta t = 0$ is rather small although the population of the excited state is the largest. With an increase in Δt, the reaction amount induced by the two-photon excitation increases up to ca. 5 ps and decreases in a few tens of ps time region. The solid line is the calculated curve with a triple-exponential function with time constants of 200 fs, 3 ps, and 12 ps. The first two time constants correspond to the rise components (increase in the amount of the two-photon induced reaction), while the time constant of 12 ps is the decay one. As was shown in Fig. 1.3, the time constant of the interconversion from 1B to 2A was 200 fs and that of 3 ps was due to the geometrical rearrangement in the 2A state. The increase of the two-photon cycloreversion reaction with these two time constants in Fig. 1.5 indicates that the excitation of the 2A state, especially the geometrically relaxed one, more efficiently induces the cycloreversion reaction than the excitation of the 1B state. The decay time constant of 12 ps corresponds to the lifetime of the 2A state, indicating that the depopulation of the initial state of the second photon absorption led to the decrease of the two-photon induced cycloreversion reaction.

The above result depending on the character of the electronic state strongly suggests that the symmetry of the electronic wave function takes an important role for the ring-opening in the higher excited state. As shown in Fig. 1.6, the excited state produced by the optically allowed transition from the A state is in the B state for which the node between the two carbon atoms is expected. On the other hand,

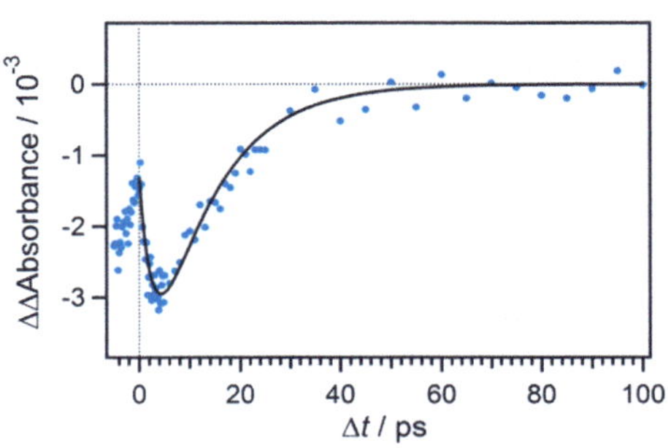

Fig. 1.5 ΔΔAbsorbance due to the stepwise two-photon cycloreversion reaction versus the time interval between the two excitation pulses, Δt. Wavelengths of the pump 1 and pump 2 pulses were 600 and 530 nm, respectively. Reprinted with permission from Ref. [18]. Copyright (2017) American Chemical Society

Fig. 1.6 Schematic illustration of the phase of wavefunctions in A and B states

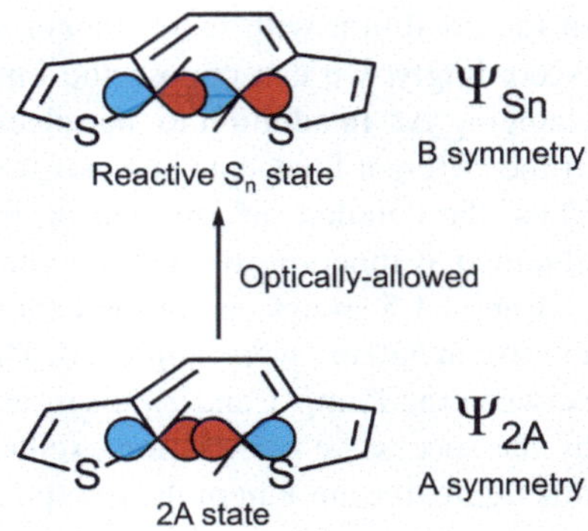

the excited state attained by the optically allowed transition from the B state is the A state where the node of the wave function between the two carbons is not plausible. Although more detailed investigation by advanced theoretical methods is necessary to derive a clear conclusion, the present experimental result strongly implies the importance of the symmetry of the wavefunction.

The second rise component in Fig. 1.5 was attributed to the geometrical rearrangement in the 2A state. This increase of the two-photon induced cycloreversion reaction may be related to the short lifetime of higher electronic state. Generally, the lifetime of higher excited states of rather large molecules is in the sub-picosecond region. In the present system, the lifetime of the higher excited state PT(c) was 300 fs [18], which is comparable to the one period of the large amplitude motion with the frequency of ca. 100 cm^{-1}. Owing to the short lifetime, molecules in higher excited states have less chance to undergo the large geometrical change before the deactivation to lower excited state. In other word, the initial geometry of the molecule in the excited Franck–Condon state takes an important role for the reaction with geometrical rearrangements, as shown in Fig. 1.7. Actually, as was shown in Fig. 1.2, large Stokes shift of PT(c) strongly suggested large geometrical change taking place in the lowest excited state. In the present case, the geometry in the relaxed 2A state is more favorable for the cycloreversion reaction in the higher excited state pumped from the 2A state. This result indicates that the Franck–Condon geometry in the higher excited state, in addition to the character of the electronic state, also takes a crucial role in the effective reaction during the short lifetime. That is, even the one-photon transition

Fig. 1.7 Schematic illustration for the relation between the subsequent reaction and the Franck–Condon geometry in the highly electronic excited state

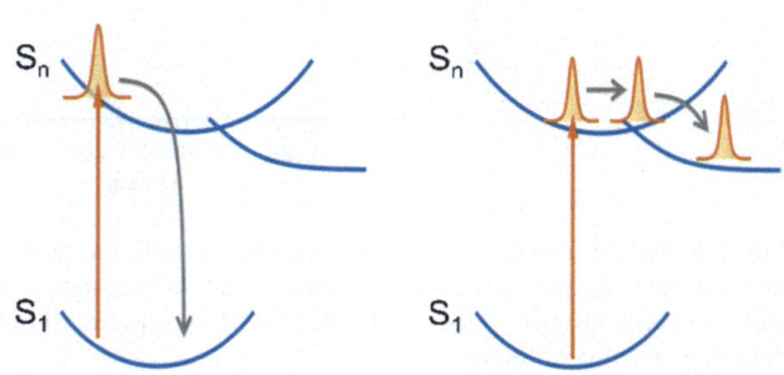

from the ground state can access to the specific electronic state, the reaction may not so efficiently occur in the vertical transition of which Franck–Condon geometry is far from the crossing point for the reaction.

The wavelength of the second laser pulse is also an important factor for the effective reaction. Figure 1.8 shows the two-photon induced cycloreversion amount as a function of Δt between the two pulses, where Pump 1 was fixed at 600 nm and various wavelengths were used for Pump 2. As was performed in Fig. 1.5, the contribution of the one-photon reaction was subtracted. The rise of the reaction amount within ca. 5 ps of Δt was observed for Pump 2 in the wavelength region of 530–680 mm, while such behavior was not detected for Pump 2 in the wavelength region longer than 730 nm and monotonous decrease of the enhancement of the reaction was observed with an increase in Δt.

Figure 1.9 shows the cycloreversion reaction yield in the high electronic state pumped up from the relaxed 2A state as a function of the wavelength of Pump 2.

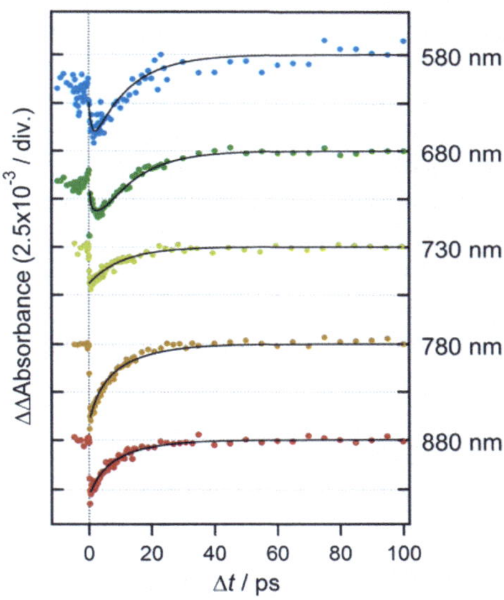

Fig. 1.8 Pump 2 wavelength dependence of ΔΔAbsorbance due to the stepwise two-photon cycloreversion reaction versus the time interval between the two excitation pulses, Δt. The wavelength of the pump 1 was 600 nm and that of the pump 2 is shown in the figure. Reprinted with permission from ref 18. Copyright (2017) American Chemical Society

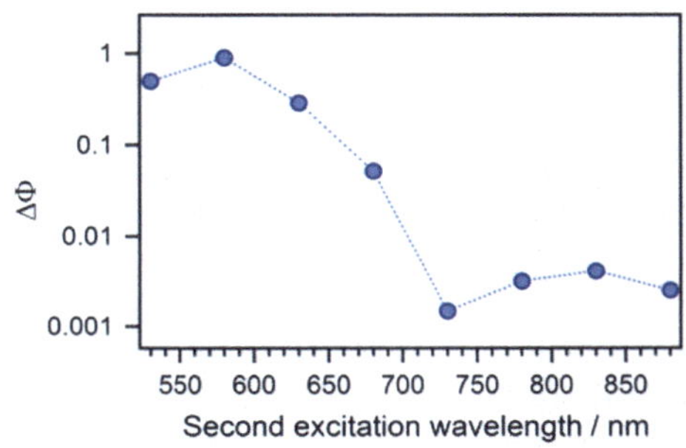

Fig. 1.9 Dependence of the reaction yields in the higher excited states on the wavelength of the second excitation pulse. Reprinted with permission from Ref. [18]. Copyright (2017) American Chemical Society

As shown in this figure, the cycloreversion yield by the second laser pulse in the wavelength region shorter than 680 nm is ca. 50–90%. The cycloreversion yield in higher excited states under the multiple excitations of PT(c) by a picosecond 532-nm laser pulse was estimated to be ca. 50% [6]. Because the picosecond laser pulse with 15-ps FWHM excites the 1B and 2A state during the pulse duration. In addition, unrelaxed 2A state is also excited in addition to the relaxed 2A state. Accordingly, the reaction yield in the high state estimated in the picosecond experiment was slightly smaller than that obtained for the relaxed 2A state in the femtosecond experiment, 70%. In anyhow, present results are almost in the same order of the picosecond pulsed excitation. On the other hand, the reaction yield is much smaller, less than 1%, in the wavelength region ≥730 nm. It is worth noting for the correlation curve with the second pulse ≥730 nm in Fig. 1.8 that the amount of the cycloreversion reaction induced by the second photon absorption showed the monotonous decrease with an increase in Δt and no selectivity depending on the initial electronic state or the geometry of the molecule for the second photon absorption was found. These results strongly suggest the contribution of the vibrationally excited state in the lowest electronically excited state, which was produced by the rapid internal conversion from the highly excited states pumped by Pump 2 in the wavelength region ≥730 nm. The excess energy of the S_1 state may assist the molecules in overcoming the activation barrier on the 2A potential energy surface, leading to the slight increase in the reaction yield of the cycloreversion. Actually, the cycloreversion reaction yield of diarylethene derivatives generally increases with increasing temperature [19].

By summarizing above results and discussion, we could conclude that the character of the electronic state, the Franck–Condon geometry of molecules in highly excited states, and the level of the final state are key factors for the marked increase of the cycloreversion reaction by the stepwise two-photon excitation.

1.2.2 Stepwise Two-Photon Cycloreversion Reactions in Fulgide Derivatives

To elucidate the generality of these reactions in 6π electron systems, we have studied the reaction dynamics of fulgide derivatives by the stepwise two-photon excitation of femtosecond laser pulses [20].

Figure 1.10 shows the photochromic reaction of F1, together with the absorption spectra of the open-ring isomer, F1(o), and closed-ring one, F1(c), in toluene solution. The fluorescence spectrum of F1(c) is also exhibited. As shown in Fig. 1.2, F1(o) has absorption bands only in the UV region while F1(c) shows the absorption bands also in the visible region. The fluorescence spectrum of F1(c) shows a very broad band with rather large Stokes shift in nonpolar solution, strongly suggesting that the relaxed excited state of F1(c) in the lowest excited state has a geometry largely different from the ground state as observed for the diarylethene derivative. The cyclization yield upon the irradiation at 366 nm and the cycloreversion yield upon the 505-nm exposure were, respectively, 0.39 and 0.11 in toluene solution at 295 K [20].

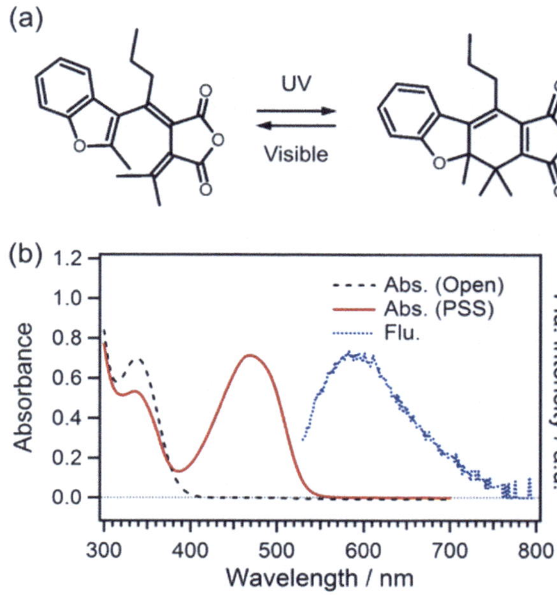

Fig. 1.10 **a** Molecular structure and photochromic reactions of F1. **b** Steady-state absorption spectrum of F1(o) and that under the photo-stationary (PSS) state under the irradiation at 330 nm (solid black line) and fluorescence spectrum of F1(c) in toluene solution. Excitation wavelength for the fluorescence measurement was 450 nm. Adapted with permission from Ref. [20]. Copyright (2018) American Chemical Society

Figure 1.11 shows the transient absorption spectra of F1(c) in toluene solution, excited with a femtosecond 500-nm laser pulse. The absorption spectrum at 0 ps shows an absorption maximum at 630 nm and a weak positive one around 700 nm. In the initial 0.5 ps following the excitation, the positive absorption band at 630 nm

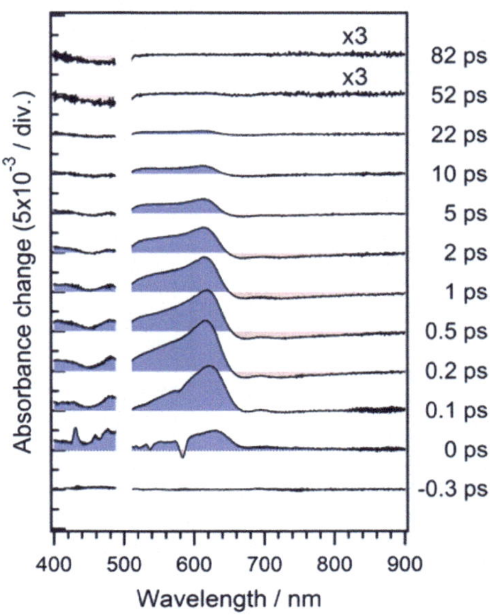

Fig. 1.11 Transient absorption spectra of F1(c) in toluene solution, excited with a femtosecond 500-nm laser pulse. Negative signal around 580 nm and the positive one around 430 nm in the spectrum at 0 ps are due to the stimulated Raman scattering. Reprinted with permission from Ref. [20]. Copyright (2018) American Chemical Society

shifts to 610 nm and the weak positive band at 720 nm evolves into a negative signal ascribable to the stimulated emission. Within a few ps following the excitation, the absorption intensity at ~520 nm relative to that at ~620 nm increases and the positive absorption bands gradually decrease together with the recovery of the negative one in the wavelength >680 nm in a few tens of ps time range. At and after ca. 50 ps following the excitation, only the weak negative signal due to the cycloreversion reaction remains in the wavelength range of 400–550 nm. From the detailed analyses of the temporal evolution of the spectra, the dynamics of F1(c) in the excited state pumped by the 500-nm absorption were summarized as the interconversion from the Franck–Condon 1B state to 2A state with the time constant of 140 fs which is followed by the geometrical rearrangement of 1.5 ps. The lifetime of the relaxed 2A state was 10 ps.

As reported previously, the stepwise two-photon excitation of fulgide derivatives by the picosecond laser pulse induced the enhancement of the cycloreversion reaction [10–12]. To precisely elucidate the mechanism, femtosecond two-pulse pump methods were applied. Figure 1.12 shows the two-photon induced reaction as a function of the time interval between the first pulse at 500 nm and second one at 480 nm. The ordinate is given as ΔΔAbsorbance as performed for Figs. 1.5 and 1.8. The positive signal at the time interval <1 ps indicates that the double-pulse excitation reduced the amount of the ring-opening reaction compared to those by the irradiation of the two pulses with Δt much longer than the excited state lifetime of F1(c). This positive signal is not attributable to the formation of F1(c) by the simultaneous two-photon absorption of F1(o) in the solution, because the excitation of the solution only with the open-ring isomer, F1(o), at $\Delta t = 0$ did not show the formation of F1(c). Accordingly, the main part of the positive signal is due to the suppression of the cycloreversion reaction by the irradiation of the second laser pulse immediately after the excitation by the first pulse. At the present stage of the investigation, it is suggested that this decrease might be due to the direct deactivation into the ground state from the higher excited state.

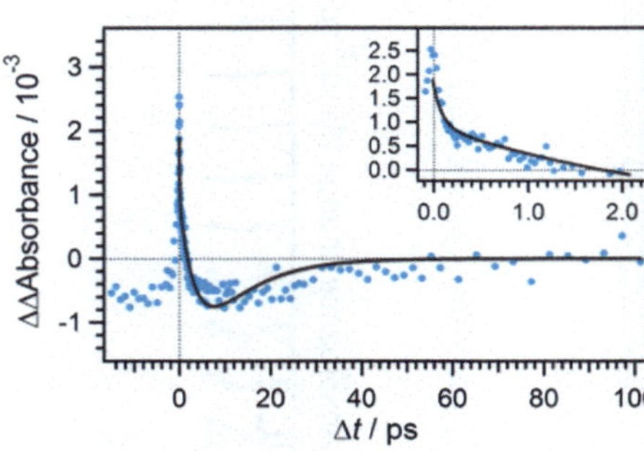

Fig. 1.12 ΔΔAbsorbance due to the stepwise two-photon cycloreversion reaction versus the time interval between the two excitation pulses, Δt. Wavelengths of the pump 1 and pump 2 pulses were set to 500 and 480 nm, respectively. Adapted with permission from Ref. [20]. Copyright (2018) American Chemical Society

With an increase in the time interval in sub-ps to several ps time region, the negative signal of ΔΔAbsorbance appears with the time constants of 100 fs and 4 ps, indicating the enhancement of the cycloreversion reaction. The former time constant of 100 fs is the same with that of the interconversion from the Franck–Condon 1B state to the subsequent 2A sate, as was shown in the previous section. On the other hand, the latter time constant, 4 ps, is slightly longer than that of the geometrical rearrangement in one-photon cycloreversion reaction dynamics. This difference might be related to the reaction probability in the higher excited state being more sensitive to the geometry. That is, the correlation curve between the reaction amount and Δt in Fig. 1.11 involves the reaction probability in the higher excited state, while the time evolution of the transient absorption in Fig. 1.11 represents the S_1 state dynamics. At and after ca. 6 ps, the signal gradually decreases due to the decay of the 2A state with the time constant of 10 ps. These results in Fig. 1.11 clearly show that the excitation of the 2A state leads to the enhancement of the cycloreversion reaction while the excitation of the 1B state results in the suppression of the cycloreversion reaction. In addition, the photoexcitation of the geometrically relaxed 2A state more efficiently leads to the cycloreversion reaction.

Figure 1.13 exhibits the relation between the reaction amount and Δt of the two pump pulses using Pump 2 with various colors. The increase of the cycloreversion reaction was observed within ca. 5 ps for the second laser pulse at 530 and 560 mm, as observed in Fig. 1.12. On the other hand, only the positive signal was observed for the second pulse at 580 nm. For the second pulses at 650 and 720 nm, the positive signal around $\Delta t = 0$ was followed by the monotonous decrease with a time constant of 10 ps, indicating that the cycloreversion reaction was suppressed by the introduction of the second laser pulse in the entire range of Δt.

Figure 1.14 shows the apparent reaction yield as a function of the wavelength of the second laser pulse. The apparent reaction yield, Φ_{P2}, was estimated to be 0.77 for Pump 2 at 480 nm, of which value is much larger than that in the S_1 state, 0.11. The apparent reaction yield decreases with the red shift of Pump 2. In the wavelength region >580 nm, the negative value was estimated. This negative value strongly

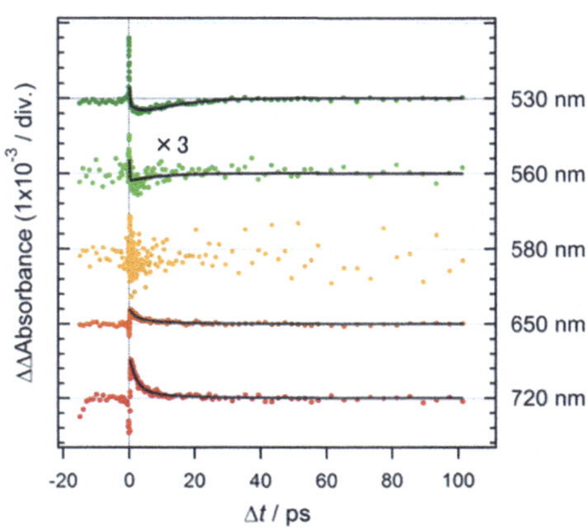

Fig. 1.13 Dependence of the stepwise two-photon cycloreversion reaction of F1(c) in toluene solution on the wavelength of the second laser pulse. The wavelengths of the first excitation pulse and probe pulse were 490 and 500 nm, respectively. Reprinted with permission from Ref. [20]. Copyright (2018) American Chemical Society

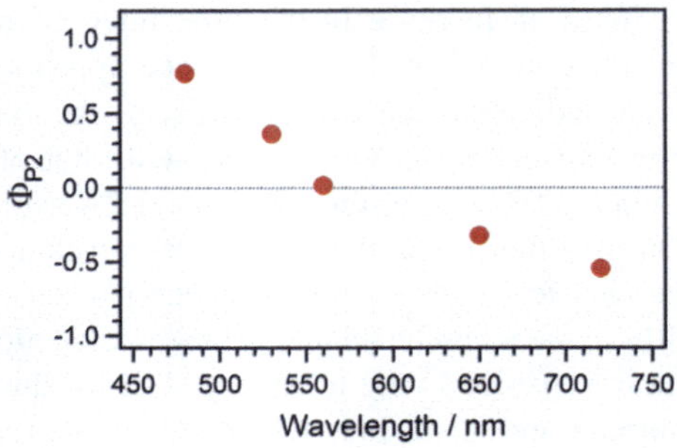

Fig. 1.14 Dependence of the apparent cycloreversion reaction yield by the stepwise two-photon excitation on the wavelength of the second excitation pulse. Reprinted with permission from Ref. [20]. Copyright (2018) American Chemical Society

suggests that the increase in the contribution of the stimulated emission leads to the decrease of the apparent reaction yield with an increase in the wavelength of Pump 2.

The reaction dynamics of F1(c) under the double-pulse excitation are summarized in Fig. 1.15. The second photon absorption from the 2A state and, in special, geometrically relaxed 2A state more effectively induces the cycloreversion reaction yield. These results are almost the same with those observed for the diarylethene derivatives. On the other hand, the rapid internal conversion from the higher excited states to the ground state may take place in the case that the 1B state in F1(c) is pumped up into the higher excited state. Moreover, the second pulse excitation in the wavelength region >580 nm remarkably suppresses the cycloreversion reaction. Actually, the transient absorbance in the wavelength region of 650–900 nm in Fig. 1.11 shows the negative signal, indicating that the stimulated emission takes an important role for the deactivation induced by the second laser light. These results indicate that the introduction of Pump 2, in principle, can induce the enhancement and suppression of the cycloreversion reaction by the control of the time interval and the color of the laser light.

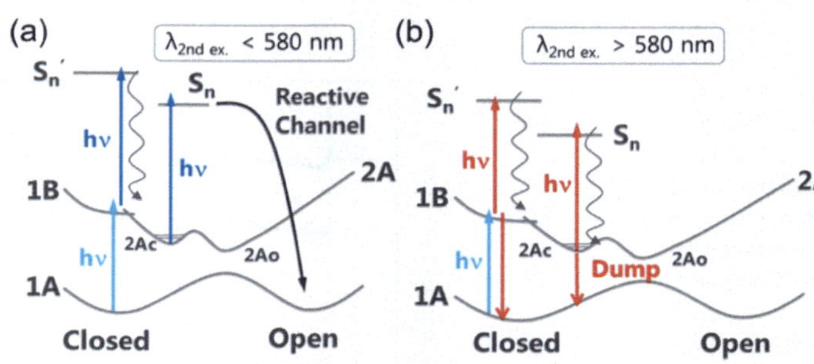

Fig. 1.15 Responses of F1(c) to the femtosecond two-color two-pulse excitation. Reprinted with permission from Ref. [20]. Copyright (2018) American Chemical Society

1.2.3 *Off-Resonant Simultaneous Two-Photon Cycloreversion Reactions*

Off-resonant simultaneous multiphoton absorption is one of representative phenomena under the photo-irradiation of high peak power lasers. Generally, simultaneous multiphoton absorption pumps the molecule into the electronic state which is optically forbidden by the one-photon absorption. The difference in the initial electronic state may lead to the different relaxation and reaction. Not only from the importance in the selection of the initial electronic state but also from the viewpoint of the laser processing, the simultaneous multiphoton absorption can provide new application of photochromic systems. Because the high peak power of lasers can induce higher-order (≥ 3) multiphoton absorption, we can induce both isomerization reactions of photochromic molecules with one-color laser light [21]. That is, the high intensity laser induces the three-photon absorption resulting in the forward reaction, and the irradiation with weak intensity induces the backward reaction via the two-photon absorption. Accordingly, we can prepare the spatial patterning of the photochromic reactions by controlling the beam profile of the one-color laser light [21]. In this section, we introduce the reaction dynamics of the closed-ring isomer of a diarylethene derivative, PT(c) in Fig. 1.2, in higher excited states attained by the simultaneous off-resonant excitation at 730 nm and discuss the detailed reactions by comparing these results with the dynamics observed by the one-photon excitation at 365 nm [22].

Before the discussion on the dynamics, we first show the excitation power dependence of transient absorbance and the relation between the time interval between the two laser pulses at 730 nm and the signal intensity in Fig. 1.16. As shown in Fig. 1.16a, the transient absorbance increases quadratically in proportion with the increase in the excitation power. This result clearly shows that the two-photon absorption is responsible for the transient signals. In Fig. 1.16b, we exhibit the correlation trace of the permanent bleaching signal as a function of time intervals between the two 730-nm excitation pulses. In this figure, the ordinate is the transient absorbance at 400 ps after the irradiation. As shown in Fig. 1.16, the maximum value of the cycloreversion reaction amount is observed at $\Delta t = 0$ with the correlation width of ca. 130-fs FWHM. Because this value is identical with the autocorrelation width of the excitation pulse monitored by SHG and the transient absorbance increases quadratically in proportion with the laser intensity, the cycloreversion reaction is induced by the *simultaneous* two-photon absorption. Efficient excitation of PT(c) is due to the large two-photon cross section, 380 GM at 730 nm [22].

Figure 1.17a shows the transient absorption spectra of PT(c) in *n*-hexane solution, excited with a femtosecond 365-nm laser pulse. The excitation at 365 nm pumps PT(c) into the excited state which is ca. 11,000 cm^{-1} higher than the relaxed S_1 state. Immediately after the excitation, an absorption peak appeared at 590 nm with broad absorption in the entire spectral window of 400–1000 nm. This 590-nm maximum is ascribable to the absorption of highly excited state (S_n state), because it was not

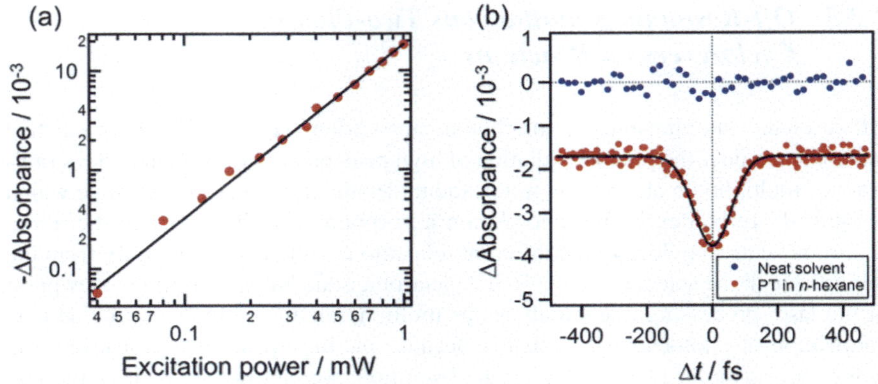

Fig. 1.16 **a** Excitation power dependence of transient absorbance changes of PT(c) in *n*-hexane solution at 2 ps after the excitation with a femtosecond laser pulse at 730 nm. The monitoring wavelength was 600 nm. **b** Correlation trace of the permanent bleaching induced by off-resonant simultaneous two-photon absorption as a function of time intervals between two 730-nm excitation pulses. A signal from neat solvent is also shown for comparison. Adapted with permission from Ref. [22]. Copyright (2017) American Chemical Society

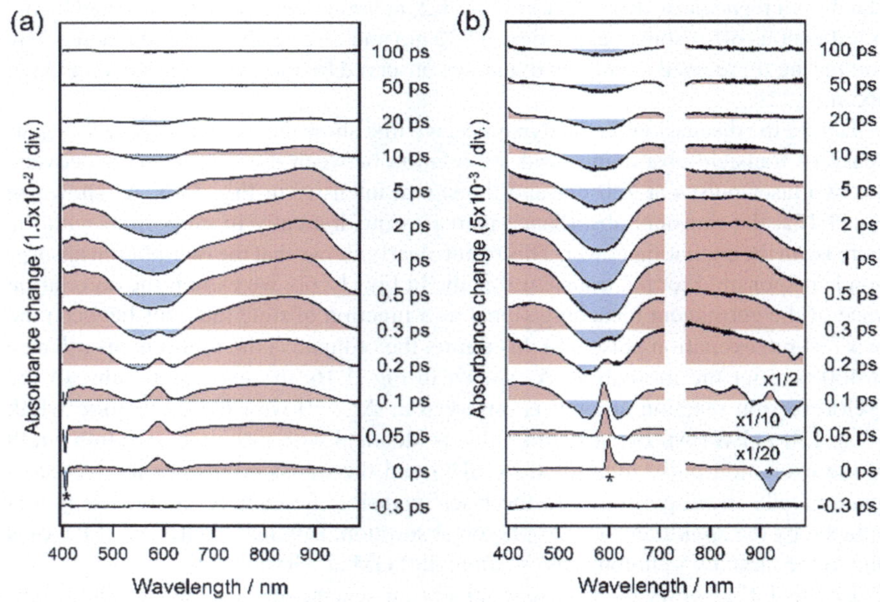

Fig. 1.17 Transient absorption spectra of PT(c) in *n*-hexane solution excited at 365 nm, (**a**), and those at 730 nm, (**b**). The spectra around 730 nm are not shown due to strong scattering of the excitation pulse at 730 nm. The spectral features denoted by asterisk are due to stimulated Raman scattering of the solvent. The excitation powers were 170 and 740 nJ for the excitation at 365 and 730 nm, respectively. Reprinted with permission from Ref. [22]. Copyright (2017) American Chemical Society

observed under the one-photon excitation in the visible region [18]. In the sub-picosecond time region, this absorption disappeared and the rise of absorption bands was observed in the whole spectral region. The spectral shape of the resultant absorption band is similar to that obtained by the visible one-photon excitation although it is broader due to larger excess vibrational energy in the 2A state [18]. These bands decrease in a few tens of ps. At 100 ps following the excitation, almost all signal disappeared and little bleaching signal remained, of which the result is consistent with the small one-photon cycloreversion reaction yield of ca. 2% under the 365-nm excitation.

On the other hand, spectral band shapes by the 730-nm off-resonant two-photon excitation are largely different from those by the 365-nm one-photon excitation. Although the signals around the time origin are affected by the coherent artifact ascribable to the response of the solvent, the spectrum at 200 fs shows absorption bands in 650–800 nm which is different from that by the 365-nm excitation. This result indicates that the excited state (S_n' state) pumped by the two-photon excitation at 730 nm is different from that by the one-photon absorption at 365 nm. The absorption band in 800–900 nm increases within 1 ps but the band shape in 650–900 nm is still different from that observed by the 365-nm excitation. This result indicates that the S_n' state is relaxed, at least in part, into the low excited state different from the 2A state. These positive absorption bands decrease into the baseline within several tens of picoseconds, and the permanent bleaching markedly remains at and after at 100 ps following the excitation. The cycloreversion reaction yield was estimated to be ca. 20% under the off-resonant two-photon excitation at 730 nm, which is 10 times larger than that by the one-photon excitation. By integrating the large reaction yield with the time evolution of the transient absorption spectra, it is strongly suggested that the enhancement of the cycloreversion occurs via some specific channels leading to the open-ring isomer without passing through the S_1/S_0 conical intersection of the low excited state produced by the one-photon absorption.

1.3 One-Color Control of Cyclization and Cycloreversion Reactions of Fluorescent Diarylethenes Under CW Visible Light Irradiation and Its Application to Ultra Longtime Single-Molecule Tracking

In this section, we introduce the one-color control of cyclization and cycloreversion reactions of fluorescent diarylethenes. Although this control uses only single color laser, the synergetic interaction between the material and the light leads to the excellent performance in the super-resolution imaging.

Super-resolution optical microscopy [23–25] is a powerful tool for the investigation of the structure of small materials and for the detection of the dynamic motion of the single fluorescent molecule with the spatial resolution beyond the diffraction limit. Single-molecule tracking (SMT) is one of the methods for the detection of the molecular motion, where the localization of fluorescence images enables to track the translational diffusion of individual molecules at very high spatial resolution (typically, 10–30 nm). The diffusional motion of molecules can probe the microscopic environments of host materials and thus provides the information on the position-depending properties of host materials. By using this method, we have investigated the microscopic properties of polymer-based materials, e.g., the evolution of the property of polymers through cross-linking [26] and photo-curing process [27] and peculiar diffusion of guest molecules depending on the distance from solid/polymer interface [28].

Although SMT is a powerful tool, two practical limitations have inhibited the tracking of large number of molecules with a long measurement time. The first one is the limited time of tracking due to the photobleaching of fluorescence molecules. For single-molecule fluorescence imaging with good signal-to-noise (*S*/*N*) ratio, photoexcitation with relatively intense laser is required. This intense excitation induces the photodegradation of fluorescence molecules typically in several tens of seconds of the light exposure. The second one is the upper limit of the concentration of guest molecules. For ensuring the single-molecule tracking based on the localization method, spatial overlap between fluorescence spots of guest molecules should be avoided during their Brownian motions. Thus, the concentration of guest molecules should be kept low. Because of these two limitations, it was difficult to cover the wide region of the specimen by longtime measurements.

The application of the fluorescence OFF–ON switching is one of the methods to overcome these limitations. In this measurement, most of the molecules in the specimen are in the fluorescent OFF state, and the number of the molecules in the fluorescent ON state is kept low to avoid the spatial overlap of fluorescence spots. The bleaching of ON state by the photodegradation or by the ON → OFF reaction is recovered by the exposure of the switching light to induce the OFF → ON reaction. Iteration of the tracking and the OFF → ON light exposure allows the single-molecule tracking of the large number of molecules for a longtime covering the wide area of the specimen.

The bleaching of ON state by the photodegradation or by the ON→OFF reaction is recovered by the exposure of the switching light to induce theOFF→ONreaction. In this measurement, the OFF → ON reaction is usually induced by the UV light, which frequently prompts the photodegradation of the dye resulting in the short measurement time. Accordingly, the new method of the OFF → ON, ON → OFF, and fluorescence excitation without UV exposure can contribute to the longtime measurement covering wide region and time-dependent change of the materials. In the following, we introduce the visible one-color control method for OFF → ON, ON → OFF, and fluorescence excitation, which was developed in this project [29].

Figure 1.18a shows the structures of the open- and closed-ring isomers fluorescent diarylethene (fDAE). The quantum yield of the cyclization (fluorescence OFF → ON) reaction in dioxane solution is 0.23, while that of the cycloreversion (fluorescence ON → OFF) is $<10^{-5}$. As Fig. 1.18b shows, the absorption band of the open-ring isomer locates in the region shorter than 450 nm, while the closed-ring

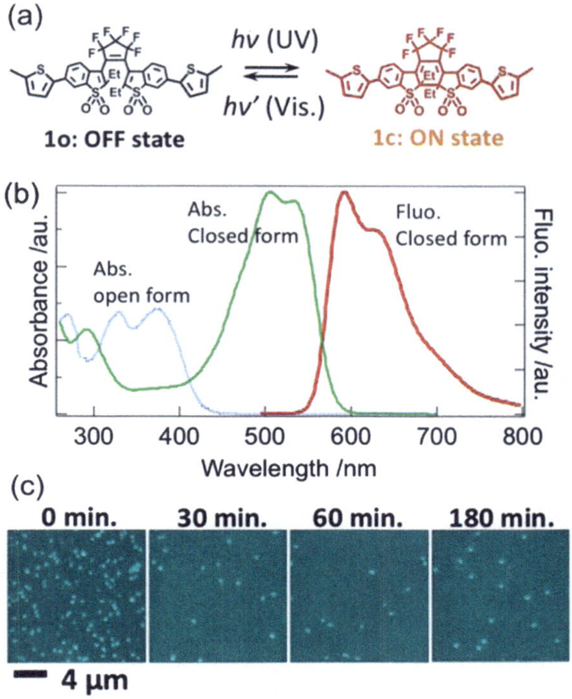

Fig. 1.18 **a** Structures of open and closed-ring isomers of fDAE. **b** Absorption and fluorescence spectra of the fDAE in 1,4-dioxane. **c** Time evolution of the single-molecule fluorescence image of fDAEs in a poly(2-hydroxyethyl acrylate) thin film. The images and plots in the figure are reproduced from Ref. [29]

isomer has absorption bands in the wavelength range of 400–600 nm and emits strong fluorescence of which quantum yield in dioxane solution is 0.78 [30].

Figure 1.18c exhibits the time evolution of fluorescence images of the single fDAEs in the poly(2-hydroxyethyl acrylate) ($Tg = 290$ K) film under CW 532-nm laser irradiation up to 180 min. The specimen was prepared with the OFF state of fDAE. At the time origin ($t = 0$ min), slightly larger number of fluorescent spots were detected, which is ascribable to the presence of the trace of the closed-ring isomer in the prepared samples. After several minutes of photo-irradiation, the number of fluorescent spots decreased to ca. 10–20 in the imaging area and remained almost constant for 3 h. Interestingly, the fluorescent spot appears in the different position in the film, indicating that the OFF → ON reaction took place by the 532-nm irradiation. Actually, the constant number of fluorescent spots is ascribed to the photo-stationary state where both ON → OFF and OFF → ON switching took place under the one-color irradiation at 532 nm. From the analysis of the band edge of the absorption spectrum of the open-ring isomer as well as the temperature effect of the appearance of new fluorescence spots, it was revealed that the new appearance of the fluorescence ON molecules is due to the absorption of the 532 nm light by the open-ring isomer through the hot band (Urbach tail [31]). Although the absorption coefficient by the Urbach tail at 532 nm is very small ($<10^{-2}$ M^{-1} cm^{-1}), it is sufficient to induce the OFF → ON reaction of several molecules in one sec to enable the longtime measurement.

Figure 1.19a shows the time evolution of superimposed trajectories of single fDAEs. With an increase in accumulation time, the number of trajectories increases and the probing area also increases. Accumulating SMT data for 63 min, whole imaging area is covered with trajectories of ca. 7000 fDAEs. Figure 1.19b shows the spatial distribution of diffusion coefficient of fDAEs of which trajectories are shown in Fig. 1.19a. Information on lateral diffusion coefficient over the whole imaging area was thus successfully obtained by the one-color photoswitching SMT. Furthermore, a time evolution of the spatial distribution of diffusion coefficient can provide the time evolution (aging) of the materials. The present SMT with one-color fluorescence switching method thus provides information on the lateral diffusion of huge number of single molecules, through which information on a host material over whole imaging area can be evaluated in terms of diffusional behaviors of single guest molecules.

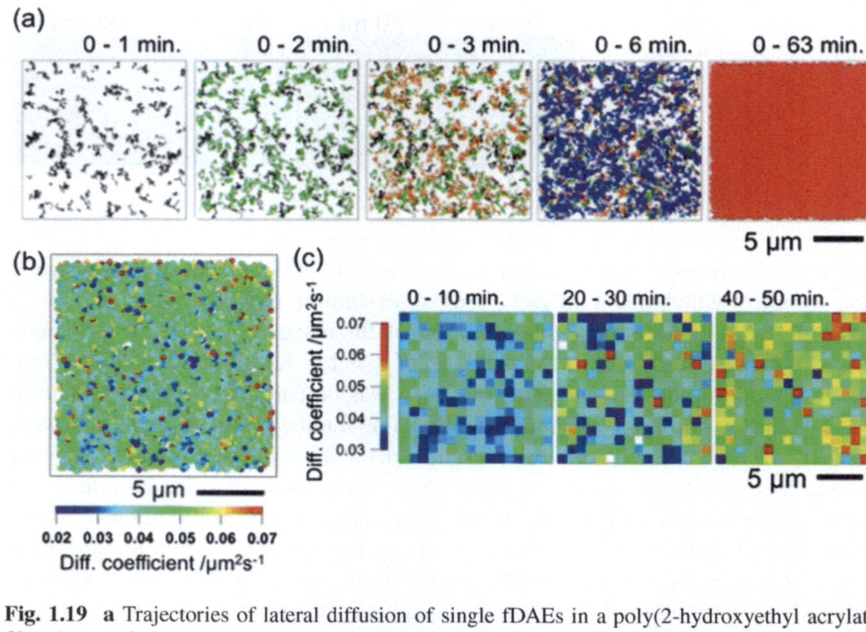

Fig. 1.19 **a** Trajectories of lateral diffusion of single fDAEs in a poly(2-hydroxyethyl acrylate) film. Accumulation times are shown in the tops of the plots. **b** Two-dimensional (2D) plot of the diffusion coefficients of fDAEs of which trajectories are shown in (**a**). **c** Time evolution of 2D distribution of diffusion coefficient obtained from the data shown in (**a**). The images and plots in the figure are reproduced from Ref. [29]

1.4 Mesoscopic Motion of Small Particles Induced by Switching of Photon Force Through Photochromic Reactions

In this section, we introduce the movement of the nano- and microparticles synchronized with the photochemical reaction under the laser trapping, which is one of the photosynergetic responses integrated with the momentum change of photon and the photochemical reaction of molecules.

A small particle under photo-irradiation experiences photon force (photon pressure) originating from the transfer of the momentum of photon to the particle [32]. This photon force is generally classified into three parts: absorption, scattering, and gradient forces. The absorption and scattering forces are, respectively, due to the momentum transfer of photons through absorption and scattering, which push a photo-irradiated particle toward the light propagation. On the other hand, the gradient force directs the spatial gradient of the light intensity. When the refractive index of a particle is larger than that of the medium, the gradient force pushes the particle to the direction along the increasing light intensity. Under the irradiation of the tightly focused laser beam, the gradient force keeps a small particle at the focal point. Because the contribution of the gradient force is generally dominant in the condition with no absorption of the particle at the wavelength of the trapping laser (off-resonant), the particle is trapped in the vicinity of the focal point. Although the absorption force is negligible in the off-resonant condition, it largely contributes in the case where the absorption of the particle is resonant to the trapping light. Hence, the change of the color of the particle may induce the change of the trapping position of the particle through the switching of the absorption force.

In the present study, we employed a photochromic diarylethene (DAE) derivative shown in Fig. 1.20a as a color switching material. In addition to the change of the absorption wavelength, photochromic reactions of this DAE are accompanied with the fluorescence switching; the open-ring isomer is non-fluorescent while the closed-ring isomer shows bright fluorescence. This fluorescence was used for tracking the position of a trapped particle [33]. Figure 1.20b shows extinction spectra of an aqueous colloid of 300-nm sized poly(methyl methacrylate) particles containing the DAE derivative (DAE-PMMA particle) for different UV irradiation time. The initial spectrum consists of the scattering of the particle and the absorption of the open-ring isomer of DAE. With an increase in UV dose, absorption band of the closed-ring isomer in the visible region increases and that of the open-ring isomer in UV region decreases, indicating that the photochromic reaction actually takes place in the PMMA particle suspended in water.

Figure 1.20c shows the time trace of the *Z*-position of the trapped DAE-PMMA particle under periodic irradiation with UV light. Reciprocating motion by several hundreds of nm, which are synchronized with the UV irradiation, was observed along the optical axis (*Z*-axis) of the trapping beam. Figure 1.20d schematically illustrates the mesoscopic motion of DAE-PMMA particle observed in Fig. 1.20c. The picture in the left-hand side shows one DAE-PMMA particle trapped in water by the CW

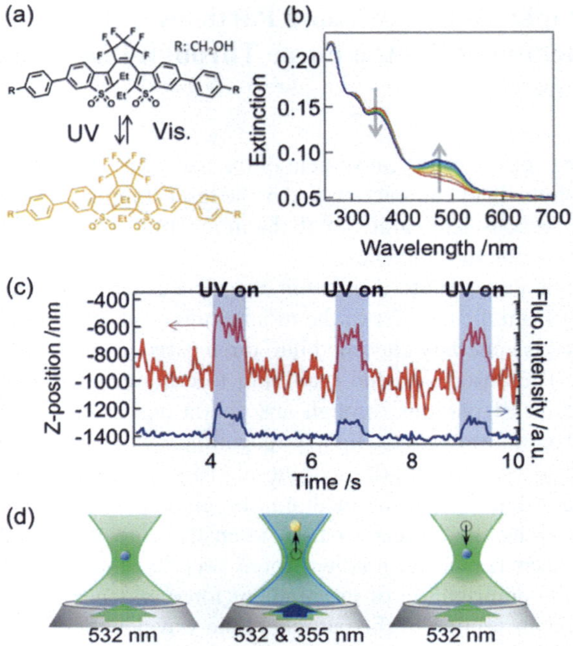

Fig. 1.20 **a** Structures of open- and closed-ring isomers of fDAE used. **b** Extinction spectra of 300-nm PMMA particles containing the fDAE in water. **c** Time trace of the *Z*-position of an optically trapped PMMA particle with the fDAEs. The fluorescence intensity of the particle is also shown at the bottom (right axis). **d** Schematic illustration of mesoscopic photomechanical motion of the single PMMA particle induced by photochromic reaction. Reprinted with permission from Ref. [33]. Copyright (2018) American Chemical Society

532-nm laser focused by an objective (X100, NA 1.35). At this stage, almost all DAE molecules in the particle are in the open-ring isomer and off-resonant to the 532-nm light. Actually, the fluorescence intensity of the particle is very weak as shown in Fig. 1.20c. Upon irradiation with CW UV laser (355 nm), the closed-ring isomer absorbing the 532-nm light is produced and, as a result, the absorption force pushes the particle toward the light propagation direction until the gradient force balances with the sum of the absorption and scattering forces. After turning the UV light OFF, the cycloreversion reaction (decolorization) is dominantly induced by the 532-nm laser light and the particle returns to the initial position, leading to the mesoscopic reciprocating motion synchronized with the colorization and decolorization of the particle.

T-type photochromic reaction was also applicable to the light-driven mesoscopic mechanical motion [34]. Figure 1.21a shows the structures and absorption spectra of colorless closed form (CF) and colored transoid-cis (TC) form of the T-type photochromic molecule (PQ) used in the present study. PQ in the colorless form turns to the colored form under the UV exposure through the photochemical reaction. The

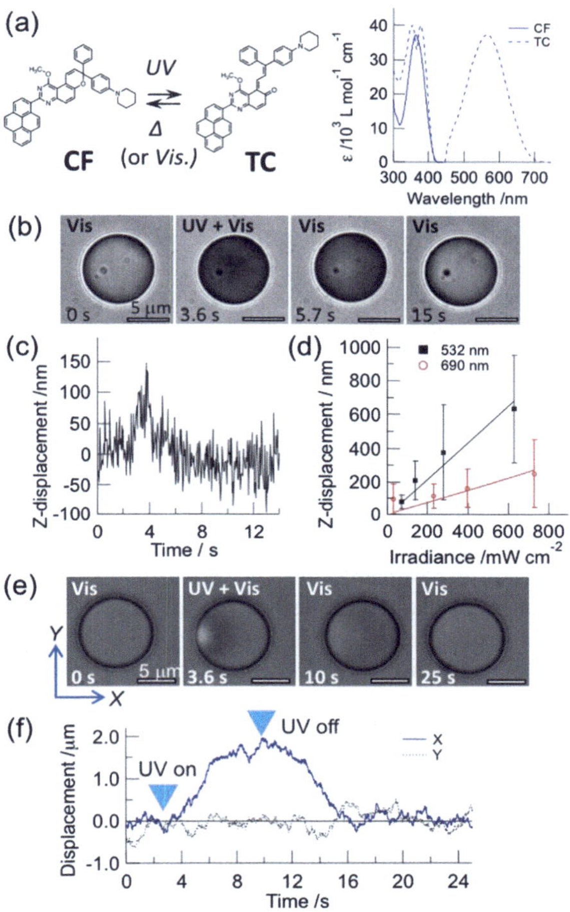

Fig. 1.21 **a** Structures and absorption spectra of colorless closed form (CF) and colored transoid-cis form (TC) of PQ. **b** Time evolution of the optical transmission image of a polymer microparticle with PQ before, during, and after UV exposure. **c** Time trace of the *Z*-position of the polymer microparticle whose images are shown in (**b**). **d** *Z*-displacement of PQ-PAE as a function of UV intensity for 532 and 690-nm trapping lasers. **e** Time evolution of the optical transmission image of a polymer microparticle with PQ before, during, and after photo-irradiation with a focused UV laser. The focal point is on the left-hand side of the particle. **f** Time trace of the *X*- and *Y*-positions of the polymer microparticle whose images are shown in (**e**). Reprinted with permission from Ref. [34]. Copyright (2018) American Chemical Society

colored form returns to the colorless form by the thermal reaction at room temperature with the time constant of 0.7 s in solution [35].

Figures 1.21b, c show the temporal evolution of the trapping behavior of the polyacrylic ester resin (PAE) microparticle containing PQ (PQ-PAE) which is trapped in water with CW 690-nm laser light. The UV light at 355 nm was turned ON at

2.0 s and OFF at 4.0 s. As was observed in the DAE-PMMA particles, the trapped PQ-PAE particle showed reciprocating motion synchronized with the photochromic reaction of PQ. The *Z*-position of the trapped PQ-PAE particle gradually increased to ca. 100 nm within ca. 1 s and the particle returned to the initial *Z*-position in >2 s. This positional shift along the *Z*-axis corresponds to the color change of the trapped particle as shown in Fig. 1.21b.

Figure 1.21d shows dependences of the *Z*-displacement of the PQ-PAE microparticles on UV dose and trapping wavelength. The *Z*-displacement monotonically increases with increasing UV dose for both trapping wavelength of 532 and 690 nm. The slope of the plot for trapping wavelength at 532 nm is much larger than that for trapping with 690-nm laser. This can be ascribed to the difference in the extinction between the two wavelengths. As shown in Fig. 1.21a, the absorption of PQ at 532 nm under UV exposure is much larger than that at 690 nm, and this difference is attributable to the wavelength dependence of the *Z*-displacement.

Not only the *Z*-direction, but also the motion on the lateral (*XY*) plane could be triggered by the locally induced photochromic reaction. Figure 1.21e shows the lateral motion of a trapped PQ-PAE microparticle induced by focusing the UV laser into an edge area of the left-hand side of the particle, where a weak emission signal was observed. This emission is presumably due to the photodegradation of PQ induced by the intense UV irradiation. The lateral position of the PQ-PAE microparticle shifted along *X*-axis by UV exposure, while the *Y*-displacement was negligible. This *X*-displacement is mainly attributed to the absorption force acting only on the left-hand side of the microparticle. After turning the UV OFF, the microparticle gradually returned to its initial position by the gradient force.

As demonstrated in this section, a new light-driven micromechanical system was developed by combining photochromic reaction and the force of light. Reciprocating motions along the *Z*-axis was achieved by switching the absorption band of microparticles. Not only along the *Z*-axis, but also lateral motion was induced by partial UV irradiation. The insight obtained in the present work will introduce a new strategy into the photomechanical control of nano/micro-objects as photosynergetic response.

1.5 Summary

We have briefly introduced our research results relating to the photosynergetic responses in the project. The femtosecond two-pulse excitation was applied also to the elucidation of the photoionization dynamics in solutions taking place with the energy lower than the ionization potential in the gas phase [36] and revealed that the solvation process of the specific electronic state regulates the ionization.

On the advanced control of the photochemical reaction by the stepwise multiphoton excitation, we extended this method to the investigation on the electron transfer reaction in higher excited states [37, 38]. In the tetraphenylporphyrin (ZnTPP)-substituted bridged imidazole dimer system [37], visible stepwise two-photon excitation of ZnTPP led to the formation of the charge separated state via the electron transfer from ZnTPP in higher exited states to the imidazole dimer, which was followed by the rapid bond cleavage in the anion radical of the imidazole dimer. After the subsequent charge recombination, the neutral radical pair with the absorption band in the visible region remains. This visible two-photon sensitization provides the new photosynergetic response of the photochemical reaction taking place only by the irradiation of the UV light [37]. Also for the protoporphyrin—CdS/ZnS core/shell nanocrystals (NCs) systems, the electron transfer from higher excited states of porphyrin pumped by the visible two-photon absorption to CdS NCs took place and successfully overcame the activation barrier with the wide bandgap ZnS shell leading to a high reduction potential [38].

It should be noted that, in the case where the initial state of the second photon absorption is long-lived such as triplet state, the high power of the excitation light source is unnecessary to induce the stepwise two-photon-gated photochemical reaction. Accordingly, it is expected that the stepwise two-photon excitation can open many kinds of photosynergetic responses in various kinds of photochemical processes such as super-resolution microscopy, laser trapping, and so forth.

Acknowledgements We deeply appreciate Professors S. Kobatake, M. Morimoto, M. Irie, Y. Yokoyama, Y. Kobayashi, K. Matsuda, J. Abe, and K. Uchida for their collaboration. The present work was supported by JSPS KAKENHI Grant Number 26107002, Grant-in-Aid for Scientific Research on Innovative Areas "Photosynergetics".

References

1. Woodward RB, Hoffman R (1969) The conservation of orbital symmetry. Angew Chem Int Ed 8:781–832
2. Irie M (2000) Diarylethenes for memories and switches. Chem Rev 100:1685–1716
3. Irie M, Fukaminato T, Matsuda K, Kobatake S (2014) Photochromism of diarylethene molecules and crystals: memories, switches, and actuators. Chem Rev 114:12174–12277
4. Yokoyama Y (2000) Fulgides for memories and switches. Chem Rev 100:1717–1739
5. Miyasaka H, Murakami M, Itaya A, Guillaumont D, Nakamura S, Irie M (2001) Multiphoton gated photochromic reaction in a diarylethene derivative. J Am Chem Soc 123:753–754
6. Murakami M, Miyasaka H, Okada T, Kobatake S, Irie M (2004) Dynamics and mechanisms of the multiphoton gated photochromic reaction of diarylethene derivatives. J Am Chem Soc 126:14764–14772
7. Ishibashi Y, Okuno K, Ota C, Umesato T, Katayama T, Mutakami M, Kobatake S, Irie M, Miyasaka H (2010) Multiphoton-gated cycloreversion reactions of photochromic diarylethene derivatives with low reaction yields upon one-photon visible excitation. Photochem Photobiol Sci 9:172–180
8. Tani K, Ishibashi Y, Miyasaka H, Kobatake S, Irie M (2008) Dynamics of cyclization, cycloreversion, and multiphoton-gated reaction of a photochromic diarylethene derivative in crystalline phase. J Phys Chem C 112:11150–11157
9. Piard J, Ishibashi Y, Saito H, Métivier R, Nakatani K, Gavrel G, Yu P, Miyasaka H (2012) Multiphoton-gated cycloreversion reaction of a photochromic 1,2-bis(hiazolyl)- perfluorocyclopentene diarylethene derivative. J Photochem Photobiol, A 234:57–65
10. Ishibashi Y, Murakami M, Miyasaka H, Kobatake S, Irie M, Yokoyama Y (2007) Laser multiphoton-gated photochromic reaction of a fulgide derivative. J Phys Chem C 111:2730–2737

11. Ishibashi Y, Katayama T, Ota C, Kobatake S, Irie M, Yokoyama Y, Miyasaka H (2009) Ultrafast laser spectroscopic study on photochromic cycloreversion dynamics in fulgide derivatives: one-photon and multiphoton-gated reactions. New J Chem 33:1409–1419
12. Miyasaka H, Ito S, Ishibashi Y (2017) Multiphoton-gated photochromic reaction induced by pulsed laser excitation. In: Yokoyama Y, Nakatani K (eds) Photon working switches. Springer, Tokyo, pp 225–235
13. Tamura H, Nanbu S, Ishida T, Nakamura H (2006) Ab initio nonadiabatic quantum dynamics of cyclohexadiene/hexatriene ultrafast photoisomerization. J Chem Phys 124:084313
14. Tamura H, Nanbu S, Ishida T, Nakamura H (2006) Laser control of reactions of photoswitching functional molecules. J Chem Phys 125:034307
15. Kondorskiy A, Nanbu S, Teranishi Y, Nakamura H (2010) Control of chemical dynamics by lasers: theoretical considerations. J Phys Chem A 114(114):6171–6187
16. Ward CL, Elles CG (2012) Controlling the excited-state reaction dynamics of a photochromic molecular switch with sequential two-photon excitation. J Phys Chem Lett 3:2995–3000
17. Ward CL, Elles CG (2014) Cycloreversion dynamics of a photochromic molecular switch via one-photon and sequential two-photon excitation. J Phys Chem A 118:10011–10019
18. Sotome H, Nagasaka T, Une K, Morikawa S, Katayama T, Kobatake S, Irie M, Miyasaka H (2017) Cycloreversion reaction of a diarylethene derivative at higher excited states attained by two-color, two-photon femtosecond pulsed excitation. J Am Chem Soc 139:17159–17167
19. Sumi T, Takagi Y, Yagi A, Morimoto M, Irie M (2014) Photoirradiation wavelength dependence of cycloreversion quantum yields of diarylethenes. Chem Commun 50:3928–3930
20. Nagasaka T, Sotome H, Yoshida Y, Yokoyama Y, Miyasaka H (2018) Cycloreversion reaction of a fulgide derivative in higher excited states attained by femtosecond two-photon pulsed excitation. J Phys Chem C 122:24987–24995
21. Mori K, Ishibashi Y, Matsuda H, Ito S, Nagasawa Y, Nakagawa H, Uchida K, Yokojima S, Nakamura S, Irie M, Miyasaka H (2011) One-color reversible control of photochromic reactions in a diarylethene derivative: Three-photon cyclization and two-photon cycloreversion by, a near-infrared femtosecond laser pulse at 1.28 μm. J Am Chem Soc 133:2621–2625
22. Sotome H, Nagasaka T, Une K, Okui C, Ishibashi Y, Kamada K, Kobatake S, Irie M, Miyasaka H (2017) J Phys Chem Lett 8:3272–3276
23. Hell SW, Wichmann J (1994) Breaking the diffraction resolution limit by stimulated emission: stimulated-emission-depletion fluorescence microscopy. Opt Lett 19:780–782
24. Betzig E, Patterson GH, Sougrat R, Lindwasser OW, Olenych S, Bonifacino JS, Davidson MW, Lippincott-Schwartz J, Hess HF (2006) Imaging intracellular fluorescent proteins at nanometer resolution. Science 313:1642–1645
25. Moerner WE, Kador L (1989) Optical detection and spectroscopy of single molecules in a solid. Phys Rev Lett 62:2535–2538
26. Ito S, Kusumi T, Takei S, Miyasaka H (2009) Diffusion processes of single fluorescent molecules in a polymer-based thin material with three-dimensional network. Chem Commun 41:6165–6167
27. Ito S, Itoh K, Pramanik S, Takei S, Miyasaka H (2009) Evaluation of diffusion coefficient in a dextrin-based photo-curable material by single molecule tracking. Appl Phys Express 2:075004
28. Ito S, Taga Y, Hiratsuka K, Takei S, Kitagawa D, Kobatake S, Miyasaka H (2015) Restricted diffusion of guest molecules in polymer thin films on solid substrates as revealed by three-dimensional single-molecule tracking. Chem Commun 51:13756–13759
29. Arai Y, Ito S, Fujita H, Yoneda Y, Kaji T, Takei S, Kashihara R, Morimoto M, Irie M, Miyasaka H (2017) One-colour control of activation, excitation and deactivation of a fluorescent diarylethene derivative in super-resolution microscopy. Chem Commun 53:4066–4069
30. Uno K, Niikura H, Morimoto M, Ishibashi Y, Miyasaka H, Irie M (2011) In situ preparation of highly fluorescent dyes upon photoirradiation. J Am Chem Soc 133:13558–13564
31. Urbach F (1953) The long-wavelength edge of photographic sensitivity and of the electronic absorption of solids. Phys Rev 92:1324
32. Ashkin A (2000) History of optical trapping and manipulation of small-neutral particle, atoms, and molecules. IEEE J Sel Top Quantum Electron 6:841–856

33. Ito S, Mitsuishi M, Setoura K, Tamura M, Iida T, Morimoto M, Irie M, Miyasaka H (2018) Mesoscopic motion of optically trapped particle synchronized with photochromic reactions of diarylethene derivatives. J Phys Chem Lett 9:2659–2664
34. Setoura K, Memon AM, Ito S, Inagaki Y, Mutoh K, Abe J, Miyasaka H (2018) Switching of radiation force on optically trapped microparticles through photochromic reactions of pyranoquinazoline derivatives. J Phys Chem C 122:22033–22040
35. Inagaki Y, Kobayashi Y, Mutoh K, Abe J (2017) A simple and versatile strategy for rapid color fading and intense coloration of photochromic naphthopyran families. J Am Chem Soc 139:13429–13441
36. Koga M, Yoneda Y, Sotome H, Miyasaka H (2019) Ionization dynamics of a phenylenediamine derivative in solutions as revealed by femtosecond simultaneous and stepwise two-photon excitation. Phys Chem Chem Phys 21:2889–2898
37. Kobayashi Y, Katayama T, Yamane T, Setoura K, Ito S, Miyasaka H, Abe J (2016) Stepwise two-photon induced fast photoswitching via electron transfer in higher excited states of photochromic imidazole dimer. J Am Chem Soc 138:5930–5938
38. Uno T, Koga M, Sotome H, Miyasaka H, Tamai N, Kobayashi Y (2018) Stepwise two-photon-induced electron transfer from higher excited states of noncovalently bound porphyrin-CdS/ZnS core/shell nanocrystals. J Phys Chem Lett 9:7098–7104

Chapter 2
Advanced Electronic Structure Theory for High-Accuracy Prediction of Higher Excited States and Its Application to Photochromic Molecules

Takeshi Yanai

Abstract In this chapter, we describe our recent development of theoretical approach to analyze photochemisty relevant electronic states (wave functions and energies) for studying photochromic molecules with a promised numerical reliability. Our approach is based on multireference (MR) theory, which is considered to be a critical treatment to properly compute complex electronic structures arising in excited states. It characterizes the wave function by explicitly expressing it as a correlated superposition of multiple electronic configurations (or Slater determinants). We mainly review our recent development of extended multi-state complete active-space second-order perturbation theory incorporated into density matrix renormalization group reference wave functions. Its illustrative applications to molecular photochromic isomerizations, diarylethene derivatives and pentaarylbiimidazole are briefly reviewed in order to show the applicability of our computational approach.

Keywords Multireference wave function theory · Density matrix renormalization group · Extended multi-state complete active-space second-order perturbation theory · Diarylethene

2.1 Introduction

In order to characterize the electronic excited states with computational methods, the multireference (MR) treatment plays an essential role. It expresses the wave function as a correlated superposition of multiple electronic configurations (or Slater determinants). In our recent work [29], we introduced an enhanced MR framework, referred to as DMRG-XMS-CASPT2 (density matrix renormalization group extended multi-state complete active-space second-order perturbation theory), that can accurately compute complex electronic structures arising in excited states. This method was

T. Yanai (✉)
Institute of Transformative Bio-Molecules (WPI-ITbM), Nagoya University, Furo-cho, Chikusa-ku, Nagoya, Aichi 464-8602, Japan
e-mail: yanait@chem.nagoya-u.ac.jp

H. Miyasaka et al. (eds.), *Photosynergetic Responses in Molecules and Molecular Aggregates*, https://doi.org/10.1007/978-981-15-5451-3_2

derived by incorporating the density matrix renormalization group (DMRG) wave functions into the multi-state reference for the XMS-CASPT2 calculations.

The CASPT2 theory is a most widely used MR method because of its low computational cost and theoretical simplicity [1, 2, 7]. In the CASPT2 calculations, the excited states of interest are first accounted for in a qualitative fashion by the complete active-space self-consistent field (CASSCF) wave functions [21, 22]. As a well-established manner, the state-averaged (SA) CASSCF treatment is used to obtain the references as a multi-root solution that provides a ground state and several low-lying excited states. Using these CASSCF states as the zeroth-order, the CASPT2 method describes the first-order perturbed wave functions and second-order perturbation energies to further account for quantitative correlation to obtain chemical accuracy, as well as in practice even qualitative correctness, including state ordering.

In order to use CASPT2 for excited states, its state-specific variant is well known to be inadequate for studying near-degenerate states, in which the strong mixing arises between the reference and other secondary CASSCF states via the second-order interactions. Another issue is a technical difficulty in determining state-specific CASSCF references of excited states because of instability of convergence causing so-called root flopping oscillation. These issues can be properly addressed by the multi-state treatment based on state-average CASSCF references, known as the MS-CASPT2 theory [7], an extension of CASPT2 formulated on basis of the quasi-degenerate perturbation theory (QDPT) [4]. It constructs an effective Hamiltonian matrix in the reference state basis under the presence of the perturbation, and its diagonalization yields the mixing of the perturbed states. This procedure allows us to properly describe the state crossing. The MS extension of MR perturbation theory was similarly developed in the MCQDPT [17, 18] and QD-NEVPT2 [3] methods. The combination of Granovsky's extended MS (XMS) method and the MS-CASPT2 method [10] was developed by Shiozaki et al. to improve the zeroth-order description. This approach (XMS-CASPT2) is able to additionally account for the off-diagonal Fock operator-based coupling [26].

2.2 Overview of DMRG-XMS-CASPT2 Theory

In this section, the multi-state (MS) extension of the CASPT2 is briefly reviewed. In the main framework of CASPT2, the first-order wave function is expanded into the internally contracted (IC) basis, which is generated by applying excitation operators to the zeroth-order reference. The MS treatment allows us to have multiple zeroth-order references resulting from the preceding state-average CASSCF calculations. The IC basis construction shown here is based on the single-state single-reference (SS-SR) contraction. As done in the state-specific CASPT2 theory, the IC configurations for the first-order space are generated solely relative to each of CAS-CI/SCF states $\{|L\rangle\}$. The SS-SR scheme is expressed with the following ansatz:

$$|\Psi_L^{(1)}\rangle = \sum_\Omega \hat{E}_\Omega |L\rangle T_{\Omega L}\,, \quad \text{(SS-SR)} \tag{2.1}$$

where $\hat{E}_\Omega$ are the spin-summed double excitation operators ($\{\hat{E}_\Omega\} = \{\hat{E}_{pq}\hat{E}_{rs}\}$), the detailed definitions of which can be consulted in Refs. [15, 16, 29]. The coefficients, $\{T_{\Omega L}\}$ (2.1), are determined by the following amplitude equations, which are solved separately for each target reference $|L\rangle$,

$$\langle L|\hat{E}_\Omega^\dagger(\hat{f} - E_L^{(0)})|\Psi_L^{(1)}\rangle + \langle L|\hat{E}_\Omega^\dagger \hat{H}|L\rangle = 0, \tag{2.2}$$

with $\forall L \in \{1, \ldots, n_{\text{state}}\}$.

Next, we turn to the construction of the effective subspace Hamiltonian,

$$H_{ML}^{\text{eff}} = H_{ML}^{\text{ref}} + \frac{1}{2}(H_{ML}^{(2)} + H_{LM}^{(2)}), \tag{2.3}$$

where the Hamiltonian matrix to the first-order is given by $H_{ML}^{\text{ref}} = \langle M|\hat{H}|L\rangle$, and the dynamic correlation arising between the perturbed states is determined by the second-order perturbation,

$$H_{ML}^{(2)} = \langle M|\hat{H}|\Psi_L^{(1)}\rangle. \tag{2.4}$$

The eigenvalues resulting from the diagonalization of $\mathbf{H}^{\text{eff}}$ are the XMS-CASPT2 energies E_P^{MS}, and the associated XMS-CASPT1 wave functions $|\Psi_P^{\text{MS}}\rangle$ are obtained using the eigenvectors R_{LP}:

$$E_P^{\text{MS}} = \sum_{ML} H_{ML}^{\text{eff}} R_{MP} R_{LP}, \tag{2.5}$$

$$|\Psi_P^{\text{MS}}\rangle = \sum_L (|L\rangle + |\Psi_L^{(1)}\rangle) R_{LP}. \tag{2.6}$$

These expressions indicate that the scaling of computational cost of (X)MS-CASPT2 with the SS-SR scheme is expressed to be linear as a function of n_{state} (i.e., $O(n_{\text{state}})$). The combination with Granovsky's extended MS (XMS) method [10, 26] results in XMS-CASPT2 theory. It uses the state rotation matrix, which can be computed to be an eigenvector matrix from the diagonalization of the generalized Fock matrix in the state basis.

In the previous study [15, 16], we developed the state-specific CASPT2 method using the DMRG-CASSCF wave function as the zeroth-order reference, as depicted in Fig. 2.1. The major task to achieve this development was the implementation to compute the first- to fourth-order reduced density matrices (k-RDMs, $k = 1, \ldots, 4$) of the active-space DMRG wave function. The combination of the DMRG and state-specific CASPT2 methods was achieved by using the RDMs of the active-space DMRG wave function in place of those arising in the tensor contraction expressions of the CASPT2 equation and energy. The RDMs are a state-specific (or single-state)

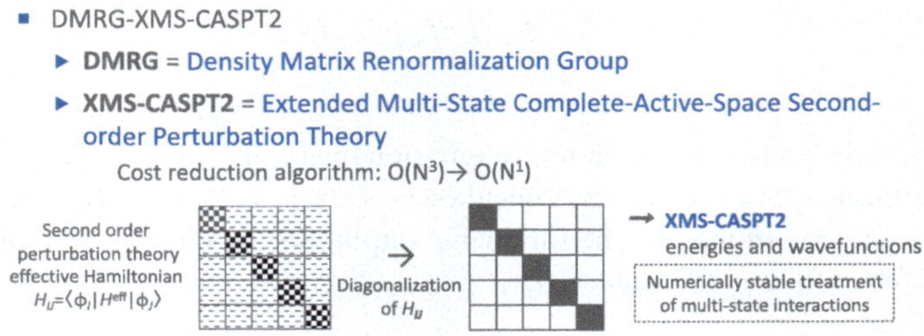

Fig. 2.1 DMRG-XMS-CASP2 is defined as use of the active-space DMRG functions as zeroth-order references of XMS-CASPT2. (DMRG-)XMS-CASPT2 constructs an effective Hamiltonian matrix in the reference state basis under the presence of the perturbation. It is diagonalized to obtain the mixing of the perturbed states

object as expressed in the following general form,

$$D^{ikm\dots}_{jln\dots}(L) = \langle \Psi^{\mathrm{DMRG}}_L | \hat{E}^{ikm\dots}_{jln\dots} | \Psi^{\mathrm{DMRG}}_L \rangle, \tag{2.7}$$

with the generator $\hat{E}^{ikm\dots}_{jln\dots} = \sum_{\sigma\tau\rho\dots} \hat{a}^\dagger_{i\sigma}\hat{a}^\dagger_{k\tau}\hat{a}^\dagger_{m\rho}\dots\hat{a}_{n\rho}\hat{a}_{l\tau}\hat{a}_{j\sigma}$, where $\sigma, \tau, \rho, \dots$ refer to the spin indices.

The 4-RDM is the $O(n^8)$ object, where n means the number of active orbitals. Thus, the computation and memory storage of its full elements become formidable when using a large active space, namely large n. In order to mitigate this complexity, we introduced an approximation to the CASPT2 framework, which is realized by replacing the 4-RDM with the approximate one built on the fly from low-order (up to third-order) RDMs using the cumulant reconstruction formula. It is referred to as the cu(4) approximation.

In Ref. [29], we extended the aforementioned state-specific DMRG-based CASPT2 to the (X)MS theory. The numerical object that is additionally required for this extension is the transition RDMs (TRDMs) of the DMRG references,

$$D^{ikm\dots}_{jln\dots}(L, M) = \langle \Psi^{\mathrm{DMRG}}_L | \hat{E}^{ikm\dots}_{jln\dots} | \Psi^{\mathrm{DMRG}}_M \rangle \quad (L \neq M). \tag{2.8}$$

The RDM (2.7) and TRDM (2.8) elements arise as building blocks of the tensor contraction form of the amplitude equation (2.2) and effective Hamiltonian matrix (2.3). As similarly done in the state-specific DMRG-CASPT2, the RDM and TRDMs supplied by the DMRG calculations are used for combining DMRG and XMS-CASPT2. The algorithm to compute TRDMs with the DMRG algorithm was developed and implemented in BLOCK [5] (see also Ref. [11]).

The SS-SR ansatz is the feasible IC basis scheme when based on DMRG references. Choosing the SS-SR scheme that can avoid the cumulant approximation to 4-TRDMs is expedient particularly when the cost of computing the pure 4-TRDMs is formidably expensive. It should be, however, noted that this feasibility comes at the price of using inconsistent subsets of the complete IC basis space for the target states and discarding their invariance with respect to the XMS rotation. The state-specific 4-RDM is required as the highest-order RDM, and thus, the cu(4) approximation is again exploited for DMRG-based (X)MS-CASPT2. In the XMS case, the approximate 4-RDM of the rotated reference $\tilde{L}$ is calculated using the rotated low-order RDMs ($\langle\tilde{L}|\hat{E}^i_j|\tilde{L}\rangle$, $\langle\tilde{L}|\hat{E}^{ik}_{jl}|\tilde{L}\rangle$, and $\langle\tilde{L}|\hat{E}^{ikm}_{jln}|\tilde{L}\rangle$) via the cumulant reconstruction formula instead of rotating the reconstructed 4-RDM.

Let us here biefly mention our computer implementation. The (X)MS-CASPT2 method with and without DMRG references was implemented into our in-house program package. Derivations and implementations of the tensor contraction form of the XMS-CASPT2 equations were achieved using the automated code generator developed in Refs. [23, 24]. The quantum chemical DMRG program BLOCK [5] was used to compute k-RDMs and k-TRDMs ($k = 1, 2, 3$) of multi-root (or multi-state) DMRG wave functions [11]. In order to accelerate the evaluation of two-electron integrals [28], the resolution-of-identity (or density-fitting) method was incorporated into the (X)MS-CASPT2 implementation. The parallel implementation was deviced by making effective use of the standard message-passing interface library and Global Arrays (GA) toolkit [19] externally linked with the Aggregate Remote Memory Copy Interface compatible library, ARMCI-MPI [6]. The memory allocations for two-electron integrals and amplitudes are distributed across nodes to the data-parallel arrays operated by the GA library. The computation of the amplitude equation (2.2) is parallelized using GA's data distribution functionality.

2.3 DMRG-XMS-CASPT2 Study of Diarylethene Derivatives

In Ref. [29], we presented an application of DMRG-XMS-CASPT2 to the multi-state calculations of the isomerization of the diarylethene derivatives, which are a well-studied type of molecular photochromic systems [13]. Two kinds of the derivatives, (**1**) the 1,2-Bis(2-methyl-5-phenyl-3-thienyl)perfluorocyclopentene [12] and (**2**) the 1,2-Bis(3-methyl-5-phenyl-2-thienyl)perfluorocyclopentene [27], denoted by N-diarylethene and I-diarylethene, respectively, for short (Fig. 2.2) were examined. They undergo a photo-switching cycloreversion transformation between two stable isomers: open-ring and closed-ring.

Geometric parameters of the open- and closed-ring isomers were derived by the geometry optimizations at the B3LYP-D3/def2-TZVPP level of theory for the singlet ground state. The intermediate structures of the isomerization were derived by interpolating the two optimized open- and closed-ring geometries. The progress of the isomerization is denoted as a function of the reaction coordinate F, ranging from 1 to 10, where the coordinates 1 and 10 correspond to the optimized open- and closed-ring isomers, respectively, denoted as i**OF** and i**CF** ($i = 1$ for N-diarylethene and $i = 2$ for I-diarylethene), respectively.

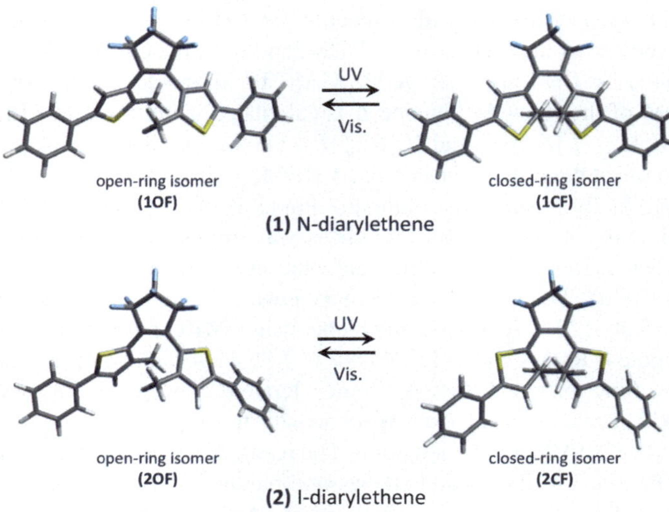

Fig. 2.2 B3LYP-D3/def2-TZVPP-optimized structures of open-ring (**OF**) and closed-ring (**CF**) isomers of (1) *N*-diarylethene and (2) *I*-diarylethene considered in this study. They undergo a ring-opening and -closing reaction upon UV and visible irradiation. Reprinted with permission from Ref. [29]. Copyright 2017 American Chemical Society

The DMRG-XMS-CASPT2 calculations were performed using the def2-SVP basis set with the DMRG-SA-CASSCF(26e,24o) reference wave functions, which accounts for the full π valence active space. Figure 2.3 shows the localized active orbitals, which were actually used as the DMRG lattice sites. The DMRG-CASSCF wave functions were computed using 256 spin-adapted renormalized basis. The four lowest-lying singlet states were considered in the multi-state XMS-CASPT2-based potential energy calculations. We employed the cu(4) approximation; the resultant approach is hereafter referred to as DMRG-cu(4)-XMS-CASPT2. It requires the computation of the pure (T)RDMs of up to third-order of the DMRG references. The XMS-CASPT2 step was carried out with an imaginary level shift [8] and an IPEA shift [9] set to $i0.1$ and 0.10 au, respectively.

The multi-state PECs of *N*-diarylethene and *I*-diarylethene obtained by the DMRG-cu(4)-XMS-CASPT2 calculations are shown in Fig. 2.4. According to the calculated oscillator strengths, the S1 and S2 states for both the diarylethene derivatives seemingly correspond to what are known as the optically allowed 1B and forbidden 2A states, respectively [13]. The calculated transition energies of the 1B state for the geometries close to **1OF**, **1CF**, **2OF**, and **2CF** are all in reasonably good agreement with the experimental absorption energies. By checking the off-diagonal elements of the effective Hamiltonian H_{ML}^{eff} (2.3), we confirmed that the avoided crossing at the reaction coordinate 7 is characterized by the mixing of the S0 (1A) and S2 (2A) states. This indicates that the characters of the S0 and S2 states intercross during the isomerization process.

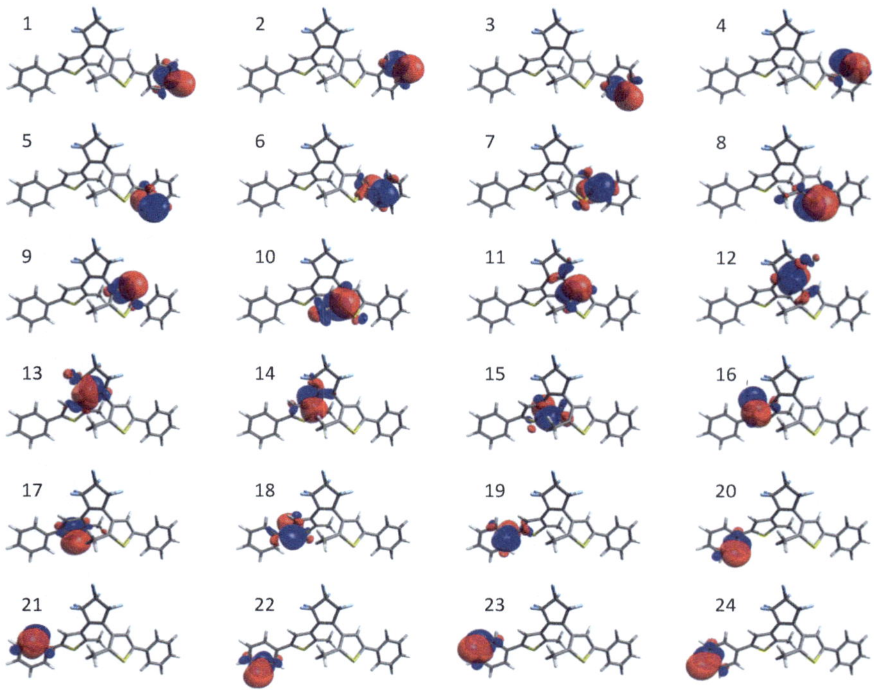

Fig. 2.3 Localized DMRG-CASSCF active orbitals of the open-ring *N*-diarylethene in the ordering on the DMRG lattice sites. Reprinted with permission from Ref. [29]. Copyright 2017 American Chemical Society

Experimentally, the ring-opening reaction of the perfluorocyclopentene diarylethene derivatives is activated under the steady-state irradiation with visible light but with an extremely low reaction yield. However, the recent laser photolysis carried out by Miyasaka et al. showed that the two-photon absorption laser treatment significantly increases its quantum yield by a factor of ca. 400 [20]. This can be explained using the calculated PECs (Fig. 2.4) as follows. The one-photon absorption upon the steady-state irradiation excites the closed isomer to the allowed S1 state. The S1 curve is not sufficiently downhill to promote the ring-opening reaction, so that it has a short lifetime. The two-photon absorption process somewhat leads to the production of the high-energy S2 state. The presence of an efficient state-crossing route that connects it to the stable ring-open isomer state (2A $\rightarrow$ 1A) seems to be highly favorable for the promotion of the reaction, which should contribute to an increase in the quantum yield. In this light, the present DMRG-cu(4)-XMS-CASPT2 results are consistent with the experimental observations; however, further detailed analysis should be carefully performed by considering structural determinations of excited-state stable states and conical intersection at the MR level of theory.

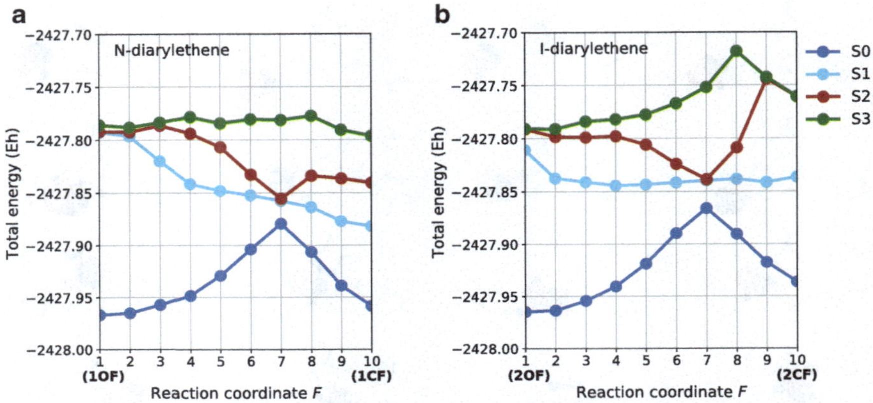

Fig. 2.4 Total energies of the S0-S3 states of **a** *N*-diarylethene and **b** *I*-diarylethene determined by the four-state DMRG-cu(4)-XMS-CASPT2 calculations with CAS(26e,24o) accounting for the full π active space. The plots are given as a function of the reaction coordinate F. $F = 1$ and $F = 10$ correspond to the DFT-optimized open-ring and closed-ring isomers, respectively. The def2-SVP basis set was used. Reprinted with permission from Ref. [29]. Copyright 2017 American Chemical Society

2.4 XMS-CASPT2 Study of Pentaarylbiimidazole (PABI)

The XMS-CASPT2 methodology was applied for illuminating the mechanism of photochromism in pentaarylbiimidazole (PABI) molecules, which was recently developed by Abe group as a novel photochromic molecular system [14]. It was shown that a mixture of diradical and quinoidal (diamagnetic) characters arises in the electronic state of the open-ring isomer and needs to be described using MR theory. While the photogenerated transient species was considered to be a single species of the biradical, the ultrafast spectroscopic technique revealed the existence of two transient isomers differing in the contributions of biradical character.

In Ref. [14], we performed quantum chemical calculations on PABI in order to understand molecular- and electronic-level details of its photochromic ring-opening and -closing reaction. In what follows, the computational setup and resultant data are presented and discussed. The density functional theory (DFT) was used to derive the one-dimensional model of the reaction pathway, determining the molecular geometries including stable states and reacting intermediates. The DFT calculations were carried out using GAUSSIAN09 program package at the CAM-B3LYP level of theory with the def2-SVP basis sets in conjunction with Grimme's van der Waals corrections6 (gd3bj option) and the broken-symmetry (or spin-polarized) treatment accounting for the open-shell singlet state of the ring-opening structure. The broken-symmetry DFT is practically suited to the geometric determination of the open form accompanied with the biradical electronic nature found in the ground state. Figure 2.5 shows the optimized molecular structures of the closed and open forms of PABI. The

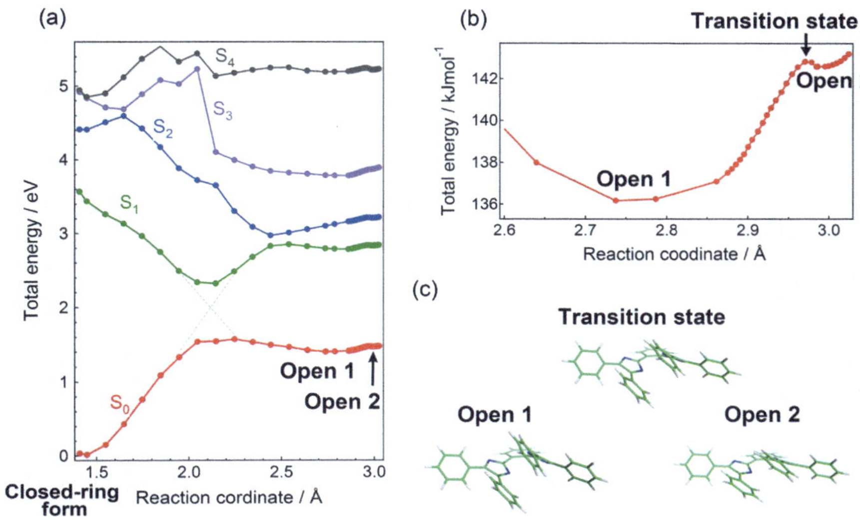

Fig. 2.5 **a** Potential energy curves of the S0 to S4 states obtained by the combinations of DFT calculations (UCAM-B3LYP/def2-SVP level of the theory) and XMS-CASPT2 calculations, **b** the magnified potential curve of the S0 state at around the open forms, and **c** optimized molecular structures of the Open 1, Open 2, and transition state of PABI. Reprinted with permission from Ref. [14]. Copyright 2017 American Chemical Society

two stable open-form structures were found, referred to as Open 1 and Open 2. The distance of C11-N18, which is to dissociate in the ring-opening reaction, is 1.448, 2.861, and 2.986 Åfor the closed form, Open 1 and Open 2, respectively. The reaction pathway was modeled using the potential energy surface (PES) scan scheme implemented in GAUSSIAN09. The scan calculations are provided the relaxed intermediate geometries along the reaction coordinate that connects the stable closed- and open-form states.

In order to examine the impact of the Grimme's van der Waals corrections on the DFT results, we again performed geometry optimization on Open 1 and Open 2 but without use of the gd3bj option. This resulted in that Open 1 and Open 2 converged to the same geometry, shown in Fig. 2.5. It has a distance of 3.007 Å for C11-N18, which is much longer than those of Open 1 and Open 2, and the distance of the stacked phenyl rings is substantially elongated by the neglect of the van der Waals corrections. This result indicates the following points: (1) the transition state (energy barrier) in the reaction path between Open 1 and Open 2 diminishes without use of van der Waals corrections; (2) the stacked phenyl rings strongly interact in an attractive way via the van der Waals interaction; (3) in a physical sense, the weak interaction arising between the stacked phenyl rings thus plays an essential role in producing the two minima of the open form in the van der Waals-corrected potential energy curve; (4) it can be alternatively viewed as causing a steric hindrance, which gives rise to the energy barrier between Open 1 and Open 2.

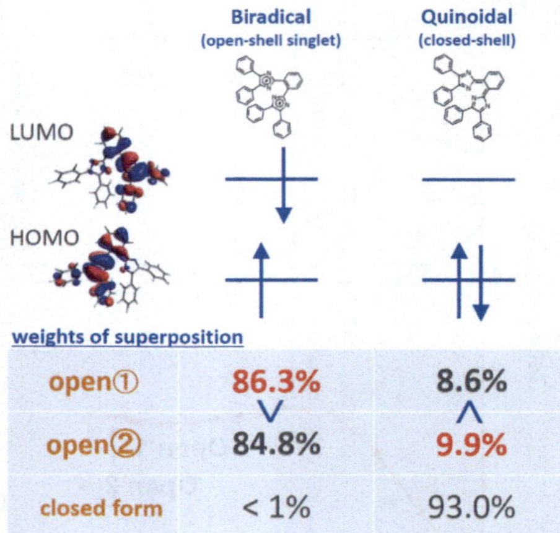

Fig. 2.6 Decomposition of the S0 state of Open 1, Open 2, and closed-ring form into a quantum superposition of biradical (open-shell singlet) and quinoidal (closed-shell) configurations, which was obtained by the multireference wave function calculations at the CASSCF level of the theory. Reprinted with permission from Ref. [14]. Copyright 2017 American Chemical Society

Using all of these DFT-scanned ground-state geometries as input structures, the single-point electronic structure calculations were performed at the level of the multireference wave function theory based on XMS-CASPT2. The def2-SVP basis sets were used for atomic basis functions. Four electrons and six orbitals, denoted (4e, 6o), were employed for the active electrons and orbitals of the CAS algorithm, respectively. We considered five states for the reference wave functions, which were determined by the preceding SA-CASSCF calculations. The imaginary level shift was set to 0.1 a.u. for the CASPT2 denominator. We used the in-house program package developed by us for the multireference calculations. The total energies of the ground-state S0 of PABI were obtained by the DFT and the XMS-CASPT2 methods; also, those of the four lowest-lying excited states Sn ($n = 1, \ldots, 4$) were determined by the XMS-CASPT2 method as a function of the reaction coordinate based on the DFT-scanned geometries. The plot of the XMS-CASPT2 potential energy curve (PEC) is presented in Fig. 2.5. At the XMS-CASPT2 level of the theory, the minima of the ground-state PEC can be located near the geometries corresponding to the closed-form, Open 1, and Open 2, respectively, and the geometry corresponding to transition state was found between Open 1 and Open 2.

Figure 2.6 shows the plots of the HOMO and LUMO-like orbitals for Open 1 and Open 2. They are the natural orbitals obtained from the state-average CASSCF calculations, which are variously occupied in the multi-configurational expansion of the wave function. The configuration analysis was performed on the reference CASSCF wave functions (Fig. 2.6) for Open 1 and Open 2. It shows that the multireference theory mainly characterizes the ground-state S0 as the open-shell singlet biradical configuration with a weight of ca. 86.3% and 84.8% for Open 1 and Open 2, respectively. It can be viewed as a charge-transfer state between the two imidazole moieties, associated with HOMO and LUMO, respectively. The rest of the description arises primarily from the closed-shell configuration, corresponding to the quinoidal structure, with a weight of 8.6% and 9.9% for Open 1 and Open 2, respectively. Overall, the biradical character is slightly more weighted toward Open 1 relative to Open 2, while the quinoidal character is less, as summarized in Fig. 2.6. The S2 state of the open form appears to closely correspond to the electronic character of the S0 state of the closed form, which was calculated to be a closed-shell state. The absorption of visible light for the open form is mainly characterized by a transition from S0 to S2 with its largest oscillator strength.

2.5 Conclusion

We have described an overview of our development of the DMRG-XMS-CASPT2 method (Ref. [29]) tailored toward highly reliable excited-state calculations of photochemical reactions including photochromic systems. The applications of this method to photochormic systems, diarylethene derivatives and pentaarylbiimidazole, have been discussed to stress the importance of multireference and multi-state treatments for achieving high accuracy in framework of quantum chemical calculations. Recently, in Ref. [25], this approach was further extended by combining the solvation theory, RISM, to account for solvation effects in excited states. This technique is important for studying bioimaging probes with photo-activity because these molecular systems are in many cases handled in solution. Earlier studies showed that they cannot be described with conventional methods such as PCM-TD-DFT, yielding results far from experimental data. Our DMRG-CASPT2 using the RISM solvation theory was shown to be powerful for calculating the photochemical properties of near-infrared bioimaging molecules in solution with higher-accuracy prediction. We hope that the establishment of these efficient computational approaches should pave the way toward deeper and further detailed understanding of the mechanism of photochromic reactions.

Acknowledgements T. Yanai greatly thanks Professors H. Miyasaka, J. Abe, Y. Shigeta, Y. Kobayashi, A. Sakamoto, H. Sotome, and K. Mutoh for their collaboration. The present work was supported by JSPS KAKENHI Grant Number 17H05274, Grant-in-Aid for Scientific Research on Innovative Areas "Photosynergetics".

References

1. Andersson K, Malmqvist PÅ, Roos BO, Sadlej AJ, Wolinski K (1990) Second-order perturbation theory with a casscf reference function. J Phys Chem 94:5483–5488
2. Andersson K, Malmqvist PÅ, Roos BO (1992) Second-order perturbation theory with a complete active space self-consistent field reference function. J Chem Phys 96:1218–1226
3. Angeli C, Borini S, Cestari M, Cimiraglia R (2004) A quasidegenerate formulation of the second order n-electron valence state perturbation theory approach. J Chem Phys 121(9):4043–4049
4. Bloch C (1958) Sur la théorie des perturbations des états liés. Nuc Phys 6:329–347
5. BLOCK verion 1.1: an implementation of the density matrix renormalization group (DMRG) algorithm for quantum chemistry. https://github.com/sanshar/Block

6. Dinan J, Balaji P, Hammond JR, Krishnamoorthy S, Tipparaju V (2012) Supporting the global arrays pgas model using mpi one-sided communication. In: 2012 IEEE 26th International Parallel and Distributed Processing Symposium, pp 739–750
7. Finley J, Malmqvist PÅ, Roos BO, Serrano-Andrés L (1998) The multi-state caspt2 method. Chem Phys Lett 288(2–4):299–306
8. Forsberg N, Malmqvist PÅ (1997) Multiconfiguration perturbation theory with imaginary level shift. Chem Phys Lett 274(1):196–204
9. Ghigo G, Roos BO, Malmqvist PÅ (2004) A modified definition of the zeroth-order hamiltonian in multiconfigurational perturbation theory (caspt2). Chem Phys Lett 396(1):142–149
10. Granovsky AA (2011) Extended multi-configuration quasi-degenerate perturbation theory: the new approach to multi-state multi-reference perturbation theory. J Chem Phys 134(21):214113
11. Guo S, Watson MA, Hu W, Sun Q, Chan GKL (2016) N-electron valence state perturbation theory based on a density matrix renormalization group reference function, with applications to the chromium dimer and a trimer model of poly(p-phenylenevinylene). J Chem Theory Comput 12(4):1583–1591
12. Irie M, Lifka T, Kobatake S, Kato N (2000) Photochromism of 1,2-bis(2-methyl-5-phenyl-3-thienyl)perfluorocyclopentene in a single-crystalline phase. J Am Chem Soc 122(20):4871–4876
13. Irie M, Fukaminato T, Matsuda K, Kobatake S (2014) Photochromism of diarylethene molecules and crystals: memories, switches, and actuators. Chem Rev 114(24):12174–12277
14. Kobayashi Y, Okajima H, Sotome H, Yanai T, Mutoh K, Yoneda Y, Shigeta Y, Sakamoto A, Miyasaka H, Abe J (2017) Direct observation of the ultrafast evolution of open-shell biradical in photochromic radical dimer. J Am Chem Soc 139(18):6382–6389
15. Kurashige Y, Yanai T (2011) Second-order perturbation theory with a density matrix renormalization group self-consistent field reference function: Theory and application to the study of chromium dimer. J Chem Phys 135(9):094104
16. Kurashige Y, Chalupský J, Lan TN, Yanai T (2014) Complete active space second-order perturbation theory with cumulant approximation for extended active-space wavefunction from density matrix renormalization group. J Chem Phys 141(17):174111
17. Nakano H (1993) Mcscf reference quasidegenerate perturbation theory with epstein-nesbet partitioning. Chem Phys Lett 207(4):372–378
18. Nakano H (1993) Quasidegenerate perturbation theory with multiconfigurational self-consistent-field reference functions. J Chem Phys 99(10):7983–7992
19. Nieplocha J, Palmer B, Tipparaju V, Krishnan M, Trease H, Aprà E (2006) Advances, applications and performance of the global arrays shared memory programming toolkit. Int J High Perform Comput Appl 20(2):203–231
20. Piard J, Ishibashi Y, Saito H, Métivier R, Nakatani K, Gavrel G, Yu P, Miyasaka H (2012) Multiphoton-gated cycloreversion reaction of a photochromic 1,2-bis(thiazolyl) perfluorocyclopentene diarylethene derivative. J Photochem Photobiol A 234:57–65 (in honour of Monique M. Martin)
21. Roos BO (1987) The complete active space self-consistent field method and its applications in electronic structure calculations. Adv Chem Phys 69:399–445
22. Ruedenberg K, Schmidt M, Gilbert M, Elbert S (1982) Are atoms intrinsic to molecular electronic wavefunctions? I. The FORS model. Chem Phys 71(1):41–49
23. Saitow M, Kurashige Y, Yanai T (2013) Multireference configuration interaction theory using cumulant reconstruction with internal contraction of density matrix renormalization group wave function. J Chem Phys 139(4):044118
24. Saitow M, Kurashige Y, Yanai T (2015) Fully internally contracted multireference configuration interaction theory using density matrix renormalization group: A reduced-scaling implementation derived by computer-aided tensor factorization. J Chem Theory Comput 11(11):5120–5131
25. Shimizu RY, Yanai T, Kurashige Y, Yokogawa D (2018) Electronically excited solute described by rism approach coupled with multireference perturbation theory: vertical excitation energies of bioimaging probes. J Chem Theory Comput 14(11):5673–5679

26. Shiozaki T, Győrffy W, Celani P, Werner HJ (2011) Communication: Extended multi-state complete active space second-order perturbation theory: energy and nuclear gradients. J Chem Phys 135(8):081106
27. Uchida K, Matsuoka T, Kobatake S, Yamaguchi T, Irie M (2001) Substituent effect on the photochromic reactivity of bis(2-thienyl)perfluorocyclopentenes. Tetrahedron 57(21):4559–4565
28. Vahtras O, Almlöf J, Feyereisen M (1993) Integral approximations for lcao-scf calculations. Chem Phys Lett 213(5):514–518
29. Yanai T, Saitow M, Xiong XG, Chalupský J, Kurashige Y, Guo S, Sharma S (2017) Multistate complete-active-space second-order perturbation theory based on density matrix renormalization group reference states. J Chem Theory Comput 13(10):4829–4840

Chapter 3
Enhanced and Selective Two-Photon Excitation of Molecular Vibronic States Using Entangled Photons

Hisaki Oka

Abstract We introduce molecular two-photon absorption and two-step excitation using quantum-mechanically entangled-photon pairs. The entangled photons possess the inherent simultaneity of photons originating from quantum correlation with energy anticorrelation, which is not shared by conventional laser light. This simultaneity enables entangled photons to achieve highly efficient two-photon absorption and two-step excitation under the condition of low light intensity. We show that ultrabroadband frequency-entangled photons can enhance the transition rates of two-photon absorption and two-step excitation more than 1000 times, compared to uncorrelated photons like laser light, and can also selectively excite a single vibrational mode in the molecular two-photon processes.

Keywords Two-photon absorption · Two-step excitation · Entangled photons

3.1 Introduction

Two-photon process is an optical nonlinear process in which two photons contribute to excitation of quantum states in materials. Generally, this process is classified into two types, namely two-photon absorption and two-step (stepwise) excitation as depicted in Fig. 3.1. In the two-photon absorption (Fig. 3.1a), the energy ω_m of intermediate state is far-off-resonant to the energy ω of an incident photon, and hence the intermediate state is not really but virtually excited only during the photon–molecule interaction. This short-lived intermediate state is therefore called "virtual state." On the other hand, in the two-step excitation (Fig. 3.1b), one photon is absorbed by the intermediate state and then another photon is absorbed by the excited state stepwisely. Generally, the energy configuration of $2\omega \approx \omega_e$ is required for the two-photon absorption, whereas $\omega \approx \omega_m$ and $\omega \approx \omega_e - \omega_m$ are sufficient for the two-step excitation. As a result, the two-photon absorption is unfavorable compared to the two-step excitation because simultaneous absorption of two photons is required.

H. Oka (✉)
Division of Engineering, Faculty of Engineering, Niigata University, Niigata 950-2102, Japan
e-mail: h-oka@eng.niigata-u.ac.jp

H. Miyasaka et al. (eds.), *Photosynergetic Responses in Molecules and Molecular Aggregates*, https://doi.org/10.1007/978-981-15-5451-3_3

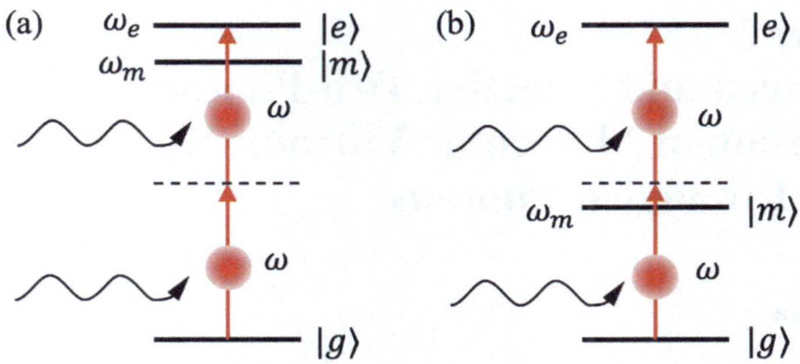

Fig. 3.1 Schematics of **a** two-photon absorption and **b** two-step excitation

Historically, two-photon absorption has been theoretically predicted by Göppert-Mayer about 80 years before [1] and is now applied to various leading-edge techniques, such as two-photon microscopy [2], three-dimensional optical storage memory [3], and coherent control of molecular processes [4]. To achieve efficient two-photon absorption in molecules, simultaneous absorption of two photons is required because two-photon absorption occurs via virtual states as mentioned above. Fortunately, thanks to recent development of optical technologies, this can be easily achieved by spatiotemporally controlling light, namely pulsing and/or focusing light, so that the photon density interacting with molecules can increase. In fact, two-photon microscopy is realized by focusing light using confocal optical system and achieves high spatial resolution microscopy. On the other hand, in molecular coherent control, ultrashort pulses are utilized in order to rapidly excite vibronic states and avoid molecular relaxation processes. For atoms and bulk materials, highly intense light might also be one way to achieve efficient two-photon absorption, however, it is not suitable for molecules because highly intense light leads to immediate deterioration and structural change of molecules. Therefore, from the view point of molecular applications, efficient two-photon absorption by low-intensity light is preferable, however, this is unrealizable in conventional laser sources.

Entangled photons provide a useful solution to the above problem. The entangled photons are a photon pair generated by nonlinear optical process, e.g., parametric down-conversion, and possess the inherent simultaneity of photons originating from quantum correlation. The studies of two-photon absorption using entangled photons have begun in the 1990s, theoretically predicted independently by Gea-Banacloche [5] and Javanainen and Gould [6]. The main target of the early studies has primarily focused on the dependence of two-photon absorption rate on incident light intensity. Generally, the two-photon absorption rate R is proportional to the square of light intensity, $R \propto I^2$. However, for nonclassical light (or entangled photons), the R dependence on I becomes linear, $R \propto I$. This indicates that the two-photons are absorbed by materials like a single photon. In fact, this linear dependence of R has been experimentally demonstrated for two-photon absorptions in Cs atoms [7] and organic molecules [8], and for sum frequency generation process in a PPKTP crystal [9]. Figure 3.2 shows the two-photon transition rate R on input photon rate $n \propto I$. The

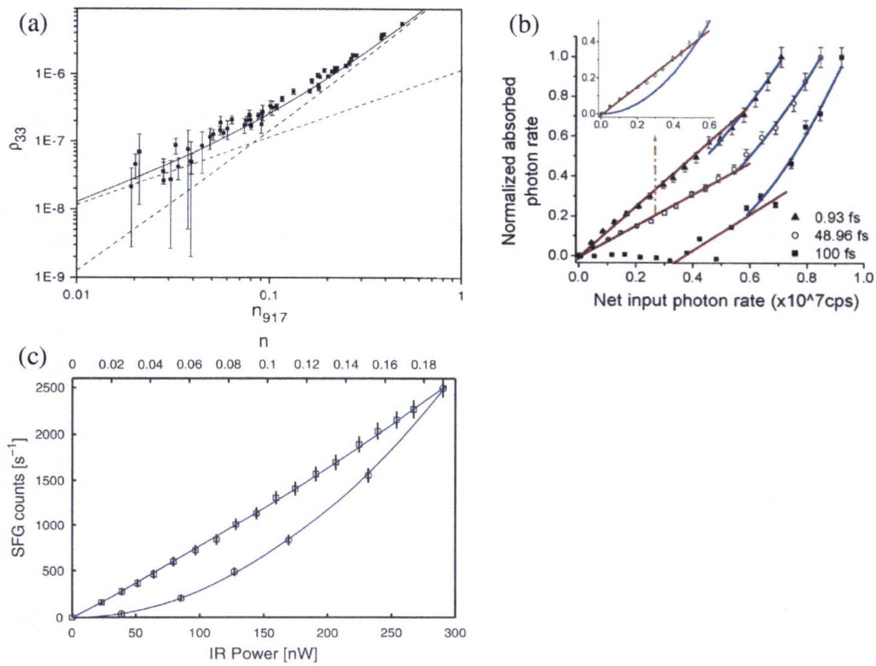

Fig. 3.2 Dependence of two-photon transition rate on input photon rate: **a** Cs atoms from Ref. [7], **b** organic molecules, reprinted with permission from Ref. [8], Copyright (2006) American Chemical Society, and **c** PPTKP crystals from Ref. [9]

linear dependence on n can be found in weak excitation region, and for lager n, the R dependence turns to quadratic. In addition, as shown in the insert of Fig. 3.2b, c, the transition rate by entangled photons (linear line) always surpasses that by classical laser light (quadratic curve). Thus, entangled photons can enhance the two-photon transition rate within the weak excitation region of small I, and hence efficient two-photon absorption can be realized under the condition of low light intensity. However, for large I, the two-photon transition rate by entangled photons becomes the same as that by classical laser. For molecular processes, highly intense light might lead to the deterioration and structural change of molecules. Therefore, from the view point of molecular applications, the two-photon absorption by entangled photons realized in low light intensity meets the purpose.

For two-step excitation by entangled photons, however, few studies have been reported so far. This is because the inherent coincidence of entangled photons is perfectly suited to the two-photon absorption process via virtual states. Intuitively, the coincidence of photons seems to be unsuitable for two-step excitation in which each of the two photons is sequentially absorbed. Within recent years, however, molecular two-step excitation utilizing entangled photons is studied with the aim of applications to molecular spectroscopy [10–12] and coherent control of molecules

[13–15]. In these applications, selective excitation of a molecular vibronic state rather than enhancement plays an important role. Since the energy sum of entangled photons is always constant even for ultrashort pulse owing to the quantum correlation with energy anticorrelation, entangled photons can achieve efficiently selective excitation of a molecular vibronic state. Thus, molecular two-step excitation utilizing entangled photons is useful for selective excitation of a molecular vibronic state.

To maximize the potential of entangled photons in both the two-photon absorption and two-step excitation processes, especially for molecular applications, ultrashort-pulsed entangled photons, in other words, ultrabroadband entangled photons are also indispensable to rapidly excite vibronic states and avoid molecular relaxation processes. In this study, we establish a theoretical framework for an efficient molecular two-photon excitation by ultrabroadband frequency-entangled photons for exactly two photons as a limiting case of low-intensity light. We show the two-photon absorption and two-step excitation of molecular vibronic states can be selectively and strongly enhanced by directly controlling the wavefunction of entangled photons.

3.2 Spatiotemporal Photon Pulse Theory for Entangled Photons

In conventional studies of two-photon absorption for molecules, the time-dependent perturbation theory is often used, in which the molecule–photon interaction is assumed to be small and external photon fields are treated as a bath so that quantum states of photons are not changed. In this study, however, we introduce the spatiotemporal photon pulse theory [16] extending input-output theory in the field of quantum optics because we consider the molecule–photon interaction in the limiting case of low-intensity light, in which exactly two photons interact with a single molecule and the wavefunction of two photons is also changed by the molecule–photon interaction.

An analytical model is depicted in Fig. 3.3, where an incident entangled-photon pulse propagates along r-axis from $r < 0$ and interacts with a molecular system located at $r = 0$. The one-dimensional input-output model can be justified by assuming the paraxial approximation. Using the natural unit of $\hbar = c = 1$, the dynamics of the entangled-photon pulse interacting with a molecule can be calculated

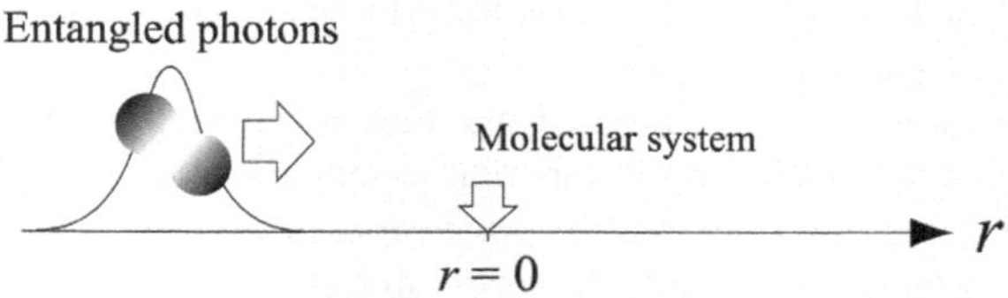

Fig. 3.3 Analytical model: one-dimensional input-output model. An incident entangled-photon pulse propagates parallel to the r-axis from $r < 0$ and locally interacts with a molecular system at $r = 0$

using the following Schrödinger equation,

$$|\psi(t)\rangle = \exp\left(-i\widehat{H}t\right)|\psi(0)\rangle,$$

where $|\psi(0)\rangle$ is the initial state at $t = 0$ and $\widehat{H}$ is the total Hamiltonian of the system. Using the photon dispersion relation of $\omega = ck = k$, the initial state $|\psi(0)\rangle$ is given by

$$|\psi(0)\rangle = \frac{1}{\sqrt{2}} \int dk \int dk' \psi_{2p}\left(k, k'\right)\hat{a}^{\dagger}(k)\hat{a}^{\dagger}(k)|0\rangle \otimes |g\rangle,$$

where $\psi_{2p}\left(k, k'\right)$ is the two-photon wavefunction and $\hat{a}^{\dagger}(k)$ is the creation operator of a photon with wavenumber k. The molecular system is assumed to be in the ground state $|g\rangle$ at $t = 0$.

To show the usefulness of entangled photons, we consider two-photon pairs as incident two-photon pulses, namely uncorrelated photons and entangled photons, for comparison. The uncorrelated photons are corresponding to classical light, given by

$$\psi_{2p}\left(k, k'\right) = \psi_{1p}(k)\psi_{1p}\left(k'\right),$$

where $\psi_{1p}(k)$ is the one-photon wavefunction. The uncorrelated photons literally have no correlation between photons and are separated into the direct product of the one-photon wavefunction, which is the characteristic of classical light. For entangled photons, we adopt two photons with energy anticorrelation, given by

$$\psi_{2p}\left(k, k'\right) = \psi_{1p}(k)\delta\left(k + k' - 2k_0\right),$$

where $\delta(\cdot)$ is the Dirac delta function and k_0 is the central wavenumber of incident photon pulse. In contrast to uncorrelated photons, the entangled photons cannot be separated into the direct product of $\psi_{1p}(k)$ and have energy anticorrelation so that the energy sum of two photons can be always $2k_0$, which is ensured by the δ function. In actual photons generated experimentally, entangled photons are not described by δ function but have a certain amount of width owing to spontaneous emission. In this study, we approximate the δ function by a Gaussian function as $\propto \exp\left[-\left(k + k' - 2k_0\right)^2/4\sigma_s^2\right]$, for simplicity. In the limiting case of $\sigma_s \rightarrow 0$, the Gaussian becomes δ function.

Figure 3.4a, b shows the two-photon joint spectra for uncorrelated and entangled photons, respectively. For ψ_{1p}, a Gaussian form with pulse width of $\sigma = k_0/20\pi$ is used and σ_s is set to $\sigma/7$. In contrast to Fig. 3.4a, ellipsoidal photon distribution can be found in Fig. 3.4b, which indicates the energy anticorrelation of two photons, conserving the total energy of $2k_0$. Figure 3.4c shows corresponding spectra of Fig. 3.4a, b. Intriguingly, one can find that two spectra are identical in spite of the presence or absence of quantum correlation. Thus, the only difference between

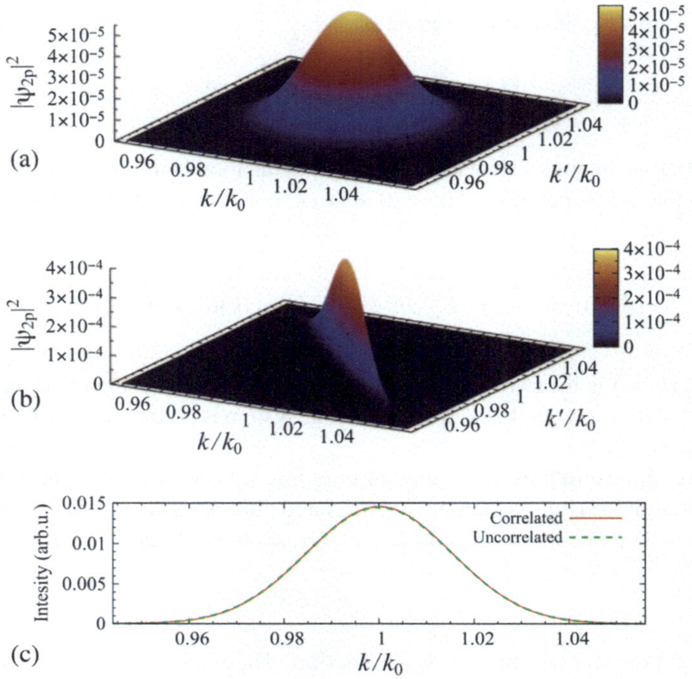

Fig. 3.4 Two-photon joint spectra: **a** uncorrelated photons, **b** entangled photons, **c** corresponding intensity spectra of (**a**) and (**b**). Reproduced from [13], with the permission of AIP Publishing

the two-photon pairs is the quantum correlation, which is distinguishable only in terms of quantum optical measurement (two-photon joint spectra), and hence the differences appearing in the calculation results in molecular two-photon absorption and two-step excitation are due to the quantum correlation.

For a molecular system, we choose a diatomic molecule with the ground state $|g\rangle$, the intermediate state $|m_\nu\rangle$, and the excited state $|e_\nu\rangle$, where ν is the mode number of molecular vibrations. For the ground state, we consider only the lowest vibrational mode because higher modes do not affect the two-photon processes at all. In addition, we approximate their adiabatic potentials as a Morse potential on the assumption of Born-Oppenheimer approximation, for simplicity. Using the natural unit of $\hbar = c = 1$, the molecular Hamiltonian can then be described as

$$\widehat{H}_{mol} = \sum_{\nu} \omega_{m_\nu} |m_\nu\rangle\langle m_\nu| + \sum_{\nu} \omega_{e_\nu} |e_\nu\rangle\langle e_\nu|,$$

where ω_m and ω_e are the eigenenergies of Morse potentials in the intermediate and excited states, respectively.

The molecule–photon interaction Hamiltonian is given by

$$\widehat{H}_{\rm int} = \int dk \sum_{\nu} \sqrt{\gamma_m F_\nu/\pi}\left(\hat{a}(k)|g\rangle\langle m_\nu| + H.c.\right)$$
$$+ \int dk \sum_{\nu,\nu'} \sqrt{\gamma_e F_{\nu\nu'}/\pi}\left(\hat{a}(k)|m_\nu\rangle\langle e_{\nu'}| + H.c.\right),$$

where F_ν and $F_{\nu\nu'}$ are the Franck–Condon factors between $|g\rangle$ and $|m_\nu\rangle$ and between $|m_\nu\rangle$ and $|e_{\nu'}\rangle$, respectively. γ_m and γ_e are the dipole relaxation rates of intermediate and excited states, respectively. The Hamiltonian for the quantized photon fields is simply given by

$$\hat{H}_{\rm photon} = \int dk \hbar k \hat{a}^\dagger(k)\hat{a}(k).$$

Consequently, the total Hamiltonian of the whole system is given by

$$\widehat{H} = \widehat{H}_{\rm mol} + \widehat{H}_{\rm photon} + \widehat{H}_{\rm int}.$$

The difference between two-photon absorption and two-step excitation is that the energy ω_m of intermediate state is either resonant or far-off-resonant to the central energy k_0 of incident photons. Therefore, we only have to modify the parameters of Morse potentials in accordance with the aim of two-photon absorption and two-step excitation.

3.3 Enhanced and Selective Two-Photon Absorption Utilizing Entangled Photons

We can now calculate the molecular dynamics driven by entangled photons. In this section, we first consider the two-phonon absorption process. Figure 3.5a shows the analytical model of molecular two-photon absorption by entangled photons, where slightly modified adiabatic potentials of $1^1\Sigma_g^+$ ($|g\rangle$), $1^1\Sigma_u^+$($|m_\nu\rangle$), and $2^1\Pi_g$ ($|e_{\nu'}\rangle$) of Na_2 are considered. The details of Morse parameters are found in Ref. [17]. The central energy k_0 of incident photons is far-off-resonant to ω_m and the sum of the central energy, $2k_0$, is resonant to the excited state with the vibrational mode of $\nu = 18$. The Franck–Condon factors of F_ν and $F_{\nu,\nu'}$ are shown in Fig. 3.5b, c. We use entangled and uncorrelated photons with pulse width of $\sigma = k_0/20\pi$ and σ_s characterizing the quantum correlation is set to $\sigma_s = \sigma/7$. The dipole relaxation rates $\gamma_m = \gamma_e = 2.5 \times 10^{-5}k_0$ are chosen, for simplicity. In addition, we assume a cold diatomic molecule and ignore vibrational phase relaxation.

Figure 3.6 shows population dynamics of excited states driven by incident entangled and uncorrelated photons, where the parameters of photons are the same for both cases expect for the quantum correlation. The horizontal axis of $r\sigma$ indicates the central position of incident photons normalized by σ^{-1}. For uncorrelated photons (Fig. 3.6a), many vibrational modes are excited as found commonly in short-pulse

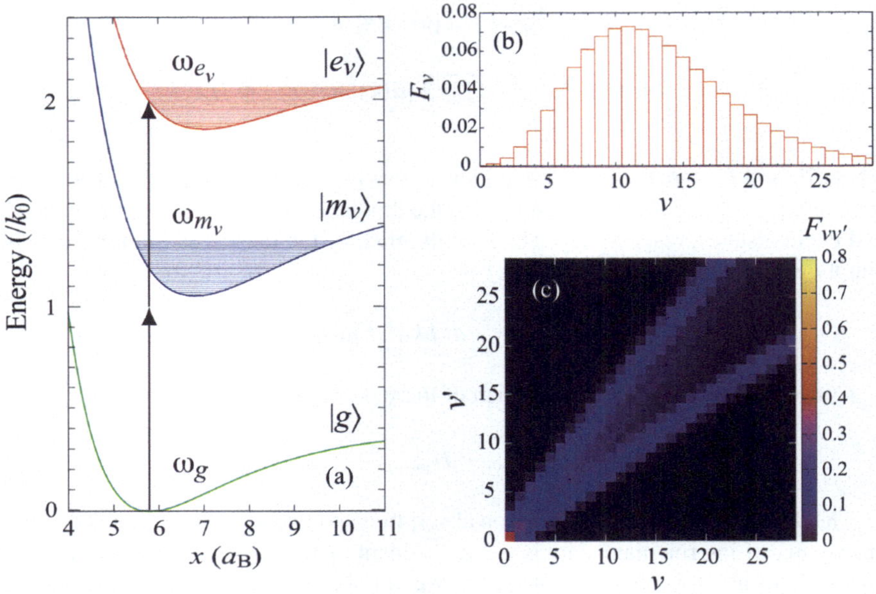

Fig. 3.5 **a** Analytical model of two-photon absorption. Franck–Condon factors: **b** for the transition between $|g\rangle$ and $|m_\nu\rangle$ and **c** for the transition between $|m_\nu\rangle$ and $|e_{\nu'}\rangle$. Reproduced from [17], with the permission of AIP Publishing

Fig. 3.6 Dynamics of excited-state population $\langle e_\nu \rangle$ for vibrational modes from $\nu = 14$ to $\nu = 22$ driven by **a** uncorrelated photons and **b** entangled photons. Reproduced from [17], with the permission of AIP Publishing

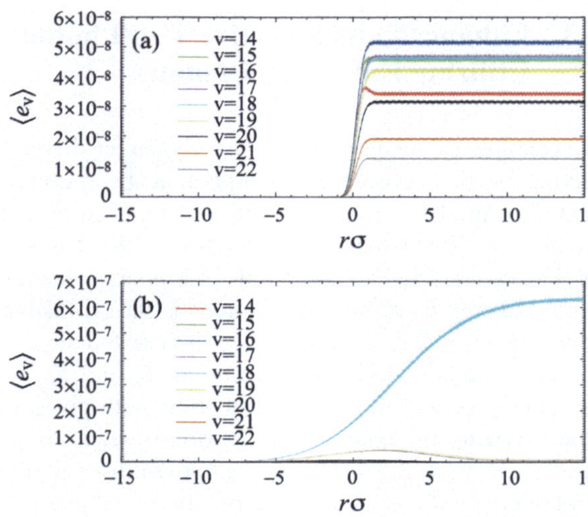

excitation of molecular vibration. Though the sum of the central energy of incident photon, $2k_0$, is resonant to the excited states with mode number $\nu = 18$, population peaks at $\nu = 16$. This is because that incident pulse is spectrally broad and the value of Franck–Condon factor for $\nu = 16$ is larger than that for $\nu = 18$. For untangled photons (Fig. 3.6b), however, the vibration mode resonant to k_0 is efficiently and selectively excited, and the excitations of other vibrational modes are strongly suppressed. The enhancement rate of the population for $\nu = 18$ is about 15 times compared to the case of uncorrelated photons and the mode selectivity $S = e_{\nu=18}/\sum_{\nu} e_{\nu}$ is about 0.97. Thus, entangled photons can concurrently achieve high selectivity and enhancement of excitation population.

By using entangled photons with broader pulse width σ and stronger quantum correlation $\sigma_s \rightarrow 0$, we can further increase the enhancement rate and mode selectivity. In fact, more than thousandfold enhancement and mode selectivity of $S \approx 0.9999$ have been theoretically predicted for atoms and molecules [17, 18] when ultrabroadband frequency-entangled photons are used. In addition, further enhancement of population excitation by utilizing plasmonic nanoantenna is also theoretically predicted [19]. Since an ultrabroadband frequency-entangled photon source is already realized [20], experimental demonstration of huge two-photon absorption of molecules might be reported in the near future.

3.4 Enhanced and Selective Two-Step Excitation Utilizing Entangled Photons

In this section, we calculate the two-step excitation of a molecule driven by entangled photons. Figure 3.7a shows the analytical model, where adiabatic potentials of $1^1\Sigma_g^+$ ($|g\rangle$), $1^1\Sigma_u^+$ ($|m_\nu\rangle$), and $2^1\Pi_g$ ($|e_{\nu'}\rangle$) of Na_2 are considered as is the case of two-photon absorption in 3.3. The details of the Morse parameters are found in Ref. [15].

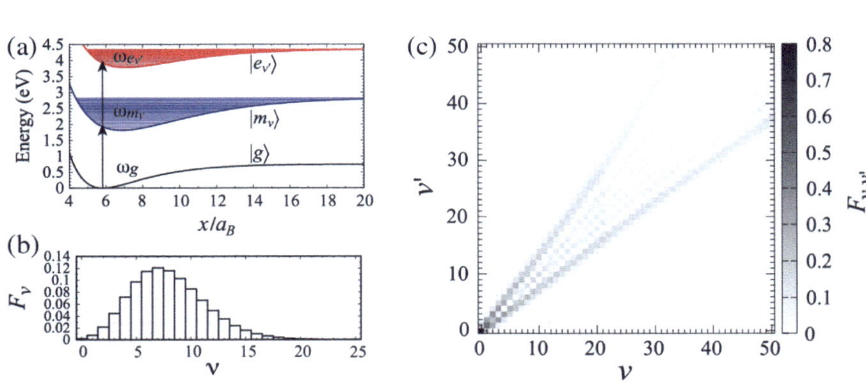

Fig. 3.7 **a** Analytical model of two-step excitation. Franck–Condon factors: **b** for the transition between $|g\rangle$ and $|m_\nu\rangle$ and **c** for the transition between $|m_\nu\rangle$ and $|e_{\nu'}\rangle$. Reproduced from Ref. [15]

The difference from Fig. 3.5a is that intermediate adiabatic potential is not modified and the central energy k_0 of incident photons is resonant to those of the intermediate vibronic states $|m_\nu\rangle$, and hence the intermediate vibrational modes are really excited. The sum of the central energy, $2k_0 \approx 4.0024$ eV, is resonant to the excited state with the vibrational mode of $\nu' = 18$ as is the case for the two-photon absorption. The Franck–Condon factors of F_ν and $F_{\nu,\nu'}$ are shown in Fig. 3.7b, c. We use entangled and uncorrelated photons with pulse width of $\sigma = 10$ THz and σ_s characterizing the quantum correlation is set to $\sigma_s = 500$ GHz. The dipole relaxation rates $\gamma_m = \gamma_e = 6$ MHz are chosen, for simplicity.

Figure 3.8 shows population dynamics of excited states driven by incident entangled and uncorrelated photons, where the parameters of photons are the same for both cases expect for the quantum correlation σ_s. The horizontal axis of $r\sigma$ indicates the central position of incident photons normalized by σ^{-1}. For uncorrelated photons (Fig. 3.8a), many vibrational modes are excited as is the case for the two-photon absorption in Fig. 3.6b. Though the sum of the central energy of incident photon, $2k_0$, is resonant to the excited states with the mode number $\nu' = 18$, the population peaks at $\nu' = 12$ in spite of the resonant excitation of the vibrational mode of $\nu' = 18$. This is because that the Franck–Condon factors $F_{\nu,\nu'=18}$ are very small for $\nu < 12$ and $F_{\nu=10,\nu'=12}$ is large, as can be seen in Fig. 3.7c. For the entangled photons (Fig. 3.8b), however, the resonant excitation of the vibrational mode of $\nu' = 18$ is selectively excited and strongly enhanced. In fact, the enhancement rate of the population for $\nu' = 18$ is about 60 times compared to the case of uncorrelated photons and the mode selectivity $S = e_{\nu'=18}/\sum_{\nu'} e_{\nu'}$ is about 0.9999 ($S = 0.0396$ for uncorrelated photons). In general, selective excitation of molecular vibronic states requires detailed Franck–Condon analysis and pulse shaping techniques. Using entangled

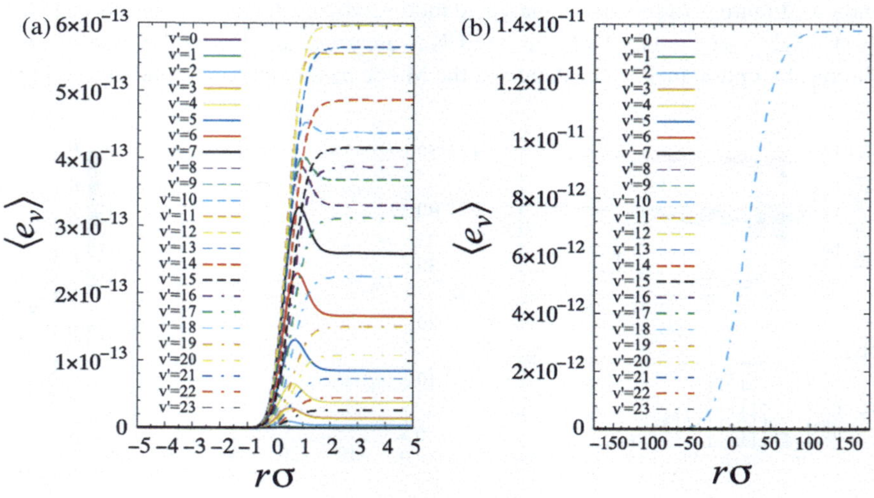

Fig. 3.8 Dynamics of excited-state population $\langle e_{\nu'} \rangle$ for vibrational modes up to $\nu' = 23$ driven by **a** uncorrelated photons and **b** entangled photons. Reproduced from Ref. [15]

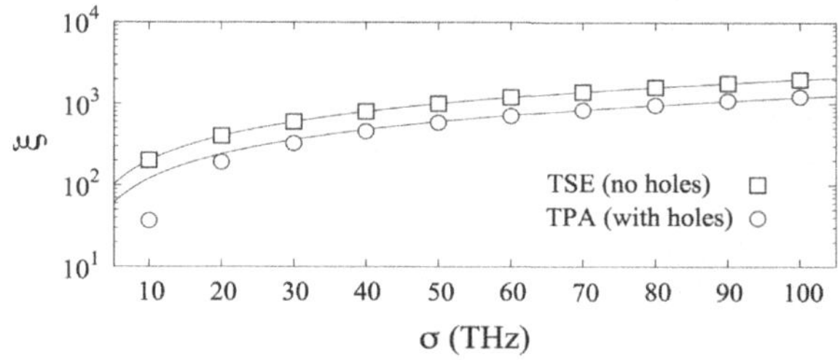

Fig. 3.9 Dependence of enhancement rate ξ on pulse width σ of incident entangled photons. Reproduced from Ref. [18]

photons, however, high selectivity and enhancement of excitation efficiency can thus be easily and concurrently achieved.

As mentioned in 3.3, the enhancement rate and mode selectivity can be further increased by using entangled photons with broader pulse width σ. Figure 3.9 shows the dependence of the enhancement rate ξ and on σ for the range from 10 to 100 THz for simple three-level molecular systems, where ζ is defined by the ratio of the population obtained from entangled photons to that obtained from uncorrelated photons. For comparison, two-photon absorption (TPA) and two-step excitation (TSE) are plotted. In both TPA and TSE, the enhancement rate ξ increases with increase in σ. For $\sigma = 100$ THz, $\xi \geq 1000$ can be achieved, especially for two-step excitation $\xi \approx 2500$. Intriguingly, the enhancement rate of TSE is always larger than TPA.

On the other hand, the mode selectivity S depends on σ_s rather than σ and increases with decrease in σ_s. In this study, however, we introduce the entanglement of formation, E, which is a well-known quantum entanglement measure, instead of σ_s to properly evaluate the dependence of S on the quantum correlation ($\sigma_s \approx 0 \Rightarrow E \approx 1$). The entanglement of formation, E, is defined as

$$E = -Tr\left[\rho' \log_{\mathrm{d}} \rho'\right] \mathrm{with} \rho' = Tr[\rho],$$

where $\rho = |\psi\rangle\langle\psi|_{\mathrm{photons}}$ is the density operator of input entangled photons. ρ' indicates the density operator partially traced for one photon, and d is the dimension of ρ. Figure 3.10 shows the dependence of the enhancement rate ζ and mode selectivity S on E for $\sigma = 50$ THz. Both ζ and S increase very gradually for $E < 0.5$, in which ζ is at most 10 and S is below 0.1. On the other hand, for $E > 0.5$, both E and S drastically increase, and in particular, ζ exceeds 1000 at $E \approx 0.83$ and S becomes nearly unity for $E \geq 0.77$. Thus, strong enhancement and high selectivity

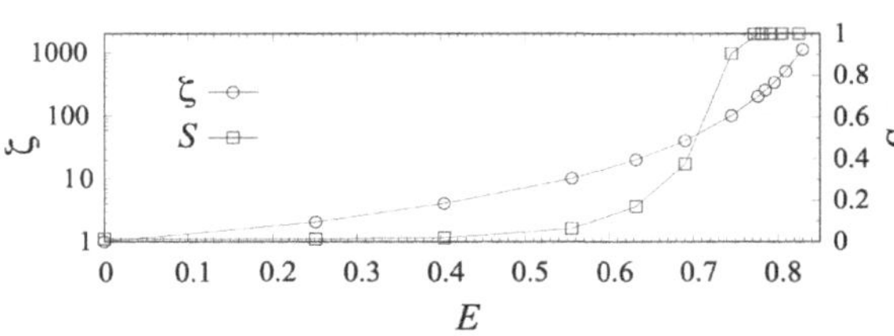

Fig. 3.10 Dependence of enhancement rate ζ and mode selectivity S on entanglement of formation E. Reproduced from Ref. [15]

are achieved for high E (in other words, small σ_s), and conversely this indicates that we cannot achieve large ζ and high S at low E even for ultrabroadband entangled photons.

For reference, entangled photons obtained from conventional parametric down-conversion exhibit $E \approx 0.55$ and therefore enhancement is at most 10 if we use the generated entangled photons directly. Our results suggest that design of entangled-photon source with high E is required in order to achieve a further strong enhancement and single vibrational-mode excitation in molecular two-photon absorption and two-step excitation.

3.5 Conclusion

We have analyzed molecular two-photon absorption and two-step excitation using entangled photons. By taking a cold diatomic molecule Na_2 as an example, we have shown that photon entanglement can strongly enhance the two-photon transition rate for both two-photon absorption (TPA) and two-step excitation (TSE) processes. The enhancement rate reaches up to more than 1000 times compared to conventional laser light under the same conditions. In particular, the TSE excitation rate for a single vibrational mode can be enhanced 2500 times by using ultrabroadband frequency-entangled photons with spectral width of $\sigma = 100$ THz. In addition, the model selectivity S of a vibrational mode nearly reaches unity for both TPA and TSE when entangled photons with strong quantum correlation (small σ_s) are used. However, if we achieve the strong enhancement and high selectivity concurrently, we have to prepare entangled photons with high quantum entanglement measure E. Entangled photons obtained from conventional parametric down-conversion exhibit $E \approx 0.55$ and our results show that the realization of strong enhancement and high selectivity requires $E \geq 0.8$. These results thus indicate that design of entangled-photon source with higher E is required in order to achieve strong enhancement and single vibrational-mode excitation in molecular two-photon absorption and two-step excitation. We hope our results facilitate the study of molecular two-photon excitation by entangled photons.

Acknowledgements The present work was supported by JSPS KAKENHI Grant Number JP17H05252, Grant-in-Aid for Scientific Research on Innovative Areas "Photosynergetics."

References

1. Göppert-Mayer M (1931) Über Elementarakte mit zwei Quantensprüngen. Ann Phys 9:273–294
2. Denk W, Strickler JH, Webb WW (1990) Two-photon laser scanning fluorescence microscopy. Science 248:73–76

3. Cumpston BH, Ananthavel SP, Barlow S, Dyer DL, Ehrlich JE, Erskine LL, Heikal AA, Kuebler SM, Lee I-YS, McCord-Maughon D, Qin J, Röckel H, Rumi M, Wu X-L, Marder SR, Perry JW (1999) Two-photon polymerization initiators for three-dimensional optical data storage and microfabrication. Nature (London) 398:51–54
4. Warren WS, Rabitz H, Dahleh M (1993) Coherent control of quantum dynamics: the dream is alive. Science 259:1581–1589
5. Gea-Banacloche J (1989) Two-photon absorption of nonclassical light. Phys Rev Lett 62:1603–1606
6. Javanainen J, Gould PL (1990) Linear intensity dependence of a two-photon transition rate. Phys Rev A 41:5088–5091
7. Georgiades NPh, Polzik ES, Edamatsu K, Kimble HJ, Parkins AS (1995) Nonclassical excitation for atoms in a squeezed vacuum. Phys Rev Lett 75:3426–3429
8. Lee D-I, Goodson T (2006) Entangled photon absorption in an organic porphyrin dendrimer. J Phys Chem B 110:25582–25585
9. Dayan B, Pe'er A, Friesem AA, Silberberg Y (2005) Nonlinear interactions with an ultrahigh flux of broadband entangled photons. Phys Rev Lett 94:043602-1-4
10. Schlawin F, Dorfman KE, Mukamel S (2016) Pump-probe spectroscopy using quantum light with two-photon coincidence detection. Phys Rev A 93:023807-1-8
11. Dorfman KE, Schlawin F, Mukamel S (2016) Nonlinear optical signals and spectroscopy with quantum light. Rev Mod Phys 88:045008-1-67
12. Schlawin F (2017) Entangled photon spectroscopy. J Phys B At Mol Opt Phys 50:203001-1-18
13. Oka H (2011) Control of vibronic excitation using quantum-correlated photons. J Chem Phys 135:164304-1-5
14. Oka H (2012) Two-photon process via internal conversion by correlated photon pair. Phys Rev A 85:013403-1-7
15. Oka H (2018) Enhanced vibrational-mode-selective two-step excitation using ultrabroadband frequency-entangled photons. Phys Rev A 97:063859-1-6
16. Oka H (2010) Real-time analysis of two-photon excitation by correlated photons: pulse-width dependence of excitation efficiency. Phys Rev A 81:053837-1-4
17. Oka H (2011) Selective two-photon excitation of a vibronic state by correlated photons. J Chem Phys 135:124313-1-6
18. Oka H (2015) Highly-efficient entangled two-photon absorption with the assistance of plasmon nanoantenna. J Phys B At Mol Opt Phys 48:115503-1-6
19. Oka H (2018) Two-photon absorption by spectrally shaped entangled photons. Phys Rev A 97:033814-1-6
20. Tanaka A, Okamoto R, Lim H-H, Subashchandran S, Okano M, Zhang L, Kang L, Chen J, Wu P, Hirohata T, Kurimura S, Takeuchi S (2012) Noncollinear parametric fluorescence by chirped quasi-phase matching for monocycle temporal entanglement. Opt Express 20:25228–25238

Chapter 4
Stepwise Two-Photon Photochromism

Jiro Abe, Yoichi Kobayashi, and Katsuya Mutoh

Abstract Stepwise two-photon absorption (2PA) is a well-known nonlinear optical phenomenon where two photons are sequentially absorbed to proceed a photophysical or photochemical reaction. The stepwise 2PA can produce a higher excited state with lower energy photons than the energy level of the excited state. Therefore, it can be an efficient tool to explore and realize nonlinear photoresponsive materials beyond Kasha's rule. Moreover, if a photogenerated chemical species is used as an intermediate state of the stepwise 2PA, the power threshold to induce the stepwise 2PA can be greatly reduced compared with that of a simultaneous 2PA. We focus on the stepwise 2PA processes related to photochromic reactions because the combination of photochromic reactions and stepwise 2PA processes is beneficial to develop unique nonlinear photoresponsive materials. We introduce several anomalous phenomena beyond one-photon photophysical and photochemical reactions.

Keywords Two-photon absorption · Nonlinear optical response · Photochromism · Imidazole dimer · Biradical · Naphthopyran

4.1 Introduction

Stark–Einstein law states that the primary act of light absorption by a molecule is a one-photon process [1]. This law can be applied in various photophysical and photochemical reactions using continuous wave (CW) light sources. However, this law would be violated when intense light sources such as laser pulses are used [2–6]. Under these conditions, nonlinear optical phenomena such as simultaneous or stepwise two- or multiple-photon absorptions may be observed. Nonlinear optical properties have several features beyond one-photon reactions, and therefore, they have

J. Abe (✉) · K. Mutoh
Department of Chemistry, School of Science and Engineering, Aoyama Gakuin University, 5-10-1 Fuchinobe, Chuo-ku, Sagamihara, Kanagawa 252-5258, Japan
e-mail: jiro_abe@chem.aoyama.ac.jp

Y. Kobayashi
Department of Applied Chemistry, College of Life Sciences, Ritsumeikan University, 1-1-1 Nojihigashi, Kusatsu, Shiga 525-8577, Japan

H. Miyasaka et al. (eds.), *Photosynergetic Responses in Molecules and Molecular Aggregates*, https://doi.org/10.1007/978-981-15-5451-3_4

been received much attention since more than a half-century ago. One of the most important nonlinear optical phenomena is a two-photon absorption (2PA) process [7, 8]. The 2PA process can be divided into two categories: simultaneous and stepwise processes. The simultaneous 2PA occurs when two photons are absorbed instantaneously via a virtual state. The simultaneous 2PA can produce the excited state with lower energy photons than the energy level of the excited state, and the probability for the excitation depends on the square of the light intensity. Therefore, the simultaneous 2PA has high spatial selectivity and can be induced deep inside of matters. Because of these characteristics, the simultaneous 2PA has been applied to biological imaging [9, 10] and three-dimensional (3D) micro-fabrications [11, 12]. However, the simultaneous 2PA requires the absorption of another photon during the interaction of the first photon and the matter (~1–2 fs) because the virtual state is not a stationary electronic state. Therefore, high-power light sources such as femtosecond and nanosecond pulse lasers are required to induce the simultaneous 2PA.

Another 2PA process is called the stepwise 2PA, which is the absorption of another photon by the photogenerated transient state. This process occurs via an actual stationary electronic state such as an excited state or a short-lived species. Therefore, the probability of the stepwise 2PA depends on the lifetime of the intermediate transient state. If the lifetime of the intermediate transient state is long enough, the power threshold to induce the stepwise 2PA is greatly reduced and the stepwise 2PA could be induced even by CW LEDs and sunlight. There are mainly three types of intermediate transient states for the stepwise 2PA reported so far: the singlet excited (S_1) state, the triplet excited (T_1) state, and the ground state of a photoproduct [13]. The first and second cases are the stepwise 2PAs via the S_1 and T_1 states to produce the higher excited singlet (S_n) and triplet (T_n) states, respectively. These two processes utilize the electronically excited states as the intermediate state of the stepwise 2PA process. The lifetimes of the S_1 states are usually nanosecond time scales, while those of T_1 states (microseconds to tens of microseconds) are usually more than a thousand times longer than those of the S_1 states in solution at room temperature due to the spin selection rule. Since the long-lived excited states have more chances to absorb another photon, the power threshold of the stepwise 2PA via the T_1 state is lower than that of the S_1 state. However, the lifetime of the T_1 state is still short to induce the stepwise 2PA with conventional CW light sources. Therefore, a pulse laser is usually necessary to promote these stepwise 2PA processes. Alternative approaches are necessary to reduce the power threshold to induce the stepwise 2PA with CW LEDs and sunlight. In the third case, the stepwise 2PA occurs via not an electronically excited state but a ground state of photogenerated transient species produced by a photochemical reaction. The lifetimes of the photogenerated transient species are usually longer than tens of nanoseconds. In some cases, photogenerated transient species can survive more than milliseconds and exist as photoproducts permanently. If we utilize these long-lived transient species as an intermediate of the stepwise 2PA, the power threshold of the stepwise 2PA can be significantly reduced.

There has been a growing interest in the development of advanced photofunctional materials whose photoresponses involve multiple photons and molecules because these materials show the photoresponses that cannot be achieved by a one-photon reaction of a single chromophore. These cooperative interactions of multiple photons and molecules are recently termed as the "photosynergetic" effects, and the understanding and utilization of these effects are becoming important research subjects. Photochromism is defined as a reversible transformation of a chemical species between two isomers having different absorption spectra induced in one or both directions by irradiation of light. If the photogenerated chemical species is thermally unstable and the back reaction to the initial structure occurs by heat (and light), this photochromism is classified as "T-type." Stepwise multiple-photon absorptions could occur when the excitation light is continuously irradiated to T-type photochromic compounds to reach the photostationary state (PSS). However, most of the photogenerated colored species of T-type photochromic compounds do not induce additional photophysical and photochemical reactions when they absorb a photon except the photodegradation. If the photogenerated colored species of photochromic compounds has a property to induce another photochemical reaction by the absorption of the second photon, this photochromic reaction becomes a reversible stepwise 2PA system [14]. The longer lifetime of the photogenerated colored species increases the probability to absorb the second photon, and therefore, reduces the power threshold of the stepwise 2PA. We have applied the fast T-type photochromic molecules to the stepwise 2PA process and succeeded in realizing the extremely low power threshold of the stepwise 2PA process, which can be initiated even by weak CW LEDs. This type of the stepwise two-photon photochromism opens up novel potential applications that cannot be achieved by conventional photofunctional materials.

4.2 Stepwise Two-Photon Photochromism of Porphyrin Substituted Bridged Imidazole Dimer Via Higher Excited States

In the electronically excited states, stepwise 2PA processes create a higher excited state. It is usually difficult to utilize higher excited states to induce photochemical reactions because of ultrafast nonradiative relaxations to the lowest excited state. However, there are several exceptions to the limitation. For example, it was reported that the photochromic reaction yields of the ring-opening reactions of diarylethene derivatives are usually at most ~1% in the one-photon process [15]. On the other hand, the ring-opening reactions are dramatically enhanced by a ten to a thousand times by the stepwise 2PA process [16]. Similar characteristic reactions by stepwise 2PA processes have been reported in fulgide, retinal, and cyanine dyes [17]. However, stepwise two-photon induced photochemical reactions via the higher excited states were reported in limited molecular frameworks within a single chromophore. To investigate the reaction from higher excited states between two chromophores, we synthesized a zinc tetraphenylporphyrin (**ZnTPP**) substituted bridged imidazole dimer (**ImD**, **ImD-ZnTPP**, Fig. 4.1) [18]. Bridged imidazole dimer is one

Fig. 4.1 Photochromism of **ImD-ZnTPP**

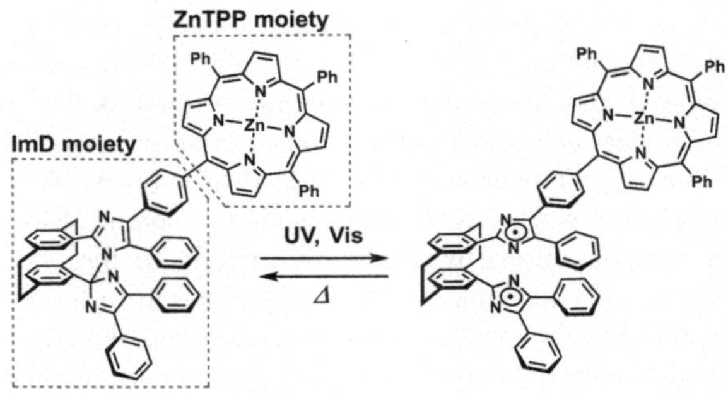

of the T-type photochromic compounds which generates an imidazolyl radical pair (biradical) upon UV light irradiation. The generated biradical quickly reverts to the initial closed form with a half-life of 33 ms in benzene solution at 298 K [19]. This fast photochromic property is promising for ophthalmic lenses, fluorescence switching [20–22], and dynamic holography [23–25]. The light intensity-dependent fast photochromic property enables to develop novel photofunctional materials such as a smart optical filter which only blocks the intense light for sensitive detectors and human eyes and a saturable absorber with reduced power thresholds. Moreover, the stepwise 2PA property will largely improve the diffraction efficiency of hologram. Therefore, the investigation of the stepwise two-photon induced photochromic reaction of **ImD-ZnTPP** is not only important for fundamental interest but also for light intensity-gated fast photochromic applications.

The steady-state absorption spectrum of **ImD-ZnTPP** is almost the superposition of those of individual **ImD** and **ZnTPP**, indicating that two chromophores do not have the electronic interaction in the ground state. At 20 ms after irradiation with a 425-nm nanosecond laser pulse, **ImD-ZnTPP** in benzene solution produces the broad absorption band over the visible light region assigned to the biradical form of the **ImD** moiety (Fig. 4.2a). The generated biradical form reverts to the initial **ImD** form with a half-life of 40 ms, which is almost identical to that of individual **ImD** (33 ms) [19]. The photochromic reaction of **ImD-ZnTPP** can be induced even with a 550–600-nm nanosecond laser pulse although the photochromic reaction of **ImD** cannot be caused by visible light irradiation. It indicates that the **ZnTPP** unit acts as a sensitizer to cause the photochromic reaction of the **ImD** unit.

The detail of the visible sensitized photochromism of **ImD-ZnTPP** was revealed by the transient absorption measurements with a 532-nm picosecond laser pulse as excitation light. Figure 4.2b shows the normalized transient absorption spectra of **ImD-ZnTPP** in benzene at 300 ps after the excitation with a picosecond 532-nm laser pulse under weak and intense excitation conditions (0.06 mJ mm^{-2} and 1.2 mJ mm^{-2}, respectively). The transient absorption spectrum of **ImD-ZnTPP** under the weak excitation condition is almost identical to that of **ZnTPP**, which is assigned to the superposition of the $S_n \leftarrow S_1$ absorption and the ground state bleaching. The spectral shape evolves in several nanoseconds time regions, and a new absorption band appeared at 840 nm, which is assigned to the T_1 state of the **ZnTPP**

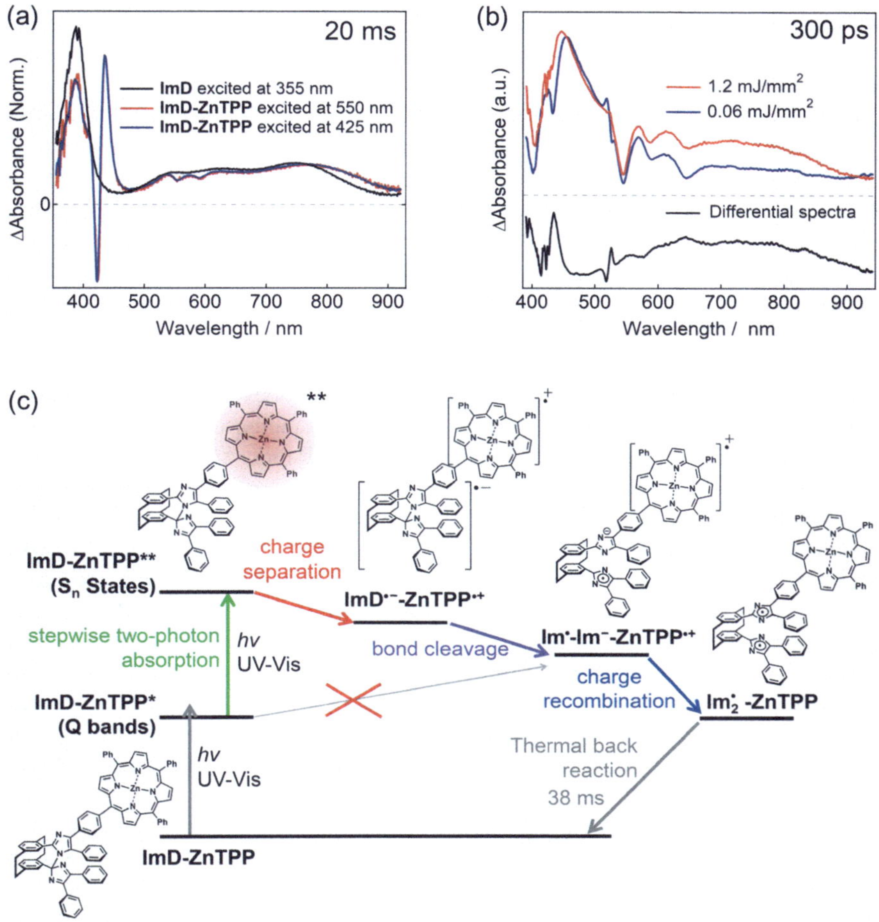

Fig. 4.2 **a** Normalized transient absorption spectra of **ImD** in benzene excited at 355 nm of the nanosecond laser pulse, and that of **ImD-ZnTPP** in benzene excited at 425 and 550 nm of the nanosecond laser pulse at 20 ms. **b** Normalized transient absorption spectra of **ImD-ZnTPP** excited at weak (blue line) and intense (red line) pump powers with a 532-nm picosecond laser pulse at 300 ps and the differential spectrum between the normalized spectra (bottom). **c** Plausible photochromic reaction pathway of **ImD-ZnTPP** upon excitation with visible light [18]

moiety. The time constant of the spectral evolution is 2.0 ns, which is almost the same as the fluorescence lifetime of free **ZnTPP**. Therefore, no effective quenching of the **ZnTPP** moiety took place in **ImD-ZnTPP** in the fluorescent state, and no photochromic reactions occured under the weak visible light excitation.

On the other hand, the transient absorption spectrum of **ImD-ZnTPP** in benzene under the intense excitation condition (1.2 mJ mm^{-2}) is quite different from that under the weak excitation condition (Fig. 4.2b). The differential spectrum between the normalized spectra at 300 ps is very similar to the transient absorption spectrum excited at 2 ms after 550-nm nanosecond laser pulse (the bottom spectrum of Fig. 4.2b), which is assigned to the biradical form of the **ImD** unit. Therefore, this result indicates that the photochromic reaction is induced by the visible light depending on the excitation intensity. The signal of the generated biradical form shows the quadratic dependence on the excitation intensity, indicating that the photochromic reaction is induced by the 2PA process. Because the photon flux density is small for the simultaneous 2PA process, this nonlinear process is most probably due to the stepwise 2PA process. Further investigations by transient absorption, electrochemical, and emission decay measurements revealed that the whole photochromic reaction pathway is written as Fig. 4.2c.

When the 550-nm light is irradiated to **ImD-ZnTPP**, the S_1 state of the **ZnTPP** unit (Q band) is formed, and the photochromic reaction does not occur from the S_1 state of the **ZnTPP** unit. When the excitation intensity increases, the stepwise 2PA occurs to produce the higher excited state of the **ZnTPP** unit. After the formation of the higher excited state of the **ZnTPP** unit, the electron is transferred to the **ImD** unit with ultrafast time scale. The formation of the charge transfer state was confirmed by detecting the cation radical of **ZnTPP**. Once the electron is injected to **ImD**, the **ImD** anion spontaneously cleaves the C–N bond between two imidazole rings, and an imidazole radical anion and an imidazolyl radical are generated [26]. The back electron transfer occurs with a time scale of 30 ps, and finally, the biradical of **ImD** is formed. The generated biradical form reverts to the initial closed form with a half-life of 40 ms. By estimating the energy level of the charge separated state between the **ZnTPP** and **ImD** units, it was revealed that the energy level of the charge separated state is lower than the higher excited state of the **ZnTPP** unit attained by the stepwise 2PA process. Thus, the charge transfer from the higher excited state is a trigger to induce the further photochromic reaction. More recently, it was demonstrated that the electron transfer occurs from higher excited states even between noncovalently bound two chromophores [27]. The stepwise 2PA process is becoming more important process to selectively produce the higher excited state and induce the advanced photochemical reactions beyond the one-photon process.

4.3 Stepwise Two-Photon Photochromism of Imidazole-Based Biphotochromic Systems

The power threshold of the stepwise 2PA process depends on the lifetime of the intermediate transient state. In organic molecular systems, because the lifetime of the T_1 state (~microseconds) is usually much longer than that of the S_1 state (~nanoseconds), the power threshold for the stepwise 2PA process can be reduced if the T_1 state is used for the intermediate state of the stepwise 2PA process instead of the S_1 state. However, it is still difficult to induce the stepwise 2PA process by weak

CW LEDs and sunlight. To further reduce the power threshold, one of the solutions is to utilize reversible stepwise photochemical reactions: photochromism [28]. T-type photochromic reactions generate a colored form upon light irradiation, and the generated colored form thermally reverts to the initial form. Because the lifetimes of the colored forms are usually much longer than microseconds, the power threshold of the stepwise 2PA process can be greatly reduced if the photogenerated colored formed are used for the intermediated state of the stepwise 2PA process. The simple molecular design to combine photochromic reactions and stepwise 2PA processes is to incorporate two photochromic units in a molecule. In this strategy, the key feature is the efficient mutual interaction between the two photochromic units to achieve the distinct photoresponses depending on the excitation intensity. Although a number of multi-switchable photochromic systems have been reported in azobenzenes, naphthopyrans, and diarylethenes, the electronic structures of the most of these derivatives were merely the superposition of the two photochromic units [29]. Besides, the switching of one photochromic unit sometimes hindered the other photochromic reaction due to the energy transfer between the two photochromic units.

Based on this idea, we developed a biphotochromic compound (**bisImD**) based on the **ImD** framework (Fig. 4.3) [30]. It is expected that **bisImD** undergoes the C–N bond cleavages upon UV light irradiation to produce two colored species: the biradical and the quinoidal forms. Because the quinoidal form, which is generated by the stepwise 2PA, has the framework of 1,4-bis-(4,5-diphenylimidazol-2-ylidene)-cyclohexa-2,5-diene (**BDPI-2Y**) [31, 32], the efficient electronic interaction is expected between the imidazolyl radicals arranged in the *para*-position of the central phenylene ring. On the other hand, the biradical form, which is generated by the one-photon absorption, has the electronic interaction originated from the face-to-face arrangement of the imidazolyl radicals, indicating that the biradical and the quinoidal forms give significantly different absorption spectra.

At millisecond time scales after the excitation with a 355-nm nanosecond laser pulse, **bisImD** shows the very similar transient absorption spectrum to that of **ImD**, which is ascribable to the biradical form of the **ImD** unit. After the decay of the

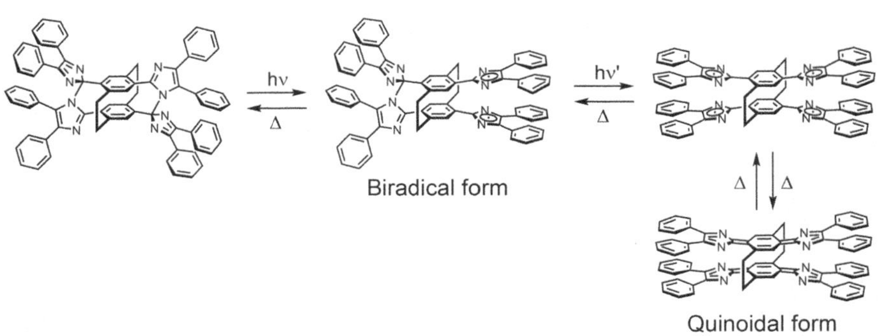

Fig. 4.3 Photochromic reaction scheme of **bisImD**

broad absorption band due to the biradical form with a half-life of 90 ms, a small absorption band remained at ~600 nm. The absorption band gradually increases with prolonged irradiation of the UV light. The spectral similarity of the absorption spectrum at around 600 nm and the Fourier-transformed infrared (FT-IR) spectrum between the photogenerated species and **BDPI-2Y** suggests that the quinoidal form was generated through the photogenerated biradical form in a stepwise manner. The absorption band of the quinoidal form gradually decreases and fades out completely within four days at room temperature.

To demonstrate the stepwise photochromic reaction via the photogenerated biradical form of **bisImD**, we conducted the nanosecond two-pulse laser flash photolysis measurements. Because the closed form of **bisImD** is not sensitive to visible light, the excitation wavelength of the second laser pulse was set to 420 nm. The transient absorption spectra after the two-pulse excitation with the delay time of 1 μs are shown in Fig. 4.4a. While the transient absorption spectrum excited only by a 355-

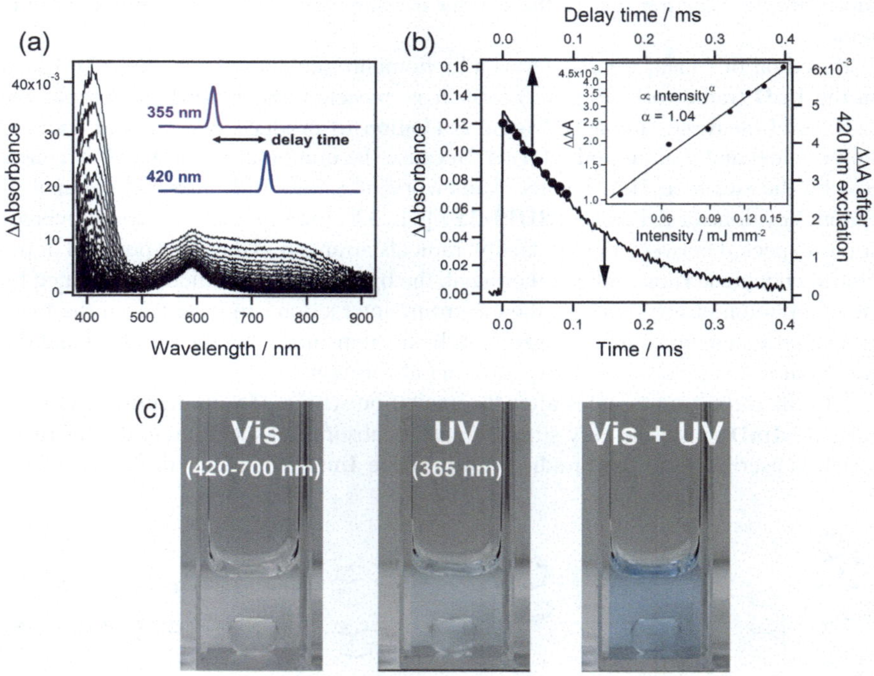

Fig. 4.4 **a** Transient absorption spectra of **bisImD** in degassed benzene after the two-pulse laser excitation (1.2×10^{-5} M, delay time; 1 μs). The excitation wavelength and intensity of two pulses were 355 and 420 nm, and 1 and 5.5 mJ/pulse. Each of spectra was recorded at 20 ms intervals. **b** ΔΔA values estimated by changing the delay time of the second 420 nm laser excitation to **bisImD** in degassed benzene, (line) the decay profile of the biradical form monitored at 400 nm, (dots) the ΔΔA values at 590 nm after the 420 nm laser excitation (1.2×10^{-5} M; delay time; 1 μs–90 ms). Inset shows the dependence of the ΔΔA values at 590 nm on the intensity of the 420 nm laser pulse. **c** Color changes of the solution of **bisImD** (293 K, 9.4×10^{-5} M) after CW light irradiation ($\lambda_{ex.} = 420$–700 nm, 500 mW; $\lambda_{ex.} = 365$ nm, 5 mW) [30]

nm nanosecond laser pulse (1 mJ/pulse) gives a small amount of the absorption band ascribable to the quinoidal form, the transient absorption spectrum excited by both a 355 and a 420-nm nanosecond laser pulses (1 and 5.5 mJ/pulse) shows the clear absorption band assigned to the quinoidal form. It clearly shows that the excitation of the biradical form of **bisImD** with a 420-nm laser pulse amplifies the formation of the quinoidal form. In contrast, the quinoidal form was not detected when the wavelength of the second laser pulse was 550 nm or 650 nm, indicating that the photocleavage reaction of the C–N bond would proceed from the higher excited state of the biradical form. We define the $\Delta\Delta A$ as the difference in the Δabsorbance values at 600 nm after the first 355 nm and the second 420 nm excitation, ΔA_{355} and ΔA_{420}, respectively ($\Delta\Delta A = \Delta A_{420} - \Delta A_{355}$). This value corresponds to the population of the quinoidal form generated by the 420 nm excitation. The $\Delta\Delta A$ decreases depending on the concentration of the biradical form with the increase in the time interval between the two excitation pulses (dot in Fig. 4.4b). The decay of the $\Delta\Delta A$ values shows good agreement with the thermal back reaction profile of the biradical form. In addition, the $\Delta\Delta A$ values linearly increase with the increase in the excitation intensity of the 420-nm laser pulse (slope = 1.04, the inset of Fig. 4.4b). Therefore, the stepwise 2PA process is involved in the photochromic reaction of **bisImD**.

This two-photon photochromic reaction can be initiated by incoherent and weak light sources (Fig. 4.4c). While a single weak incoherent light (5 mW and 355-nm light or a halogen lamp, 420–700 nm as white light) does not cause the two-photon induced photochromism, the irradiation of both lights at the same time clearly enhances the reaction in spite that they are weak and incoherent light sources. It clearly shows that the power threshold for the stepwise photochromic reaction is extremely low (~mW/cm^2) as compared to those of conventional nonlinear optical processes (>MW/cm^2). The efficient stepwise photochromic reaction of **ImD** was realized based on the two key features: (i) the short-lived transient species of T-type photochromic compounds as an intermediate state for the stepwise two-photon photochemical reactions and (ii) the strong electronic coupling between the short-lived photogenerated transient chromophores.

The concept of the stepwise photochromic reactions of **bisImD** can be expanded to other radical-dissociation type photochromic compounds such as pentaarylbi-imidazole (**PABI**) [33] and phenoxyl-imidazolyl radical complex (**PIC**) [34, 35], which are recently developed radical-dissociation type photochromic compounds. It is important to note that the half-lives of the thermal back reactions of **PIC** can be tuned from tens of nanoseconds to tens of seconds by slight modifications of the molecular framework. Especially, the substitution of the phenyl ring at the 2-position of the imidazole ring to the thiophene ring (**TPIC**) dramatically decelerates the thermal back reaction most probably due to the donor effect of the thiophene ring. Based on **PIC** and **TPIC** frameworks, we designed and synthesized stepwise two-photon induced photochromic compound derivatives: ***p*-bisPIC**, ***m*-bisPIC**, and bis(thiophene-coupled **PIC**) (**bisTPIC**) (Fig. 4.5) [36]. The photogenerated biradical species of **PIC** has a through-bond spin-spin interaction between the imidazolyl and phenoxyl radicals at the *ortho*-position of the phenylene ring. In the similar manner,

Fig. 4.5 Molecular structures of **bisPIC** derivatives

the one-photon reaction of **bisPIC** derivatives generate the biradical species which has a through-bond interaction at the *ortho*-position of the phenylene moiety. On the other hand, the quinoidal form generated by the stepwise two-photon induced reaction has both interactions between the radicals at *ortho*- and *para*-positions of the phenylene moiety. Therefore, the biradical and quinoidal forms show significantly different colors and the rate of the thermal back reaction.

BisPIC derivatives show the similar stepwise two-photon photochromic behaviors to that of **bisImD**, although the half-lives of their biradical and quinoidal forms are quite different from those of **bisImD**. That is, under the weak excitation condition, the one-photon induced photochromic reaction proceeds and generates a biradical form. On the other hand, under the intense excitation condition, the stepwise photochromic reaction proceeds from the biradical form and generates a tetetraradical form. The generated tetraradical form quickly equilibrates toward the quinoidal form. For example, in **bisTPIC**, under the weak excitation condition with a 355-nm nanosecond laser pulse (0.05 mJ mm^{-2}), a broad transient absorption spectrum assigned to the biradical form was observed (black line of Fig. 4.6a). On the other hand, the increase in the excitation intensity (2.23 mJ mm^{-2}) gives another spectral component at 520 nm most probably assigned to the quinoidal form. While the half-life of the thermal back reaction of the biradical form of **bisTPIC** is 3.2 ms, that of the light intensity-dependent component is 0.99 s (Fig. 4.6b). The half-lives of the biradical and quinoidal forms are tabulated in Table 4.1.

The mechanism of the stepwise two-photon photochromic reaction of **TPIC** was further revealed by time-resolved FT-IR absorption spectroscopy with a 355-nm nanosecond laser pulse. Since the carbonyl group of the phenoxyl moiety gives the characteristic and intense signal, this signal can be used as a marker band for identifying the biradical and quinoidal forms. Under the weak excitation condition (0.5 mJ/pulse), a characteristic intense peak was observed at 1560 cm^{-1} ascribable to the C–O stretching vibrational mode. The signal monotonically decays in millisecond time scales. On the other hand, under the intense excitation condition (4.0 mJ/pulse), the initial intense peak assigned to the C–O stretching vibrational mode was quickly converted to the other peak at 1618 cm^{-1}. This high frequency-shifted peak was assigned to the quinoidal form generated by the stepwise photochromic reaction. The tetraradical species was difficult to detect in these systems due to the low concentration. While the lifetimes of the biradical and quinoidal forms of bisTPIC are

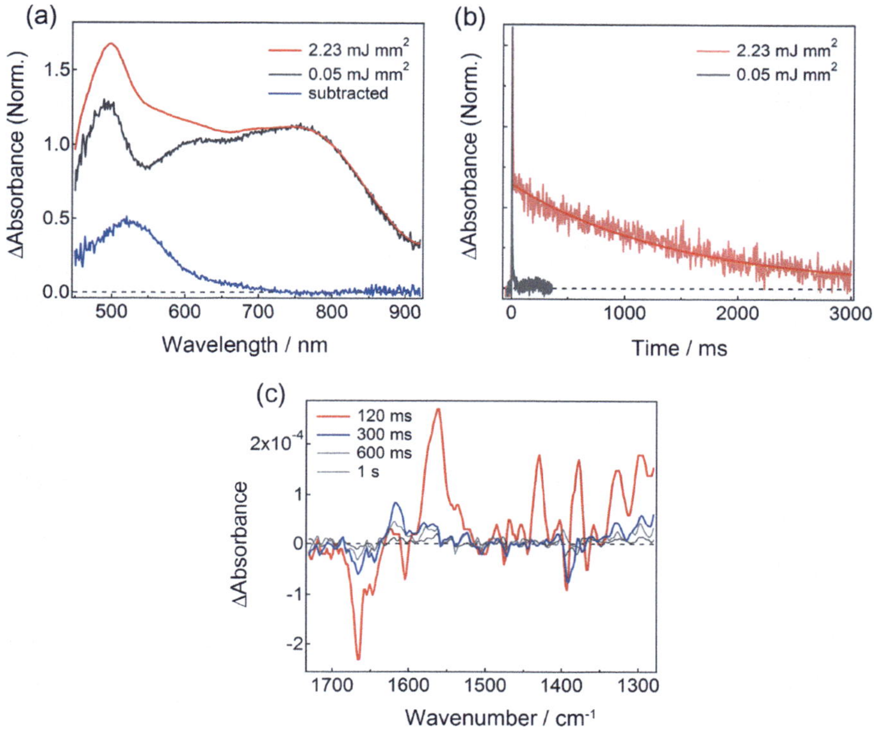

Fig. 4.6 **a** Transient absorption spectra and **b** dynamics of **bisTPIC** in benzene at different excitation intensity of the 355-nm nanosecond laser pulse. **c** Time evolution of the time-resolved FT-IR spectra of **bisTPIC** in CD_2Cl_2 after irradiation of intense laser pulse (4.0 mJ/pulse) [36]

Table 4.1 Half-lives of the transient species of PIC and **bisPIC** derivatives in benzene at 298 K

	PIC	*p*-bisPIC (μs)	*m*-bisPIC (μs)	bisTPIC
BR	0.25 μs	0.40	0.65	3.2 ms
Q	–	8.0	3.6	0.99 s

much shorter than those of bisImD, the stepwise two-photon photochromism of **bisTPIC** can be induced by CW light sources. Therefore, these fast stepwise two-photon photochromic reactions are promising for practical fast-responsive nonlinear photoresponsive materials with the reduced power threshold.

4.4 Stepwise Negative Photochromism of Binaphthyl-Bridged Phenoxyl-Imidazolyl Radical Complex

While remarkable examples of photochromic molecules in various research fields have been reported, UV light has been frequently used as a switching light for artificial photochromic molecules. In contrast, the utilization of visible light has been expected to regulate biochemical functions and functional materials due to the high energy of UV light which causes degradation and mutagenesis of cells. Therefore, the development of visible or near-infrared (NIR) light responsive molecules has been extensively reported [37–43]. However, the increase in the photosensitivity to visible and NIR light causes an unexpected photochemical reaction by weak illuminating or background light sources because typical photochemical reactions are the linear photoresponsive process on excitation light intensity. Specific nonlinear or orthogonal reactions depending on light intensity or wavelength are of particular importance to activate photoswitches in complex systems involving intermolecular networks of photofunctional molecules [44–47]. Thus, the development of visible light-responsive photochromic molecules showing excitation intensity dependence is an important subject for applications of photochromic molecules to advanced photoswitches in systems chemistry.

From this circumstance, we designed and synthesized a stepwise negative photochromic molecule (**bisBN-PIC**, Fig. 4.7) consisting of two negative photochromic units in a molecule. The binaphthyl-bridged phenoxyl-imidazolyl radical complex (**BN-PIC**) is one of the fast-switchable negative photochromic molecules in which the most thermally stable colored form (C) photoisomerizes to the metastable colorless form (CL) upon visible light irradiation [48, 49]. Because the electronic coupling between the two fast photochromic units is effective to realize nonlinear stepwise photochemical reactions, the combination of a couple of fast-switchable negative photochromic molecules will be a basic idea to design organic molecules for the selective photochemical response to incident visible light condition.

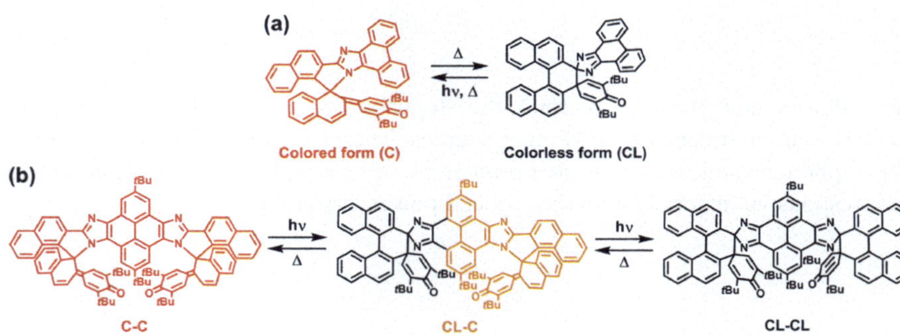

Fig. 4.7 Negative photochromic reactions of **a BN-PIC** and **b bisBN-PIC**

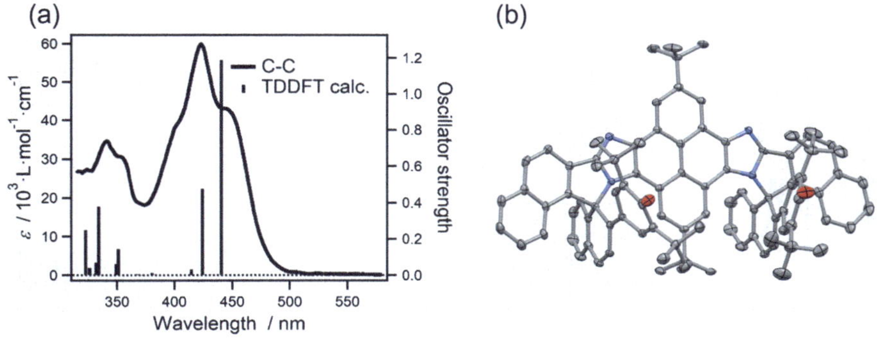

Fig. 4.8 **a** UV-vis absorption spectrum of **C-C** and **b** molecular structure of **C-C** obtained by the X-ray crystallography

The molecular structure of **bisBN-PIC** revealed by X-ray crystallography and the absorption spectrum in benzene are shown in Fig. 4.8. The absorption band at 450 nm is assigned to the π–π* transition on the two cyclohexadienone units by the time-dependent density functional theory (TDDFT) calculation (MPW1PW91/6-31G(d) level of the theory). In contrast, the absorption band at 424 nm is attributable to the charge transfer transition from the electron-donating pyrenyl unit to the bridging naphthalene unit, suggesting the effective conjugation between the pyrenyl unit and the two **BN-PIC** units in the most stable **C-C** state due to the planar structure.

The two-photon induced stepwise negative photochromic reaction was investigated by UV-vis absorption spectroscopy in detail. The absorption spectrum shows stepwise changes upon 405-nm CW laser irradiation (Fig. 4.9). The absorption band of **C-C** in the visible region initially decreases accompanied by the shift of the absorption maximum from 422 to 445 nm upon intense 405-nm light irradiation. The intensity of the absorption band further decreased with the two isosbestic points at 336 and 362 nm by subsequent CW light irradiation. Finally, the solution became colorless at the PSS. The thermal recovery process of the absorption spectrum after 405-nm light irradiation was proceeded in a stepwise manner (Fig. 4.9b). The global analysis for this stepwise spectral change revealed that this thermal recovery process is well described by a sequential model with three components. The isolated evolution associated spectra are in good agreement with the TDDFT calculated spectra for **C-C**, **CL-C,** and **CL-CL** (Fig. 4.9c). From the Eyring analysis of the rate constants for the stepwise thermal back reactions, the activation-free energy barriers ($\Delta G^{\ddagger}$) of the thermal back reaction process for **CL-C** to **C-C** and that for **CL-CL** to **CL-C** were estimated to be 75.4 kJ/mol and 71.3 kJ/mol, respectively.

The excitation light intensity dependence of the stepwise negative photochromism was shown in Fig. 4.10. The weak laser irradiation (less than 0.1 mW) causes only the one-photon reaction from **C-C** to **CL-C**, indicating that the concentration of **CL-CL** is not accumulated because of the fast thermal back reaction of **CL-CL**. In contrast, the photoisomerization to **CL-CL** was observed upon intense laser irradiation, and the conversion ratio to **CL-CL** reached almost 100% by irradiation with 260-mW

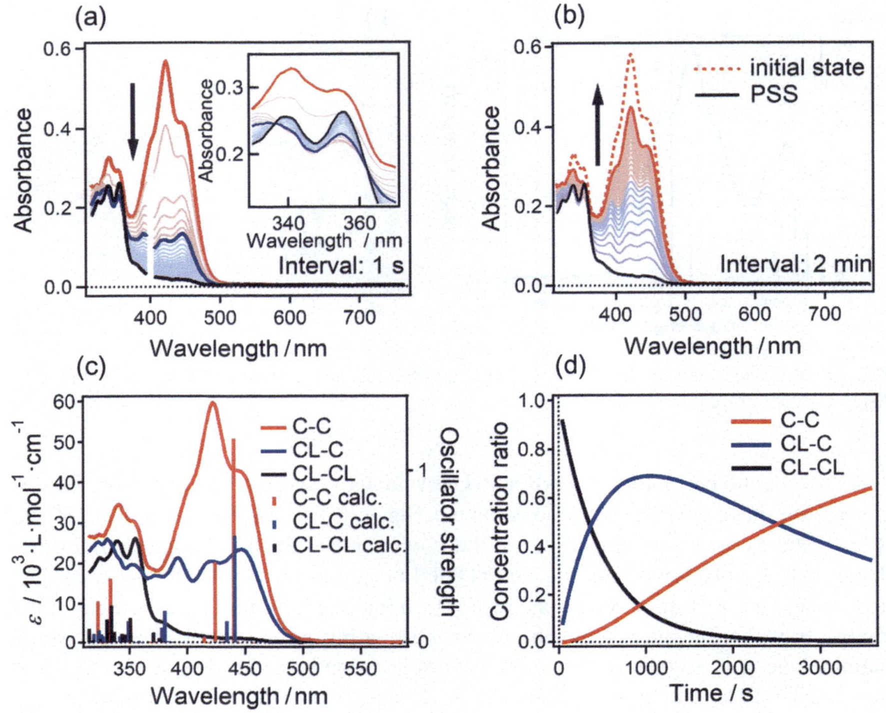

Fig. 4.9 Temporal changes of the UV-vis absorption spectrum of **bisBN-PIC** in degassed toluene (1.0×10^{-5} M) at 245 K, **a** recorded under irradiation with a 405 nm CW laser (260 mW), and **b** recorded after turning off the irradiation. **c** UV-vis absorption spectra of **C-C**, **CL-C,** and **CL-CL**. The calculated spectra (MPW1PW91/6-31G(d) level of the theory) are shown by the perpendicular lines. **d** Time evolutions of the concentrations of **C-C**, **CL-C,** and **CL-CL** obtained from the global analysis

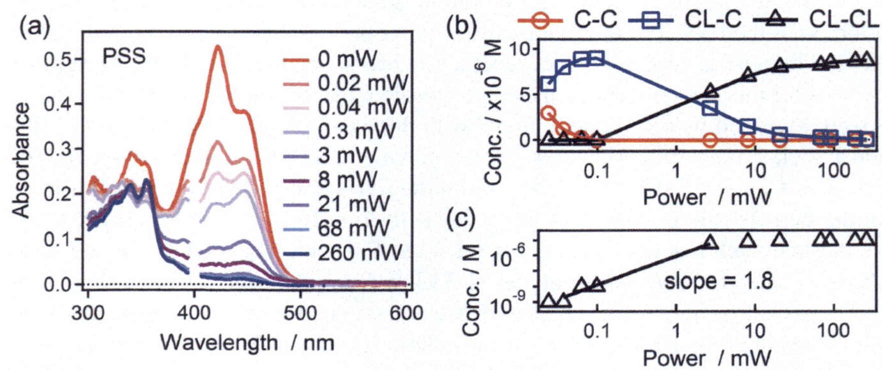

Fig. 4.10 **a** Dependence of the UV-vis absorption spectra of **bisBN-PIC** (in Ar-bubbled toluene (9.2×10^{-6} M), at 245 K) at the PSS on the incident CW laser intensity (0–260 mW, 405 nm). **b** Concentrations of **C-C**, **CL-C,** and **CL-CL** at the PSS. **c** Concentrations of **CL-CL** at the PSS

laser. The slope of the concentration of **CL-CL** against the incident laser power was estimated to be 1.8. This indicates that **bisBN-PIC** generates colorless **CL-CL** in a stepwise manner and shows the nonlinear photoresponse to the incident light intensity (Fig. 4.10c).

In conclusion of this section, we demonstrated the stepwise negative photochromism of **bisBN-PIC** which consists of electronically coupled two fast-switchable negative photochromic units. **bisBN-PIC** nonlinearly responds to incident light intensity. The two transient species **C-C** (initial state) and **CL-C** (one-photon product) have absorption bands at around 450 nm which is attributable to the π–π^* transition on the cyclohexadienone unit of **C**. In contrast, **CL-CL** generated by the two-photon negative photochromic reaction has no absorption band in the visible light region. Because the thermal back reaction rate from **CL-CL** to **CL-C** is 10-times faster than that from **CL-C** to **C-C**, the concentration of **CL-CL** is not accumulated upon weak light irradiation, leading to the excitation intensity-dependent photochromism in which only the intense light irradiation makes the solution of **bisBN-PIC** transparent. This visible light intensity-dependent negative photochromism will give an important strategy to design the organic molecule applicable to nonlinear optical materials such as saturable absorber.

4.5 Stepwise Photochromism of Bisnaphthopyrans Exhibiting Excitation Intensity-Dependent Color Change

Our strategy to realize stepwise photochromic reactions depending on the excitation intensity is not limited to the radical complex system. Naphthopyran derivatives are an important class of T-type photochromic molecules for practical applications such as photochromic lenses because of their high colorability and tunable spectrokinetic properties [50–54]. The pyran ring of naphthopyran shows the ring-opening reaction upon UV light irradiation to form the two colored open-ring isomers, *transoid-cis* (TC) and *transoid-trans* (TT) forms (Fig. 4.11). Although the slow thermal back reaction of the TT form was one of the issues to be solved for real-time switching, we have recently succeeded to decrease the generation amount of the TT form. The 8*H*-pyrano[3,2-*f*]quinazoline (8*H*-PQ) reported by Pozzo and co-workers is an attractive molecular framework to accelerate the thermal back reaction of the TC form [55]. We recently demonstrated that the introduction of an alkoxy group to the 1-position of 8*H*-PQ effectively reduces the generation amount of the undesired long-lived TT form because of the C–H···O intramolecular hydrogen bond in the TC form [56].

In this study, we developed a bisnaphthopyran system showing excitation light intensity-dependent photochromism. We designed and synthesized bisnaphthopyran derivatives bridged by a dihydroanthracene moiety, **bis-NP,** and **bis-PQ**, which are composed of 3*H*-naphtho[2,1-*b*]pyran (3*H*-NP) and 8*H*-PQ units, respectively

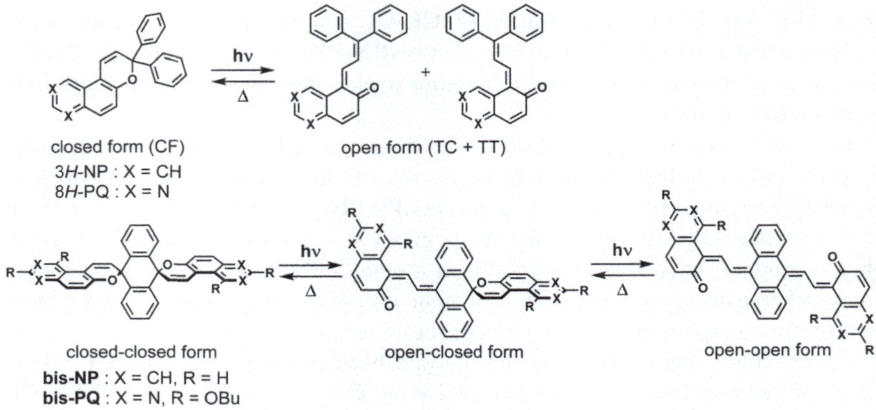

Fig. 4.11 Photochromic reaction schemes of 3*H*-NP, 8*H*-PQ, **bis-NP,** and **bis-PQ**

(Fig. 4.11). The two pyran units are perpendicularly connected to the central dihydroanthracene moiety through a spiro carbon of each pyran rings. Because the two pyran units are electronically separated in the closed-closed form and the closed-open form, the energy transfer from the closed form to the open form can be negligible. Moreover, the substitution of a *n*-butoxy group to the 1-position of the *8H*-PQ units can increase the thermal decoloration rate of the transient open-ring form and reduces the generation ratio of the long-lived TT form [56]. Therefore, **bis-PQ** is a potential candidate to realize fast-switchable stepwise photochromism by controlling excitation light conditions.

The X-ray crystallographic analysis of the crystal of **bis-NP** revealed that the two pyran units are arranged perpendicularly against the central dihydroanthracene unit. Figure 4.12 shows the UV-vis absorption spectra of **bis-NP**, **bis-PQ,** and 3*H*-NP in

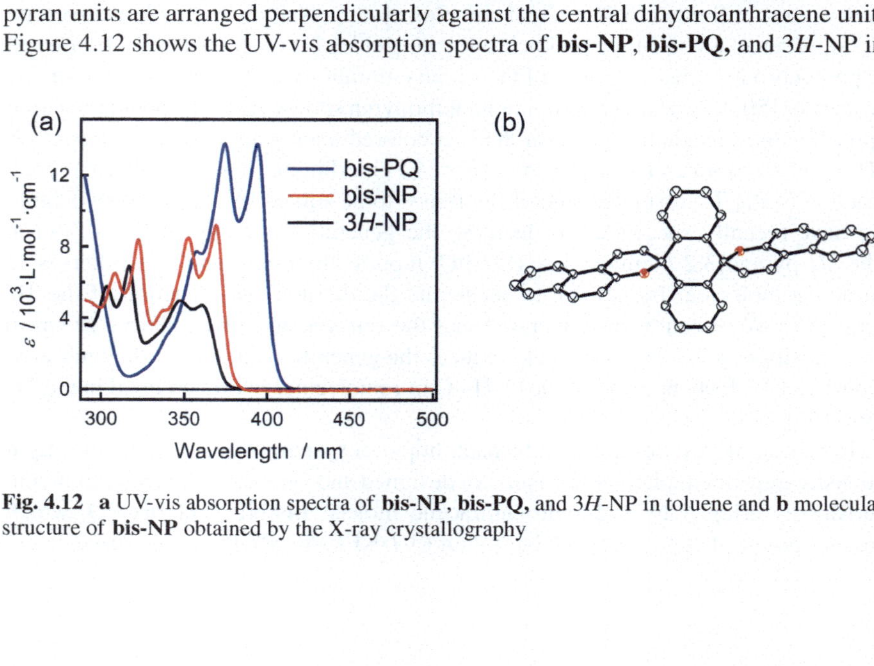

Fig. 4.12 **a** UV-vis absorption spectra of **bis-NP**, **bis-PQ,** and 3*H*-NP in toluene and **b** molecular structure of **bis-NP** obtained by the X-ray crystallography

toluene at 298 K. The similar absorption spectral shape and the almost twice larger molar extinction coefficient of **Bis-NP** in UVA region, compared with that of 3*H*-NP, indicate the two pyran units of **bis-NP** which are electronically independent. In contrast, the large red-shift of the absorption bands of **bis-PQ** was observed. This is presumably because of the introduction of the electron-donating *n*-butoxy groups to the *8H*-PQ units. The time-resolved absorption spectroscopy was conducted to investigate the excitation light intensity and the temperature dependence of the stepwise two-photon induced photochromic properties of **bis-NP** and **bis-PQ**. Figure 4.13a shows the absorption spectra of **bis-NP** at the PSS under CW UV light irradiation at 298 K. While an absorption band at 450 nm was observed upon weak UV light irradiation (30 mW), the other band at 528 nm appeared at 298 K with increasing the excitation light intensity to 120 mW. This absorption band was clearly increased by 120-mW UV light irradiation by decreasing the temperature to 263 K (Fig. 4.13b). This new absorption band of 528 nm is attributable to the open-open form of **bis-NP** produced by the stepwise 2PA process. The significant red-shift relative to the open-closed form is due to the extended π-conjugation over the two *o*-quinone monomethide units arranged in a plane in the open-open form. Because the thermal back

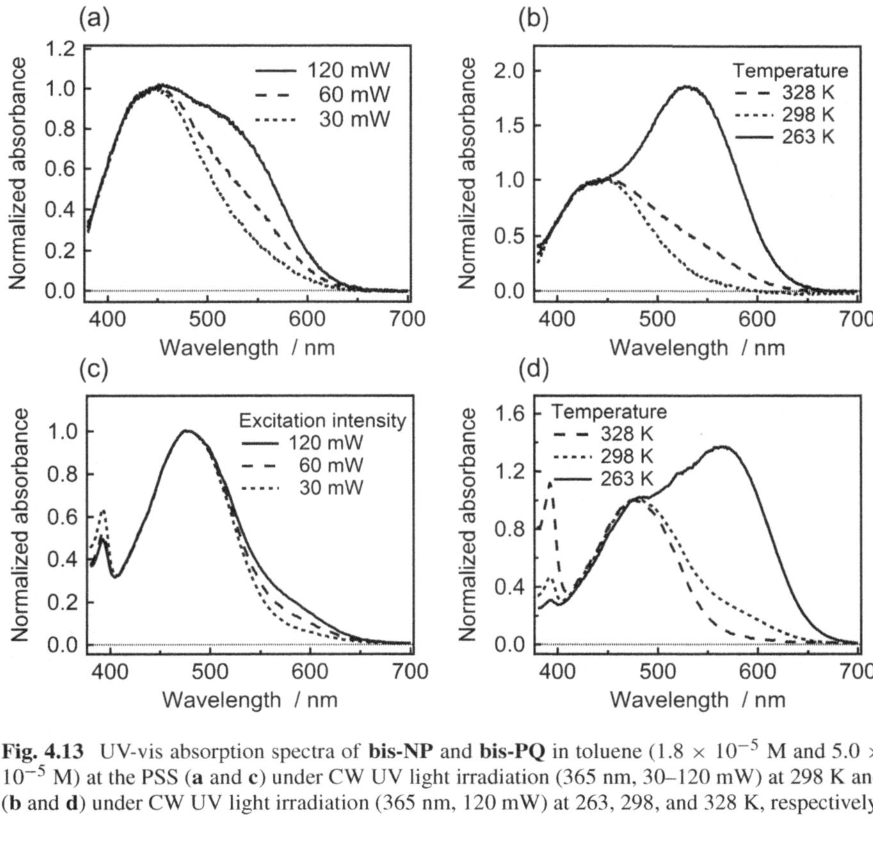

Fig. 4.13 UV-vis absorption spectra of **bis-NP** and **bis-PQ** in toluene (1.8×10^{-5} M and 5.0×10^{-5} M) at the PSS (**a** and **c**) under CW UV light irradiation (365 nm, 30–120 mW) at 298 K and (**b** and **d**) under CW UV light irradiation (365 nm, 120 mW) at 263, 298, and 328 K, respectively

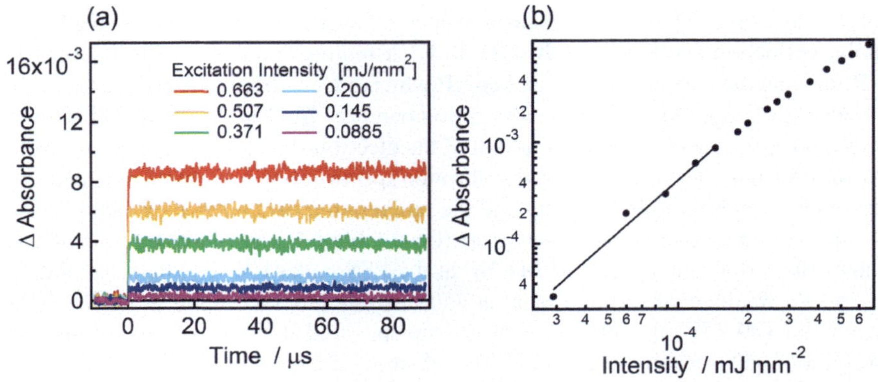

Fig. 4.14 **a** Time profiles of the Δabsorbance at 610 nm of **bis-PQ** in toluene (5.0×10^{-5} M) upon 355 nm laser excitation (pulse width = 5 ns, pulse energy = 0.092–2.1 mJ) at 298 K. **b** Double logarithmic plots of the average of the Δabsorbance at 610 nm of **bis-PQ**

reaction rate of the open-closed form of **bis-NP** is sufficiently decreased to absorb an additional photon by decreasing the temperature, the stepwise photochromic reaction of **bis-NP** was efficiently observed upon intense UV light irradiation at low temperature. Even in the case of **bis-PQ**, the similar dependences on the excitation intensity and temperature were observed as shown in Fig. 4.13c, d.

The dependence of the generation amount of the open-open form on the excitation intensity was investigated. The Δabsorbance at 610 nm, which corresponds to the amount of the open-open form after 355-nm laser pulse irradiation, shows a quadratic dependence on the intensity of the incident 355 nm laser pulse (Fig. 4.14), indicating the photochromic reaction of **bis-PQ** in a stepwise manner. Thus, it is concluded that the open-closed form generated by one-photon reaction of the initial closed-closed form shows the further photochemical ring-opening reaction, resulting in the generation of the open-open form.

We demonstrated the distinct color change of the **bis-PQ** in toluene solution depending on the intensity of incident light. Figure 4.15 shows the photochromic reaction of the toluene solution upon 365-nm light irradiation. The color of the film changes from colorless to orange by the irradiation of weak CW light (365 nm, 15 mW). In the whole irradiated area, the orange color monotonically disappeared after 365-nm light irradiation. This thermal decoloration process is attributable to the thermal back reaction of the open-closed form. In contrast, high-power UV light irradiation induces the different color changes between the central region and the edge region in the irradiated spot. While the central region turns dark reddish-purple upon UV light irradiation, the edge region shows the same color change to orange with that upon weak UV light irradiation. The central reddish-purple colored spot sequentially returned to the initial colorless via the orange state while the edge of the spot shows uniform bleaching of the orange color. This indicates that the open-open form was generated by the sequential 2PA process only in the spot irradiated with

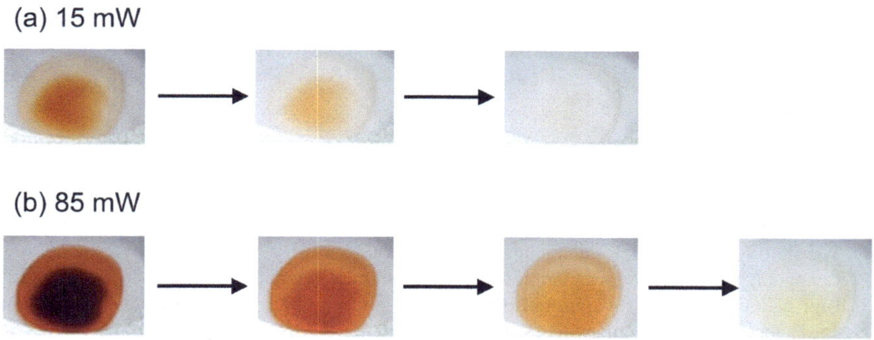

Fig. 4.15 Intensity-dependent color change of **bis-PQ** in toluene upon UV light irradiation (365 nm, **a** 15 mW and **b** 85 mW) at 283 K

intense UV light. This threshold system showing the color change depending on excitation light intensity was achieved by the nonlinear photoresponsive nature and the effective acceleration of the thermal back reaction of **bis-PQ**.

In conclusion of this section, the novel bispyran derivatives, **bis-NP** and **bis-PQ**, displaying the two-photon induced stepwise photochromic reaction are developed. The two-photon reaction generates the electronically coupled open-ring pair (open-open form) with planar conformation. The color of the open-open form is drastically red-shifted compared with that of the one-photon product (open-closed form) because of the extended π-conjugation. Furthermore, **bis-PQ** shows the large excitation light intensity dependence of the stepwise photochromism because the thermal fading speed of the TC form is accelerated by the introduction of alkoxy groups to the 1-position of the 8*H*-PQ unit. The novel molecular design for fast photoswitching will give an attractive insight for the future development of advanced photochromic compounds.

4.6 Conclusion

We described the unique photoresponsive systems developed by combining stepwise 2PA and photochromism. The efficient stepwise two-photon photochromic reactions were achieved by the combination of the fast photochromism with other optical properties such as the biradical-quinoid valence isomerization. If the lifetime of the photogenerated transient species is longer than microseconds but shorter than seconds, the power threshold of the stepwise 2PA process can be clearly observed. Thus, these biphotochromic systems composed of the fast T-type photochromic chromophores realize the photosynergetic effect: the cooperative interaction involving two photons and two photochromic units. The biphotochromic molecules enabling the stepwise 2PA processes can be a novel platform for photofunctional systems realizing the photosynergetic effects.

Acknowledgements This work was supported by JSPS KAKENHI Grant Numbers JP26107010, grant-in-aid for Scientific Research on Innovative Areas "Photosynergetics."

References

1. Wardle B (2009) Principles and applications of photochemistry. Wiley
2. Lengyel BA (1962) Lasers generation of light by simulated emission. Wiley
3. Bret G, Gires F (1964) Giant-pulse laser and light amplifier using variable transmission coefficient glasses as light switches. Appl Phys Lett 4:175–176
4. Geusic JE, Marcos HM, Van Uitert LG (1964) Laser oscillations in Nd-doped yttrium aluminum, yttrium gallium and gadolinium garnets. Appl Phys Lett 4:182–184
5. Moulton P (1986) Spectroscopic and laser characteristics of Ti:Al_2O_3. J Opt Soc Am B 3:125–133
6. Spence D, Kean P, Sibbett W (1991) 60-fsec pulse generation from a self-mode-locked Ti:sapphire laser. Opt Lett 16:42–44
7. He GS, Tan LS, Zheng Q, Prasad PN (2008) Multiphoton absorbing materials: molecular designs, characterizations, and applications. Chem Rev 108:1245–1330
8. Pawlicki M, Collins HA, Denning RG, Anderson HL (2009) Two-photon absorption and the design of two-photon dyes. Angew Chem Int Ed 48:3244–3266
9. König K (2000) Multiphoton microscopy in life sciences. J Microsc 200:83–104
10. Helmchen F, Denk W (2005) Deep tissue two-photon microscopy. Nat Methods 2:932–940
11. Kawata S, Kawata Y (2000) Three-dimensional optical data storage using photochromic materials. Chem Rev 100:1777–1788
12. Kawata S, Sun HB, Tanaka T, Takada K (2001) Finer features for functional microdevices. Nature 412:697–698
13. Ferreira KN, Iverson TM, Maghlaoui K, Barber J, Iwata S (2004) Architecture of the photosynthetic oxygen-evolving center. Science 303:1831–1838
14. Dürr H, Bouas-Laurent H (1990) Photochromism: molecules and systems. Elsevier B.V., Amsterdam
15. Irie M, Fukaminato T, Matsuda K, Kobatake S (2014) Photochromism of diarylethene molecules and crystals: memories, switches, and actuators. Chem Rev 114:12174–12277
16. Yokoyama Y, Nakatani K (2017) Photon-working switches. Springer, Japan, Tokyo
17. Kobayashi Y, Mutoh K, Abe J (2018) Stepwise two-photon absorption processes utilizing photochromic reactions. J Photochem Photobiol C Photochem Rev 34:2–28
18. Kobayashi Y, Katayama T, Yamane T, Setoura K, Ito S, Miyasaka H, Abe J (2016) Stepwise two-photon induced fast photoswitching via electron transfer in higher excited states of photochromic imidazole dimer. J Am Chem Soc 138:5930–5938
19. Kishimoto Y, Abe J (2009) A fast photochromic molecule that colors only under UV light. J Am Chem Soc 131:4227–4229
20. Mutoh K, Sliwa M, Abe J (2013) Rapid fluorescence switching by using a fast photochromic [2.2]paracyclophane-bridged imidazole dimer. J Phys Chem C 117:4808–4814
21. Mutoh K, Sliwa M, Fron E, Hofkens J, Abe J (2018) Fluorescence modulation by fast photochromism of a [2.2]paracyclophane-bridged imidazole dimer possessing a perylene bisimide moiety. J Mater Chem C 6:9523–9531
22. Mutoh K, Miyashita N, Arai K, Abe J (2019) Turn-on mode fluorescence switch by using negative photochromic imidazole dimer. J Am Chem Soc 141:5650–5654
23. Ishii N, Kato T, Abe J (2012) A real-time dynamic holographic material using a fast photochromic molecule. Sci Rep 2:819
24. Ishii N, Abe J (2013) Fast photochromism in polymer matrix with plasticizer and real-time dynamic holographic properties. Appl Phys Lett 102:163301

25. Kobayashi Y, Abe J (2016) Real-time dynamic hologram of a 3D object with fast photochromic molecules. Adv Opt Mater 4:1354–1357
26. Mutoh K, Nakano E, Abe J (2012) Spectroelectrochemistry of a photochromic [2.2]paracyclophane-bridged imidazole dimer: clarification of the electrochemical behavior of HABI. J Phys Chem A 116:6792–6797
27. Uno T, Koga M, Sotome H, Miyasaka H, Tamai N, Kobayashi Y (2018) Stepwise two-photon-induced electron transfer from higher excited states of noncovalently bound porphyrin-CdS/ZnS core/shell nanocrystals. J Phys Chem Lett 9:7098–7104
28. Kobayashi Y, Mutoh K, Abe J (2016) Fast photochromic molecules toward realization of photosynergetic effects. J Phys Chem Lett 7:3666–3675
29. Fihey A, Perrier A, Browne WR, Jacquemin D (2015) Multiphotochromic molecular systems. Chem Soc Rev 44:3719–3759
30. Mutoh K, Nakagawa Y, Sakamoto A, Kobayashi Y, Abe J (2015) Stepwise two-photon-gated photochemical reaction in photochromic [2.2]paracyclophane-bridged bis(imidazole dimer). J Am Chem Soc 137:5674–5677
31. Mayer U, Baumgartel H, Zimmermann H (1966) Über biradikale, chinone und semichinone der imidazolyl-reihe. Angew Chem 78:303
32. Kikuchi A, Iwahori F, Abe J (2004) Definitive evidence for the contribution of biradical character in a closed-shell molecule, derivative of 1,4-bis-(4,5-diphenylimidazol-2-ylidene)cyclohexa-2,5-diene. J Am Chem Soc 126:6526–6527
33. Yamashita H, Abe J (2014) Pentaarylbiimidazole, PABI: an easily synthesized fast photochromic molecule with superior durability. Chem Commun 50:8468–8471
34. Yamashita H, Ikezawa T, Kobayashi Y, Abe J (2015) Photochromic phenoxyl-imidazolyl radical complexes with decoloration rates from tens of nanoseconds to seconds. J Am Chem Soc 137:4952–4955
35. Ikezawa T, Mutoh K, Kobayashi Y, Abe J (2016) Thiophene-substituted phenoxyl-imidazolyl radical complexes with high photosensitivity. Chem Commun 52:2465–2468
36. Mutoh K, Kobayashi Y, Yamane T, Ikezawa T, Abe J (2017) Rate-tunable stepwise two-photon-gated photoresponsive systems employing a synergetic interaction between transient biradical units. J Am Chem Soc 139:4452–4461
37. Siewertsen R, Neumann H, Buchheim-stehn B, Herges R, Näther C, Renth F, Temps F (2009) Highly efficient reversible Z−E photoisomerization of a bridged azobenzene with visible light through resolved $S_1(n\pi^*)$ absorption bands. J Am Chem Soc 131:15594–15595
38. Beharry AA, Sadovski O, Woolley GA (2011) Azobenzene photoswitching without ultraviolet light. J Am Chem Soc 133:19684–19687
39. Bléger D, Schwarz J, Brouwer AM, Hecht S (2012) *o*-Fluoroazobenzenes as readily synthesized photoswitches offering nearly quantitative two-way isomerization with visible light. J Am Chem Soc 134:20597–20600
40. Fukaminato T, Hirose T, Doi T, Hazama M, Matsuda K, Irie M (2014) Molecular design strategy toward diarylethenes that photoswitch with visible Light. J Am Chem Soc 136:17145–17154
41. Bléger D, Hecht S (2015) Visible-light-activated molecular switches. Angew Chem Int Ed 54:11338–11349
42. Petermayer C, Dube H (2018) Indigoid photoswitches: visible light responsive molecular tools. Acc Chem Res 51:1153–1163
43. Hemmer JR, Page ZA, Clark KD, Stricker F, Dolinski ND, Hawker CJ, Read De Alaniz J (2018) Controlling dark equilibria and enhancing donor-acceptor stenhouse adduct photoswitching properties through carbon acid design. J Am Chem Soc 140:10425–10429
44. Bochet CG (2000) Wavelength-selective cleavage of photolabile protecting groups. Tetrahedron Lett 41:6341–6346
45. Lerch MM, Hansen MJ, Velema WA, Szymanski W, Feringa BL (2016) Orthogonal photoswitching in a multifunctional molecular system. Nat Commun 7:12054
46. Ashkenasy G, Hermans TM, Otto S, Taylor AF (2017) Systems chemistry. Chem Soc Rev 46:2543–2554

47. Yonekawa I, Mutoh K, Kobayashi Y, Abe J (2018) Intensity-dependent photoresponse of biphotochromic molecule composed of a negative and a positive photochromic unit. J Am Chem Soc 140:1091–1097
48. Hatano S, Horino T, Tokita A, Oshima T, Abe J (2013) Unusual negative photochromism via a short-lived imidazolyl radical of 1,1′-binaphthyl-bridged imidazole dimer. J Am Chem Soc 135:3164–3172
49. Yamaguchi T, Kobayashi Y, Abe J (2016) Fast negative photochromism of 1,1′-binaphthyl-bridged phenoxyl-imidazolyl radical complex. J Am Chem Soc 138:906–913
50. Lenoble C, Becker RS (1986) Photophysics, photochemistry and kinetics of photochromic 2H-pyrans and chromenes. J Photochem 33:187–197
51. Crano JC, Guglielmetti R (1999) Organic photochromic and thermochromic compounds, vol 1. Plenum Press, New York
52. Kim SH (2006) Functional dyes. Elsevier, Amsterdam
53. Sousa CM, Pina J, de Melo JS, Berthet J, Delbaere S, Coelho PJ (2011) Preventing the formation of the long-lived colored transoid-trans photoisomer in photochromic benzopyrans. Org Lett 13:4040–4043
54. Bergamini G, Silvi S (2016) Applied photochemistry, when light meets molecules. Springer International Publishing, Switzerland
55. Pozzo JL, Lokshin VA, Guglielmetti R (1994) New photochromic 2,2-diphenyl-[2H]-chromenes annellated with nitrogenated six-membered ring. Mol Cryst Liq Cryst Sci Technol, Sect A 246:75–78
56. Inagaki Y, Kobayashi Y, Mutoh K, Abe J (2017) A simple and versatile strategy for rapid color fading and intense coloration of photochromic naphthopyran families. J Am Chem Soc 139:13429–13441

Chapter 5
Suppression of Internal Conversions from Pseudo-Degenerate Excited Electronic States

Wataru Ota and Tohru Sato

Abstract We describe the relationship between the rate constant of internal conversion and vibronic coupling constant (VCC) based on the crude-adiabatic approximation. Vibronic coupling density (VCD) is introduced to clarify the origin of vibronic couplings from the local picture. The control of vibronic couplings from pseudo-degenerate excited electronic states gives the suppression of internal conversions. We discuss the fluorescence via higher triplets (FvHT) mechanism observed in the organic light-emitting diodes (OLEDs) of 1,4-bis(10-phenylanthracene-9-yl)benzene (BD1) used as a fluorescent dopant and the aggregation-induced enhanced emission (AIEE) of 1,2-bis(pyridylphenyl)ethene (CNPPE).

Keywords Vibronic coupling density · Pseudo-degeneracy · Internal conversion · Fluorescence via higher triplets mechanism · Organic light-emitting diodes · Aggregation-induced enhanced emission

5.1 Introduction

Vibronic couplings, the interactions between vibrational and electronic motions, induce internal conversions between electronic states. The suppression of internal conversions is achieved by the control of vibronic couplings, which can provide a design principle for novel light-emitting molecules. Vibronic coupling density (VCD) defined by the spatial density distribution of a vibronic coupling constant (VCC) is developed to elucidate the origin of vibronic couplings from the local picture [1, 2]. In this chapter, using the concept of VCD, we show that pseudo-degeneracy of

W. Ota · T. Sato (✉)
Fukui Institute for Fundamental Chemistry, Kyoto University, Sakyo-ku, Kyoto 606-8103, Japan
e-mail: tsato@scl.kyoto-u.ac.jp

Department of Molecular Engineering, Graduate School of Engineering, Kyoto University, Nishikyo-ku, Kyoto 615-8510, Japan

T. Sato
Unit of Elements Strategy Initiative for Catalysts & Batteries, Kyoto University, Nishikyo-ku, Kyoto 615-8510, Japan

H. Miyasaka et al. (eds.), *Photosynergetic Responses in Molecules and Molecular Aggregates*, https://doi.org/10.1007/978-981-15-5451-3_5

electronic states can give the suppression of internal conversions between electronic excited states against Kasha's rule [3]. This is applicable to designing novel emitting molecules for organic light-emitting diodes (OLEDs) in which the fluorescence via higher triplets (FvHT) mechanism [4–6] operates and elucidating the origin of the aggregation-induced enhanced emission (AIEE).

5.2 Internal Conversion and Vibronic Coupling Density

Consider a transition from initial vibronic state $|\Phi_{mi}(\mathbf{r}, \mathbf{Q})\rangle$ associated with electronic m and vibrational mi states to finial vibronic state $|\Phi_{mi}(\mathbf{r}, \mathbf{Q})\rangle$ where $\mathbf{r} = (\mathbf{r}_1, \ldots, \mathbf{r}_i, \ldots, \mathbf{r}_N)$ is a set of N electronic configurations and $\mathbf{Q} = (Q_1, \ldots, Q_\alpha, \ldots Q_M)$ is a set of M mass-weighted normal coordinates. From the Fermi's golden rule, the rate constant of internal conversion is given by [7]

$$k^{\mathrm{IC}}_{n\leftarrow m}(T) = \frac{2\pi}{\hbar} \sum_{ij} P_{mi}(T) \left| \left\langle \Phi_{nj}(\mathbf{r}, \mathbf{Q}) | H' | \Phi_{mi}(\mathbf{r}, \mathbf{Q}) \right\rangle \right|^2 \delta\left(E_{mi} - E_{nj}\right), \tag{5.1}$$

where $P_{mi}(T)$ is the Boltzmann distribution function of $|\Phi_{mi}(\mathbf{r}, \mathbf{Q})\rangle$ at temperature T, and E_{mi} and E_{nj} are the eigenvalues of $|\Phi_{mi}(\mathbf{r}, \mathbf{Q})\rangle$ and $|\Phi_{nj}(\mathbf{r}, \mathbf{Q})\rangle$, respectively. In the crude-adiabatic approximation (CA), the vibronic state is represented as the product of vibrational and electronic states fixed at nuclear configuration $\mathbf{R}_0$ [8]:

$$|\Phi_{mi}(\mathbf{r}, \mathbf{Q})\rangle = |\chi_{mi}(\mathbf{Q})\rangle |\Psi_m(\mathbf{r}; \mathbf{R}_0)\rangle . \tag{5.2}$$

$\mathbf{R}_0$ is generally chosen as an equilibrium nuclear configuration. Using the Herzberg–Teller expansion, the matrix element of the interaction Hamiltonian is expressed as

$$\left\langle \Phi_{nj}(\mathbf{r}, \mathbf{Q}) | H' | \Phi_{mi}(\mathbf{r}, \mathbf{Q}) \right\rangle = \sum_{\alpha} V_{nm,\alpha} \left\langle \chi_{nj}(\mathbf{Q}) | Q_\alpha | \chi_{mi}(\mathbf{Q}) \right\rangle, \tag{5.3}$$

where $V_{mn,\alpha}$ is the off-diagonal VCC defined by

$$V_{mn,\alpha} = \left\langle \Psi_m(\mathbf{r}; \mathbf{R}_0) \middle| \left(\frac{\partial \widehat{H}(\mathbf{r}, \mathbf{R})}{\partial Q_\alpha} \right)_{\mathbf{R}_0} \middle| \Psi_n(\mathbf{r}; \mathbf{R}_0) \right\rangle. \tag{5.4}$$

Here, $\widehat{H}(\mathbf{r}, \mathbf{R})$ is a molecular Hamiltonian, and $\mathbf{R}$ is a set of nuclear configurations. $V_{n,\alpha} := V_{nn,\alpha}$ is called the diagonal VCC. As a result, the rate constant of internal conversion in the CA representation is written as [9]

$$k^{\mathrm{IC}}_{n\leftarrow m}(T) = \frac{2\pi}{\hbar} \sum_{ij} P_{mi}(T) \sum_{\alpha\beta} V_{nm,\alpha} V_{mn,\beta}$$

$$\times \langle \chi_{nj}(\mathbf{Q})|Q_\alpha|\chi_{mi}(\mathbf{Q})\rangle\langle \chi_{mi}(\mathbf{Q})|Q_\beta|\chi_{nj}(\mathbf{Q})\rangle\delta(E_{mi} - E_{nj}). \tag{5.5}$$

It should be noted that this expression is different from the rate constant in the Born–Oppenheimer (BO) representation [10, 11] in which the electronic states depend on nuclear configuration $\mathbf{R}$ rather than fixed at $\mathbf{R}_0$.

Suppose that $|\chi_{mi}(\mathbf{Q})\rangle$ is represented as the product of vibrational states of a single mode

$$|\chi_{mi}(\mathbf{Q})\rangle = \prod_\gamma |n_{mi,\gamma}(Q_\gamma)\rangle, \tag{5.6}$$

then ignoring the Duschinsky effect the vibrational part is expressed as [9]

$$\langle \chi_{mi}|Q_\alpha|\chi_{nj}\rangle = \langle n_{mi,\alpha}|Q_\alpha|n_{nj,\alpha}\rangle \prod_{\gamma\neq\alpha} \langle n_{mi,\gamma}|n_{nj,\gamma}\rangle, \tag{5.7}$$

where

$$\langle n_{mi,\alpha}|Q_\alpha|n_{nj,\alpha}\rangle = \frac{\hbar}{2\pi}\left(\sqrt{n_{nj,\alpha}+1}\langle n_{mi,\alpha}|n_{nj,\alpha}+1\rangle + \sqrt{n_{nj,\alpha}}\langle n_{mi,\alpha}|n_{nj,\alpha}-1\rangle\right) \tag{5.8}$$

and the Franck–Condon (FC) overlap integral is [12]

$$\langle n_{mi,\alpha}|n_{nj,\alpha}\rangle = \sqrt{\frac{n_{mi,\alpha}!n_{nj,\alpha}!}{2^{n_{mi,\alpha}+n_{nj,\alpha}}}}e^{-\frac{1}{4}g_{n,\alpha}^2} \times \sum_{l=0}^{\min[n_{mi,\alpha},n_{nj,\alpha}]} (-1)^{n_{mi,\alpha}-l}2^l \frac{g_{n,\alpha}^{n_{mi,\alpha}+n_{nj,\alpha}-2l}}{l!(n_{mi,\alpha}-l)!(n_{nj,\alpha}-l)!}. \tag{5.9}$$

Here, $g_{n,\alpha}$ is the dimensionless diagonal VCC defined by

$$g_{n,\alpha} = \frac{V_{n,\alpha}}{\sqrt{\hbar\omega_{n,\alpha}^3}}, \tag{5.10}$$

where $\omega_{n,\alpha}$ is the angular frequency of vibrational mode α. In general, the rate constant of internal conversion decreases with the diagonal and off-diagonal VCCs. The control of the diagonal and off-diagonal VCCs is achieved by the analysis of the VCD.

The VCD is developed to clarify the local picture of VCC, which is given by the integrand of the VCC [1, 2]:

$$V_{mn,\alpha} = \int d\boldsymbol{x}\,\eta_{mn,\alpha}(\boldsymbol{x}), \tag{5.11}$$

where $\boldsymbol{x} = (x, y, z)$ is a three-dimensional spatial coordinate. The diagonal $\eta_{n,\alpha}(\boldsymbol{x}) := \eta_{nn,\alpha}(\boldsymbol{x})$ and off-diagonal VCD are expressed as

$$\eta_{n,\alpha}(\boldsymbol{x}) = \Delta\rho_{nm}(\boldsymbol{x}) \times v_\alpha(\boldsymbol{x}), \tag{5.12}$$

$$\eta_{mn,\alpha}(\boldsymbol{x}) = \rho_{mn}(\boldsymbol{x}) \times v_\alpha(\boldsymbol{x}). \tag{5.13}$$

Here, $\Delta\rho_{nm}(\boldsymbol{x})$ is the electron density difference between $|\Psi_n(\mathbf{r}; \mathbf{R}_0)\rangle$ and $|\Psi_m(\mathbf{r}; \mathbf{R}_0)\rangle$:

$$\Delta\rho_{nm}(\boldsymbol{x}) = \left\langle \Psi_n(\mathbf{r}; \mathbf{R}_0)|\hat{\rho}(\boldsymbol{x})|\Psi_n(\mathbf{r}; \mathbf{R}_0)\right\rangle - \left\langle \Psi_m(\mathbf{r}; \mathbf{R}_0)|\hat{\rho}(\boldsymbol{x})|\Psi_m(\mathbf{r}; \mathbf{R}_0)\right\rangle, \tag{5.14}$$

$|\Psi_m(\mathbf{r}; \mathbf{R}_0)\rangle$ is generally chosen as the reference electronic state at the equilibrium nuclear configuration. $\hat{\rho}(\boldsymbol{x})$ is the electron density operator defined by

$$\hat{\rho}(\boldsymbol{x}) = \sum_{ij}\sum_{\sigma\tau} \hat{c}^\dagger_{i\sigma}\hat{c}_{j\tau}\psi^*_{i\sigma}(\boldsymbol{x})\psi_{j\tau}(\boldsymbol{x}),$$

where $\hat{c}^\dagger_{i\sigma}$ and $\hat{c}_{j\tau}$ are creation and annihilation operators, and $\psi^*_{i\sigma}(\boldsymbol{x})$ and $\psi_{j\tau}(\boldsymbol{x})$ are spatial orbitals. Here, i and j are the orbital indices, and σ and τ are the spin indices. $\rho_{mn}(\boldsymbol{x})$ is the overlap density between $|\Psi_m(\mathbf{r}; \mathbf{R}_0)\rangle$ and $|\Psi_n(\mathbf{r}; \mathbf{R}_0)\rangle$:

$$\rho_{mn}(\boldsymbol{x}) = \left\langle \Psi_m(\mathbf{r}; \mathbf{R}_0)|\hat{\rho}(\boldsymbol{x})|\Psi_n(\mathbf{r}; \mathbf{R}_0)\right\rangle. \tag{5.15}$$

$v_\alpha(\boldsymbol{x})$ is the potential derivative with respect to vibrational mode α given by

$$v_\alpha(\boldsymbol{x}) = \left(\frac{\partial u(\boldsymbol{x})}{\partial Q_\alpha}\right)_{\mathbf{R}_0}, \tag{5.16}$$

where $u(\boldsymbol{x})$ is the electron-nucleus attractive potential acting on a single electron. Thus, the vibronic couplings are understood from the electronic term $\Delta\rho_{nm}(\boldsymbol{x})$ or $\rho_{mn}(\boldsymbol{x})$ and vibrational term $v_\alpha(\boldsymbol{x})$. Since the manipulation of vibrational term is rather difficult, the vibronic couplings are controlled through the electronic term. It should be noted that the disappearance of $\Delta\rho_{nm}(\boldsymbol{x})$ and $\rho_{mn}(\boldsymbol{x})$ leads to the decrease of the diagonal and off-diagonal VCCs, respectively, which contributes to the suppression of internal conversions between electronic states. The transition dipole moment also depends on $\rho_{mn}(\boldsymbol{x})$ [9]. Therefore, the disappearance of $\rho_{mn}(\boldsymbol{x})$ leads to the suppression of both radiative and non-radiative transitions. The concept of VCD has been applied not only to a theoretical design of light-emitting molecules [13–16] but also to carrier-transporting molecules [17–20] and chemical reactivity [21–24].

5.3 Fluorescence Via Higher Triplets in Organic Light-Emitting Diodes

The use of triplet excitons generated by electrical excitation is required for achieving highly efficient OLEDs. Thermally activated delayed fluorescence (TADF) has gathered attention as a mechanism for the use of triplet excitons; that is, the reverse intersystem crossing (RISC) from T_1 to S_1 is thermally activated by decreasing the energy difference between T_1 and S_1 ($\Delta E_{S_1-T_1}$) using the HOMO-LUMO separation [25, 26]. However, the OLED dopants using triplet excitons in spite of large $\Delta E_{S_1-T_1}$ has been reported. 1,4-bis(10-phenylanthracene-9-yl)benzene (BD1) is one of examples where $\Delta E_{S_1-T_1}$ are calculated to be more than 1.0 eV [27], which is extremely large for T_1 excitons to thermally overcome. Therefore, the RISC from higher triples than T_1 is expected. We propose a fluorescence via higher triplets (FvHT) as a mechanism for the use of high triplet excitons [4–6]. The excitons with long lifetime are possible if radiative and non-radiative transitions from the high to low triplets are sufficiently suppressed. In the following, we showed using BD1 that the suppression of the transitions between triplets is achieved by the cancelation of the overlap densities between these states because of the pseudo-degenerate electronic states [5].

BD1 with *X*-*Y*-*X* structure belongs to D_2 symmetry. Figure 5.1 shows the frontier orbitals and orbital levels of BD1 at the S_0 optimized structure. Since the frontier orbitals are delocalized over the anthracene moieties *X* in *X*-*Y*-*X*, the NHOMO and HOMO as well as the LUMO and NLUMO are pseudo-degenerate. These orbitals are approximately expressed as

$$\psi_{\text{NHO}} = \tfrac{1}{2}(\phi_{\text{HO}}(L) - \phi_{\text{HO}}(R)), \tag{5.17}$$

$$\psi_{\text{HO}} = \tfrac{1}{2}(\phi_{\text{HO}}(L) + \phi_{\text{HO}}(R)), \tag{5.18}$$

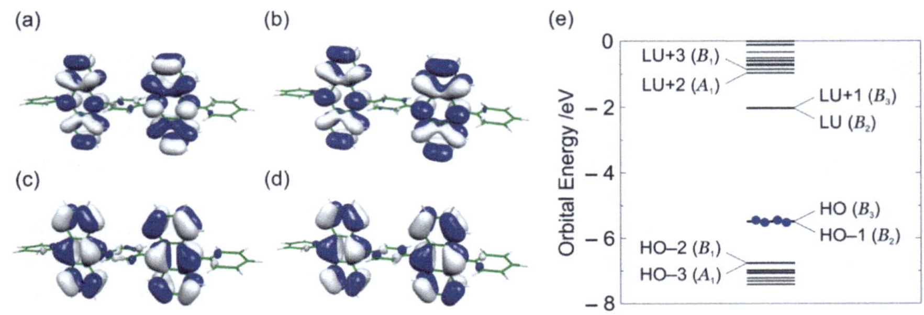

Fig. 5.1 **a** LUMO, **b** NLUMO, **c**, HOMO, and **d** NHOMO of BD1. Isosurface values are 2.0 × 10^{-2} a.u. White region is positive while blue region is negative. **e** Orbital levels of BD1. Reprinted from Ref. [4]

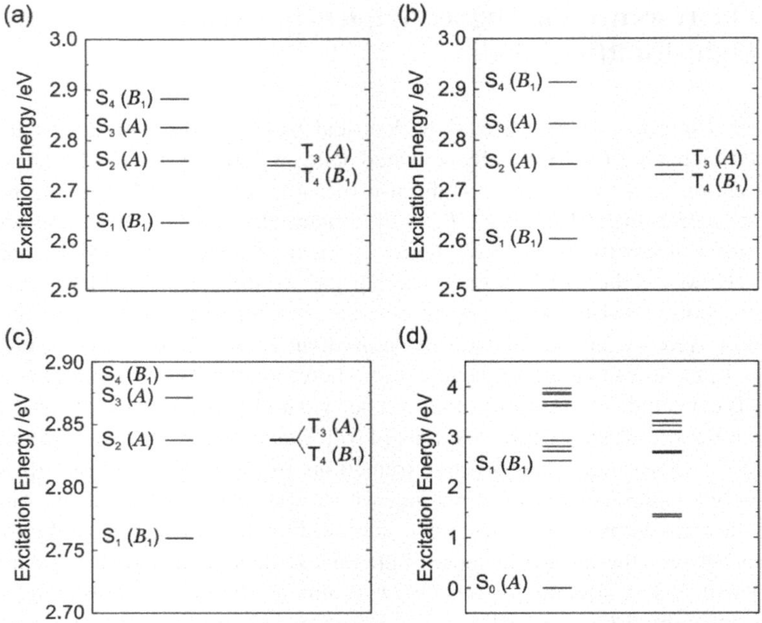

Fig. 5.2 Energy levels of excited states at **a** T_3, **b** T_4, **c** S_2, and **d** S_1 optimized structures. Reprinted from Ref. [4]

$$\psi_{\rm LU} = \frac{1}{2}(\phi_{\rm LU}(L) - \phi_{\rm LU}(R)), \tag{5.19}$$

$$\psi_{\rm NLU} = \frac{1}{2}(\phi_{\rm LU}(L) + \phi_{\rm LU}(R)), \tag{5.20}$$

where $\phi_{\rm HO/LU}$ is the HOMO/LUMO of the left (L) and right (R) anthracene moieties.

Figure 5.2 shows the excited states of BD1 at the T_3, T_4, S_2, and S_1 optimized structures. $\Delta E_{S_2-T_3}$ at the T_3 optimized structure is 0.8 meV, which is small enough for triplet excitons to thermally convert into singlet excitons. However, the RISC from T_3 to S_2 is symmetry-forbidden because the symmetry of both T_3 and S_2 are A. On the other hand, the RISC from T_4 with B_1 symmetry to S_2 is symmetry-allowed. In addition, $\Delta E_{S_2-T_4}$ at the T_4 optimized structure is only 21 meV. Therefore, the RISC from T_4 to S_2 is considered to be possible if the transitions from T_4 to T_1 and T_2 are suppressed.

Figure 5.3 shows the overlap densities between excited states and between excited and ground states. The overlap densities of T_4–T_1 and T_4–T_2 exhibit small distributions, which indicate that the transitions from T_4 to T_1 and T_2 are suppressed because the off-diagonal VCCs between these states become small. In contrast, the overlap densities of T_3–T_4 and S_2–S_1 exhibit large distributions, and the transition probabilities between these states are large. Therefore, the T_3 and S_2 excitons are promptly converted into T_4 and S_1 excitons, respectively. The magnitude of the overlap density

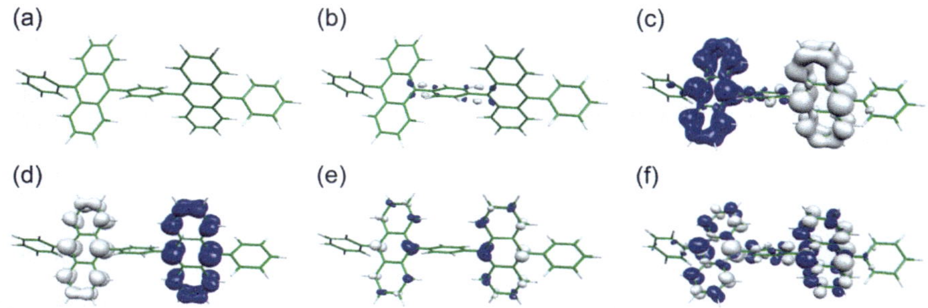

Fig. 5.3 Overlap densities of **a** $T_4@T_4$–$T_2@T_4$, **b** $T_4@T_4$–$T_1@T_4$, **c** $T_3@T_3$–$T_4@T_3$, **d** $S_2@S_2$–$S_1@S_2$, **e** $S_2@S_2$–$S_0@S_2$, and **f** $S_1@S_1$–$S_0@S_1$. White region is positive while blue region is negative. Isosurface values are 1.0×10^{-3} a.u. Reprinted from Ref. [4]

of S_1–S_0 is moderate so that the internal conversion from S_1 to S_0 is suppressed while the radiative transition is allowed.

The TD-DFT wave functions of T_1, T_2, and T_4 are approximately represented as [5]

$$|\Psi_{T_1}\rangle \approx c\left(|\Phi_{\mathrm{HO}}^{\mathrm{LU}}\rangle + |\Phi_{\mathrm{NHO}}^{\mathrm{NLU}}\rangle\right), \tag{5.21}$$

$$|\Psi_{T_2}\rangle \approx c\left(|\Phi_{\mathrm{HO}}^{\mathrm{NLU}}\rangle + |\Phi_{\mathrm{NHO}}^{\mathrm{LU}}\rangle\right), \tag{5.22}$$

$$|\Psi_{T_3}\rangle \approx c\left(|\Phi_{\mathrm{HO}}^{\mathrm{LU}}\rangle - |\Phi_{\mathrm{NHO}}^{\mathrm{NLU}}\rangle\right), \tag{5.23}$$

where c is the CI coefficient, and $1/\sqrt{2}$ when the contributions of the orbitals other than the frontier ones are ignored. Since NHOMO/HOMO and LUMO/NLUMO are pseudo-degenerate, T_1 and T_2 are pseudo-degenerate. The overlap density of T_4–T_1 is expressed as

$$\rho_{T_4-T_1} \approx c^2\left(|\psi_{\mathrm{NHO}}|^2 - |\psi_{\mathrm{HO}}|^2 + |\psi_{\mathrm{LU}}|^2 - |\psi_{\mathrm{NLU}}|^2\right). \tag{5.24}$$

Thus, $\rho_{T_4-T_1}$ is canceled because $|\psi_{\mathrm{NHO}}|^2$ and $|\psi_{\mathrm{HO}}|^2$ as well as $|\psi_{\mathrm{LU}}|^2$ and $|\psi_{\mathrm{NLU}}|^2$ exhibit almost the same distributions owing to the pseudo-degeneracy of the frontier orbitals. The overlap density of T_4–T_2 is expressed as

$$\rho_{T_4-T_2} \approx \left(c^2 - c^2\right)\psi_{\mathrm{NHO}}\psi_{\mathrm{HO}} + \left(c^2 - c^2\right)\psi_{\mathrm{LU}}\psi_{\mathrm{NLU}}. \tag{5.25}$$

In this case, $\rho_{T_4-T_2}$ exhibits a small distribution attributed to the cancelation of the CI coefficients. The TD-DFT wave functions of S_1 and S_2 are represented as

$$|\Psi_{S_1}\rangle \approx |\Phi_{\mathrm{HO}}^{\mathrm{LU}}\rangle, \tag{5.26}$$

$$|\Psi_{S_2}\rangle \approx c\left(|\Phi_{\mathrm{HO}}^{\mathrm{NLU}}\rangle - |\Phi_{\mathrm{NHO}}^{\mathrm{LU}}\rangle\right). \tag{5.27}$$

As a result, the overlap densities of S_1–S_0 and S_2–S_1 are expressed as

$$\rho_{S_1-S_0} \approx \psi_{\mathrm{HO}}\,\psi_{\mathrm{LU}}, \tag{5.28}$$

$$\rho_{S_2-S_1} \approx c(\psi_{\mathrm{NLU}}\,\psi_{\mathrm{LU}} - \psi_{\mathrm{NHO}}\,\psi_{\mathrm{HO}}). \tag{5.29}$$

Therefore, $\rho_{S_1-S_0}$ and $\rho_{S_2-S_1}$ exhibit large distributions because the cancelations of the overlap densities do not occur. Thus, in conclusion, the pseudo-degeneracy plays an important role in suppressing the transitions between excited states through the disappearance of the overlap densities, which enables the RISC from higher triplets than T_1 to singlets.

5.4 Role of Pseudo-Degeneracy on Aggregation-Induced Enhanced Emission

AIEE is a phenomenon in which fluorescent dyes exhibit strong luminescence in the aggregation phase [28, 29]. Understanding the mechanism of AIEE is required for the applications to the OLEDs in solid phase. A cyano-substituted 1,2-bis(pyridylphenyl)ethene (CNPPE) shows the AIEE behavior in solid phase; that is, the fluorescence quantum yield increases from 0.002 in CH_2Cl_2 solution to 0.72 in solid phase following the significant decrease of the rate constant of non-radiative transitions from >1.0 × 10^{10} s^{-1} in solution to 5.0 × 10^7 s^{-1} in solid [30]. In the following, we discussed the role of the pseudo-degeneracy on the suppression of the non-radiative transitions in solid phase.

Figure 5.4 shows the crystal structure of the CNPPE solid. The CNPPE in solid phase is modeled as a dimer with a cofacial configuration where this dimer is expected to have large intermolecular interaction energies. The vibronic couplings of the dimer are compared with that of a single molecule as a model in solution phase. The monomer belongs to C_1 symmetry while the dimer to C_i symmetry. The ground and excited states are computed at the CAM-B3LYP/6-31G(d,p) and TD-CAM-B3LYP/6-31G(d,p) levels of theory, respectively, using the Gaussian 16 [31]. In this study, the optimization of excited states is not performed for simplicity.

Table 5.1 lists the excited states of the dimer at the S_0 optimized structure. Since S_1 (A_g) is symmetry-forbidden and S_2 (A_u) is symmetry-allowed transitions, a photon absorption occurs in S_2. Although the emission occurs from the lowest excited state according to Kasha's rule [3], the emissions from S_2 are considered to be possible if the transitions from S_2 to S_1 are suppressed. Figure 5.5 shows the frontier orbitals and their orbital levels. The frontier orbitals are delocalized over molecules indicating the excimer formation, which suggests that in solid phase the absorption and fluorescence

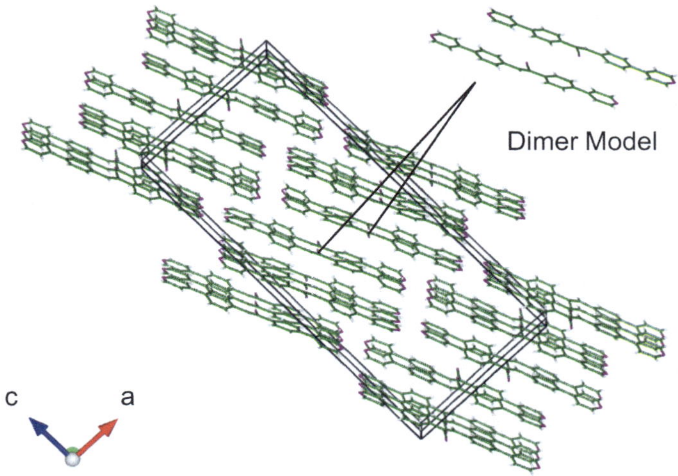

Fig. 5.4 Crystal structure of CNPPE solid taken from Ref. [30]. The CNPPE solid is modeled as a dimer with a cofacial configuration

Table 5.1 Excited states of the CNPPE dimer at the S_0 optimized structure. f denotes the oscillator strength

State	Excitation energy		f	Major configurations
	eV	nm		(CI coefficient)
$S_1(A_g)$	3.717	333.5	0	HO-1 → LU + 1 (0.4519)
				HO → LU (0.5174)
$S_2(A_u)$	3.806	325.74	2.7058	HO-1 → LU (0.4646)
				HO → LU + 1 (0.5061)

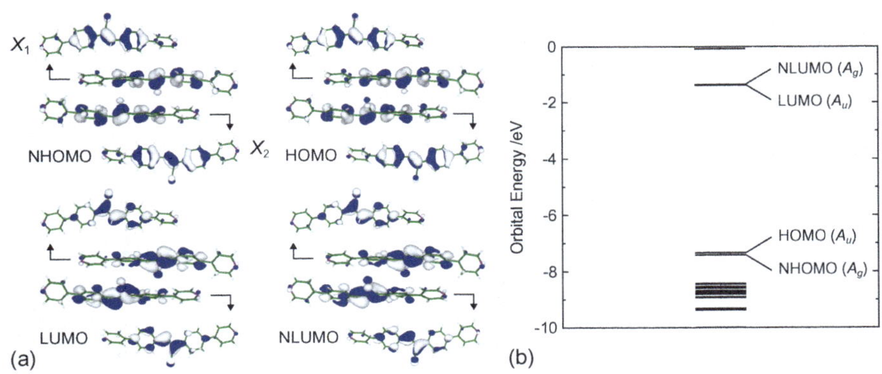

Fig. 5.5 **a** Frontier orbitals and **b** orbital levels of the dimer at the S_0 optimized structure. White region is positive while blue region is negative. Isosurface values of the frontier orbitals are 3.0×10^{-2} a.u.

occur within a dimer or larger polymer rather than a monomer. The NHOMO and HOMO as well as the LUMO and NLUMO are pseudo-degenerate because of the delocalized electronic states. These orbitals are approximately represented as

$$\psi_{\mathrm{NHO}} \approx \frac{1}{\sqrt{2}}(\phi_{\mathrm{HO}}(X_1) + \phi_{\mathrm{HO}}(X_2)), \tag{5.30}$$

$$\psi_{\mathrm{HO}} \approx \frac{1}{\sqrt{2}}(\phi_{\mathrm{HO}}(X_1) - \phi_{\mathrm{HO}}(X_2)), \tag{5.31}$$

$$\psi_{\mathrm{LU}} \approx \frac{1}{\sqrt{2}}(\phi_{\mathrm{LU}}(X_1) - \phi_{\mathrm{LU}}(X_2)), \tag{5.32}$$

$$\psi_{\mathrm{NLU}} \approx \frac{1}{\sqrt{2}}(\phi_{\mathrm{LU}}(X_1) + \phi_{\mathrm{LU}}(X_2)), \tag{5.33}$$

where $\phi_{\mathrm{HO/LU}}(X_1/X_2)$ is the HOMO/LUMO of X_1/X_2 consisting of the dimer. Figure 5.6 (a), (b), and (c) show the electron density differences of S_1-S_0, S_2-S_0, and S_2-S_1, respectively. The electron density differences of S_1-S_0 and S_2-S_0 exhibit similar distributions, and S_1 and S_2 are pseudo-degenerate. As a result, the electron density difference of S_2-S_1 is canceled, which contributes to the suppression of the internal conversion from S_2 to S_1 via the decrease of the diagonal VCCs. Figure 5.6 (d) and (e) shows the overlap densities of S_2-S_0, and S_2-S_1, respectively. The overlap density of S_2-S_0 exhibits a large distribution. In contrast, the overlap density of S_2-S_1 exhibits a small distribution, which contributes to the suppression of the internal conversion from S_2 to S_1 via the decrease of the off-diagonal VCCs. The overlap

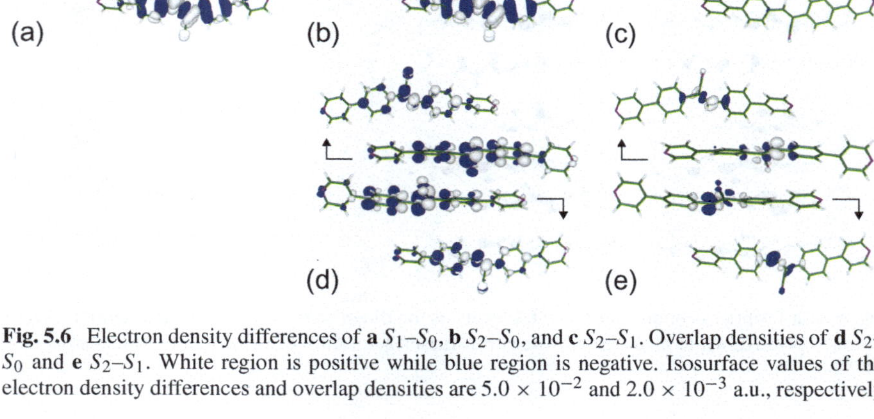

Fig. 5.6 Electron density differences of **a** S_1–S_0, **b** S_2–S_0, and **c** S_2–S_1. Overlap densities of **d** S_2–S_0 and **e** S_2–S_1. White region is positive while blue region is negative. Isosurface values of the electron density differences and overlap densities are 5.0×10^{-2} and 2.0×10^{-3} a.u., respectively

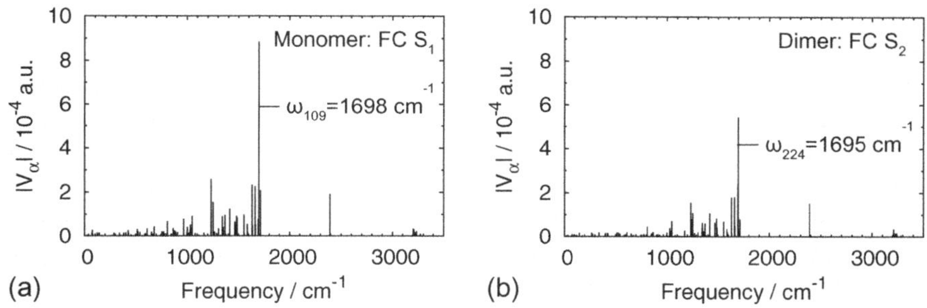

Fig. 5.7 Diagonal VCCs of **a** the monomer in the FC S_1 state and **b** the dimer in the FC S_2 state. The largest VCC is observed for the vibrational mode 109 in the monomer and 224 in the dimer

densities of S_2-S_0 and S_2-S_1 are approximately expressed as

$$\rho_{S_2-S_0} \approx c(\psi_{NHO}\psi_{LU} + \psi_{HO}\psi_{NLU}), \tag{5.34}$$

$$\rho_{S_2-S_1} \approx 2c^2(\psi_{LU}\psi_{NLU} - \psi_{NHO}\psi_{HO}), \tag{5.35}$$

where $c \approx 1/2$ is the CI coefficient. The sum of $\psi_{NHO}\psi_{LU}$ and $\psi_{HO}\psi_{NLU}$ in $\rho_{S_2-S_0}$ is not canceled. On the other hand, the difference of $\psi_{LU}\psi_{NLU}$ and $\psi_{NHO}\psi_{HO}$ in $\rho_{S_2-S_1}$ is canceled because of the pseudo-degenerate frontier orbitals. Therefore, in this pseudo-degenerate system, the fluorescence from S_2 expects to be possible because of the cancelation of the electron density difference and overlap density between S_2 and S_1.

Figure 5.7 compares the diagonal VCCs of the monomer in the FC S_1 state and the dimer in the FC S_2 state. The reducible representation of vibrational modes for the monomer belonging to C_I symmetry is

$$\Gamma_{vib}(C_1) = 129A, \tag{5.36}$$

where all the vibrational modes give the non-zero diagonal VCCs. Namely, all the vibrational modes are vibronic active. In contrast, the reducible representation of the dimer belonging to C_i symmetry is

$$\Gamma_{vib}(C_i) = 132A_g + 132A_u, \tag{5.37}$$

Since the vibronic active mode for the diagonal VCC is the totally symmetric A_g mode, the dimer has 132 vibronic active modes. Therefore, the number of vibronic active modes is almost the same as in monomer and dimer, although the number of the vibrational modes of the dimer is twice as large as that of the monomer. This results indicate that the high symmetry is preferable to the reduction of the VCCs.

The diagonal VCCs of the dimer are smaller than those of the monomer. For example, the largest VCC of the monomer is 8.849×10^{-4} a.u. While that of the dimer is 5.429×10^{-4} a.u. The ratio of the VCCs of dimer to monomer is 0.614, which is approximately equal to $1/\sqrt{2} \approx 0.707$. This is because the electron density difference of the dimer is delocalized over molecules [32] as shown in Fig. 5.6. Since the spatial integration of the electron density difference is zero by definition, the value of the electron density difference of X_1 (or X_2) consisting the dimer is half of that of the monomer. In addition, the value of the potential derivative of X_1 is $1/\sqrt{2}$ times that of the monomer because of the normalized condition of vibrational modes. As a result, the diagonal VCD of X_1 expressed as the product of the electron density difference and potential derivative is $1/(2\sqrt{2})$ times that of the monomer. Therefore, the spatial integration of the diagonal VCD of the dimer gives the diagonal VCCs that is $1/\sqrt{2}$ times smaller than those of the monomer. The deviation from $1/\sqrt{2}$ occurs because the molecules in the dimer do not have the same structures as the monomer. The rate constant of the internal conversion strongly depends on the diagonal VCCs, and the slight decrease of the diagonal VCCs reduces the rate constant in several orders [9]. Thus, the internal conversion from S_2 to S_0 in the dimer is more suppressed than that from S_1 to S_0 in the monomer originating from the decrease of the diagonal VCCs, which contributes to the appearance of the AIEE behavior.

Consequently, the pseudo-degenerate delocalized electronic states for the dimer suppress the internal conversions from S_2 to S_1 and S_2 to S_0. It should be noted that this does not occur when the electronic states are localized on a single molecule. In the present case, the electronic states are delocalized because the dimer in the CNPPE solid has C_i symmetry. The electronic states could be delocalized even when a crystal structure has no site symmetry, though the site symmetry in solid phase is preferable to realize such an excited electronic states. To confirm the delocalization of adiabatic wave functions, the geometry optimizations of the excited states should be performed. We have found that the adiabatic wave functions of the dimer are delocalized so that the vibronic couplings are reduced [33]. This suggests that the origin of the AIEE observed in the CNPPE solid is the excimer formation in which the vibronic couplings are reduced to suppress the internal conversions because of the pseudo-degenerate electronic states.

5.5 Conclusion

We discussed the role of pseudo-degenerate delocalized electronic states on suppressing the internal conversions between excited states using the VCD theory. The pseudo-degeneracy gives rise to the cancelations of the electron density differences and overlap densities resulting in the decrease of the diagonal and off-diagonal VCCs, respectively. This enables the fluorescence using the RISC from high triplet excited states and the AIEE in cofacial configurations. The control of vibronic couplings utilizing the pseudo-degeneracy can provide a new design principle for OLED dopants and AIEE dyes.

Acknowledgements This study was supported by JSPS KAKENHI Grant Number JP17H05259 in Scientific Research on Innovative Areas "Photosynergetics", JSPS KAKENHI Grant Number JP18K05261 in Scientific Research (C), and Element Strategy Initiative of MEXT Grant Number JPMXP0112101003. The computations were partly performed at Supercomputer System, Institute for Chemical Research, Kyoto University, Academic Center for Computing and Media Studies (ACCMS), Kyoto University, and Research Center for Computational Science, Okazaki.

References

1. Sato T, Tokunaga K, Tanaka K (2008) Vibronic coupling in naphthalene anion: vibronic coupling density analysis for totally symmetric vibrational modes. J Phys Chem A 112:758–767
2. Sato T, Uejima M, Iwahara N, Haruta N, Shizu K, Tanaka K (2013) Vibronic coupling density and related concepts. J Phys Conf Ser 428:012010
3. Kasha M (1950) Characterization of electronic transitions in complex molecules. Disc Faraday Soc 9:14–19
4. Sato T, Hayashi R, Haruta N, Pu YJ (2017) Fluorescence via reverse intersystem crossing from higher triplet states in a bisanthracene derivative. Sci Rep 7:4820
5. Sato T (2015) Fluorescence via reverse intersystem crossing from higher triplet states. J Comput Chem Jpn 14:189–192
6. Pu YJ, Satake R, Koyama Y, Otomo T, Hayashi R, Haruta N, Katagiri H, Otsuki D, Kim D, Sato T (2019) Absence of delayed fluorescence and triplet–triplet annihilation in organic light emitting diodes with spatially orthogonal bianthracenes. J Mater Chem C 7:2541–2547
7. Lin SH (1966) Rate of interconversion of electronic and vibrational energy. J Chem Phys 44:3759–3767
8. Azumi T, Matsuzaki K (1977) What does the term "vibronic coupling" mean? Photochem Photobiol 25:315–326
9. Uejima M, Sato T, Yokoyama D, Tanaka K, Park JW (2014) Quantum yield in blue-emitting anthracene derivatives: vibronic coupling density and transition dipole moment density. Phys Chem Chem Phys 16:14244–14256
10. Niu Y, Peng Q, Shuai Z (2008) Promoting-mode free formalism for excited state radiationless decay process with Duschinsky rotation effect. Sci China, Ser B: Chem 51:1153–1158
11. Niu Y, Peng Q, Deng C, Gao X, Shuai Z (2010) Theory of excited state decays and optical spectra: application to polyatomic molecules. J Phys Chem A 114:7817–7831
12. Hutchisson E (1930) Band Spectra intensities for symmetrical diatomic molecules. Phys Rev 36:410
13. Uejima M, Sato T, Tanaka K, Kaji H (2014) Enhancement of fluorescence in anthracene by chlorination: vibronic coupling and transition dipole moment density analysis. Chem Phys 430:47–55
14. Uejima M, Sato T, Detani M, Wakamiya A, Suzuki F, Suzuki H, Fukushima T, Tanaka K, Murata Y, Adachi C, Kaji H (2014) A designed fluorescent anthracene derivative: theory, calculation, synthesis, and characterization. Chem Phys Lett 602:80–83
15. Kameoka Y, Uebe M, Ito A, Sato T, Tanaka K (2014) Fluorescent triphenylamine derivative: theoretical design based on reduced vibronic coupling. Chem Phys Lett 615:44–49
16. Uebe M, Ito A, Kameoka Y, Sato T, Tanaka K (2015) Fluorescence enhancement of nonfluorescent triphenylamine: a recipe to utilize carborane cluster substituents. Chem Phys Lett 633:190–194
17. Sato T, Shizu K, Kuga T, Tanaka K, Kaji H (2008) Electron–vibration interactions in carrier-transport material: vibronic coupling density analysis in TPD. Chem Phys Lett 458:152–156
18. Shizu K, Sato T, Tanaka K, Kaji H (2010) Electron–vibration interactions in triphenylamine cation: why are triphenylamine-based molecules good hole-transport materials? Chem Phys Lett 486:130–136

19. Shizu K, Sato T, Tanaka K, Kaji H (2010) A boron-containing molecule as an efficient electrontransporting material with low-power consumption. Appl Phys Lett 97:142111
20. Shizu K, Sato T, Ito A, Tanaka K, Kaji H (2011) Theoretical design of a hole-transporting molecule: hexaaza [1 6] parabiphenylophane. J Mater Chem 21:6375–6382
21. Sato T, Iwahara N, Haruta N, Tanaka K (2012) C_{60} bearing ethylene moieties. Chem Phys Lett 531:257–260
22. Haruta N, Sato T, Iwahara N, Tanaka K (2013) Vibronic couplings in cycloadditions to fullerenes. J Phys Conf Ser 428:012003
23. Haruta N, Sato T, Tanaka K (2014) Regioselectivity in multiple cycloadditions to fullerene c60: vibronic coupling density analysis. Tetrahedron 70:3510–3513
24. Kojima Y, Ota W, Teramura K, Hosokawa S, Tanaka T, Sato T (2019) Model building of metal oxide surfaces and vibronic coupling density as a reactivity index: regioselectivity of CO_2 adsorption on Ag-loaded Ga_2O_3. Chem Phys Lett 715:239–243
25. Endo A, Ogasawara M, Takahashi A, Yokoyama D, Kato Y, Adachi C (2009) Thermally activated delayed fluorescence from Sn4+–porphyrin complexes and their application to organic light emitting diodes-a novel mechanism for electroluminescence. Adv Mater 21:4802–4806
26. Adachi C (2014) Third-generation organic electroluminescence materials. Jpn J Appl Phys 53:060101
27. Hu JY, Pu YJ, Satoh F, Kawata S, Katagiri H, Sasabe H, Kido J (2014) Bisanthracene-based donor–acceptor-type light-emitting dopants: Highly efficient deep-blue emission in organic light-emitting devices. Adv Funct Mater 24:2064–2071
28. Hong Y, Lam JWY, Tang BZ (2009) Aggregation-induced emission: phenomenon, mechanism and applications. Chem Commun 29:4332–4353
29. Hong Y, Lam JWY, Tang BZ (2011) Aggregation-induced emission. Chem Soc Rev 40:5361–5388
30. Nishio S, Higashiguchi K, Matsuda K (2014) The effect of cyano substitution on the fluorescence behavior of 1,2-bis(pyridylphenyl)ethene. Asian J Org Chem 3:686–690
31. Frisch MJ, Trucks GW, Schlegel HB, Scuseria GE, Robb MA, Cheeseman JR, Scalmani G, Barone V, Petersson GA, Nakatsuji H, Li X, Caricato M, Marenich AV, Bloino J, Janesko BG, Gomperts R, Mennucci B, Hratchian HP, Ortiz JV, Izmaylov AF, Sonnenberg JL, Williams-Young D, Ding F, Lipparini F, Egidi F, Goings J, Peng B, Petrone A, Henderson T, Ranasinghe D, Zakrzewski VG, Gao J, Rega N, Zheng G, Liang W, Hada M, Ehara M, Toyota K, Fukuda R, Hasegawa J, Ishida M, Nakajima T, Honda Y, Kitao O, Nakai H, Vreven T, Throssell K, Montgomery Jr JA, Peralta JE, Ogliaro F, Bearpark MJ, Heyd JJ, Brothers EN, Kudin KN, Staroverov VN, Keith T, Kobayashi R, Normand J, Raghavachari K, Rendell A, Burant JC, Iyengar SS, Tomasi J, Cossi M, Millam JM, Klene M, Adamo C, Cammi R, Ochterski JW, Martin RL, Morokuma K, Farkas O, Foresman JB, Fox DJ (2016) Gaussian 16, Revision B. 01. Gaussian Inc., Wallingford, CT
32. Shizu K, Sato T, Tanaka K (2013) Inverse relationship of reorganization energy to the number of π electrons from perspective of vibronic coupling density. J Comput Chem Jpn 12:215–221
33. Ota W, Takahashi K, Higashiguchi K, Matsuda K, Sato T (2020) Origin of aggregation-induced enhanced emission: role of pseudo-degenerate electronic states of excimers formed in aggregation phases. J Mater Chem C. https://doi.org/10.1039/C9TC07067B

Chapter 6
Advanced Function Control of Photochemical Reactions Using Mesoscopic Structures

Keisuke Imaeda and Kohei Imura

Abstract Strong light fields induced on mesoscopic structures have the potential to pave the way to develop novel excitation schemes that cannot be realized using conventional light fields. In this chapter, we use various scanning near-field optical microscopy techniques to obtain near-field spatial and spectral characterizations of metal mesoscopic structures. We also describe the enhancement of molecular optical responses by mesoscopic structures.

Keywords Mesoscopic structure · Plasmon resonance · Scanning near-field optical microscope

6.1 Introduction

Recent advances in chemical synthesis and theoretical calculations have enabled us to prepare advanced functional molecules for a variety of applications. However, our understanding of the optical responses of these molecules has been mostly limited by the dipolar approximation [1, 2]. The spatial structure of the optical field is neglected in this approximation, resulting in selection rules involving a dipolar operator and the wave functions of initial and final states. For instance, the optical transition from an initial to a final state is forbidden when both states have the same symmetry. This restriction can be overcome when the spatial scale of the optical field is comparable to or smaller than the subwavelength scale because the optical field of excitation light cannot be decoupled to obtain the selection rule. That is, the selection rule is modified under a localized field excitation [3–7]. Thus, forbidden transitions become optically accessible, leading to advanced functions of molecular systems. For this reason, localized optical fields have the potential to overcome conventional photophysical and photochemical restrictions that often originate from selection rules under plane wave irradiation.

K. Imaeda · K. Imura (✉)
Department of Chemistry and Biochemistry, School of Advanced Science and Engineering, Waseda University, Shinjuku, Tokyo 169-8555, Japan
e-mail: imura@waseda.jp

H. Miyasaka et al. (eds.), *Photosynergetic Responses in Molecules and Molecular Aggregates*, https://doi.org/10.1007/978-981-15-5451-3_6

In order to overcome the dipolar approximation, we induced localized optical fields in mesostructures with dimensions ranging from several tens of nanometers to a few micrometers. Mesoscopic structures can confine light fields tightly on subwavelength scales to generate nanoscale light fields in close proximity to the structures. This light confinement capability paves the way for the development of novel excitation schemes beyond conventional light-matter interactions. For example, confined fields possess larger electric field gradients and wavevectors than plane waves: these characteristics surpass conventional frameworks and can potentially induce ladder-type transitions, intraband transitions, and indirect band-gap transitions. In addition, mesoscopic structures amplify the electromagnetic field of light, which can boost a variety of molecular optical responses, such as nonlinear optical responses and magnetic dipole responses [8–11]. Furthermore, mesoscopic structures can modulate the electronic states of molecules via near-field interactions and strong coupling schemes, leading to the active control of molecular functionalities [12–17].

To utilize localized fields for photophysical and photochemical processes, a detailed understanding of optical fields is absolutely indispensable. For this purpose, visualization of the optical field distribution on mesoscopic structures is highly desirable. However, the spatial scale of mesoscopic structures is much smaller than the diffraction limit of light, and thus a conventional optical microscope cannot visualize these fields in real space. The scanning near-field optical microscope (SNOM) is a powerful tool for the nanoscale optical imaging of near-field distributions of mesoscopic structures [18, 19]. We used the SNOM in conjunction with a variety of spectroscopic methods to develop nano-spectral imaging techniques to elucidate the near-field spatial features of mesoscopic structures. We also explored the near-field interactions between molecules and mesoscopic optical fields. We demonstrated that these localized optical fields can be used to induce novel photophysical and photochemical processes beyond the conventional light-matter interaction. In this chapter, we describe the development of advanced near-field imaging methods, the characterization of localized fields and plasmons excited in mesostructures, and applications of optical fields to photophysical and photochemical processes.

6.2 Development of Advanced Near-Field Imaging Methods

The optical characteristics of mesostructures are closely related to the elementary excitations induced by electromagnetic fields of light. For example, unique optical characteristics of noble metal mesostructures are derived from the coherent oscillation of conduction electrons, which is known as surface plasmon resonance. Plasmon resonance confines free propagating light into a subwavelength space to generate intense optical fields in the vicinity of mesostructures [20–23]. To exploit these plasmonic fields as a nanoscale light source, the spatial structures of the fields have to be elucidated. So far, we have visualized the spatial patterns of the plasmonic fields induced on mesostructures of various shapes using an aperture-type SNOM [18]. In most of the studies with the aperture-type SNOM, the transmitted light or

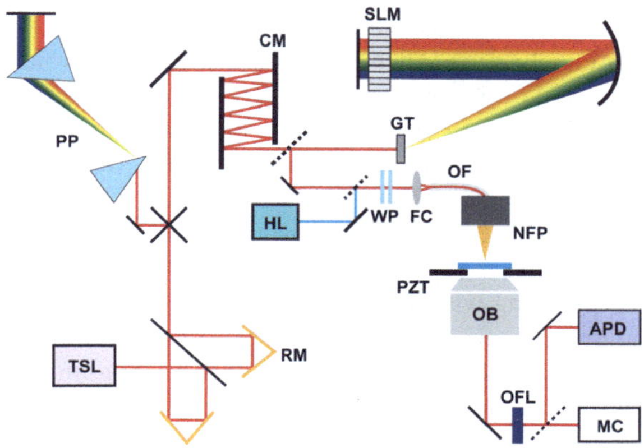

Fig. 6.1 Schematic illustration of an ultrafast scanning near-field optical microscope. TSL: mode-locked Ti:sapphire laser, RM: reflecting mirror, PP: prism pairs, CM: chirp mirrors, GT: grating, SLM: spatial light modulator, HL: halogen lamp, WP: wave plates, FC: fiber coupler, OF: optical fiber, NFP: near-field fiber probe, PZT: piezo-driven stage, OB: objective lens, OFL: optical filter, MC: monochromator, APD: avalanche photodiode. Reprinted with permission from [31]. Copyright 2016 American Chemical Society

photoluminescence (PL) from the sample is acquired for near-field optical imaging [24–26]. These spectroscopic imaging methods allow the direct observation of the steady-state spatial features of the elementary excitations from the visible to the near-infrared spectral region.

The optical characteristics of a mesostructure are also strongly related to the dynamic behavior of the elementary excitations, and hence the detailed characterization of the spatial and temporal features of the elementary excitations is extremely important [27–29]. Time-resolved near-field measurement is a promising method for the direct observation of the spatio-temporal behavior of the elementary excitations in a mesostructure. However, simultaneous achievement of high spatial and temporal resolution has remained highly challenging until recently. Figure 6.1 shows a schematic of an ultrafast aperture SNOM [30, 31]. In the aperture-type SNOM, a near-field optical fiber probe is used to generate a near-field light source, and the output from a mode-locked Ti:sapphire laser (central wavelength ~800 nm, repetition rate 80 MHz, pulse width <20 fs) is coupled to the near-field fiber probe. The dispersion from the optical fiber broadens the temporal width of the incident ultrashort pulse, thereby significantly degrading the temporal resolution at the aperture of the near-field probe. To overcome this problem, we pre-compensated the dispersion of the optical fiber by using a prism pair (PP), chirp mirrors (CM), and a spatial light modulator (SLM). We measured the second harmonic generation (SHG) autocorrelation on a β-barium borate (BBO) crystal and confirmed that a pulse width of ~17 fs was recovered at the aperture of the near-field probe, demonstrating that high spatial and temporal resolution are attained simultaneously.

As previously mentioned in this section, transmission or photoluminescence spectroscopy is conventionally utilized in the aperture-type SNOM, which has been challenging for the near-field optical imaging of opaque or non-luminescent samples. To overcome this difficulty, we developed two novel near-field imaging methods: dye-assisted luminescent imaging [32] and near-field reflection imaging [33]. In the former method, we measured the two-photon fluorescence signal from dye molecules to visualize the near-field distribution of a non-luminescent mesostructure. The emission intensity of emitters, such as dye molecules and quantum dots, is dramatically enhanced by coupling with the near-fields of mesostructures [34–36]. Therefore, the detection of enhanced fluorescence from molecules near the mesostructures can realize near-field luminescent imaging even for non-luminescent samples. Figure 6.2a shows a scanning electron microscope (SEM) image of a hexagonal gold mesoplate (edge length ~400 nm, thickness ~30 nm). We performed a near-field two-photon excitation measurement on the mesoplate by using a mode-locked Ti:sapphire laser as a light source. Figure 6.2b shows a near-field two-photon excitation image of the hexagonal mesoplate. As is evident from this figure, two-photon

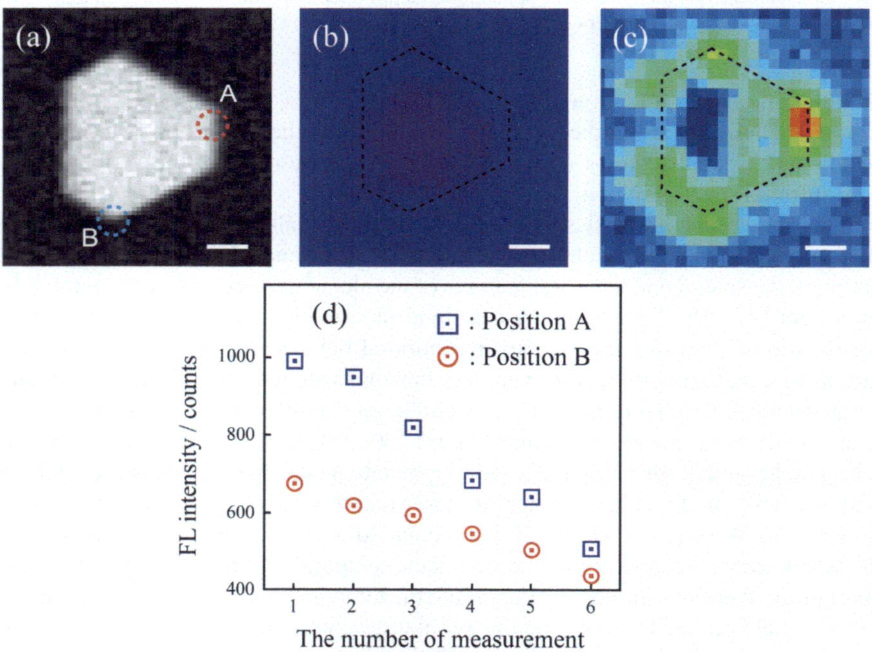

Fig. 6.2 **a** A SEM image of a hexagonal gold mesoplate (side length ~400 nm, thickness ~30 nm). **b** Near-field two-photon excitation image of the mesoplate. **c** Near-field two-photon fluorescence image of the R6G dispersed mesoplate. Excitation wavelength ~800 nm. Black dotted lines represent the approximate shapes of the mesoplate. Scale bars: 200 nm. **d** Measurement time dependence of the fluorescence intensity at the positions A and B in (**a**). Reprinted with permission from [32]. Copyright 2016 Elsevier

PL was not observed for the mesoplate. This observation indicates the mesoplate is non-luminescent because of a low PL quantum yield. To obtain a near-field luminescent image of a non-luminescent sample, we dispersed rhodamine 6G (R6G) solution on the sample substrate by spin-coating and then obtained a near-field two-photon excitation fluorescence image, as shown in Fig. 6.2c. Unlike Fig. 6.2b, bright spots can clearly be observed in Fig. 6.2c. This result indicates that the fluorescence of the R6G molecules was induced via a two-photon excitation process. We also found that two-photon fluorescence was locally enhanced on the mesoplate and consequently a unique spatial pattern was observed. We repeated the two-photon excitation measurements on the same mesoplate several times and examined the position dependency of the photobleaching on the mesoplate. The fluorescence intensity taken at positions A and B in Fig. 6.2a are plotted in Fig. 6.2d as a function of the number of measurements. As shown in the figure, the fluorescence intensity declines as the number of measurements increases, indicating the photobleaching proceeds on the mesoplate. We also found that the photobleaching rate at the position A is faster than that in position B, suggesting that the strongly enhanced near-fields at position A locally accelerate the photobleaching rate of the dye molecules. This result supports that near-fields on mesostructures can modify the optical response of molecular systems with subwavelength spatial resolution.

In the latter method, we developed a near-field reflection imaging method to visualize near-field spatial characteristics for opaque mesostructures [33]. In this method, we adopted the illumination-collection (I-C) mode of the aperture-type SNOM, wherein the sample is locally illuminated through the aperture of the near-field probe, and the reflected light from the sample is locally collected through the same aperture. This method can be used to characterize the near-field of an opaque sample, but it is very difficult to detect the reflection signal from the sample beyond the large background originating from the reflection at the end face of the near-field probe. To improve the signal-to-noise ratio in the I-C mode, we used the phase-stepping modulation technique [37–39], in which the distance between the tip of the near-field probe and the sample surface is modulated sinusoidally to selectively retrieve the near-field reflection signals from the sample [33]. Figure 6.3 a, b show near-field

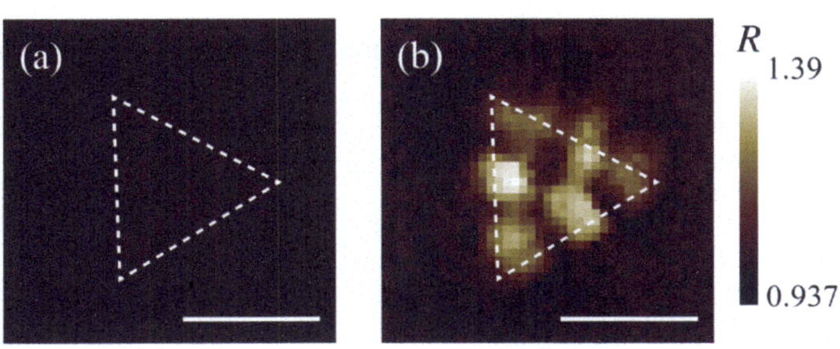

Fig. 6.3 **a** Near-field reflection images of the triangular silver mesoplate (edge length ~600 nm, thickness ~25 nm) **a** without and **b** with the phase-stepping modulation, respectively. Scale bars: 500 nm. Reprinted with permission from [33]. Copyright 2017 American Chemical Society

reflection images of the triangular silver mesoplate (edge length ~600 nm, thickness ~25 nm) obtained without and with phase-stepping modulation, respectively. In Fig. 6.3a, the reflection signals from the sample are buried in the large background, and thus the near-field spatial pattern of the mesoplate is not observable. On the other hand, in Fig. 6.3b, the near-field reflection signals are substantially recovered on the mesoplate, and the unique spatial pattern is clearly visualized inside the mesoplate. From theoretical calculations, the observed spatial pattern is assignable to the plasmon mode resonantly excited in the triangular mesoplate. These results imply that the phase-stepping modulation technique is very useful for improving the signal-to-noise ratio under the I-C mode of the SNOM. The present results also prove that the developed near-field reflection imaging method is a powerful tool for the near-field characterization of elementary excitations in opaque mesostructures.

6.3 Near-Field Characterization of Mesoscopic Structures

In general, the macroscopic optical response of a bulk material can be described by the dielectric function of the constituent material [40]. On the other hand, the optical response of the mesostructure is also strongly dependent on its geometrical shape, since the shape provides the boundary condition of the spatial mode of the elementary excitations induced in the mesostructure [41]. Therefore, the spatial mode patterns of elementary excitations vary enormously depending on the geometrical shapes of the mesostructures. For example, in the case of a one-dimensional mesostructure, such as a metal nanowire (Fig. 6.4a, left), a periodical oscillating pattern is resonantly excited along the long axis of the nanowire [24, 25, 42–45]. This mode pattern can be verified theoretically by calculating the electric field distribution using an electromagnetic simulation, as shown in Fig. 6.4b. It is also widely known that a geometrically complemental screen made of the same metal as the nanowire, called

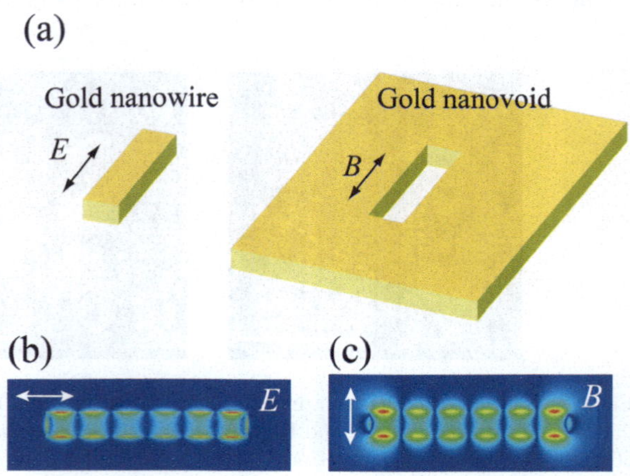

Fig. 6.4 **a** Schematic illustrations of a gold nanowire (left) and gold nanovoid (right). Black arrows represent the directions of the electric and magnetic fields induced in the nanowire and nanovoid. **b** Simulated electric field distribution of a gold nanowire. **c** Simulated magnetic field distribution of a gold nanovoid. White arrows indicate the polarization directions of the excitation light

a rectangular nanovoid (Fig. 6.4a, right), induces electromagnetic modes similar to those of the nanowire. Babinet's principle in electromagnetism states that the spectral feature of the transmittance of a metal nanowire exhibits inverse with respect to the transmittance of the metal nanovoid [40, 46]. This principle also states that the electric field distributions induced in the metal nanowire coincide with the magnetic field distributions inside the metal nanovoid when the structures are illuminated by complementary polarized light. Figure 6.4c shows the simulated magnetic field distribution inside the metal nanovoid. The spatial pattern in Fig. 6.4c is in good agreement with that observed in Fig. 6.4b. This agreement theoretically demonstrates that Babinet's principle is applicable to a plasmonic system. This result also supports that strong magnetic fields inside the nanovoid can boost magnetic dipole responses in a molecular system.

To utilize the strong electromagnetic fields in metal nanowires and nanovoids in practical applications, the optical complementarity in the near-field region needs to be elucidated. For this purpose, we examined the genuine near-field transmission spectra of a metal nanowire and nanovoid with same dimensions [47]. An ordinary near-field transmission spectrum measured by the aperture-type SNOM contains both near- and far-field components because the near-field optical fiber probe generates near- and far-field components of light. Therefore, the far-field component has to be removed from the ordinary near-field transmission spectrum to evaluate the genuine near-field extinction spectrum. Figure 6.5 shows the genuine near-field extinction spectra of the gold nanowire (red) and nanovoid (blue). The spectral features of the gold nanowire in the figure are reasonably consistent with the inverted spectral features of the nanovoid for wavelengths range longer than 620 nm. This result implies that the Babinet's principle is applicable even in the near-field optical region. It can also

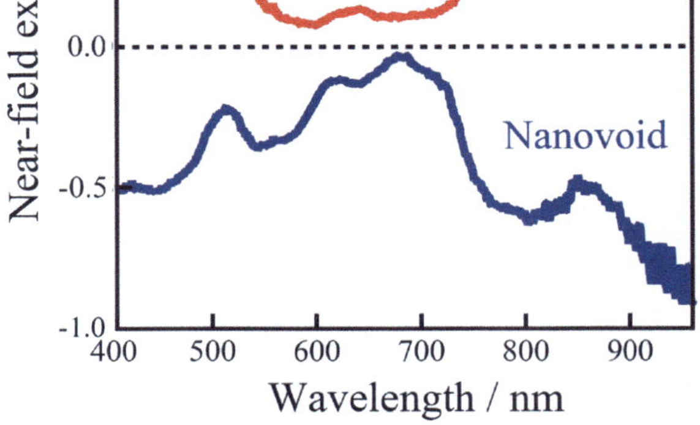

Fig. 6.5 Near-field extinction spectra of a gold nanowire and nanovoid obtained by subtracting the far-field transmission spectra from the near-field ones [47]

be seen that spectral complementarity is not feasible for wavelengths range shorter than 620 nm, which can be attributed to the interband transition of gold.

As depicted in Fig. 6.4b, the spatial mode patterns excited in one-dimensional mesostructures are systematically described by the Fabry–Pérot modes of a one-dimensional cavity resonator [44]. By contrast, the spatial mode patterns excited in two-dimensional mesoplates are much more complex because of the higher dimensionality [48–51]. So far, several research groups have reported high-resolution imaging of the spatial patterns of plasmon modes excited in metal mesoplates using scanning transmission electron microscopy (STEM) combined with electron energy loss spectroscopy (EELS) [52–59]. Although these results provide crucial information on the spatial features of plasmon modes, visualization by optical microscopy is highly desirable for photophysical and photochemical applications. Recently, we have successfully visualized the spatial patterns of plasmon modes induced in mesoplates of various shapes by adopting the near-field optical imaging method and revealed that the spatial patterns observed in the near-field images can be comprehensively interpreted based on the eigenfunctions of a particle confined inside the mesoplates [60]. Figure 6.6a shows an SEM image of a triangular gold mesoplate (edge length ~810 nm, thickness ~30 nm). We obtained the near-field spectral and spatial characteristics of the mesoplate by performing near-field transmission measurements using a halogen lamp as a light source. Figure 6.6b shows the near-field extinction spectrum taken near the apex of the mesoplate. Several plasmon resonance peaks can clearly be observed in the visible to near-infrared region of this spectrum. We obtained near-field transmission images by mapping the transmitted light intensity at the resonances, as shown in Fig. 6.6d–f. The bright areas in these images represent the reduction of the transmitted light intensity, indicating high excitation probability of plasmons [25, 60]. As seen in these images, a variety of spatial patterns can be observed inside the mesoplate, depending on the observation wavelengths. To elucidate the physical origin of the observed spatial patterns, we calculated the eigenfunctions and eigenenergies of the mesoplate by solving the Schrödinger equation for a particle confined in a two-dimensional triangular potential well [61]. Figure 6.6g–i show the square moduli of the calculated eigenfunctions (which are henceforth referred to as eigenmodes). The bright areas of the eigenmodes correspond to regions where the existence probability of a particle is high, that is, the bright spots of the calculated eigenmodes correspond to those in the near-field images. As seen in Fig. 6.6d–i, the spatial patterns of the calculated eigenmodes are evidently in good agreement with those in the near-field transmission images. This agreement implies that the spatial patterns in the near-field transmission images can be interpreted as plasmonic standing waves confined inside the boundary of the mesoplate. We also found a linear correlation between the observed plasmon resonance energies and the calculated eigenenergies, as shown in Fig. 6.6c. This means that the plasmon resonance energy can be qualitatively evaluated by calculating the eigenenergies.

We also adopted the group theory to classify the irreducible representations of each eigenfunction [61]. A triangular mesoplate on a glass substrate belongs to the C_{3v} point group shown in Table 6.1. The eigenmodes in Fig. 6.6g, i exhibit the irreducible representation E, that is, these modes have the same symmetrical characteristics as the

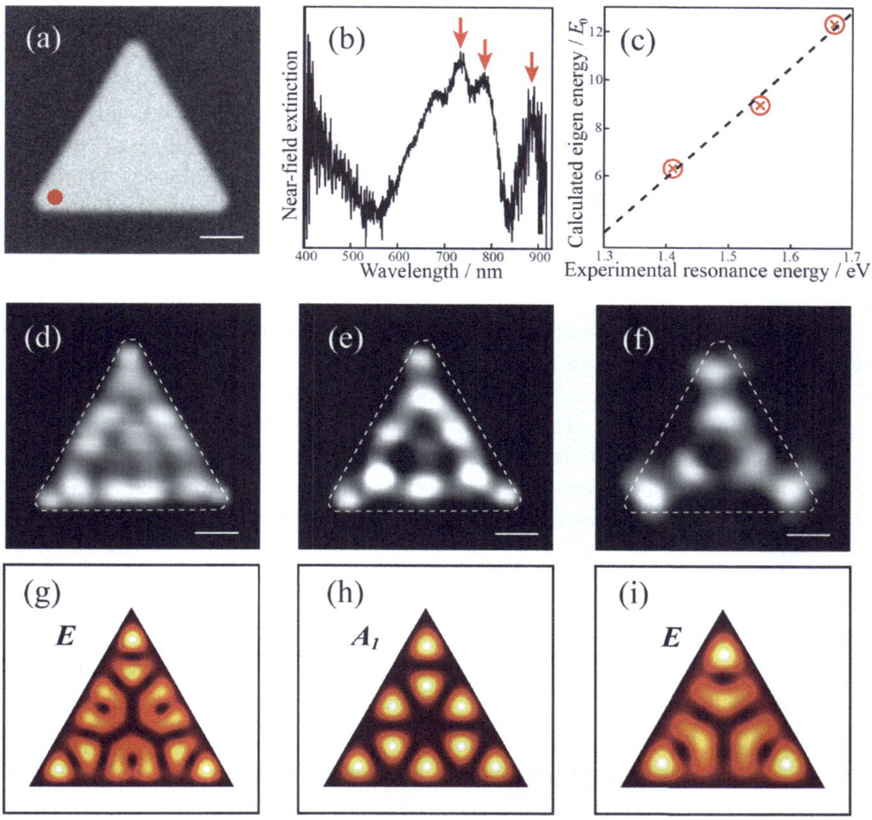

Fig. 6.6 **a** A SEM image of the triangular gold mesoplate (edge length ~810 nm, thickness ~30 nm). **b** Near-field extinction spectrum taken at the red point in (**a**). **c** Relation between the experimental resonance energy and the calculated eigenenergy. E_0 is the eigenenergy of the lowest eigenmode. **d–f** Near-field transmission images of the mesoplate. Observation wavelength: ~740 nm for (**d**), ~800 nm for (**e**), and ~880 nm for (**f**). White dotted lines represent the approximate shapes of the mesoplate. Scale bars: 200 nm. **g–i** Square moduli of the calculated eigenfunctions. Eigenenergy: $12.33E_0$ for (**g**), $9E_0$ for (**h**), and $6.33E_0$ for (**i**). The irreducible representations are shown for each eigenfunction. Reprinted with permission from [60]. Copyright 2018 American Chemical Society

Table 6.1 Character table for the C_{3v} point group

C_{3v}	E	$2C_3$	$3\sigma_v$		
A_1	1	1	1	z	$x^2 + y^2, z^2$
A_2	1	1	−1	R_z	
E	2	−1	0	(x, y); (R_x, R_y)	($x^2 - y^2$, xy); (xz, yz)

x- and y-vector components, and hence can be assigned as in-plane polarized modes. These eigenmodes are optically allowed by the in-plane electric field components, and thus can be excited by polarized light from normal incidence. On the other hand, the eigenmode with the irreducible representation of A_1 is assigned as the out-of-plane polarized mode. According to the conventional optical selection rule, this eigenmode is only accessible by an out-of-plane electric field component. Therefore, the eigenmode in Fig. 6.6h is forbidden for normal plane wave incidence. In the case of the aperture-type SNOM, both in-plane and out-of-plane electric field components are generated in the vicinity of the aperture [62], and thereby, both out-of-plane and in-plane modes are experimentally observable in the near-field images.

We also performed near-field transmission measurements on the hexagonal gold mesoplate shown in Fig. 6.7a [63]. The near-field transmission image observed at the resonance peak exhibits a periodically oscillating pattern, as shown in Fig. 6.7b. To clarify the origin of the observed patterns, we calculated the square moduli of the eigenfunctions of a particle confined in a two-dimensional hexagonal potential well. Figure 6.7c, d show the square moduli of the eigenfunctions (eigenenergies: $5.25E_0$ for (c) and $6.66E_0$ for (d)). From the group theory analysis, the eigenmodes in Fig. 6.7c, d can be assigned to out-of-plane and in-plane modes, respectively. By comparing the spatial patterns in Fig. 6.7b–d, we found that the spatial pattern in the near-field image (Fig. 6.7b) is partially consistent with the eigenmode patterns in Fig. 6.7c, d. This result suggests that the spatial pattern in Fig. 6.7b cannot be assigned to a single eigenmode, but is a superposition of two eigenmodes. Here, we emphasize that the polarization directions of the two eigenmodes shown in Fig. 6.7c,

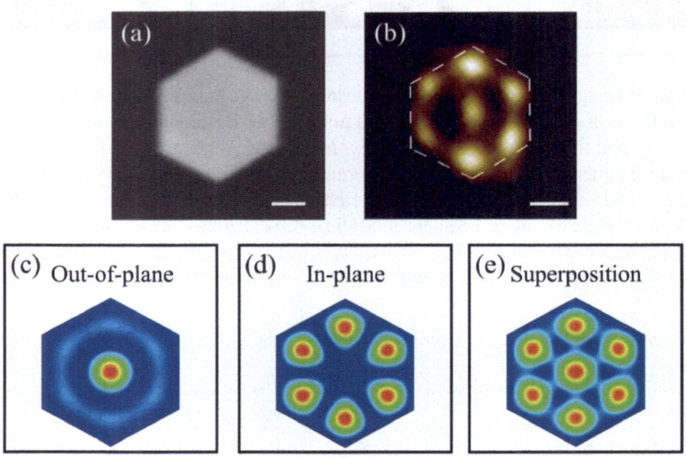

Fig. 6.7 **a** A SEM image of the hexagonal gold mesoplate. **b** Near-field transmission image observed near 900 nm. White dotted line represents the approximate shape of the mesoplate. Scale bar: 200 nm. **c**, **d** Square moduli of eigenfunction calculated for a hexagonal potential well. Eigenenergy: **c** 5.25 E_0 and **d** 6.66 E_0. These functions are assignable to the out-of plane and in-plane modes, respectively. **e** Spatial superposition of the eigenmodes shown in (**c**) and (**d**). Reprinted with permission from [63]. Copyright 2019 American Chemical Society

d are perpendicular, and hence the two eigenmodes do not coherently interact with each other. From this consideration, we calculated the spatial superposition of the eigenmodes, as shown in Fig. 6.7e. The spatial pattern of this image is consistent with that of the near-field transmission image (Fig. 6.7b), which means the two eigenmodes are spectrally overlapped, and consequently, the spatial superposition of these modes is experimentally observed.

We also investigated the steric near-field distributions induced on the hexagonal mesoplate by using three-dimensional near-field optical microscopy [63]. In this method, two-dimensional near-field transmission measurements are conducted at various distances d between the near-field probe tip and the sample surface. Figure 6.8a shows the near-field transmission images measured at distances d = 20, 40, and 90 nm. The spatial features in these images depend on the near-field probe tip-sample distance d. In the center of the mesoplate, the extinction spot blurs as the distance d increases. On the other hand, periodical extinction patterns along the side of the mesoplate are clearly observable regardless of the distance d. These results indicate that the steric near-field distribution varies with the observation position on the mesoplate. To obtain further insight into the steric near-field distribution of the mesoplate, we obtained a series of near-field transmission images of the mesoplate by incrementally varying d from 20 to 90 nm. In Fig. 6.8b, the extinction intensities taken at the apex and the center on the mesoplate are plotted against the distance d. The vertical scale of this figure is logarithmic. In this figure, the extinction intensity at the apex of the mesoplate decays exponentially with a single component. However, the extinction at the center decays with two components. We fitted these plots with

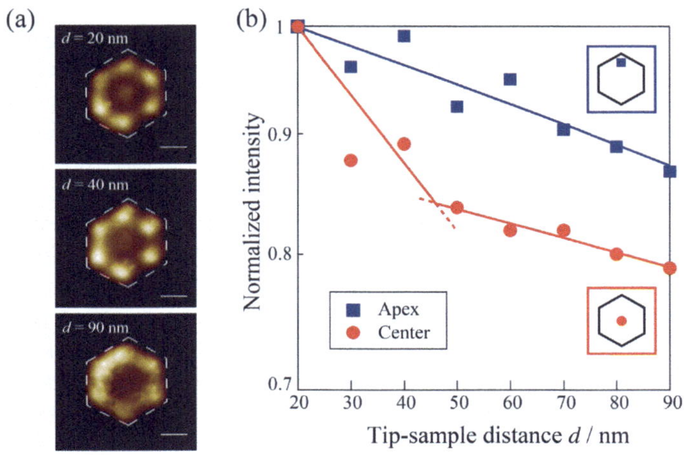

Fig. 6.8 **a** Near-field transmission images of the hexagonal gold mesoplate observed at the near-field probe tip-sample distances $d = 20, 40$, and 90 nm. Scale bars: 200 nm. Observation wavelength: 900 nm. **b** Tip-sample distance dependence of the extinction intensity taken near the center (red) and apex (blue) on the mesoplate. The vertical axis is plotted on a logarithmic scale. Reprinted with permission from [63]. Copyright 2019 American Chemical Society

exponential functions that are shown as solid lines in Fig. 6.8b. From the figure, we found that the decay lengths are strongly dependent on the positions on the mesoplate: the decay length at the apex on the mesoplate is estimated to be 520 nm, whereas the decay lengths at the center of the mesoplate are 160 and 700 nm. The position dependency of the decay lengths can be interpreted in terms of the spatial patterns of the eigenmodes. As mentioned above, the spatial features of the near-field transmission image are consistent with the spatial superposition of the two eigenmodes shown in Fig. 6.7c, d. From the spatial distribution of these eigenmodes, we found that the excitation probability of these eigenmodes are strongly dependent on the positions on the mesoplate: the out-of-plane eigenmode shown in Fig. 6.7c is predominantly excited at the center of the mesoplate, whereas the in-plane mode in Fig. 6.7d is predominantly excited near the apices of the mesoplate. Considering this position dependency of the excitation probability of the eigenmodes, the observed results in Fig. 6.8b indicates that the decay length of the in-plane mode is longer than that of the out-of-plane mode. This fact implies that the out-of-plane mode in Fig. 6.7c confines light fields more tightly compared with the in-plane mode. Therefore, the near-field distribution induced by the out-of-plane plasmon mode is more promising for the realization of advanced function control of molecular systems.

6.4 Time-Resolved Near-Field Measurements on Metal Mesostructures

The steady-state spatial patterns of plasmons in metal mesostructures have been studied extensively using a variety of high-resolution microscopic imaging methods [24, 25, 42–45, 48–59]. However, few studies have been conducted on the spatio-temporal dynamics of plasmons due to the ultrashort plasmon lifetimes of several femtoseconds [64, 65]. Since plasmon lifetime (dephasing time) is directly related to the enhancement of the local field inside the mesostructure [66–68], a deeper understanding of the dynamic behavior of plasmons is highly desirable for practical application of plasmonic fields. We observed real-time plasmon dynamics using the ultrafast near-field optical microscope shown in Fig. 6.1. This apparatus can achieve high spatial and temporal resolutions simultaneously, enabling the space- and time-resolved imaging of plasmons induced in metal mesostructures. First, we explored plasmon dynamics in a one-dimensional gold nanorod [69]. Figure 6.9a shows a near-field two-photon excitation image of a chemically synthesized gold nanorod (length ~240 nm, width ~40 nm). In this image, two bright spots are clearly observed along the long axis of the nanorod, indicating the resonant excitation of a second order plasmon mode. We performed time-resolved two-photon induced photoluminescence (TPI-PL) autocorrelation measurements by changing the delay time between the two excitation pulses. The red dotted curve in Fig. 6.9b shows the TPI-PL autocorrelation trace taken near the edge of the nanorod. We also measured the SHG autocorrelation trace on a BBO crystal, as shown by the blue curve in

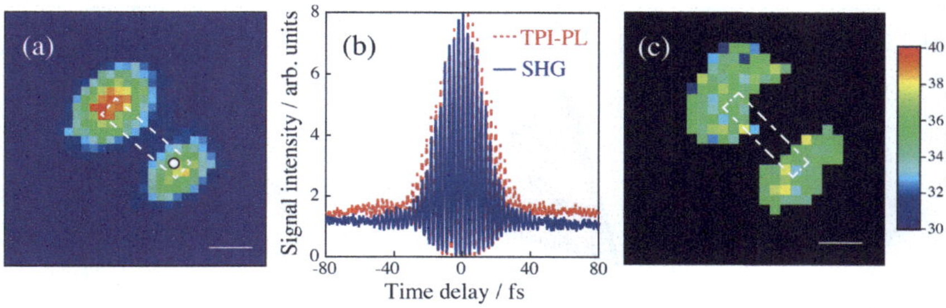

Fig. 6.9 **a** Near-field two-photon excitation image of the gold nanorod (length ~240 nm, width ~40 nm). The dotted line represents the approximate shape of the nanorod. **b** (Red dotted curve) Time-resolved TPI-PL autocorrelation trace taken at the marked position in (**a**). (Blue curve) Time-resolved SHG autocorrelation trace taken on a BBO crystal. **c** TPI-PL correlation width map on the gold nanorod. The black areas indicate the areas where the correlation widths are not evaluated due to low signal intensity. Scale bars: 100 nm. Reprinted with permission from [69]. Copyright 2015 American Chemical Society

Fig. 6.9b. In Fig. 6.9b, the observed TPI-PL correlation width for the gold nanorod (~35 fs) is broader than the SHG autocorrelation trace (~31 fs). The broadening of the correlation width on the nanorod originates from the dephasing process of plasmons. From a Fourier transformation analysis on the TPI-PL autocorrelation trace, a plasmon dephasing time of the gold nanorod was evaluated to be 5 ± 1 fs. We measured the spatial distribution of the plasmon dephasing time by performing time-resolved TPI-PL measurements at every scanned point on the nanorod and obtained a TPI-PL correlation width map, as shown in Fig. 6.9c. In the image, the TPI-PL correlation width is almost constant regardless of the measurement position on the nanorod, indicating that the plasmon dephasing time is constant over the nanorod. This result indicates that locally excited plasmons are rapidly delocalized throughout the nanorod.

We also investigated the spatio-temporal behavior of plasmons induced in a two-dimensional mesoplate [70]. As we described in the previous section, in-plane and out-of-plane polarized plasmon modes are resonantly excited in a two-dimensional mesoplate. Since these plasmon modes have different light confinement capabilities, the plasmon dephasing time should also depend on the polarization direction of the plasmons. To substantiate this prediction, we performed time-resolved near-field measurements on a triangular gold mesoplate by simultaneously exciting in-plane and out-of-plane plasmon modes. Figure 6.10a is a near-field transmission image of the triangular gold mesoplate (side length ~660 nm, thickness ~45 nm) taken near 800 nm. In this image, three extinction spots can be clearly visualized along each side of the mesoplate. The observed near-field distribution is qualitatively reproduced by the spatial superposition of the in-plane (Fig. 6.10b) and out-of-plane (Fig. 6.10c) plasmon modes, as shown in Fig. 6.10d. This agreement indicates that the in-plane and out-of-plane polarized plasmon modes are spectrally overlapped, and consequently, the spatial superposition of these modes is experimentally observed

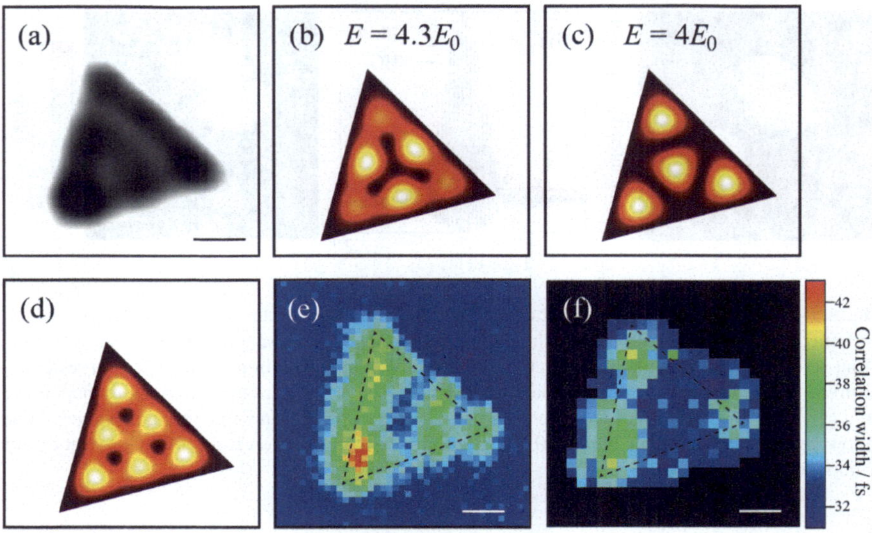

Fig. 6.10 **a** Near-field transmission image of the triangular gold mesoplate (side length ~660 nm, thickness ~45 nm). Observation wavelength ~800 nm. **b**, **c** Square moduli of the eigenfunction calculated for a particle confined inside a two-dimensional triangular potential well. The eigenenergy of (**b**, **c**) are $4.3E_0$ and $4E_0$. (Here, E_0 represents the eigenenergy of the lowest eigenfunction.) These eigenfunctions are assignable to the in-plane and out-of-plane modes, respectively. **d** Spatial superposition of the eigenmodes in (**b**, **c**). **e** Near-field two-photon excitation image of the mesoplate. **f** Correlation width map of the mesoplate. Scale bars: 200 nm. Black dotted lines represent the approximate shape of the triangular mesoplate. Reprinted with permission from [70]. Copyright 2018 American Chemical Society

in the near-field transmission image (Fig. 6.10a). Figure 6.10e shows a near-field two-photon excitation image of the mesoplate. In this image, three bright spots are observable along each side of the mesoplate. This spatial feature is consistent with that on the one observed in the near-field transmission image in Fig. 6.10a, that is, the two eigenmodes shown in Fig. 6.10b, c are excited simultaneously by ultrashort pulses. To visualize the spatial distribution of the plasmon dephasing time, we obtained time-resolved TPI-PL autocorrelation traces at every point on the mesoplate and mapped the TPI-PL correlation width, as shown in Fig. 6.10f. From this image, we found that broadened correlation widths dominate near the apices of the triangular mesoplate. This spatial feature signifies that the plasmon dephasing time near the apices is longer than that measured at other regions on the mesoplate. We also found that the spatial pattern of the correlation width map is considerably different from that of the static near-field two-photon excitation image shown in Fig. 6.10e. The disagreement between the spatial patterns is interpreted in terms of the spatial distributions of the eigenmodes. As explained above, the spatial pattern in Fig. 6.10e is composed of in-plane and out-of-plane eigenmodes shown in Fig. 6.10b, c. The spatial distributions of these modes show that the in-plane eigenmode (Fig. 6.10b) is dominantly excited at the center of the side of the mesoplate, whereas the out-of-plane mode (Fig. 6.10c) is dominantly excited at the apices of the mesoplate. Considering this position-dependent excitation probability of the eigenmodes, the spatial pattern in Fig. 6.10f can be considered to be similar to the spatial distribution of the out-of-plane mode (Fig. 6.10c). This result indicates that the out-of-plane mode has a longer dephasing time than the in-plane mode. Since the plasmon dephasing time is significantly related to the temporal confinement of light, our finding suggests that the out-of-plane plasmon mode is more strongly enhanced in the near-field optical region.

6.5 Enhancement of Optical Responses of Molecules by Mesoscopic Structures

To elucidate the interaction between locally enhanced fields on metal mesostructures and molecular systems, we explored plasmon-enhanced photochemical reactions on assemblies of gold nanorods and poly-diarylethene (DE) [71]. Longitudinal plasmon resonance induced on a gold nanorod amplifies the electric fields in the vicinity of the end-edges, as shown in Fig. 6.11a. The averaged field enhancement near the surface (within 5 nm from the surface) is evaluated to be 4.6 from an electromagnetic simulation. To investigate the influence of the plasmon-enhanced fields on the DE cycloreversion reaction, gold nanorod-poly(DE) assemblies were prepared by mixing chemically synthesized gold nanorods with a poly(DE) tetrahydrofuran (THF) solution. To excite the DE cycloreversion reaction, the sample solution was irradiated using a mode-locked Ti:sapphire laser (wavelength: 780 nm, power: 40 mW–1.4 W). We measured the time dependence of the absorption spectra of the sample solution to monitor the photochemical reaction yields of the poly(DE), as

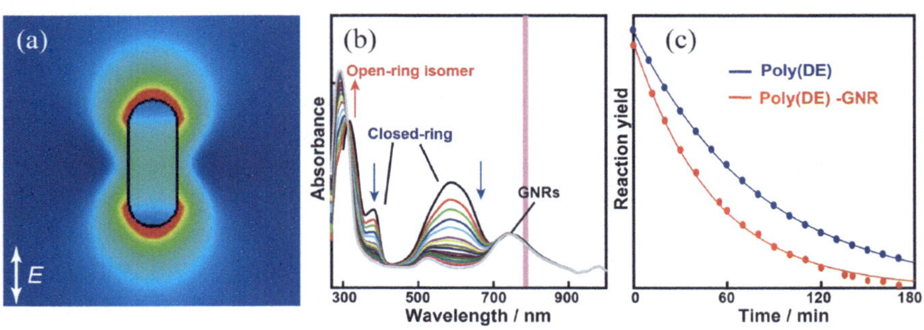

Fig. 6.11 **a** Electric field distribution simulated for a gold nanorod (GNR). Excitation wavelength: 785 nm. White arrow indicates the polarization direction of the incident light. **b** Time evolution of absorption spectra of gold nanorod capped with poly(diarylethene). Red vertical line represents the excitation wavelength. **c** Time evolution of the absorption for the neat poly(diarylethene) and gold nanorods with poly(diarylethene) at 590 nm. Solid lines are results of the exponential fit. Reprinted with permission from [71]. Copyright 2019 Chemical Society of Japan

shown in Fig. 6.11b. The black curve in the figure represents the absorption spectrum of the sample before laser irradiation. The absorption band centered at 740 nm is assignable to the longitudinal plasmon mode in the gold nanorod. On the other hand, the three peaks observed near 310, 370, and 540 nm are attributed to the resonance bands of poly(DE). As depicted by the arrows in Fig. 6.11b, the absorption intensities at 370 and 540 nm decrease as the laser irradiation time increases. We also found that a new absorption band appears near 290 nm and becomes stronger with increasing laser irradiation time. These spectral changes imply that the cycloreversion reaction of poly(DE) proceeds by near-infrared laser irradiation. To examine the influence of the gold nanorod on the reaction, we performed the same measurements on the neat poly(DE) solution. The red and blue plots in Fig. 6.11c are time-traces of the absorption intensities at 540 nm measured for the assembled and neat poly(DE), respectively. In this figure, the absorption intensity decays exponentially in both the assembled and neat samples, indicating that the cycloreversion reaction proceeds even in the neat poly(DE) solution. We evaluated the reaction rates by fitting the time-traces with single exponential functions shown as solid curves in Fig. 6.11c. From the analysis, the reaction rate constants of the assembled and neat poly(DE) samples are determined to be $k_{\text{GR-DE}} = 0.020\ \text{min}^{-1}$ and $k_{\text{DE}} = 0.012\ \text{min}^{-1}$. We also found that a reaction ratio between the assembled and neat poly(DE) samples is constant regardless of the incident laser power. This fact indicates that thermal reaction does not occur in the present study. These results demonstrate that the plasmonic fields induced in the gold nanorods accelerate the one-photon induced cycloreversion reaction of poly (DE).

Strong interactions between plasmonic fields on mesostructures and molecules in the vicinity of the structures can amplify various optical signals, including fluorescence and Raman scattering [34–36, 72–74]. This striking property has potential applications in a wide range of fields, such as high-sensitive sensors and biological imaging. For the practical implementation of plasmon-enhanced light-molecule interactions, rational design, and construction of wide-scale plasmonic substrates are highly desirable. The metallic nanohole array is a representative example of a plasmonic substrate and has attracted enormous attention for its unique optical characteristics, such as extraordinary light transmission and amplification of local fields [75–79]. To elucidate the interaction between molecules and plasmonic fields on a metallic nanohole array, we investigate the fluorescence characteristics of dye molecules dispersed on a gold nanohole array using the SNOM [80]. Figure 6.12a shows an SEM image of the gold nanohole array. From this image, the diameter and periodicity of the nanohole arrays were estimated to be 240 and 480 nm, respectively. An atomic force microscope (AFM) was used to measure the topography, and the depth of the nanohole was determined to be 40 nm. To avoid quenching by the gold, a thin SiO_2 film (20 nm) was deposited on the sample surface. A drop of methanol solution containing R6G molecules (10^{-4} mol/dm^3) was spin-coated onto the sample for the fluorescence measurements. We used a continuous wave laser (532 nm) to excite the fluorescence of R6G molecules. Figure 6.12b shows near-field fluorescence spectra taken at the red and blue points shown in Fig. 6.12a. We found that spectral shapes in Fig. 6.12b are different from that of the original R6G molecules.

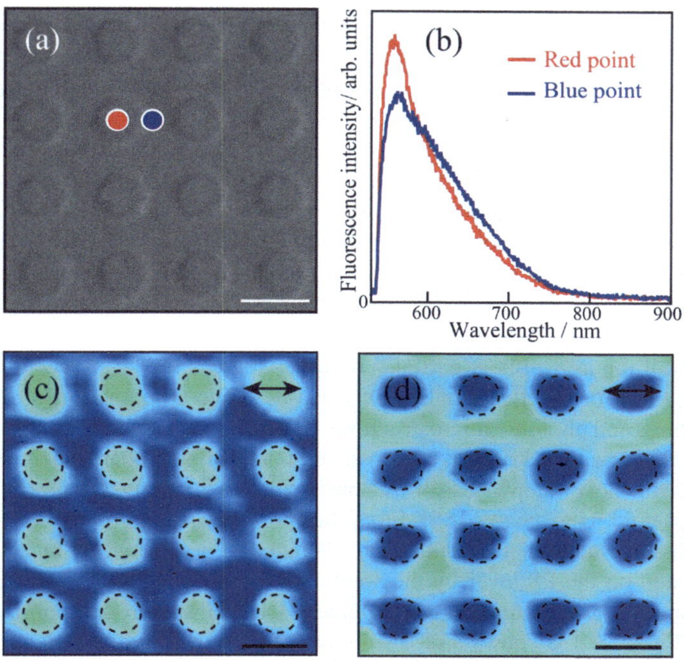

Fig. 6.12 **a** A SEM image of the gold nanohole arrays. **b** Near-field fluorescence spectra from R6G molecules observed at the red and blue points in (**a**). **c**, **d** Near-field fluorescence excitation images of the plasmonic chip. Observation wavelength: 550–580 nm for (**c**) and 635–665 nm for (**d**). Black arrows represent the polarization directions of the incident light. Black dotted circles indicate the position of the nanoholes. Scale bars: 400 nm. Reprinted with permission from [80]. Copyright 2019 Chemical Society of Japan

This result indicates that the spectral profile of the fluorescence is modulated by the optical characteristics of the gold nanohole arrays. We also found from Fig. 6.12b that the spectral profile of the fluorescence varies with the observation position on the nanohole array. To examine the position dependency of the fluorescence spectral modulation, we mapped the fluorescence intensity over a given spectral range and obtained near-field fluorescence excitation images of the sample. Figure 6.12c is a near-field fluorescence excitation image at 550–580 nm. Dotted circles indicate the approximate shapes of the nanoholes. The fluorescence excitation probability in this image is locally enhanced at the nanoholes. To elucidate the origin of the observations, we simulated the electric field distribution of the sample and found that intense fields are generated inside the nanoholes. These results indicate that strong electric fields induced at nanoholes boost the fluorescence excitation probability of the R6G molecules. We also obtained near-field fluorescence excitation images by mapping the fluorescence intensity from 635 to 665 nm, as shown in Fig. 6.12d. In this image, the fluorescence is predominantly excited outside the nanoholes, which is entirely different from the observation in Fig. 6.12c. These results indicate that the spectral features and excitation rate of molecular fluorescence are locally modulated by the sample position. This fact indicates that mesoscopic structures can be used for the spatially selective modulation of optical functions of molecular systems at the subwavelength scale.

Although the metallic nanohole array has been exploited in various optical applications, its operation spectral range is relatively narrow. To increase the versatility of plasmonic fields for a wide variety of applications, the fabrication of plasmonic substrates with a wide spectral range is highly desirable. Two-dimensional closed-packed assembly of nanoparticles exhibits broad spectral features. However, the local field enhancement of the assembly is relatively low because spatially confined fields are substantially delocalized over the assembly [81]. Hence, the state-of-the-art application of the assembly is partially limited, whereas the large preparation area is advantageous for practical applications [82–84]. The combination of plasmonic nanostructures and photonic microstructures is the most promising means of simultaneously realizing a broad optical response and intense field enhancement. Recently, we proposed a novel stepwise self-assembly scheme for polystyrene (PS) microspheres and gold nanospheres [85]. First, the PS microspheres (diameter ~500 nm) were self-assembled on a glass substrate. Then, a thin gold film (thickness ~20 nm) was deposited on the PS layer by the sputtering method. Finally, an aqueous solution of dispersed gold nanospheres (diameter ~94 nm) was dropped on the sample substrate. Figure 6.13a shows an SEM image of the fabricated sample. As seen in this image, gold nanosphere trimers are selectively located at the hollow sites of the PS assembly. In this sample, not only in-plane plasmonic coupling inside the trimers but also out-of-plane coupling between the gold nanospheres and gold-coated PS assembly are induced, and thus giant optical field enhancements are involved. In addition, the photonic mode couples the optical fields inside the gold-coated PS assembly. Since these coupling schemes operate in concert with each other, the fabricated structure attains a very broad spectral response and high local field enhancement simultaneously. We measured the two-photon photoluminescence of the sample

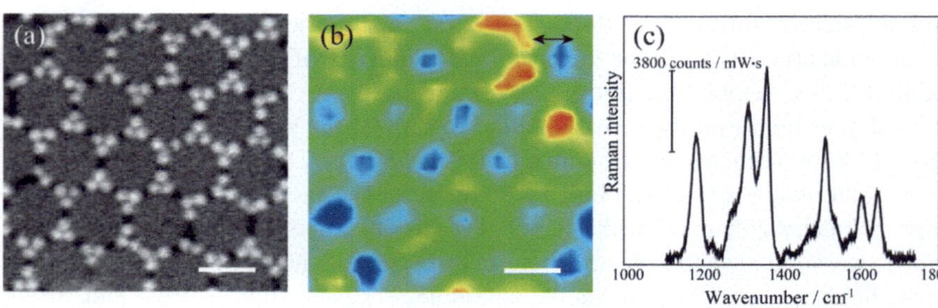

Fig. 6.13 **a** A SEM image of gold nanoparticles on the gold-coated polystyrene microsphere assembly. **b** Two-photon excitation image of the assembly. Black arrows indicate the incident polarization direction of the excitation light. Scale bars: 500 nm. **c** Raman scattering spectrum from R6G molecules excited at 785 nm. Incident power: 0.4 mW. Exposure time: 3 s. Reprinted with permission from [85]. Copyright 2016 American Chemical Society

using a confocal optical microscope and near-infrared ultrashort pulses (wavelength: 800 nm, repetition rate: 80 MHz, pulse width <100 fs). The two-photon excitation image in Fig. 6.13b reveals that the two-photon excitation probability is locally enhanced on the sample. This result directly indicates that locally enhanced fields are generated on the assembly. To investigate the influence of the enhanced fields on the optical response of the molecules, we dispersed R6G molecules on the fabricated sample and excite Raman scattering from the molecules using a CW laser (785 nm) as the excitation source. Figure 6.13c shows the Raman scattering spectrum from the R6G molecules on the sample. We compared the observed Raman signal intensity with that on the bare glass substrate and found that the Raman enhancement factor reaches more than 10^8. We also carried out Raman scattering measurements with 633 and 532 nm CW laser sources and found that the Raman enhancement for these excitations is comparable to that achieved at 785 nm, indicating that the fabricated structure exhibits intense optical fields with a broad spectral response. These characteristics are indispensable for the amplification of optical responses in molecular systems.

6.6 Conclusion

In this chapter, we described the development of advanced near-field imaging methods and their applications to the spatio- and temporal-characterization of plasmons excited in mesoscopic structures. The excited plasmons in these structures exhibit periodic wavy features that depend on the size and shape of the structures and the observed wavelength. The observed spatial features are attributed to the plasmonic standing waves generated in the structures from theoretical simulations. We demonstrate that out-of-plane and in-plane plasmons are excited in mesoplates and these modes can be used to amplify and modulate the optical responses of molecules in a space-resolved manner. To utilize both in-plane and out-of-plane modes, we fabricated gold nanoparticle trimer assemblies on gold-capped hexagonal structures. The assemblies provide large optical field amplification, with maintaining a broad spectral response. Plasmons excited in gold nanorods and plasmonic nanohole arrays were also utilized to accelerate photochemical processes and the spectral profile of molecular fluorescence. These observations clearly demonstrate that the near-field interactions between plasmons and molecular systems modulate the electronic structure of the molecules and provide novel excitation schemes beyond the conventional light-matter interaction. The localized optical fields generated on the mesoscopic structures can produce novel photophysical and photochemical processes and thus are promising for the advanced function control of materials. The use of novel excitation schemes with coherent control techniques enables not only the exploration of novel fundamental areas in physical chemistry but also the development of advanced future applications, such as ultrafast optical communication and computational techniques.

Acknowledgements We are grateful to the collaborators who contributed to this research project, in particular, Prof. H. Okamoto, Prof. H. Misawa, Prof. K. Ueno, Prof. K. Tawa, Prof. S. Kobatake, Prof. T. Iida, Prof. Y. Nishiyama, Dr. H. Mizobata, Mr. T. Matsuura, Ms. T. Uchida, Mr. S. Noda, and Mr. S. Hasegawa. The present work was supported by JSPS KAKENHI Grant Number 26107001, 26107003, 26620018, 15K21725, JP16H04100, Grant-in-Aid for Scientific Research on Innovative Areas "Photosynergetics".

References

1. Turro NJ, Ramamurthy V, Scaiano JC (2009) Principles of molecular photochemistry. University Science Books, USA
2. Demtröder W (2014) Laser spectroscopy 1 basic principles. Springer, Heidelberg
3. Yang N, Tang Y, Cohen AE (2009) Spectroscopy in sculpted fields. Nano Today 4:269–279
4. Iida T, Ishihara H (2009) Unconventional control of excited states of a dimer molecule by a localized light field between metal nanostructures. Phys Sat Sol A 206:980–984
5. Iida T, Aiba Y, Ishihara H (2011) Anomalous optical selection rule of an organic molecule controlled by extremely localized light field. App Phys Lett 98:053108
6. Jain PK, Ghosh D, Baer R, Rabani E, Paul Alivisatos A (2012) Near-field manipulation of spectroscopic selection rules on the nanoscale. Proc Natl Acad Sci USA 109:8016–8019
7. Takase M, Ajiki H, Mizumoto Y, Komeda K, Nara M, Nabika H, Yasuda S, Ishihara H, Murakoshi K (2013) Selection-rule breakdown in plasmon-induced electronic excitation of an isolated single-walled carbon nanotube. Nature Photon 7:550–554
8. Tsuboi Y, Shimizu R, Shoji T, Kitamura N (2009) Near-infrared continuous-wave light driving a two-photon photochromic reaction with the assistance of localized surface plasmon. J Am Chem Soc 131:12623–12627
9. Murazawa N, Ueno K, Mizeikis V, Juodkazis S, Misawa H (2009) Spatially selective nonlinear photopolymerization induced by the near-field of surface plasmons localized on rectangular gold nanorods. J Phys Chem C 113:1147–1149
10. Wu B, Ueno K, Yokota Y, Sun K, Zeng H, Misawa H (2012) Enhancement of a two-photon-induced reaction in solution using light-harvesting gold nanodimer structures. J Phys Chem Lett 3:1443–1447
11. Volpe G, Noack M, Aćimović SS, Reinhardt C, Quidant R (2012) Near-field mapping of plasmonic antennas by multiphoton absorption in poly(methyl methacrylate). Nano Lett 12:4864–4868
12. Hakala TK, Toppari JJ, Kuzyk A, Pettersson M, Tikkanen H, Kunttu H, Törmä P (2009) Vacuum Rabi splitting and strong-coupling dynamics for surface-plasmon polaritons and rhodamine 6G molecules. Phys Rev Lett 103:053602
13. Zengin G, Johansson G, Johansson P, Antosiewicz TJ, Käll M, Shegai T (2013) Approaching the strong coupling limit in single plasmonic nanorods interacting with J-aggregates. Sci Rep 3:3074
14. Tame MS, McEnery KR, Özdemir SK, Lee J, Maier SA, Kim MS (2013) Quantum plasmonics. Nat Phys 9:329–340
15. Törmä P, Barnes WL (2015) Strong coupling between surface plasmon polaritons and emitters: a review. Rep Prog Phys 78:013901
16. Pelton M, David Storm S, Leng H (2019) Strong coupling of emitters to single plasmonic nanoparticles: exciton-induced transparency and Rabi splitting. Nanoscale 11:14540–14552

17. Hertzog M, Wang M, Mony J, Börjesson K (2019) Strong light-matter interactions: a new direction within chemistry. Chem Soc Rev 48:937–961
18. Okamoto H, Imura K (2013) Visualizing the optical field structures in metal nanostructures. J Phys Chem Lett 4:2230–2241
19. Neuman T, Alonso-González P, Garcia-Etxarri A, Schnell M, Hillenbrand R, Aizpurua J (2015) Mapping the near fields of plasmonic nanoantennas by scattering-type scanning near-field optical microscopy. Laser Photonics Rev 9:637–649
20. Barnes WL, Dereux A, Ebbesen TW (2003) Surface plasmon subwavelength optics. Nature 424:824–830
21. Schuller JA, Barnard ES, Cai W, Jun YC, White JS, Brongersma ML (2010) Plasmonics for extreme light concentration and manipulation. Nat Mater 9:193–204
22. Novotny L, van Hulst N (2011) Antennas for light. Nature Photon 5:83–90
23. Giannini V, Fernández-Domínguez AI, Heck SC, Maier SA (2011) Plasmonic nano-antennas: fundamentals and their use in controlling the radiative properties of nanoemitters. Chem Rev 111:3888–3912
24. Imura K, Nagahara T, Okamoto H (2004) Plasmon mode imaging of single gold nanorods. J Am Chem Soc 126:12730–12731
25. Imura K, Nagahara T, Okamoto H (2005) Near-field optical imaging of plasmon modes in gold nanorods. J Chem Phys 122:154701
26. Imura K, Nagahara T, Okamoto H (2005) Near-field two-photon-induced photoluminescence from single gold nanorods and imaging of plasmon modes. J Phys Chem B 109:13214–13220
27. Piatkowski L, Accanto N, van Hulst NF (2016) Ultrafast meets ultrasmall: controlling nanoantennas and molecules. ACS Photon 3:1401–1414
28. Dąbrowski M, Dai Y, Petek H (2017) Ultrafast microscopy: imaging light with photoelectrons on the nano-femto scale. J Phys Chem Lett 8:4446–4455
29. Beane G, Devkota T, Brown BS, Hartland GV (2019) Ultrafast measurements of the dynamics of single nanostructures: a review. Rep Prog Phys 82:016401
30. Imaeda K, Imura K (2013) Optical control of plasmonic fields by phase-modulated pulse excitations. Opt Express 21:27481–27489
31. Imaeda K, Imura K (2016) Raman activity and dynamics of plasmons on a rough gold film studied by ultrafast scanning near-field optical microscopy. In: Ozaki Y, Schatz GC, Graham D, Itoh T (eds) Frontiers of plasmon enhanced spectroscopy. ACS symposium series 1246, vol 2. pp 121–137
32. Imaeda K, Imura K (2016) Dye-assisted visualization of plasmon modes excited in single gold nanoplates. Chem Phys Lett 646:179–184
33. Mizobata H, Hasegawa S, Imura K (2018) Development of aperture-type near-field re-flection spectroscopy and its application to single silver nanoplates. J Phys Chem C 121:11733–11738
34. Anger P, Bharadwaj P, Novotny L (2006) Enhancement and quenching of single-molecule fluorescence. Phys Rev Lett 96:113002
35. Kühn S, Håkanson U, Rogobete L, Sandoghdar V (2006) Enhancement of single-molecule fluorescence using a gold nanoparticle as an optical nanoantenna. Phys Rev Lett 97:017402
36. Li J-F, Li C-Y, Aroca RF (2017) Plasmon-enhanced fluorescence spectroscopy. Chem Soc Rev 46:3962–3979
37. Smith RJ, Somekh MG, Sharples SD, Pitter MC, Harrison I, Rossignol C (2008) Parallel detection of low modulation depth signals: application to picosecond ultrasonics. Meas Sci Technol 19:055301
38. Smith RJ, Light RA, Sharples SD, Johnston NS, Pitter MC, Somekh MG (2010) Multichannel, time-resolved picosecond laser ultrasound imaging and spectroscopy with custom complementary metal-oxide-semiconductor detector. Rev Sci Instrum 81:024901
39. Smith RJ, Light RA, Johnston N, Sharples S, Pitter MC, Somekh MG (2010) Parallel detection in laser ultrasonics. J Phys: Conf Ser 214:012006
40. Born M, Wolf E (2002) Principle of optics. Cambridge Univ Press, Cambridge
41. Boudarham G, Kociak M (2012) Modal decompositions of the local electromagnetic density of states and spatially resolved electron energy loss probability in terms of geometric modes. Phys Rev B 85:245447

42. Ditlbacher H, Hohenau A, Wagner D, Kreibig U, Rogers M, Hofer F, Aussenegg FR, Krenn JR (2005) Silver nanowires as surface plasmon resonators. Phys Rev Lett 95:257403
43. Vesseur EJR, de Waele R, Kuttge M, Polman A (2007) Direct observation of plasmonic modes in Au nanowires using high-resolution cathodoluminescence spectroscopy. Nano Lett 7:2843–2846
44. Dorfmüller J, Vogelgesang R, Weitz RT, Rockstuhl C, Etrich C, Pertsch T, Lederer F, Kern K (2009) Fabry-Pérot resonances in one-dimensional plasmonic nanostructures. Nano Lett 9:2372–2377
45. Rossouw D, Couillard M, Vickery J, Kumacheva E, Botton GA (2011) Multipolar plasmonic resonances in silver nanowire antennas imaged with a subnanometer electron probe. Nano Lett 11:1499–1504
46. Jackson JD (1999) In: Classical electrodynamics, 3rd ed. Wiley
47. Mizobata H, Ueno K, Misawa H, Okamoto H, Imura K (2017) Near-field spectroscopic properties of complementary gold nanostructures: applicability of Babinet's principle in the optical region. Opt Express 25:5279–5289
48. Rang M, Jones AC, Zhou F, Li Z-Y, Wiley BJ, Xia Y, Raschke MB (2008) Optical near-field mapping of plasmonic nanoprisms. Nano Lett 8:3357–3363
49. Awada C, Popescu T, Douillard L, Charra F, Perron A, Yockell-Lelièvre H, Baudrion A-L, Adam P-M, Bachelot R (2012) Selective excitation of plasmon resonances of single Au triangles by polarization-dependent light excitation. J Phys Chem C 116:14591–14598
50. Viarbitskaya S, Teulle A, Marty R, Sharma J, Girard C, Arbouet A, Dujardin E (2013) Tailoring and imaging the plasmonic local density of states in crystalline nanoprisms. Nat Mater 12:426–432
51. Imura K, Ueno K, Misawa H, Okamoto H, McArthur D, Hourahine B, Papoff F (2014) Plasmon modes in single gold nanodiscs. Opt Express 22:12189–12199
52. Nelayah J, Kociak M, Stéphan O, García de Abajo FJ, Tencé M, Henrard L, Taverna D, Pastoriza-Santos I, Liz-Marzán LM, Colliex C (2007) Mapping surface plasmons on a single metallic nanoparticle. Nat Phys 3:348–353
53. Nelayah J, Gu L, Sigle W, Koch CT, Pastoriza-Santos I, Liz-Marzán LM, van Aken PA (2009) Direct imaging of surface plasmon resonances on single triangular silver nanoprisms at optical wavelength using low-loss EFTEM imaging. Opt Lett 34:1003–1005
54. Nelayah J, Kociak M, Stéphan O, Geuquet N, Henrard L, García de Abajo FJ, Pastoriza-Santos I, Liz-Marzán LM, Colliex C (2010) Two-dimensional quasistatic stationary short range surface plasmons in flat nanoprisms. Nano Lett 10:902–907
55. Gu L, Sigle W, Koch CT, Ögüt B, van Aken PA, Talebi N, Vogelgesang R, Mu J, Wen X, Mao J (2011) Resonant wedge-plasmon modes in single-crystalline gold nanoplatelets. Phys Rev B 83:195433
56. Schmidt F-P, Ditlbacher H, Hohenester U, Hohenau A, Hofer F, Krenn JR (2012) Dark plasmonic breathing modes in silver nanodisks. Nano Lett 12:5780–5783
57. Schmidt FP, Ditlbacher H, Hofer F, Krenn JR, Hohenester U (2014) Morphing a plasmonic nanodisk into a nanotriangle. Nano Lett 14:4810–4815
58. Kawasaki N, Meuret S, Weil R, Lourenço-Martins H, Stéphan O, Kociak M (2016) Extinction and scattering properties of high-order surface plasmon modes in silver nanoparticles probed by combined spatially resolved electron energy loss spectroscopy and cathodoluminescence. ACS Photon 3:1654–1661
59. Campos A, Arbouet A, Martin J, Gérard D, Proust J, Plain J, Kociak M (2017) Plasmonic breathing and edge modes in aluminum nanotriangles. ACS Photon 4:1257–1263
60. Imaeda K, Hasegawa S, Imura K (2018) Imaging of plasmonic eigen modes in gold triangular mesoplates by near-field optical microscopy. J Phys Chem C 122:7399–7409
61. Li W-K, Blinder SM (1985) Solution of the Schrödinger equation for a particle in an equilateral triangle. J Math Phys 26:2784–2786
62. Betzig E, Chichester RJ (1993) Single molecules observed by near-field scanning optical microscopy. Science 262:1422–1425

63. Matsuura T, Imaeda K, Hasegawa S, Suzuki H, Imura K (2019) Characterization of overlapped plasmon modes in a gold hexagonal plate revealed by three-dimensional near-field optical microscopy. J Phys Chem Lett 10:819–824
64. Mårsell E, Losquin A, Svärd R, Miranda M, Guo C, Harth A, Lorek E, Mauritsson J, Arnold CL, Xu H, L'Huillier A, Mikkelsen A (2015) Nanoscale imaging of local few-femtosecond near-field dynamics within a single plasmonic nanoantenna. Nano Lett 15:6601–6608
65. Sun Q, Yu H, Ueno K, Kubo A, Matsuo Y, Misawa H (2016) Dissecting the few-femtosecond dephasing time of dipole and quadrupole modes in gold nanoparticles using polarized photoemission electron microscopy. ACS Nano 10:3835–3842
66. Sönnichsen C, Franzl T, von Plessen G, Feldmann J, Wilson O, Mulvaney P, Wilk T (2002) Drastic reduction of plasmon damping in gold nanorods. Phys Rev Lett 88:077402
67. Stockman MI (2011) Nanoplasmonics: past, present, and glimpse into future. Opt Exp 19:22029–22106
68. Hanke T, Cesar J, Knittel V, Trügler A, Hohenester U, Leitenstorfer A, Bratschitsch R (2012) Tailoring spatiotemporal light confinement in single plasmonic nanoantennas. Nano Lett 12:992–996
69. Nishiyama Y, Imaeda K, Imura K, Okamoto H (2015) Plasmon dephasing in single gold nanorods observed by ultrafast time-resolved near-field optical microscopy. J Phys Chem C 119:16215–16222
70. Imaeda K, Hasegawa S, Imura K (2018) Static and dynamic near-field measurements of high-order plasmon modes induced in a gold triangular nanoplate. J Phys Chem Lett 9:4075–4081
71. Noda S, Hasegawa S, Hamada H, Kobatake S, Imura K (2019) Plasmon enhanced optical response of diarylethene molecules absorbed on gold nanorods. Chem Lett 48:537–540
72. Stiles PL, Dieringer JA, Shah NC, Van Duyne RP (2008) Surface-enhanced Raman spectroscopy. Annu Rev Anal Chem 1:601–626
73. Sharma B, Fernanda Cardinal M, Kleinman SL, Greeneltch NG, Frontiera RR, Blaber MG, Schatz GC, Van Duyne RP (2013) High-performance SERS substrates: advanced and challenges. MRS Bull 38:615–624
74. Reguera J, Langer J, Jime´nez de Aberasturi D, Liz-Marzán (2017) Anisotropic metal nanoparticles for surface enhanced Raman scattering. Chem Soc Rev 46:3866–3885
75. Ebbesen TW, Lezec HJ, Ghaemi HF, Thio T, Wolff PA (1998) Extraordinary optical transmission through sub-wavelength hole arrays. Nature 391:667–669
76. Martín-Moreno L, García-Vidal FJ, Lezec HJ, Pellerin KM, Thio T, Pendry JB, Ebbesen TW (2001) Theory of extraordinary optical transmission through subwavelength hole arrays. Phys Rev Lett 86:1114–1117
77. Kwak E-S, Henzie J, Chang S-H, Gray SK, Schatz GC, Odom TW (2005) Surface plasmon standing waves in large-area subwavelength hole arrays. Nano Lett 5:1963–1967
78. Rindzevicius T, Alaverdyan Y, Sepulveda B, Pakizeh T, Käll M, Hillenbrand R, Aizpurua J, García de Abajo FJ (2007) Nanohole plasmons in optically thin gold films. J Phys Chem C 111:1207–1212
79. Imaeda K, Minoshima W, Tawa K, Imura K (2019) Direct visualization of near-field distributions on a two-dimensional plasmonic chip by scanning near-field optical microscopy. J Phys Chem C 123:10529–10535
80. Imaeda K, Tawa K, Imura K (2019) Nanoscopic visualization of fluorescence excitation probability on two-dimensional periodical gold nanohole arrays. Chem Lett 48:1119–1121
81. Shimada T, Imura K, Hossain MK, Okamoto H, Kitajima M (2008) Near-field study on correlation of localized electric field and nanostructures in monolayer assembly of gold nanoparticles. J Phys Chem C 112:4033–4035
82. Scanlon MD, Smirnov E, Stockmann TJ, Peljo P (2018) Gold nanofilms at liquid-liquid interfaces: an emerging platform for redox electrocatalysis nanoplasmonic sensors and electrovariable optics. Chem Rev 118:3722–3751
83. Yang G, Ivanov IN, Ruther RE, Sacci RL, Subjakova V, Hallinan D, Nanda J (2018) Electrolyte solvation structure at solid-liquid interface probed by nanogap surface-enhanced Raman spectroscopy. ACS Nano 12:10159–10170

84. Lu X, Huang Y, Liu B, Zhang L, Song L, Zhang J, Zhang A, Chen T (2018) Light-controlled shrinkage of large-area gold nanoparticle monolayer film for tunable SERS activity. Chem Mater 30:1989–1997
85. Uchida T, Yoshikawa T, Tamura M, Iida T, Imura K (2016) Multiple resonances induced by plasmonic coupling between gold nanoparticle trimers and hexagonal assembly of gold-coated polystyrene microspheres. J Phys Chem Lett 7:3652–3658

Chapter 7
Plasmon-Associated Control of Chemical Reaction at Nanometer Scale

Shuichi Toyouchi, Tomoko Inose, Kenji Hirai, and Hiroshi Uji-i

Abstract In this chapter, we introduce multiphoton photochromic reversible reaction in a diarylethene derivative at nanometer scale much beyond the diffraction limit. For this, we have utilized nonlinear plasmonics along a plasmonic waveguide, particularly on the chemically synthesized silver nanowire. Propagating near-field surface plasmons excited by near-infrared femtosecond laser pulses successfully excite diarylethene derivatives surrounding a silver nanowire. Three-photon cyclization and two-photon cycloreversion reaction induced through the propagating surface plasmon polaritons were visualized by means of fluorescence imaging. We demonstrated one-color near-infrared laser reversible control by tuning the laser power.

Keywords Multiphoton photochromic reaction · Fluorescent diarylethene · Silver nanowire · Plasmonic waveguiding · Nonlinear optical effects · Second-harmonic generation

7.1 Introduction (Plasmonic Waveguiding on Silver Nanowire)

The plasmonic waveguide effect arises from the strong coupling of light to collective electron motions and allows for the confinement and propagation of optical fields on sub-diffraction limit sized metallic structures. It has the potential to miniaturize photonic and optoelectronic devices to the nanoscale. Among various metal nanostructures for plasmonic waveguiding, chemically synthesized silver nanowires

S. Toyouchi · H. Uji-i (✉)
Department of Chemistry, KU Leuven, Heverlee, Flemish-Brabant 3001, Belgium
e-mail: hiroshi.ujii@es.hokudai.ac.jp; hiroshi.ujii@kuleuven.be

S. Toyouchi
e-mail: toyo.toyouchi@kuleuven.be

T. Inose · K. Hirai · H. Uji-i
Research Institute for Electronic Science (RIES), Hokkaido University, Sapporo, Hokkaido 011-0020, Japan

H. Miyasaka et al. (eds.), *Photosynergetic Responses in Molecules and Molecular Aggregates*, https://doi.org/10.1007/978-981-15-5451-3_7

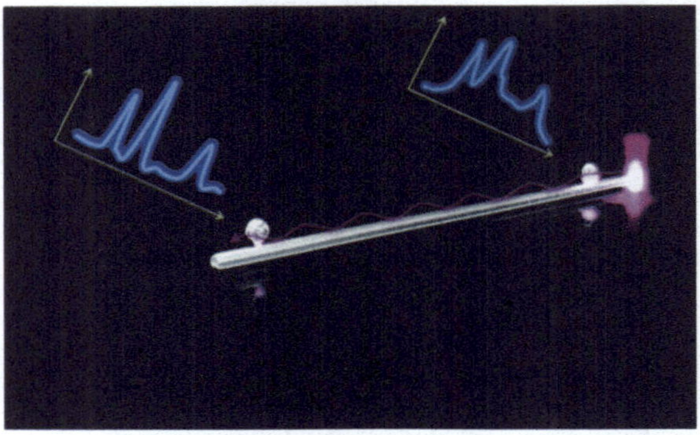

Fig. 7.1 The concept of remote spectroscopy on sub-diffraction limit diameter metallic nanowires. In this case, SERS hotspots at points of nanoparticle adsorption along the nanowire are excited by SPP propagating along the wire. Separating the excitation laser focus from the detection volume reduces the spectroscopic background and photo-degradation of the sample. Figure adapted from Ref. [4]

(AgNWs) are especially promising with their atomically smooth surfaces, and low absorption at visible and near-infrared spectral ranges, affording them excellent plasmonic properties [1–3]. AgNWs are also excellent tools for nanoscopy, especially as they can allow for the separation in space of the points of optical excitation and probing (Fig. 7.1). Remote excitation of surface-enhanced Raman scattering (RE-SERS) [4, 5] and single-molecule fluorescence [6], have all been developed in this context. RE-SERS has even been applied inside single live cells, such that AgNWs become plasmonic endoscopes [7].

In this present chapter, we first introduce our pioneering remote excitation spectroscopic works using AgNW one-photon (linear) plasmonic waveguiding (Sect. 7.2). Then we discuss AgNW's nonlinear optical response under near-IR femtosecond laser irradiation and expend the AgNW plasmonic waveguide into nonlinear regime (Sect. 7.3). Finally, we demonstrate one-color reversible photochromic reaction in a diarylethene derivative using the AgNW nonlinear remote excitation scheme (Sect. 7.4).

7.2 Remote Excitation of One-Photon Raman/Fluorescence

In this section, we introduce two pioneering works for remote excitation spectroscopy using AgNW plasmonic waveguiding, namely remote excitation of SERS [4] and single-molecule fluorescence [6]. Since our initial report in 2009 [4], the use of metallic nanowires for remote spectroscopy has expanded dramatically, and recent reviews have detailed the experiments showing dual remote excitation of RE-SERS on coupled nanowires, remote sensing of Raman optical activity, remote excitation of SPP-induced catalysis, live-cell endoscopy, and much more [8, 9].

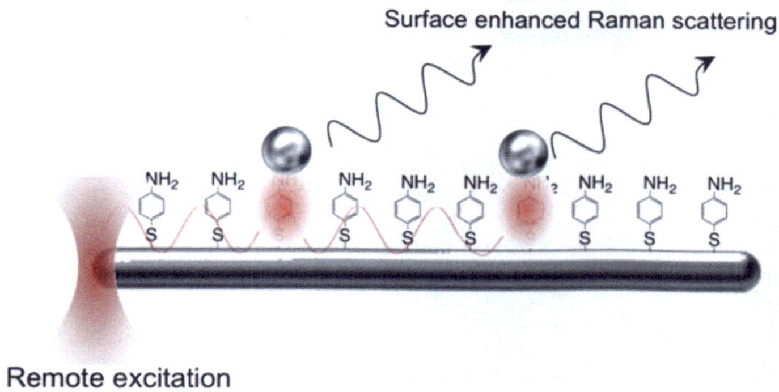

Fig. 7.2 A schematic representation of remote excitation of SERS hotspots at points of NPs adsorption on a 4-ATP/AgNW. Figure adapted from Ref. [4]

Remote Excitation of Surface-Enhanced Raman Spectroscopy For the first demonstration on remote excitation of SERS, AgNW surface was functionalized with Raman reporter molecules, 4-aminothiophenol (4-ATP) (Fig. 7.2). The 4-ATP/AgNWs were then dispersed onto glass substrates and all non-bound 4-ATP rinsed away. Then, silver nanoparticles (AgNPs) were spin-cast on the sample such that a few NPs were bound to each AgNW. In this way, 4-ATP molecules would be likely positioned at the junction between the AgNW and the adsorbed NP (Fig. 7.2).

Two excitation configurations were employed, wide-field excitation in which the whole field of view was excited by a 632.8 nm laser, and focused excitation with the laser focused at the left end of the AgNW to launch propagating SPPs. Figure 7.3a shows an optical transmission image of a NPs/4-ATP/AgNW with diameter of ~100 nm and length of ~19 μm, and that displayed seven SERS hotspots when excited both by wide-field and focused excitation at one end as shown in Fig. 7.3b and c, respectively. Remote excitation SERS hotspots under the focused excitation were clearly observed even more than 5 μm away from the excitation end. Under remote excitation, the intensities of the SERS hotspots decayed in a roughly exponential fashion as a function of distance along the AgNW and could be fitted with the following Eq. (7.1).

$$I(x) = I_0 \exp(-x/L) \tag{7.1}$$

Here, x is the distance from the SERS hotspot to the focused excitation end and L is the decay length. For the AgNW in Fig. 7.3, a value for L of 3.4 ± 1.6 μm was extracted. Although the analysis is rudimentary and ignores damping due to SPP leakage at the multiple hotspots, the L values extracted are similar to the characteristic propagation lengths observed for SPP propagation in AgNWs of similar dimensions and measured with excitation at 800–830 nm (~3 μm) [10–12].

The background typically observed in SERS spectroscopy is due to Raman signal from carbon contamination, enhanced fluorescence, or image-molecule/electronic

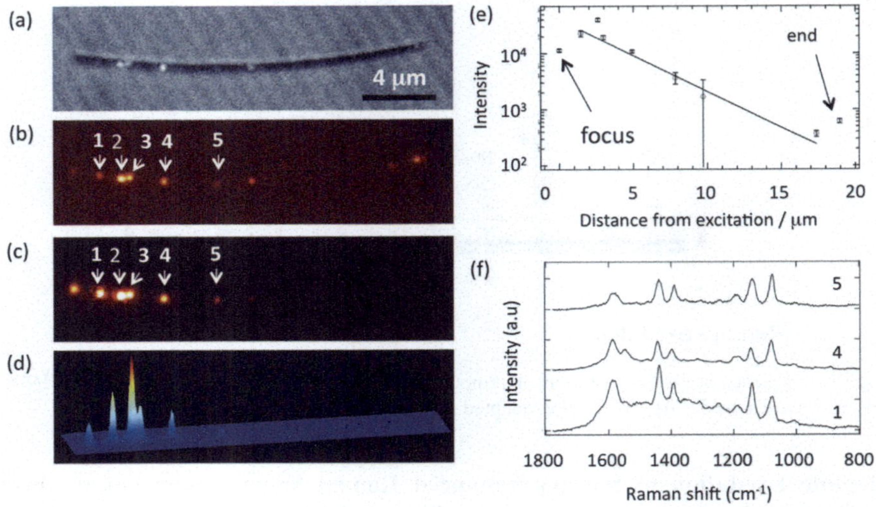

Fig. 7.3 **a** White light optical transmission image, **b** Raman image under wide-field excitation, and Raman image under focused excitation at left end of the wire in **c** 2-D, and **d** 3-D. Multiple hotspots are numbered 1–5 as indicated by arrows. **e** RE-SERS intensity at each hotspot as a function of distance from the excited end of the AgNW. **f** 4-ATP spectra at the hotspots indicated the background is reduced as the distance between the hotspot and the point of excitation increases. Figure adapted from Ref. [4]

continuum coupling [13]. It is worth mentioning that the RE-SERS spectrum has much less background compared to typical SERS under direct excitation (Fig. 7.3f). This is because the excitation volume for remote excitation is reduced to the sub-diffraction limit dimensions of the SPP mode volume and is an example of the fundamental advantages of the RE-SERS technique.

Remote Excitation Single-Molecule Fluorescence Detection In 2012, we published the first studies that attempted to use fluorescence localization of emitters in the vicinity of metallic nanostructures in order to visualize nanostructures [14]. However, it was immediately apparent that the super-resolution reconstructed imaging based on molecule localization to sub-diffraction limit certainty by fitting the point spread function (PSF) of its emission with a 2-D Gaussian [15] did not perfectly reproduce the dimensions of the nanostructures as shown by electron microscopy. As the laser excitation overlapped the detection volume, scattering and broad emission from the metallic nanostructure itself made it difficult to resolve the single-molecule fluorescence spot above the background. Some of the PSF appeared elongated rather than Gaussian, but due to the high background, it was difficult to make a more accurate assessment of the true form of the PSF.

To gain a clearer understanding of the nature of molecular fluorescence PSFs in the vicinity of metallic nanostructures, we turned once again to remote spectroscopy using AgNW to avoid the problems of high background due to direct excitation [6]. In this study, we employed stochastic switching for fluorescence localization, so-called dSTORM technique together with remote excitation [16, 17]. Fluorescent dyes were attached to the AgNW surface via a bulky bio-conjugation to prevent direct contact between fluorophore and metal (Fig. 7.4a). The dye Alexa647 was chosen as the label since excitation light intensity can modify its rate of stochastic switching between fluorescent and non-fluorescent states in appropriate buffer conditions [18].

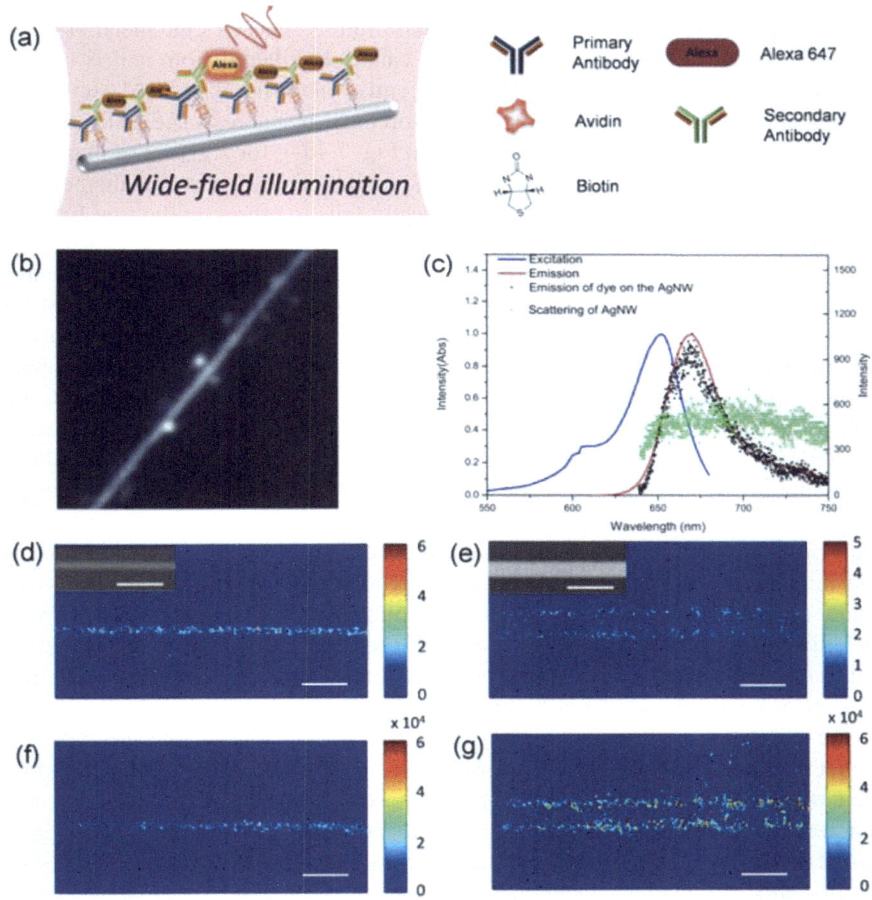

Fig. 7.4 **a** Schematic of the binding of Alexa647 to AgNWs. **b** Fluorescence image of an Alexa647-bound AgNW under wide-field excitation. **c** Absorption (blue curve) and emission (red curve) of Alexa647 in bulk solution. The fluorescence spectrum of an Alexa647-bound AgNW (black dots) and a bare AgNW (greens dots) is also included. **d–g** High resolution reconstructed images of AgNWs following wide-field excitation assuming that each emission spot is due to a single Alexa647 molecule localized at its centroid. **d, e** Localized single-molecule density maps for 110 and 250 nm diameter nanowires, respectively as shown in SEM images inset. **f, g** Localized single-molecule intensity maps. Figure reproduced with permission from Ref. [6]

In Fig. 7.4b, a fluorescence image of an Alexa647-labeled AgNW under wide-field excitation at 632.8 nm is shown. The fluorescence spots appear most clearly as satellites along the AgNW, which also exhibit significant background emission (Fig. 7.4b). The large background obscured the single-molecule emission so that it had to be assumed that each fluorescent spot was due to a single molecule at its centroid. A time-averaged emission spectrum from an Alexa647-bound AgNW clearly matches that of Alexa647 in buffer solution (Fig. 7.4c). In contrast, the time-averaged spectrum from an Alexa-free AgNW is broad, typical of metallic nanostructures, and attributed to Raman scattering and/or luminescence of the metal or other impurities (Fig. 7.4c). This broad emission cannot be filtered from and is clearly of a similar order of magnitude to the Alexa emission during wide-field excitation, a fundamental problem for the precise measurement of fluorescence PSFs. Super-resolution reconstructed images are shown in Fig. 7.4d–g for AgNWs of diameters 110 and 250 nm, respectively. The reconstruction imaging suggests the diameter of the 110 nm nanowire is 205 nm and completely fails to represent the true form of the nanowire with 250 nm diameter.

When the same experiment is done utilizing remote excitation, much clearer images of the fluorescence PSFs emerge (Fig. 7.5). They appear as bright spots clearly discernable against the weak background emission, demonstrating once again the benefit of using remote excitation for reducing spectroscopic background. Importantly, some groups of spots clearly blink on and blink off again at the same time, indicating that they are actually multi-spot emission PSFs due to a single Alexa647 molecule. Finite-difference time-domain (FDTD) calculations confirm that these multi-spot PSFs are particularly sensitive to the fluorophores position and orientation of adsorption on the nanowire, and dependent further on the dimensions of the nanowire itself. While this is problematic for accurate super-resolution reconstruction imaging of metallic nanowires, it highlights the need for rigorous checks to ensure correct PSF fitting for molecules near metallic nanoparticles generally. With such knowledge, the complex PSFs could be used predictively in far-field imaging assays of nanoparticle size/shape, or of molecular adsorption site/orientation, given some prior knowledge of the NP sample.

These experiments show once again the advantage of using remote spectroscopy on metallic NWs for minimizing background emission/scattering. In this case, remote excitation revealed a complex array of multi-spot emission PSFs for single molecules that were otherwise obscured during direct excitation. The results also suggest a fundamental paradox for the goal of taking super-resolution optical images of metallic NPs. Normally the technique is used to assess the size of an unknown object of interest. Unfortunately, for molecules near metallic NPs, it is not possible to know the form of the PSF without some a priori knowledge of the NP dimensions. This work has consequences also for counting (for instance) single-molecule catalytic turn over events on metallic NPs using fluorescence imaging.

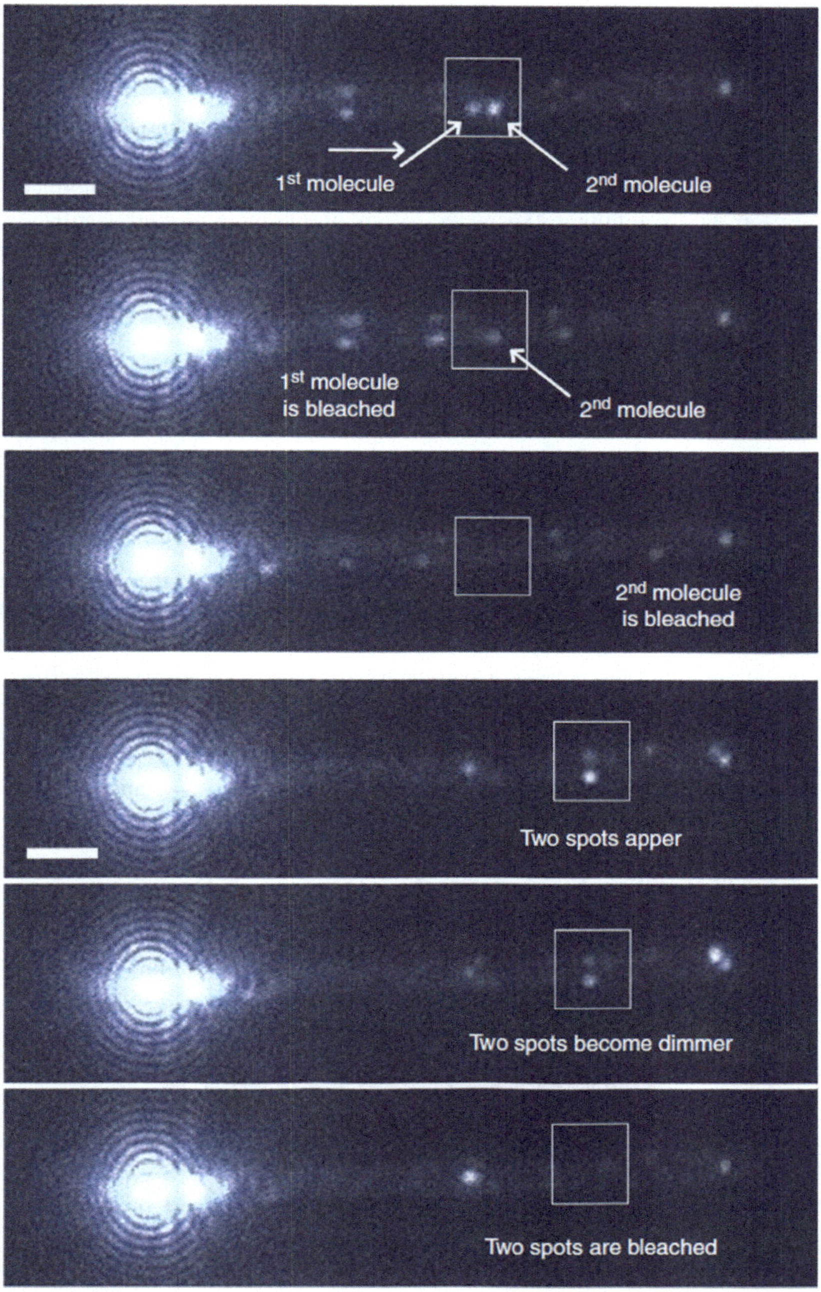

Fig. 7.5 Fluorescence images of Alexa647-bound AgNWs under remote excitation at the left end of the wire at different times. The synchronous blinking on and off of multiple emission spots are indicated. Figure reproduced with permission from Ref. [6]

7.3 Direct/Remote Excitation of Second-Harmonic Generation, Multiphoton Fluorescence

Recently nonlinear optics (NLO) has merged with plasmonics, known as nonlinear plasmonics, shed light on new technologies and applications [19]. Nonlinear plasmonics have been regarded as one of the important fields in modern nanophotonics. In this section, we are aiming to expend the AgNW plasmonic waveguiding effect from linear to nonlinear regimes. Since AgNWs serve as excellent plasmonic waveguide due to their well-defined crystal structure and atomically smooth surface [4, 6, 7], we investigated the possibility of remote excitation of NLO, as analogous to one-photon SERS and fluorescence discussed in the previous section.

Direct Excitation of NLO First, we discuss second-order nonlinear optical responses under direct excitation, second-harmonic generation (SHG) (Fig. 7.6a) [20], on AgNWs under polarized femtosecond NIR pulsed laser irradiation (820 nm, 120 fs, 80 MHz). Figure 7.6 presents position and excitation polarization dependence of SHG on an AgNW with a diameter of ~150 nm and length of ~10 μm laying on glass surface. Under excitation with a polarization parallel to the longitudinal axis of an AgNW (p-polarization), high SHG signal was observed at the apex parts (the red line), while much lower SHG has been detected at the middle part of the AgNW (the black line). SHG signal intensity at apexes is approximately 8 times higher than that in the middle of the AgNW. This position dependence is clearly visualized by

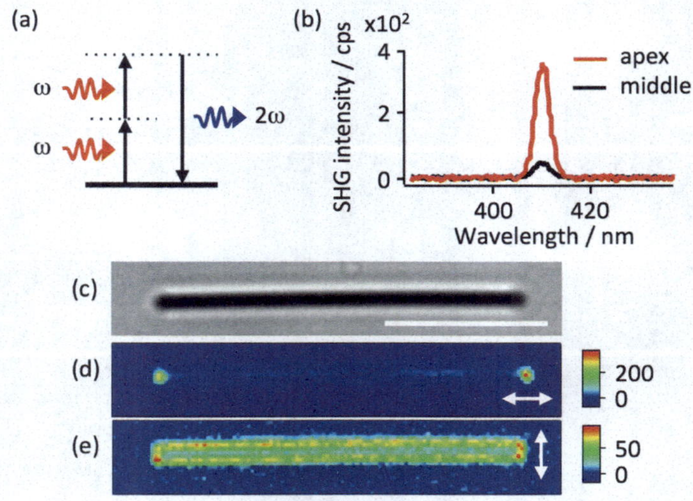

Fig. 7.6 **a** Energy diagram of second-harmonic generation (SHG). **b** SHG spectra obtained from a silver nanowire excited at an apex (red) and middle (black) part of an AgNW with 820 nm p-polarized excitation light. An optical transmission image **c**, SHG intensity map under p-polarized **d** and s-polarized excitation light **e** of an AgNW with direct excitation configuration, respectively. Scale bar is 5 μm

constructing an SHG intensity map, in which p-polarized excitation light focus was scanned over an AgNW (Fig. 7.6d). In contrast to this, almost no position dependence with very low SHG efficiency was confirmed with excitation polarized perpendicular to the longitudinal axis (s-polarization) as shown in Fig. 7.6e. This position and polarization dependence can be explained by the excitation efficiency of SPPs along an AgNW as well as symmetry selectivity of NLO. p-polarized light efficiently couple with SPPs mainly at apexes but much less at the middle part of AgNWs due to the momentum matching condition between far-field light and SPPs, suggesting higher and lower SHG efficiency at apexes and the middle part of AgNWs, respectively. In addition to this, SHG requires a break of centro-symmetry because it is second-order nonlinear optical effect. AgNW apexes seemingly work as a symmetry breaking point for p-polarized excitation light, resulting in a higher yield of SHG. In contrast, the middle part is almost symmetry for p-polarized light causing significantly lower SHG efficiency. In the case of s-polarized excitation, transverse SPPs are excited at the entire AgNW and its surface could be centro-symmetry breakpoint. This could be the reason for relatively weak SHG at the entire part of the AgNW.

This centro-symmetry scenario was further confirmed by comparing it with third-harmonic generation (THG) (Fig. 7.7a). THG doesn't require centro-symmetry break and is able to be generated in bulk because it is third-order NLO. Figure 7.7b displays a comparison of SHG and THG intensity maps on an AgNW. Indeed THG map shows less position dependence while SHG intensity strongly depends on the position, indicating that centro-symmetry plays an important role in NLO in addition to SPPs excitation efficiency.

Remote Excitation of NLO Remote excitation of SHG [20] has been demonstrated by focusing 820 nm fs laser light at one end of AgNW (indicated as "in" in Fig. 7.8) while out-coupling light at the distal end was monitored ("out" in Fig. 7.8). After propagating SPPs excited at an AgNW end reach the distal end, the SPPs should localize at an apex of the end. In case that the localized plasmon intensity is high enough, SHG at 410 nm should be remotely generated as schematically illustrated in Fig. 7.8a. As shown in Fig. 7.8c, SHG was indeed observed at the right end of AgNW when the left end was excited, vice versa. The spectrum measurement proves that second-harmonic was generated at both ends, while nothing was detected in the middle part of the AgNW (Fig. 7.8d). This result indicates that SHG can be

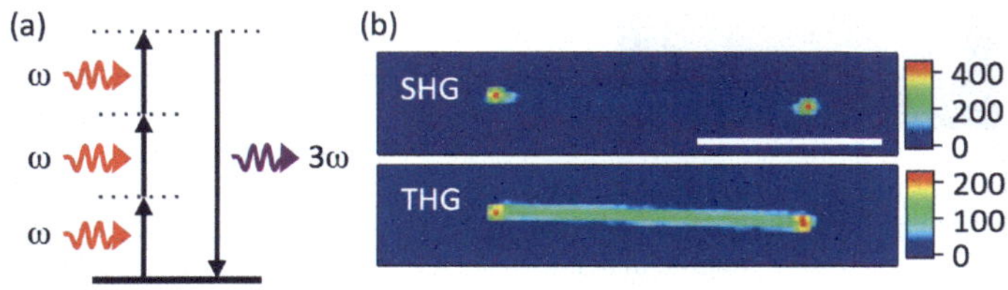

Fig. 7.7 **a** Energy diagram of third-harmonic generation (THG). **b** SHG and THG intensity map on an AgNW excited at 1164 nm with p-pol. Scale bar is 5 μm

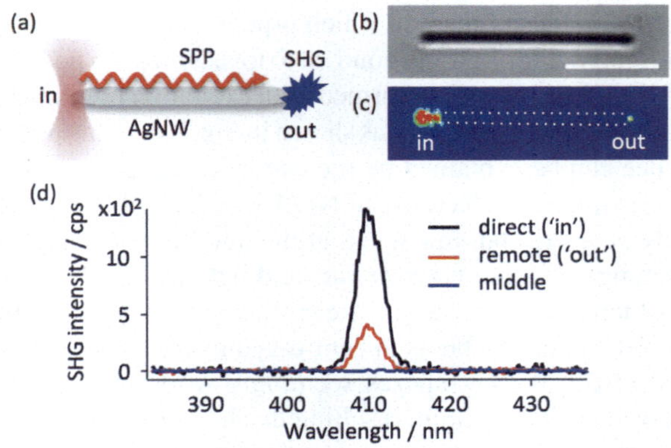

Fig. 7.8 **a** Schematic illustration of remote excitation SHG. An optical transmission image **b** and SHG image under p-polarized focused excitation **c** of an AgNW, respectively. Scale bar is 5 μm. **d** SHG spectra taken at left end (black line, direct excitation, "in"), at right end (red line, remote excitation, "out"), and body part of the AgNW (blue line, remote excitation, body), respectively

excited by surface plasmon localized at AgNW ends but not while propagating along an AgNW. Since 410 nm light hardly propagates over micrometers along AgNWs, the observed SHG at the distal end is not SHG generated at and propagated from the left end. Instead, it is most likely that propagating 820 nm SPPs was localized and induced SHG at the distal end. This remote excitation SHG via propagating SPPs could provide pulsed point light source at nanometer scale for super-resolution spectroscopy and microscopy.

As a point light source application, a similar experiment was conducted on AgNWs embedded in a polymer layer doped with fluorescence dye (Rhodamine 6G). A fluorescence spot was observed at the distal end as shown in Fig. 7.9b, proving that remote excitation of SHG (or multiphoton absorption) could serve as a point light

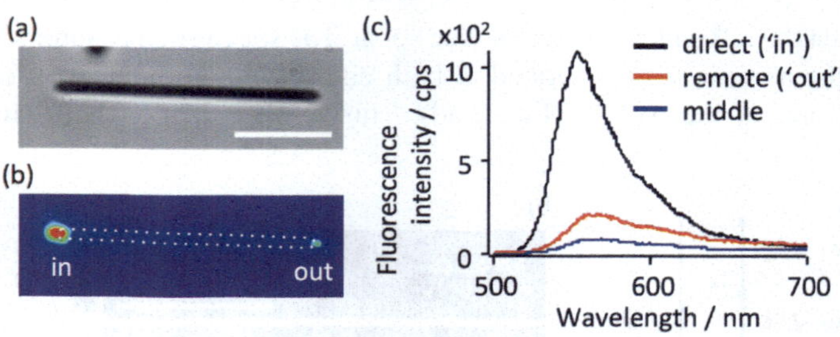

Fig. 7.9 An optical transmission image **a** and fluorescence image **b** of an AgNW embedded Rhodamine 6G/PVA polymer matrix under focused excitation at left end. Scale bar is 5 μm. **c** Rhodamine 6G fluorescence spectra taken at the left end (black line, direct excitation, "in"), the right end (red line, remote excitation, "out"), and middle part (blue line, middle) of the AgNW, respectively

source. Fluorescence spectra taken at the excitation end, body part, and distal end are shown in Fig. 7.9c. Unlike SHG experiments, fluorescence signals could be observed both at the end and middle part. This result indicates that Rhodamine 6G molecules at body parts are most likely excited via 2-photon absorption of 820 nm, instead of one-photon absorption of SHG light, 410 nm. This site dependence on nonlinear interactions between dipole molecules and plasmonic waveguide provides a unique opportunity for control of photochemical reactions along AgNWs.

7.4 Multiphoton Photochromic Reaction on Silver Nanowire

Diarylethene (DAE) derivatives have attracted great attention as thermally irreversible but photochemically reversible (so-called as P-type) photochromic molecules with excellent thermal stability and light durability. Many studies on DAE photochromism have been reported not only for application as photo-functional materials but also for a basic understanding of photochemical reactions [21]. In this section, we aim to nonlinear optically control the photochromic reaction in a fluorescent DAE derivative at the nanometer scale by using the AgNW plasmonic waveguide, as shown in a conceptual illustration in Fig. 7.10.

Here, AgNWs are embedded in polyvinyl alcohol (PVA) film doped with a fluorescent DAE derivative, 1, 2-bis(2-ethyl-6-phenyl-1-benzothiophene-1,1-dioxide-3-yl)perfluorocyclopentene. fDAE is pro-fluorescent molecule; the open form has absorption bands below 400 nm and non-fluorescent, while the closed-form absorbs

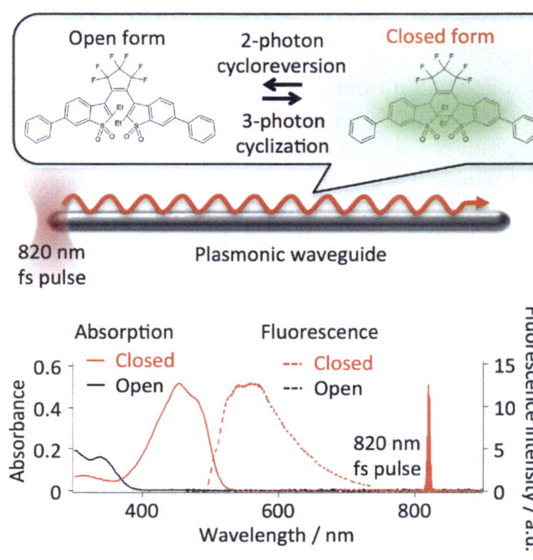

Fig. 7.10 Upper; schematic illustration of the multiphoton photochromic reaction on AgNW plasmonic waveguide. Lower; absorption and fluorescence spectra of fDAE in IPA solution and PVA polymer matrix, respectively

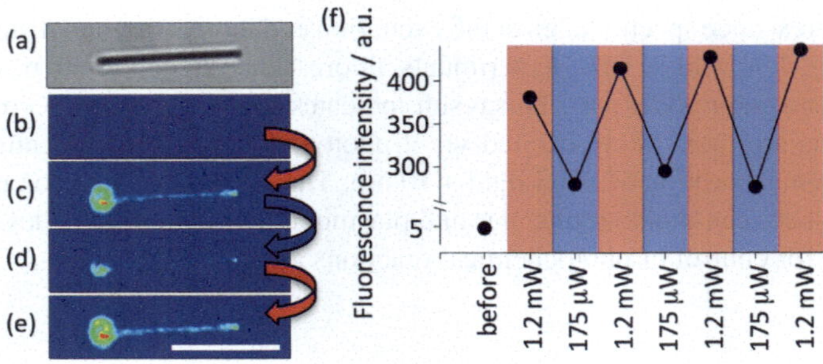

Fig. 7.11 **a** An optical transmission image of an AgNW. (**b–e**) Wide-field fluorescence images of the AgNW excited at 488 nm. Before (**b**), after sequential remote excitation with high power (1.2 mW, 1 s) **c**, **e** and with low power (175 μW, 3 min) (**d**). Scale bar is 5 μm. **f** Fluorescence intensity at middle part of the AgNW before and after the sequential remote excitations

around 450 nm and emits fluorescence around 550 nm (Fig. 7.10). Since 820 nm is quite far from the fDAE absorption bands for both closed and open form (Fig. 7.10), multiphoton photochromic reactions are required, namely 3-photon cyclization and 2-photon cycloreversion reaction.

For remote excitation of reaction, p-polarized NIR fs laser pulse (820 nm, 120 fs, 80 MHz, 1.2 mW, 1 s) was focused at the left end. Before and after the remote excitation, wide-field fluorescence images were taken to check cyclization (Fig. 7.11b and c). The images show that the fluorescence signal recovered not only at the left end but also along the AgNW, indicating that open form of the fDAE undergoes three-photon cyclization reaction to closed-form along the AgNW through propagating SPPs.

For further confirmation of three-photon cyclization, we conducted a similar experiment NIR laser having s-polarization. It is known that s-polarization light can launch much less propagating SPPs. In contrast to p-polarization, s-polarization laser indeed didn't induce the fluorescence recovery along AgNW. From the polarization dependence on the fluorescence recovery, we concluded that the 3-photon cyclization reaction of fDAE was induced via propagating SPPs.

We demonstrate that the photostationary state (PSS) between closed and open forms can be controlled by merely tuning one-color NIR laser power. After the fluorescence recovery with high power excitation (1.2 mW, 1 s, Fig. 7.11b–c), NIR laser was again focused at the same apex with lower power (175 μW) for 3 min. A fluorescence image captured after the low power irradiation shows fluorescence bleaching as shown in Fig. 7.11d, indicating that closed fluorescent form of the fDAE undergoes two-photon cycloreversion reaction to open non-fluorescence form. The bleaching can be explained by that three-photon cyclization reaction is suppressed while two-photon cycloreversion reaction becomes dominant under low power irradiation, and resulting in PSS shifting toward open form. The fluorescence recovery and bleaching can be repeated several times as shown in Fig. 7.11f. Since the propagating SPPs are near-field and localize at nanometer scale from the AgNW surface, this photochromic reaction of fDAE also localized at nanometer scale. To our best knowledge, this is the first experimental demonstration of nanometer-scale control of photochromic reaction without the influence of far-field light.

Miyasaka and co-workers have reported similar one-color reversible multiphoton photochromic reactions in amorphous film of a DAE derivative [22]. In their report, the ratio of the number of open and closed-form molecules, N_O and N_C, respectively was discussed as follows. Time dependencies of the open and closed-form at the PSS are represented by the following equations.

$$-\frac{\partial N_O}{\partial t} = N_O I^n \delta^{(n)} \Phi_{O\to C} - N_C I^m \delta^{(m)} \Phi_{C\to O} = 0 \tag{7.2}$$

$$-\frac{\partial N_{\mathrm{C}}}{\partial t} = N_{\mathrm{C}} I^m \delta^{(m)} \Phi_{\mathrm{C}\to\mathrm{O}} - N_{\mathrm{O}} I^n \delta^{(n)} \Phi_{\mathrm{O}\to\mathrm{C}} = 0 \tag{7.3}$$

Here, I is the intensity of the laser pulse. $\delta^{(n)}$ and $\delta^{(m)}$ are the n- and m-photon absorption cross-sections for the open and the closed-form, respectively. $\Phi_{\mathrm{O}\to\mathrm{C}}$ and $\Phi_{\mathrm{C}\to\mathrm{O}}$ are the cyclization and cycloreversion quantum yields. From Eqs. (7.2) and (7.3) and the relation $N_O + N_C = N_{total}$, Eq. 7.4 can be obtained.

$$\frac{N_{\mathrm{C}}}{N_{\mathrm{O}}} = \frac{N_{\mathrm{C}}}{N_{\mathrm{total}} - N_{\mathrm{C}}} = \frac{\Phi_{\mathrm{O}\to\mathrm{C}}}{\Phi_{\mathrm{C}\to\mathrm{O}}} \times \frac{\delta^{(n)}}{\delta^{(m)}} \times I^{(n-m)} \tag{7.4}$$

By considering that n and m would be 3 and 2, respectively, the relation between N_C/N_O and I would be linear dependence on excitation intensity as the following.

$$\frac{N_{\mathrm{C}}}{N_{\mathrm{O}}} \propto I \tag{7.5}$$

They experimentally confirmed the relation using absorbance of the open and closed-form. In our study, we measured fluorescence intensity instead of absorbance, and open form is non-fluorescent. Therefore, measured fluorescence intensity, *F(I)* is directly related to the number of closed-form. From Eqs. (7.3), *F(I)* can be represented as the following.

$$F(I) \propto N_{\mathrm{C}} = (N_{\mathrm{total}} - N_{\mathrm{C}}) \times \frac{\Phi_{\mathrm{O}\to\mathrm{C}}}{\Phi_{\mathrm{C}\to\mathrm{O}}} \times \frac{\delta^{(n)}}{\delta^{(m)}} \times I^{(n-m)} \tag{7.6}$$

By assumption that N_{total} is extremely large to N_C, $N_{total} - N_C = N_{total}$, *F(I)* can be also represented as linear equation to *I*.

$$F(I) \propto N_{\mathrm{total}} \times \frac{\Phi_{\mathrm{O}\to\mathrm{C}}}{\Phi_{\mathrm{C}\to\mathrm{O}}} \times \frac{\delta^{(n)}}{\delta^{(m)}} \times I^{(n-m)} \propto I \tag{7.7}$$

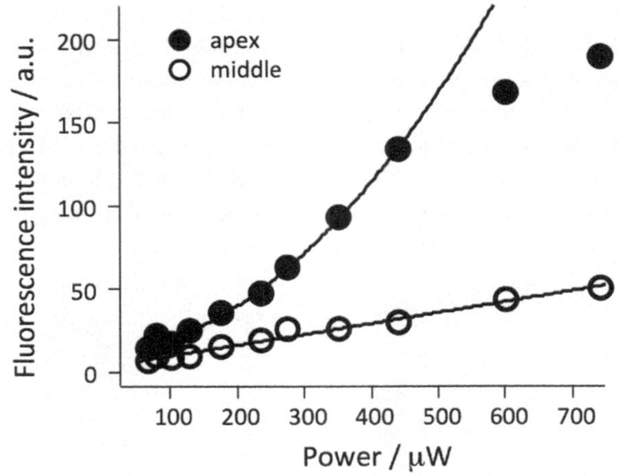

Fig. 7.12 Fluorescence intensity recovery as a function of NIR fs laser power

The assumption is supported by a fact, that fluorescence intensity obtained when fDAE is fully isomerized to closed-form by irradiation of 365 nm UV light for 30 min is typically 20 times higher than the intensity after the fluorescence recovery.

The relation between *F(I)* and *I* was experimentally examined and plotted in Fig. 7.12. It was found that the fluorescence recovery in the middle part of AgNW (circle) was indeed linearly proportional to the laser power. The fluorescence intensity is fitted with a power equation $F(I) = A * I^{B} + C$. At the middle part, the B value was calculated to be around 1.1, suggesting that the PSS between open and closed-form was attained and could be tuned by merely changing NIR power.

Interestingly, the fluorescence recovery at distal end of AgNW, however, showed nonlinear dependence to the laser power (Fig. 7.12, black filled circle). The fluorescence intensity was also fitted with the power equation, and B was calculated to be around 1.9. This result suggests that photochemical interactions between fDAE and AgNW plasmonic properties are different at middle and apex, and the two-photon cycloreversion reaction is suppressed somehow at the apexes of AgNW. We propose a possible explanation for the site-specific photochemical behavior by considering (1) NLO effect on AgNW under fs pulsed laser irradiation discussed in the previous session, (2) difference on cycloreversion quantum yields via one-photon and two-photon absorption.

Figure 7.13 describes possible energy diagrams of the NLO fDAE reactions at the middle and apex part of AgNWs. At middle part of AgNW, open form of fDAE absorbs three photons of 820 nm light and is excited into an excited state (S_n). Cyclization reaction takes place from S_n to closed-form and turns on fluorescence. While the closed-form fDAE absorbs two photons of 820 nm light and is excited into an excited state (S_1'). Cycloreversion reaction takes place from S_1' to open form and turns off fluorescence. Here we assumed that the cycloreversion quantum yield via two-photon excitation was high enough that the PSS can be attained between cycloreversion and cyclization. Although the cycloreversion quantum yield is known to be usually quite

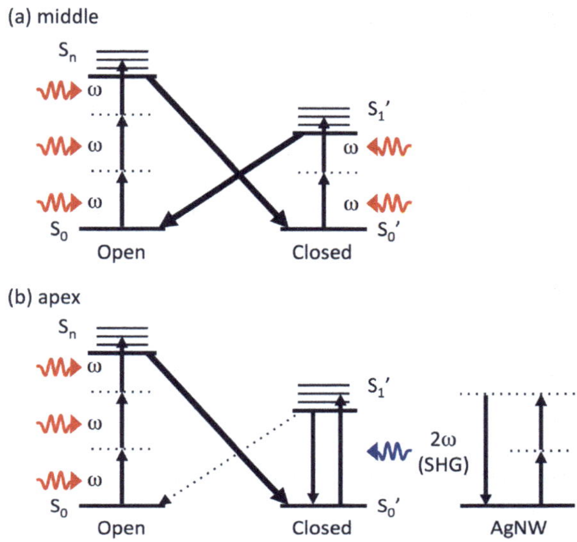

Fig. 7.13 Energy diagrams of multiphoton photochromic reaction in fDAE at middle (**a**) and apex part (**b**)

small (~10^{-5}), Miyasaka and co-workers recently reported that multiphoton excitation enhances the quantum yield of cycloreversion reaction in a DAE derivative (~10 ~10^4 times) [23]. Generally speaking, the photoexcited state (S_n) generated by light absorption is quickly relaxed to the lowest excited state S_1 regardless of the excited state (known as Kasha's rule) [24]. Therefore, photochemical processes are regarded as less sensitive by the difference on excitation scheme. However, there are increasing reports on, in some photochemical reactions, the reaction yield from the excited states (dark states, S_n') generated via multiphoton excitation, which is not generated via ordinary one-photon excitation, is greatly different from the yield from S_1. They have been clarified as the anti-Kasha photochemical reactions [25–27].

On the other hand, at apex of ANW, cyclization reaction from open form could be the same at body part. Cycloreversion reaction, however, would be different from middle part. At apex part, as discussed in the previous section, SH can be generated efficiently. The SHG having energy of 410 nm light can excite the closed-form fDAE molecules via one-photon absorption. Since the quantum yield of cycloreversion reaction via one-photon excitation is quite low, the excited fDAE molecules are relaxed to ground state (S_0') rather than undergoing cycloreversion reaction. Therefore, the SHG mediated "quasi" one-photon excitation of the closed molecules works as an anti-filter and blocks cycloreversion reaction.

7.5 Conclusion

We have expended the AgNW's plasmonic waveguide effect from linear to nonlinear regime, namely remote excitation SHG has been demonstrated as a promising nanometer-scale pulsed point light source. We hope this tool can open a new field of remote excitation "nonlinear" spectroscopy. In this chapter, we further demonstrated the one-color reversible photochromic reactions in a fDAE derivative by using "non-linear" remote excitation. By interacting with propagating SPPs on AgNW, fDAE molecules undergo both three-photon cycloreversion and two-photon cyclization reactions. The PSS of fDAE between closed and open forms can be controlled by merely tuning one-color NIR laser power. The "nonlinear" remote excitation allows reducing excitation volume to the sub-diffraction limit dimensions and provides a unique platform for control of photochemical reaction along AgNWs.

Acknowledgements This work was supported by JSPS KAKENHI (JP17H05458, 17H03003, 18H01948, and 18K19085). The Nakatani Foundation to T. I, The Fund for Scientific Research-Flanders (FWO, G0D4519N and G081916N), KU Leuven Internal Funds (C14/15/053, C14/19/079), and the JSPS Core-to-Core Program A. to H. U. are also acknowledged. The work was partially supported by the RIES International Exchange Program of "Dynamic Alliance for Open Innovation Bridging Human, Environment and Materials" from MEXT and the Photo-excitonix Project at Hokkaido University.

References

1. Graff A, Wagner D, Ditlbacher H, Kreibig U (2005) Silver nanowires. Eur Phys J D 34:263–269
2. Wei H, Xu H (2012) Nanowire-based plasmonic waveguides and devices for integrated nanophotonic circuits. Nanophotonics 1:155–169
3. Xiong X, Zou CL, Ren XF, Liu AP, Ye YX, Sun FW, Guo GC (2013) Silver nanowires for photonics applications. Laser Photon Rev 1–19
4. Hutchison JA, Centeno SP, Odaka H, Fukumura H, Hofens J, Uji-i H (2009) Subdiffraction limited, remote excitation of surface enhanced Raman scattering. Nano Lett 9:995–1001
5. Fang Y, Wei H, Hao F, Nordlander P, Xu H (2009) Remote-excitation surface-enhanced Raman scattering using propagating Ag nanowire plasmons. Nano Lett 9:2049–2053
6. Su L, Lu G, Kenens B, Rocha S, Fron E, Yuan H, Chen C, Dorpe PV, Roeffaers BJ, Mizuno H, Hofkens J, Hutchison JA, Uji-i H (2015) Visualization of molecular fluorescence point spread functions via remote excitation switching fluorescence microscopy. Nat. Commun. 6:1–9
7. Lu G, De Keersmaecker H, Su L, Kenens B, Rocha S, Fron E, Chen C, Van Dorpe P, Mizuno H, Hofkens J, Hutchison JA, Uji-i H (2014) Live-cell SERS endoscopy using plasmonic nanowire waveguides. Adv Mater 26:5124–5128
8. Halas NJ, Lal S, Chang W-S, Link S, Nordlander P (2011) Plasmons in strongly coupled metallic nanostructures. Chem Rev 111:3913–3961
9. Huang Y, Fang Y, Zhang Z, Zhu L, Sun M (2014) Nanowire-supported plasmonic waveguide for remote excitation of surface-enhanced Raman scattering. Light Sci Appl 3:e199
10. Krenn JR, Weeber J-C (2004) Surface plasmon polaritons in metal stripes and wires. Philos Trans R Soc Lond Ser A 362:739–756
11. Ditlbacher H, Hohenau A, Wagner D, Kreibig U, Rogers M, Hofer F, Aussenegg FR, Krenn JR (2005) Silver nanowires as surface plasmon resonators. Phys Rev Lett 95:257403

12. Sanders AW, Routenberg DA, Wiley BJ, Xia Y, Dufresne ER, Reed MA (2006) Observation of plasmon propagation, redirection, and fan-out in silver nanowires. Nano Lett 6:1822–1826
13. Ru EL, Etchegoin P (2008) Principles of surface-enhanced Raman spectroscopy: and related plasmonic effects. Elsevier Science
14. Lin H, Centeno SP, Su L, Kenens B, Rocha S, Sliwa M, Hofkens J, Uji-i H (2012) Mapping of surface-enhanced fluorescence on metal nanoparticles using super-resolution photoactivation localization microscopy. Chem Phys Chem 13:973–981
15. Orrit M (2014) Nobel prize in chemistry: celebrating optical nanoscopy. Nat Photon 8:887–888
16. Rust MJ, Bates M, Zhuang X (2006) Sub-diffraction-limit imaging by stochastic optical reconstruction microscopy (STORM). Nat Methods 3:793–796
17. Schüttpelz M, Wolter S, van de Linde S, Heilemann M, Sauer M (2010) dSTORM: realtim subdiffraction-resolution fluorescence imaging with organic fluorophores. In: Proceedings of SPIE, vol 7571. pp 75710 V–7
18. van de Linde S, Loschberger A, Klein T, Heidbreder M, Wolter S, Heilemann M, Sauer M (2011) Direct stochastic optical reconstruction microscopy with standard fluorescent probes. Nat Protoc 6:991–1009
19. Kauranen M, Zayats AV (2012) Nonlinear plasmonics. Nat Photon 6:737–748
20. Inose T, Toyouchi S, Lu G, Umemoto K, Tezuka Y, Lyu B, Masuhara A, Fron E, Fujita Y, Hirai K, Uji-i H (2019) Water-mediated polyol synthesis of pencil-like sharp silver nanowires suitable for nonlinear plasmonics. Chem Common 55:11630–11633
21. Irie M, Fukaminato T, Matsuda K, Kobatake S (2014) Photochromism of diarylethene molecules and crystals: Memories, switches, and actuators. Chem Rev 114:12174
22. Mori K, Ishibashi Y, Matsuda H, Ito S, Nagasawa Y, Nakagawa H, Uchida K, Yokojima S, Nakamura S, Irie M, Miyasaka H (2011) One-color reversible control of photochromic reactions in a diarylethene derivative: three-photon cyclization and two-photon cycloreversion by a near-infrared femtosecond laser pulse at 1.28 μm. J Am Chem Soc 133:2621
23. Sotome H, Nagasaka T, Une K, Okui C, Ishibashi Y, Kamada K, Kobatake S, Irie M, Miyasaka H (2017) Efficient cycloreversion reaction of a diarylethene derivative in higher excited states attained by off-resonant simultaneous two-photon absorption. J Phys Chem Lett 8:3272
24. Kasha M (1950) Characterization of electronic transitions in complex molecules. Discuss Faraday Soc 9:14
25. de Klerk JS, Szemik-Hojniak A, Ariese F, Gooijer C (2007) Intramolecular proton-transfer processes starting at higher excited states: a fluorescence study on 2-butylamino-6-methyl-4-nitropyridine N-oxide in nonpolar solutions. J Phys Chem A 111:5828
26. Tseng HW, Shen JY, Kuo TY, Tu TS, Chen YA, Demchenko AP, Chou PT (2016) Excited-state intramolecular proton-transfer reaction demonstrating anti-Kasha behavior. Chem Sci 7:655
27. Sotome H, Nagasaka T, Une K, Morikawa S, Katayama T, Kobatake S, Irie M, Miyasaka H (2017) Cycloreversion reaction of a diarylethene derivative at higher excited states attained by two-color, two-photon femtosecond pulsed excitation. J Am Chem Soc 139:17159

Chapter 8
Modulations of Electronic States in Plasmonic Strong Coupling Systems and Their Application to Photochemical Reaction Fields

Kosei Ueno

Abstract The modulation of electronic or transition states plays an important role in improving the selectivity and reactivity of chemical reactions. In recent years, the strong coupling between an optical mode and excitons or molecular vibrational modes has received attractive attention as a physical phenomenon for controlling the activation energy of chemical reactions by modulating the electronic state or the vibrational state. In this study, we investigated the spectral properties of hybrid states formed by the strong coupling between plasmon and molecular/intermolecular vibrational modes in the infrared wavelength region. The strong coupling between plasmon and molecular excitonic states have been also studied for demonstrating how to confirm the formation of hybrid states. We developed a highly efficient photochemical reaction field by using modal strong coupling systems between plasmons and different optical modes.

Keywords Strong coupling · Modulation of electronic states · Infrared localized surface plasmon · Modal strong coupling

8.1 Introduction

Strong coupling has received considerable attention as a physical phenomenon modulating molecular electronic states based on quantum mechanical interactions with optical modes such as microcavity, plasmon, and so on [1, 2]. If the electronic state is modulated by the strong coupling, it is expected to change the reactivity of molecules. Therefore, strong coupling works as a catalyst for promoting photochemical reactions. On this basis, we have explored strong coupling between localized surface plasmon resonance (LSPR) and molecular excitonic states by using J-aggregate of molecules having a relatively larger dipole moment [3, 4]. Ebbesen and co-workers recently demonstrated the strong coupling between microcavity and

K. Ueno (✉)
Department of Chemistry, Faculty of Science, Hokkaido University, Kita-Ku, Sapporo, Hokkaido 060-0810, Japan
e-mail: ueno@sci.hokudai.ac.jp

H. Miyasaka et al. (eds.), *Photosynergetic Responses in Molecules and Molecular Aggregates*, https://doi.org/10.1007/978-981-15-5451-3_8

molecular vibrational modes in the infrared wavelength range and the possibility of modulating chemical reaction activities by the vibrational strong coupling [5, 6]. We also study the spectrum modulation of LSPR band in the infrared wavelength region in the presence of molecules [7]. In this chapter, we describe spectrum modulations due to the interaction between infrared LSPR and molecular/intermolecular vibrational modes and demonstrate how to confirm the formation of hybrid states using the strong coupling regime between LSPR and molecular excitonic states. Finally, an application to a highly efficient photochemical reaction field using modal strong coupling systems between plasmons and different optical modes is described.

8.2 Modulation of Infrared LSPR Bands by Interaction with Molecular/Intermolecular Vibrational Modes

Figure 8.1a shows the extinction spectra of gold (Au) nanorods with various rod lengths. A typical example of a scanning electron microscope (SEM) image of Au nanorods fabricated by electron beam lithography and lift-off methods on a sapphire substrate is shown in the inset of Fig. 8.1b. Not only a dipole resonance band but also a hexapole resonance band can be clearly seen relatively at the shorter wavelength region of the dipole resonance band in each spectrum [8]. Importantly, the peak wavelength of each dipole resonance band shows a red-shift and linearly increased with the rod length as shown in Fig. 8.1b.

When the Au rod length increases by several tens of micrometers, the LSPR band shows a similar response to the spectral properties of Au nanorods even in the far-infrared wavelength region. Figure 8.2a shows the extinction spectra of Au rod structures fabricated on a silicon substrate. The length of the Au rod is 25, 35, 50, and 100 μm. The inset in Fig. 8.2b shows a SEM image of Au rods whose length is

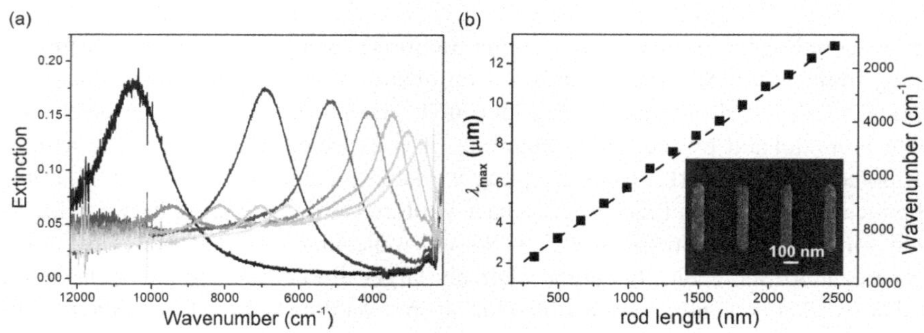

Fig. 8.1 **a** Extinction spectra of Au nanorods with a different rod length; 140, 280, 420, 560, 700, 850, and 990 nm from left, respectively. Both the width and the thickness of each rod are 40 nm. **b** The rod length dependence of the peak wavelength. The inset shows an SEM image of Au nanorods with a rod length of 700 nm Adapted with permission from [7], Copyright 2015 Optical Society of America

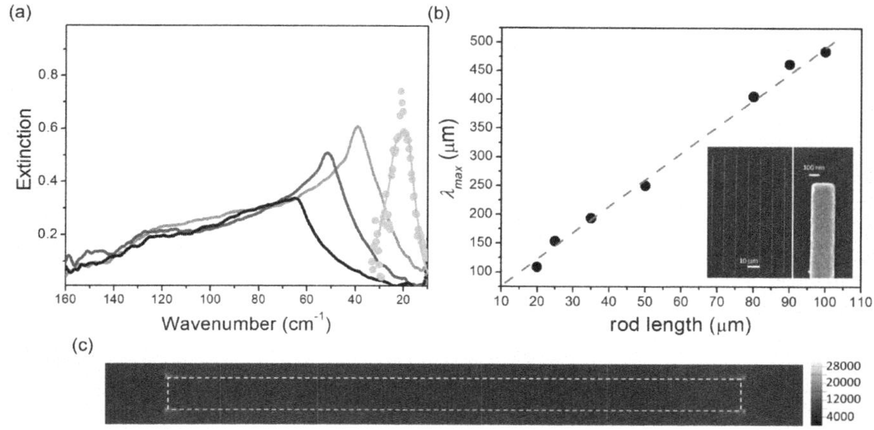

Fig. 8.2 **a** Extinction spectra of Au rods with several lengths; 25, 35, 50, and 100 μm from left. The width and thickness of each rod are 200 and 40 nm, respectively. **b** The rod length dependence of the peak wavelength. The inset shows an SEM image of Au rods with a rod length of 100 μm. **c** The near-field intensity distribution of the Au rod with a length of 40 μm simulated by an FDTD method Adapted with permission from [7], Copyright 2015 Optical Society of America

100 μm. As analogous to Fig. 8.1a, not only dipole resonance band but also higher order bands can be seen in the shorter wavelength region and the dipole resonance band shows a red-shift with the rod length. The rod length dependence of the dipole resonance band is shown in Fig. 8.2b. Even in the far-infrared wavelength region, a linear relationship between the peak wavelength and the rod length can be seen. The near-field intensity distribution simulated by the finite-difference time-domain (FDTD) method is shown in Fig. 8.2c. The enhancement factor as high as 28,000 is obtained at the four corners of the Au rod with a length of 40 μm [7].

The near-field intensity of Au rod is relatively higher than that of Au nanorod in the near-infrared wavelength region. Therefore, strong electromagnetic interaction with molecular/intermolecular vibrational modes is expected. Figure 8.3a shows an extinction spectrum of L-sodium glutamate pellets with a thickness of 0.9 mm. The obvious lattice vibrational mode is clearly observed at about 23 cm^{-1}. Interestingly, the spectrum is split into two bands when the peak wavelength of the vibrational mode overlaps with the LSPR band. Notice that the spectral modulation was obtained only by dropping the aqueous solution of L-sodium glutamate (20 mmol dm^{-3}) and air-dried on the Au rod structured substrate. The spectral modulation might be obtained by electromagnetic interaction between LSPR and the vibrational mode of the amino acid derivative. It is speculated that the electromagnetic interaction results in the Rabi splitting based on the strong coupling or the Fano dip around 23 cm^{-1} based on the weak coupling regime [7].

Ebbesen and his co-workers elucidated the vacuum Rabi splitting based on a strong coupling between the molecular vibrational mode and the Fabry–Pérot micro-cavity mode [9]. The spectral properties of polyvinyl acetate (PVAc) employed in

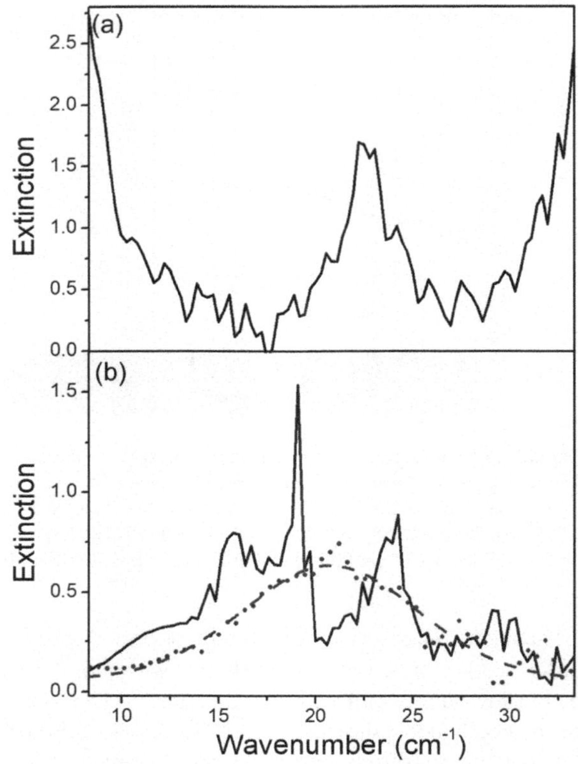

Fig. 8.3 **a** Extinction spectrum of L-sodium glutamate pellet with a thickness of 0.9 mm, **b** Extinction spectra of Au rods with a length of 100 μm (broken line) and that with L-sodium glutamate (solid line) Adapted with permission from [7], Copyright 2015 Optical Society of America

the study of microcavity have been explored also by using LSPR as an optical mode in the infrared wavelength region. Au nanochain structures have been fabricated on a silicon substrate whose plasmon resonance band exists in the mid-infrared wavelength region. An SEM image of the typical Au nanochain structure is shown in Fig. 8.4a. The Au nanochain is constituted from 4 or 5 nanoblocks and each nanoblock's size is $100 \times 100 \times 40\ \text{nm}^3$. The length of Au nanochains was controlled by the number of nanoblocks (4 or 5) and the overlapping length between adjacent nanoblocks. Figure 8.4b shows the extinction spectra of Au nanochains with a different chain length. In the case of the Au nanochain structure, interestingly, the higher order plasmon resonance bands are suppressed and the distinct dipole resonance band can be seen [8, 10]. The dipole resonance band shows a monotonous red-shift with a length of Au nanochains. The noise at $2370\ \text{cm}^{-1}$ and the dip around $1280\ \text{cm}^{-1}$ can be assigned by CO_2 in air and an interference with the Si–O–Si asymmetry stretching vibrational mode on the substrate surface, respectively. Then, PVAc acetone solution (2 wt%) was spin-coated on the Au nanochain structured substrate under the conditions of 1000 rpm for 5 s and 4000 rpm for 90 s. The thickness of PVAc film was estimated to be 100 nm on the Au nanochains structured substrate from the cross-sectional SEM image.

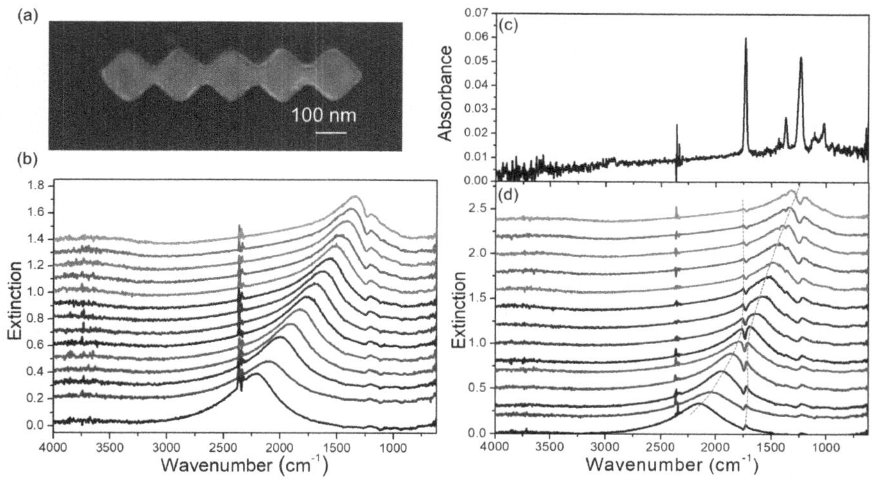

Fig. 8.4 **a** A SEM image of the typical Au nanochain structure. **b** Extinction spectra of Au nanochains with a different chain length; 601, 636, 672, 707, 742, 778, 813, 848, 884, 919, 954, 990, 1025, 1061 nm from the bottom. **c** Absorption spectrum of the PVAc film deposited on a silicon substrate in the mid-infrared wavelength region. **d** Extinction spectra of PVAc-coated Au nanochains with a different chain length; 601, 636, 672, 707, 742, 778, 813, 848, 884, 919, 954, 990, 1025, 1061 nm from the bottom

Figure 8.4c indicates absorption spectrum of PVAc film deposited on a silicon substrate in the mid-infrared wavelength region. A distinct C=O stretching vibrational mode can be seen at 1740 cm^{-1} while CH_3 and C–O–C asymmetry stretching vibrational mode near the Si–O–Si asymmetry stretching vibrational mode was also observed at 1370 cm^{-1} and 1240 cm^{-1}, respectively. Importantly, extinction spectra of PVAc-coated Au nanochains as shown in Fig. 8.4d is modulated around the wavenumbers of 1740 and 1240 cm^{-1}. Here, we focus on the spectrum dip at 1740 cm^{-1} because the spectral dip cannot be seen in Fig. 8.4b. It is noteworthy that there appear to be two peaks across 1740 cm^{-1} or there appears to be a spectral dip at 1740 cm^{-1}. When guidelines are added to the two peaks in Fig. 8.4d, it can be seen that the two guidelines show an anti-crossing behavior. Figure 8.5 shows a dispersion curve which was made by plotting the two peak energies of the spectrum in Fig. 8.4d to the bare LSPR peak wavenumber (Fig. 8.4b). An anti-crossing behavior which is a characteristic of strong coupling can be seen. By fitting the plot using a harmonic oscillator model, the vacuum Rabi splitting energy ($\hbar\Omega_R$) can be estimated to be 12 meV, which is slightly smaller than that obtained by using a Fabry–Pérot microcavity mode in the former study [9]. Almost similar results measured by using microcavities as an optical mode were obtained even by using the LSPR modes.

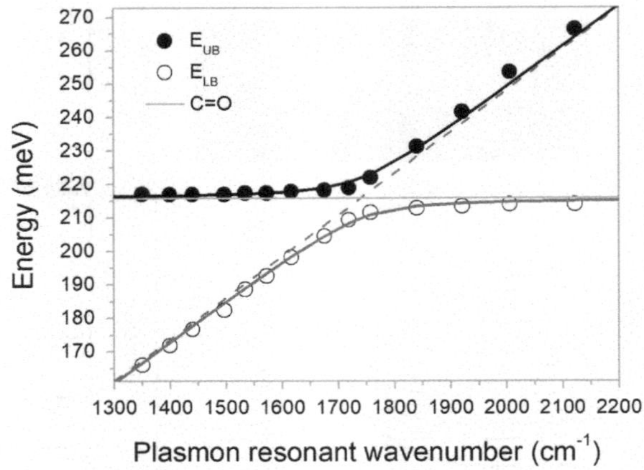

Fig. 8.5 A dispersion curve which was made by plotting the two peak energies of the spectrum in Fig. 8.4d to the bare LSPR resonant wavenumber (Fig. 8.4b)

8.3 Confirmation of Hybrid States by Excitation Spectrum Measurements

Only from the extinction spectroscopy, it is difficult to distinguish between the strong and weak couplings. One of the analysis methods is to evaluate the Rabi splitting energy from the dispersion curve and the spectrum widths before the coupling because the Rabi splitting energy should be larger than the average of spectrum widths of both optical and molecular modes. However, the easiest way to confirm the formation of hybrid states is by measuring excitation spectrum because excitation spectrum elucidates the existence of hybrid states (upper and lower branch polaritons) [11–14].

To investigate the concept, aluminum (Al) nanostructures (nanodisks) were fabricated on a glass substrate by electron beam lithography and lift-off methods [15]. Figure 8.6a shows an SEM image of Al nanodisks with a diameter (d) of 200 nm. The thickness of Al nanodisks was set at 30 nm. The period is slightly larger than twice the nanodisk diameter to avoid near-field interactions. Figure 8.6b exhibited the extinction spectra of Al nanodisks with various diameters from 90 to 200 nm by a 5 nm difference (total 23 spectra). A monotonous spectrum shift to the longer wavelength has successfully been observed with increasing the diameter of Al nanodisks.

Tetraphenylporphyrin tetrasulfonic acid hydrate (TPPS) J-aggregates methanol solution (3 mmol dm^{-3}) was spin-coated under the conditions of 3000 rpm for 90 s on the Al nanodisks-loaded glass substrate. The thickness of the TPPS J-aggregate film was estimated to be 20 nm. Figure 8.6c shows the extinction spectra of Al nanodisks ($d = 115$ nm) with a TPPS J-aggregate film. As a reference, extinction spectra of the bare LSPR band and the TPPS J-aggregate film without Al nanodisks are also inserted in the spectrum by black broken and solid lines. Because the LSPR band of Al disks ($d = 115$ nm) shows a red-shift by depositing the TPPS J-aggregate film due to the change of surrounding dielectric constants, the LSPR band and the Soret band of TPPS J-aggregate might accord with each other. The spectrum was

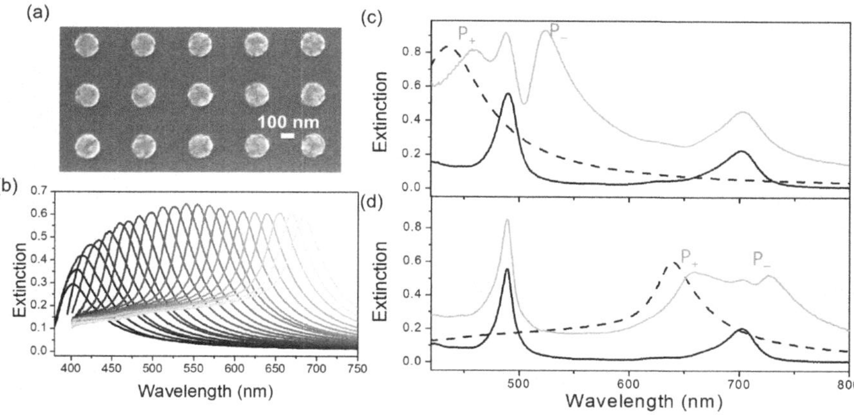

Fig. 8.6 **a** A SEM image of Al nanodisks with a diameter of 200 nm. **b** Extinction spectra of Al nanodisks with various diameters from 90 to 200 nm by a 5 nm difference (total 23 spectra). **c** Extinction spectrum of Al nanodisks ($d = 115$ nm) with a TPPS J-aggregate film on the Al nanodisks (gray solid line). The black broken line shows the bare LSPR band without the TPPS J-aggregate film. The black solid line indicates an extinction spectrum of a TPPS J-aggregate film on the glass substrate. **d** Extinction spectrum of Al nanodisks ($d = 180$ nm) with a TPPS J-aggregate film on the Al nanodisks (gray solid line). The black broken line shows the bare LSPR band without the TPPS J-aggregate film. The black solid line indicates an extinction spectrum of a TPPS J-aggregate film on the glass substrate Adapted with permission from [6], Copyright 2016 American Chemical Society

clearly split into two bands which are corresponding to upper and lower branch states although the uncoupled one can be also found as a peak at the same position of the Soret band of TPPS J-aggregate. On the other hand, Fig. 8.6d shows extinction spectra of Al nanodisks ($d = 180$ nm) with a TPPS J-aggregate film. The spectrum was clearly split into two bands because of the LSPR band and Q-band of TPPS J-aggregate accord with each other [6].

Dispersion curves were made by plotting both peak energies of upper and lower branch states to the plasmon resonant wavenumber as shown in Fig. 8.7a. Anti-crossing behaviors can be observed not only around the Soret band wavelength region but also the Q-band wavelength region, which is attributed to the formation of two hybrid states based on the strong coupling between LSPRs and Soret/Q-bands of TPPS J-aggregate as shown in the energy diagrams of hybrid states formed by the strong coupling between LSPRs and Soret/Q-bands of TPPS J-aggregate (Fig. 8.7b). Note that the Rabi splitting energies attributed to the strong coupling between Soret band and LSPR and that between Q-band and LSPR are estimated to be 300 meV and 180 meV, respectively. The difference between these two Rabi splitting energies is caused by the difference of the dipole moment between Soret and Q-bands because the Rabi splitting energies are dependent on the dipole moment of excitonic state and electric-field of LSPR [6].

To confirm the formation of hybrid states, excitation spectra have been measured. Figure 8.8 shows the extinction and excitation spectra of hybrid states around from

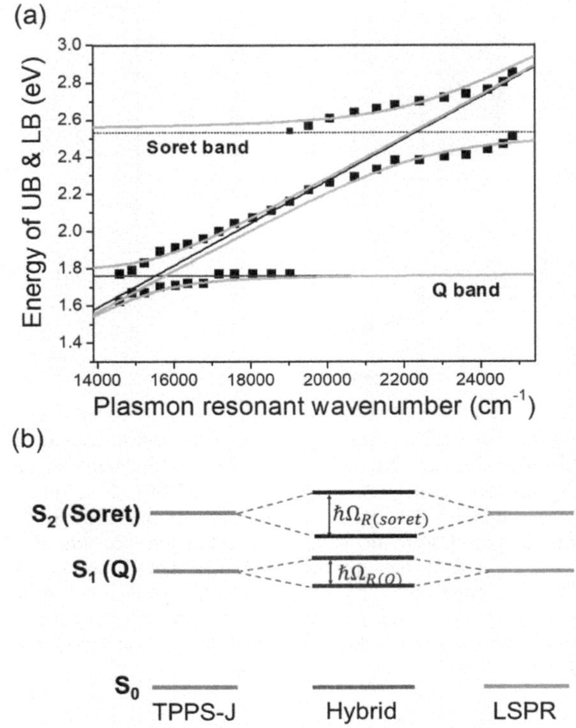

Fig. 8.7 **a** Energy diagrams of hybrid states formed by the strong coupling between LSPRs and Soret/Q-bands of TPPS J-aggregate. **b** Dispersion curves made by plotting peak energies of upper and lower branch states to the plasmon resonant wavenumber Adapted with permission from [6], Copyright 2016 American Chemical Society

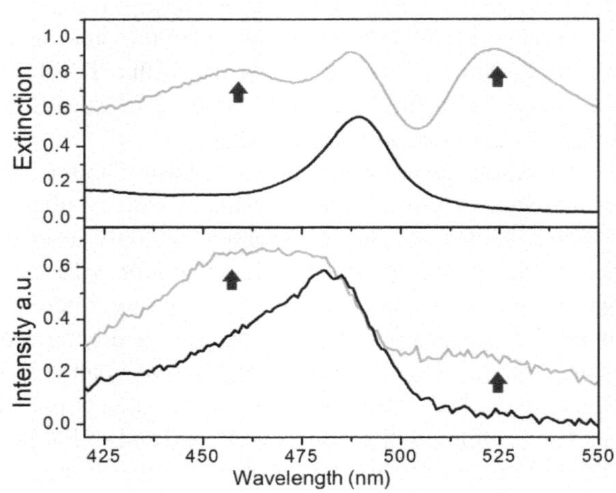

Fig. 8.8 Extinction (upper) and excitation (lower) spectra of hybrid states around from 450 to 530 nm wavelength which is corresponding to the Soret band of TPPS J-aggregate (gray solid lines). Black solid lines show extinction (upper) and excitation (lower) spectra of TPPS J-aggregate Adapted with permission from [6], Copyright 2016 American Chemical Society

420 to 550 nm wavelength which is corresponding to the Soret band of TPPS J-aggregate. The monitor wavelengths for the excitation spectra were set at 718 nm. It is emphasizing that two shoulder peaks derived from upper and lower branches of the hybrid states, which are corresponding to the wavelengths of the hybrid states in the extinction spectrum, can be observed. In the Q-band region, a similar response was observed (The data are not shown here). Therefore, the excitation spectrum provides strong evidence that a hybrid state is really formed in the strong coupling regime [6].

8.4 Modal Strong Coupling Systems that Enhances Photochemical Reactions

Metallic nanostructures showing LSPRs have received considerable attention as a light-harvesting optical antenna for light-energy conversion systems such as solar cells as well as artificial photosynthesis [16–19]. To construct efficient light-harvesting optical antennae, an optimization of the structural design is one of the most important research topics in the plasmon-induced light-energy conversions. Coupled plasmonic systems such as nanogap heptamer, dolmen, metal-insulator-metal nanostructures, and so on are promising as a photoelectrode design because of their strong near-field enhancement and the wide wavelength responsibility [12–14]. Here, the modal strong coupling between the waveguide mode and LSPR is employed to investigate whether the photocurrent response extends over a wide wavelength range and is promoted by near-field enhancement in the plasmon-induced photocurrent generation using gold nanostructured titanium dioxide (TiO_2) photoelectrodes. The near-field spectrum and the photocurrent action spectrum are compared to elucidate the effect of near-field enhancement on the photocurrent generation.

TiO_2 photoelectrodes having periodic gold nanogratings (AuNGs) with a different pitch size have been fabricated by deposition of TiO_2 on a glass substrate with a thickness of 250 nm using an atomic layer deposition (ALD), and subsequent electron beam lithography and lift-off methods. A conventional photoelectrochemical measurement using three-electrodes was performed for the measurement of photocurrent generation efficiencies. The AuNGs/TiO_2 photoelectrode, platinum wire, and a saturated calomel electrode (SCE) were employed as working, counter and reference electrodes, respectively. An aqueous Ar–gas–bubbled $KClO_4$ (0.1 mol/dm_3) solution was used as a supporting electrolyte solution. A plasmon-induced water oxidation as a half-reaction of the water-splitting was explored.

Figure 8.9a shows an SEM image of AuNGs/TiO_2 photoelectrode with a pitch size of 300 nm. Extinction spectra of the AuNGs/TiO_2 photoelectrode with a different pitch size are shown in Fig. 8.9b. There is only one peak corresponding to the LSPR band of periodic AuNGs with 200 and 225 nm pitch sizes. Starting with a pitch size of 250 nm, three peaks can be observed, and the peaks show a spectral shift with increasing the pitch size. The two outer peaks can be assigned as coupled modes

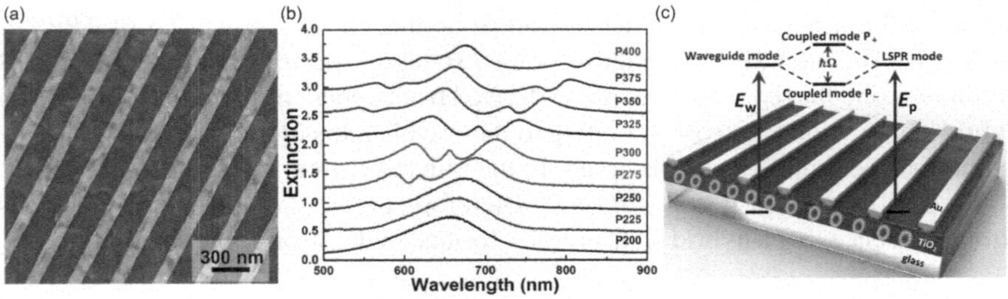

Fig. 8.9 **a** A typical SEM image of the AuNGs/TiO_2 structure with a 300 nm pitch size. Extinction spectra of the AuNGs/TiO_2 photoelectrode with various pitch sizes from 200 to 400 nm by a 25 nm difference (total 9 spectra). The inset number shows each pitch size. **c** A schematic illustration of the periodic AuNGs/TiO_2 photoelectrode and the formation of the hybrid states based on the modal strong coupling between the waveguide mode and LSPR. P_+ and P_- are energy levels of upper and lower branches of the hybrid states, and $\hbar\Omega$ indicates the splitting energy of the hybrid states Adapted with permission from [20], Copyright 2017 American Chemical Society

(P_+ (upper branch) and P_- (lower branch)) deriving from the symmetric waveguide mode E_w and LSPR mode E_p as shown in the energy diagram of hybrid states based on the modal strong coupling between the waveguide mode and LSPR (Fig. 8.9c). The center peak can be considered as a coupling mode between the asymmetric waveguide mode and the waveguide-LSPR modal strong coupling system [20].

Internal quantum efficiency (IQE) spectra and simulated near-field spectra of the AuNGs/TiO_2 photoelectrodes with 300 and 350 nm pitch sizes are shown in Fig. 8.10. The near-field intensity in the spectra was calculated by monitoring at the interface between the AuNGs and the TiO_2 film by the electromagnetic simulations using a finite-difference time-domain (FDTD) method. It was clearly elucidated that IQE spectrum has successfully reproduced the near-field spectrum under the modal strong coupling conditions. This indicates that the photocurrent response extended

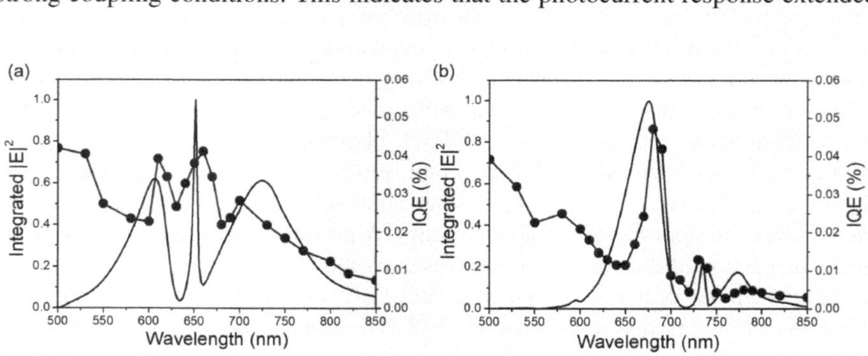

Fig. 8.10 IQE spectra (plot & line) and simulated near-field spectra (solid line) of the AuNGs/TiO_2 photoelectrodes with 300 (**a**) and 350 nm (**b**) pitch sizes, respectively Adapted with permission from [20], Copyright 2017 American Chemical Society

over a wide wavelength range utilizing the coupled plasmonic systems and near-field enhancement effects promoted the plasmon-induced water oxidation because IQE values increased responding to the near-field spectra [20].

8.5 Conclusion

We described the spectral properties of LSPR with molecules in the infrared wavelength region. A distinct spectrum modulation could be observed by the electromagnetic interactions between LSPR and molecular/intermolecular vibrational modes. Excitation spectrum measurements are crucial to distinguish the coupling regimes between strong and weak couplings, which was confirmed by the strong coupling system between LSPR and molecular excitonic states. The modal strong coupling between LSPR and the other optical mode is promising to control the photochemical reaction because it was demonstrated that the modal strong coupling between LSPR and the waveguide mode enhances the plasmon-induced water oxidation at the wavelengths of hybrid states. Actually, we have successfully demonstrated the modal strong coupling between LSPR and the Fabry–Pérot cavity mode promoted the water-splitting reaction at the wavelengths of hybrid states [21]. The dephasing time of LSPR can be also controlled by the modal strong coupling between LSPR and the long-lived optical mode such as a propagating surface plasmon polariton [22]. Thus, the strong coupling is promising to modulate electronic and transition states of materials for controlling the photochemical reactions.

Acknowledgements The author is much grateful to Professors H. Misawa, K. Sasaki, T. Oshikiri, Q. Sun, X. Shi, and K. Imura, Drs. J. Li, H. Yu, and J. Guo for the fruitful discussions, collaborations, and experiments. The present work was supported by JSPS KAKENHI Grant Numbers JP15H01073, JP17H05245, JP18H05205, JP19H02737, JP19H04667, the Nanotechnology Platform and the Photo-excitonix Project in Hokkaido University.

References

1. Houdré R, Stanley RP, Oesterle U, Ilegems M, Weisbuch C (1993) Room temperature exciton-photon Rabi splitting in a semiconductor microcavity. J Phys IV 3:51–58
2. Dintinger J, Klein S, Bustos F, Barnes WL, Ebbesen TW (2005) Strong coupling between surface plasmon-polaritons and organic molecules in subwavelength hole arrays. Phys Rev B 71:035424-1-5
3. Hao Y-W, Wang H-Y, Jiang Y, Chen Q-D, Ueno K, Wang W-Q, Misawa H, Sun H-B (2011) Hybrid states dynamics of gold nanorods/dye J-aggregate under strong coupling. Angew Chem Int Ed 50:7824–7828
4. Li J, Ueno K, Uehara H, Guo J, Oshikiri T, Misawa H (2016) Dual strong couplings between TPPS J-aggregates and aluminum plasmonic states. J Phys Chem Lett 7:2786–2791

5. Thomas A, George J, Shalabney A, Dryzhakov M, Varma SJ, Moran J, Chervy T, Zhong X, Devaux E, Genet C, Hutchison JA, Ebbesen TW (2016) Ground-state chemical reactivity under vibrational coupling to the vacuum electromagnetic field. Angew Chem Int Ed 55:11462–11466
6. Thomas A, Lethuillier-Karl L, Nagarajan K, Vergauwe RMA, George J, Chervy T, Shalabney A, Devaux E, Genet C, Moran J, Ebbesen TW (2019) Tilting a ground-state reactivity landscape by vibrational strong coupling. Science 363:615–619
7. Ueno K, Nozawa S, Misawa H (2015) Surface-enhanced terahertz spectroscopy using gold rod structures resonant with terahertz waves. Opt Exp 23:28584–28592
8. Ueno K, Sun Q, Mino M, Itoh T, Oshikiri T, Misawa H (2016) Surface plasmon optical antennae in the infrared region with high resonant efficiency and frequency selectivity. Opt Exp 24:17728–17737
9. Shalabney A, George J, Hutchison J, Pupillo G, Genet C, Ebbesen TW (2015) Coherent coupling of molecular resonators with a microcavity mode. Nat Commun 6:5981-1-6
10. Ueno K, Juodkazis S, Mizeikis V, Ohnishi D, Sasaki K, Misawa H (2007) Inhibition of multipolar plasmon excitation in periodic chains of gold nanoblocks. Opt Exp 15:16527–16539
11. Murata N, Hata R, Ishihara H (2015) Crossover between energy transparency resonance and Rabi splitting in antenna–molecule coupled systems. J Phys Chem C 119:25493–25498
12. Yu H, Sun Q, Ueno K, Oshikiri T, Kubo A, Matsuo Y, Misawa H (2016) Exploring coupled plasmonic nanostructures in the near field by photoemission electron microscopy. ACS Nano 10:10373–10381
13. Yu H, Sun Q, Yang J, Ueno K, Oshikiri T, Kubo A, Matsuo Y, Gong Q, Misawa H (2017) Near-field spectral properties of coupled plasmonic nanoparticle arrays. Opt Exp 25:6883–6894
14. Ueno K, Yang J, Sun Q, Aoyo D, Yu H, Oshikiri T, Kubo A, Matsuo Y, Gong Q, Misawa H (2019) Control of plasmon dephasing time using stacked nanogap gold structures for strong near-field enhancement. Appl Mater Today 14:159–165
15. Lecarme O, Sun Q, Ueno K, Misawa H (2014) Robust and versatile light absorption at near infrared wavelengths by plasmonic aluminum nanorods. ACS Photon 1:538–546
16. Ueno K, Oshikiri T, Zhong Y, Shi X, Misawa H (2015) Plasmon-induced artificial photosynthesis. Interface Focus 5:20140082-1-9
17. Ueno K, Oshikiri T, Murakoshi K, Inoue H, Misawa H (2015) Plasmon-enhanced light energy conversion using gold nanostructured oxide semiconductor photoelectrodes. Pure Appl Chem 87:547–555
18. Ueno K, Oshikiri T, Misawa H (2016) Plasmon-induced water splitting using metallic nanoparticle-loaded photocatalysts and photoelectrodes. ChemPhysChem 17:199–215
19. Ueno K, Oshikiri T, Sun Q, Shi X, Misawa H (2018) Solid-state plasmonic solar cells. Chem Rev 118:2955–2993
20. Guo J, Ueno K, Yang J, Shi X, Li J, Sun Q, Oshikiri T, Misawa H (2017) Exploring the near-field of strongly coupled waveguide-plasmon modes by plasmon-induced photocurrent generation using a gold nanograting-loaded titanium dioxide photoelectrode. J Phys Chem C 121:21627–21633
21. Shi X, Ueno K, Oshikiri T, Sun Q, Sasaki K, Misawa H (2018) Enhanced water splitting under modal strong coupling conditions. Nat Nanotechnol 13:953–958
22. Yang J, Sun Q, Ueno K, Shi X, Oshikiri T, Misawa H, Gong Q (2018) Manipulation of the dephasing time by strong coupling between localized and propagating surface plasmon modes. Nat Commun 9:4858-1-8

Chapter 9
Photosynergetic Effects on Triplet–Triplet Annihilation Upconversion Processes in Solid Studied by Theory and Experiments

Kenji Kamada, Ryuma Sato, Toshiko Mizokuro, Hirotaka Kito-Nishioka, and Yasuteru Shigeta

Abstract Utilization of high density of triplets is important to overcome the dissipation of excitation energy. Triplet–triplet annihilation (TTA) has attracted attention to overcome the conventional limitation for the application of photon upconversion (UC). Synergetic approaches of experimental and computational studies on TTA have been made for efficient TTA-UC in the solid state. Binary solid fabricated by the rapid-drying casting from mixed solution enabled efficient triplet sensitization. Structural modification of emitter molecules has enhanced the performance of TTA-UC in solid. Computational studies clarified the importance of the dimensionality of the exciton diffusion process and orbital overlap in the TTA process. Various extended studies such as solid TTA-UC by near-infrared (NIR) excitation and sensitizer-fixed approach on a solid surface are also presented.

Keywords Singlet fission · Triplet–triplet annihilation · Upconversion · Casting · Binary solid · Triplet–triplet energy transfer · Exciton migration · Marcus theory · Random walk models

K. Kamada
NMRI, Kansai Center, National Institute of Advanced Industrial Science and Technology (AIST), Ikeda, Osaka 563-8577, Japan
e-mail: k.kamada@aist.go.jp

R. Sato
Center for Biosystems Dynamics Research, RIKEN, 6-2-3 Furuedai, Suita, Osaka 565-0874, Japan

T. Mizokuro
RIAEP, Tsukuba Center, National Institute of Advanced Industrial Science and Technology (AIST), Tsukuba, Ibaragi 305-8565, Japan

H. Kito-Nishioka · Y. Shigeta (✉)
Center for Computational Sciences, University of Tsukuba, Tsukuba, Ibaraki 305-8577, Japan
e-mail: shigeta@ccs.tsukuba.ac.jp

H. Miyasaka et al. (eds.), *Photosynergetic Responses in Molecules and Molecular Aggregates*, https://doi.org/10.1007/978-981-15-5451-3_9

9.1 Introduction

9.1.1 Background of the Multiple Spin Conversion Processes in Excited States

In recent years, there is a growing need for efficient use of excited states of relatively large molecules for obtaining unique optical functions and energy/material conversions. Photoexcitation and the use of its excitation energy have been governed by a "one-photon absorption and single-molecule response" basis. This causes the limitation of the usage because of rapid annihilation processes between the molecules, even if many numbers of excited molecules are generated in condensed systems. Thus, the annihilation processes have been considered to be harmful to efficient photoluminescence and photo-induced chemical reactions for many years. However, the annihilation processes and the intermolecular interaction in the excited state has the potential to break the limitation beyond the conventional viewpoint and can employ as a photosynergetic process. For example, it is known that there are the complex excitation processes in which multiple molecules including excited states cooperatively change its spin state such as *triplet–triplet annihilation* (TTA) and *singlet fission* (SF), which are the reverse processes each other as

$$^3M^* + {}^3M^* \underset{\text{SF}}{\overset{\text{TTA}}{\rightleftharpoons}} {}^1M^* + {}^1M$$

where ^1M, $^1M^*$, and $^3M^*$ are the ground, singlet and triplet excited states of the molecule, M. TTA is involved in *upconversion* (UC) for converting from low energy photons to a high-energy photon, and SF is involved in the generation of long-lived triplet excitons by photoexcitation. These processes are now of particular interest from the viewpoint of improving the conversion efficiency of solar cells and can be a good example of the photosynergetic effect.

In Photosynergetic Project, the authors and coworkers aim at investigating TTA-UC and SF to understand the basic mechanism of them and obtain the guideline for a molecular design suitable for these processes by both theoretical and experimental approaches. For SF process, theoretical works clarified that rational quantum design of efficient materials based on diradical character [1], we have proposed (i) a novel class of highly efficient SF systems called "pancake-bonded systems" [2], where electronic couplings for both SF and charge mobility are found to be maximized simultaneously in the face-to-face configuration, (ii) rational design guidelines for singly-/doubly-bridged SF/TTA chromophores [3], and (iii) optimal packing forms of oligorylenes [4]. Furthermore, we have clarified the size and structure dependences of SF dynamics for molecular aggregates and have discussed dynamics design principles using the quantum master equation method, including vibronic couplings [5–7]. As experimental work, we combined SF and two-photon absorption (TPA), multiple-photon excitation, as a demonstration of the usage of higher excited states [8]. On the other hand, for TTA-UC, we have experimentally and theoretically studied the

TTA process with new molecules in solution and condensed solid, particularly binary solid as dense solid consisting of chromophores. Since the pages are limited in this chapter, in the following parts, we will describe the results obtained for TTA-UC in solid. We will start with the general background of TTA-UC.

9.1.2 Basic Mechanism of Triplet–Triplet Annihilation Upconversion (TTA-UC)

As described before, TTA, also called as Triplet Fission (TF), is the reverse process of SF. Once, it was known as an annoying process that quenches triplet state useful for various photochemical reactions. TTA also caused delayed fluorescence (DF), i.e., P-type (pyrene-type) DF, which has also been known since the early 1960s [9] including triplet–sensitized DF [10]. The DF, however, had been generally recognized as a fancy and rare event of the photophysical process until Castellano and coworkers found the potential of the process for low-intensity photon upconversion (UC) [11].

TTA-based UC, i.e., TTA-UC, is triplet–sensitized DF; thus, two chromophores, sensitizer (S) and emitter (E), are used. Its basic mechanism is illustrated with the energy diagram in Fig. 9.1. The role of sensitizer is (1) to absorb incident photon and be populated to its singlet excited state ($^1S^*$), (2) to generate the triplet excited state ($^3S^*$) via intersystem crossing (ISC), and (3) to transfer the energy to emitter by triplet–triplet energy transfer (TET). On the other hand, emitter plays a role of TTA and DF; the triplets ($^3E^*$) generated by TET are accumulated in the system by continuous excitation and then interact each other by TTA to populate the excited

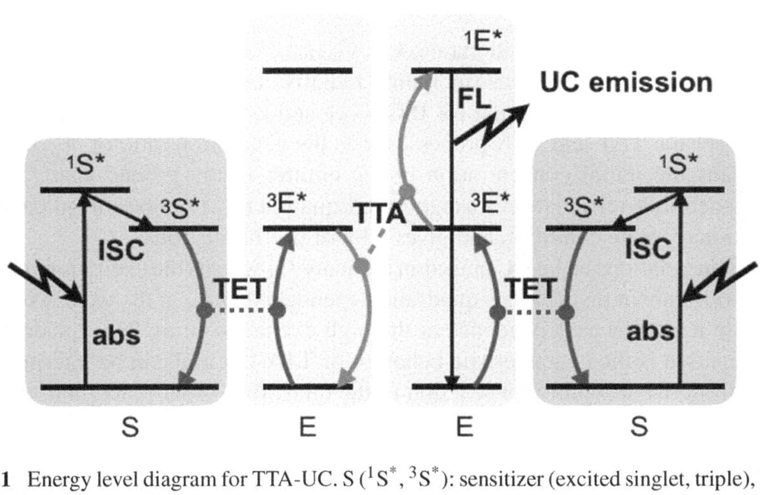

Fig. 9.1 Energy level diagram for TTA-UC. S ($^1S^*$, $^3S^*$): sensitizer (excited singlet, triple), E ($^1E^*$, $^3E^*$) emitter (excited singlet, triplet), abs: absorption, ISC: intersystem crossing, TET: triplet–triplet energy transfer; TTA: Triplet–triplet annihilation, FL: fluorescence

singlet ($^1E^*$), which emits a photon that has higher energy than that of incident photon as fluorescence (FL). Thus, the UC emission is obtained as DF. Of the four sequential processes (ISC, TET, TTA, FL) consisting of the overall TTA-UC process after excitation.

The quantum yield of UC emission (Φ_{UC}), one of the important parameters of TTA-UC, can be defined as the number of photons emitted per photon absorbed. This definition gives 0.5 as the theoretical maximum. However, another definition, where the value is multiplied by a factor of 2, is often used to scale to unity for full conversion. Here we use the latter convention (i.e., $\Phi_{UC} = 1$ for full conversion). Under this convention, Φ_{UC} can be represented as the product of the quantum efficiencies of the four-element processes:

$$\Phi_{UC} = \Phi_{ISC}\Phi_{TET}\Phi_{TTA}\Phi_{FL} \tag{9.1}$$

In this equation, Φ_{ISC} and Φ_{FL} depend on the molecular property of sensitizer and emitter, respectively. On the other hand, Φ_{TET} and Φ_{TTA} are based on bimolecular processes (TET and TTA) and thus depend not only on their molecular properties but also their mutual distance, orientation, and others that relate intermolecular interaction.

Based on the sequence of the reaction scheme shown in Fig. 9.1, the coupled rate equations for the excited species for TTA-UC,

$$\begin{cases} d[^1S^*]/dt = k_{ex} - (k_{1s} + k_{ISC})[^1S^*] \\ d[^3S^*]/dt = k_{ISC}[^1S^*] - (k_{3s} + k_{TET}[^1E]_0)[^3S^*] \\ d[^3E^*]/dt = k_{TET}[^3S^*][^1E]_0 - k_{3E}[^3E^*] - k_{TTA}[^3E^*]^2 \\ d[^1E^*]/dt = fk_{TTA}[^3E^*]^2 - k_{1E}[^1E^*] \end{cases} \tag{9.2}$$

are obtained. Here, k_{ex} is excitation rate; $k_{1S}, k_{3S}, k_{3E}, k_{1S}, k_{ISC}$ are the first-order rate constants for de-excitation rates (including radiative and non-radiative) of, respectively, $^1S^*$, $^3S^*$, $^3E^*$, $^1E^*$ and that for ISC. k_{TET} and k_{TTA} are the second-order rate constants for the TET and TTA processes. f is the generation ratio of $^1E^*$ by TTA; $[^1E]_0$ means the initial concentration of the emitter in the ground state, and the change under photo-irradiation was ignored. Equation (9.2) can be solved under the photostationary-state condition and gives a lot of useful relations.

One of the relations is the UC emission intensity I_{UC} versus the excitation intensity I_{ex}. It is well known that I_{UC} has quadratic dependence to I_{ex} at the weak excitation limit while it has linear dependence at the high excitation limit. This quadratic-to-linear transition is the characteristic behavior of TTA-UC and can be interpreted as the shift from the second-order reaction to the quasi-first-order reaction at the high concentration of $^3E^*$. From the solution of Eq. (9.2), this relation is given as [12],

$$I_{UC} = K\left[1 + \left(1 - \sqrt{1 + 4I_{ex}/I_{th}}\right)/(2I_{ex}/I_{th})\right]I_{ex} \tag{9.3}$$

where K is the instrumental constant. This equation agreed perfectly with that derived from the formulation for the relative UC quantum yield reported by Murakami [13] and gives the quadratic-to-linear change. The threshold intensity I_{th} defined as the excitation intensity at the crossing point between two extrapolated lines of the quadratic at low intensity and linear at high intensity [14]. I_{th} governs the quadratic-to-linear transition, and low I_{th} means that the TTA-UC system can operate at low excitation intensity. Conventionally, one can obtain I_{th} by extrapolation of the two slopes in the double logarithmic plot of I_{UC} against I_{ex}. However, he/she can obtain in a more reproducible manner by cure fitting with Eq. (9.2).

For the derivation of Eq. (9.9) from Eq. (9.2), I_{th} is defined as,

$$I_{th} = J(\varepsilon_{ex}[S]_0 \Phi_{ISC} \Phi_{TET})^{-1} \left(k_{3E}^2 / k_{TTA}\right) \tag{9.4}$$

where $J = N_A(hc/\lambda_{ex})/(2 \times 10^3 \ln 10)$ and N_A is Avogadro's number, hc/λ_{ex} is photon energy of excitation. Equation (9.4) shows that I_{th} can be reduced by increasing ε_{ex} $[S]_0$, the product of molar extinction coefficient of sensitizer at the excitation wavelength and the molar concentration of sensitizer, i.e., Φ_{ISC} and Φ_{TET}. This means the efficient generation of $^3E^*$ is important to reduce I_{th}, which can be achieved by strong absorption to generate $^1S^*$, and efficient conversion and transfer of energy form $^1S^*$ to $^3S^*$ to $^3E^*$via ISC and TET processes.

9.1.3 Importance of Bimolecular Processes in TTA-UC

TET and TTA are bimolecular processes and strongly depend on samples in which state and distribution of molecules are different.

In narrow meaning, the TTA process is the spin conversion process of a molecular pair, $^3E^* + {}^3E^* \rightarrow {}^1S^* + {}^1S$. However, it often used in broad meaning where the processes before two emitter triplets encounter are also included. In the solution system, this is governed by molecular diffusion, and the rate of TTA is diffusion-controlled. TTA immediately occurs after they encountered compared to the time scale of molecular diffusion. This may be reasonable because TTA is a spin-allowed process as molecular pair as SF. It was reported that SF can occur at the timescale of picosecond to nanosecond. Thus, the reverse process can occur at the same timescale at a favorable condition. Molecular diffusion is usually well described by the Stokes-Einstein relation and thus depends on the viscosity of solvent and the molecular volume of solute. The viscosity does not only affect molecular diffusion but also does the TTA process via a cage-forming process. The solvent molecules may form the solvation cage surrounding the encounter complex of the TTA process. The stability of the solvent cage may be affected by viscosity and affect the equilibrium between TTA and SF.

Different mechanism governs the process for condensed solid, which consists only of chromophores. In such solid, each molecule cannot move in solid state, but the energy can move by transferring to neighbor molecules by TET of the same molecular species, i.e, triplet exciton migration. This is a necessary process for two emitter triplets to encounter and to cause TTA, which has been known to occur in the molecular crystal [15]. Triplet exciton migration may be slower than that of TTA and be an important process governing the efficiency of overall (or wide meaning) TTA process.

TET is another bimolecular process and a critically important process in TTA-UC. In the solution system, the process is mostly diffusion-controlled like TTA, and the triplet level matching between sensitizer and emitter is a remaining essential factor. In solid system, on the other hand, direct contact between sensitizer and emitter with keeping the energy-level matching is critically important. The details will be discussed in the following sections.

9.2 Experimental Studies on Triplet–Triplet Annihilation Photon Upconversion in Solid and Related Systems

9.2.1 TTA-UC in Binary Solid

TTA-UC has been developed in solution systems because molecular diffusion can mediate TET and TTA processes during the lifetime of their triplets. However, they have the inconvenience of handling and require hermetic sealing because they are volatile, flammable, and easily dissolving oxygen that quenches triplets. Thus, solid systems are suitable for device applications. There are the approaches of the diluted solid systems, such as doped-polymer and gel; however, molecular diffusion mediated TET and TTA in these systems are slow because of the high viscosity, leading low overall efficiency. Thus, the solid system with condensed chromophore of the emitter is preferable media because triplet exciton migration can mediate the processes, allowing the realization of fast and efficient bimolecular processes.

From this point of view, binary crystals can be an ideal medium for solid TTA-UC because the speed of migration in a crystal can be higher than in other dense media thanks to the crystal regularity and dense packing. As TTA-UC medium, binary solids, however, has a problem of segregation of the sensitizer from the emitter matrix [14, 16]. Segregation forms the dimer of sensitizers. The triplet level of the dimer is lower than that of sensitizer, which interferes with TET. High excitation intensities in the order of kW/cm^2, was reported to induce TTA-UC in binary solids [17].

In order to overcome the segregation problem in binary solid, we developed a simple method called *rapid-drying casting* [12]. The key idea is crystallizing host matrix of emitter before guest molecule of sensitizer aggregating. We choose Pt-octaethylporphyrin (PtOEP) as sensitizer and 9,10-diphenyl-anthracene (DPA) as emitter (Fig. 9.2a) because the PtOEP-DPA system has been used in many works and can be a good benchmark. Moreover, DPA has very high fluorescence (FL) QY, $\Phi_{FL} = 0.95$, in crystals [18]. The binary crystal (PtOEP:DPA) were fabricated by drop casting from mixed solution on a glass slide at the aerated condition. For the

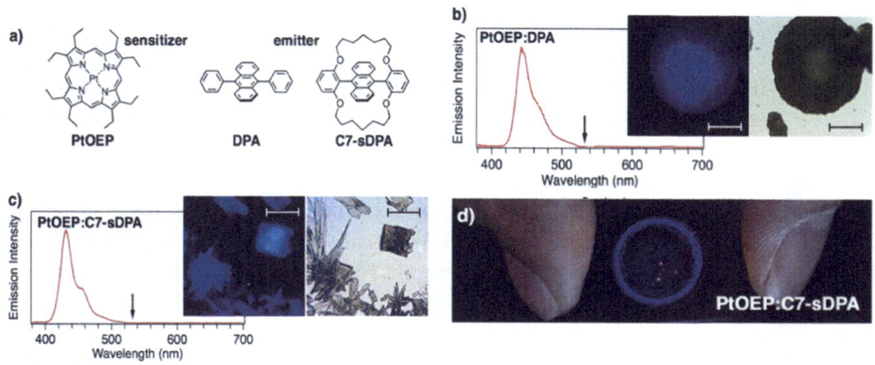

Fig. 9.2 **a** Chemical structures. Transmission and emission images of microparticles and the emission spectra of binary solid of **b** PtOEP:DPA and **c** PtOEP:C7-sDPA. Bar stand for 50 μm. The arrows in the spectra shows the excitation wavelength (532 nm). **d** Centimeter-scale upconverted emission image of the PtOEP:C7-sDPA cast sample illuminated by a diode laser (~1 mW/cm^2). Reproduced from [12]—Published by The Royal Society of Chemistry

rapid-drying casting, we paid attention to the following points: (1) the concentration of emitter was close to that of its saturation; (2) the concentration of sensitizer is much lower than that of emitter (around 1:1000 in molar ratio of sensitizer to emitter); and (3) solvent has low boiling point and good wettability to glass surface. By dropping the mixed solution from a pipette, a single drop of the solution was spread widely on the slide and formed a thin liquid film, which quickly evaporated from the round rim. The circular thin film of the binary solid with "coffee stain" rim with a diameter of ca. 1 cm was obtained.

In the rim, there were some round particles (10–100 μm) having spherulite-like structures as an aggregation of needle-like microcrystals. By cw-excitation of the particles at 532 nm (matching to the peak wavelength of the Q-band of PtOEP) under an optical microscope, we found that the round microparticles showed the clear blue emission whose spectrum matched to FL spectrum of DPA (Fig. 9.2b)—UC emission. The excitation intensity was only 4 W/cm^2 at the sample surface. By applying the simple fabrication technique of rapid-drying casting, the excitation intensity can be reduced by three orders of magnitudes than the previous reports [17].

The same strategy was applied to another derivative of DPA, which has the loop-like alkoxy chain protecting the anthracene core (C7-sDPA, Fig. 9.2a). C7-sDPA also has high FL-QY of $\Phi_f = 0.85$ in crystal and shows excellent photostability compared to DPA because of the loop [18]. By casting the mixed solution of PtOEP and C7-sDPA, rectangular- and petal-shaped crystals, showing strong UC emission were obtained (Fig. 9.2c). Not all particles were luminescent, but they were formed with good reproducibility. Centimeter-scale illumination by laser diode gave excitation intensity of ~1 mW/cm^2, at the level of sunlight (for 532 ± 5 nm). At this excitation intensity, the binary solid of PtOEP:C7-sDPA showed clear UC emission (Fig. 9.2d), which shows the potential of these fabrication methods. The ring-like emission shows the UC-luminescent particles are located in the rim.

The I_{ex}-dependence of I_{UC} and Φ_{UC} at the saturation intensity (*i.e.*, at linear-response regime) of single microparticles were investigated by using a micro-spectroscopic setup because those UC-luminescent particles were mixed with a lot of non-luminescent particles. Both binary solids of PtOEP:DPA and PtOEP:C7-sDPA showed the quadratic-to-linear transition of the dependence, characteristic to TTA-UC, and well reproduced by Eq. (9.3). Meanwhile, I_{th} was largely different (Fig. 9.3a); the observed range of I_{th} for the microparticles of PtOEP:C7-sDPA was two or three orders of magnitudes lower than those of PtOEP:DPA, as expected by the demonstration shown in (Fig. 9.2d). Since both systems used the same sensitizer and roughly the same concentrations, the difference originates from Φ_{TET} and the ratio of rate constants of the emitter, $(k_{3E})^2/k_{TTA}$ from Eq. 9.4. On the other hand, Φ_{UC} at the saturation intensity was higher for PtOEP:C7-sDPA than PtOEP:DPA (Fig. 9.3b) regardless of the environments. This means that oxygen cannot penetrate and quench the triplets for the short time scale, and the scattering of microparticle did not affect the Φ_{UC} measurement significantly. It should be noted here that the Φ_{UC} value for PtOEP:C7-sDPA was above 10% and nearly 20% in center. This is one of highest values in solids and closed to the values of solution.

The absorption spectroscopy of microparticles on Q-band of PtOEP clarified that PtOEP was dispersed as a monomer in microparticle of PtOEP:C7-sDPA while it

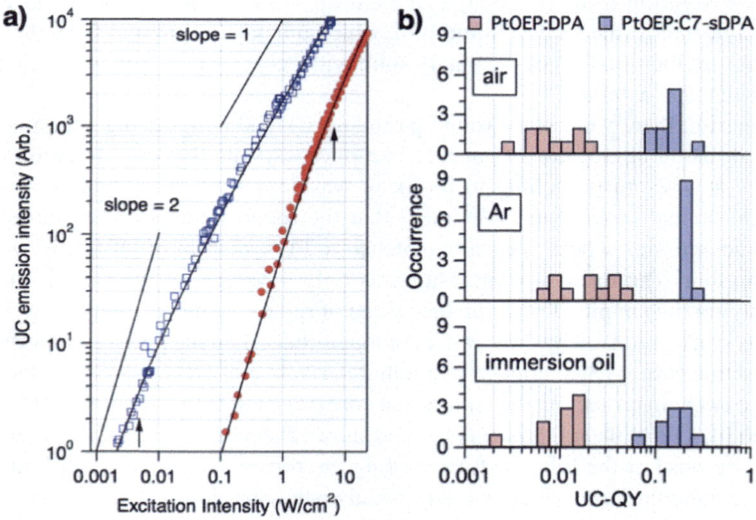

Fig. 9.3 **a** Examples of the excitation intensity dependence of the UC emission of single grain of binary solid of PtOEP:DPA (red filled circle) and PtOEP:C7-sDPA (blue open square) with the curve fits with Eq. (3.3) (black curve). The arrows indicate I_{th} obtained by the fits. **b** Histogram of UC-QY of single microparticles of PtOEP:DPA (red) and PtOEP:C7-sDPA (blue) under different environment. Reproduced from [12]—Published by The Royal Society of Chemistry

still aggregated in some degree in those of PtOEP:DPA. These results were supported by the high magnification spectrographic images of the upconverted emission of the microparticles. Microparticles of PtOEP:C7-sDPA have uniformly emitted the UC emission while that of PtOEP:DPA had some bright cores of UC emission. Thus, aggregation of sensitizer was prevented by the rapid-drying casting technique, but the degree was different between the emitters and affected the rate of the TET and the efficiency. Another interesting feature was that triplet diffusion length C7-sDPA in the crystal was estimated to be ~230 nm from the decay of the UC emission. This value is much longer than the nanometer-sized grain structure (20–50 nm) of the microparticle of PtOEP:C7–sDPA and can also enhance TTA process by increasing the encounters.

9.2.2 Binary Solid with Various Alkyl-Chain-Attached DPAs

As shown in the previous section, the rapid-drying casting technique enables efficient TTA-UC in binary solid by preventing the segregation of sensitizer and accelerating the TET process. The addition of loop-like alkoxy groups in C7-sDPA may contribute to enhance the compatibility with PtOEP. To understand the effect of the introduction of alkyl chain to sensitizer molecule, we fabricated binary solids and investigated their TTA-UC properties with the four octhyl derivatives of DPA and C7-sDPA at the different substituted position as emitter (Fig. 9.4).

The same conditions as before were applied for the rapid-drying casting with the four different emitters. The concentration ratio of sensitizer to emitter was 1:1000, and PtOEP was used as the sensitizer for all. Microcrystals were obtained by the casting for all samples except one with **1a**. Most of the microcrystals were found to have a spherulite-like structure by observation with crossed polarizers. The microcrystals of binary solids showed UC emission under low intensity (4 W/cm^2), cw-excitation at 532 nm under aerated condition. On the other hand, the sample with **1a** gave microdroplets on a glass plate. Compound **1a** has a low melting point below

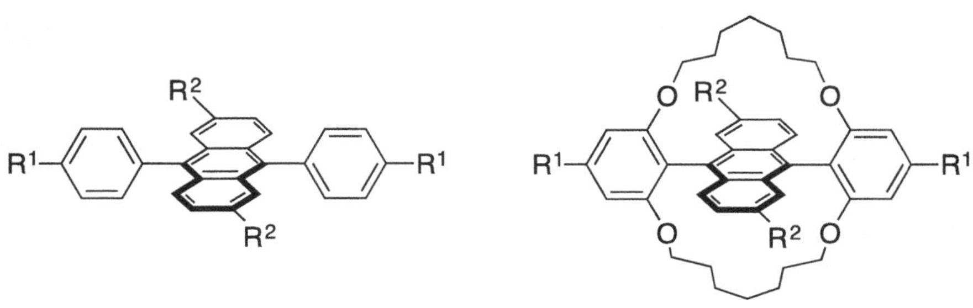

Fig. 9.4 Chemical structures of the alkyl-extended derivatives of DPA (left, **1p**: $R^1 = C_8H_{17}$, $R^2 = H$ and **1a**: $R^1 = H$, $R^2 = C_8H_{17}$) and C7-sDPA (right, **2p**: $R^1 = C_8H_{17}$, $R^2 = H$, and **2a**: $R^1 = H$, $R^2 = C_8H_{17}$)

80 °C and may easily melt by mixing with not only PtOEP but also impurities of oligomers in THF solvent. The microdroplets also showed UC emission under air like the microcrystals of the other three samples. This may be due to the high viscosity of the liquid. The I_{th} and Φ_{UC} at the saturation intensity were also characterized for all samples obtained by the casting. The values of I_{th} for the samples were between those of PtOEP:C7-sDPA and PtOEP:DPA and tended to be lower for C7-sDPA derivatives (**2p** and **2a**) than for DPA derivatives (**1p** and **1a**). The binary solid with **2a** found to have a comparable value of Φ_{UC} with PtOEP:C7-sDPA, but the others were in the order of a few percent. These results suggested that C7-sDPA and its derivatives (**2p** and **2a**) tended to show better TTA-UC performance through the low I_{th} and relatively high Φ_{UC} at the saturation intensity.

9.2.3 TTA-UC in Binary Solid from Near-Infrared

The rapid-drying casting technique was also applied for fabricating the binary solid that enables to convert light from near-infrared (NIR) to visible wavelengths [19]. Extending the excitation wavelength to the NIR region is important for the application such as solar cell. However, NIR-to-Vis conversion by TTA-UC in condensed solids is still very limited [20, 21]. We chose PdTPTAP (Fig. 9.5) as an NIR sensitizer absorbing 785 nm and rubrene as an emitter with the emission peak at 570 nm), allowing NIR-to-vis upconversion.

By the rapid-drying casting, the binary solid of PdTPTAP:rubrene was fabricated. This gave many spherical microparticles mainly in the rim of the casted

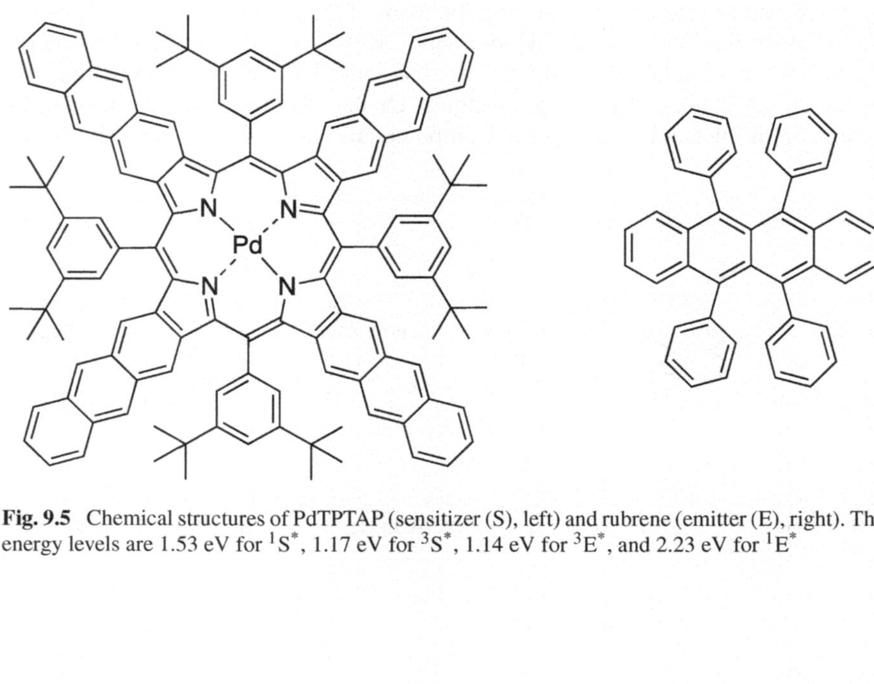

Fig. 9.5 Chemical structures of PdTPTAP (sensitizer (S), left) and rubrene (emitter (E), right). The energy levels are 1.53 eV for $^1S^*$, 1.17 eV for $^3S^*$, 1.14 eV for $^3E^*$, and 2.23 eV for $^1E^*$

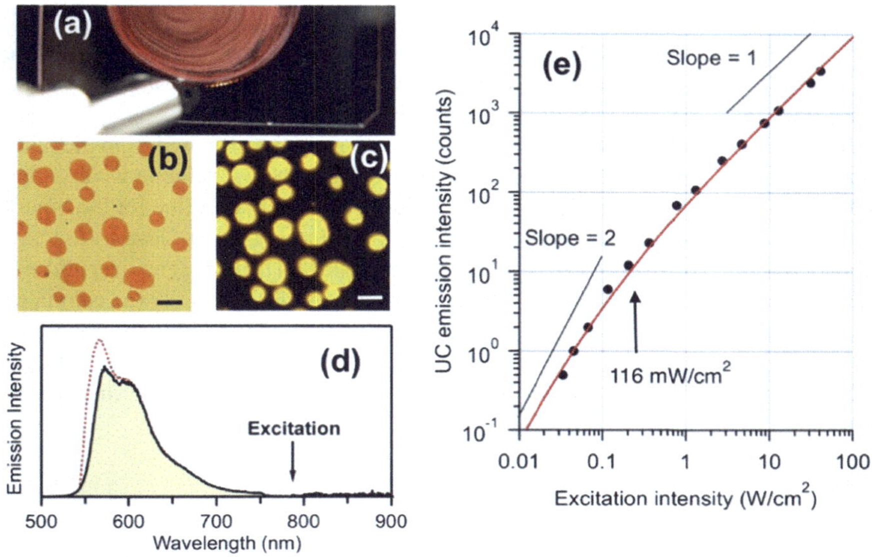

Fig. 9.6 Binary solid of PdTPTAP:rubrene fabricated by solution casting: **a** appearance of the sample illuminated by 785 nm light from the optical fiber. Microscopic images of **b** transmission and **c** UC emission in the rim of the sample. Bars are 50 μm. **d** UC (filled yellow) and fluorescence (red dotted) emission spectra. **e** I_{ex}-dependence of I_{UC} (black dot) with theoretical fit by Eq. (3.3) (red curve). Reprinted with permission from [19] Copyright 2019. American Chemical Society

region (Fig. 9.6b). These particles were confirmed to be amorphous by polarization microscopy. By cw-excitation at 785 nm (7 W/cm^2), the microparticles showed bright yellow emission under air (Fig. 9.6a, c) with the emission spectrum of individual particles matched to the fluorescence spectrum of rubrene (Fig. 9.6d). Depression in the short wavelength edge is due to reabsorption. The I_{th} was found to be in the order of 100 mW/cm^2 by measuring the I_{ex}-dependence of I_{UC} (Fig. 9.6e). Again, the quadratic-to-linear transition in the dependence proofed that the yellow emission was due to TTA-UC.

For more than 40 individual particles, Φ_{UC} at the saturate intensity (7 W/cm^2) was measured. The obtained value was $\Phi_{UC} = 0.005 \pm 0.001$ under air and the almost the same value under air. Based on Eq. (9.1), a determining process of the Φ_{UC} value can be clarified by investigating of efficiencies of individual processes.

Femtosecond transient absorption spectroscopy of PdTPTAP unveiled that ISC occurs with lifetime 2.9 ps, and the ISC rate is two-orders of magnitude faster than other deactivation processes of $^1S^*$, giving the estimation of $\Phi_{ISC} > 0.99$. Next, the TET process of the binary solid was investigated by measuring the rise time of the UC emission. It should be noted that the UC emission intensity is proportional to the square of the concentration of emitter triplet, $I_{UC} \propto [^3E^*]^2$ (from Eq. (9.2)); thus, the risetime of $(I_{UC})^{1/2}$ corresponds to the decay lifetime of $^3S^*$. The observed risetime was found to be much faster than 10 ns. The conservative estimation of the lifetime was also obtained to be shorter than $\tau = 2\ \mu s$ from a precise analysis of the UC emission. These estimations of lifetimes are much faster than the natural lifetime of $^3S^*$, $\tau_0 = 16.2\ \mu s$ determined by nanosecond transient absorption spectroscopy. Thus, $\Phi_{ISC} = 1-\tau/\tau_0 = 0.88$ or larger.

Since there is no direct method to characterize Φ_{TTA}, it is estimated from the other values. The remaining efficiency is Φ_{FL} of the emitter. Φ_{FL} of rubrene is very high and close to unity; however, it is very low in solid. The obtained value for the binary solid of the cast sample was $\Phi_{FL} = 0.015$. This value is almost close to the cast sample without sensitizer ($\Phi_{FL} = 0.015$), suggesting PdTPTAP sensitizer did not quench the emission of rubrene. With these values and Eq. (9.1), it was estimated $\Phi_{TTA} = 0.36$. Thus, the most determining process was found to be FL of rubrene. The value of Φ_{TTA} was comparable to the reported highest value for rubrene in solution [22]; nevertheless, further studies are needed to explorer the room to improve in condensed solid.

9.2.4 Fixing Sensitizer Molecule on the Surface of Nanopore

The rapid-drying casting technique has been applied to prevent the segregation of sensitizer molecules on the fabrication of solid. An alternative approach to prevent segregation was explored. The method is fixing sensitizer molecules on the solid surface. It was reported that, for example, porphyrin derivatives could be fixed on the surface of clay nanosheets without aggregation [23]. Thus, a similar approach can be adopted. As the platform to fix sensitizer, here, nanoporous glass was employed. Nanoporous glass (NPG) has 3D-network of nanopores, enabling extremely large pore surface area per volume. Thus, thick interaction volume like bulk-hetero-junction can be possible.

To fix the sensitizer, we choose Pt(II) coproporphyrin I (PtCP), a tetracarboxylic-acid derivative of PtOEP, as a sensitizer. The carboxylic acids of PtCP reacted with the amino groups at the surface of the nanopores to fix it. NPG was treated with the silane-coupling agent with the amino group to cover the surface with the amino groups prior to the fixation (Fig. 9.7a–b). To fix the sensitizer, a piece of the amino-treated NPG was dipped or refluxed in PtCP solution of THF. The piece of NPG was colored in pink, and its absorption spectrum well agreed with PtCP in THF solution after subtraction of the scattering due to the nanoporous structure. This showed that PtCP was fixed in its monomeric form on the surface of the nanopore.

This PtCP-fixed NPG was showed strong red emission assigned to the phosphorescence of PtCP under deaerated condition filled with Ar gas. Interestingly, it was found that the phosphorescence QY (~0.25) was enhanced by compared to that in THF solution (~0.16) for the same loading concentration (75 μM) as shown in Fig. 9.7c. The high phosphorescence QY of PtCP fixed in NPG was unchanged by increasing the loading concentration by a factor of ten and then decreased for further increase. The phosphorescence decay of the samples fixed in NPG was single exponential with the lifetime of 95 μs, which is nearly twice that in THF solution

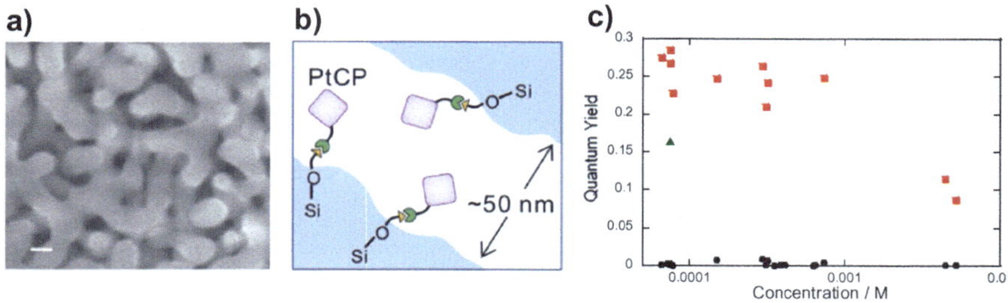

Fig. 9.7 **a** SEM image of nanoporous glass (bar: 50 nm). **b** Schematic drawing of fixing the sensitizer (PtCP) to the pore surface of nanoporous glass. **c** Phosphorescene QY of PtCP-fixed nanoporous glass (red square) and in THF (green triangle) under deoxygenated by Ar-bubbling, together with that of PtCP-fixed nanoporous glass under air (black circle) (Reproduced from [24]—Published by The Royal Society of Chemistry)

(48 μs), both at 75 μM. The analysis showed that the non-radiative deactivation was suppressed by fixing the sensitizer in NPG. These results clarified that excited triplet $^3S^*$ of the sensitizer was not quenched by fixing on the surface of the nanopore and rather maintained by suppressing the non-radiative deactivation pass probably due to limited freedom of the molecular motion. This causes the enhanced phosphorescence in NPG with the higher QY and the longer lifetime [24].

This successful result of fixing the sensitizer on the solid surface in NPG without losing the function encouraged us to use it for TTA-UC [25]. First, we prepared the toluene solution of DPA, and the piece of PtCP-fixed NPG was immersed in it. The blue UC emission was successfully observed by cw-excitation at 532 nm, suggesting TET occurs from PtCP covalently bonded on the surface of the solid to DPA diffused into the nanopore (~50 nm), and TTA occurred in the confined space of nanopore. The Φ_{UC} at a saturation excitation intensity (0.13) was found to close to that in solution (0.15, toluene), suggesting that both TET and TTA processes can proceed in nanopore and its surface as the same level in bulk solution. We further studied the phosphorescence decay and the UC emission rise of the sample. Careful analysis of the correspondence between the decay lifetime and rise time clarified that fast and slow TET's existed in the sample of PtCP-fixed NPG immersed in the DPA solution. This may suggest that the sensitizer molecules were fixed on the nanopore surface in two different fashions. Moreover, the all-solid TTA-UC system was demonstrated by filling DPA in the nanopore of PtCP-fixed NPG.

9.2.5 *TTA of Alkyl Strapped DPA*

As shown in some of the previous sections, DPA derivatives with loop-like alkoxy groups (Fig. 9.8), so-called strapped DPA, such as C7-sDPA ($n = 7$) has

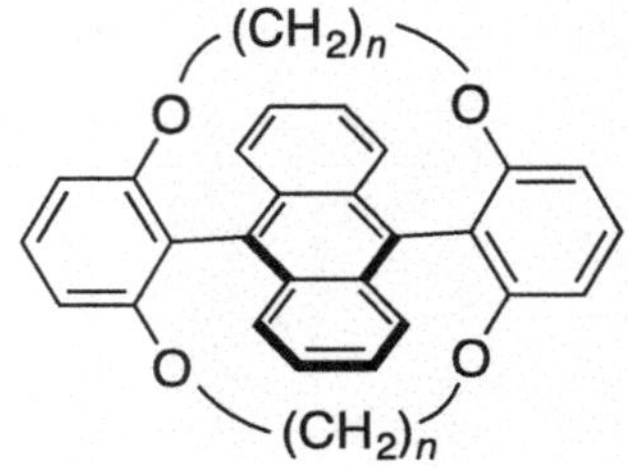

Fig. 9.8 Chemical structure of C*n*-sDPA

interesting property as the emitter of TTA-UC. In relatively dilute solution ([sensitizer]:[emitter] = 2 μM: 200 μM; sensitizer is PtOEP), C7-sDPA showed brighter UC emission than DPA for whole range of excitation intensity (up to 100 mW/cm^2). The theoretical limit of Φ_{UC} at the infinite excitation intensity ($\Phi_{UC,\,inf}$) of the sample was estimated to be 0.24 with C7-sDPA while 0.12 for DPA. These estimations were based on the curve fitting to the equation,

$$\Phi_{UC}(I_{ex}) = \Phi_{UC,inf}\left[1 + \left(1 - \sqrt{1 + 4I_{ex}/I_{th}}\right) \Big/ (2I_{ex}/I_{th})\right] \quad (9.5)$$

which is the source equation of Eq. (9.3).

The obtained $\Phi_{UC,\,inf}$ can be factorized as Eq. (9.1). With the knowledge of Φ_{ISC} and Φ_{FL}, and the concentration dependence of emitter on Φ_{TET} with the Stern-Volmer relation, the singlet generation ratio of TTA process was obtained. The obtained singlet generation ratio for all samples was higher than 0.11 = 1/(1 + 3 + 5), which is predicted by the simple spin statistic of the triplet pairs (giving singlet, triplets, and quintets without their equilibriums). Interestingly, the generation ratio was higher for C7-sDPA than that of unsubstituted DPA and showed concentration dependence. That for C7-sDPA was found to increase as increasing the concentration while DPA showed the opposite behavior. Derivatives of other chain lengths were also studied (*n* = 6 and 8). C6-sDPA showed the same behavior as C7-sDPA, while the concentration dependence was not obtained for C8-sDPA due to its solubility problem. It is not easy to interpret the trends of the singlet generation ratio of the TTA process, but the extensive results of theoretical calculation gave interesting suggestions to these systems [26, 27].

9.3 Theoretical Studies on Triplet–Triplet Annihilation Photon Upconversion in Solid State

In the previous section, we explained the experimental results of TTA-UC in the solid state and the related topics in detail. In this section, we give the theoretical background of the energy transfer rate constants and encounter probability of triplet excitons to understand the underlying mechanism of TTA-UC processes [28]. For the former issue, we adopt Marcus theory of electron transfer, where all the parameters appearing in this theory are evaluated using quantum chemical calculations for monomers or dimers extracted from the crystal structure. For the latter, triplet exciton migration and thus the encounter probability are modeled by random walk process. Using the experimental condition and observation, we focus on the dimensionality of the exciton migration for the encounter probability.

9.3.1 Theoretical Background of Energy Transfer Rate Constant Represented by Marcus Theory

Dominant TTA processes are categorized into two types. One is the direct change from the T_1–T_1 pair state to the S_1–S_0 pair state, and the other is the indirect (sequential) change from T_1–T_1 state to S_1–S_0 via charge-separated states. On the other hand, triplet–triplet energy transfer (TTET) between two emitters in crystal occurs under the Dexter mechanism, where an overlap of molecular orbitals between two neighboring emitters determines the energy transfer efficiency. Note here that we use the abbreviation of TTET to distinguish this process (i.e., between the emitters) and the triplet energy transfer (TET) from the sensitizer to the emitter in Fig. 9.1. We here estimate the rate constant of energy transfer processes concerning TTA-UC in solid, k_{ET}s, for each process by the Marcus theory [29],

$$k_{ET} = \frac{2\pi}{\hbar}|T_{DA}|^2 \frac{1}{\sqrt{4\pi\lambda k_B T}} \exp\left[-\frac{(\Delta G + \lambda)^2}{4\lambda k_B T}\right], \tag{9.6}$$

where T_{DA}, ΔG, and λ are the electronic coupling matrix elements between a donor (D) and an acceptor (A) of the electron, the driving force, and the reorganization energy, respectively. $\hbar$, k_B, and T is the Dirac's constant, the Boltzmann constant, and the temperature, respectively.

In the TTA process, the rate constants of several elementary processes should be evaluated to yield a total TTA rate constant. For the direct TTA process, we estimated the electronic coupling matrix elements using a fourfold way [30–32], which is a general diabatization scheme based on diabatic molecular orbitals (DMOs) obtained by a threefold way density criterion and a configurational uniformity [33] by limiting an active space. To construct a model Hamiltonian, we use only six singlet states for the dimer, S_0S_0, T_1T_1, S_1S_0, S_0S_1, CA, and AC. Here S_0S_0 and T_1T_1 indicate that both monomers are either the ground state or the triplet state. S_1S_0 and S_0S_1 indicate that one monomer is the singlet excited state, and the other is the ground state (i.e., local single-exciton). CA and AC are charge-separated states, where one monomer is anion (A), and the other is cation (C). We use the reference molecular orbitals (MO) obtained by the restricted Hartree–Fock (RHF) calculations followed by the threefold way (CASSCF(4,4)/6–31 g(d)) and obtained the reference DMOs. Then, we set dominant CSFs which, are divided into six groups (S_0S_0, T_1T_1, S_1S_0, S_0S_1, CA, and AC) and performed the fourfold way diabatization scheme. These monomer DMOs are used to generate the diabatic configurations [34].

After determination of diabatic states, we estimated the electronic coupling matrix element using the following equation [35];

$$T_{\mathrm{DA}} = \langle \mathrm{T_1T_1}|\hat{H}|\mathrm{S_1S_0}\rangle - \sum_{X=\mathrm{AC,CA}} \frac{\langle \mathrm{T_1T_1}|\hat{H}|X\rangle\langle X|\hat{H}|\mathrm{S_1S_0}\rangle}{\langle X|\hat{H}|X\rangle - E_{\mathrm{tun}}} \tag{9.7}$$

where $|\mathrm{T_1T_1}>$, $|\mathrm{CA}>$, $|\mathrm{AC}>$, and $|\mathrm{S_1S_0}>$ are the electronic structures of the triplet–triplet pair, the cation-anion pair, the anion–cation pair, and the singlet excited-ground pair states, respectively. These pair state wave functions are obtained from the fourfold way as the diabatic states. $\hat{H}$ is the total Hamiltonian of the system, and E_{tun} is the average of the energies of $\mathrm{T_1T_1}$ and $\mathrm{S_1S_0}$. The first term on the right-hand side is the direct term (i.e., $\mathrm{T_1T_1} \rightarrow \mathrm{S_1S_0}$), and the second and third terms on the right-hand side are the super-exchange terms (i.e., $\mathrm{T_1T_1} \rightarrow \mathrm{CA} \rightarrow \mathrm{S_1S_0}$ and/or $\mathrm{T_1T_1} \rightarrow \mathrm{AC} \rightarrow \mathrm{S_1S_0}$). Note here that the super-exchange mechanism includes the virtual charge-separated states (CA and/or AC).

For the TTET process, we also adopted the Marcus theory and estimated the electronic coupling matrix element using the fragment difference approaches [36–42]. In this study, we adopted the fragment excitation difference (FED) scheme, which is one of the simplest fragment difference approaches, with the configuration interaction singles (CIS/6−31 + g(d)) method. We performed FED scheme calculations using a modified version of GAMESS [ver. Dec 5, 2014] [43].

The reorganization energy, λ, was estimated by the sum of internal, λ^{intra}, and external, λ^{inter}, contributions as [44, 45]

$$\lambda = \lambda^{\mathrm{intra}} + \lambda^{\mathrm{inter}}, \tag{9.8}$$

To estimate λ^{intra}, we used four-point method. On the other hand, λ^{inter} was approximated by using a Born-Hush approach [44, 46, 47]. Note here that λ^{inter} for the TTET process is zero since there is no charge change upon the energy transfer [48, 49].

9.3.2 TTA and TTET Reaction Time for DPA and C7-SDPA in Crystal

These protocols in hand, we have implemented the in-house codes for calculating all required terms and evaluated both the TTA and TTET reaction times for DPA and C7-sDPA, which are estimated from the kinetic constants as $\tau_{\mathrm{TTA}} = 1/k_{\mathrm{TTA}}$ and $\tau_{\mathrm{TTET}} = 1/k_{\mathrm{TTET}}$, respectively. Figure 9.9 show the non-redundant structures of neighboring molecule pairs extracted from the crystal structures and the corresponding reaction times among the monomers. It is easily seen that the TTA reaction times of DPA are ns time scale (or much less for a specific direction), while the TTET ones are μs time scale (or more). This means that the TTET process is the time-limiting step among several elementary steps in the TTA-UC process, and TTA immediately occurs when

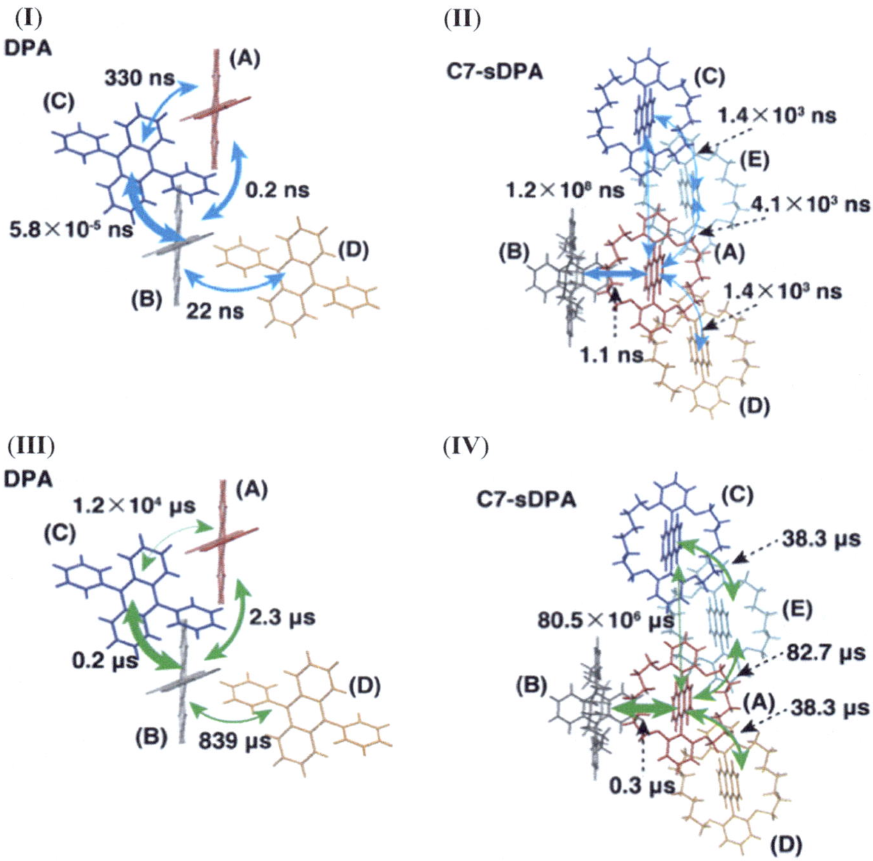

Fig. 9.9 TTA reaction time, $\tau_{\mathrm{TTA}} = 1/k_{\mathrm{TTA}}$, [ns] for DPA (**I**) and C*n*-sDPAs (**II**). Here, red, gray, blue, orange, and cyan colors represent monomer A, B, C, D, and E respectively. Corresponding TTET reaction time, $\tau_{\mathrm{TTET}} = 1/k_{\mathrm{TTET}}$, [μs] for DPA (**III**) and C*n*-sDPAs (**IV**). Reprinted with permission from [28] Copyright 2019. American Chemical Society

two triplet excitons locate at any neighboring site because of $\tau_{\mathrm{TTA}} << \tau_{\mathrm{TTET}}$. Another finding is that there exist preferable directions both for TTA and TTET. For DPA, TTA between B and C monomers are so fast (fs time scale), and others are ns time scale and comparable with the minimum time of TTA in C7-sDPA. Among four directions, the ratio between the minimum reaction time and the second minimum one is about 1:10. It means that the triplet energy transfer occurs quasi-two-dimensionally.

For C7-sDPA, the situation is different from DPA. One of the TTA reaction times of C7-sDPA (A–B pair) is ns time scale, while the others are much longer than it. This behavior is also found for the TTET process. However, comparing to DPA, the minimum reaction time of the TTET process is shorter than those of the TTA reaction times except for that in the A–B pair. This indicates that TTA occurs only when two excitons locate neighboring sites along the A–B pair direction. It means that the triplet energy transfer occurs one-dimensionally.

Judging only from the TTA reaction times, the performance of DPA in the TTA process is superior to C7-sDPA, which contradicts with the experimental finding. Compared to the TTA process, the difference in the minimum reaction time for TTET between DPA and C7-sDPA is small, i.e., 0.2 μs for DPA and 0.3 μs for C7-sDPA. However, this result does not also explain why DPA exhibits less efficient TTA-UC than C7-sDPA. These results suggest that other mechanisms are needed for considering TTA-UC efficiency. One possible reason is the difference in the lifetime of triplet exciton, τ_{triplet}. τ_{triplet} for DPA and C7-sDPA are 54 and 140 μs, respectively. TTET occurs at most $n_{\text{TTET}} = 335$ (510) times, estimated from $n_{\text{TTET}} = \tau_{\text{triplet}}/\min(\tau_{\text{TTET}})$. This means that the triplet exciton possibly encounters more often in C7-sDPA than in DPA. The encounter probability is directly related to the quantum yield and emission intensity. Thus, in order to measure the TTA-UC performance between DPA and C7-sDPA quantitatively, we analyze the encounter probability of triplet excitons by referring both the differences in the dimensionality, the TTET reaction times, and the experimentally measured triplet state lifetime.

9.3.3 Random Walk Models for the Encounter of Triplet Excitations

Before formulating the encounter probability, we must consider the experimental conditions. The diffusion length L_{D} of triplet exciton is approximately estimated as $L_{\text{D}} = \sqrt{R^2 n_{\text{TTET}}/2m}$, where R and m are the averaged distance between neighboring emitters and the dimensionality. The diffusion length is estimated as 60.1Å for DPA and 71.3Å for C7-sDPA, respectively. In this sense, C7-sDPA encounters more likely than DPA. However, the situation is not so simple. Since the concentration of the triplet sensitizer is 1/1000 of that of DPAs, the average distance between two sensitizers is approximately 10 times longer than the lattice constant of DPAs for each direction, 91–211 Å for DPA and 149–154 Å for C7-sDPA, which clearly exceed both L_{D} values. This implies that it is likely that two excitons are generated from the same sensitizer with a lag time, τ_{lag}, which is defined as the mean recovery time to the initial state of the sensitizer after TTET to the emitter. The normalized lag time is defined as $n_{\text{lag}} = \tau_{\text{lag}}/\min(\tau_{\text{TTET}})$. Hereafter, we first consider the encounter probability of two excitons as a function of n_{TTET} and n_{lag}. Then, we go back to the encounter probability of the triplet excitons, which are generated independently from two different sensitizers.

Here, we consider a rough estimation of an encounter probability of two excitons by using uniform random walk models, where we assume that the TTET between two emitters is treated as a stochastic problem. For the one-dimensional (1D) systems, the probability that one finds an exciton at the position i after n steps, $P_i(n)$, is given by

$$P_i(n) = {}_nC_{(n+i)/2}2^{-n}, \tag{9.9}$$

where n and i are often assumed to be the even integer number. To estimate the encounter probability of two excitons starting both from the same triplet sensitizer, we evaluate the following equation:

$$R_{1D}(n_{TTET}, n_{rag}) = \sum_{i \le |n_{TTET} - n_{rag}|} P_i(n_{TTET}) P_{i\pm1}(n_{TTET} - n_{rag}), \tag{9.10}$$

where n_{TTET} and n_{lag} are the number of TTET before the spontaneous emission from the emitter and that of the lag time that the sensitizer absorbs the light and then the intersystem crossing occurs in the sensitizer, respectively. Note here that $\pm$ in the above equation means a sum of the $i + 1$ and $i - 1$ terms.

Since the distribution function of the random walk is asymptotically given by the normal distribution (Gaussian approximation) as

$$P(x, n) = \frac{1}{\sqrt{2\pi n}} \exp\left(-\frac{x^2}{2n}\right), \tag{9.11}$$

The encounter probability of two excitons are given by the overlap of two distributions S_{1D} as

$$\begin{aligned} R_{1D}(n_{TTET} \cdot n_{lag}) &= 2\int dx\, P(x, n_{TTET}) P(x, n_{TTET} - n_{lag}) \\ &= \frac{2}{\sqrt{2\pi(2n_{TTET} - n_{lag})}} \equiv 2S_{1D} \end{aligned} \tag{9.12}$$

To validate this approximation, we numerically evaluated the equations Eqs. (9.10) and (9.12) as a function of n_{lag} with $n_{TTET} = 200$ in Fig. 9.10 (left). As seen in this figure, the Gaussian approximation works quite well. Thus, we also adopt the Gaussian approximation to the more complex quasi-two-dimensional (2D) diffusion problem.

The distribution for the 2D anisotropic random walk is given as

$$P(\mathbf{r}, n, \delta) = \frac{1}{n\pi\sqrt{1-\delta^2}} \exp\left(-\frac{x^2}{n(1+\delta)} - \frac{y^2}{n(1-\delta)}\right), \tag{9.13}$$

where a degree of anisotropy, $\delta \ge 0$, is described by the TTET reaction times of the x and y directions ($\tau_{TTET,x}$ and $\tau_{TTET,y}$) as

$$\delta = \frac{\tau_{TTET,y} - \tau_{TTET,x}}{\tau_{TTET,y} + \tau_{TTET,x}} \tag{9.14}$$

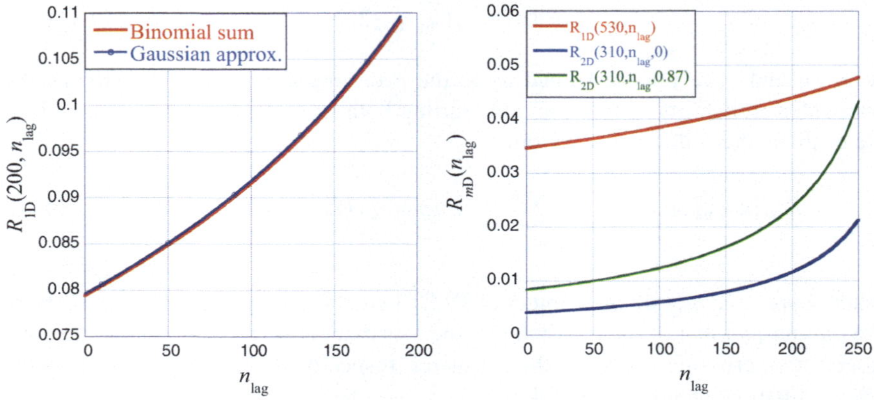

Fig. 9.10 Encounter probability of two excitons using one- and two-dimensional isotropic random walk models. (left) a comparison of the binomial sum formula and the Gaussian approximation to the one-dimensional random walk, where $n_{TTET} = 200$. (right) a comparison of magnitude between one- and two-dimensional random walk, where $n_{TTET} = 530$ for the one-dimensional (red) and $n_{TTET} = 310$ for the two-dimensional problem (blue for the isotropic and green for the anisotropic cases), respectively, based on the estimate from the triplet exciton lifetime. Reprinted with permission from [28] Copyright 2019. American Chemical Society

Note here that the factor 1/2 is missing, and the degree of anisotropy is included in the above equation owing to the fact that the numbers of the trials for x and y directions are $n^*(1 + \delta)/2$ and $n^*(1-\delta)/2$, respectively. In analogy to Eq. (9.12), the 2D encounter probability that one particle is found at a position $(x,y) = (i, j)$ and other at either $(i \pm 1, j \pm 1)$ simultaneously and they are summed over the entire space, is approximated by

$$\begin{aligned} R_{2D}\left(n_{TTET} \cdot n_{lag}\right) &= \sum_{i \le |\Delta n|} \sum_{j \le |\Delta n - |i||} P_{i,j}(n_{TTET}) P_{i\pm1,j\pm1}(\Delta n) \\ &\sim 4 \int d\mathbf{r} P(\mathbf{r}, n_{TTET}, \delta) P(\mathbf{r}, \Delta n, \delta) \\ &= \frac{4}{\sqrt{1-\delta^2}\left(2n_{TTET} - n_{lag}\right)\pi} \equiv 4S_{2D}, \end{aligned} \tag{9.15}$$

where $P_{i,j}(n)$ is the probability of the anisotropic 2D random walk problem (not shown in detail) and $\Delta n = n_{TTET} - n_{lag}$. Note here again that $\pm$ in the above equation means a sum of $(i + 1, j + 1)$, $(i + 1, j-1)$, $(i-1, j + 1)$, and $(i + 1, j-1)$ terms.

To compare the encounter probabilities for 1D and quasi-2D TTETs for C7-sDPA and DPA, respectively, the magnitude of the R_{1D} and R_{2D} are depicted as a function of n_{lag} (and $\delta = 0$ and 0.87 for the homogeneous 2D and the quasi-2D cases, respectively) in Fig. 9.10 (right). Even if the degree of the anisotropy is introduced the magnitude of R_{1D} is always lager than that of R_{2D} when n_{lag} (i.e., the lag time, τ_{lag}) is small enough compared to n_{TTET} (i.e., the triplet lifetime $\tau_{Triplet}$) indicating that the encounter probability of the quasi-2D energy migration tends to be less than the 1D one for this specific region.

Next, we consider the case that two excitons are generated at the different sensitizers independently (the distance of them is assumed to be the separation of two sensitizers, L_S), and they encounter after the migration. The encounter probability can again be approximated by the Gaussian with a different center and a (normalized) duration time, $\tau_{duration}$ ($n_{duration} = \tau_{duration}/\tau_{TTET}$) between two different sensitizers to two different emitters, $P(x-L_S, n_{TTET}-n_{duration})$. Note that the duration time is completely random compared to the lag time. In this case, the encounter probability, R_{mD} (m: dimension), is approximately proportional to $S_{mD}\exp[-\alpha\,(L_S/a)^2]$, where the exponent ($\alpha > 0$) is a function of n_{TTET} and $n_{duration}$, and a is the distance between the neighboring emitters, $a << L_S$. This means that the encounter probability is generally smaller than the values estimated above when L_S is distant enough. Thus, the obtained results for the single-site case here seems to be an upper bound for the encounter probability. However, it is noticed here that the quasi-two-dimensional diffusion case might have higher probability than one-dimensional case when the concentration of the sensitizer is high enough since the TTA in DPA occurs all neighboring sites, while that in C7-sDPA does not always.

9.4 Conclusion

In this chapter, we mainly described experimental and theoretical studies on TTA-UC processes, an important application of the TTA process. Usage of the TTA process is one of the key issues of Photosynergetics to overcome the conventional limitation of efficient usage of photoexcitation. Especially, we focused solid-state system of TTA-UC because it is interesting in views of fundamental studies and device applications. Using the rapid-drying casting method, we have achieved the TTA-UC in solid as efficient as that in the solution for green-to-blue conversion by using PtOEP:DPA (or its derivative). The drop casting method improved TET in solid by achieving molecular dispersion of the sensitizer molecules. By changing the emitter molecule from DPA to C7-sDPA, a strapped derivative, I_{th} for the microparticles was reduced by two or three orders of magnitudes, and Φ_{UC} was enhanced up to 10–20%.

To explain these experimental facts, we have established the theoretical protocol to estimate kinetic constants (inverse of reaction times) for the elementary processes concerning about TTA-UC in crystal. The results indicated that the TTET of exciton migration is the time-limiting step compared to the spin conversion process of TTA. Using random walk models of TTET, we concluded that not only the time scale of TTET but also the difference in dimensionality of the TTET direction, the lifetime of the triplet excitons, and the lag (recovery) time of the sensitizer after TTET to the emitter play an important role in determining the encounter probability, i.e., quantum yield and emission intensity of TTA-UC.

We have extended the achievement by the rapid-drying technique to NIR excitation. By using the binary solid of PdTPTAP: rubrene, NIR-to-vis TTA-UC was successfully demonstrated. The quantum efficiency of each elementary process was elucidated by various spectroscopies, including time-resolved UC emission and transient absorption spectroscopies. Very fast TET was found in solid, and the bottleneck step of the total QY was clarified. Moreover, a novel approach was tested by bonding sensitizer on a solid surface of nanopore to prevent the aggregation of sensitizer. The non-radiative deactivation of the triplet was suppressed by fixing on the surface. TET from sensitizer on the solid surface to emitter diffusing in solvent was found to occur, leading TTA-UC.

These results are open a new page in the usage of the excitation energy of triplets, especially in solid medium. The importance of manipulating triplets and the efficient usage in solid will be increased in the future. These works will be a part of the foundation in further studies.

Acknowledgements We deeply appreciate to Professors M. Nakano, Y. Kitagawa, R. Kishi, K. Kobayashi, S. Hirata, M. Vacha, H. Sotome, H. Miyasaka, D. Kuzuhara, H. Yamada for their collaboration. The present work was supported by JSPS KAKENHI Grant Number 26107004, Grant-in-Aid for Scientific Research on Innovative Areas "Photosynergetics."

References

1. Ito S, Nagami T, Nakano M (2018) Molecular design for efficient singlet fission. J Photochem Photobiol, C 34:85–120
2. Ito S, Nagami T, Nakano M (2017) Singlet fission in pancake-bonded systems. Phys Chem Chem Phys 19:5737–5745
3. Ito S, Nagami T, Nakano M (2017) Rational design of doubly-bridged chromophores for singlet fission and triplet–triplet annihilation. RSC Adv 7:34830–34845
4. Nagami T, Ito S, Kubo T, Nakano M (2017) Intermolecular packing effects on singlet fission in oligorylene dimers. ACS Omega 2:5095–5103
5. Ito S, Nagami T, Nakano M (2015) Density analysis of intra- and intermolecular vibronic couplings toward bath engineering for singlet fission. J Phys Chem Lett 6:4972–4977
6. Nakano M, Ito S, Nagami T, Kitagawa Y, Kubo T (2016) Quantum master equation approach to singlet fission dynamics of realistic/artificial pentacene dimer models: relative relaxation factor analysis. J Phys Chem C 120:22803–22815
7. Nakano M (2019) Quantum master equation approach to singlet fission dynamics in pentacene ring-shaped aggregate models. J Chem Phys 150: 234305 (18 page)
8. Garoni E, Zirzlmeier J, Basel BS, Hetzer C, Kamada K, Guldi DM, Tykwinski RR (2017) Two-photon absorption in pentacene dimers: the importance of the tether using upconversion as an indirect route to singlet fission. J Am Chem Soc 139:14017–14020
9. Birks JB (1970) Photophysics of aromatic molecule. Wiley, London
10. Parker CA, Hatchard CG (1962) Sensitized anti-Stokes delayed fluorescence. Proc Chem Soc 386–387
11. Islangulov RR, Kozlov DV, Castellano FN (2005) Low power upconversion using MLCT sensitizers. Chem Commun 1:3776–3778
12. Kamada, K, Sakagami Y, Mizokuro T, Fujiwara Y, Kobayashi K, Narushima K, Hirata S, Vacha M (2017) Efficient triplet–triplet annihilation upconversion in binary crystalline solids fabricated by solution casting and operated in air. Mater Horiz 4:83–87; ibid. (2018) 5:1219
13. Murakami Y (2011) Photochemical photon upconverters with ionic liquids. Chem Phys Lett 516:56–61

14. Monguzzi A, Tubino R, Hoseinkhani S, Campione M, Meinardi F (2012) Low power, non-coherent sensitized photon up-conversion: modelling and perspectives. Phys Chem Chem Phys 14:4322–4332
15. Sternlicht H, Nieman GC, Robinson GW (1963) Triplet–triplet annihilation and delayed fluorescence in molecular aggregates. J Chem Phys 38:1326–1335
16. Campione M, Fumagalli E, Raimondo L, Monguzzi A, Meinardi F, Sassella A (2011) Control of π–π interactions in epitaxial films of platinum(II) octaethyl porphyrin. Chem Mater 23:832–840
17. Goudarzi H, Keivanidis PE (2014) Triplet–triplet annihilation-induced up-converted delayed luminescence in solid-state organic composites: monitoring low-energy photon up-conversion at low temperatures. J Phys Chem C 118:14256–14265
18. Fujiwara Y, Ozawa R, Onuma D, Suzuki K, Yoza K, Kobayashi K (2013) Double alkylene-strapped diphenylanthracene as a photostable and intense solid-state blue-emitting material. J Org Chem 78:2206–2212
19. Abulikemu A, Sakagami Y, Heck C, Kamada K, Sotome H, Miyasaka H, Kuzuhara D, Yamada H (2019) Solid state, near-infrared to visible photon upconversion via triplet–triplet annihilation of a binary system fabricated by solution casting. ACS Appl Mater Interf 11:20812–20819
20. Wu M, Congreve DN, Wilson MWB, Jean J, Geva N, Welborn M, Van Voorhis T, Bulović V, Bawendi MG, Baldo MA (2016) Solid-state infrared-to-visible upconversion sensitized by colloidal nanocrystals. Nat Photon 10:31–34
21. Amemori S, Sasaki Y, Yanai N, Kimizuka N (2016) Near-infrared-to-visible photon upconversion sensitized by a metal complex with spin-forbidden yet strong S_0-T_1 absorption. J Am Chem Soc 138:8702–8705
22. Cheng YY, Khoury T, Clady RGCR, Tayebjee MJY, Ekins-Daukes NJ, Crossley MJ, Schmidt TW (2010) On the efficiency limit of triplet–triplet annihilation for photochemical upconversion. Phys Chem Chem Phys 12:66–71
23. Takagi S, Eguchi M, Tryk DA, Inoue H (2006) Porphyrin photochemistry in inorganic/organic hybrid materials: Clays, layered semiconductors, nanotubes, and mesoporous materials. J Photochem Photobio C 7:104–126
24. Mizokuro T, Abulikemu A, Sakagami Y, Jin T, Kamada K (2018) Enhanced phosphorescence properties of Pt-porphyrin derivative fixed on the surface of nano-porous glass. Photochem Photobio Sci 17:622–627
25. Mizokuro T, Abulikemu A, Suzuki K, Sakagami Y, Nishii R, Jin T, Kamada K (2020) Triplet-triplet annihilation upconversion through triplet energy transfer at a nanoporous solid-liquid interface. Phys Chem Chem Phys, accepted.
26. Sato R, Kitoh-Nishioka H, Yanai T, Shigeta Y (2017) Theoretical analyses on triplet–triplet annihilation process of 9,10-diphenylanthracene in solution. Chem Lett 46:873–875
27. Sato R, Kitoh-Nishioka H, Kamada K, Mizokuro T, Kobayashi K, Shigeta Y (2018) Does inactive alkyl chain enhance triplet–triplet annihilation of 9,10-diphenylanthracene derivatives. J Phys Chem C 122:5334–5340
28. Sato R, Kitoh-Nishioka H, Kamada K, Mizokuro T, Kobayashi K, Shigeta Y (2018) Synergetic effects of triplet–triplet annihilation and triplet–triplet energy transfer processes on photon up-conversion in crystalline systems. J Phys Chem Lett 9:6638–6643
29. Marcus RA, Sutin N (1985) Electron transfer in chemistry and biology. Biochem Biophys Acta 811:265–322
30. Nakamura H, Truhlar DG (2001) The direct calculation of diabatic states based on configurational uniformity. J Chem Phys 115:10353–10372
31. Nakamura H, Truhlar DG (2002) Direct diabatization of electronic states by the fourfoldway. II. Dynamical correlation and rearrangement processes. J Chem Phys 117:5576–5593
32. Yang KR, Xu X, Truhlar DG (2013) Direct diabatization of electronic state by the fourfold-way: including dynamical correlation by multi-condiguration quasidegenerate perturbation theory with complete active space self-consistent-field diabatic molecular orbitals. Chem Phys Lett 573:84–89

33. Atchity GJ, Ruedenberg K (1997) Determination of diabatic states through enforcement of configuration uniformity. Theor Chem Acc 97:47–58
34. Zeng T, Hoffmann R, Ananth N (2014) The low-lying electronic states of pentacene and their roles in singlet fission. J Am Chem Soc 136:5755–5764
35. Berkelbach TC, Hybertsen MS, Reichman DR (2013) Microscopic theory of singlet exciton fission. II. Application to pentacene dimers and the role of superexchange. J Chem Phys 138:114103
36. Voityuk AA, Rosch N (2002) Fragment charge difference method for estimating donor-acceptor electronic couping: application to DNA π-Stacks. J Chem Phys 117:5607–5616
37. Voityuk AA (2005) Estimates of electronic coupling for excess electron transfer in DNA. J Chem Phys 123:034903
38. You Z-Q, Hsu C-P, Fleming GR (2006) Triplet–triplet energy transfer coupling: theory and calculation. J Chem Phys 124:044506
39. Hsu C-P, You Z-Q, Chen H-C (2008) Characterization of the short-range coupling in excitation energy transfer. J Phys Chem C 112:1204–1212
40. You Z-Q, Hsu C-P (2010) The fragment spin difference scheme for triplet–triplet energy transfer coupling. J Chem Phys 133:074105
41. Curutchet C, Voityuk AA (2012) Distance dependence of triplet energy transfer in water and organic solvents: a QM/MD study. J Phys Chem C 116:22179–22185
42. You Z-Q, Hsu C-P (2014) Theory and calculation for the electronic coupling in excitation energy transfer. Int J Quantum Chem 114:102–115
43. Schmidt MW, Baldridge KK, Boatz JA, Elbert ST, Gordon MS, Jensen JH, Koseki S, Matsunaga N, Nguyen A, Su S et al (1993) General atomic and molecular electronic structure system. J Comp Chem 14:1347–1363
44. Renaud N, Sherratt PA, Ratner MA (2013) Mapping the relation between stacking geometries and singlet fission yield in a class of organic crystals. J Phys Chem Lett. 4:1065–1069
45. Tavernier HL, Fayer MD (2000) Distance dependence of electron transfer on DNA: the role of the reorganization energy and free energy. J Phys Chem B 104:11541–11550
46. van Hal PA, Knol J, Langeveld-Voss BMW, Neskers SCJ, Hummelen JC, Janssen RA (2000) Photoinduced energy and electron transfer in fullerence-oligothiophene-fullerene triads. J Phys Chem A 104:5974–5988
47. Guo J, Liang Y, Xiao S, Szarko JM, Sprung M, Mukhopadhyay MK, Wang J, Yu L, Chen X (2009) Structure and dynamics correlations of photoinduced charge separation in rigid conjugated liner donor-acceptor dyads towards photovoltaic applications. New J Chem 33:1497–1507
48. Nayyar IH, Batista ER, Tretiak S, Saxena A, Smith DL, Martin RL (2011) Localization of electronic excitations in conjugated polymers studies by DFT. J Phys Chem Lett 2:566–571
49. Yost SR, Hontz E, Yeganeh S, Voorhis TV (2012) Triplet versus singlet energy transfer in organic semiconductors: the tortoise and the hare. J Phys Chem C 116:17369–17377

Chapter 10
Hot Carrier Transfer and Carrier Manipulation of Semiconductor Nanocrystals

Naoto Tamai and Sadahiro Masuo

Abstract Hot carrier transfer of semiconductor nanocrystals (NCs) plays an important role for solar energy conversion. In this chapter, effects of quantum confinement of colloidally synthesized semiconductor NCs on hot carrier transfer and the carrier transfer mechanism are discussed on the basis of state-selective excitation of femtosecond transient absorption spectroscopy and initial bleach yield of band-edge state. The role of phonon emission from higher excited states on hot carrier transfer in quantum-confined NCs is revealed. In addition, carrier manipulation of a single semiconductor NC by plasmonic nanostructures is demonstrated with single particle spectroscopy. The distance dependence between a single semiconductor quantum dot (QD) and a plasmonic nanostructure on luminescence intensity and lifetime of a single semiconductor QD is discussed in terms of the electromagnetic enhancement of absorption and luminescence and energy transfer quenching by the plasmonic nanostructure.

Keywords Hot carrier transfer · Semiconductor nanocrystal · Quantum confinement · State-selective transient absorption spectroscopy · Phonon emission · Carrier manipulation · Single particle spectroscopy · Energy transfer · Plasmonic nanostructure

10.1 Introduction

Colloidally synthesized semiconductor nanocrystals (NCs) have unique optical and physical properties resulting from the strong confinement of electron and hole, which can be used as promising materials such as photocatalysis and photovoltaics. One

N. Tamai (✉)
Department of Chemistry, School of Science and Technology, Kwansei Gakuin University, Sanda, Hyogo 669-1337, Japan
e-mail: tamai@kwansei.ac.jp

S. Masuo
Department of Applied Chemistry for Environment, School of Science and Technology, Kwansei Gakuin University, Sanda, Hyogo 669-1337, Japan
e-mail: masuo@kwansei.ac.jp

H. Miyasaka et al. (eds.), *Photosynergetic Responses in Molecules and Molecular Aggregates*, https://doi.org/10.1007/978-981-15-5451-3_10

of the typical quantum-confined materials is semiconductor quantum dots (QDs), in which electron and hole are confined along three-dimensional (3D) directions, showing the size-tunable absorption and luminescence spectra originating from the discrete electronic states [1], and highly efficient multiple exciton generation (MEG) [2–4] by one-photon absorption. One- and two-dimensional quantum-confined semiconductors are known as nanoplatelets (NPLs) and nanorods (NRs), respectively, as illustrated in Fig. 10.1. The relaxation rate of excited electron and hole with excess energy (hot electron, hot hole) in semiconductor NCs is strongly dependent on the dimensionality of quantum confinement.

Concerning the discrete electronic states of 3D confined CdSe QDs, it is well known that the energy spacing between 1S(e) and 1P(e) states is ten times larger than the LO phonon energy [1, 5]. Thus, the hot electron higher than 1P(e) state in QDs relaxes to the band-edge 1S(e) state by different processes from the bulk semiconductors such as Auger cooling process that is the energy transfer from hot electron to hole, and energy transfer to surface ligands [6–9]. Kambhanpati's group examined the size dependence of intraband transition rate in CdSe QDs by the state-selective excitation experiments and found that the dominant hot carrier relaxation processes from 1P(e) to 1S(e) were due to Auger cooling [8]. According to these previous studies, the rate of intraband relaxation from 1P(e) to 1S(e) state can be regulated by the wavefunction separation between electron and hole as well as surface ligands. On the other hand, the intraband relaxation of hot carrier in CdSe NPLs with 1D confined quantum well structure is less than 100 fs that is much faster than that of CdSe QDs [10]. For the conventional solar cells using bulk semiconductors, the hot carrier relaxes to the band-edge state immediately after the excitation by efficient phonon emission through continuous electronic states, which leads to the well-known Shockley–Queisser limit giving the maximum conversion efficiency of ~32% [11]. By utilizing the excess energy of hot carriers, the maximum efficiency of solar energy conversion could reach up to ~67% in a theoretical calculation [12]. Thus, the relationship between the dimensionality of quantum confinement in semiconductor NCs (QDs, NRs, NPLs) and the efficiency of hot carrier extraction by attaching acceptors is very important issue for constructing the next-generation solar cells. Firstly, we show the experimental methods to evaluate the efficiency of hot carrier extraction by using a state-selective excitation technique of femtosecond transient absorption

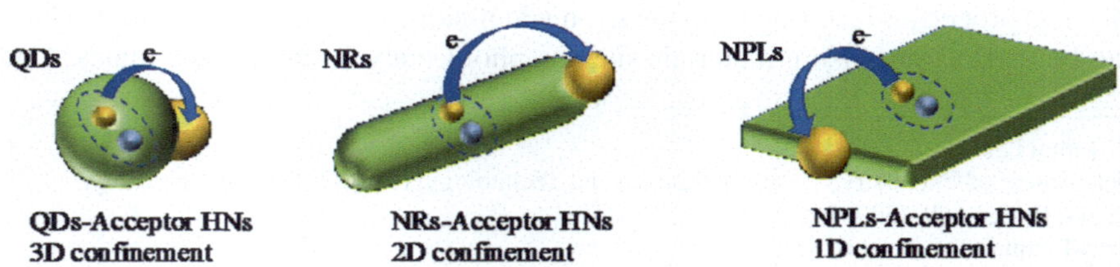

Fig. 10.1 Structures of quantum-confined semiconductor NCs from 3D to 1D confinements of carriers: quantum dots (QDs), nanorods (NRs), and nanoplatelets (NPLs) and their hybrid nanostructures (HNs) attached with acceptors for carrier extraction

spectroscopy and the initial bleach yield of the band-edge state of semiconductor NCs. The effect of quantum confinement of semiconductor NCs on the extraction of hot carriers by acceptor molecules and metal nanoparticles (NPs) (Fig. 10.1) will be discussed.

Given the situation that the size of metal NPs becomes larger, plasmonic field appears near NPs called plasmonic nanostructures. Multiple carriers or excitons can be generated in semiconductor NCs by the interaction with plasmonic nanostructures. The generated multiple excitons disappear within a few ps ~ hundred ps by Auger recombination (AR) processes [13–18], so the extraction of multiple excitons is also important for the application to solar energy conversion. Secondly, the manipulation of a single and multiple excitons in a single QD is demonstrated by the distance regulation between plasmonic nanostructures and a single QD. The competition between electromagnetic enhancement of absorption and luminescence and energy transfer quenching by a single plasmonic nanostructure will be discussed.

10.2 Analytical Methods of Hot Carrier Extraction from Semiconductor NCs—Acceptor Hybrid System

In transient absorption spectrum, the bleach dynamics of the band-edge state of semiconductor NCs can be interpreted in terms of the state-filling [19]. A representative energy diagram of CdSe QDs excited at *i* state higher than 1P(e) is shown in Fig. 10.2. The 1S(e) state is occupied by the electron relaxed from higher excited *i* state, so the rise time of 1S bleach can be observed with the intraband relaxation rate of k_1. The rate constant k_1 includes the sequential relaxation processes in QDs from $i \rightarrow$ 1P and 1P $\rightarrow$ 1S, which can be approximately expressed as one process ($i \rightarrow$ 1S) with the rate constant of intraband relaxation. In QD-acceptor HNs, providing that the hot electron is transferred from the higher excited state to acceptor with the rate constant of k_{HET}, 1S bleach dynamics (absorbance change, -ΔOD(1S)) can be represented by the following equation convolved with the instrumental response function (IRF),

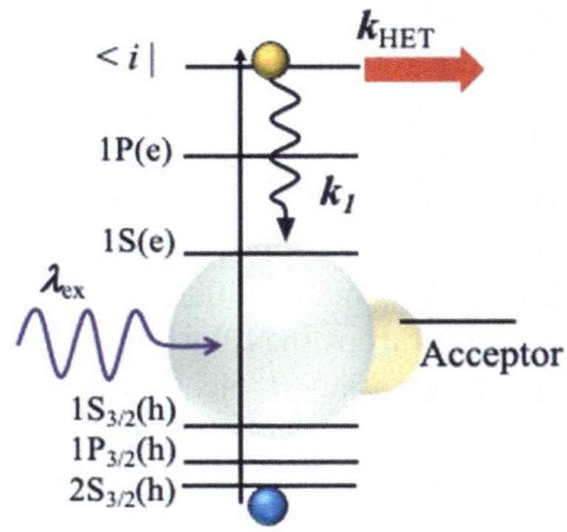

Fig. 10.2 Energy diagram of CdSe QDs—acceptor system excited at *i* state. Hot electron transfer (k_{HET}) from higher excited state to the acceptor competes with the intraband relaxation (k_1)

where $[i]_0$ is the initial concentration of the i state and ⊗ indicates the convolution. This equation is based in the long lifetime of 1S(e) state. In the existence of band-edge electron transfer from 1S(e) state with the rate constant of k_{ET}, the decay component of $exp(-k_{ET}\ t)$ should be introduced instead of 1 in Eq. (10.1).

$$-\Delta \mathrm{OD}(1S) \propto [1S(e)] = \frac{k_1}{k_1 + k_{\mathrm{HET}}}[i]_0\left[1 - \exp\left(-(k_1 + k_{\mathrm{HET}})^t\right)\right] \otimes \mathrm{IRF} \quad (10.1)$$

The 1S initial bleach amplitude of QDs in transient absorption spectra, defined as $B_{QDs} = \Delta OD(1S)_{QDs}/OD(\lambda_{ex})_{QDs}$ at a maximum absorbance change just after the excitation, is an another important parameter to evaluate the hot carrier transfer, where $OD(\lambda_{ex})_{QDs}$ is the absorbance of QDs at the excitation wavelength. The ratio of the initial bleach amplitude of QD-acceptor HNs ($B_{HNs} = \Delta OD(1S)_{HNs}/OD(\lambda_{ex})_{HNs}$) against B_{QDs}, $B_{HNs}/B_{QDs} = k_1/(k_1 + k_{HET})$, gives the hot carrier transfer yield (Φ_{HET}) by the following Eq. (10.2).

$$\Phi_{\mathrm{HET}} = \frac{k_{\mathrm{HET}}}{k_1 + k_{\mathrm{HET}}} = 1 - \frac{B_{\mathrm{HNs}}}{B_{\mathrm{QDs}}} \quad (10.2)$$

In addition, Φ_{HET} can be calculated directly from the rise time analysis of 1S bleach dynamics in Eq. (10.1), which should be compared with the value estimated from Eq. (10.2). If the disagreement between both values exists and Φ_{HET} calculated from 1S bleach amplitude is smaller than the value by rise time analysis, another factor should be considered. In the case of $B_{HNs} < B_{QDs}$ even in the excitation of 1S state, in which no hot carrier is generated, ultrafast electron transfer faster than the excitation pulse width ($k_{ET} \gg$ (pulse width)$^{-1}$) should be considered as a possible mechanism. This will be discussed in Sect. 10.10.5 for CdSe QDs-Au HNs.

10.3 Electron Transfer in One-Dimensional Quantum-Confined System: CdSe NPL-Acceptor HNs

Colloidal CdSe NPLs, typical 1D quantum-confined materials as similar to quantum wells, were synthesized by Ithurria and Dubertret in 2008, with atomic layer precision of the order of several monolayers in thickness direction [20]. NPLs exhibit bright and tunable luminescence with narrow line widths even at room temperature and large absorption cross sections [21]. The strong confinement of electrons and holes exists only in the thickness direction, and the absorption and emission spectra are almost independent of the lateral size of NPLs. The quantum confinement with large shape anisotropy in NPLs lifts the light/heavy-hole degeneracy in balance band while the quasi-continuous electronic states in conduction band, which will make the large difference in carrier relaxation and extraction features from those of 3D confined semiconductor QDs.

For carrier extraction from NPLs, we prepared CdSe NPLs-molecular acceptor, methyl viologen (MV^{2+}), HNs to make clear the hot and band-edge electron transfer by analyzing the formation of radical cation (MV^{+}), as well as the decay dynamics of the band-edge state of NPLs [10]. CdSe NPLs attached with Au NPs (acceptor) were also examined to analyze the hot electron transfer.

Figure 10.3a illustrates the absorption spectra and a typical TEM image of CdSe NPLs (21.7 × 6.7 nm, 4 monolayer thickness) and NPL–MV^{2+} HNs. The average number of adsorbed MV^{2+} to a CdSe NPL is ~80. Two narrow absorption bands observed at 510 and 479 nm can be assigned to electronic transitions from heavy-hole (HH) and light-hole (LH) valence bands to the conduction band edge, respectively. The band-edge luminescence of the CdSe NPLs has a peak at 511 nm with very narrow width (FWHM ~8 nm) at room temperature and an average lifetime of 1~2 ns depending on the lateral size of NPLs. In CdSe NPL–MV^{2+} HNs, luminescence of CdSe NPLs is strongly quenched, suggesting the existence of band-edge and/or hot electron transfer from CdSe NPLs to MV^{2+} as shown in the energy diagram in Fig. 10.3b.

Transient absorption spectra of CdSe NPL–MV^{2+} HNs excited at 400 nm are illustrated in Fig. 10.4a. Very fast relaxation of the HH and LH bleach signals was observed for CdSe NPL–MV^{2+} HNs. In addition, positive transient absorption band appears in near-IR region (~1060 nm) for CdSe NPLs, which shows faster decay for NPL–MV^{2+} HNs as similar to the HH bleach. By adding hole and electron acceptor experiments, HH band dynamics of CdSe NPLs is mainly originated from the excited electrons. Thus, the fast relaxation at HH band of CdSe NPL–MV^{2+} HNs represents the electron transfer from the band-edge state of CdSe NPLs to MV^{2+}. The electron transfer from CdSe NPLs to MV^{2+} can be confirmed by the absorption spectrum of MV^{+} radical cation with a peak at 620 nm [22] as shown in Fig. 10.4b.

For CdSe NPL–MV^{2+} HNs, the rise time of HH band can be expressed as the inverse of the sum of rate constants of intraband relaxation k_1 and hot electron

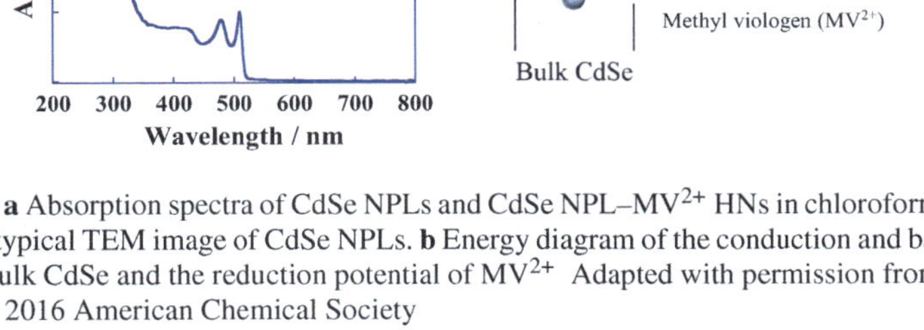

Fig. 10.3 **a** Absorption spectra of CdSe NPLs and CdSe NPL–MV^{2+} HNs in chloroform. The inset of (**a**) is a typical TEM image of CdSe NPLs. **b** Energy diagram of the conduction and balance band edges of bulk CdSe and the reduction potential of MV^{2+} Adapted with permission from Ref. [10]. Copyright 2016 American Chemical Society

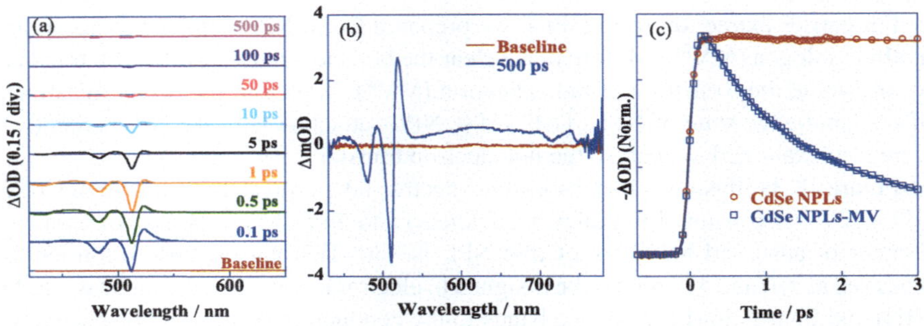

Fig. 10.4 **a** Transient absorption spectra of CdSe NPL–MV^{2+} HNs, **b** transient absorption spectrum of CdSe NPL–MV^{2+} HNs observed at 500 ps, and **c** transient absorption dynamics of HH band bleach in CdSe NPLs and CdSe NPL–MV^{2+} HNs. The samples were excited at 400 nm Adapted with permission from Ref. [10]. Copyright 2016 American Chemical Society

transfer k_{HET}. The rise dynamics of CdSe NPL–MV^{2+} HNs at HH band is almost the same (~70 fs) as the CdSe NPLs as clearly shown in Fig. 10.4c. The rate of hot electron relaxation (~60–70 fs) in CdSe NPLs is much faster than that in CdSe QDs and NRs irrespective of lateral sizes of NPLs. In addition, the initial bleach amplitudes of HH band of both systems, $B_{NPL-MV2+}$ and B_{NPL}, are also very similar [10]. From these results, the hot electron transfer does not occur in CdSe NPL–MV^{2+} HNs. This is in contrast with CdSe QDs or CdSe NRs, where the hot electron transfer occurs in these NCs as discussed in Sects. 10.4 and 10.5.

We observed two types of electron transfer from band-edge state of CdSe NPL to MV^{2+}. As shown in Fig. 10.5a, sub ps and ps scale rise components were detected in the formation of MV^{+}, in accordance with the two decay components of HH band. To

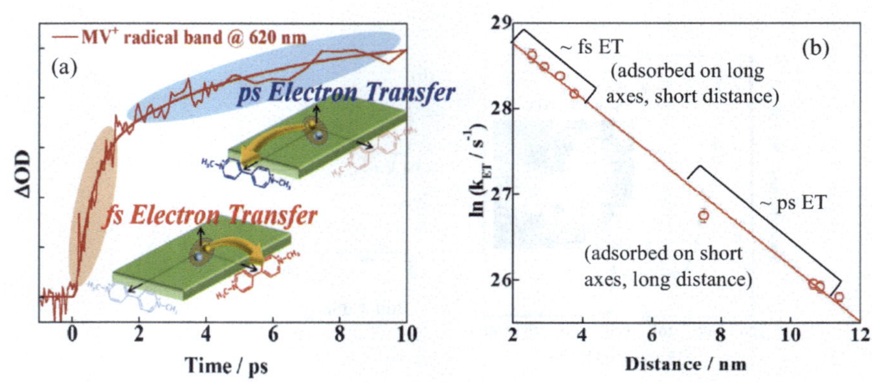

Fig. 10.5 Transient absorption dynamics of **a** CdSe NPL–MV^{2+} HNs observed at 620 nm (MV^{2+}), **b** electron transfer rates k_{ET} as a function of distance from the center of CdSe NPLs. A good correlation was obtained between k_{ET} and distance, with $k_{ET}(d) = k_0 \exp(-\beta d)$, β ~ 0.33 nm^{-1} Adapted with permission from Ref. [10]. Copyright 2016 American Chemical Society

reveal the origin of two different electron transfer rates, we examined the lateral size dependence on the rate of electron transfer dynamics by using four kinds of CdSe NPLs. The rates of fast and slow electron transfer processes became longer as the lateral dimension of NPLs increased (τ_{fast}: from 370 to 580 fs, and τ_{slow}: from 2.4 to 6.2 ps). The electronic coupling strength, $|\overline{H}(E)|$, between the CdSe NPLs and MV^{2+} should be different at the different NPLs faces since the shape of CdSe NPLs is highly anisotropic, with a large lateral size (short axis of 5.1–7.5 nm and long axis of 15.0–22.8 nm) and small thickness (4 monolayers, ~1.2 nm), resulting in the difference of wavefunction penetration depending on the face. Therefore, the fast electron transfer is probably originated from adsorbed MV^{2+} on long axes of CdSe NPLs faces, and the slow electron transfer is due to MV^{2+} adsorbed on short axes of CdSe NPLs faces, since the wavefunction penetration of electrons along with the long axes of NPLs is expected to be much larger than that of the short axes. If MV^{2+} were adsorbed on the plane of CdSe NPLs, $|\overline{H}(E)|$ would not change with the lateral size because all the CdSe NPLs have the same thickness, and the electron transfer rate should be similar irrespective of the lateral size. In addition, it is known that the faces perpendicular to the thickness direction are well passivated by carboxylic acid groups because of the termination with cadmium atoms [23]. Furthermore, MV^{2+} tends to adsorb to the Se-terminated anionic surface, and thus, the adsorption of MV^{2+} is not so easy on the face perpendicular to the thickness direction. By considering the wavefunction penetration and electronic coupling, $|\overline{H}(E)|$, the observed fast and slow electron transfer from the CdSe NPLs to MV^{2+} can be assigned to electron transfer at the faces of the long and short axes in the lateral dimension, respectively (Fig. 10.5a). These rate constants are plotted against the half-length of the short and long axes in the lateral dimension of CdSe NPLs, as illustrated in Fig. 10.5b. A good correlation was obtained between two electron transfer rates and the length of the short and long axes, showing the well-known exponential behavior of $k_{ET}(d) = k_0 \exp(-\beta d)$, where d is the distance between donor and acceptor (half-length of the lateral dimension, or the distance from the center of NPLs), k_0 is the preexponential factor, and β characterizes the exponential distance dependence of the transfer (β ~0.33 nm^{-1}). This correlation also supports the importance of wavefunction penetration and the electronic coupling with MV^{2+}. This electronic coupling of CdSe NPL-MV^{2+} HNs is relatively weak as compared with that of CdSe QD-MV^{2+} HNs, since the ultrafast electron transfer was observed from CdSe QDs to MV^{2+} within 100 fs [24]. Because of the fast intraband relaxation due to efficient phonon emission in quasi-continuous conduction band of NPLs and the weak electronic coupling, the hot electron transfer does not occur from CdSe NPLs to MV^{2+}.

Au NPs can be used to extract electrons from CdSe NPLs. As shown in Fig. 10.6a, small-sized Au NPs (~1.1 nm) were attached with four monolayer CdSe NPLs (lateral size of 5.8×20.8 nm^2). Transient absorption dynamics at heavy-hole wavelength (~510 nm) of both CdSe NPLs and CdSe NPL-Au HNs were illustrated in Fig. 10.6b. The heavy-hole band dynamics of CdSe NPL-Au HNs could be fitted by a sum of exponential functions with the lifetime of ~0.9 ps and ~4 ps components, which can be assigned to the electron transfer from the band-edge state of CdSe NPLs to Au NPs. The rates of electron transfer are slower than those for CdSe QD-Au HNs as

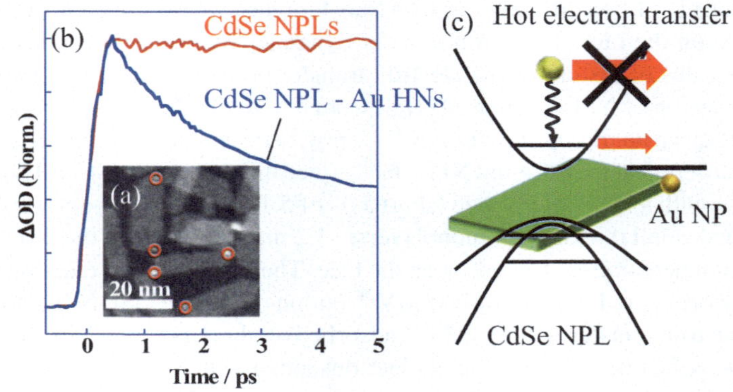

Fig. 10.6 **a** A typical STEM image of CdSe NPL-Au HNs with Au NP diameter of ~1.1 nm (red circle). The average lateral size of CdSe NPLs is 5.8 × 20.8 nm^2. **b** Transient absorption dynamics of CdSe NPLs and CdSe NPL–Au HNs observed at heavy-hole bleach wavelength (~510 nm). The samples were excited at 400 nm. **c** Schematic illustration of electron transfer in CdSe NPL-Au HNs

shown in Sect. 10.5. The slower electron transfer rate in NPLs is likely due to the difference of electronic coupling constant. CdSe NPLs have much larger lateral size as compared to the diameter of CdSe QDs (~4.2 nm), leading to the weak electronic coupling in CdSe NPL-Au HNs. In addition, the rise time of the band-edge state of CdSe NPL-Au HNs is almost the same as the CdSe NPLs, as illustrated in Fig. 10.6b. The initial bleach yield of CdSe NPL-Au HNs ($B_{\text{NPL-Au}}$) is also similar to that of CdSe NPLs (B_{NPL}). These results suggest that the hot electron transfer from higher excited states does not exist in CdSe NPL-Au HNs as illustrated in Fig. 10.6c.

10.4 Electron Transfer in Two-Dimensional Quantum-Confined System: CdSe NR-Acceptor HNs

The electron transfer dynamics from CdSe NRs (4.0 nm × 14.0 nm, inset of Fig. 10.7a) to Au NPs attached to the tip of NRs has been examined by femtosecond transient absorption and single particle spectroscopy [25]. Figure 10.7a shows absorption spectra of CdSe NRs and CdSe NR-Au HNs A, B, and C with the average Au NPs diameter of 1.5, 1.6, and 2.2 nm for HNs A, B, and C. Absorption bands of CdSe NRs at 590 nm, 550 nm, and 480 nm can be assigned to the transitions of $1\sigma_e$-$1\sigma_h$ (1Σ), $2\sigma_e$-$2\sigma_h$ (2Σ), and $1\pi_e$-$1\pi_h$ (1Π), respectively, by considering the electronic structure of CdSe NRs [26]. Luminescence of CdSe NRs is efficiently quenched with increasing Au NPs diameter, indicating the existence of electron transfer from CdSe NRs to Au NPs. By analyzing the transient absorption dynamics of CdSe NR-Au HNs, the lifetimes of 1Σ bleach are 2–4 ps and become faster with increasing Au NPs diameter. The faster decay is due to the electron transfer from $1\sigma_e$ state of NRs

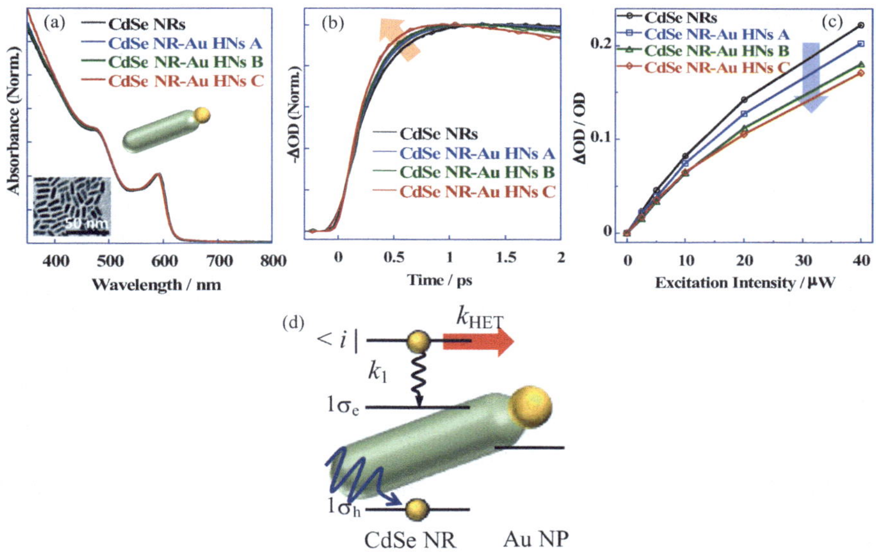

Fig. 10.7 **a** Absorption spectra of CdSe NRs (inset: TEM image, 4.0 × 14.0 nm^2) and CdSe NR-Au HNs (Au diameter: 1.5, 1.6, 2.2 nm for *A*, *B*, and *C*). **b** 1Σ bleach dynamics and **c** 1Σ initial bleach yields of CdSe NRs and CdSe NR-Au HNs. The excitation wavelength is 400 nm. **d** Schematic illustration of electron transfer in CdSe NR-Au HNs Adapted with permission from Ref. [25]. Copyright 2013 Royal Society of Chemistry

to Au NPs. The electron transfer rate in CdSe NR-Au HNs is slower than that in CdSe QD-Au HNs (Sect. 10.5), which is mainly due to the weak electronic coupling between the electronic wavefunction penetrated into the long axis (14.0 nm, larger than the exciton Bohr radius ~5.6 nm) of CdSe NRs and Au NPs attached to the tip.

Rise dynamics and initial bleach amplitude of 1Σ bleach band of CdSe NR-Au HNs excited at 400 nm are illustrated in Figs. 10.7b, c, respectively. With increasing the diameter of Au NPs, the rise time of 1Σ bleach band becomes faster as clearly shown in Fig. 10.7b (0.30 ps for NRs only, 0.27 to 0.23 ps from HNs A to C). This tendency is consistent with the initial bleach amplitude of 1Σ band as shown in Fig. 10.7c, in which the bleach amplitude of HNs A, B, and C decreased by 10, 19, and 23%, respectively, as compared to that of CdSe NRs. Both results of rise time and initial bleach amplitude support the existence of hot electron transfer from higher excited state of CdSe NRs to Au NPs as illustrated in Fig. 10.7d. By analyzing the rise times of CdSe NR-Au HNs and CdSe NRs, the hot electron transfer rates were estimated to be ~2.7, ~2.0, and ~1.0 ps for CdSe NR-Au HNs A, B, and C. In addition, the hot electron transfer yield (Φ_{HET}) of CdSe NR-Au HNs obtained by the rise time analyses are 0.10, 0.14, and 0.23 in CdSe NR-Au HNs A, B, and C, respectively. These yields are in good agreement with the values obtained from the initial bleach amplitude of 1Σ band in CdSe NR-Au HNs. Thus, the ultrafast electron transfer faster than the pulse width ($\ll$ 100 fs) does not exist in CdSe NR-Au HNs, which is in contrast with CdSe QD-Au HNs as shown in Sect. 10.5. This difference is originated from the relatively weak electronic coupling between CdSe NRs and Au NPs even though the existence of hot electron transfer.

10.5 Electron Transfer in Three-Dimensional Quantum-Confined System: CdSe QD-Acceptor HNs

For 3D confined system of PbS QD-Au HNs, we observed ultrafast electron transfer and ps scale hole transfer from the band-edge state of PbS QDs to Au NPs by state-selective femtosecond transient absorption spectroscopy [26]. For CdSe/CdS core/shell QD-Au HNs, the electron transfer from higher excited and band-edge states of core/shell QDs to Au NPs was reported, although the excitation wavelength dependence and the consistency between the rise time of bleach dynamics and the initial bleach amplitude has not been examined yet [27]. We synthesized and characterized CdSe QD-Au HNs with different-sized Au NPs, and the mechanism of electron transfer from CdSe QDs to Au NPs was revealed by state-selective transient absorption spectroscopy [28].

Absorption spectra of CdSe QDs (diameter 4.2 ± 0.3 nm) and CdSe QD-Au HNs are illustrated in Fig. 10.8a. Absorption spectra of CdSe QD-Au HNs slightly shift to the shorter wavelength and become broader with increasing the diameter of Au NPs. The broad absorption spectra of CdSe QD-Au HNs are originated from both strong electronic coupling and absorption of small Au NPs (*A* ~1.3 nm, *B* ~2.0 nm) attached to CdSe QDs. The band-edge luminescence of CdSe QDs is strongly quenched in

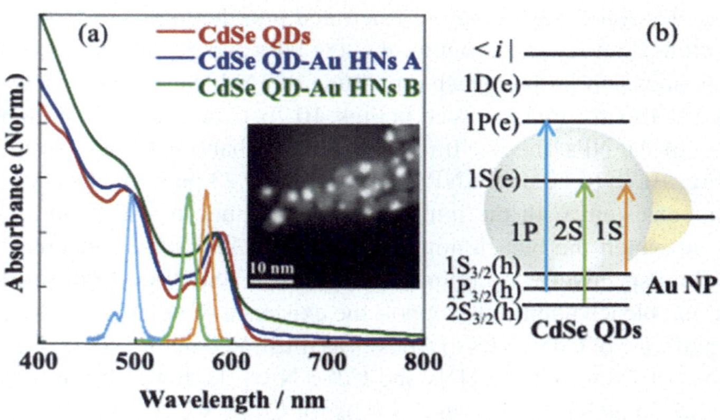

Fig. 10.8 **a** Absorption spectra of CdSe QDs (diameter: 4.2 ± 0.3 nm) and CdSe QD-Au HNs *A*, *B* (Au diameter: *A* = 1.3 ± 0.4 nm, *B* = 2.0 ± 0.4 nm, inset: TEM image of *B*). Femtosecond pulse spectra for 1P (496 nm), 2S (555 nm), and 1S (574 nm) excitation are also shown. **b** Schematic energy diagram of CdSe QD-Au HNs Adapted with permission from Ref. [28]. Copyright 2019 American Chemical Chemistry

CdSe QD-Au HNs, suggesting the existence of efficient electron transfer from CdSe QDs to Au NPs. State-selective transient absorption spectroscopy was used to reveal the electron transfer mechanism from CdSe QDs to Au NPs, in which HNs were excited at 574, 555, and 496 nm as well as 400 nm, corresponding to the selective excitation of 1S(e)-1$S_{3/2}$(h) (1S), 1S(e)-2$S_{3/2}$(h) (2S), and 1P(e)-1$P_{3/2}$(h) (1P) as illustrated in Fig. 10.8b, respectively.

The transient absorption spectra of CdSe QD-Au HNs excited at 400 nm are shown in Fig. 10.9a. The spectral features of QD-Au HNs are similar with CdSe QDs. However, the fast relaxation (750 fs (60%) for *A*, 450 fs (75%) for *B*) of transient absorption signal can be observed in CdSe QD-Au HNs, which corresponds to the band-edge electron transfer from 1S(e) of CdSe QDs to Au NPs. Another difference of the spectral feature is the existence of long-lived positive absorption at ~610 nm in CdSe QD-Au HNs originating from the charge separation in HNs. In previous studies, the positive absorption longer than 1S bleach band with a short lifetime (<1 ps) was attributed to the spectral shift of 1S transition due to the biexciton effect between hot carrier and 1S exciton generated by the probe pulse [6]. On the other hand, the positive absorption band was also observed by the charge separation from semiconductor NCs to the outside because of the electric field-induced exciton peak shift by a Stark effect [26]. The long-lived positive absorption at ~610 nm observed in transient absorption spectra of CdSe QD-Au HNs suggests the charge separation between CdSe QDs and Au NPs.

The important differences between CdSe QDs and HNs are faster rise time of 1S bleach dynamics and low 1S bleach yield for CdSe QD-Au HNs as illustrated in Figs. 10.9b and c. By using the rise time at 1S bleach band of CdSe QDs (~290 fs ($1/k_1$)) and the both of CdSe QD-Au HNs *A* and *B* (~150 fs ($1/(k_1 + k_{HET})$)), the rate of hot electron transfer ($1/k_{HET}$) and the yield (Φ_{HET}) in both CdSe QD-Au HNs *A* and *B* were calculated to be ~310 fs and ~0.48, respectively. On the other hand, the ratios of 1S bleach yields in CdSe QD-Au HNs *A* and *B* against CdSe QDs (B_{HNs}/B_{QDs}) calculated from Fig. 10.9c were ~0.31 and ~0.15, respectively. These values are much smaller than the estimated values from 1-Φ_{HET} (=0.52). The decrease of initial 1S

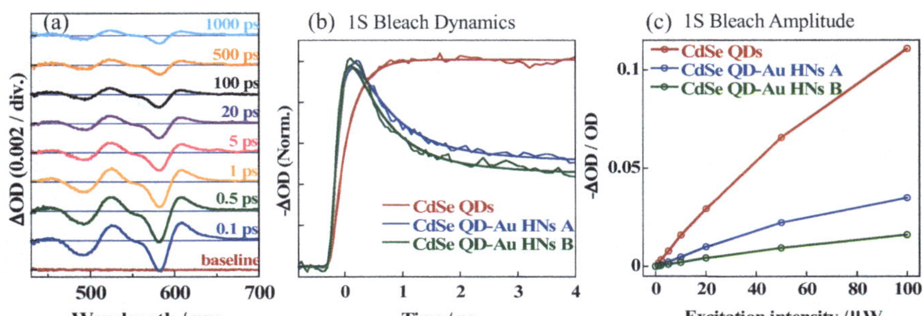

Fig. 10.9 **a** Transient absorption spectra of CdSe QD-Au HNs A, **b** 1S bleach dynamics, and **c** 1S bleach amplitude of CdSe QDs and CdSe QD-Au HNs, excited at 400 nm Adapted with permission from Ref. [28]. Copyright 2019 American Chemical Chemistry

bleach amplitude observed in our experiments is probably due to ultrafast electron transfer faster than the instrument response function from the 1S(e) state of CdSe QDs to Au NPs.

The 1S bleach yields (B_{HNs}/B_{QDs}) of CdSe QD-Au HNs *A* and *B* with 1P and 1S excitation are plotted against the excitation intensity as illustrated in Fig. 10.10. In 1S excitation experiments (Fig. 10.10b), B_{HNs}/B_{QD} of CdSe QD-Au HNs *A* and *B* is estimated to be ~0.71 and ~0.28, respectively. The lower value of B_{HNs}/B_{QD} for 1S excitation is due to the existence of ultrafast electron transfer from 1S(e) to Au NPs, and the yields (Φ_{UET}) in CdSe QD-Au HNs *A* and *B* can be calculated to be ~0.29 and ~0.72, respectively. In addition, B_{HNs}/B_{QD} of both HNs *A* and *B* for 1P excitation is lower as compared with those in 1S excitation. From rise time analyses, Φ_{HET} of both HNs *A* and *B* under 1P excitation was estimated to be ~0.50 as similar to 400 nm excitation (~0.48). Thus, the lower B_{HNs}/B_{QD} under 1P excitation is originated from both hot electron transfer and ultrafast electron transfer from 1S(e). In addition, B_{HNs}/B_{QDs} for 1P excitation is similar to that for 400 nm excitation, suggesting that the hot electron transfer mainly occurs from 1P(e) state of CdSe QDs to Au NPs even though the hot electron is generated at higher excited state than 1P(e). B_{HNs}/B_{QDs} of CdSe QD-Au HNs *A* and *B* for 1P excitation can be calculated by the products of (1-Φ_{HET}) and (1-Φ_{UET}). The calculated B_{HNs}/B_{QDs} for HNs *A* and *B* is ~0.36 and ~0.14, respectively, which are in good agreement with the observed B_{HNs}/B_{QDs} (0.36 and 0.15 for *A* and *B*) [28]. This result suggests the importance of hot electron transfer from 1P(e) and ultrafast electron transfer from 1S(e) in CdSe QD-Au HNs as illustrated in Fig. 10.10c. Phonon bottle neck in the relaxation from 1P(e) to 1S(e) and strong electronic coupling originated from the wavefunction penetration in 3D quantum-confined CdSe QDs attached with acceptors play an important role for these unique properties of electron transfer.

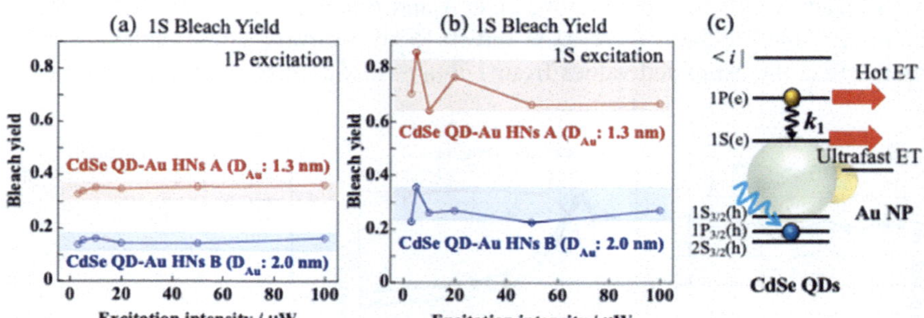

Fig. 10.10 **a** 1S bleach yields of CdSe QD-Au HNs A and B under 1P excitation, **b** 1S bleach yields of CdSe QD-Au HNs A and B under 1S excitation, and **c** schematic illustration of hot electron transfer from 1P(e) and ultrafast electron transfer from 1S(e) in CdSe QDs Adapted with permission from Ref. [28]. Copyright 2019 American Chemical Chemistry

10.6 Control of Multiexciton Dynamics Using Plasmonic Nanostructures

Colloidal semiconductor nanocrystal QDs are one of the most attractive fluorophores because of the size-dependent emission wavelengths, narrow emission line width, and high photodurability. A crucial property of the QDs is the simultaneous existence of multiple excitons (MX) in a single QD. By utilizing the MX, the efficiency of the optoelectronic devices can be considerably increased. Furthermore, multiphoton emission from triexciton (TX) and biexciton (BX) states can behave as correlated photon pairs, i.e., quantum entangled photons, to realize quantum information technologies. However, when MX are produced in a single QD, nonradiative Auger recombination (AR) efficiently occurs, i.e., the MX decay to a single exciton (SX) by AR, and thus, the excitons are wastefully consumed by AR. AR also caused the emission blinking behavior which is called Auger ionization. Therefore, AR suppression has been extensively studied for effective use of the excitons. On the other hand, AR promotes single-photon emission from a single QD, i.e., photon antibunching in the QD emission, because the SX as a result of AR can emit a single photon even when MX are generated in a single QD. The single-photon emission is also important emission property of a single QD for quantum information technologies. Therefore, it is crucial to control the MX dynamics and subsequent emission photon statistics for the applications employing QDs.

We previously reported that the emission photon statistics, i.e., multiphoton emission and a single-photon emission from a single QD, can be modified by interactions of the QD with plasmonic nanostructures, i.e., metallic nanostructures (MNSs) [29–36]. After that, similar enhancements of multiphoton emission were reported using several QD-MNS systems [37–54]. Two possibilities have been discussed as the mechanism for the enhancement of the BX emission. One mechanism is the enhancement of the BX emission rate by the MNS, which is same as our previous results. Another is the quenching of SX emission by the MNS, i.e., a decrease in the quantum yield of SX emission rather than an actual increase in the quantum yield of BX emission. Thus, the interaction of QD with MNS, particularly the influence of the MNS on the MX dynamics, is unrevealed. To elucidate the influence, a single QD-MNS system in which both the spectral overlap and the distance are fully controlled would be used because the interaction between the QD and MNS strongly depends on the spectral overlap and the distance. However, the investigation using such systems has not been reported. Additionally, one would have to directly observe changes in the emission behavior of a single QD that accompany the interaction with an MNS. To achieve the direct observation, we used two atomic force microscopy (AFM) techniques.

10.6.1 Multiexciton Control of Single QDs Using Nanosized AuCube

As a first AFM technique, AFM manipulation of a cubic-shaped gold nanoparticle (AuCube) was employed [33]. A transmission electron microscopy (TEM) image of the AuCube is shown in Fig. 10.11a. The edge length of the AuCube was estimated to be 87.0 ± 2.4 nm. An extinction spectrum of the AuCube is shown in Fig. 10.11b with the absorption and photoluminescence (PL) spectra of CdSe/ZnS core/shell QD in toluene. The localized surface plasmon resonance (LSPR) band of the AuCube overlaps with both the absorption and PL spectra of the QD. This overlap means that both excitation and relaxation (i.e., radiative and nonradiative) processes of the QD can be modified by the AuCube to select the excitation wavelength. In this work, a ps-pulsed laser of 405 nm was used as an excitation light source. Under this excitation condition, the LSPR could not be efficiently excited by the laser (Fig. 10.11b). Therefore, the enhancements of the radiative and nonradiative processes were mainly considered. By using the well-defined AuCube, we can simply discuss the mechanism of the interaction between QD and MNS because the wavelength of the LSPR band is not varied in each AuCube.

Figure 10.12 shows one example of AFM manipulation of an AuCube. In these AFM images, single QDs and an AuCube are visualized. The AuCube approached

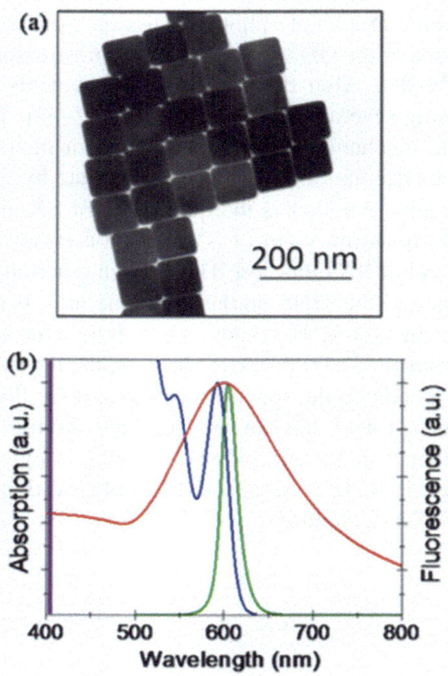

Fig. 10.11 **a** Transmission electron microscopy (TEM) image of mono-dispersed cubic Au nanoparticles (AuCubes). **b** Absorption (blue) and PL (green) spectra of a CdSe/ZnS QD in dispersed solution, an extinction spectrum of the AuCube in dispersed solution (red). The excitation wavelength (405 nm) is marked as a purple vertical line Adapted with permission from Ref. [33]. Copyright 2016 American Chemical Society

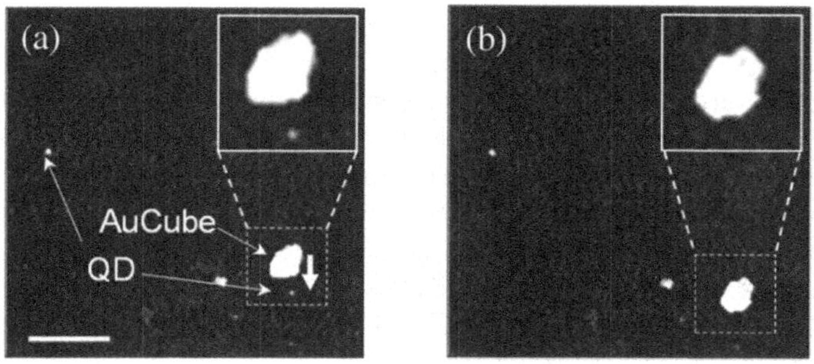

Fig. 10.12 AFM images of single QDs and an AuCube on a glass coverslip before (**a**) and after (**b**) the AFM manipulation of an AuCube. The scale bar in (**a**) represents 500 nm

the single QD by pushing the AuCube in the direction of the arrow (Fig. 10.12a). After approach of the AuCube near the QD, the single QD and the AuCube overlapped (Fig. 10.12b). The center-to-center distance between the QD and AuCube was estimated to be 5 nm, indicating that the QD overlapped with the AuCube or might be attached to the side of the AuCube. After measuring the emission behavior of the single QD, the AuCube was separated from the single QD by the manipulation. And then, the emission behavior of the QD was measured again.

The emission behaviors of the single QD accompanying the AFM manipulation of the AuCube are shown in Fig. 10.13. Before the manipulation of the AuCube, the maximum PL count from the single QD was estimated to be ca. 45 counts/ms (Fig. 10.13a). The photon correlation histogram obtained from a single QD is shown in Fig. 10.13d. The center peak at a delay time of 0 ns was lower than other peaks.

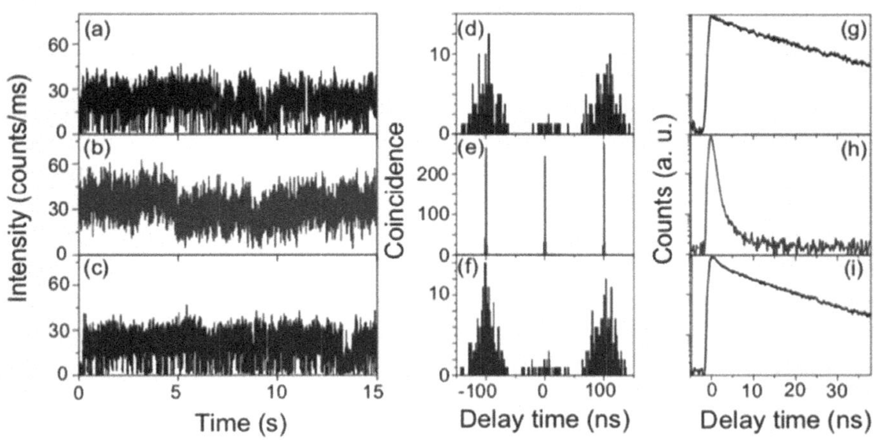

Fig. 10.13 Time traces of the PL intensity (**a**–**c**), photon correlation histograms (**d**–**f**), and PL decay curves (**g**–**i**) detected from a single QD before (**a**, **d**, **g**) and after (**b**, **e**, **h**) the approach of the AuCube, and after the AuCube was pushed away (**c**, **f**, **i**)

The photon correlation histogram indicates that the probability of the single-photon emission increases when the center peak at delay time 0 ns is close to zero. To analyze the photon correlation histogram quantitatively, the second-order correlation function, $g^{(2)}(0)$, was defined as the ratio of the center peak to the average number of the other peaks. Thus, the $g^{(2)}(0)$ value means that the probability of a single-photon emission increases when $g^{(2)}(0)$ is close to zero. In the case of emission from a single QD, the $g^{(2)}(0)$ value corresponds to Φ_{BX}/Φ_{SX}, where Φ_{BX} and Φ_{SX} are quantum yields of the BX and SX emission, respectively. In the case of the photon correlation histogram shown in Fig. 10.13d, the $g^{(2)}(0)$ value was calculated to be 0.14, indicating that the isolated single QD exhibited single-photon emission, i.e., photon antibunching. PL decay curve of the single QD (Fig. 10.13g) was well fitted using a single exponential decay function with a lifetime of 29.6 ns. These are typical PL behavior of the single QD itself, which were dramatically modified by the approach of the AuCube. The PL behavior of the single QD after the manipulation of the AuCube is shown in Fig. 10.13b, e, and h. The maximum PL count from the single QD increased from ca. 45 counts/ms to ca. 60 counts/ms. In the photon correlation histogram, the center peak dramatically increased, and the $g^{(2)}(0)$ values increased to 0.97. This increase in the $g^{(2)}(0)$ values indicated that the contribution of the BX emission from the single QD was 6.9 times higher than that of the single QD before the approach of the AuCube. In Fig. 10.13h, the PL was decayed rapidly, indicating the radiative and/or nonradiative processes of the QD which was modified by the AuCube. By the approach of the AuCube to the single QD, the PL intensity and the contribution of the BX emission increased 1.3 times and 6.9 times, respectively, with a shortening of the PL lifetime. The PL behavior of the single QD after separating the AuCube is shown in Fig. 10.13c, f, and i. The PL intensity, photon antibunching behavior, and decay curve were returned to original behavior, i.e., the emission behavior of the single QD before the approach of the AuCube. The importance of the present results is that the modification of the PL behavior was directly observed accompanying the interaction between a single QD and an AuCube. To discuss the mechanism of the modification, the enhancement factors of the radiative and nonradiative rates were estimated from the modified PL intensity and the PL lifetime. As a result, the enhancement factors for the radiative and nonradiative rates were estimated to be 95 and 54 times, respectively [33]. These values indicated that the proposed two mechanism, i.e., both the enhancement of the BX emission rate and the quenching of the SX state, result in an increase in the contribution of the BX emission upon the interaction of the QD with the AuCube. The present results directly demonstrated that the emission photon statistics and PL intensity from a single QD can be tuned by controlling the exciton relaxation processes using MNS.

10.6.2 Multiexciton Control of Single QDs Using AFM AgTip

The interaction between QDs and MNS is based on dipole-dipole interaction. Thus, the interaction should be strongly influenced by the distance with a nanometer-scale between the QD and MNS. In the case of above AFM manipulation, however, control of the distance between the single QD and the AuCube with nanometer-scale precision was quite hard. Thus, we utilized a silver-coated AFM cantilever (AgTip) as an MNS to achieve the distance control. By setting the AgTip on an AFM system as a cantilever, distance between a QD and an AgTip can be controlled with the spatial resolution of a Piezo actuator unit, i.e., nanometer-scale precision. In addition, using the AgTip, which exhibits a well-defined LSPR band, both the distance and the spectral overlap can be controlled [34].

The AgTip was prepared by sputtering Ag on an Si cantilever. Figure 10.14 shows the scanning electron microscopy (SEM) images of the Si cantilever (a) and the AgTip (b). In image (b), the coating of the cantilever by Ag was confirmed. The tip radius of the AgTip was estimated to be 20–30 nm from the SEM images. The scattering spectrum of the AgTip is shown with the absorption and PL spectra of CdSe/ZnS QD in Fig. 10.14c. The LSPR band of the AgTip fully overlaps the absorption spectrum and slightly overlaps the PL spectrum of the QD. In this work, 405 and 465 nm lasers were used as the excitation light source. The LSP of the AgTip could be generated under the 465 nm excitation. Thus, the excitation rate of the QD could be enhanced

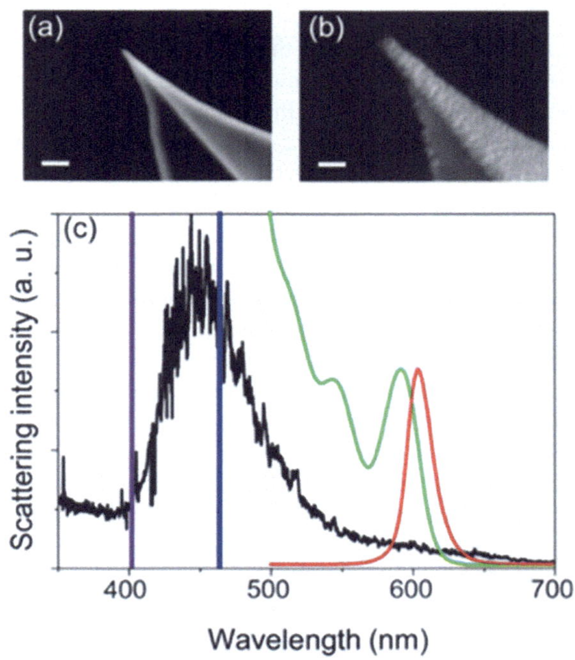

Fig. 10.14 Scanning electron microscopy (SEM) images of the Si cantilever (**a**) and an AgTip prepared by sputtering Ag (**b**). The scale bars in the images represent 100 nm. **c** A scattering spectrum of the AgTip (black line), the absorption (green line), and PL (red line) spectra of CdSe/ZnS QD dispersed in toluene. The excitation wavelengths of 405 and 465 nm were indicated by the vertical purple and blue lines Adapted with permission from Ref. [34]. Copyright 2016 American Chemical Society

by the electric field of the LSP under 465 nm excitation because of the overlap between the LSPR band and the absorption spectrum of the QD. On the other hand, the relaxation rate of the QD could also be enhanced by the AgTip independent of the excitation wavelength because of the spectral overlap between the PL spectrum and LSPR band. To generate the electric field of z-direction on the AgTip, a radially polarized laser beam was introduced to the microscope.

The PL and AFM images of the single QDs dispersed on a glass coverslip are summarized in Fig. 10.15. In the PL images, individual QDs exhibit a double-lobed PL intensity pattern, which is characteristics of radially polarized beam excitation. The full-width at half-maximum (fwhm) of the center lobe (Fig. 10.15a) was estimated to be 220 nm (Fig. 10.15b). The PL and AFM images of the same area as image (a) were then measured again under 405 nm excitation with approaching the AgTip (Fig. 10.15c, d). The center lobe of each PL spot was vanished by approaching the AgTip (Fig. 10.15c), which was also confirmed by the cross section of the PL spot shown in Fig. 10.15g. This result indicates that the PL from the individual QDs was quenched, namely the nonradiative decay rate of the QD was enhanced by approaching the AgTip. On the other hand, in the PL image under 465 nm excitation with approaching the AgTip (Fig. 10.15e), the center lobe of the PL spots was not vanished. The fwhm of the center lobe was smaller (97 nm, Fig. 10.15h) than the cross section (Fig. 10.15b) obtained under 405 nm excitation without approaching the AgTip. These results indicated that the PL was not quenched in the case of 465 nm excitation, and the spatial resolution of the PL image was increased by the approach of the AgTip.

The representative emission behavior of a single QD depending on the distance between a single QD and the AgTip (z-distance) is summarized in Fig. 10.16. In the time traces of the PL intensity (Fig. 10.16a–e), the PL intensity decreased with decreasing the z-distance. In the PL decay curves (Fig. 10.16k–o), the curves was shortened with decreasing z-distance. The decrease in the PL intensity with shortening lifetime depending on the z-distance clearly indicated that the PL of the QD was quenched, namely the nonradiative decay rate of the QD was enhanced by the AgTip due to the resonance energy transfer from the QD to the AgTip. In the photon correlation histogram (Fig. 10.16f–j), the center peak gradually increased with decreasing the z-distance. The $g^{(2)}(0)$ before the approach of the AgTip was estimated to be 0.09 (Fig. 10.16f). This value increased to 0.17 for $z = 10$ nm (Fig. 10.16g) and 0.85 for $z = 6$ nm (Fig. 10.16h). In the case of $z = 2$ nm (Fig. 10.16i), the photon correlation histogram could not be constructed because of the low PL intensity. After retracting the AgTip (Fig. 10.16j), the $g^{(2)}(0)$ value returned to the original value of 0.08. These results indicated that single-photon emission from the single QD (before the approach of the AgTip) changed to multiphoton emission with the approach of the AgTip, i.e., the emission photon statistics of the QD can be controlled by the resonance energy transfer. This control of the emission photon statistics can be interpreted as follows. In the case of 405 nm excitation, no excitation enhancement occurs because the LSP is not generated on the AgTip. On the other hand, enhancement of the nonradiative rate (quenching) by the energy transfer occurs through the spectral overlap between the PL and LSPR band. Thus, we conclude that, in the case of the QD-AgTip system,

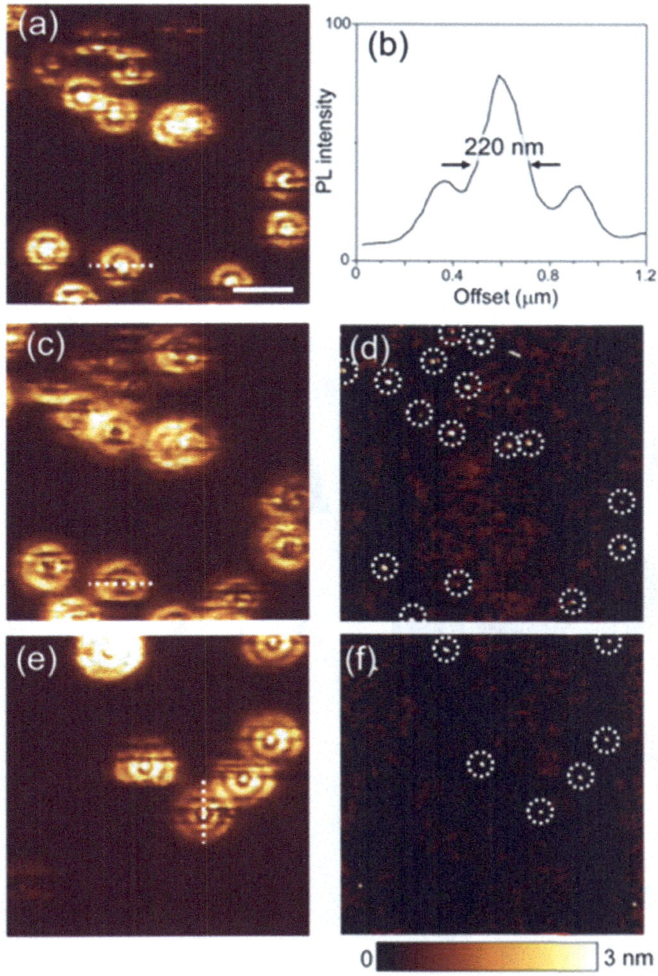

Fig. 10.15 **a** A PL image of single QDs obtained under 405 nm excitation without the approach of the AgTip. **b** A cross section of the single PL spot indicated by the dotted line in image (**a**). **c, d** A PL image of single QDs obtained from the same area as image (**a**) under 405 nm excitation with the approach of AgTip. **d** An AFM image obtained from the same area as images (**a**, **c**). **e** A PL image of single QDs obtained under 465 nm excitation with an approach of the AgTip. **f** An AFM image obtained from the same area as images (**e**). **g, h** Cross sections of the single PL spots indicated by the dotted lines in images (**c**) and (**e**), respectively. The size of all images is 5 μm × 5 μm. The scale bar in image (**a**) represents 1 μm Adapted with permission from Ref. [34]. Copyright 2016 American Chemical Society

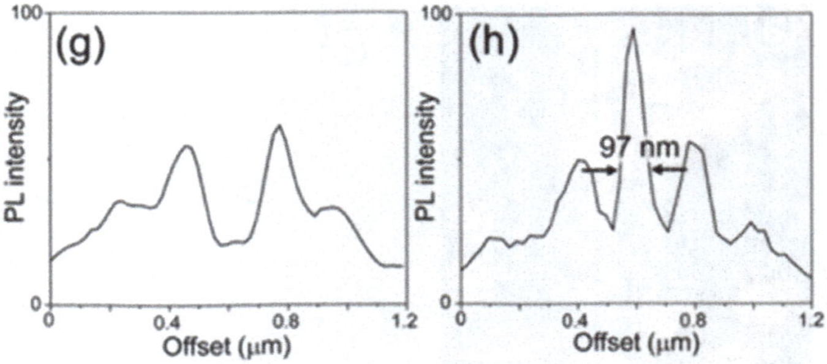

Fig. 10.15 (continued)

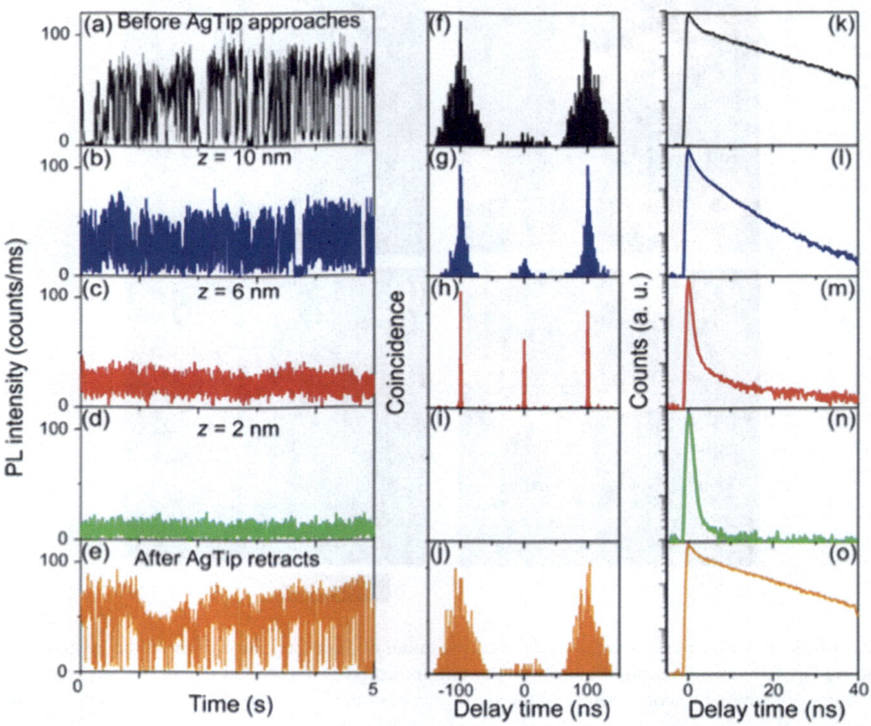

Fig. 10.16 Time traces of the PL intensity (**a–e**), photon correlation histograms (**f–j**), and PL decay curves (**k–o**) detected from the same single QD depending on the z-distance under 405 nm excitation: (**a**, **f**, **k**) before the approach of the AgTip; (**b**, **g**, **l**) $z = 10$ nm; (c, h, m) $z = 6$ nm; (**d**, **i**, **n**) $z = 2$ nm; and (**e**, **j**, **o**) after the AgTip was retracted Adapted with permission from Ref. [34]. Copyright 2016 American Chemical Society

the efficient quenching of the SX of the QD is the mechanism for the increase in the probability of the multiphoton emission.

The representative emission behavior of the single QD observed under 465 nm excitation with approaching the AgTip is shown in Fig. 10.17. In the case of 465 nm excitation, enhancement of the excitation rate of the single QD was expected because the LSP of the AgTip could be generated under 465 nm excitation. In addition, the enhancement of the nonradiative decay rate also should occur as in the case of the 405 nm excitation due to the spectral overlap between the PL and the LSPR. Actually, in contrast to the 405 nm excitation, the PL intensity from the single QD increased with a decrease in *z*-distance as shown in Fig. 10.17a–c. When the *z*-distance further

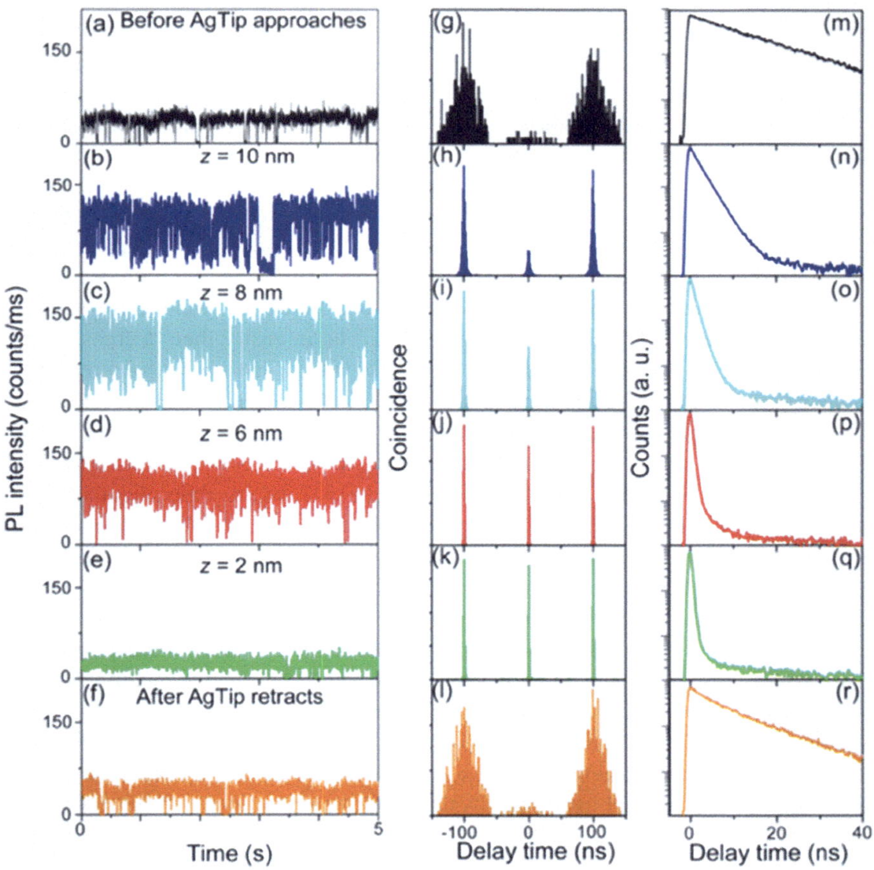

Fig. 10.17 Time traces of the PL intensity (**a–f**), photon correlation histograms (**g–l**), and PL decay curves (**m–r**) detected from the same single QD depending on the *z*-distance under 465 nm excitation: (**a, g, m**) before the approach of the AgTip; (**b, h, n**) $z = 10$ nm; (**c, i, o**) $z = 8$ nm; (**d, j, p**) $z = 6$ nm; (**e, k, q**) $z = 2$ nm; and (**f, l, r**) after the AgTip was retracted Adapted with permission from Ref. [34]. Copyright 2016 American Chemical Society

decreased (Fig. 10.17d, e), the intensity decreased. After the AgTip was retracted, the PL intensity returned to the same value as before the approach of the AgTip. In the photon correlation histograms (Fig. 10.17g–l), the center peak gradually increased with decreasing the z-distance, which was same as in the case of the 405 nm excitation. Thus, the probability of multiphoton emission increased with decreasing z-distance. The decay curves (Fig. 10.14m–r) were also shortened with decreasing the z-distance. Interestingly, when the z-distance was 10 nm, the PL intensity increased with a slight increase in the center peak. As the z-distance further decreased, the center peak increased, i.e., the probability of multiphoton emission increased, with increasing the PL intensity ($z = 8$ and 6 nm). These results demonstrated that the control of single-photon and multiphoton emission associated with the increased PL intensity could be achieved through nanometer-scale control of the z-distance. Above emission behavior observed under 465 nm excitation can be interpreted as follows. The PL intensity increased by the enhancement of the excitation rate because the LSP could be generated on the AgTip under 465 nm excitation. Furthermore, the quenching of SX also occurred as in the case of the 405 nm excitation, which resulted in an increase in multiphoton emission with a shortening of PL lifetime. The mechanism of the emission behavior is further discussed below.

The observed enhancements of the PL intensity and $g^{(2)}(0)$ value as a function of the z-distance built from average values of ten single QD under 465 nm excitation are summarized in Fig. 10.18. In Fig. 10.18a, the PL intensity increased with decreasing z-distance, and the intensity reached the maximum value, i.e., enhancement of 2.3 times on average was obtained near $z \approx 10$ nm. The intensity then decreased with decreasing the z-distance. In Fig. 10.18b, the $g^{(2)}(0)$ value increased with decreasing z-distance below 10 nm. The enhancement of 10.5 times was obtained as the maximum value at $z \approx 2$ nm. To reveal the mechanism of the modification of the emission behavior induced by the approach of the AgTip, theoretical analysis was applied to the enhancement of the PL intensity (Fig. 10.18a) [54]. In this theoretical analysis, the enhancement of the excitation rate by the LSP of the AgTip and the quenching of the PL by resonance energy transfer from the single QD to the AgTip were considered to fit the PL enhancement shown in Fig. 10.18a. The red and blue lines in Fig. 10.18a exhibit the theoretical curves obtained by assuming $\Phi^0_{\mathrm{PL}} = 30$ and 40%, respectively, with tip radius $a = 20$ nm. The theoretical curves well reproduced the experimental results using reasonable Φ^0_{PL} and a values. This result supports that the observed PL enhancement can be explained by the combination of the excitation rate enhancement by LSP of the AgTip and the quenching of SX by the resonance energy transfer from the QD to the AgTip. Therefore, the observed increase in multiphoton emission can be explained by the quenching of the SX due to the energy transfer. The $g^{(2)}(0)$ value increased with decreasing the z-distance (Fig. 10.18b), because the rate of energy transfer increases with decreasing z-distance. This quenching occurred independent of the excitation wavelength because the spectral overlap between PL of the QD and LSPR of the AgTip is important for the quenching. Thus, the increase in multiphoton emission was observed under both 405 and 465 nm excitation. In the case of 465 nm excitation, enhancement of the excitation rate also occurred. Hence, the PL intensity increased with decreasing z-distance, and then, the PL intensity decreased with

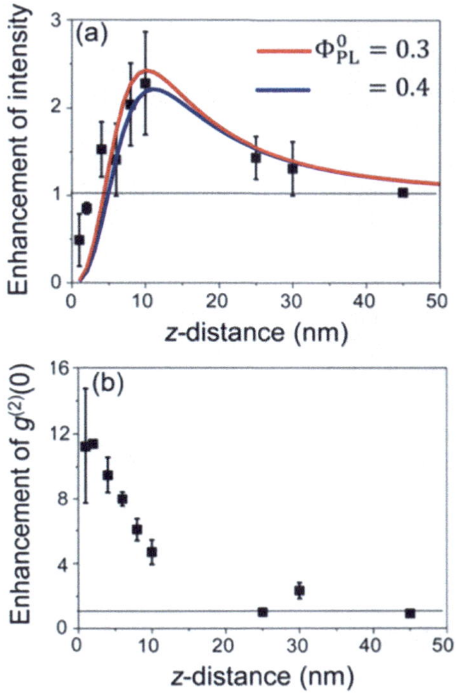

Fig. 10.18 Observed enhancements of QD emission by approaching the AgTip as a function of the z-distance. **a** PL intensity, I/I_0, **b** $g^{(2)}(0)$ value, $g^{(2)}(0)/g^{(2)}(0)^0$. The superscript 0 indicates the values before the approach of the AgTip. These figures were built from the average results of ten single QD measurements under 465 nm excitation. The red and blue lines in (**a**) show the theoretical curves obtained by assuming PL quantum yield (Φ^0_{PL}) = 30% and 40%, respectively, with tip radius a = 20 nm Adapted with permission from Ref. [34]. Copyright 2016 American Chemical Society

further decreasing the distance because of the increase in the energy transfer rate. On the other hand, in the case of 405 nm excitation, the PL intensity simply decreased with decreasing z-distance due to no enhancement of the excitation rate.

As shown above, we directly demonstrated that the PL behavior of QDs can be controlled by the interaction with MNSs. Based on the above results, we can conclude that the PL intensity can be controlled by the enhancements of the excitation rate and radiative rate, and multiphoton emission can be enhanced by the radiative rate and the quenching of SX. Furthermore, these enhancements strongly depend on the distance between the QD and MNS. Therefore, we can modify the PL behavior of QDs on demand by the combination of these enhancements and the distance.

10.7 Conclusion

The effects of quantum confinement of colloidally synthesized semiconductor NCs (QDs, NRs, and NPLs) on hot carrier transfer from higher excited states and band-edge carrier transfer were described. The state-selective femtosecond transient absorption spectroscopy and the initial bleach yield of band-edge state of NCs-acceptor HNs are important tools to reveal the carrier transfer mechanism. 1D quantum-confined system, CdSe NPLs with quantum well structure, shows a rapid intraband relaxation (60–70 fs) due to efficient phonon emission and weak electron-hole coupling, in which the carrier extraction from higher excited states is not so efficient. 3D quantum-confined system, CdSe QDs, shows a phonon bottle neck and strong electron-hole coupling, which is suitable to extract the hot carrier from higher excited states. The wavefunction penetration leading to the strong electronic coupling strength between CdSe QDs and Au NPs plays an important role for ultrafast electron transfer. In addition, the manipulation of single and multiexcitons of a single QD was demonstrated by using plasmonic nanostructures or metallic nanostructures (MNSs) based on AFM and single particle spectroscopy. The electromagnetic enhancement of absorption and luminescence of a single QD and energy transfer quenching by MNSs are key factors to understand the strong distance dependence between a single QD and a MNS.

Acknowledgements we deeply appreciate Professors K. Sasaki, T. Teranishi, H. Fujiwara, R. Sato, Drs T. Okuhata, T. Katayama, and L. Wang for their collaboration. The present work was supported by JSPS KAKENHI Grant numbers JP17H05254, 26390023 and 26107005 in grant-in-aid for Scientific Research on Innovative Areas "Photosynergetics."

References

1. Klimov VI (2000) Optical nonlinearities and ultrafast carrier dynamics in semiconductor nanocrystals. J Phys Chem B 104:6112–6123
2. Schaller RD, Klimov VI (2004) High efficiency carrier multiplication in PbSe nanocrystals: implications for solar energy conversion. Phys Rev Lett 92:186601
3. Kobayashi Y, Udagawa T, Tamai N (2009) Carrier multiplication in CdTe quantum dots by single—photon timing spectroscopy. Chem Lett 38:830–831
4. McGuire JA, Joo J, Pietryga JM, Schaller RD, Klimov VI (2008) New aspects of carrier multiplication in semiconductor nanocrystals. Acc Chem Res 41:1810–1819
5. Klimov VI (2007) Spectral and dynamical properties of multiexcitons in semiconductor nanocrystals. Ann Rev Phys Chem 58:635–673
6. Kambhampati P (2011) Hot exciton relaxation dynamics in semiconductor quantum dots: radiationless transitions on the nanoscale. J Phys Chem C 115:22089–22109
7. Kambhampati P (2011) Unraveling the structure and dynamics of excitons in semiconductor quantum dots. Acc Chem Res 44:1–13
8. Cooney RR, Sewall SL, Dias EA, Sagar DM, Anderson KEH, Kambhampati P (2007) Unified picture of electron and hole relaxation pathways in semiconductor quantum dots. Phys Rev B 75:245311
9. Guyot – Sionnest P, Wehrenberg B, Yu D (2005) Intraband relaxation in CdSe nanocrystals and the strong influence of the surface ligands. J Chem Phys 123:074709
10. Okuhata T, Tamai N (2016) Face-dependent electron transfer in CdSe nanoplatelet-methyl viologen complexes. J Phys Chem C 120:17052–17059
11. Shockley W, Queisser HJ (1961) Detailed balance limit of efficiency of p–n junction solar cells. J Appl Phys 32:510–519
12. Ross RT, Nozik AJ (1982) Efficiency of hot–carrier solar energy converters. J Appl Phys 53:3813–3818
13. Kobayashi Y, Pan L, Tamai N (2009) Effects of size and capping reagents on biexciton auger recombination dynamics of CdTe quantum dots. J Phys Chem C 113:11783–11789

14. Kobayashi Y, Tamai N (2010) Size-dependent multiexciton spectroscopy and moderate temperature dependence of biexciton auger recombination in colloidal CdTe quantum dots. J Phys Chem C 114:17550–17556
15. Sagarzazu G, Kobayashi Y, Murase N, Ping Y, Tamai N (2011) Auger recombination dynamics in hybrid silica-coated CdTe nanocrystals. Phys Chem Chem Phys 13:3227–3230
16. Kobayashi Y, Nishimura T, Yamaguchi H, Tamai N (2011) Effect of surface defects on auger recombination in colloidal CdS quantum dots. J Phys Chem Lett 2:1051–1055
17. Hyeon-Deuk K, Kobayashi Y, Tamai N (2014) Evidence of phonon-assisted auger recombination and multiple exciton generation in semiconductor quantum dots revealed by temperature-dependent phonon dynamics. J Phys Chem Lett 5:99–105
18. Wang L, Tian Y, Okuhata T, Tamai N (2015) Charge transfer dynamics and auger recombination of CdTe/CdS core/shell quantum dots. J Phys Chem C 119:17971–17978
19. Klimov VI, McBranch DW, Leatherdale CA, Bawendi MG (1999) Electron and hole relaxation pathways in semiconductor quantum dots. Phys Rev B 60:13740
20. Ithurria S, Dubertret B (2008) Quasi 2D colloidal CdSe platelets with thicknesses controlled at the atomic level. J Am Chem Soc 130:16504–16505
21. Ithurria S, Tessier MD, Mahler B, Lobo RPSM, Dubertret B, AlL Efros (2011) Colloidal nanoplatelets with two–dimensional electronic structure. Nature 10:936–941
22. Vuckovic M, Mentus SV, Janata E, Milosavljevic BH (2001) Fast dimerisation of the triparaquat radical dication. Phys Chem Chem Phys 3:4310–4315
23. Lhuillier E, Pedetti S, Ithurria S, Nadal B, Heuclin H, Dubertret B (2015) Two-dimensional colloidal metal chalcogenides semiconductors: synthesis, spectroscopy, and applications. Acc Chem Res 48:22–30
24. Matylitsky VV, Dworak L, Breus VV, Basché T, Wachtveitl J (2009) Ultrafast charge separation in multiexcited CdSe quantum dots mediated by adsorbed electron acceptors. J Am Chem Soc 131:2424–2425
25. Sagarzazu G, Inoue K, Saruyama M, Sakamoto M, Teranishi T, Masuo S, Tamai N (2013) Ultrafast dynamics and single particle spectroscopy of Au–CdSe nanorods. Phys Chem Chem Phys 15:2141–2152
26. Okuhata T, Kobayashi Y, Nonoguchi Y, Kawai T, Tamai N (2015) Ultrafast Carrier transfer and hot carrier dynamics in PbS–Au hybrid nanostructures. J Phys Chem C 119:2113–2120
27. Dana J, Partha M, Ghosh HN (2017) Hot-electron transfer from semiconductor domain to metal domain in CdSe@CdS{Au} nano-heterostructure. Nanoscale 9:9723–9731
28. Okuhata T, Katayama T, Tamai N (2020) Ultrafast and hot electron transfer in CdSe QD–Au hybrid nanostructures. J Phys Chem C 124:1099–1107
29. Masuo S, Naiki H, Machida S, Itaya A (2009) Photon statistics in enhanced fluorescence from a single CdSe/ZnS quantum dot in the vicinity of silver nanoparticles. Appl Phys Lett 95:193106
30. Naiki H, Masuo S, Machida S, Itaya A (2011) Single-photon emission behavior of isolated CdSe/ZnS quantum dots interacting with the localized surface plasmon resonance of silver nanoparticles. J Phys Chem C 115:23299–23304
31. Masuo S, Tanaka T, Machida S, Itaya A (2012) Photon antibunching in enhanced photoluminescence of a single CdSe/ZnS nanocrystal by silver nanostructures. J Photochem Photobio A 237:24–30
32. Naiki H, Masuhara A, Masuo S, Onodera T, Kasai H, Oikawa H (2013) Highly controlled plasmonic emission enhancement from metal-semiconductor quantum dot complex nanostructures. J Phys Chem C 117:2455–2459
33. Masuo S, Kanetaka K, Sato R, Teranishi T (2016) Direct observation of multiphoton emission enhancement from a single quantum dot using AFM manipulation of a cubic gold nanoparticle. ACS Photonics 3:109–116
34. Takata H, Naiki H, Wang L, Fujiwara H, Sasaki K, Tamai N, Masuo S (2016) Detailed observation of multiphoton emission enhancement from a single colloidal quantum dot using a silver-coated AFM tip. Nano Lett 16:5770–5778
35. Naiki H, Oikawa H, Masuo S (2017) Modification of emission photon statistics from single quantum dots using Metal/SiO_2 core/shell nanostructures. Photochem Photobio Sci 16:489–498

36. Naiki H, Uedao T, Wang L, Tamai N, Masuo S (2017) Multiphoton emission enhancement from a single colloidal quantum dot using SiO_2-coated silver nanoparticles. ACS Omega 2:728–737
37. Yuan CT, Yu P, Ko HC, Huang J, Tang J (2009) Antibunching single-photon emission and blinking suppression of CdSe/ZnS quantum dots. ACS Nano 3:3051–3056
38. Mallek-Zouari I, Buil S, Quelin X, Mahler B, Dubertret B, Hermier JP (2010) Plasmon assisted single photon emission of CdSe/CdS nanocrystals deposited on random gold film. Appl Phys Lett 97:053109
39. Vion C, Spinicelli P, Coolen L, Schwob C, Frigerio JM, Hermier JP (2010) Maitre A E (2010) Controlled modification of single colloidal CdSe/ZnS nanocrystal fluorescence through interactions with a gold surface. Opt Express 18:7440–7455
40. Canneson D, Mallek-Zouari I, Buil S, Quelin X, Javaux C, Mahler B, Dubertret B, Hermier JP (2011) Strong Purcell effect observed in single thick-shell CdSe/CdS nanocrystals coupled to localized surface plasmons. Phys Rev B 84:245423
41. Leblanc SJ, McClanahan MR, Jones M, Moyer PJ (2013) Enhancement of multiphoton emission from single CdSe quantum dots coupled to gold films. Nano Lett 13:1662–1669
42. Park YS, Ghosh Y, Chen Y, Piryatinski A, Xu P, Mack NH, Wang HL, Klimov VI, Hollingsworth JA, Htoon H (2013) Super-poissonian statistics of photon emission from single CdSe-CdS core-shell nanocrystals coupled to metal nanostructures. Phys Rev Lett 110:117401
43. Park YS, Ghosh Y, Xu P, Mack NH, Wang HL, Hollingsworth JA, Htoon H (2013) Single-nanocrystal photoluminescence spectroscopy studies of plasmon–multiexciton interactions at low temperature. J Phys Chem Lett 4:1465–1470
44. Yuan CT, Wang YC, Cheng HW, Wang HS, Kuo MY, Shih MH, Tang J (2013) Modification of fluorescence properties in single colloidal quantum dots by coupling to plasmonic gap modes. J Phys Chem C 117:12762–12768
45. Dey S, Zhou YD, Tian XD, Jenkins JA, Chen O, Zou SL, Zhao J (2015) An Experimental andTheoretical Mechanistic study of biexciton quantum yield enhancement in single quantum dots near gold nanoparticles. Nanoscale 7:6851–6858
46. Wang F, Karan NS, Nguyen HM, Ghosh Y, Hollingsworth JA, Htoon H (2015) Coupling single giant nanocrystal quantum dots to the fundamental mode of patch nanoantennas tthrough fringe field. Sci Rep 5:14313
47. Wang F, Karan NS, Nguyen HM, Ghosh Y, Sheehan CJ, Hollingsworth JA, Htoon H (2015) Correlated structural-optical study of single nanocrystals in a gap-bar antenna: effects of plasmonics on excitonic recombination pathways. Nanoscale 7:9387–9393
48. Wang F, Karan NS, Nguyen HM, Mangum BD, Ghosh Y, Sheehan CJ, Hollingsworth JA, Htoon H (2015) Quantum optical signature of plasmonically coupled nanocrystal quantum dots. Small 11:5028–5034
49. Dey S, Zhao J (2016) Plasmonic effect on exciton and multiexciton emission of single quantum dots. J Phys Chem Lett 7:2921–2929
50. Hoang TB, Akselrod GM, Mikkelsen MH (2016) Ultrafast room-temperature single photon emission from quantum dots coupled to plasmonic nanocavities. Nano Lett 16:270–275
51. Dey S, Zhou Y, Sun Y, Jenkins JA, Kriz D, Suib SL, Chen O, Zou S, Zhao J (2018) Excitation wavelength dependent photon anti-bunching/bunching from s single quantum dots near gold nanostructures. Nanoscale 10:1038–1046
52. Mundoor H, Sheetah GH, Park S, Ackerman PJ, Smalyukh II, van de Lagemaat J (2018) Tuning and switching a plasmonic quantum dot "sandwich" in a nematic line defect. ACS Nano 12:2580–2590
53. Krivenkov V, Goncharov S, Samokhvalov P, Sanchez-Iglesias A, Grzelczak M, Nabiev I, Rakovich YP (2019) Enhancement of biexciton emission due to long- range interaction of single quantum dots and gold nanorods in a thin-film hybrid nanostructure. J Phys Chem Lett 10:481–486
54. Bharadwaj P, Novotny L (2007) Spectral dependence of single molecule fluorescence enhancement. Opt Express 15:14266–14274

Chapter 11
The Confinement and Migration of Charge-Carriers in Lead Halide Perovskites

Sushant Ghimire, Lata Chouhan, and Vasudevanpillai Biju

Abstract With the growing success of halide perovskites as functional materials in solar cells, LEDs, and lasers, studies are progressing to reveal their optical and charge-carrier properties. This chapter describes the consequences of charge-carrier generation, stabilization, and recombination in perovskite nanocrystals, quantum dots, and their assemblies. The strong confinement of charge-carriers in perovskite quantum dots causes photoluminescence blinking with distinct ON and OFF events, which is due to photocharging and ultrafast Auger nonradiative recombination. Such a blinking with long OFF periods suppresses superoxide generation and oxidation of perovskites. When perovskite nanocrystals and quantum dots are closely-packed into superlattices, the carrier confinement is broken due to the narrowing of inter-particle energy levels and the formation of minibands that allow for carrier migration. This results in unexpectedly delayed photoluminescence at low intensities of excitation light, whereas at high intensities of excitation light, the ultrafast radiative recombination of charge-carriers occurs. These properties of quantum dots, nanocrystals, and assemblies of perovskites are important to be considered during the construction of devices such as solar cells, LEDs, and lasers.

Keywords Halide perovskites · Nanocrystals · Quantum dots · Perovskite assemblies · Single particle studies

---

Sushant Ghimire and Lata Chouhan: Equal contribution.

S. Ghimire · L. Chouhan · V. Biju
Graduate School of Environmental Science, Hokkaido University, N10, W5, Sapporo, Hokkaido 060-0810, Japan
e-mail: ghimiresushant@eis.hokudai.ac.jp

V. Biju (✉)
Research Institute for Electronic Science, Hokkaido University, N20, W10, Sapporo, Hokkaido 001-0020, Japan
e-mail: biju@es.hokudai.ac.jp

H. Miyasaka et al. (eds.), *Photosynergetic Responses in Molecules and Molecular Aggregates*, https://doi.org/10.1007/978-981-15-5451-3_11

11.1 Introduction

Halide perovskites are exciting materials for light-harvesting [1–3] and light-emitting [4–6] applications. Their success foots on the large absorption coefficient, high charge-carrier mobility, high photoluminescence (PL) quantum yield (QY), and cost-effective synthesis [1–6]. Further, the emission color from halide perovskites can cover the entire UV-visible-near infrared spectrum, which emanates from different band-gaps induced by changing the halogen compositions in the material [4–6]. Interestingly, these properties of halide perovskite are maintained irrespective of their size. Such attributes make halide perovskites highly promising semiconductor materials that can compete with the existing silicon technology for solar cells and quantum dot (QD) technology for light-emitting devices and displays. In addition to the chemical and physical stability of perovskites, the generation, stabilization, and binding energy of charge-carriers in these materials are also fundamental aspects to be considered for such device applications.

Charge-carrier dynamics in halide perovskites are complicated by the coexistence of bound electron-hole pairs (excitons) and free charge-carriers [7–11]. The coulombic binding between an electron and a hole in an exciton can be strong or weak, depending upon the exciton binding energy (E_b). Weakly and strongly bound excitons in semiconductors are shown schematically in Fig. 11.1. Weakly bound excitons are called Mott-Wannier excitons (Fig. 11.1a), where low E_b results in the larger delocalization of electron-hole pair, resulting in exciton Bohr radius (R_B) much greater than the lattice constant (a) of the material. On the other hand, strongly bound

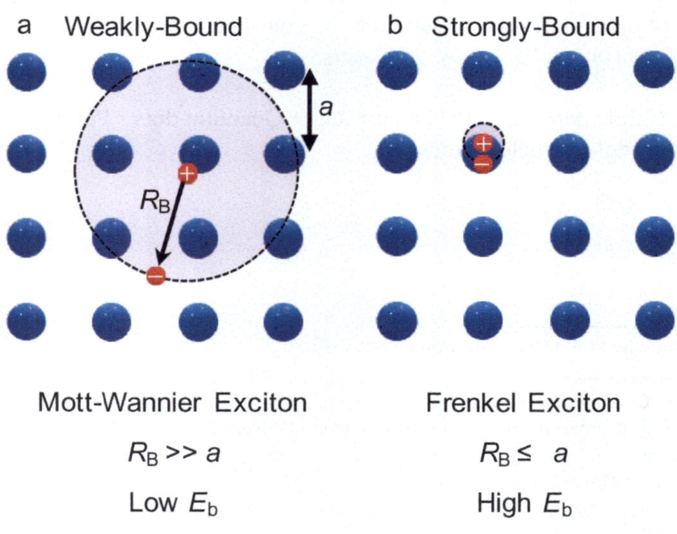

Fig. 11.1 Weakly and strongly bound excitons in semiconductors. **a** Mott-Wannier type and **b** Frenkel type excitons, where R_B is exciton Bohr radius, a is lattice constant, and E_b is exciton binding energy

exciton with high E_b is called Frenkel exciton whose R_B is much smaller than the a of the material (Fig. 11.1b). In halide perovskites, E_b lies in between those of Mott-Wannier and Frenkel excitons and is either less than or comparable to the thermal energy. Hence, free charge-carriers dominate in the films and bulk single crystals of these materials. Studies have shown that the free charge-carriers are the consequence of dissociation of excitons when the associated binding energy is overcome by the thermal energy [8–12]. The free charge-carrier property of halide perovskites is associated with long carrier diffusion lengths and lifetimes, as observed in the case of bulk films and single crystals [2, 3]. Nevertheless, the carrier lifetime varies in the range of sub-nanoseconds to microseconds in different types of halide perovskites, depending upon their size, dimensionality, and composition [2, 3, 8–15].

Exciton binding energies in perovskite nanocrystals (PNCs) are higher than the corresponding bulk crystal. For example, E_b of 320 meV is reported for $MAPbBr_3$ (MA = methylammonium, $CH_3NH_3^+$) nanocrystals, which is more than three times higher (84 meV) than that for the corresponding bulk crystal [16]. Therefore, quantum confinement effects are observed in PNCs as well, when their physical size becomes smaller than the corresponding R_B values [13, 15]. Additionally, the quantum confinement effects in PNCs are depended upon not only the size but also the dimensionality. For instance, blue-shift in absorption and PL spectra, which is associated with the increase in optical band-gap, is observed in the case of perovskite nanoplatelets when the layer thickness decreases below R_B [15]. Despite the quantum confinement effects in PNCs, when compared to the conventional chalcogenide QDs, slow cooling and efficient extraction of hot carriers are reported in their colloidal solutions [17]. The intrinsic phonon bottleneck and Auger heating effects at low and high carrier densities, respectively, in PNCs account for such observations. Apart from this, the excellent optoelectronic properties of PNCs, which are comparable to the conventional QDs, are maintained even at the weak quantum confinement regime. While R_B in lead halide perovskites lies in the range of 2–7 nm [13, 15, 16, 18, 19], most of the studies demonstrating the light-emitting [4, 6] and photovoltaic [17, 18] applications of PNCs show the nanocrystal size either only slightly smaller or larger than the corresponding R_B value. This suggests that, compared to the chalcogenide QDs, the application of PNCs in light-harvesting and light-emitting device is diverse and immense.

11.2 Photoluminescence Blinking

In quantum-confined semiconductor nanocrystals or QDs, the photogenerated electrons and holes are strongly correlated. Carrier dynamics in such strongly confined QDs is governed by the discrete energy states showing sharp excitonic band in the absorption spectrum, fast radiative recombination of charge-carriers, carrier multiplication or multi-exciton generation, and increased Auger recombination [20]. Moreover, the PL properties of these semiconductor nanocrystals at their single particle level are often intricated by the stochastic emission behavior with a sequence of bright

'ON' and dark 'OFF' events known as PL intermittency or blinking. This is attributed to the random charging by photoionization and neutralization or discharging of the QD. The blinking behavior in QDs is often described as A-type and B-type [21]. In A-type blinking, the 'OFF' state is associated with the decrease in PL lifetime, whereas the carrier-recombination rates are not affected by the 'OFF' or the 'ON' states in B-type blinking. The short PL lifetime observed during the 'OFF' state in A-type blinking is attributed to the formation of a charged excitonic state called a trion (X^-) which promotes the mono-exponential nonradiative Auger-type carrier recombination in QDs. On the other hand, B-type blinking mostly depends on the activation and deactivation of the traps or its recombination centers which leads to the 'ON' and 'OFF' states, respectively.

Apart from the QDs, diverse stochastic emission behavior has been observed in the case of perovskite single nanocrystals, microcrystals, and thin films [22–27]. Some perovskite single-emitters have shown blinking with two-state fluctuations between 'ON' and 'OFF' events [24, 25, 27], which is similar to the case of chalcogenide QD and is ascribed to the process of photocharging and neutralization. Semiconductor nanocrystals or QDs show various interesting optoelectronic properties in the isolated forms. On the other hand, multi-state PL blinking has also been reported from the nanocrystals, microcrystals, and films of lead halide perovskite [22, 23, 26]. An early report on PL blinking in $MAPbI_3$ PNCs and microcrystals has shown that the chemical and structural properties of material are responsible for the multi-state stochastic emission behavior [22]. This is explained through two mechanisms, namely the quenching-site and the emitting-site model, which is shown in Fig. 11.2. In emitting-site model, as shown in Fig. 11.2a, the emitters are at the either end of the nanocrystal which randomly get charged and discharged leading to the 'ON' and 'OFF' states, respectively. Quenching-site model (Fig. 11.2b) considers the whole nanocrystal as an emissive site, where the emission localization position and center of the nanocrystals lie at the same point. Quenchers in quenching-site model act as traps which are present at the either end of the nanocrystal and deactivate the electrons and holes during the diffusion, resulting in the 'OFF' state. An interesting behavior of PL blinking in lead halide perovskites is that, unlike lead chalcogenide QDs, the multi-state PL fluctuations in these semiconductor materials is suppressed under high-intensity of excitation, which on the other hand becomes more intense at low-intensity of excitation. Figure 11.3a shows the excitation laser intensity-dependent PL blinking in $MAPbI_3$ PNCs at the ambient environment. The suppression of PL blinking in such perovskite samples at high-intensity excitation is attributed to the decrease in efficiency of active quenchers to trap large number of charges. Alternately, more emitting sites are generated at high-intensity excitation, resulting in the averaged PL intensity over many emitting sites. Nevertheless, as shown in Fig. 11.3a(i) and b(i), photobleaching cannot be ruled out at high-intensity excitation under the ambient atmosphere [22, 25, 27].

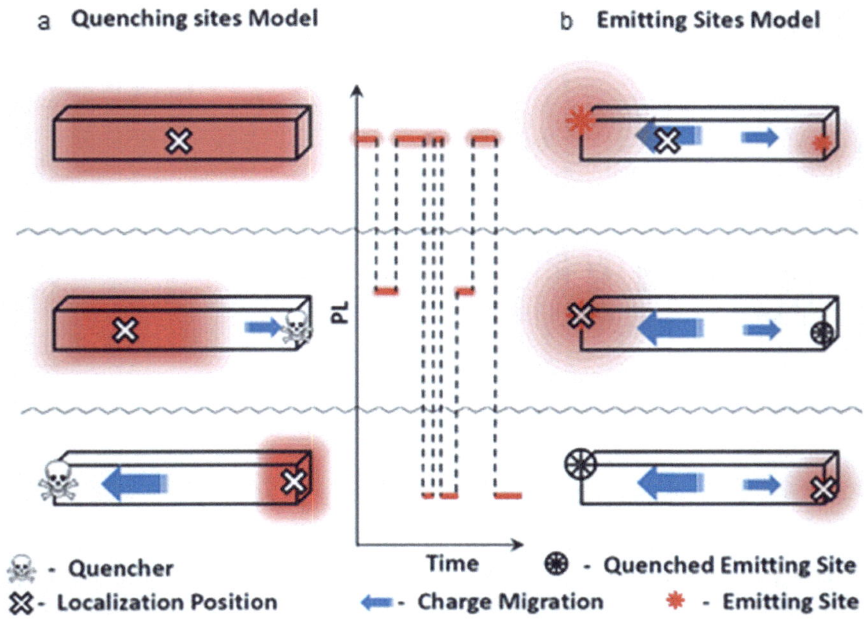

Fig. 11.2 Models of PL blinking in lead halide perovskites: **a** Quenching-site model and **b** emitting-site model. Reproduced with permission from Ref. [22]. Copyright 2015, American Chemical Society

11.3 External Factors Affecting the Photoluminescence

Importantly, different environmental factors such as moisture, nitrogen, oxygen, polymer, and light have shown variable results on the PL properties of perovskites at their single particle level [24, 27]. The PL blinking trajectories of $MAPbI_3$ nanocrystals at different local environment are shown in Fig. 11.3b. For example, different roles of oxygen on the PL properties of lead halide perovskites have been reported for different types of perovskite samples [22, 25, 27]. While PL brightening was observed as the consequence of photoinduced trap-filling in the case of $MAPbI_3$ PNCs in presence of oxygen [22], PL bleaching was reported in $CsPbI_3$ perovskite QDs under the similar environment [27]. Recently, a study carried out at the single particle level in $MAPbI_3$ PNCs under air, argon, and polymer matrix (Fig. 11.3b) reveals its PL degradation mechanism [25]. The PNCs degrade or photobleach under the air atmosphere, whereas the PL intensity remains constant under the argon and the polymer atmosphere. The photostability of a PNC under the argon or the polymer atmosphere is the result of low reactivity of oxygen with the nanocrystal during its 'ON' state. Interestingly, as shown in Fig. 11.3b(iv), a recovery of PL intensity level is observed even at the air atmosphere after the long 'OFF' event, suggesting that there

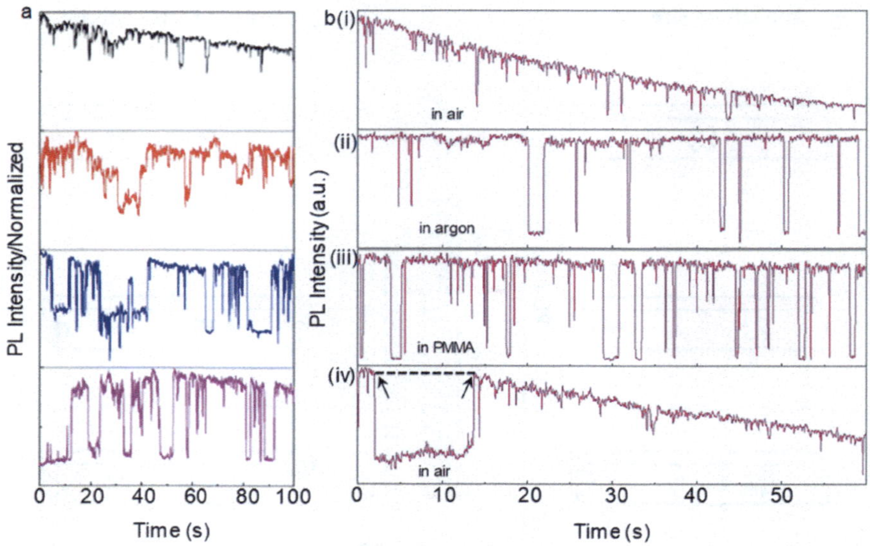

Fig. 11.3 Excitation intensity- and local environment-dependent PL blinking in lead halide perovskites. **a** PL blinking in $MAPbI_3$ PNCs under increasing intensity of the excitation laser: (bottom) 0.01 to (top) 0.5 Wcm^{-2}. **b** PL blinking in $MAPbI_3$ PNCs (i, iv) in air (ii) under continuous argon purging, and (iii) in polymer matrix Reproduced with permission: (**a**) from Ref. [22]. Copyright 2015, American Chemical Society. (**b**) From Ref. [25]. Copyright 2019, Wiley-VCH

was no degradation of PNCs during the long 'OFF' duration that involves nonradiative recombination. Alternately, the PL degradation at the air atmosphere when the PNCs were more frequently in the 'ON' state or with the long-living 'ON' duration can be ascribed to the oxidation of nanocrystals by the generation of superoxide, which is discussed schematically in Fig. 11.4. Here, the neutral and ionized states are responsible for the 'ON' and 'OFF' events in the PL blinking trajectories of PNCs. When the nanocrystal is in excited state, it can either oxidize and transfer electron to the oxygen present in the atmosphere generating superoxide or get ionized. The generated superoxide can take the vacancies of the anion and deprotonate the methyl ammonium cation, thereby slowly degrading the PNCs. On the other hand, when the nanocrystal is in the ionized state, nonradiative recombination becomes faster than the electron transfer, and the reactivity of oxygen is suppressed during the long 'OFF' duration. In such ON-OFF dynamics and oxidation/antioxidation processes, the excitation light intensity plays a crucial role on the density and fate of charge-carriers. Apart from this, the confinement or diffusion of locally generated multiple charge-carriers depends upon the particle-particle interaction.

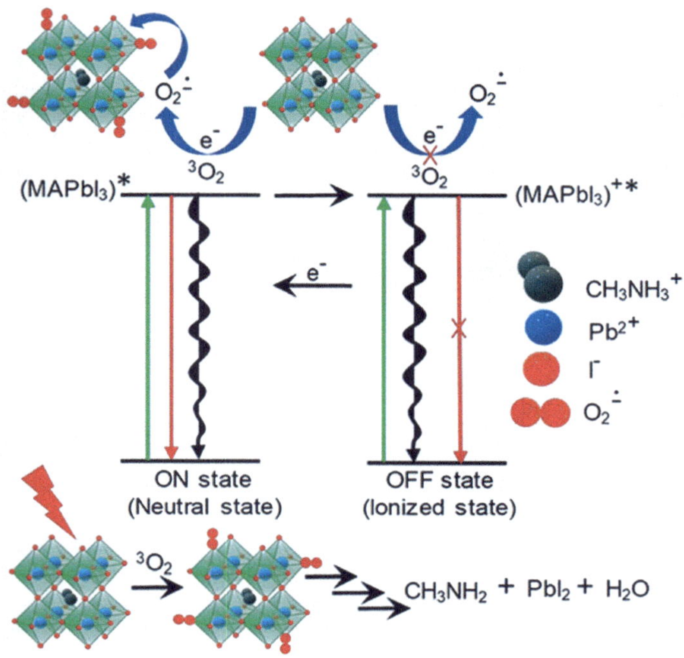

Fig. 11.4 Oxidation in perovskite. Schematic representation of the PL blinking and the degradation of the $MAPbI_3$ PNCs. Reproduced with permission from Ref. [25]. Copyright 2019, Wiley-VCH

11.4 Carrier Migration in Perovskite Assemblies

When semiconductor nanocrystals or QDs are assembled into a superlattice, their individual properties transform to produce more unique optical, electronic, and thermal properties. This arises from the interaction between the individual particles in the assembly [28–30]. In a superlattice, when the QDs are very close to each other, the interaction between their electronic wavefunctions increases. At a point, when the inter-particle spacing is too short, the discrete energy states of individual nanocrystals or QDs split to form minibands. As the result, long-range energy and charge transport in these superlattices through these bands become possible [30]. Strong electronic coupling and the formation of minibands as a function of inter-particle distance in semiconductor nanocrystal superlattice is shown schematically in Fig. 11.5. Such properties of QD superlattice are highly applicable in electronic, opto-electronic and thermoelectric devices [30, 31]. Since the pioneering work of Bawendi and coworkers on chalcogenide QD superlattices [28], many researchers have carried out a successful coupling of QDs and migration of energy or excitons among them in their two- or three-dimensional superlattice structures [31–33]. Exciton diffusion is observed in a highly ordered CdSe QD superlattice with center-to-center distance less than 7 nm between the adjacent QDs [32]. In such QD superlattices, the exciton

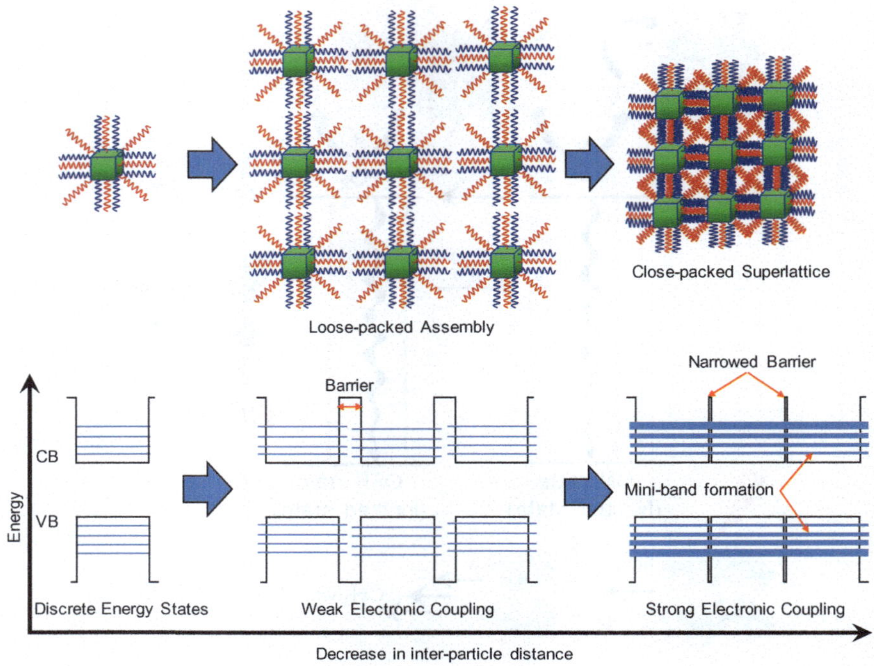

Fig. 11.5 Scheme showing strong electronic coupling and the formation of minibands in semiconductor nanocrystal superlattice

diffusion length as longer as 125 nm is reported. A faster exciton diffusion among highly interacting QDs can result in a faster exciton decay, which is attributed to the larger number of traps or defect sites on different QDs that the excitons may encounter within their lifetime. Besides, a highly ordered QD superlattice can result lower PLQY when compared to the disordered films and colloidal QD samples. This is ascribed to the faster energy transfer among closely-packed QDs. Similarly, directional transport of excitons is observed in long-range ordered and strongly coupled binary nanocrystal superlattice of PbSe consisting of two different sizes [33]. The surface-to-surface inter-particle spacing in such QD array can be reduced further by ligand exchange which results in the ultrafast exciton decay within few picoseconds, showing strong electronic coupling between QDs.

In a QD superlattice, the coupling energy between the individual QDs is much smaller than the exciton binding energy. Therefore, the excitons do not dissociate into free charge-carriers in the superlattice, rather they diffuse or hop and undergo ultrafast recombination, or the nonradiative energy transfer takes place among the closely-packed QDs. On the other hand, if the inter-particle distance between the QDs is very small (for example, center-to-center distance <1 nm), it induces sintering and straining of QDs, leading to increased disorder [29, 32]. Owing to these limitations of QD solids, a superlattice of semiconductor nanocrystals where long-range diffusion

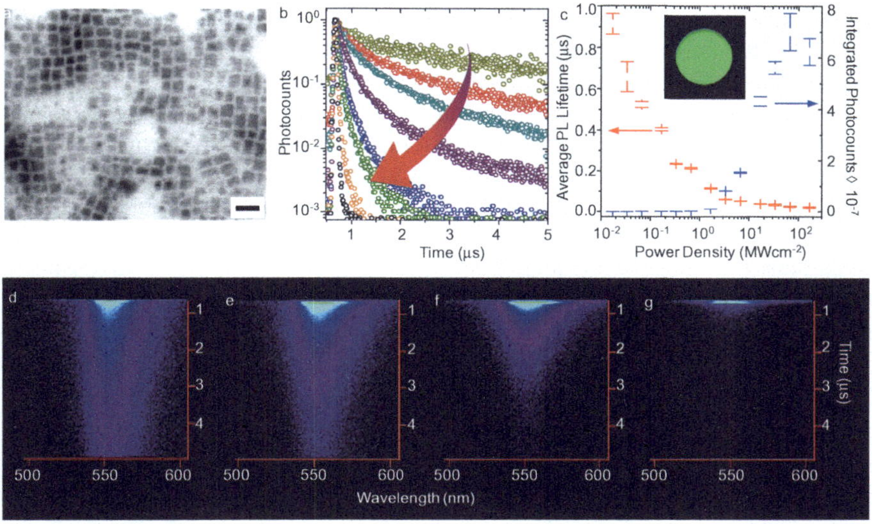

Fig. 11.6 Excitation intensity-dependent PL and charge-carrier properties of PNC films. **a** TEM image of $MAPbBr_3$ PNCs. The scale bar is 20 nm. **b** PL decay profiles of a $MAPbBr_3$ PNC film under increasing intensity of excitation laser: (top) 0.0175 to (bottom) 170 $MWcm^{-2}$. **c** Plots of the PL lifetime and the number of photons emitted as a function of excitation laser power for a $MAPbBr_3$ PNC film. Inset: $MAPbBr_3$ PNC film under UV light. **d–g** Temporally and spectrally resolved photocount maps of a $FAPbBr_3$ PNC film with an increase in excitation laser power: **d** 0.17, **e** 1.7, **f** 17, **g** 170 $MWcm^{-2}$. Reproduced with permission from Ref. [34]. Copyright 2019, American Chemical Society

of free charge-carriers, which is associated with delayed recombination, is largely required for their efficient applications in solar cells; and the organic-inorganic hybrid lead halide perovskites are promising candidates for this purpose.

Organic-inorganic hybrid PNCs (Fig. 11.6a) when assembled into close-packed film (Fig. 11.1c inset) show unexpectedly delayed PL (Fig. 11.6b, c, and d) under low-intensity excitation [34, 35], which are comparable to the free carrier recombination in single microcrystals and bulk films of perovskites [2, 3, 7, 14]. For instance, a PL lifetime >900 ns is observed for $MAPbBr_3$ PNC film [34], which becomes much longer (>4 μs) in the case of $FAPbBr_3$ (FA = formamidinium,$CHN_2H_3^+$) perovskite nanocrystal PNC film [35]. Under the increasing excitation intensity, the radiative recombination becomes extremely fast, resulting in amplified emission. The excitation intensity-dependent PL decay profiles of $MAPbBr_3$ PNC film and the temporally and spectrally resolved photocount maps of $FAPbBr_3$ PNC film are shown in Fig. 11.6b, and d–g, respectively. The fast PL lifetimes at higher intensities of excitation are associated with the exponential rise in the number of photons emitted by the PNC film, which is discussed in Fig. 11.6c. On the other hand, a fast PL is obvious in the case of isolated PNCs in their colloidal solution, irrespective of the intensity of excitation. These observations point toward the long-range migration of excitonically unbound charge-carriers at low-intensity excitation and spatial confinement of these charge-carriers at higher intensities in PNC films. The amplified emission can be assigned to the radiative recombination of high-density electrons and holes that are generated and spatially confined among the close-packed PNCs in the film [34, 35].

The dynamics of charge-carriers generated in a PNC film at low- and high-intensity excitations is shown schematically in Fig. 11.7a and b, respectively. At high-intensity excitation, manifold charge-carriers are generated among different PNCs in the irradiated area (Fig. 11.7b). On the other hand, the concentration of photogenerated charge-carriers is less at low-intensity excitation, which reduces the carrier-carrier interactions and allows for the long-range diffusion in the film (Fig. 11.7a). Owing to such a long-range carrier diffusion, PNC films excited with low-intensity laser show

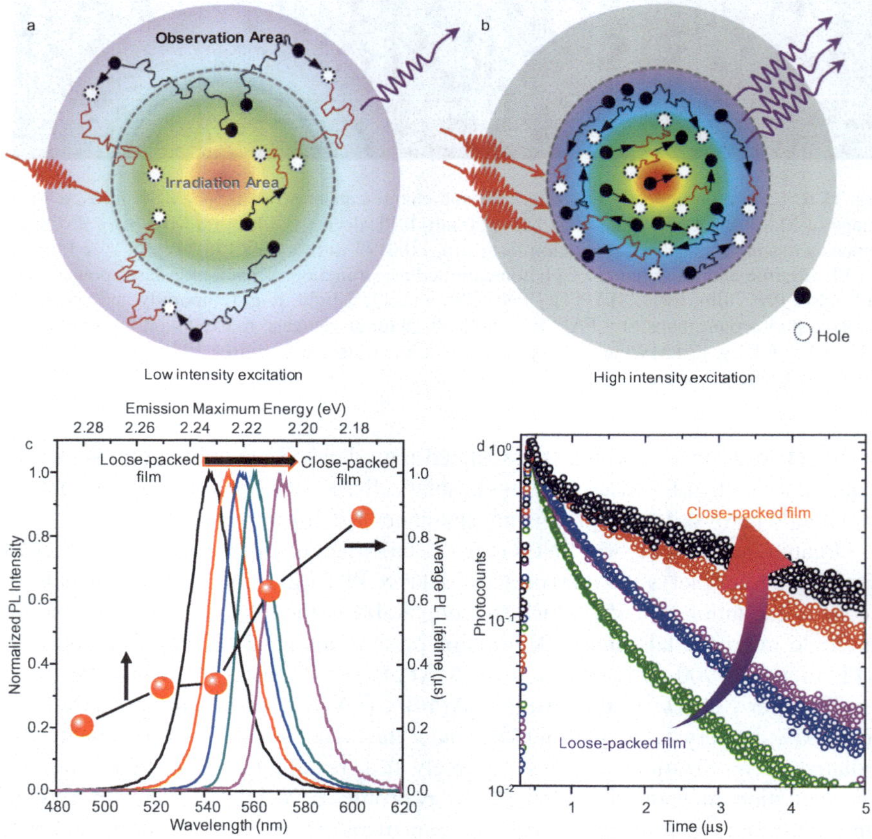

Fig. 11.7 Diffusion and recombination of charge-carriers in a PNC film **a**, **b** Scheme of charge-carrier diffusion and recombination in a PNC film under **a** low-intensity and **b** high-intensity excitations. **c** PL spectra of $FAPbBr_3$ PNC films with different PNC density. The corresponding PL lifetime values are shown in the line-symbol plot. **d** PL decay profiles of $FAPbBr_3$ PNC films with different PNC density. Reproduced with permission from Ref. [34]. Copyright 2019, American Chemical Society

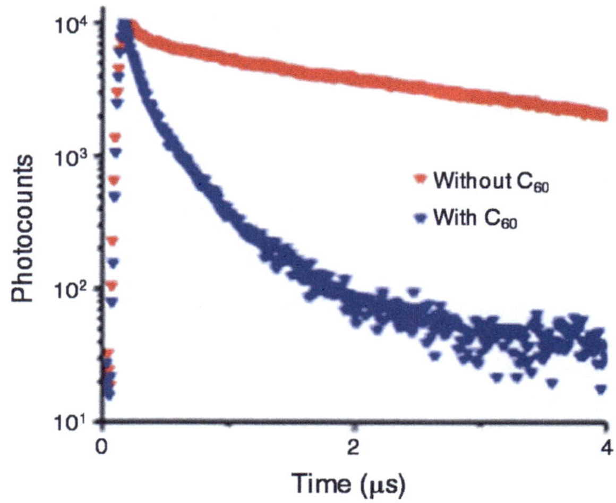

Fig. 11.8 PL decay profiles of a PNC film excited with low-intensity laser (0.0175 μJ/pulse) (red) before and (blue) after doping with C_{60}. Reproduced with permission from Ref. [35]. Copyright 2019, Royal Society of Chemistry

unexpectedly delayed PL. On the other hand, the diffusion of photogenerated charge-carriers is spatially confined by increasing the intensity of excitation laser, which is the result of an increase in the carrier concentration. Accordingly, the rate of radiative recombination within the irradiated area controls over the diffusion, providing amplified emission.

The degree of charge-carrier migration across PNCs in their close-packed assembly (or film) depends upon their packing density [34]. As seen in Fig. 11.7c, the PL spectra of a loose-packed PNC film is blue-shifted, whereas that of a close-packed film is red-shifted. Additionally, as shown in Fig. 11.7c and d, a red-shifted PL spectrum is associated with long PL lifetime. The low-energy photons which are associated with the delayed emission in PNC film point toward the diffusion of charge-carriers through closely-spaced energy states during which some energy is lost, resulting in long-lived low-energy photons. Such delayed PL, as a result of long-range migration of photogenerated charge-carriers, in close-packed assembly of PNCs is a promising property for the harvesting of charge-carriers in high-efficiency solar cells. Indeed, an appreciable rate of electron transfer ($3.3 \times 10^6\ s^{-1}$) is achieved from the photoexcited close-packed assembly of PNCs to fullerene (C_{60}), revealed by the fast PL decay of the PNC film doped with C_{60}, when compared to a pristine film without C_{60}, as shown in Fig. 11.8 [35].

11.5 Conclusions

The optical and electronic properties of individual QDs modify when they are arranged into superlattices. The overlapping of QDs electronic wavefunctions in the lattice facilitates the long-range diffusion or hopping of excitons. Due to the high exciton binding energy in strongly quantum-confined systems, the diffusing excitons do not dissociate into free charge-carriers. Thus, the exciton recombination becomes ultrafast, and the isolated semiconductor nanocrystals or QDs show PL blinking. In contrast, metal halide perovskite nanocrystals and QDs arranged into superlattices exhibit long-range carrier migration and delayed radiative recombination. This is analogous to the long-range diffusion of free charge-carriers in bulk perovskites, suggesting a weak confinement of photogenerated electron-hole pairs. The formation of new inter-particle energy states in superlattices of PNCs assist in the diffusion of coulombically uncorrelated electrons and holes. Such freely diffusing electrons are effectively harvested using C_{60}. Therefore, superlattices of PNCs are promising structures for the development of high-efficiency solar cells.

Acknowledgements S. G. acknowledges the MEXT scholarship for doctoral research, and L. C. acknowledges the JICA scholarship for doctoral research. V. B. acknowledges financial support (Grants 15H01099 and 17H05243) from MEXT under the JSPS Grant-in-Aid for Scientific Research on Innovative Areas "Photosynergetics."

References

1. Kojima A, Teshima K, Shirai Y, Miyasaka T (2009) Organometal halide perovskites as visible-light sensitizers for photovoltaic cells. J Am Chem Soc 131:6050–6051
2. Dong Q, Fang NY, Shao Y, Mulligan P, Qiu J, Cao L, Huang J (2015) Electron-hole diffusion lengths > 175 μm in solution-grown $CH_3NH_3PbI_3$ single crystals. Science 347:967–970
3. deQuilettes DW, Vorpahl SM, Stranks SD, Nagaoka H, Eperon GE, Ziffer ME, Snaith HJ, Ginger DS (2015) Impact of microstructure on local carrier lifetime in perovskite solar cells. Science 348:683–686
4. Zhang F, Zhong H, Chen C, Wu X, Hu X, Huang H, Han J, Zou B, Dong Y (2015) Brightly luminescent and color-tunable colloidal $CH_3NH_3PbX_3$ (X = Br, I, Cl) quantum dots: potential alternatives for display technology. ACS Nano 9:4533–4542
5. Kim Y-H, Cho H, Heo JH, Kim T-S, Myoung N, Lee C-L, Im SH, Lee T-W (2015) Multicolored organic/inorganic hybrid perovskite light-emitting diodes. Adv Mater 27:1248–1254
6. Yakunin S, Protesescu L, Krieg F, Bodnarchuk MI, Nedelcu G, Humer M, De Luca G, Fiebig M, Heiss W, Kovalenko MV (2015) Low-threshold amplified spontaneous emission and lasing from colloidal nanocrystals of caesium lead halide perovskites. Nat Commun 6:8056
7. Saba M, Cadelano M, Marongiu D, Chen F, Sarritzu V, Sestu N, Figus C, Aresti M, Piras R, Lehmann AG, Cannas C, Musinu A, Quochi F, Mura A, Bongiovanni G (2014) Correlated electron-hole plasma in organometal perovskites. Nat Commun 5:5049
8. Yang Y, Yang M, Li Z, Crisp R, Zhu K, Beard MC (2015) Comparison of recombination dynamics in $CH_3NH_3PbBr_3$ and $CH_3NH_3PbI_3$ perovskite films: influence of exciton binding energy. J Phys Chem Lett 6:4688–4692
9. Jha A, Duan H-G, Tiwari T, Nayak PK, Snaith HJ, Thorwart M, Miller RJD (2018) Direct observation of ultrafast exciton dissociation in lead iodide perovskite by 2D electronic spectroscopy. ACS Photonics 5:852–860
10. Phuong LQ, Yamada Y, Nagai M, Maruyama N, Wakamiya A, Kanemitsu Y (2016) Free carriers versus excitons in $CH_3NH_3PbI_3$ perovskite thin films at low temperatures: charge transfer from the orthorhombic phase to the tetragonal phase. J Phys Chem Lett 7:2316–2321
11. Blancon J-C, Tsai H, Nie W, Stoumpos CC, Pedesseau L, Katan C, Kepenekian M, Soe CMM, Appavoo K, Sfeir MY, Tretiak S, Ajayan PM, Kanatzidis MG, Even J, Crochet JJ, Mohite AD (2017) Extremely efficient internal exciton dissociation through edge states in layered 2D perovskites. Science 355:1288–1292

12. Miyata A, Mitioglu A, Plochocka P, Portugall O, Wang JT, Stranks SD, Snaith HJ, Nicholas RJ (2015) Direct measurement of the exciton binding energy and effective masses for charge carriers in organic-inorganic tri-halide perovskites. Nat Phys 11:582–587
13. Butkus J, Vashishtha P, Chen K, Gallaher JK, Prasad SKK, Metin DZ, Laufersky G, Gaston N, Halpert JE, Hodgkiss JM (2017) The Evolution of quantum confinement in $CsPbBr_3$ perovskite nanocrystals. Chem Mater 29:3644–3652
14. Shi D, Adinolfi V, Comin R, Yuan M, Alarousu E, Buin A, Chen Y, Hoogland S, Rothenberger A, Katsiev K, Losovyj Y, Zhang X, Dowben PA, Mohammed OF, Sargent EH, Bakr OM (2015) Low trap-state density and long carrier diffusion in organolead trihalide perovskite single crystals. Science 347:519–522
15. Levchuk I, Osvet A, Tang X, Brandl M, Perea JD, Hoegl F, Matt GJ, Hock R, Batentschuk M, Brabec CJ (2017) Brightly luminescent and color-tunable formamidinium lead halide perovskite $FAPbX_3$ (X = Cl, Br, I) colloidal nanocrystals. Nano Lett 17:2765–2770
16. Zheng K, Zhu Q, Abdellah M, Messing ME, Zhang W, Generalov A, Niu Y, Ribaud L, Canton SE, Pullerits T (2015) Exciton binding energy and the nature of emissive states in organometal halide perovskites. J Phys Chem Lett 6:2969–2975
17. Li M, Bhaumik S, Goh TW, Kumar MS, Yantara N, Grätzel M, Mhaisalkar S, Mathews N, Sum TC (2017) Slow cooling and highly efficient extraction of hot carriers in colloidal perovskite nanocrystals. Nat Commun 8:14350
18. Li M, Begum R, Fu J, Xu Q, Koh TM, Veldhuis SA, Grätzel M, Mathews N, Mhaisalkar S, Sum TC (2018) Low threshold and efficient multiple exciton generation in halide perovskite nanocrystals. Nat Commun 9:4197
19. Tanaka K, Takahashi T, Ban T, Kondo T, Uchida K, Miura N (2003) Comparative study on the excitons in lead-halide-based perovskite-type crystals $CH_3NH_3PbBr_3$ $CH_3NH_3PbI_3$. Solid State Commun 127:619–623
20. Ghimire S, Biju V (2018) Relations of exciton dynamics in quantum dots to photoluminescence, lasing, and energy harvesting. J Photochem Photobiol C 34:137–151
21. Galland C, Ghosh Y, Steinbrück A, Sykora M, Hollingsworth JA, Klimov VI, Htoon H (2011) Two types of luminescence blinking revealed by spectroelectrochemistry of single quantum dots. Nature 479:203–207
22. Tian Y, Merdasa A, Peter M, Abdellah M, Zheng K, Ponseca CS Jr, Pullerits T, Yartsev A, Sundström V, Scheblykin IG (2015) Giant photoluminescence blinking of perovskite nanocrystals reveals single-trap control of luminescence. Nano Lett 15:1603–1608
23. Gerhard M, Louis B, Camacho R, Merdasa A, Li J, Kiligaridis A, Dobrovolsky A, Hofkens J, Scheblykin IG (2019) Microscopic insight into nonradiative decay in perovskite semiconductors from temperature-dependent luminescence blinking. Nat Commun 10:1698
24. Park Y-S, Guo S, Makarov NS, Klimov VI (2015) Room temperature single-photon emission from individual perovskite quantum dots. ACS Nano 9:10386–10393
25. Chouhan L, Ghimire S, Biju V (2019) Blinking beats bleaching: the control of superoxide generation by photo-ionized perovskite nanocrystals. Angew Chem Int Ed 58:4875–4879
26. Li C, Zhong Y, Luna CAM, Unger T, Deichsel K, Gräser A, Köhler J, Köhler A, Hildner R, Huettner S (2016) Emission enhancement and intermittency in polycrystalline organolead halide perovskite films. Molecules 21:1081
27. Yuan G, Ritchie C, Ritter M, Murphy S, Gómez DE, Mulvaney P (2018) The degradation and blinking of single $CsPbI_3$ perovskite quantum dots. J Phys Chem C 122:13407–13415
28. Murray CB, Kagan CR, Bawendi MG (1995) Self-organization of CdSe nanocrystallites into three-dimensional quantum dot superlattices. Science 270:1335–1338
29. Collier CP, Vossmeyer T, Heath JR (1998) Nanocrystal superlattices. Annu Rev Phys Chem 49:371–404
30. Kagan CR, Murray CB (2015) Charge transport in strongly coupled quantum dot solids. Nat Nanotechnol 10:1013–1026
31. Harman TC, Taylor PJ, Walsh MP, LaForge BE (2002) Quantum dot superlattice thermoelectric materials and devices. Science 297:2229–2232

32. Yoon SJ, Guo Z, dos Santos Claro PC, Shevchenko EV, Huang L (2016) Direct imaging of long-range exciton transport in quantum dot superlattices by ultrafast microscopy. ACS Nano 10:7208–7215
33. Wu Y, Li S, Gogotsi N, Zhao T, Fleury B, Kagan CR, Murray CB, Baxter JB (2017) Directional carrier transfer in strongly coupled binary nanocrystal superlattice films formed by assembly and *in situ* ligand exchange at a liquid-air interface. J Phys Chem C 121:4146–4157
34. Ghimire S, Chouhan L, Takano Y, Takahashi K, Nakamura T, Yuyama K, Biju V (2019) Amplified and multicolor emission from films and interfacial layers of lead halide perovskite nanocrystals. ACS Energy Lett. 4:133–141
35. Ghimire S, Nair VC, Muthu C, Yuyama K, Vacha M, Biju V (2019) Photoinduced photoluminescence enhancement in self-assembled clusters of formamidinium lead bromide perovskite nanocrystals. Nanoscale 11:9335–9340

Chapter 12
Plasmon-Induced Carrier Transfer for Infrared Light Energy Conversion

Masanori Sakamoto, Zichao Lian, and Toshiharu Teranishi

Abstract The infrared (IR) region of the solar spectrum is a source of untapped potential energy. Conversion of IR-light to electrical energy or fuel would provide a plentiful energy source for modern society. Plasmonic energy conversion holds the key to conversion of solar energy through materials that efficiently absorb photons of the desired wavelength, including the IR region. Herein, we investigated plasmon-induced carrier transfer from plasmonic heavily doped semiconductor nanocrystals in a wide bandgap semiconductor to convert IR-light to energy. We discovered that efficient hot-carrier transfer proceeds from a heavily doped semiconductor nanocrystal to a wide bandgap semiconductor upon excitation of a localized surface plasmon resonance band. In addition, this material system achieved photocatalytic H_2 evolution based on excitation at long wavelengths of the solar spectrum (i.e., 2500 nm). The apparent quantum yield of photocatalytic H_2 evolution based on a catalyst with a plasmonic heavily doped semiconductor as a light-harvesting material represents a highly efficient conversion of IR energy to fuel. The relationship between the plasmon-induced carrier dynamics and photocatalytic activity paves the way for use of this undeveloped low-energy light as a solar energy resource.

Keywords Infrared light · Localized surface plasmon resonance · Photocatalyst · Nanocrystal · Hydrogen evolution · Solar fuel generation

12.1 Introduction

Solar light is an important energy resource supporting life on Earth and many human activities. The parts of the solar spectrum used for solar energy conversion, including photosynthesis and artificial photosynthesis, are limited to the ultraviolet (UV), visible, and a limited region of the near-infrared (NIR) light (700–1000 nm) [1]. IR-light, which accounts for almost half of all solar energy, particularly IR-light at wavelengths longer than 1000 nm, including shortwave IR (SWIR: 1400–3000 nm)

M. Sakamoto (✉) · Z. Lian · T. Teranishi
Department of Chemistry, Graduate School of Science, Kyoto University, Gokasho, Uji, Kyoto 611-0011, Japan
e-mail: sakamoto@scl.kyoto-u.ac

H. Miyasaka et al. (eds.), *Photosynergetic Responses in Molecules and Molecular Aggregates*, https://doi.org/10.1007/978-981-15-5451-3_12

[2], represents a vast source of untapped energy. Developing artificial IR-light-to-fuel conversion systems would open up this resource for applications. However, despite extensive research efforts, conversion of IR-light to chemical energy remains challenging. Although certain narrow bandgap semiconductors, such as HgCdTe and InGaAs are known to be effective IR-light-absorbing materials, the high toxicity, low electromotive force, and poor stability of these materials have restricted their applications as photocatalysts [3, 4].

Localized surface plasmon resonance (LSPR) is a phenomenon related to collective oscillation of free carriers in nanomaterials, which can be applied to overcome the limitations of harvesting solar energy in the IR region [5–16]. Recently, copper chalcogenide nanocrystals (NCs) have drawn interest as a novel series of compound semiconductors. These materials show tunable hole-based LSPR absorption in the near-IR (NIR) region and have attracted much attention as candidates for IR-responsive photocatalysts (Fig. 12.1) [6–8, 12, 13, 16]. Because the carrier density (~10^{21}–10^{22} cm^{-3}) of heavily doped semiconductors is low relative to that of metal NCs (~10^{23} cm^{-3}), their LSPR response lies in the NIR to mid-infrared (MIR) regions. This property is understood from the equation:

$$\omega_p = \sqrt{\frac{Ne^2}{\varepsilon_0 m}}$$

where ω_p is the bulk plasma frequency, N is the free carrier density, e is the electron charge, ε_0 is the dielectric constant of vacuum, and m is the free carrier effective mass [12]. Owing to the low carrier density compared with that of metal NCs, heavily doped semiconductor NCs show LSPR in the IR region.

Although plasmonic materials have excellent light-harvesting ability, low conversion efficiency is a drawback. The low efficiency of plasmonic energy conversion is mainly attributed to ultrafast-relaxation of hot carriers and ultrafast annihilative recombination at the heterointerface between plasmonic materials and carrier

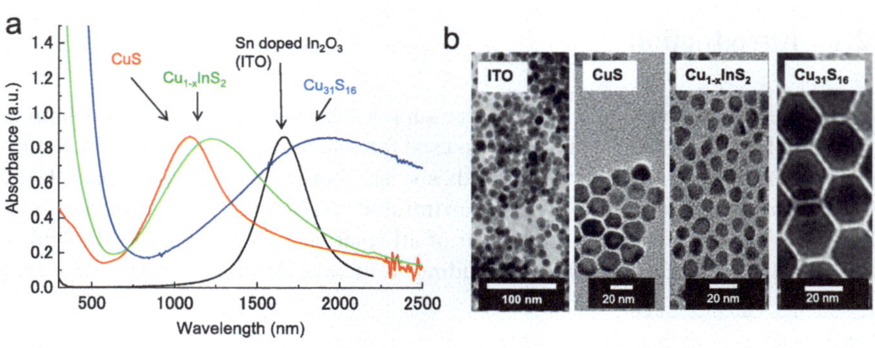

Fig. 12.1 **a** Absorption spectra of typical heavily doped semiconductor NCs. **b** TEM images of heavily doped semiconductor NCs

acceptors [9]. These disadvantages have restricted the use of plasmonic materials as light energy conversion materials. Thus, a better understanding of this hot-carrier dynamics at the interface is essential to achieve highly efficient IR-responsive photocatalyst.

A heterostructure based on CdS/Cu_7S_4 nanocrystals (HNCs) is an ideal system for spectroscopic tracing of the LSPR-induced carrier transfer from the plasmonic Cu_7S_4 phase to the CdS phase. Furthermore, the LSPR band of Cu_7S_4 NCs covers almost all the IR region of the solar spectrum. In the present work, we elucidated the LSPR-induced behavior of hot carriers in plasmonic Cu_7S_4 NCs and CdS/Cu_7S_4 HNCs by time-resolved transient absorption spectroscopy. Transient absorption spectroscopy enabled us the directly observe carriers in photo-excited NCs.

Here, we investigated the LSPR-induced hot-carrier dynamics of CdS/Cu_7S_4 HNCs with the use of femtosecond transient absorption measurement [13]. Through detailed investigations of the carrier dynamic of CdS/Cu_7S_4 HNCs we identified efficient electron transfer from the Cu_7S_4 phase to the CdS phase. Notably, CdS/Cu_7S_4 has long-lived charge separation (>273 μs), which has not been observed in plasmon-induced carrier-injection systems. On the basis of the ultralong lifetime of the charge separation, the CdS/Cu_7S_4 HNCs achieved an efficient photocatalytic H_2 evolution reaction (HER). The CdS/Cu_7S_4 HNCs exhibited an exceedingly high apparent quantum yield (AQY) of 3.8% at 1100 nm, which exceeds the current record for photocatalytic HER under IR-light irradiation. Furthermore, we demonstrated that our novel system can convert energy from the solar spectrum up to long wavelengths (i.e., 2000–2500 nm).

We revealed that the high catalytic activity is related to efficient hot-electron injection and long-lived charge separation at the plasmonic p–n heterojunction of the CdS/Cu_7S_4 HNCs. This behavior is unlike a conventional Schottky junction at the heterointerface of plasmonic metal/semiconductor NCs. Because ultrafast charge recombination is a major drawback of all plasmonic energy conversion systems, the observed mechanism proposed here will change the consensus on LSPR-induced energy conversion and highlight the great advantages of high hot-carrier transfer efficiency and long-lived charge separation.

12.2 Synthesis and Characterization of IR-Responsive Photocatalyst

The CdS/Cu_7S_4 HNCs were synthesized through a seeded growth reaction of disk-shaped Cu_7S_4 NCs and subsequent partial cation exchange of the Cu_7S_4 NCs with Cd^{2+} cations (Fig. 12.2). Figure 12.2a shows transmission electron microscope (TEM) images of monodisperse disk-shaped Cu_7S_4 NCs (size: 16.2 ± 0.9 nm, thickness: 3.5 ± 0.4 nm). After the cation exchange reaction, CdS phases were formed on the Cu_7S_4 NCs to give dimer structures (Fig. 12.2b). The X-ray diffraction (XRD) patterns in Fig. 12.1c clearly show that the CdS/Cu_7S_4 HNCs comprised

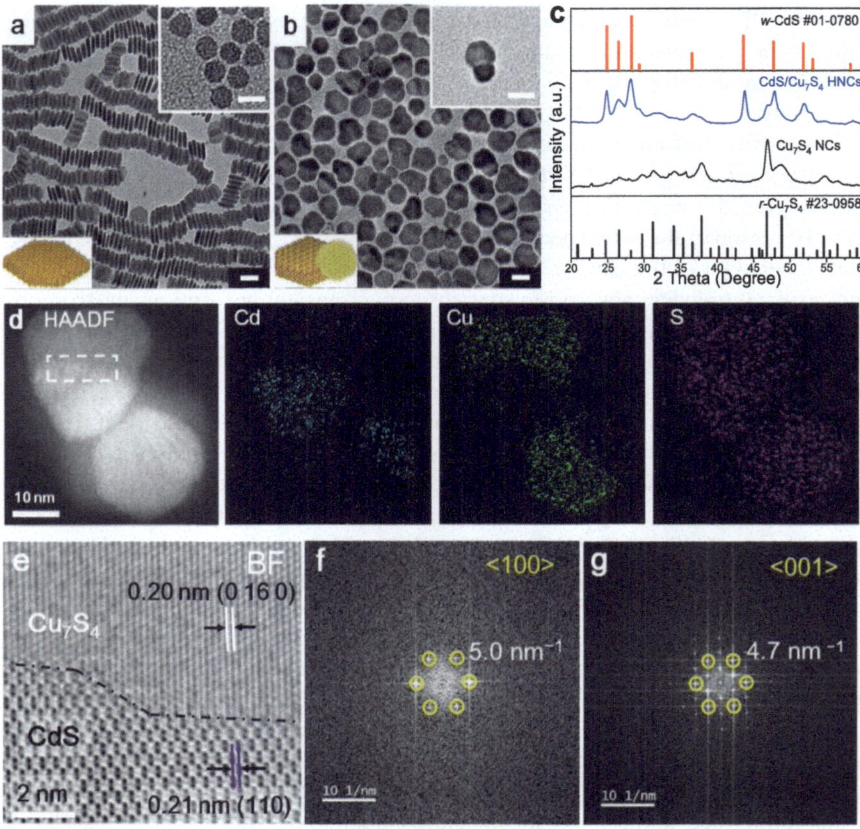

Fig. 12.2 Representative TEM images of (**a**) disk-shaped Cu_7S_4 NCs [inset: Cu_7S_4 NCs (upper right), schematic representation (lower left)], (**b**) CdS/Cu_7S_4 HNCs [inset: a typical CdS/Cu_7S_4 HNC (upper right), schematic representation (lower left)]. Scale bars are 20 nm. (**c**) XRD patterns of Cu_7S_4 NCs and CdS/Cu_7S_4 HNCs. (**d**) HAADF-STEM-EDS elemental mapping images of CdS/Cu_7S_4 HNCs. (**e**) BF-STEM image of the heterointerface of a single CdS/Cu_7S_4 HNC indicated by a dashed white rectangle of a HAADF-STEM image in (**d**). (**f**, **g**) FFT patterns of the Cu_7S_4 phase from <100> direction at the upper region of (**e**) and the CdS phase from <001> direction at the lower region of (**e**), respectively. Reprinted with permission from *J Am. Chem. Soc.* **2019**, *141*, 2446–2450. Copyright 2019 American Chemical Society

triclinic roxbyite Cu_7S_4 (r-Cu_7S_4, Joint Committee on Power Diffraction Standards (JCPDS) no. 23-0958) and wurtzite CdS (w-CdS, JCPDS no. 01-0780) phases with a Cd/Cu molar ratio of 49:51, as estimated by X-ray fluorescence (XRF) spectroscopy. High-angle annular dark-field (HAADF) scanning TEM (STEM) and STEM-energy dispersive X-ray spectrometry (EDS) elemental mapping (Fig. 12.2d) also indicated the formation of heterodimers composed of Cu_7S_4 and CdS phases. The bright-field (BF)-STEM image revealed that the heterointerface between the r-Cu_7S_4 and w-CdS phases had a small lattice mismatch of 5% (Fig. 12.2e). The lattice fringes of 0.20 and

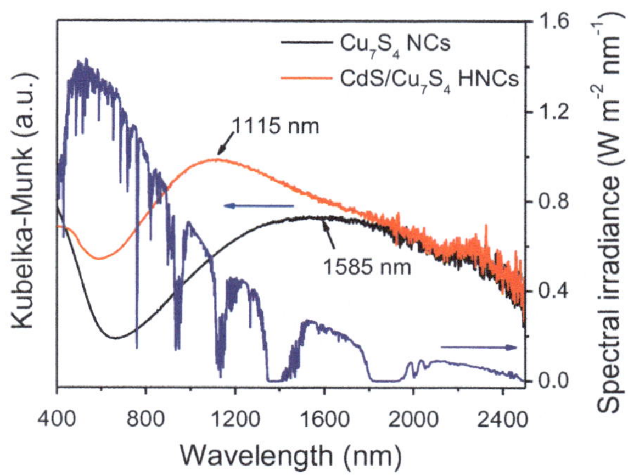

Fig. 12.3 Diffuse reflectance spectra of Cu_7S_4 NCs and CdS/Cu_7S_4 HNCs and AM1.5 solar spectrum. Reprinted with permission from *J. Am. Chem. Soc.* **2019**, *141*, 2446–2450. Copyright 2019 American Chemical Society

0.21 nm (in Fig. 12.2e), are assigned to *r*-Cu_7S_4 (0160) and *w*-CdS (110), respectively, which agrees with the results of the fast Fourier transform (FFT) patterns (Fig. 12.2f, g).

The ground state diffuse reflectance spectra of Cu_7S_4 NCs and CdS/Cu_7S_4 HNCs with LSPR peaks at 1585 and 1115 nm, respectively, are shown in Fig. 12.3. The blue-shifted LSPR peak of the CdS/Cu_7S_4 HNCs compared with that of the Cu_7S_4 NCs might be attributed to differences in size, shape, and/or dielectric environment [16]. The LSPR band of Cu_7S_4 almost covered the whole IR region of the solar spectrum.

12.3 Transient Absorption Measurements and IR-Light-Induced Plasmonic Hot-Electron Injection

We investigated LSPR-induced carrier transfer in the CdS/Cu_7S_4 HNCs from the *r*-Cu_7S_4 phase to the *w*-CdS phase. Transient absorption (TA) measurements were conducted to reveal the mechanism of the IR-induced carrier dynamics over the CdS/Cu_7S_4 HNCs. The band diagrams of Cu_7S_4 [14, 17] and CdS [18, 19] are shown in Fig. 12.4. The conduction band edge of the CdS NCs was accessed based on hot carriers generated by LSPR excitation of the Cu_7S_4 phases.

The time-resolved shift of TA spectra (TAS) of Cu_7S_4 NCs and CdS/Cu_7S_4 HNCs after selective excitation of the LSPR band by a 1300-nm laser are shown in Fig. 12.5. Upon excitation of plasmonic Cu_7S_4 NCs, a broad absorption peak was observed at 540 nm. A similar transient absorption was reported by Burda et al. They investigated the TAS of Cu_7S_4 upon excitation of an interband transition of Cu_7S_4 NCs by visible

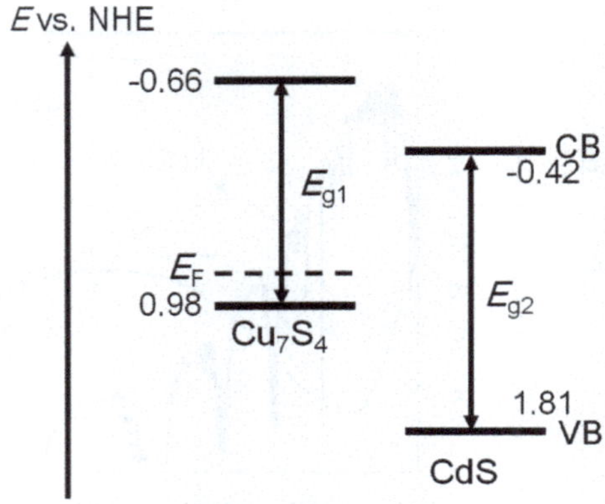

Fig. 12.4 Schematic illustration of the band alignment of the CdS/Cu_7S_4 HNC. E_F = Fermi level, E_{g1} = 1.64 eV (Cu_7S_4 optical energy bandgap), E_{g2} = 2.23 eV (CdS optical energy bandgap)

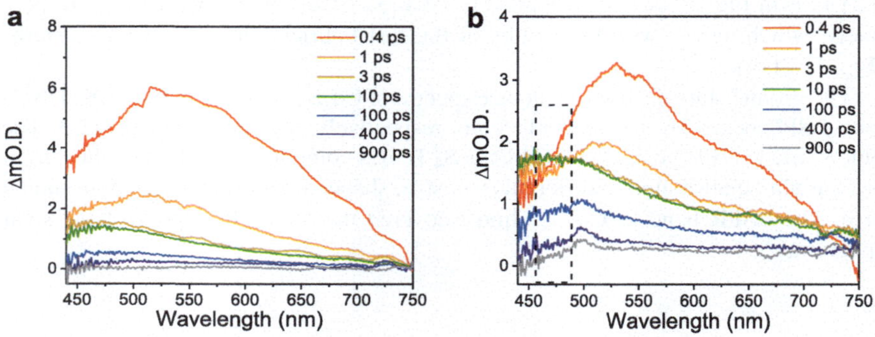

Fig. 12.5 Time-resolved TAS of (**a**) Cu_7S_4 NCs and (**b**) CdS/Cu_7S_4 HNCs in chloroform upon 1300-nm laser excitation. Dashed rectangle in (**b**) indicates a bleaching feature of the CdS phases. Reprinted with permission from *J. Am. Chem. Soc.* **2019**, *141*, 2446–2450. Copyright 2019 American Chemical Society

laser light and attributed this feature to holes trapped in the Cu_7S_4 NCs [19]. Interestingly, in the present system, we observed a similar TAS although the excitation band of the Cu_7S_4 NCs was the LSPR in NIR region. This result suggests that an IR LSPR-induced hole trapping process exists in the Cu_7S_4 NCs. Because the TAS was not affected by the excitation laser power, the non-linear optical phenomenon or a many-body effect does not contribute to the observed hole trapping process.

Notably, the TAS of CdS/Cu_7S_4 HNCs showed a similar broad absorption with a dip at approximately 450–500 nm. We assigned the dip in the TAS to CdS exciton bleaching caused by state filling [20] derived from electron transfer from the Cu_7S_4 to CdS phases under 1300-nm laser excitation. The dipping feature was observed even after 900 ps of 1300-nm laser excitation.

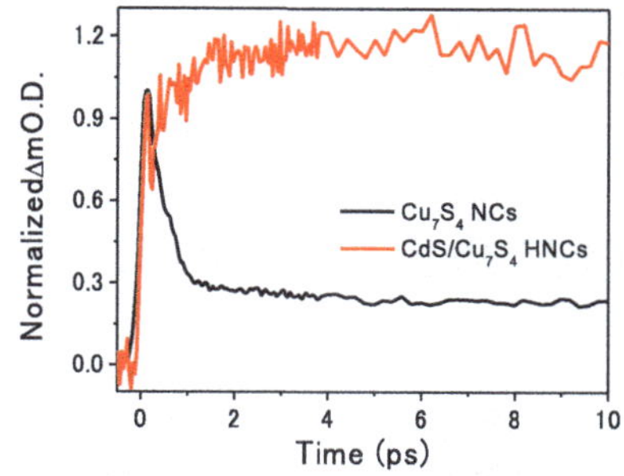

Fig. 12.6 Kinetic profiles of Cu_7S_4 NCs and CdS/Cu_7S_4 HNCs in chloroform at 460 nm in the ps region upon 1300 nm laser excitation. Reprinted with permission from *J. Am. Chem. Soc.* **2019**, *141*, 2446–2450. Copyright 2019 American Chemical Society

Kinetic traces of Cu_7S_4 NCs and CdS/Cu_7S_4 HNCs at 460 nm are shown in Fig. 12.6. Upon excitation of the LSPR band of the Cu_7S_4 phase, we observed the TA derived from trapped holes of Cu_7S_4 NCs, which decayed with fast and slow components (0.40 and 16 ps, respectively), indicating that multiple trapping process, as proposed by Burda and coworkers [19], was induced by the excitation of LSPR.

Conversely, the kinetic profile of the CdS/Cu_7S_4 HNCs at 460 nm showed only a rising component corresponding to the decay of state filling of CdS. This result indicates that electron transfer caused by state filling of CdS was completed within the laser pulse (<100 fs). This ultrafast electron transfer, within 100 fs, suggests a ballistic hot-carrier transfer process. We estimated the recovery rate of the bleaching of CdS phases from the results of TAS in the ns region (Fig. 12.7). Because the recovery of bleaching (i.e., the rising component) of CdS reflects a charge recombination process, we estimated the lifetime of interphase charge separation to be 273 μs (Fig. 12.7b),

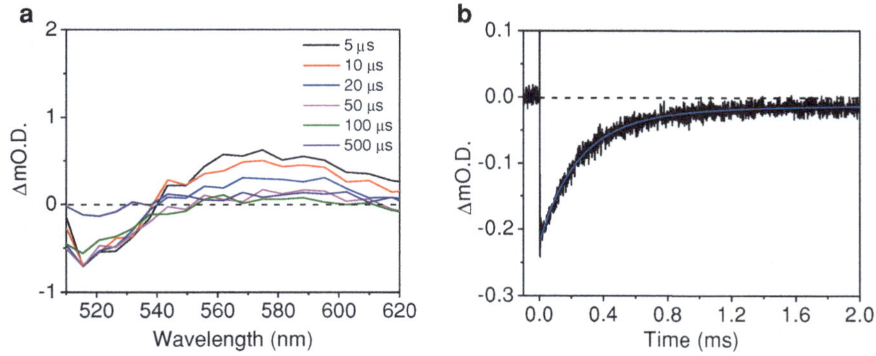

Fig. 12.7 (**a**) Time-resolved spectral changes of the CdS/Cu_7S_4 HNCs in the visible region upon excitation at 800 nm. The bleaching feature of CdS was observed even 100 μs after excitation by the 800-nm laser at the NIR-LSPR band. (**b**) Kinetic profile of CdS/Cu_7S_4 HNCs probing at 520 nm in the μs region upon the excitation of the LSPR band. Blue line is the best fit. Reprinted with permission from *J. Am. Chem. Soc.* **2019**, *141*, 2446–2450. Copyright 2019 American Chemical Society

which is exceptionally long compared with that of typical plasmonic hot-carrier transfer systems (several tens of ps) [21].

12.4 Photocatalytic H_2 Evolution

The photocatalytic H_2 evolution reaction (HER) activity under IR-light irradiation (λ > 800 nm, i.e., only exciting the LSPR) of Cu_7S_4 or CdS/Cu_7S_4 HNCs was evaluated in an aqueous solution containing H_2PtCl_6 as a H_2 evolution cocatalyst precursor and Na_2S–Na_2SO_3 as a hole scavenger.

For photocatalytic H_2 generation measurements, we used a home-made online-flow sampling system with a quartz reactor connected to a gas chromatograph (GC-2004, Shimadzu, with argon as a carrier gas) and a thermal conductivity detector (TCD). To measure of photocatalytic activity, a 10.0-mg portion of the photocatalyst was dispersed in 100 mL of an aqueous solution containing 0.25 M Na_2S and 0.35 M Na_2SO_3 as sacrificial reagents, and 27 μL 1 wt% H_2PtCl_6 aqueous solution was then added to the solution. Before IR-light irradiation, the system was stirred under an argon flow to remove oxygen. The reaction temperature was maintained at 20 °C. A 300-W Xe lamp (Aegle lamp house R300-3 J, Cermax, Excelitas Technology) equipped with a UV–visible cut-off filter (λ > 800 nm, power density: 50 mW cm^{-2}) was used as the irradiation light source.

The Cu_7S_4 NCs exhibited no H_2 evolution activity. On the contrary, the CdS/Cu_7S_4 HNCs showed high HER activity under IR-light irradiation (Fig. 12.8a). The average photocatalytic H_2 evolution rate of the CdS/Cu_7S_4 HNCs was 14.7 μmol h^{-1} g^{-1}. Notably, the mixture of Cu_7S_4 NCs and CdS NCs exhibited no H_2 evolution activity

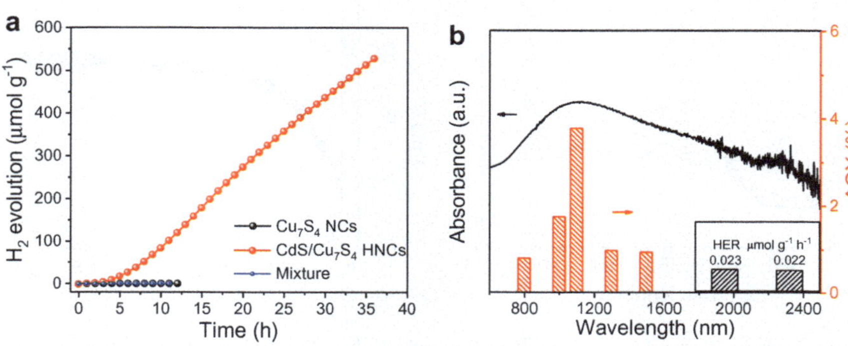

Fig. 12.8 **a** Temporal photocatalytic H_2 evolution over Cu_7S_4 NCs, CdS/Cu_7S_4 HNCs and a mixture of Cu_7S_4 NCs and CdS NCs under IR illumination (λ > 800 nm) in a Na_2S–Na_2SO_3 aqueous solution (0.25 M Na_2S and 0.35 M Na_2SO_3, pH = 12). **b** Absorption spectrum and AQY for HER of the CdS/Cu_7S_4 HNCs under the monochromic light (6 mW cm^{-2}). Reprinted with permission from *J. Am. Chem. Soc.* **2019**, *141*, 2446–2450 Copyright 2019 American Chemical Society

under IR irradiation. Consequently, we conclude that the photo-induced carrier transfer from the *r*-Cu_7S_4 phase to the *w*-CdS phase in a CdS/Cu_7S_4 HNC is the key step contributing to the HER activity.

Remarkably, the CdS/Cu_7S_4 HNCs showed no notable changes in their catalytic activity after 36 h, indicating high stability of the CdS/Cu_7S_4 HNCs. We found that the amount of evolved H_2 increased with illumination time after a substantial induction period (Fig. 12.8) [22]. We attribute the induction period of the HER under IR-light irradiation to in situ deposition of Pt-cocatalyst, which increased the photocatalytic activity over time. When the Pt-cocatalyst was photodeposited on the CdS/Cu_7S_4 HNCs beforehand, the time-dependent photocatalytic HER showed no induction period.

The apparent quantum yield (AQY) was measured under the above photocatalytic reaction conditions with the use of various monochromic wavelength bandpass filters. The light intensity was measured with a power meter. The AQY was calculated according to the following equation:

$$\begin{aligned} \mathrm{AQY} &= \frac{\text{number of reacted electrons}}{\text{number of incident photons}} \times 100\% \\ &= \frac{\text{number of evolved } \mathrm{H_2} \text{ molecules} \times 2}{\text{number of incident photons}} \times 100\% \end{aligned}$$

For wavelength-dependent AQY measurements of the photocatalytic HER, we analyzed the photoreacted gas in a quartz cell (10 mm) with a gas chromatograph (GC-2004, Shimadzu, with argon as a carrier gas). A 300-W xenon lamp was irradiated onto the sample through different wavelength filters for 4 h.

The AQYs obtained at several monochromic light wavelengths were consistent with the LSPR spectrum of the CdS/Cu_7S_4 HNCs, indicating that the present photocatalytic reaction was caused by LSPR excitation (Fig. 12.8). The estimated AQY at 1100 nm was 3.8%, which is considerably greater than the previously reported record. The damping of the AQYs from 1300 to 1500 nm is likely caused by strong absorption of light by water. Notably, we observed H_2 evolution at 0.023 and 0.022 μmol g^{-1} h^{-1} from the 1615–2280 and 2093–2547 nm regions (center power density: 6 mW cm^{-2}), respectively. The Cu_7S_4 NCs showed no catalytic activity even under visible light irradiation.

12.5 Mechanism of IR-Induced Photocatalytic H_2 Evolution

Finally, on the basis of the TA measurements, we summarize the mechanism of IR-induced photocatalytic HER in Fig. 12.9. First, excitation of the Cu_7S_4 LSPR band in CdS/Cu_7S_4 HNCs generated hot carriers in the Cu_7S_4 phase. Hot electrons generated in the Cu_7S_4 phases were then injected into the CdS phases through the

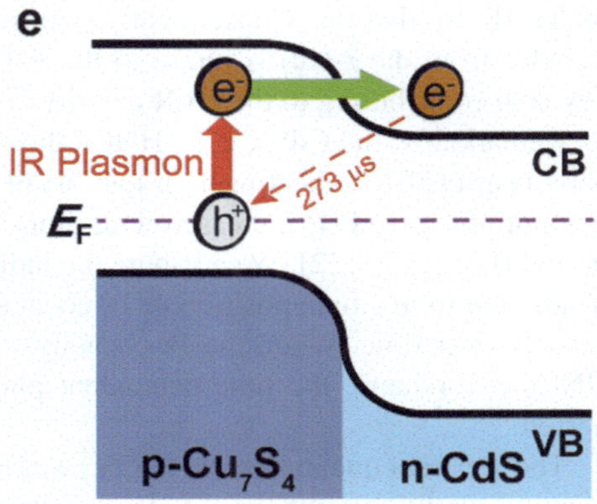

Fig. 12.9 Hot-electron injection at the plasmonic p–n heterojunction upon IR-light-plasmon excitation. Reprinted with permission from *J. Am. Chem. Soc.* **2019**, *141*, 2446–2450. Copyright 2019 American Chemical Society

p–n heterojunction within the fs-laser pulse. Hot electrons directly injected into the conduction band of CdS phases reacted with H_2O to generate H_2. Conversely, holes were consumed by sacrificial agents. Ultrafast charge recombination is a critical drawback for conventional plasmonic energy conversion. We believe that the key to achieving high AQY in the present system is the extraordinary long-lived charge separation. As mentioned above, hot-electron injection occurs at the p–n heterojunction and is unlike the behavior at a conventional Schottky-interface. At the p–n heterojunction, the p–n boundary creates an electric field, which promotes charge separation. Plasmonic hot-carrier transfer at the p–n heterojunction formed by p-type plasmonic Cu_7S_4 and n-type CdS breaks the limit of plasmonic energy conversion to realize highly efficient photocatalytic solar fuel generation.

12.6 Conclusion

In summary, we successfully synthesized CdS/Cu_7S_4 HNCs as a photocatalyst for H_2 evolution, with record-breaking performance under IR-light irradiation. The AQY of the photocatalytic H_2 evolution reached 3.8% at 1100 nm, which exceeds the highest performance of previously reported IR-responsive photocatalysts. TA measurements revealed that both efficient hot-electron injection and ultralong-lived charge separation (>273 μs) at the p–n heterojunction of the HNCs are key issues to enable extraordinarily high catalytic activity.

As ultrafast charge recombination is a major drawback of all plasmonic energy conversion systems, we anticipate that LSPR-induced carrier transfer at p–n heterojunction will remove the limitations of conventional plasmon-induced energy conversion. Our results pave the way for developing novel and efficient solar fuel generation systems based on untapped solar energy in IR region and might also enable the development of plasmonic sensors and detectors responsive to the IR-light.

Acknowledgements We appreciate useful collaborations with Dr. Z. Lian, Dr. J. J. M. Vequizo, Dr. C. S. K. Ranasinghe, Prof. A. Yamakata, Dr. T. Nagai, Dr. K. Kimoto, Prof. Y. Kobayashi, and Prof. N. Tamai. This work was supported by The Ministry of Education, Culture, Sports, Science, and Technology (MEXT)/Japan Society for the Promotion of Science (JSPS) Kakenhi as follows: for Scientific Research B (Grant No. 18H01827), for Scientific Research on Innovative Areas (Grant No. JP17H05257 (Photosynergetics). This work was also supported by NIMS microstructural characterization platform as a program of "Nanotechnology Platform" of MEXT, Japan.

References

1. Kubacka A, Fernández-García M, Colón G (2012) Advanced nanoarchitectures for solar photocatalytic applications. Chem Rev 112:1555–1614
2. Carr JA et al (2018) Shortwave infrared fluorescence imaging with the clinically approved near-infrared dye indocyanine green. Proc Nat Acad Sci 115:4465–4470
3. Cox A (1992) In: Bryce-Smith D, Gilbert A (eds) Photochemistry, vol. 23. The Royal Society of Chemistry, London, pp 471–478
4. Tahara H, Sakamoto M, Teranishi T, Kanemitsu Y (2018) Quantum coherence of multiple excitons governs absorption cross-sections of PbS/CdS core/shell nanocrystals. Nat Commun 9:3179
5. Lian Z et al (2018) Near infrared light induced plasmonic hot hole transfer at a nano-heterointerface. Nat Commun 9:2314
6. Sakamoto M et al (2019) Clear and transparent nanocrystals for infrared-responsive carrier transfer. Nat Commun 10:406
7. Kanehara M, Koike H, Yoshinaga T, Teranishi T (2009) Indium tin oxide nanoparticles with compositionally tunable surface plasmon resonance frequencies in the near-IR region. J Am Chem Soc 131:17736–17737
8. Furube A, Du L, Hara K, Katoh R, Tachiya M (2007) Ultrafast plasmon-induced electron transfer from gold nanodots into TiO_2 nanoparticles. J Am Chem Soc 129:14852–14853
9. Han C, Quan Q, Chen HM, Sun Y, Xu YJ (2017) Progressive design of plasmonic metal–semiconductor ensemble toward regulated charge flow and improved vis–NIR-driven solar-to-chemical conversion. Small 13:1602947
10. Zhang Z et al (2017) A Nonmetal plasmonic Z-scheme photocatalyst with UV- to NIR-driven photocatalytic protons reduction. Adv Mater 29:1606688
11. Zheng Z, Tachikawa T, Majima T (2014) Single-particle study of Pt-modified Au nanorods for plasmon-enhanced hydrogen generation in visible to near-infrared region. J Am Chem Soc 136:6870–6873
12. Luther JM, Jain PK, Ewers T, Alivisatos AP (2011) Localized surface plasmon resonances arising from free carriers in doped quantum dots. Nat Mater 10:361–366
13. Lian Z et al (2018) Plasmonic p–n junction for infrared light to chemical energy conversion. J Am Chem Soc 141:2446–2450
14. Yuan Q et al (2017) Noble-metal-free Janus-like structures by cation exchange for Z-scheme photocatalytic water splitting under broadband light irradiation. Angew Chem 56:4206–4210
15. Cui J et al (2015) Near-infrared plasmonic-enhanced solar energy harvest for highly efficient photocatalytic reactions. Nano Lett 15:6295–6301
16. Chen L, Sakamoto M, Sato R, Teranishi T (2015) Determination of a localized surface Plasmon resonance mode of Cu_7S_4 nanodisks by Plasmon coupling. Faraday Discuss 181:355–364
17. Choi Y, Beak M, Yong K (2014) Solar-driven hydrogen evolution using a $CuInS_2$/CdS/ZnO heterostructure nanowire array as an efficient photoanode. Nanoscale 6:8914–8918
18. Wu K, Chen J, McBride JR, Lian T (2015) Efficient hot-electron transfer by a plasmon-induced interfacial charge-transfer transition. Science 349:632–635

19. Lou Y et al (2003) Evaluation of the photoinduced electron relaxation dynamics of $Cu_{1.8}S$ quantum dots. Phys Chem Chem Phys 5:1091–1095
20. Lian Z et al (2018) Durian-shaped CdS@ZnSe core@mesoporous-shell nanoparticles for enhanced and sustainable photocatalytic hydrogen evolution. J Phys Chem Lett 9:2212–2217
21. Furube A, Hashimoto S (2017) Insight into plasmonic hot-electron transfer and plasmon molecular drive: new dimensions in energy conversion and nanofabrication. NPG Asia Mater 9:e454
22. Du P et al (2008) Bi- and terpyridyl platinum(II) chloro complexes: molecular catalysts for the photogeneration of hydrogen from water or simply precursors for colloidal platinum? J Am Chem Soc 130:5056–5058

Chapter 13
Controlling Optical Properties of Multinary Quantum Dots for Developing Novel Photoelectrochemical Reactions

Tsukasa Torimoto and Tatsuya Kameyama

Abstract We introduce solution-phase preparation of multinary QDs composed of less-toxic Ag-III-VI-based semiconductors and control of their physicochemical properties. The energy gaps (E_gs) of multinary QDs could be adjusted by the chemical composition as well as by the particle size. The photochemical properties, including photoluminescence (PL), photocatalysis, and photocurrent generation, of prepared QDs, are discussed in terms of the E_g, the non-stoichiometric chemical composition, and the particle morphology. The PL peak was controlled from visible to near-IR wavelength regions by varying the chemical composition of QDs, in which the peak width of Ag-In-Ga-S QDs was remarkably narrowed by the removal of deep defect levels via tuning of non-stoichiometry and the surface condition. A nonlinear photoresponse induced by hot hole transfer was observed by visible light irradiation to near-IR-light-responsive Zn-Ag-In-Te QDs. The findings will provide new insights into design and fabrication of novel QD-based devices.

Keywords Quantum dot · Multinary semiconductor · Photoluminescence · Photocatalyst · Photocurrent generation · Hot hole transfer

13.1 Introduction

Quantum dots (QDs) are semiconductor nanocrystals with sizes of less than ca. 10 nm that show the quantum size effect. Their physicochemical properties are intermediate between those of molecules and bulk materials and strongly depend on their particle size and shape. Many efforts have been devoted to the development of high-efficient light energy conversion systems consisting of QDs as components, such as photocatalysts and solar cells, because QDs have high absorption coefficients in the visible and near-IR wavelength regions and their E_gs are tunable depending on the nanocrystal size [1, 2]. Recently, unique photoresponses originating from the

T. Torimoto (✉) · T. Kameyama
Department of Materials Chemistry, Graduate School of Engineering, Nagoya University, Chikusa-Ku, Nagoya 464-8603, Japan
e-mail: torimoto@chembio.nagoya-u.ac.jp

H. Miyasaka et al. (eds.), *Photosynergetic Responses in Molecules and Molecular Aggregates*, https://doi.org/10.1007/978-981-15-5451-3_13

quantum confinement in QDs have been reported. For example, the relaxation times of hot carriers in highly excited states to band-edge states were remarkably increased by the discrete energy levels of QDs, because of the phonon bottleneck effect [3]. This increased the probability of hot carrier transfer [4, 5]. Furthermore, multiple exciton generation (MEG) in a single QD is a nonlinear phenomenon caused by the absorption of one photon having an energy of more than twice the E_g [6–8]. The MEG efficiency was reported to become much larger in QDs than in bulk materials [6]. Thus, although strategies to efficiently extract photogenerated carriers from QDs without energy loss have not been developed yet, these nonlinear phenomena have attracted much attention as a means for increasing the efficiency of solar light energy conversion.

So far, binary QDs, such as PbS, PbSe, CdS, and CdSe, have been utilized for developing novel photocatalysts and photovoltaics because synthetic methods for obtaining high-quality QDs have been established. Although the use of these conventional binary QDs has enabled fabrication of systems having relatively high efficiencies, their use in practical applications has been severely restricted due to the large content of highly toxic elements, such as Cd and Pb. On the other hand, group I-III-VI-based multinary semiconductors consisting of elements with less toxicity, such as $AgInS_2$, $CuInS_2$, and their solid solutions with ZnS, have been more and more intensively investigated to develop thin-film solar cells or visible light-driven photocatalysts. Since the colloidal syntheses were reported to produce highly photoluminescent QDs of these materials, these multinary QDs have also been receiving much attention as alternatives to highly toxic binary ones [9–12]. In this chapter, we introduce solution-phase synthetic methods to obtain high-quality QDs consisting of Ag-III-VI semiconductors and their solid solutions with Zn-VI semiconductors. The prepared QDs had optical properties that were tunable by changing the chemical composition and morphology of particles. Furthermore, their photoelectrochemical responses were investigated with irradiation of visible light.

13.2 Tunable Optical Properties of Solid Solution QDs

13.2.1 Photoluminescence Properties of $AgInS_2$-Based QDs

High-quality multinary QDs could be colloidally synthesized by methods similar to these used for the preparation of binary QDs, though the difference in reactivities of two or more metal salts used as precursors often resulted in non-stoichiometry of QDs. We prepared highly photoluminescent multinary QDs composed of a ZnS-$AgInS_2$ solid solution, $(AgIn)_xZn_{2(1\text{-}x)}S_2$ (ZAIS), by a heating up method [13]. The reaction of corresponding metal acetates with thiourea was carried out at 250 °C in oleylamine (OLA) containing 1-decanethiol (DDT), in which the ratio of Ag^+: In^{3+}: Zn^{2+} was fixed to x: x: $2(1-x)$. A decrease of the x value in the preparation resulted in an increase of the Zn fraction in the obtained spherical ZAIS QDs. Furthermore, the increase in the DDT content decreased the particle size: The average size (d_{ave}) of ZAIS QDs prepared with $x = 0.5$ was increased from 3.9 to 8.5 nm with a reduction in the DDT content from 8.7 to 0 vol.%, and the particle composition was almost constant at Ag:In:Zn = 0.25:0.30:0.45, being slightly In-rich and Zn-deficient. Thus, the present method enabled independent control of the size and chemical composition of spherical ZAIS QDs.

Figure 13.1 shows the absorption and PL spectra of ZAIS QDs with different chemical compositions and sizes. ZAIS QDs exhibited absorption spectra in a wide visible light region, the onset wavelength being blue-shifted with a decrease in d_{ave} (Fig. 13.1a) or with a decrease in the x value (Fig. 13.1c). These results indicated that the E_g of ZAIS QDs could be two-dimensionally controlled by changes in the particle size and chemical composition. A broad PL peak, the full width at half-maximum intensity (FWHM) of which was ca. 100 ~ 200 nm, was observed in the PL spectra shown Fig. 13.1b and d for each kind of ZAIS QDs. The broadness of the peak width arose from donor–acceptor pair (DAP) radiative recombination of trapped carriers, not from band-to-band recombination of charge carriers, that is, band-edge recombination. The PL peak wavelength was also shifted to a shorter wavelength with an increase in the E_g, indicating that the PL color was widely tunable from deep red to green. The highest PL quantum yield (QY) of 79% for 630-nm emission was obtained with ZAIS QDs of 5.7 nm in d_{ave} prepared with $x = 0.5$.

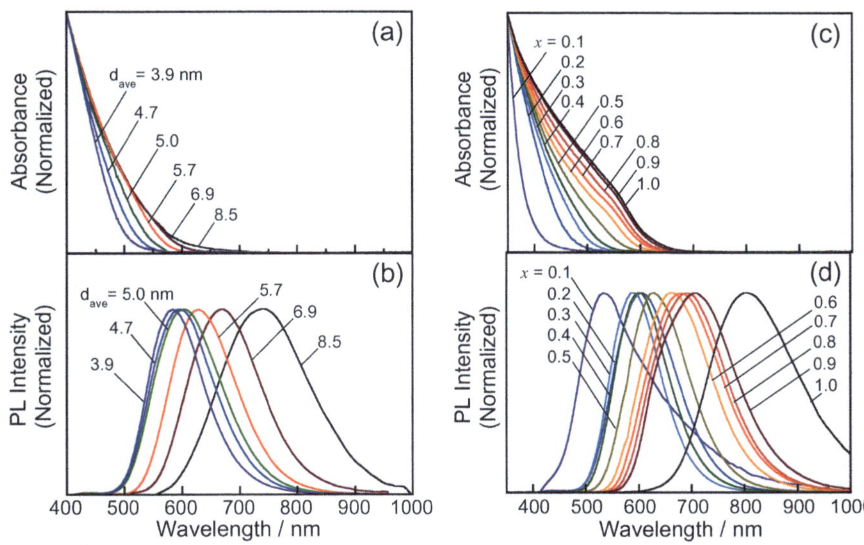

Fig. 13.1 Absorption spectra (**a**, **c**) and PL spectra (**b**, **d**) of spherical ZAIS QDs in chloroform. Samples are (**a**, **b**) ZAIS QDs of $x = 0.5$ with various sizes and (**c**, **d**) ZAIS QDs of ca. 5.5 nm in d_{ave} with different x values. Reprinted with permission from ref. [13]

As aforementioned, ZAIS QDs exhibited a very broad PL peak owing to energy levels of defect sites in the bandgap. Although the PL QY of ZAIS QDs was considerably high, the lack of monochromaticity of PL was recognized as a major problem in applications to biological imaging systems and LCD devices. Therefore, in order to narrow the PL peak width, we investigated the influence of defect sites formed in $AgInS_2$ (AIS) QDs on their PL properties. AIS QDs with different Ag fractions, that is, with a non-stoichiometric composition, were prepared by changing the Ag/(Ag + In) ratio in precursors [14]. The obtained AIS QDs had similar d_{ave} values of 3.6 ~ 4.3 nm. With a decrease in the Ag/(Ag + In) ratio from 0.60 to 0.30, the composition of AIS QDs changed from stoichiometric to Ag-deficient, indicating the formation of a large amount of defect sites in QDs, such as Ag^+ vacancies and antisites of In^{3+} on Ag^+ sites. Although the absorption spectra shown in Fig. 13.2a were similar regardless of the Ag/(Ag + In) ratio, the PL spectra significantly varied depending on the Ag/(Ag + In) ratio in the preparation (Fig. 13.2b): The broad defect-site PL peak was blue-shifted with a decrease in the Ag/(Ag + In) ratio, and then a band-edge

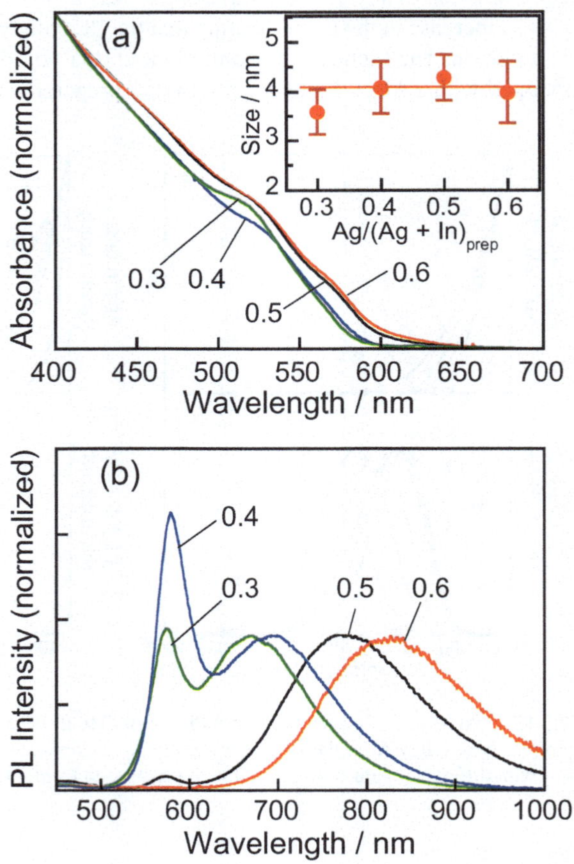

Fig. 13.2 Absorption spectra (**a**) and photoluminescence spectra (**b**) of AIS QDs having different Ag fractions. Each number in the figure represents the Ag/(Ag + In) ratio in the preparation. The inset in panel a shows the average size (d_{ave}) (solid circles) with the standard deviation (error bars) as a function of Ag/(Ag + In) in the preparation. Reprinted with permission from ref. [14]

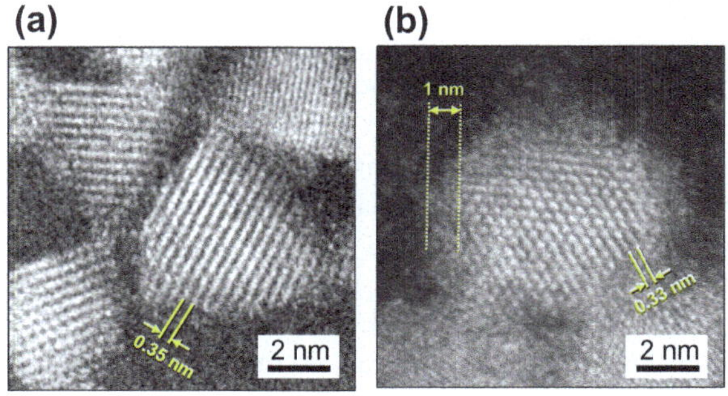

Fig. 13.3 High-resolution HAADF-STEM images of AIS (**a**) and AIS@GaS_x QDs (**b**). Reprinted with permission from Ref. [14]

PL peak appeared on the shorter wavelength side of the broad peak for Ag-deficient AIS QDs prepared with Ag/(Ag + In) of less than 0.40. The band-edge peak had a much smaller width than that of the defect-site broad peak, in which the former peak intensity was 1.8-times larger than that of the latter peak for AIS QDs with Ag/(Ag + In) = 0.40. It has been theoretically reported for $CuInSe_2$ by Zhang et al. [15] that pairing two Cu-vacancies (V_{Cu}) with In substituting for Cu (In_{Cu}) raised the deep levels of In_{Cu} to the conduction-band minimum, resulting in the defect sites not being able to act as harmful carrier traps. This seemed to be true of the present case of AIS QDs.

QDs have a large surface-to-volume ratio, and a considerable amount of defect sites are possibly formed on their surface. Among the various strategies reported for eliminating surface defect sites and enhancing PL intensity, surface coating with wide gap semiconductors, such as ZnS, is one of the most effective methods, with resulting QDs being a core-shell structure with type-I heterojunctions. However, since ZnS easily formed a solid solution with $AgInS_2$ with heat treatment, as shown, for example, in Fig. 13.1, we chose gallium sulfide as a shell material [16]. Heat treatment of AIS QDs with $Ga(acac)_3$ and thiourea in OLA at 300 °C produced core-shell-structured particles [14]. Figure 13.3 shows HAADF-STEM images of AIS QDs used as a core and GaS_x-coated AIS QDs (AIS@GaS_x). AIS QDs were polygonal nanocrystals having clear lattice fringes with spacing of 0.35 nm, being in good agreement with the (012) plane of an orthorhombic $AgInS_2$ crystal structure. In contrast, surface coating with a GaS_x layer slightly changed the core shape to a rounded one and then produced an amorphous thin shell with a thickness of ca. 1 nm on each core. A clear lattice fringe assignable to the (002) plane of an orthorhombic $AgInS_2$ structure was also observed in the core (Fig. 13.3b), suggesting that the diffusion of Ga^{3+} into the AIS core was negligibly small in the present conditions. The obtained AIS@GaS_x QDs predominantly exhibited a narrow band-edge PL peak. Thus, it was found that the remaining broad PL peak observed for Ag-deficient AIS QDs as shown in Fig. 13.2b originated from the defect sites on the surface of QDs used as a core, most of which could be removed by GaS_x coating.

The position of the band-edge emission peak can be controlled by changing the E_g of QDs. It is well known that solid solutions are formed between different kinds of I-III-VI semiconductors, giving an E_g that depends on their chemical composition. We tried to modify the E_g of AIS QDs by Ga doping, because the bulk E_g of $AgGaS_2$, 2.5 eV, was larger than that of $AgInS_2$, 1.8 eV. QDs of Ag-In-Ga-S (AIGS) were synthesized at 300 °C by reacting metal precursors of Ag(OAc), $In(acac)_3$, and $Ga(acac)_3$ with elemental sulfur as an S^{2-} precursor. Thus-obtained AIGS QDs were spherical or polygonal, the d_{ave} of which was in the range of 2.9 ~ 4.5 nm. The GaS_x coating on AIGS QDs enlarged the particle size by 0.4 ~ 1 nm, indicating the formation of core-shell-structured AIGS@GaS_x QDs. The absorption spectra of AIGS@GaS_x QDs (Fig. 13.4a) were blue-shifted with a decrease in the In/(In + Ga) ratio, that is, with an increase in the Ga fraction in the AIGS cores. This indicated that the E_g of AIGS cores was tunable between 2.1 and 2.6 eV by changing the Ga/In ratio. The as-prepared AIGS QDs exhibited peaks of both broad defect-site PL and narrow band-edge PL. However, the surface coating of AIGS cores with GaS_x shell layers predominantly decreased the intensity of defect-site PL peaks and then sharp

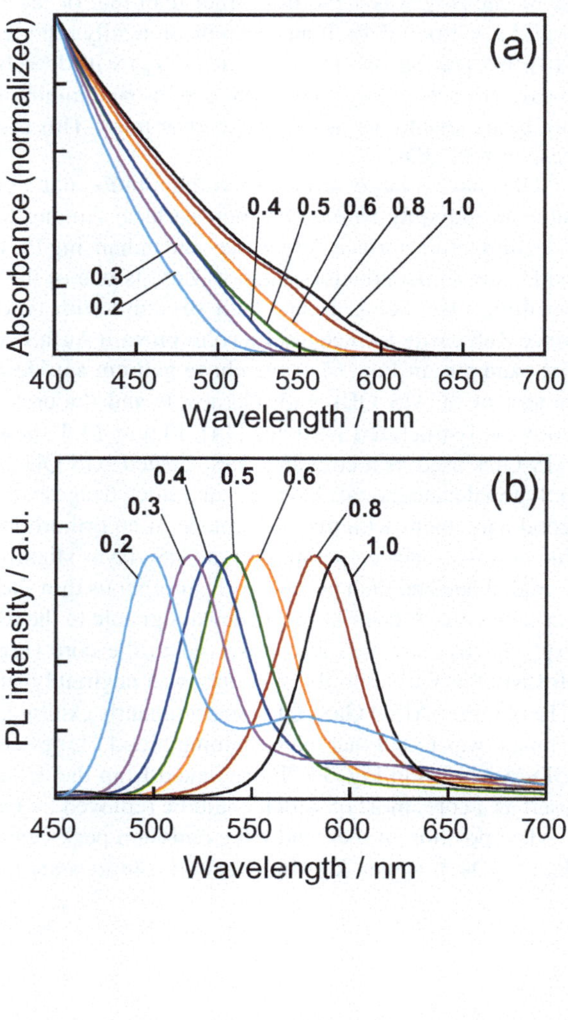

Fig. 13.4 Absorption spectra (**a**) and PL spectra (b) of AIGS QDs surface-coated with a GaS_x shell. The number in each panel represents the In/(In + Ga) ratio in the preparation. Reprinted with permission from Ref. [14]

band-edge peaks clearly appeared as shown in Fig. 13.4b. The FWHMs of band-edge emission peaks were ca. 40 nm (180 meV), being much narrower than those of defect-site emission peaks, 100 ~ 200 nm (Figs. 13.1 and 13.2). The position of the PL peak was also blue-shifted from 590 to 500 nm with an increase in the E_g of QDs. The PL QYs of AIGS@GaS_x QDs were relatively high, the optimal value being 28% for 530-nm band-edge emission with AIGS@GaS_x QDs prepared with In/(In + Ga) = 0.40.

PL peaks of $AgInS_2$-based QDs were narrowed by removing defect sites causing broad PL peaks both via precise tuning of the reaction conditions and via surface coating with a GaS_x shell. However, the thus-obtained band-edge emission peak still had a relatively large width for the application to optical devices. Single-particle spectroscopy enables clarification of the origins of PL peak broadening. We found that the band-edge emission peaks from single AIS QDs were much narrower than those observed in the ensemble measurement, in which the peak widths (FWHM) were ca. 20 nm (73 meV), while the peak position fluctuated greatly from particle to particle between ca. 600 nm (2.0 eV) and ca. 550 nm(2.3 eV). These results indicated that the ensemble PL peaks were inhomogeneously broadened, probably due to the size and composition variations of individual QDs. Thus, we can conclude that the band-edge PL peak in the ensemble measurement of $AgInS_2$-based QDs will be potentially further narrowed by reducing their inhomogeneity.

13.2.2 Controllable PL Peak of Zn-Ag-in-Te QDs in the Near-IR Region

Tunable optical properties in the near-IR wavelength region are useful for developing novel QD-based photovoltaic systems. We successfully prepared near-IR-light-responsive QDs of high quality by making a solid solution between ZnTe and $AgInTe_2$, $(AgIn)_xZn_{2(1-x)}Te_2$ (ZAITe) [17], in place of multinary metal sulfides. ZAITe has a widely tunable E_g because bulk E_gs of $AgInTe_2$ and ZnTe are 1.0 and 2.3 eV, respectively. The metal precursors of corresponding metal acetates were reacted with trioctylphosphine telluride in DDT at 180 °C for 180 min with the ZAITe composition being modified by changing the metal precursor ratio, Ag:In:Zn = x:x:2(1−x) (0.25 < x<1). The thus-obtained ZAITe QDs had a rod-like structure as shown in Fig. 13.5a and 13.5b, the dimensions of which were ca. 16 ~ 18 nm in length and ca. 4 ~ 5 nm in width, regardless of the x values. Each QD was a single crystal grown along the c-axis of a wurtzite structure without a grain boundary.

As shown in Fig. 13.5c, the ZAITe QDs had clear exciton peaks in the absorption spectra, the peak position of which was in the near-IR wavelength region and was shifted from 965 to 710 nm with an increase in the Zn fraction, that is, with a decrease in the x value. The E_g of ZAITe QDs, determined from the absorption onset, was enlarged from 1.2 to 1.6 eV by increasing the Zn content. Furthermore, band-edge emission peaks appeared at around the corresponding absorption onsets. The PL

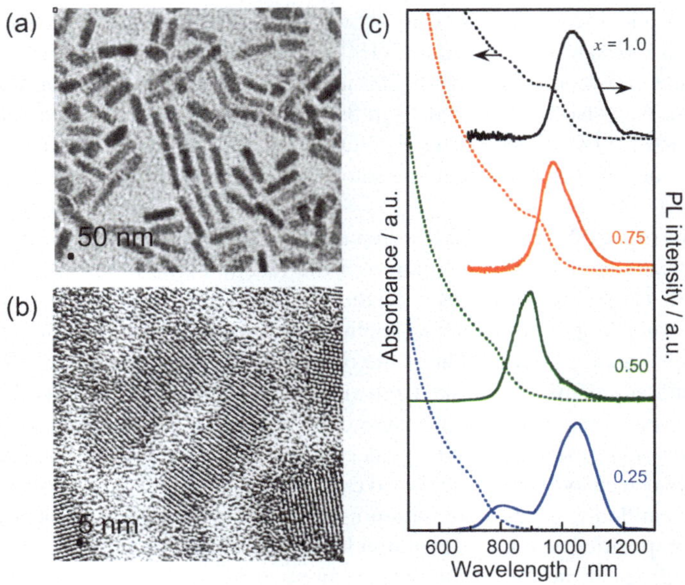

Fig. 13.5 (**a**, **b**) TEM images of ZAITe QDs prepared with $x = 0.5$. Panel b is a high-resolution image of panel a. (**c**) Absorption and PL spectra of ZAITe QDs prepared with different x values. Reprinted with permission from ref. [17]

peak width (FWHM) was almost constant at 140 ~ 170 meV, being comparable with those of CdTe and PbS QDs, ca. 50 ~ 100 meV, except for the case of QDs with $x =$ 0.25 showing a broad defect-site PL in addition to the narrow band-edge PL.

Consequently, we clearly demonstrated fascinating optical properties of less-toxic Ag-III-VI-based multinary QDs, that is, Zn-Ag-In-S, Ag-In-Ga-S, and Zn-Ag-In-Te QDs, that were controllable by their chemical composition, being different from conventional binary QDs.

13.3 Visible Light-Driven Photocatalytic H_2 Evolution with ZAIS QDs

The electronic energy structure of these multinary QDs can be designed by tuning the chemical composition as well as by changing their size. Figure 13.6 shows the potentials of the conduction band edge (E_{CB}) and the valence band edge (E_{VB}) of multinary QDs of ZnX-AgInX_2 (ZAIX, X = S, Se, and Te) solid solution QDs, determined by photoemission yield spectroscopy in air. The E_{CB} levels of individual kinds of solid solution were shifted more negatively with an increase in the E_g due to the increase of the Zn fraction. On the other hand, the E_{VB} levels of ZAIS and ZAITe were shifted more positively with an increase in the E_g, but the E_{VB} of ZAISe

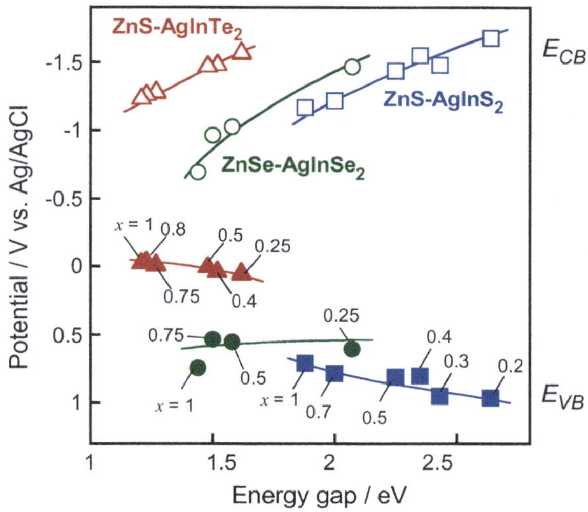

Fig. 13.6 Potentials of E_{CB} (open symbols) and E_{VB} (solid symbols) of QDs composed of ZnX-AgInX$_2$ (X = S (squares), Se (circles), or Te (triangles)) solid solution. The numbers in the figure indicate the x values in the formula of $(AgIn)_xZn_{2(1-x)}X_2$. The data were obtained from refs. [13, 18, 19]

was constant or slightly negatively shifted. Furthermore, the E_{VB} level of ZAIX was positively shifted in the order of X = Te, Se, and S when a comparison was made at a constant x value.

Recently, QDs have been attractive materials for photocatalysts because of their tunable optical properties in a wide wavelength range of the solar spectrum. However, there have been only a few reports on photocatalytic activities of multinary I-III-VI-based QDs [20, 21]. As shown in Fig. 13.6, the E_{CB} levels of ZAIS QDs were sufficiently negative to reduce H^+ in an aqueous solution to H_2. Thus, we investigated photocatalytic H_2 evolution with the use of ZAIS QDs in the presence of S^{2-} as a sacrificial agent [13]. Figure 13.7a shows the time course of H_2 evolution of spherical ZAIS QDs with different sizes, which were prepared with $x = 0.5$. With elapse of light irradiation time, the amount of H_2 evolved monotonously increased, regardless of the particle size. From the initial slope of the increase in the amount of H_2 evolved, the H_2 evolution rate ($R(H_2)$) was determined, as shown in Fig. 13.7b. A volcano-type dependence was observed between $R(H_2)$ and d_{ave} of ZAIS QDs with each x value: The photocatalytic activity tended to increase with a decrease in the d_{ave}, and then the optimal value of $R(H_2)$ for each x value was obtained at a similar d_{ave}, ca. 4.2 ~ 5.5 nm. Furthermore, the change in chemical composition of ZAIS greatly affected the $R(H_2)$ value when a comparison was made at a constant d_{ave}. We could reasonably explain these tendencies from the points of negative shift of E_{CB} and enlargement of E_g. The ZAIS QDs had a larger E_g with a decrease in the d_{ave} or with a decrease in the x value, resulting in a negative shift of E_{CB}, which caused an increase in the driving force for H^+ reduction. However, excessive enlargement of the E_g remarkably decreased the number of photons absorbed by ZAIS QDs, deteriorating the $R(H_2)$.

The shape of ZAIS QDs also influenced their photocatalytic activity. We successfully prepared ZAIS QDs having anisotropic shapes, rod-like shapes, and rice seed-like shapes via solution-phase synthesis with two-step heat treatment [22]. The $R(H_2)$

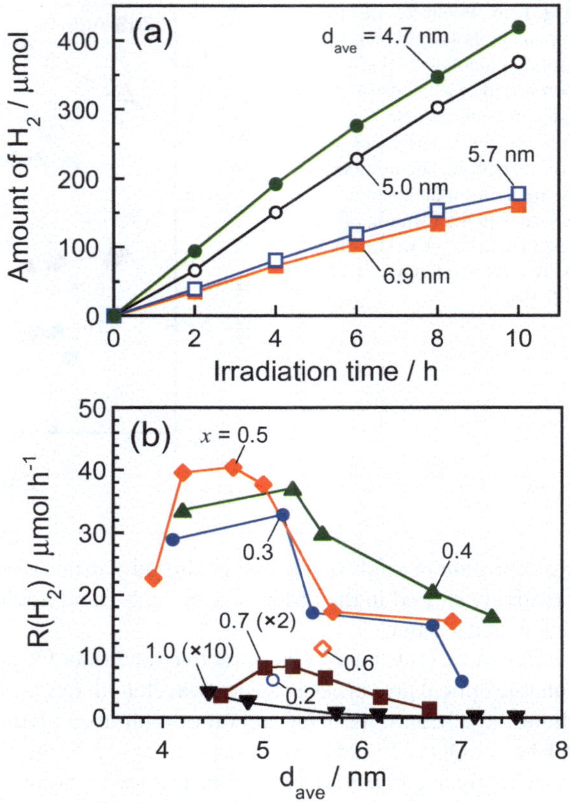

Fig. 13.7 Time courses of H_2 evolution by irradiation of spherical ZAIS QDs with different d_{ave} values. The QDs were prepared with the condition of $x = 0.5$. (**b**) Relationship between $R(H_2)$ and d_{ave} of ZAIS QDs. The x values used in the preparation are shown in the panel. Reprinted with permission from Ref. [13]

value obtained with ZAIS QDs of similar x values of 0.35 ~ 0.45 was enlarged in the order of rice (size: ca. 9 × ca. 16 nm), sphere (diameter: ca. 5.5 nm), and rod (size: 4.6 × 27 nm) QDs. The apparent quantum yield for H_2 evolution was determined to be 5.9% with rod-shaped ZAIS QDs. Furthermore, the formation of a type-II heterojunction in a particle was advantageous for enhancing the photocatalytic activity of ZAIS QDs, in which photogenerated electrons could be effectively separated from holes. Heat treatment of rod-shaped ZAIS QDs in the presence of precursors enabled epitaxial growth of ellipsoidal ZAIS domains on both of their termini [23]. The resulting QDs were dumbbell-shaped ones of two ellipsoidal nanocrystals (ca. 4 ~ 6 nm in width × 7 ~ 11 nm in length) connected by a nanorod (ca. 4 nm in width × 16 ~ 23 nm in length). Since the Zn fraction in the ellipsoidal parts was smaller than that in the rod part, a type-II heterojunction was formed between the rod and tip parts, as expected from the energy levels shown in Fig. 13.6. Photogenerated electrons could be trapped in the ellipsoidal tip parts, while holes were delocalized over the whole particle. Dumbbell-shaped ZAIS QDs exhibited much higher photocatalytic activity for H_2 evolution than the photocatalytic activities of original rod-like QDs or free ellipsoidal ZAIS QDs with similar composition and size. Thus, it was clarified that precise control of the chemical composition, particle size, shape, and heterojunction was necessary to design efficient QD photocatalysts composed of multinary semiconductors.

13.4 Nonlinear Photoelectrochemical Responses of ZAITe QD Film Electrodes

As mentioned in the introductory part, semiconductor QDs have been reported to exhibit higher probabilities to induce nonlinear photoresponses, such as hot carrier transfer and MEG by the absorption of a high-energy single photon, due to the increased relaxation times of hot carriers in highly excited states. These can potentially enlarge the efficiencies of light energy conversion systems. Here, we clarified the nonlinear behavior of photocurrent generation for photoelectrodes composed of ZAITe QDs, in which hot carriers were generated with visible light irradiation because of their small E_g in the near-IR wavelength region [19].

Rod-shaped ZAITe QDs (Fig. 13.5) were immobilized on ITO substrates by an alternative layer-by-layer deposition technique using 1,2-ethanedithiol as the crosslinking agent. Figure 13.8 shows AFM images of ZAITe QD films prepared with $x = 0.75$. The substrate was not completely covered with a QD film prepared with one deposition cycle, the thickness of which was ca. 10 nm. In contrast, the five deposition cycles enabled the formation of ZAITe QD multilayer films that completely covered the electrode surface.

Photoelectrochemical measurements of ZAITe QD monoparticle films were carried out in an aqueous solution containing $Eu(NO_3)_3$ as an electron scavenger. A cathodic photocurrent was observed with irradiation as shown in Fig. 13.9a, being similar to bulk p-type semiconductor photoelectrodes. However, the photocurrent onset potential of the ZAITe QD film with $x = 0.75$ was remarkably varied by the excitation photon energy of monochromatic light and was more positive than the E_{VB} of QDs used, -0.05 V vs. Ag/AgCl (Fig. 13.6), the degree being enlarged with an

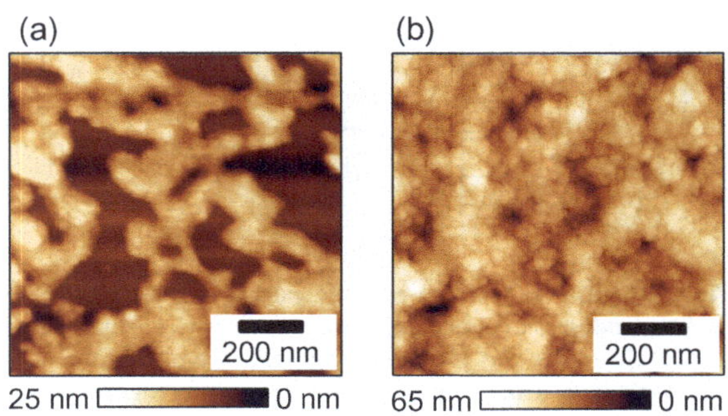

Fig. 13.8 AFM images of ZAITe QD films prepared on substrates with 1 deposition cycle (**a**) and 5 deposition cycles (**b**). The ZAITe QDs used were prepared with $x = 0.75$. Reprinted with permission from ref. [19]

increase in the photon energy. Such behavior was not observed with bulk semiconductors. An action spectrum of the photocurrent with a ZAITe QD monoparticle film revealed that the onset photon energy of photocurrent generation agreed with that of the absorption spectrum of QDs used, but the spectrum outline was considerably deviated from that of the absorption spectrum, especially at photon energies larger than ca. 2.5 eV, where the incident photon-to-current efficiencies (IPCE) were much larger than those obtained with irradiation of lower-energy photons.

One of the possible mechanisms for the observed nonlinear photoresponses seemed to be the MEG in a ZAITe QD. To explore this possibility, we measured ultrafast carrier dynamics in ZAITe QDs. It was found that the absorption of a single 3.1-eV photon did not produce a biexciton in a ZAITe QD, though the ratio of irradiated photon energy to the E_g of QDs, $h\nu/E_g$, was more than 2.4. Thus, we could conclude that the MEG did not occur in a ZAITe QD and thus was not the origin of nonlinear photoresponses.

Figure 13.9c shows the relationship between the onset potential of the cathodic photocurrent and the energy of photons irradiated to ZAITe QD monoparticle film electrodes. The onset potentials were comparable to the E_{VB} values of ZAITe QDs in the case of photon energy lower than ca. 2.1 eV, regardless of the composition of ZAITe QDs used. However, the onset potential was more positively shifted with irradiation of photons with energy higher than ca. 2.4 eV, indicating that the oxidation

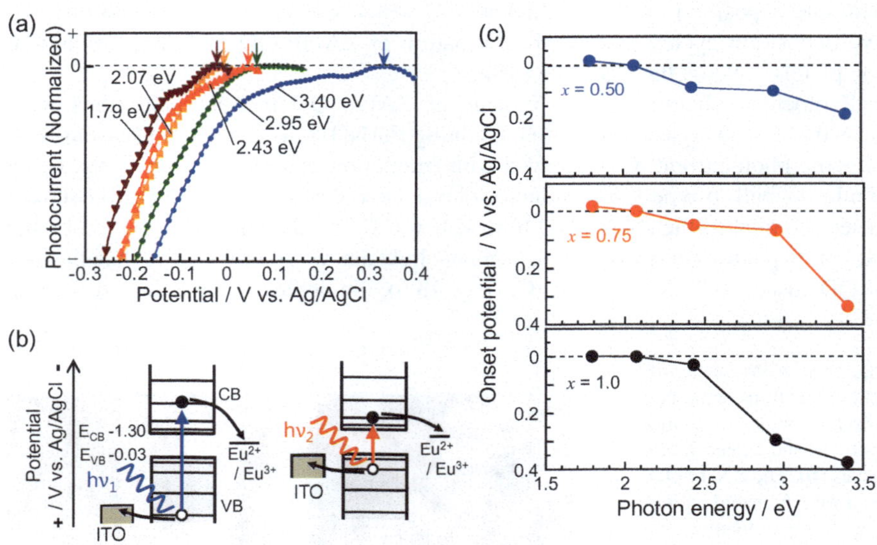

Fig. 13.9 (**a**) Photocurrent–potential curves of a ZAITe QD monoparticle film prepared with $x = 0.75$ depending on the excitation photon energy. The number of deposition cycles of ZAITe QDs on ITO electrodes was 1. (**c**) Schematic illustration of hot hole injection from ZAITe QDs into ITO electrodes with irradiation of photons with different energies ($h\nu_1 > h\nu_2$). (**c**) Photocurrent onset potential of ZAITe QD monoparticle film electrodes as a function of excitation photon energy. The x value used is indicated in each panel. Reprinted with permission from Ref. [19]

power of ZAITe QDs was variable depending on the irradiated photon energy. This behavior could be reasonably explained by the hot hole transfer from ZAITe QDs into ITO electrodes as schematically illustrated in Fig. 13.9b, the probability of which was increased for hot holes at more highly excited states.

The difference between the onset potential and the E_{VB} was at most ca. 0.37 V as shown in Fig. 13.9c, being much smaller than that between the E_g of ZAITe QDs used and the excitation photon energy. This indicated that a large part of the excess energy of hot holes was dissipated as heat before the hole transfer to collecting electrodes. Furthermore, the onset potential shift with irradiation of 3.40-eV photons became larger with a decrease in the E_g of ZAITe QDs, suggesting that ZAITe QDs having a lower E_{CB} (Fig. 13.6) could produce hot holes at more highly excited states. It should be noted that a shift of the photocurrent onset potential was not observed for ZAITe QD multilayer films prepared with 5 deposition cycles, regardless of the particle composition: The onset potential agreed well with the E_{VB} of ZAITe QDs, indicating that hot holes were relaxed to the valence band bottom during the transfer between QDs in multilayer films.

13.5 Conclusion

We described solution-phase preparation of multinary QDs composed of less-toxic Ag-III-VI-based semiconductors, enabling precise control of their chemical composition, size, shape, and heterostructure. Multinary QDs had E_gs that are tunable by both the chemical composition and particle size. Controlling photoluminescence properties of multinary QDs by the chemical composition is a significant advantage in comparison with conventional binary QDs. A band-edge PL peak was successfully observed by tuning the non-stoichiometric composition of Ag-In-Ga-S QD cores and by surface coating with a GaS_x shell layer, the peak width of which was much smaller than that of a defect-site emission peak. Irradiation to Zn-Ag-In-S QDs enabled photocatalytic H_2 evolution, the activity being modulated by the particle composition and morphology. A nonlinear photoelectrochemical response was observed by visible light irradiation to near-IR-light-responsive QDs composed of a Zn-Ag-In-Te solid solution, in which a cathodic photocurrent was generated at a potential more positive than the E_{VB} of QDs. Further precise control of the photochemical properties will be achieved by reducing the inhomogeneity in both the chemical composition and the particle size of individual multinary QDs. Our findings and the underlying mechanism will be essential for the development of novel QD-based devices such as bio-imaging devices, luminescent devices, photocatalysts, and solar cells.

Acknowledgements The present work was supported by JSPS KAKENHI Grant Numbers JP15H01082 and JP17H05254 in Scientific Research on Innovative Areas "Photosynergetics" and JP16H06507 in Scientific Research on Innovative Areas "Nano-Material Optical-Manipulation."

References

1. Kovalenko MV, Manna L, Cabot A, Hens Z, Talapin DV, Kagan CR, Klimov VI, Rogach AL, Reiss P, Milliron DJ, Guyot-Sionnnest P, Konstantatos G, Parak WJ, Hyeon T, Korgel BA, Murray CB, Heiss W (2015) Prospects of nanoscience with nanocrystals. ACS Nano 9:1012–1057. https://doi.org/10.1021/nn506223h
2. Pietryga JM, Park YS, Lim JH, Fidler AF, Bae WK, Brovelli S, Klimov VI (2016) Spectroscopic and device aspects of nanocrystal quantum dots. Chem Rev 116:10513–10622. https://doi.org/10.1021/acs.chemrev.6b00169
3. Klimov VI, Mikhailovsky AA, McBranch DW, Leatherdale CA, Bawendi MG (2000) Mechanisms for intraband energy relaxation in semiconductor quantum dots: the role of electron-hole interactions. Phys Rev B 61:13349–13352. https://doi.org/10.1103/PhysRevB.61.R13349
4. Tisdale WA, Williams KJ, Timp BA, Norris DJ, Aydil ES, Zhu XY (2010) Hot-electron transfer from semiconductor nanocrystals. Science 328:1543–1547. https://doi.org/10.1126/science.1185509
5. Wang YF, Wang HY, Li ZS, Zhao J, Wang L, Chen QD, Wang WQ, Sun HB (2014) Electron extraction dynamics in CdSe and CdSe/CdS/ZnS quantum dots adsorbed with methyl viologen. J Phys Chem C 118:17240–17246. https://doi.org/10.1021/jp5024789
6. Nozik AJ, Beard MC, Luther JM, Law M, Ellingson RJ, Johnson JC (2010) Semiconductor quantum dots and quantum dot arrays and applications of multiple exciton generation to third-generation photovoltaic solar cells. Chem Rev 110:6873–6890. https://doi.org/10.1021/cr900289f
7. Semonin OE, Luther JM, Choi S, Chen HY, Gao JB, Nozik AJ, Beard MC (2011) Peak external photocurrent quantum efficiency exceeding 100% via MEG in a quantum dot solar cell. Science 334:1530–1533. https://doi.org/10.1126/science.1209845
8. Hanna MC, Beard MC, Nozik AJ (2012) Effect of solar concentration on the thermodynamic power conversion efficiency of quantum-dot solar cells exhibiting multiple exciton generation. J Phys Chem Lett 3:2857–2862. https://doi.org/10.1021/jz301077e
9. Torimoto T, Kameyama T, Kuwabata S (2014) Photofunctional materials fabricated with chalcopyrite-type semiconductor nanoparticlescomposed of $AgInS_2$ and its solid solutions. J Phys Chem Lett 5:336–347. https://doi.org/10.1021/jz402378x
10. Torimoto T (2017) Nanostructure engineering of size-quantized semiconductor particles for photoelectrochemical applications. Electrochemistry 85:534–542. https://doi.org/10.5796/electrochemistry.85.534
11. Xu GX, Zeng SW, Zhang BT, Swihart MT, Yong KT, Prasad PN (2016) New generation cadmium-free quantum dots for biophotonics and nanomedicine. Chem Rev 116:12234–12327. https://doi.org/10.1021/acs.chemrev.6b00290
12. Stroyuk O, Raevskaya A, Gaponik N (2018) Solar light harvesting with multinary metal chalcogenide nanocrystals. Chem Soc Rev 47:5354–5422. https://doi.org/10.1039/c8cs00029h
13. Kameyama T, Takahashi T, Machida T, Kamiya Y, Yamamoto T, Kuwabata S, Torimoto T (2015) Controlling the electronic energy structure of $ZnS\text{-}AgInS_2$ solid solution nanocrystals for photoluminescence and photocatalytic hydrogen evolution. J Phys Chem C 119:24740–24749. https://doi.org/10.1021/acs.jpcc.5b07994
14. Kameyama T, Kishi M, Miyamae C, Sharma DK, Hirata S, Yamamoto T, Uematsu T, Vacha M, Kuwabata S, Torimoto T (2018) Wavelength-tunable band-edge photoluminescence of nonstoichiometric Ag-In-S nanoparticles via Ga^{3+} doping. ACS Appl Mater Interfaces 10:42844–42855. https://doi.org/10.1021/acsami.8b15222
15. Zhang SB, Wei SH, Zunger A, Katayama-Yoshida H (1998) Defect physics of the $CuInSe_2$ chalcopyrite semiconductor. Phys Rev B 57:9642–9656. https://doi.org/10.1103/PhysRevB.57.9642
16. Uematsu T, Wajima K, Sharma DK, Hirata S, Yamamoto T, Kameyama T, Vacha M, Torimoto T, Kuwabata S (2018) Narrow band-edge photoluminescence from $AgInS_2$ semiconductor nanoparticles by the formation of amorphous III-VI semiconductor shells. NPG Asia Mater 10:713–726. https://doi.org/10.1038/s41427-018-0067-9

17. Kameyama T, Sugiura K, Ishigami Y, Yamamoto T, Kuwabata S, Okuhata T, Tamai N, Torimoto T (2018) Rod-shaped Zn-Ag-In-Te nanocrystals with wavelength-tunable band-edge photoluminescence in the near-IR region. J Mater Chem C 6:2034–2042. https://doi.org/10.1039/c7tc05624a
18. Kameyama T, Douke Y, Shibakawa H, Kawaraya M, Segawa H, Kuwabata S, Torimoto T (2014) Widely controllable electronic energy structure of ZnSe-AgInSe$_2$ solid solution nanocrystals for quantum-dot-sensitized solar cells. J Phys Chem C 118:29517–29524. https://doi.org/10.1021/jp508769f
19. Kameyama T, Sugiura K, Kuwabata S, Okuhata T, Tamai N, Torimoto T (2019) Enhanced photoelectrochemical properties of Zn-Ag-In-Te nanocrystals with high energy photon excitation. Chem Nano Mat 5:1028–1035. https://doi.org/10.1002/cnma.201900241
20. Regulacio MD, Han MY (2016) Multinary I-III-VI$_2$ and I$_2$-II-IV-VI$_4$ semiconductor nanostructures for photocatalytic applications. Acc Chem Res 49:511–519. https://doi.org/10.1021/acs.accounts.5b00535
21. Coughlan C, Ibanez M, Dobrozhan O, Singh A, Cabot A, Ryan KM (2017) Compound copper chalcogenide nanocrystals. Chem Rev 117:5865–6109. https://doi.org/10.1021/acs.chemrev.6b00376
22. Torimoto T, Kamiya Y, Kameyama T, Nishi H, Uematsu T, Kuwabata S, Shibayama T (2016) Controlling shape anisotropy of ZnS-AgInS$_2$ solid solution nanoparticles for improving photocatalytic activity. ACS Appl Mater Interfaces 8:27151–27161. https://doi.org/10.1021/acsami.6b10408
23. Kameyama T, Koyama S, Yamamoto T, Kuwabata S, Torimoto T (2018) Enhanced photocatalytic activity of Zn-Ag-In-S semiconductor nanocrystals with a dumbbell-shaped heterostructure. J Phys Chem C 122:13705–13715. https://doi.org/10.1021/acs.jpcc.8b00255

Part II
Development of High-Order Photofunctions Based on Synergetic Action of Photoresponsive Molecules

Chapter 14
Photosynergetic Enhancement of Photosensitivity of Photochromic Terarylenes

Tsuyoshi Kawai

Abstract Photochromic molecules are one of the major targets of research among various scopes of photosynergetic approaches on molecular photochemistry. In this chapter, the authors describe progress of photosynergetic enhancement of photoreactivity of photochromic substances beyond classic photochemical stoichiometry based on one molecule/one photon hypothesis. Catalytic activity of photoproduct is one of the key concepts to achieve this and the classic concept of photoacid generators, PAGs, was far expanded for photo-Lewis Acid Generators, PLAGs. Cascade reactions involving photochromic molecules are further maximized with terarylene derivatives and amplified radio-chromic nature was rationally demonstrated, opening critical scope for the future of molecular photochemistry.

Keywords Photochromism · Photosynergetics · Photoisomerization · Terarylene · Diarylethene

14.1 Introduction

Photo-induced reactions of various molecules are widely observed in various natural and artificial systems. Vision systems in monomania are, for example generally based on the photoisomerization of retinal in the host protein such as opsin [1–3]. One of their typical properties is their capability for fine tunings on sensitive wavelength and photochemical sensitivity based on the protein-retinal interaction. Photosensitivity is a more general figure of merits for most of artificial photoreactive substances and materials. As the photo-induced reaction of molecules is practically promoted with a remote way and can be spatially confined in the photo-irradiated area, they are widely used in industrial photoresist polymer systems and photo-curable resin systems [4]. While biological and medical aspects have become pronounced targets, such as phototherapy and photopharmacology [5], industrial photoreactive substances are still considered to be of the most practical and social importance. In these systems,

T. Kawai (✉)
Division of Materials Science, Nara Institute of Science and Technology, NAIST, F410-8916-5, Takayama, Ikoma, Nara 630-0192, Japan
e-mail: tkawai@ms.naist.jp

H. Miyasaka et al. (eds.), *Photosynergetic Responses in Molecules and Molecular Aggregates*, https://doi.org/10.1007/978-981-15-5451-3_14

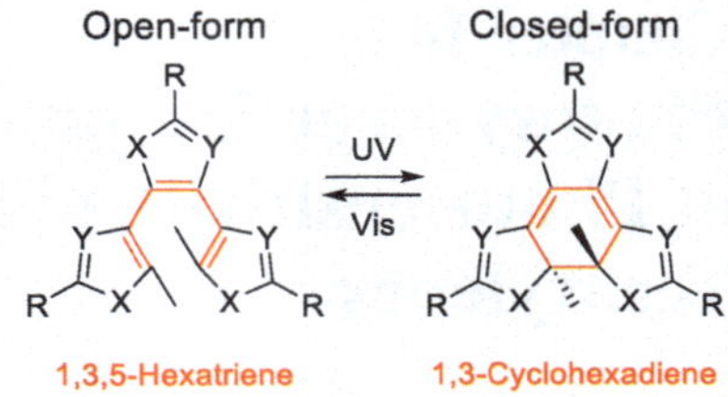

Scheme 14.1 General structure of photochromic terarylenes

most photo-triggered reactions are initiated by the photochemically generated reactive substances such as radicals, acids, and bases. While most industrial photoreactive substances undergo irreversible reactions, some substances undergo fatigue resistive reversible photoisomerization cycles for many times. Diarylethene has been recognized as one of the most fatigue resistive photochromic organic molecules, which shows reversible 6π-photocyclization and cyclo-reversion reactions [6, 7]. It also displays specific thermal stability at the ring-closed form, which is supported by the partially forbidden thermal cycloreversion process along the electronic ground-state potential surface and relatively large thermal activation energy. Triangle terarylene (Scheme 14.1) which involves an additional aromatic bridging unit also displays well-defined 6π-photocyclization like diarylethene [8, 9]. Although it is not large modification on the diarylethene structure, the introduction of central aromatic unit offers a marked variety of combinations of aromatic units and flexibility of molecular design for unique photofunctional properties. In this article, the author reviews recent progress on photochromic terarylenes and their photosynergetic enhanced photosensitivity.

14.2 Terarylene as Typical Expansion of Diarylethene

After the dramatic success of diarylethene mostly delivered by Prof. Irie with his coworkers, a considerable number of publications have been published on photoresponsive molecular systems based on diarylthenes, such as liquid crystals, molecular amorphous films, organic semiconductors, gels, membranes, metal-coordination compounds and polymers [10, 11]. These studies clearly presented significant adaptability to various molecular systems and promising future applications of diarylethenes. Chemical modification on diarylethene core structures has also been extensively studied. For example, Tien et al. have introduced eight-aromatic leg units into a porphyrin core to form a new photochromic molecule [12]. Professor Krayushkin and coworker systematically synthesized 1,2-diaryl-thiophene and related substances [13]. Independently, the author and coworkers synthesized terarylene derivatives as photo-switching unit for π-conjugation connection. Wavelength of UV-vis optical absorption band of these compounds in the open- and closed-ring forms are rationally modulated with the π-conjugation expansion, which is roughly evaluated with the bond alternation as shown in Scheme 14.2 [8]. The terary-

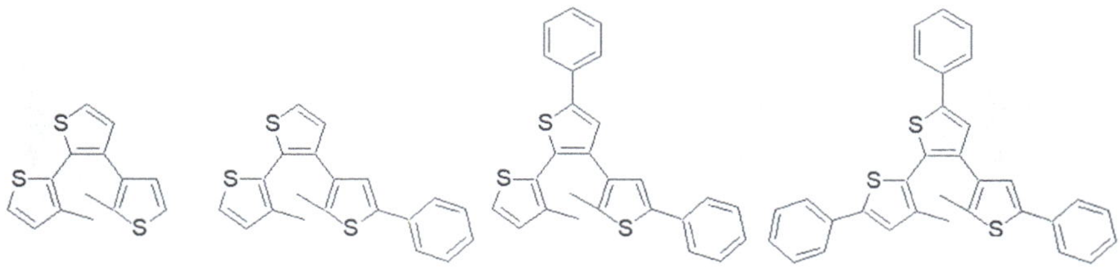

Scheme 14.2 Examples of terarylenes with different p-conjugation length

lene structure has thus been suggested to have additional flexibility in controlling various photofunctionality because of expanded π-conjugation.

The additional aromatic unit in the terarylene structure was expected to stabilize its ring-open form isomer and enhance the enthalpy increase upon ring cyclization. Therefore, spontaneous thermal cycloreversion reaction from the ring-closed form was supposed to be firster than the diarylethene. Since the decay lifetime of the colored state of diarylethene is enough long (can be 10^6 years) [10, 11], diarylethenes are recognized as "P-type" one. We have then expected tunability in the fading lifetime of terarylenes. It was practically modulated in the range of several years to several seconds, in which the decay kinetics was especially enhanced with four-phenylacetyl units (Scheme 14.3) [14]. Their specifically short lifetime of closed form was attributed mostly not to the activation energy but to the apparent frequency factor in the Arrhenius-type kinetics. The spontaneous reverse kinetics seems beneficial for molecular-based switching systems.

The photochromic ring-cyclization of terarylenes is considered to proceed in the conrotatory motion among the hexatriene structure as like as diarylethene. In this sense, their ground-state structure of quasi-C2 symmetry is expected to be photoactive, which is typically called as "anti-parallel" structure for diarylethene. Indeed, some crystalline diarylethenes which are fixed in the "anti-parallel" conformation exhibit almost 100% of photochemical quantum yield of ring-cyclization [10, 11]. Conformational equilibrium of terarylene should be more complex than diarylethene as its non-symmetric central hetero-aromatic units. However, those with higher stability and larger population ratio in the quasi-C2 symmetric structure should promote relatively high photochemical quantum yield. In the typical examples shown in Scheme 14.4, specific intramolecular tethering interaction between specific atoms,

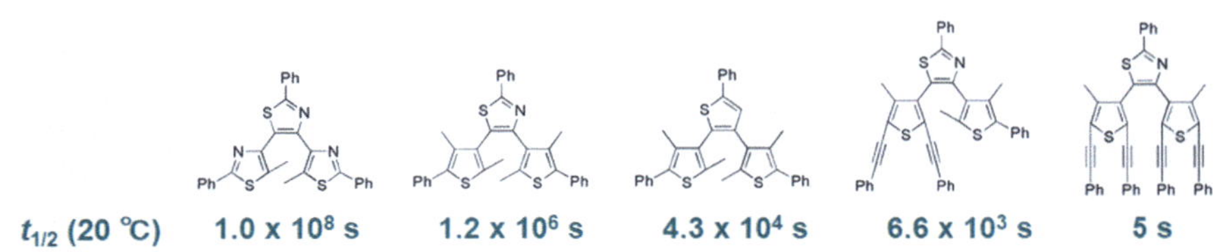

Scheme 14.3 Photochromic terarylenes with fast-fading dynamics

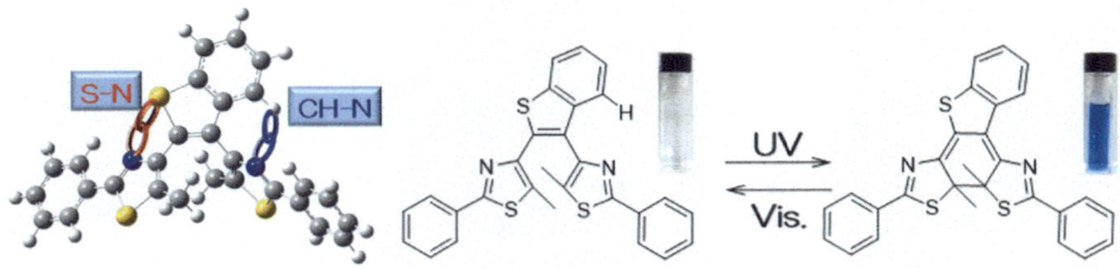

Scheme 14.4 Photochromic terarylene showing "Photon-quantitative Reaction"

such as S/N and CH/N should selectively stabilize their quasi-C_2 symmetric conformation. This terarylene of photochemical quantum yield as high as 98%, which was distinguished as "photon-quantitative reaction" [15]. This implies almost 100% of molecules are in the reactive conformation in solution phase. Such conformational control has been typically achieved in the crystal state of diarylethenes as mentioned above. Since excited-state reaction of diarylethene proceeds in several picoseconds or less, their conformation should be maintained during the photoexcitation. The compounds of non-reactive conformation should have less or almost no chance to proceed to the ring-cyclization reaction after conformational isomerization. We could thus control conformational equilibrium and enhanced photochemical sensitivity by means of chemical modification at the central unit of terarylenes.

The authors further verified this conformation control approach by means of guest-binding terarylene with enhanced photosensitivity in methanol (Scheme 14.5). Because of the hydrogen-binding pocket, the reactive conformation was stabilized in methanol showing enhanced photochemical sensitivity [16]. This conformational-control approach was also expanded for photochromic foldamers based on oligo-thiazole structure, where the intramolecular S/N interaction promotes stable helicate structure [17]. The foldamer helicity was rationally controlled by the chiral units at the ends of oligo-thiazole wire, which was justified by systematic DFT calculations. The photocyclization reaction with almost 100% diastereo-selectivity was achieved in this approach. Specific photochemical modulation of circularly polarized luminescence was demonstrated in a chiral tetra-thiazole derivative with two pyrene units at both ends [18]. Di-nuclear Eu(III) complex based on the tetra-thiazole foldamar

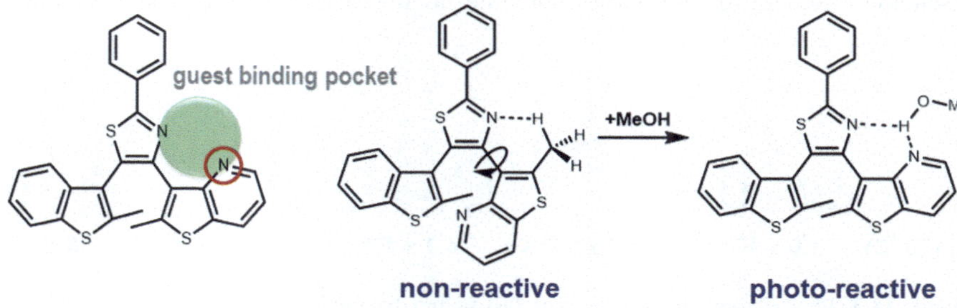

Scheme 14.5 Terarylene with a guest-binding pocket

was also studied [19]. Both of them displayed specific circularly polarized luminescence in the O-form and its diminish in C-form because of suppressed stereospecific interaction between two emitting units at both ends of tetra-thiazole. This reversible photo-switching nature associates with configurational conversion between sp2 and sp3 at the reaction center carbon atoms.

14.3 Photoacid Generators Based on Terarylenes

Among various studies on photo-switching applications of diarylethenes, some successful attempts have been reported on photo-reversible modulation of catalytic activity [20–24]. They may offer remote control of chemical reactions in a limited space and on-demand reaction triggering at the ambient temperature. The authors have studied terarylene backbone for new photoacid generator, PAG. PAG is one of the most frequently used industrial chemicals in photolithography technology and photocuring processes, which forms Brfnsted acid after photo-decomposition reaction [25, 26]. Especially, non-ionic PAGs are widely used for the industrial process in VLSI fabrication, in which they are demanded to be compatible with non-ionic polymers and monomers [27]. The non-ionic PAGs are usually composed of sulfonyl ester and sulfonyl imide, which decompose under UV light irradiation forming sulfonic acids. In order to trigger polymerization of epoxy monomers, sufficiently strong acids such as triflic acids and other superacids are also desired. Since most industrial PAGs have photochemical quantum yield of about 30% or less, it is still worth for challenging to explore non-ionic PAGs with higher photosensitivity.

The authors have studied a tearylene with H– and CH_3O– units at the reaction center carbon atoms instead of CH_3– units, which exhibited photo-induced release of CH_3OH affording condensed aromatic substance [28, 29]. This dissociation reaction has been assigned to the E1-like elimination reaction, in which methoxy anion dissociates from the ring-closed form isomer, and H^+ successively dissociates from the carbocation intermediate. For developing new efficient PAGs based on terarylenes, we have introduced H– and acid-ester group, CH_3SO_3– (Scheme 14.6) [30]. After the photocyclization reaction, an anion, $CH_3SO_3^-$ dissociates from the sp^3 carbon atom forming a sp^2 carbocation. This should be followed by the release of H^+ providing the Bronsted acid, CH_3SO_3H. Interestingly, this new PAG releases the whole CH_3SO_3H without external source of H^+, which is generally desired for

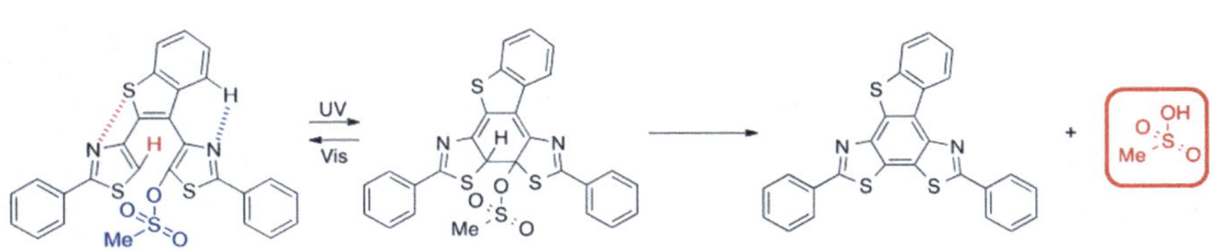

Scheme 14.6 Reaction scheme of S-PAG based on terarylenes

precious PAGs to form Brϕnsted acids. Because of this specific nature, we called this new PAG as the self-contained PAG, S-PAG. The new S-PAG was indeed capable to use as the photo-initiator for cationic epoxy polymerization reaction. Various kinds of Brϕnsted acids can be released such as carboxylic acid and sulfonic acid [31, 32]. In the S-PAG for triflic acid which is a typical "super acid", we stably introduced CF_3SO3- unit in the open form on a naphthalene unit [33]. The ring-cyclized form quantitatively releases acid affording the condensed aromatic form. The overall quantum yield generating the acid is thus determined with the initial photocyclization reaction and can be as high as 71%, which is highest among the neutral PAGs so far reported. Another advantage of the S-PAG is absence of radical intermediate, which is commonly involved in the usual PAGs. The extended π-conjugation system in terarylene seems supportive of the quantitative release of acid via the carbocation intermediate stabilized over their π-conjugation.

More recently, the author and coworkers expanded this S-PAG concept into photo-induced Lewis Acid Generator, PLAG, in which Lewis acid is reliably formed under UV light irradiation instead of Brϕnsted acid [34]. Scheme 14.7 depicts the concept of PLAG, which affords stable carbocation, C^+-form, as a Lewis acid catalyst under UV light irradiation. The carefully characterized C^+ form showed life-time of about 100 days and demonstrated to trigger the cationic polymerization of epoxy monomers, in which end-terminal unit was interestingly protected with the C^+ unit. This clearly demonstrates the concept of PLAG affording the C^+-form as a Lewis acid catalyst. Moreover, we successfully demonstrated to trigger Mukaiyama-Aldol reaction with this PLAG, which cannot be initiated by any Brϕnsted acids [35, 36]. The product of the Mukaiyama-Aldol reaction was importantly obtained as silyl-ester, which is fragile toward Brϕnsted acid. This again clearly indicated that PLAG promotes Lewis acid but not Brϕnsted acid. This was the first example of photo-triggered one despite more than 5000 reports have been published on Mukaiyama-Aldol reaction and related topics. The conventional Lewis acid catalysts such as metal chlorides and their coordination compounds mostly desire careful handling because of ambient instability. The new PLAG is fairly stable under ambient condition but promotes highly reactive carbocation Lewis acid with 50% of photochemical quantum yield. PLAGs are expected to be compatible with cationic photo-polymerization of epoxy and vinyl ether monomers having protected functional groups such as silyl esters, which are cleavable with Brϕnsted acid after the photo-polymerization.

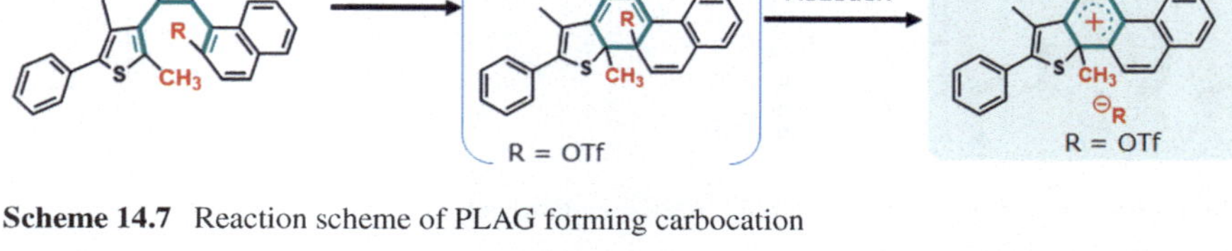

Scheme 14.7 Reaction scheme of PLAG forming carbocation

14.4 New Domino Reaction for Photosynergetic Approach

The mammalian vision systems are based on ionic transport through the photo-switchable ion-channels at membrane and capable of amplifying light signal over 1:1 ratio between numbers of ions and photons. Artificial systems such as avalanche photodiode (APD) and Photo-Multiplying Tube (PMT) detectors also have significant amplification nature. Tokumaru et al. and Irie et al. have independently reported photoisomerization reaction with photochemical quantum yield higher than 100% [37, 38]. Both systems of these advanced research are based on triplet state energy transfer between substances. Recently Garcia-Garibay et al. have reported specific quantum-chain reactions in single crystalline state. They have reported specific photodecarbonylation reaction of diarylcyclopropenones in the crystalline state with photochemical quantum yield higher than 300%, which is based on singlet energy migration [39]. Since all these photo-amplifying systems are based on irreversible photoreactions and it was still challenging to explore those of reversible photochemical systems. The author has reported a specific chain-like reaction in diarylethene, where one-electron electrochemical oxidation of C-form resulted in isomerization of several C-forms to O-forms with chain-like reaction [40]. After Branda et al. have reported the same phenomenon by means of chemical oxidation, Matsuda proposed a clear concept of spontaneous isomerization reaction in the oxidized and even reduced states which are supported with suppressed activation energy between two isomers [41, 42]. These studies are summarized as the reaction scheme shown in Scheme 14.8. The first oxidation of C-form generates cation radical of C-form and the C^+-cation radical proceeds to isomerization reaction to O^+-cation radical. Since O^+-cation radical has sufficiently high oxidizing capability against another C-form for recovering C^+-cation radical for propagating chain-like reactions. The chain-propagation of this system was supposed to be much pronounced in some molecules with longer lifetime of cation radicals. Since most radical cations can have longer lifetime than the triplet and the singlet excited states, chain propagation can be much pronounced in the present systems in comparison with previous studies.

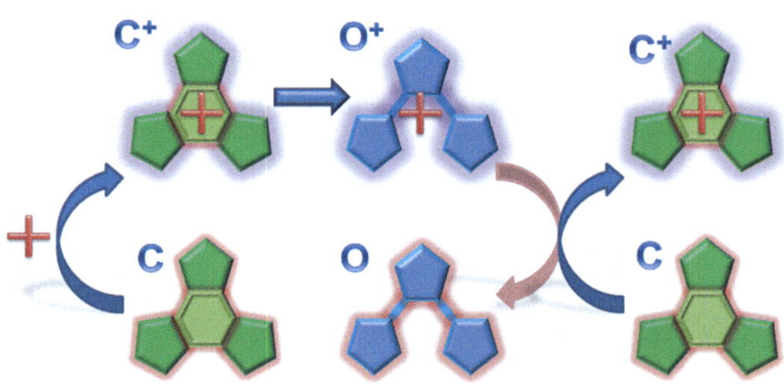

Scheme 14.8 Schematic illustration of chain-like cascade reaction of terarylenes

The author and coworkers have expected marked enhancement of chain propagation in terarylenes, because of their expanded π-conjugation systems supporting stable cation radicals. The authors have indeed studied on some terarylene derivatives and found efficient chain-like isomerization reaction in terthiazole [43]. Their kinetics were systematically studied with the stopped-flow analysis. The author and coworker expanded the terarylene derivatives to the compound shown in Scheme 14.9 [44]. The phenyl units at the reaction center carbon atoms are considered to be involved in extended π-conjugation system in the O^- and the O^+-forms but not in the C^- and the C^+-forms. Therefore, the introduction of these phenyl groups seems to enhance the relative stability of the O^+-form against the C^+-form and to make the unimolecular isomerization reaction favorable, which is apparently the rate-determining step of in chain-propagation process. Indeed, this new compound proceeds the propagation of chain-like reaction for more than 1000 times. Once the oxidant of about 0.001 equivalent to the C-form isomer was added, the whole molecules isomerized into the O-form. Specific quasi-zero-order decay kinetics of C-form was observed in this chemical oxidation, suggesting the rate-determining step is unimolecular reaction but not bi-molecular reaction. The apparent zero-order rate constant was roughly proportional to the amount of initial oxidant. These photochromic substances display reversible photochromic color changes and additional electrochromic nature with extremely large charge sensitivity with an amplification factor over 1000. For instance, we may expect future light shading material based on these molecules, in which coloration is induced by the ambient UV light, and electro-bleaching can be induced with a very small amount of external current charge.

More recently, the author and coworkers have demonstrated photo-induced trigger of oxidative chain reaction. The photochemical reactivity and oxidative cascade amplification capability of were studied for compounds shown in Scheme 14.10 [45]. The closed forms demonstrated fading reaction under UV light illumination in chloroform. This reaction was considered to be triggered by the electron detach from the singlet state excited of the substances or charge separation from the CT excited state, in which chloroform molecules operate as photo-oxidizing electron-scavenging substance. The photo-triggered fading reaction proceeds efficiently in the presence of oxygen. Because of the chain-like reaction amplification, the authors have observed surprisingly large photochemical fading quantum yield, apparently about 3300%. This seems a typical example of photosynergetic enhancement of photochemical sensitivity of organic substances, which is one of highest photosensitivity

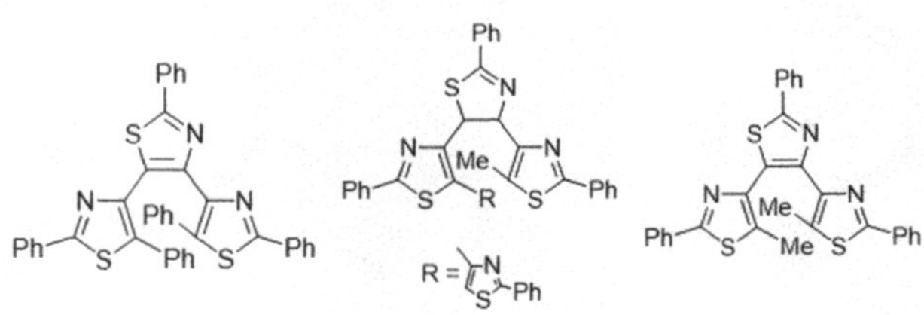

Scheme 14.9 Terarylenes of highly efficient cascade reactions

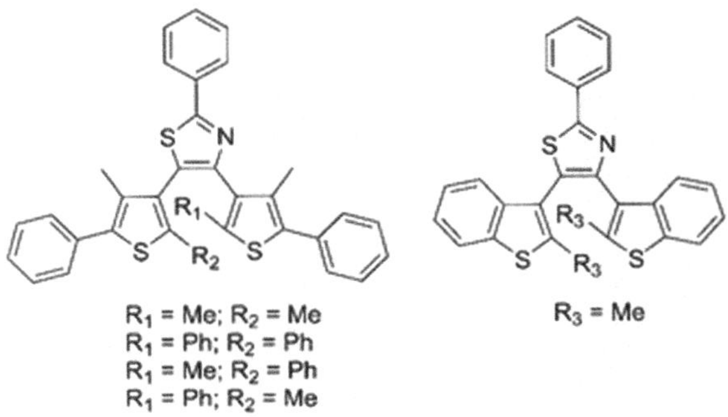

Scheme 14.10 Molecular structures of terarylenes for cascade reaction study

among organic photochemical color-changing reactions. The author has thus motivated to apply this photo-triggered chain-like reaction to the radio-detection process. That is, the C-form/chloroform solution was irradiated with an X-ray. Systematic bleaching reaction was clearly observed after irradiation of X-ray for 6 s. The degree of bleaching was roughly proportional to the irradiation dose amount of X-ray. Lowest detection limit for the X-ray dose was evaluated to about 0.3 mGy, which is more than 100 times higher in comparison with previous organic radio-chromic molecules [46–48]. Although there can be several different chemical processes upon irradiation of X-ray competitively, significant cascade amplification should be pronounced for the oxidation reaction forming cation.

14.5 Conclusion

Photochromic molecules have been widely studied for the color and light controlling materials. They have been also expected as dynamic photo-switching molecules and optical recording materials. Recent intensive and systematic studies on various photochromic substances, especially on diarylethenes and related substances, promoted considerable progress in their photosensitivity and provided highly-photosensitive molecules. Indeed, their photosensitivity is considerably higher than most of the irreversible photoreactive substances, which are widely used in industrial, medical, and analytical purposes. The author and colleagues have retracted in some cases the discriminating scope for diarylethene derivatives as reversible systems and exploited for efficient new PAGs and even PLAGs. Especially, the PLAG system is expected to alter many occasions of chemical reactions. Most of the previous catalytic Lewis acids are composed of metal halides of moisture sensitivity, such as $SnCl_4$, $TiCl_4$, $AnCl_3$, $ZnCl_2$, etc. The new PLAGs are expected to be compatible with some of Lewis acid catalytic reactions but inert under ambient conditions before illumination of UV light. In addition to this advantage in usability, the photochemical quantum yield of PLAG (QY = 50%) is considerably higher than the most of PAGs.

On the other hand, the chain-like ring-cycloreversion reactions seem to be a spectacle example of amplification molecular system. The chain-propagation number is not obviously limited by the reaction thermodynamics but by competitive kinetics of chain-propagation dynamics with the spontaneous decay of radicals. Control on radical lifetime and appropriate design of reaction space may offer the ultimate challenging fields.

Acknowledgements We deeply appreciate to Professors M. Irie, H. Miyasaka, J. Abe, K. Matsuda, N. Nakatani, R. Metivier T. Nakashima, for their collaboration. The present work was supported by JSPS KAKENHI Grant Number 26107002, Grant-in-Aid for Scientific Research on Innovative Areas "Photosynergetics".

References

1. Hille B (2001) Ionic channels of excitable membranes. Oxford University Press Inc
2. Kliger DS, Menger EL (1975) Special issue on the chemistry of vision. Acc Chem Res 8:81–84
3. Kim JE, Tauber MJ, Mathies RA (2001) Wavelength dependent cis-trans isomerization in vision. Biochemistry 40:13774–13778
4. Fouassier JP, Lalevee J (2012) Photoinitiators for polymer synthesis-scope, reactivity, and efficiency. Wiley-VCH, Weinheim
5. Velema WA, Szymanski W, Feringa BL (2014) Photopharmacology: beyond proof of principle. J Am Chem Soc 136:62178–62191
6. Irie M, Mohri M (1988) Thermally irreversible photochromic systems. Reversible photocyclization of diarylethene derivatives. J Org Chem 53:803–808
7. Irie M, Yokoyama Y, Seki T (2013) New frontiers in photochromism. Tokyo, Springer Japan
8. Kawai T, Iseda T, Irie M (2004) Photochromism of triangle terthiophene derivatives as molecular re-router. Chem Commun 72–73
9. Nakashima T, Atsumi K, Kawai S, Nakagawa T, Hasegawa Y, Kawai T (2007) Photochromism of thiazole-containing triangle terarylenes. Eur J Org Chem 2007:3212–3218
10. Irie M (2000) diarylethenes for memories and switches. Chem Rev 100:1685–1716
11. Irie M, Fukaminato T, Matsuda K, Kobatake S (2014) Photochromism of diarylethene molecules and crystals: memories, switches, and actuators. Chem Rev 114:12174–12277
12. Tian H, Chen B, Tu H, Müllen K (2002) Novel bisthienylethene-based photochromic tetraazaporphyrin with photoregulating luminescence. Adv Mater 14:918
13. Krayushkin MM, Ivanov SN, Martynkin AY, Lichitsky BV, Dudinov AA, Uzhinov BM (2002) Photochromic dihetarylethenes. 17. Synthesis and photochromic properties of dithienylethenes containing new heterocyclic bridging fragments. Russ Chem Bull 50:116–121
14. Kawai S, Nakashima T, Atsumi K, Sakai T, Harigai M, Imamoto Y, Kamikubo H, Kataoka M, Kawai T (2007) Novel photochromic molecules based on 4,5-dithienyl thiazole with fast thermal bleaching rate. Chem Mater 19:3479–3483
15. Fukumoto S, Nakashima T, Kawai T (2011) Photon- quantitative reaction of a dithiazolylarylene in solution. Angew Chem Int Ed 50:1565–1568
16. Nakashima T, Fujii R, Kawai T (2011) Regulation of folding and photochromic reactivity of terarylenes through a host–guest interaction. Chem Eur J 17:10951–10957
17. Nakashima T, Yamamoto K, Kimura Y, Kawai T (2013) Chiral photoresponsive tetrathiazoles that provide snapshots of folding states. Chem Eur J 19:16972–16980
18. Hashimoto M, Nakashima T, Kuno J, Yamada M, Kawai T (2018) Dynamic modulation of circularly polarized luminescence in photoresponsive assemblies. Chem Nano Mat 4:815–820

19. Hashimoto M, Nakashima T, Yamada M, Yuasa J, Rapenne G, Kawai T (2018) Hierarchical emergence and dynamic control of chirality in a photoresponsive dinuclear complex. J Phys Chem Lett 9:2151–2157
20. Peters MV, Goddard R, Hecht S (2006) On the illusive nature of o-formylazobenzenes: exploiting the nucleophilicity of the azo group for cyclization to indazole derivatives. J Org Chem 71:7840–7845
21. Gurke J, Budzak S, Chmidt BM, Jacquemin D, Hecht S (2018) Efficient light-induced pKa modulation coupled to base-catalyzed photochromism. Angew Chem Int Ed 57:4797–4801
22. Stoll RS, Hecht S (2010) Artificial light-gated catalyst systems. Angew Chem Int Ed 49:5054–5075
23. Lemieux V, Spantulescu MD, Baldridge KK, Branda NR (2008) Modulating the lewis acidity of boron using a photoswitch. Angew Chem Int Ed 47:5034–5037
24. Kaur B, Raza R, Stashick MJ, Branda NR (2019) Using light to control the inhibition of karstedt's catalyst. Org Chem Front 6:1253–1256
25. Crivello JV, Reichmanis E (2014) Photopolymer materials and processes for advanced technologies. Chem Mater 26:533–548
26. Zivic N, Kuroishi PK, Dumur F, Gigmes D, Dove AP, Sardon H (2019) Recent advances and challenges in the design of organic photoacid and photobase generators for polymerizations. Angew Chem Int Ed 58:10410–10422
27. Martin CJ, Rapenne G, Nakashima T, Kawai T (2018) Recent progress in development of photoacid generators. J Photochem Photobiol C Photochem Rev 34:41–51
28. Nakagawa H, Kawai S, Nakashima T, Kawai T (2009) Synthesis and photochemical reactions of photochromic terarylene having a laving methoxy group. Org Lett 11:1475–1478
29. Nakagawa H, Nakashima T, Kawai T (2012) Subsequent chemical reactions of photochromic 4,5-dibenzothienylthiazoles. Eur J Org Chem 24:4493–4500
30. Nakashima T, Tsuchie K, Kanazawa R, Li R, Iijima S, Galangau O, Nakagawa H, Mutoh K, Kobayashi Y, Abe J, Kawai T (2015) Self-contained photoacid generator triggered by photocyclization of Ttiangle terarylene backbone. J Am Chem Soc 137:7023–7026
31. Li R, Nakashima T, Kanazawa R, Galangau O, Kawai T (2016) Efficient self-contained photoacid generator system based on photochromic terarylenes. Chem Eur J 22:16250–16257
32. Galangau O, Delbaere S, Ratel-Ramond N, Rapenne G, Li R, Calupitan JPDC, Nakashima T, Kawai T (2016) Dual photochemical bond cleavage for a diarylethene-based phototrigger containing both methanolic and acetic sources. J Org Chem 81:11282–11290
33. Li R, Nakashima T, Kawai T (2017) A self-contained photoacid generator for super acid based on photochromic terarylene. Chem Commun 53:4339–4341
34. Mizutsu R, Asato R, Martin CJ, Yamada M, Nishikawa Y, Katao S, Yamada M, Nakashima T, Kawai T (2019) Photo-lewis acid generator based on radical-free 6π photo-cyclization reaction. J Am Chem Soc (in press). https://doi.org/10.1021/jacs.9b11821
35. Mukaiyama T, Banno K, Narasaka K (1974) New cross-aldol reactions. J Am Chem Soc 96:7503–7509
36. Frias M, Cieslik W, Fraile A, Rosado-Abon A, Garrido-Castro AF, Yuste F, Aleman J (2018) Development and application of asymmetric organocatalytic Mukaiyama and vinylogous Mukaiyama-type reactions. Chem Eur J 24:10906–10933
37. Irie M, Yorozu T, Yoshida K, Hayashi (1977) Radiation induced racemization of 1,1'-binaphthyl in tetrahydrofuran and toluene. J Phys Chem 81:973–976
38. Arai T, Karatsu T, Sakuragi H, Tokumaru K (1983) "One-way" photoisomerization between cis- and trans-olefin. A novel adiabatic process in the excited state. Tetrahedron Lett 24:2873–2876
39. Kuzmanich G, Gard MN, Garcia-Garibay MA (2009) Photonic amplification by a singlet-state quantum chain reaction in the photodecarbonylation of crystalline diarylcyclopropenones. J Am Chem Soc 131:11606–11614
40. Koshido T, Kawai T, Yoshino K (1995) Optical and electrochemical properties of cis-1,2-dicyano-1,2-bis(2,4,5-trimethyl-3-thienyl)ethane. J Phys Chem 99:6110–6114
41. Peters A, Branda NR (2003) Electrochromism in photochromic dithienylcyclopentenes. J Am Chem Soc 125:3404–3405

42. Moriyama Y, Matsuda K, Tanifuji N, Irie S, Irie M (2005) Electrochemical cyclization/cycloreversion reactions of diarylethenes. Org Lett 7:3315–3318
43. Nakashima T, Kajiki Y, Fukumoto S, Taguchi M, Nagao S, Hirota S, Kawai T (2012) Efficient oxidative cycloreversion reaction of photochromic dithiazolythiazole. J Am Chem Soc 134:19877–19883
44. Calupitan JPDC, Nakashima T, Hashimoto Y, Kawai T (2016) Fast and efficient oxidative cycloreversion reaction of a π-extended photochromic terarylene. Chem Eur J 22:10002–10008
45. Asato R, Martin CJ, Calupitan JP, Mizutsu R, Nakashima T, Okada G, Noriaki Kawaguchi N, Yanagida T, Kawai T (2019) Photosynergetic amplification of radiation input: from efficient UV induced cycloreversion to sensitive X-ray detection. Chem Sci to be published
46. Irie S, Kim MS, Kawai T, Irie M (2004) The radiation-induced coloration of amorphous photochromic dithienylethene films. Bull Chem Soc Jpn 77:1037–1040
47. Kinashi K, Miyashita Y, Ishida K, Ueda Y (2012) Radiochromic film containing spiropyran dye for dosimetry of low energy X-rays. J Phys Org Chem 25:427–430
48. Kinashi K, Miyamae Y, Nakamura R, Sakai W, Tsutsumi N, Yamane H, Hatsukano G, Ozaki M, Jimbo K, Okabe T (2015) Chem Commun 51:11170–11173

Chapter 15
Creation of Molecularly Integrated Multi-responsive Photochromic Systems

Yasushi Yokoyama

Abstract In this chapter, we describe: (1) the use of photochromic acid-generating spiropyran in (i) controlling the absorption wavelength of bisphenanthrolinodiarylethene and (ii) the on/off switching of photochromism in a diarylethene; (2) highly enantioselective light-induced chirality transfer from human serum albumin to diarylethenes and (3) the short-step assembly of (i) thermally irreversible stealth photochromic arylbutadienes and (ii) thermally reversible photochromic teraryls by way of "click" Huisgen reactions catalyzed by Ru(I) complexes.

Keywords Diarylethene · Spiropyran · Arylbutadiene · Teraryl · Human serum albumin · McMurry coupling · Huisgen click reaction

15.1 Introduction

Photochromism is an interesting field of research since irradiation of photochromic materials not only induces changes of color but also various useful properties [1–5]. Of such photochromic materials, diarylethenes are among the most investigated [6–8] due to their high thermal as well as photochemical durability, facile synthetic accessibility and wide range of switching abilities of their color and properties.

Here, we have described our results mainly related to diarylethenes and their derivatives obtained from the JSPS Kakenhi project on "Photosynergetics" led by Professor Miyasaka of Osaka University. When the project started in 2014, our main objective was to construct multifunctional photochromic systems which are composed of more than one chemical species. The components of such a system must work in a synergetic manner by photoirradiation to act as a photoswitch for such novel functions and properties strongly related to photochromic phenomena (Sect. 15.2) as well as for the transfer of chiral information from its environment

Y. Yokoyama (✉)
Graduate School of Engineering Science, Department of Chemistry and Life Science, Yokohama National University, Yokohama, Kanagawa 240-8501, Japan
e-mail: yokoyama-yasushi-wp@ynu.ac.jp

H. Miyasaka et al. (eds.), *Photosynergetic Responses in Molecules and Molecular Aggregates*, https://doi.org/10.1007/978-981-15-5451-3_15

(Sect. 15.3). As a secondary objective, we aimed to establish short-step synthetic pathways for the photochromic molecules in order to allow a more facile process for the construction of complex systems (Sect. 15.4).

15.2 Photon-Working Conjugate of Photochromic Acid-Generating Spiropyran and Diarylethenes

15.2.1 Fine-Tuning of the Absorption Wavelength of Bisphenanthrolinodiarylethene [9]

A thermally reversible and water-soluble photochromic spiropyran **1mc** was shown in a 2012 paper by Liao et al. [10]. It takes the merocyanine form with the phenol moiety instead of the phenolate at its thermal equilibrium due to the existence of a sulfonic acid group. Upon visible light irradiation (ca. 450 nm), it changes to spiropyran form **1sp** in which the phenolic proton moves to the indoline nitrogen. We found this compound to be of great interest so that its use in organic solvents was investigated. We then designed a spiropyran **2mc** by introducing a *t*-butyl group. A fluorine atom was also introduced on the indoline moiety to enhance the acidity of **2sp** (Chart 15.1). Its absorption spectral change by visible light irradiation and thermal back reaction are shown in Fig. 15.1.

According to the reported data of similar compounds, the pK_a values of **2mc** and **2sp** are estimated to be ca. 10 [11] and ca. 4 [12] in aqueous solutions, respectively. Therefore, during the photochromic reactions, the acidity will change by *ca.* 10^6 times.

In our previous research, we found that a diarylethene **3o** (Chart 15.1) changed its photochromic properties dramatically when it forms a complex with Cu(I) (Lewis acid) [13] or a salt with oxalic acid (Brønsted acid) [14] in solution. When four equivalents of CF_3SO_3H was added to the chloroform solution of **3c** as a strong acid, the absorption maximum wavelength shifted to longer wavelengths as large as

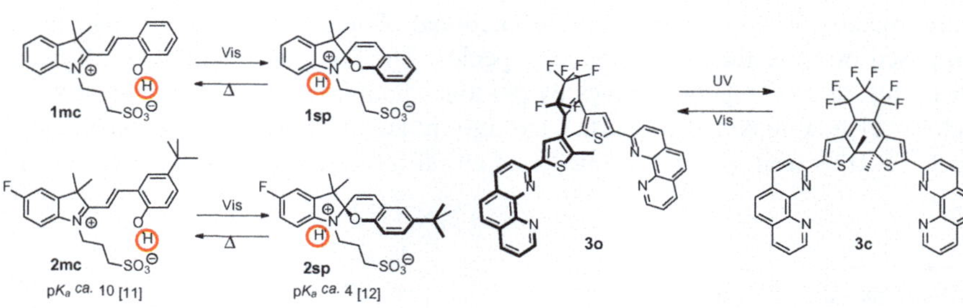

Chart 15.1 Thermally reversible photochromic strong acid-generating spiropyrans **1** and **2** and thermally irreversible photochromic bisphenanthrolino-bisthienylethene **3**

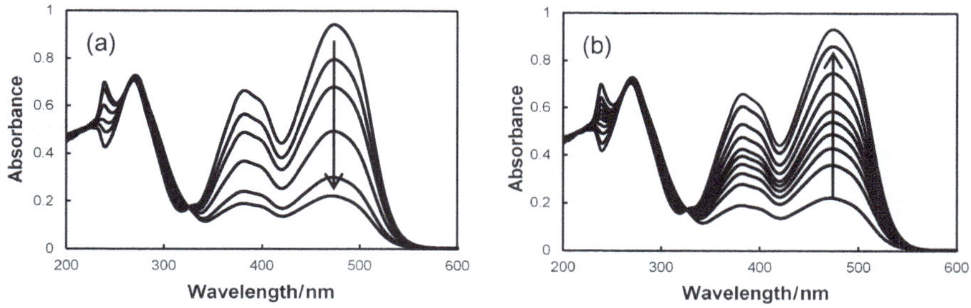

Fig. 15.1 The thermally reversible photochromism of **2mc** in chloroform at room temperature (ca. 20 °C). **a** Irradiated with 450-nm light. Concentration: 6.4×10^{-5} mol dm^{-3}, optical path length: 1 cm, light source: 450 nm (0.42 mW/cm^2), irradiation time/min: 0, 0.5, 1, 2, 4, 12. **b** Thermal back reaction. Time interval/min: 0, 3, 6, 9, 12, 15, 21, 30, 50, 180. From [9] Copyright © 2016 by Wiley, Inc. Reprinted by permission of Wiley, Inc.

46 nm (from 638 to 684 nm) with a change in the color of the solution from blue to green (Fig. 15.2).

Since the pK_a value of protonated phenanthroline is around 4.86 [15] which is between the pK_a values of **2mc** and **2sp**, we envisaged that the phenolic proton in **2mc** should move to the phenanthroline moieties of **3c** instead of to the indoline nitrogen of **2sp** when **2mc** coexists with **3c** in the solution and **2sp** is generated from **2mc** by visible light irradiation. One concern was whether visible light for the generation of **2sp** would also change **3c** to **3o**. Fortunately, as shown in Fig. 15.2, the molar absorption coefficients of both **3c** and **3c**-$(H^+)_2$ between 450 and 500 nm are small. We therefore carried out the all-optical fine-tuning of the absorption band position by visible light irradiation to the **2mc** and **3c** mixture solution. The results are shown in Fig. 15.3.

Since **2mc** and **3o** are photochromic, the former thermally reversible and the latter thermally irreversible, we generated distinctive color states using these molecules.

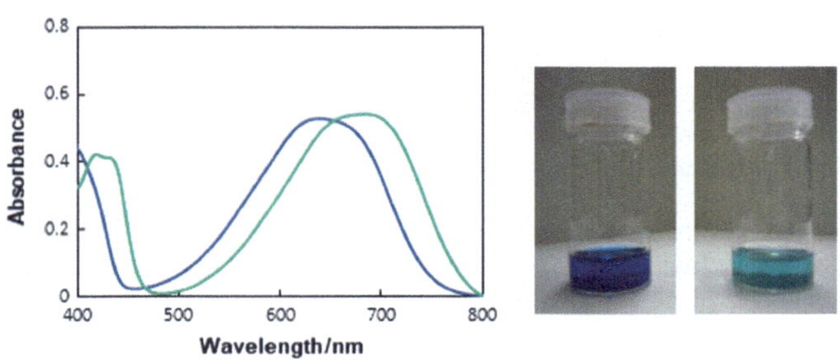

Fig. 15.2 Absorption spectra of **3c** and **3c**-$(H^+)_2$ at the photostationary state of 366-nm irradiation (**3c**:**3o** = >98:<2) in chloroform. Blue line (left photograph): **3c** (3.5×10^{-5} mol dm^{-3}). Green line (right photograph): **3c** (3.5×10^{-5} mol dm^{-3}) in the presence of 4 eq CF_3SO_3H (1.4×10^{-4} mol dm^{-3}). From [9] Copyright © 2016 by Wiley, Inc. Reprinted by permission of Wiley, Inc.

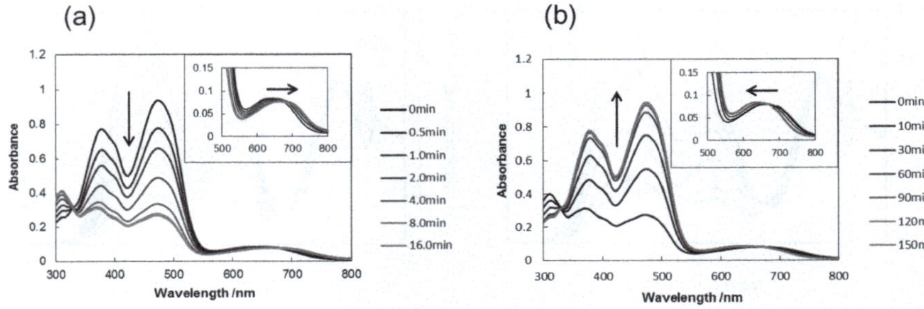

Fig. 15.3 Shift of absorption band of **3c** in the 550–800 nm range by the photochromic reaction between **2mc** and **2sp** in chloroform. **a** Irradiation of 450-nm light to induce the change from **2mc** to **2sp**. **b** Thermal reaction to induce the change from **2sp** to **2mc** at r.t. (ca. 20 °C). Concentration: **3c** 2.40×10^{-5} mol dm^{-3}, **2mc** 3.84×10^{-4} mol dm^{-3} (16 eq). From [9] Copyright © 2016 by Wiley, Inc. Reprinted by permission of Wiley, Inc.

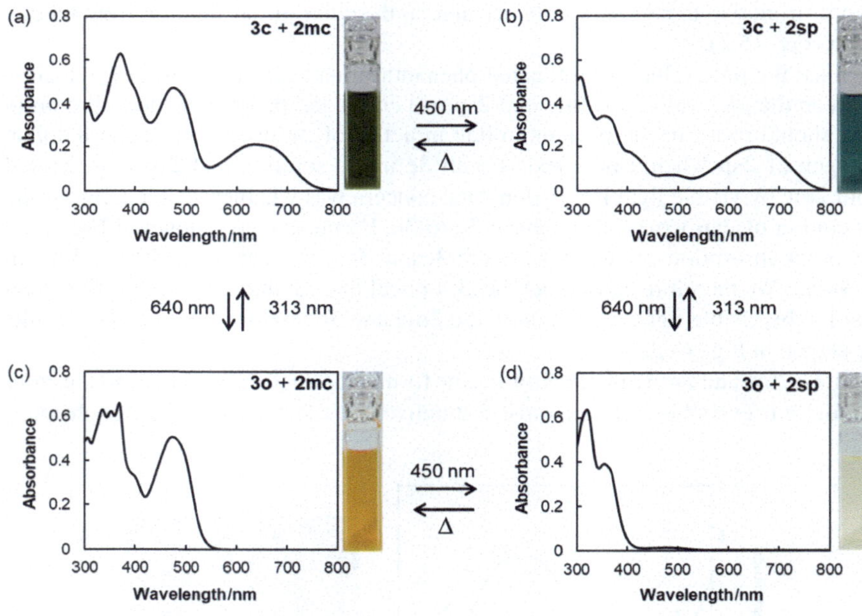

Fig. 15.4 Four states of the **3**-and-**2** system generated by light irradiation and thermal treatment in chloroform. **a 3c** and **2mc**, dark green; **b 3c** and **2sp**, dark cyan, 450-nm light irradiation of (**a**); **c 3o** and **2mc**, orange, generated by 640-nm light irradiation of (**a**); **d 3o** and **2sp**, faint yellow, generated by 450-nm light irradiation of (**c**). Other conversion methods are described in the figure. Concentration: **3** 1.2×10^{-4} mol dm^{-3}, **2** 4.8×10^{-4} mol dm^{-3}. Optical path length: Spectra 2 mm, photographs 20 mm. From [9] Copyright © 2016 by Wiley, Inc. Reprinted by permission of Wiley, Inc.

Figure 15.4 shows the four different colors generated only by alternating the irradiation light wavelength and application of heat.

Thus, we have synthesized a photochromic and strong acid generator spiropyran **2mc** which works in organic solvents. By making use of the large difference in the pK_a values for **2mc** and **2sp**, we have succeeded in controlling the continuous shift of the absorption bands of **3c** by light irradiation and thermal treatment. It is well known that when photochromism induces a change in color of the solution or material by photoirradiation, it occurs between two "remote" wavelengths. The system we showed in this section exhibits "continuous" change of the absorption band by photoirradiation. It is the "fine-tuning" of the absorption band position of a compound controlled by the photochromism of another compound. Since this system uses two different photochromic compounds, four distinctive states with regard to the absorption spectra can be generated independently just by light irradiation and thermal treatment [9].

15.2.2 All-Optical On/Off Switching of the Photochromic Reaction of Diarylethene [16]

The construction of all-optical control for the position of the absorption band of a photochromic diarylethene was successfully carried out by photochromism of the spiropyran which produces a highly acidic proton upon visible light irradiation. Simulaneously, we synthesized a unique diarylethene **4o** (Chart 15.2) which did not show photochromism upon UV irradiation when dissolved in acetonitrile (MeCN) due to the twisted intramolecular charge transfer (TICT) [17]. When an excess amount (10–1000 eq) of CF_3CO_2H was added, it recovered its photochromic nature (Fig. 15.5) since the strong electron-donating ability of the dimethylamino group was lost by protonation to its nitrogen atom.

Since the pK_a value of **4o**-H^+ was estimated to be about 4.0 [18], it was uncertain that if **2sp** (pK_a about 4.0 [12]) was present in the same solution, the proton would

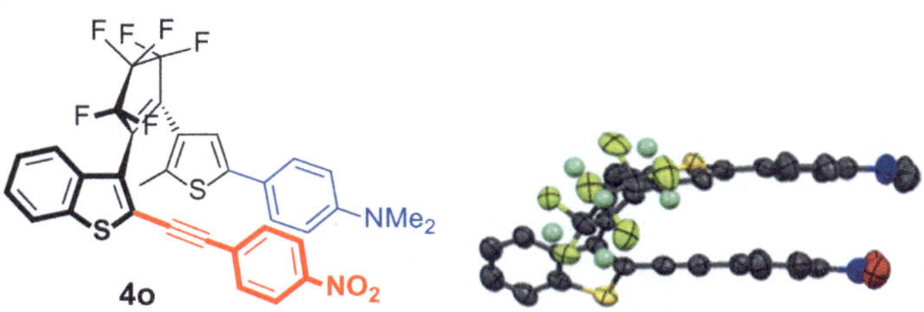

Chart 15.2 Diarylethene **4o** and its ORTEP drawing obtained by X-ray crystallographic analysis (shown with 50% probability of thermal ellipsoids). Fluorine atoms of the secondary conformer are shown as green balls. Hydrogen atoms were omitted for clarity. Reprinted with permission from [16]. Copyright 2016 American Chemical Society

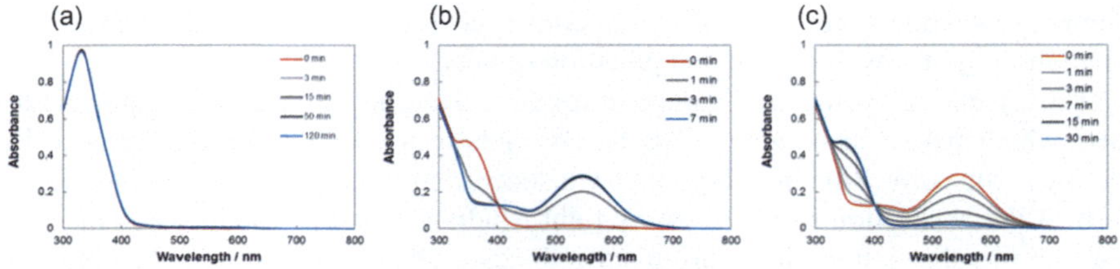

Fig. 15.5 Absorption spectral change of **4o** by photoirradiations in MeCN. Concentration: 2.05 × 10^{-5} mol dm^{-3}. Light intensity: 366 nm 0.400 mW cm^{-2}, 578 nm 2.20 mW cm^{-2}. **a** Irradiation with 366 nm light to **4o**; **b** irradiation with 366 nm light to the mixture of **3o** and ca. 1000 eq. of trifluoroacetic acid; **c** successive irradiation with 578 nm light to the photostationary state solution of (**b**). Reprinted with permission from [16]. Copyright 2016 American Chemical Society

be transferred from **2sp** to **4o**. However, we still attempted the use of **2mc** as the photochromic acid generator by irradiation of 450 nm light to produce **4o**-H^+. It first had to be proved that before 450 nm light irradiation, no photochromism of **4o** to **4c** occurs by UV irradiation. If 366 nm light was used for UV irradiation, it could change **2mc** to **2sp**; then, the latter would give a proton to **4o**, and **4c**-H^+ would be formed during 366 nm light irradiation. However, we found that irradiation of 313 nm light produces little **2sp** from **2mc**. Consequently, the small amount of **2sp** formed can be trapped if small amounts of alkylamine are initially added.

We envisaged the following scenario: Even though irradiation of 313 nm light is carried out on the mixture of **4o**, **2mc** and the small amount of triethylamine in MeCN, it would not yield any **4c**. Irradiation of 450 nm light to change **2mc** to **2sp** followed by 313 nm light irradiation would generate **4c**-H^+ which would revert to **4o**-H^+ by visible light irradiation. While the last transformation from **4c**-H^+ to **4o**-H^+ has already been proved by experiments with CF_3CO_2H, our interest was to see whether **4c** would revert to **4o** by visible light irradiation. Thus, further experiments were carried out, and the spectral changes are shown in Fig. 15.6.

As shown in Fig. 15.6a, the mixture of an equimolar amount of **4o** and **2mc** and 10 mol% of triethylamine in MeCN was irradiated with 313 nm for 5 min. The small decrease in the absorbance at around 440 nm indicates the change from **2mc** to **2sp**, implying the generation of a small amount of strong acid. As expected, 10 mol% of triethylamine worked very well to trap the acid so that no **4c**-H^+ was formed. Successive irradiation of 450 nm light changed **2mc** to **2sp**, as evidenced by the entire disappearance of the absorption band around 450 nm. Irradiation of 313 nm light to this solution induced the photochromic coloration of **4o**-H^+ to **4c**-H^+ (Fig. 15.6b). When the solution was placed in the dark, **2sp** changed to **2mc** gradually, as shown by the increase in the absorbance at around 450 nm in Fig. 15.6c, while no change was observed for the absorption band in the visible region. Successive irradiation of the solution by 578 nm light caused the change from **4c** to **4o**.

We thus succeeded in assembling an all-optical lock/unlock system of a thermally irreversible photochromic system composed of **4o**, **2mc** and triethylamine in MeCN (Chart 15.3) [16].

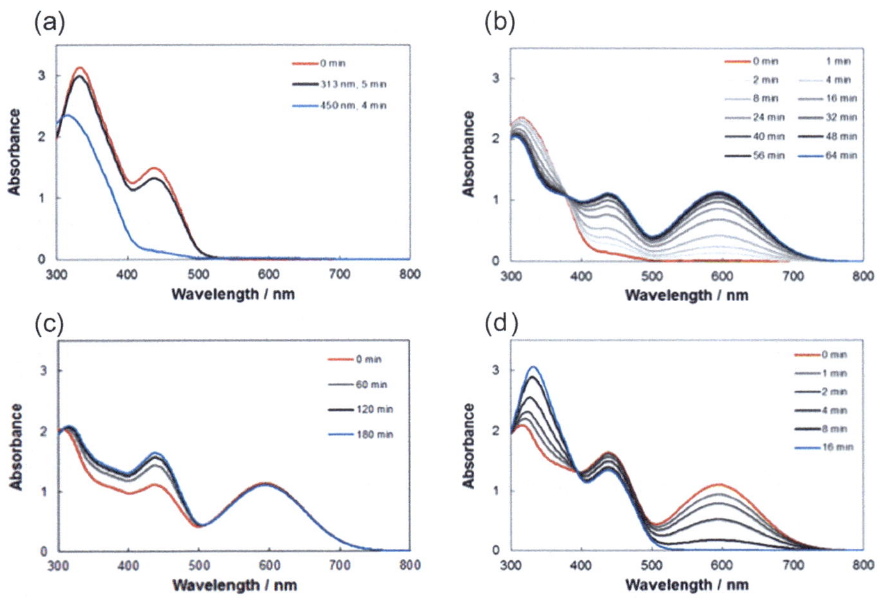

Fig. 15.6 Absorption spectral changes of **4o** and **2mc** in MeCN in the presence of triethylamine at room temperature. **a** Irradiation of 313 nm light (5 min), then 450 nm light (4 min); **b** successive irradiation to the solution of (**a**) with 313 nm light; **c** thermal change of **2sp** to **2mc**; **d** irradiation of 578 nm light to the solution of (**c**). Concentration: **4o** 2.50×10^{-5} mol dm^{-3}, **2mc** 2.50×10^{-5} mol dm^{-3}, Et_3N 2.50×10^{-6} mol dm^{-3}. Light intensity: (a) 313 nm 0.400 mW cm^{-2}; (b) 450 nm 0.420 mW cm^{-2}; (c) 313 nm 0.400 mW cm^{-2}; (d) 578 nm 2.20 mW cm^{-2}. Reprinted with permission from [16]. Copyright 2016 American Chemical Society

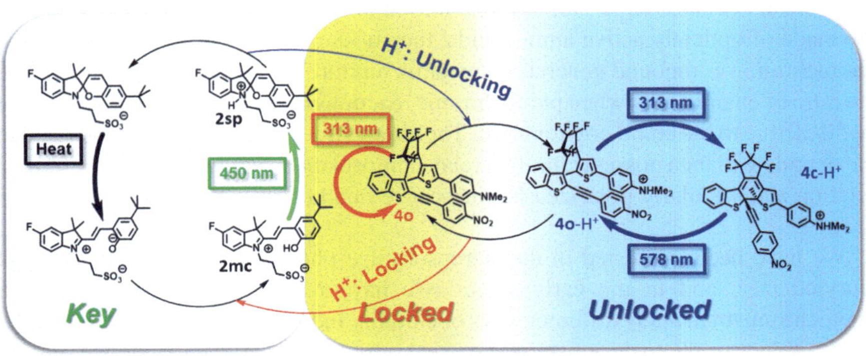

Chart 15.3 Schematic diagram of all-optical lock/unlock system of thermally irreversible photochromic **4o** with thermally reversible photochromic **2mc** as the key. Adapted with permission from [16]. Copyright 2016 American Chemical Society

15.2.3 Section Conclusion

Spiropyran **2** takes the merocyanine form **2mc** in organic solvents due to the existence of the sulfonic acid moiety in the molecule. Despite the zwitterioninc structure, **2mc** is yet soluble in MeCN because of the t-butyl group. Upon visible light irradiation at 450 nm, **2mc** changes to the strong acid **2sp** which transfers a proton to the stronger base such as the phenanthroline in **3c** and the dimethylamino group in **4o**, giving rise to dramatic changes in the principal properties of these photochromic compounds. Namely, the absorption maximum wavelength of **3c** shifted continuously [9], and the firmly locked photochromic ability of **4o** was unlocked by visible light irradiation [16].

15.3 Light-Induced Chirality Transfer from a Peptide Molecule Placed in a Quasi-artificial Medium to Diarylethenes

15.3.1 Photochromic Molecules in Peptides

Photochromic molecules are gaining much more attention than ever because they can be used in biological media as switching units for functions and imaging probes [19]. When such molecules are used in water, they should be water soluble, otherwise they will form aggregates. When those molecules are attached or incorporated in a peptide such as human serum albumin (HSA), they are water soluble, though the size of the probe itself becomes large [20, 21]. In addition to water solubility, because peptides are made of optically active amino acids, they offer chiral environments. When the photochromic compound generates a racemic mixture upon photochromic reaction, the ratio of enantiomers when photochromic reactions occur in the peptide would not be 50/50 due to the chiral environment. The photochromic molecules incorporated in the peptide will then possess chirality-related information such as CD signals, optical rotation and circularly polarized luminescence in addition to their usual properties [22].

We have been interested in the stereoselective photochromism of fulgides and diarylethenes, and in the early stages, we have reported on diastereoselective photochromism in many different ways of chirality induction [23–27]. Subsequently, enantioselective photochromism was carried out in HSA [28, 29]. Here, we present our results showing that the fine-tuning of the reaction media surrounding HSA plays an important role in attaining highly enantioselective photochromic ring closure of diarylethenes [29].

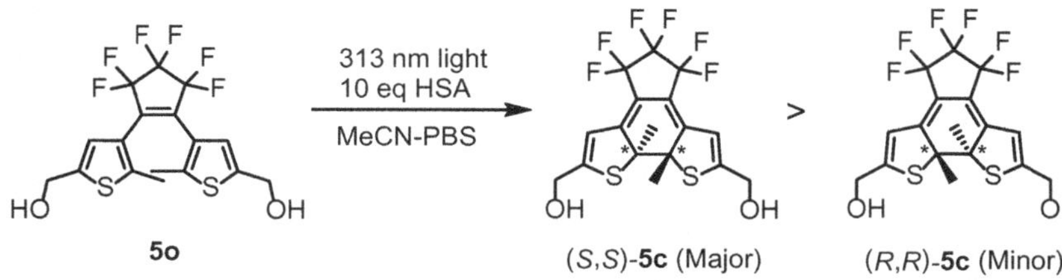

Chart 15.4 Enantioselective photochromic ring closure of diarylethene **5o**

15.3.2 First Enantioselective Photochromism of Diarylethenes in HSA [28]

Diarylthene **5o** (Chart 15.4) was first employed to be incorporated in 10 eq HSA in phosphate buffer solution (PBS) by adding its MeCN solution (the final ratio of MeCN/PBS was 1/99 v/v). The mixture solution was then kept in the dark at room temperature for 24 h (i.e., incubation) to secure complete incorporation in the hydrophobic pockets within HSA [30]. The enantiomer excess (ee) value after 313-nm light irradiation was 63%. When the solution was kept at −4 °C for 24 h before UV irradiation, the ee increased to 71%.

The absolute stereochemistry of the major **5c** generated in HSA was determined by X-ray crystallographic analysis of the bis(4-chlorobenzoyloxy) derivative of the optically resolved **5c** to be (*S*,*S*)-**5c** [31]. HPLC (column: Daicel OD-H) analysis showed that it moves slower than the other enantiomer and exhibited a negative Cotton effect in the CD spectrum in the visible region [28].

15.3.3 Highly Enantioselective Photochromic Ring Closure of Diarylethenes in HSA Attained in Quasi-artificial Media [29]

The enantioselectivity of the diarylethenes with different substituents shown in Chart 15.5 was then examined. Although experiments on **6o**, a dicarboxylic acid with

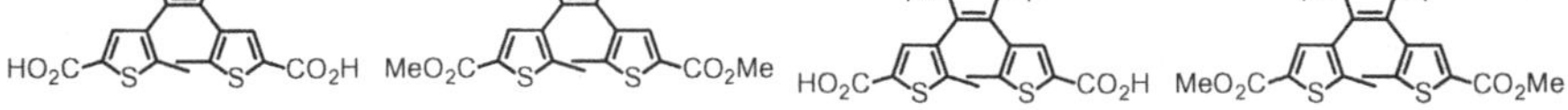

Chart. 15.5 Diarylethenes used in the experiments in HSA with different media compositions

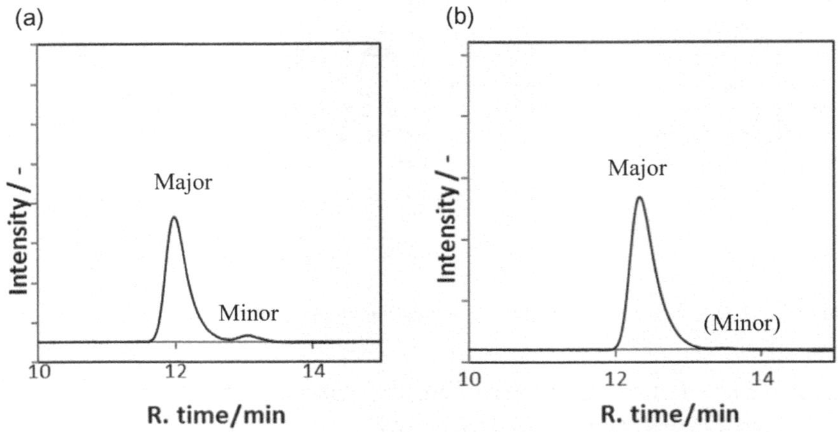

Fig. 15.7 HPLC chart of **9c**: **a** incubated at 25 °C, **b** incubated at −4 °C. HPLC column: Daicel OD-3, 2.1 mm × 150 mm. Eluent: hexane—2-propanol (99.5/0.5 v/v). Flow rate: 0.5 mL/min. Detection wavelength: 585 nm. Reproduced from Ref. [29] with permission from the Royal Society of Chemistry

cyclopentene in place of hexafluorocyclopentene, was not successful since **6c** was not released from HSA after photoirradiation, its methyl ester **7o** led to interesting results.

Using our usual experimental procedures **7o** gave 28% ee. However, when the composition of the reaction medium was changed from MeCN/PBS = 1/99 to 5/95, the ee value increased to 50%. The ee values were then examined by changing the composition of MeCN in PBS. Surprisingly, the ee value of **7c** formed by photochromism increased to 80% when the composition of MeCN was 15%. At 20%, however, it decreased.

The effect of MeCN to the ee values of photochromic ring closure could also be applied to other diarylethenes. Although it did not work effectively for **5o**, it worked well for diarylethenes **8o** and **9o** possessing the perfluorocyclopentene unit. In particular, the ee value of **9c** increased dramatically from 27% when MeCN was 1% to as high as 94% when MeCN was 15%.

When the incubation temperature for 24 h was lowered, the ee values increased further. For **9c**, the minor enantiomer could hardly be detected by HPLC (Fig. 15.7). These results are summarized in Table 15.1 [29].

15.3.4 Determination of the Absolute Stereochemistry of the Major Enantiomers Formed in HSA [29]

Our next interest was to determine the absolute stereochemistry of the major closed forms generated in HSA. We have already determined the major enantiomer of **5c**

Table 15.1 Enantioselectivity of diarylethenes in HSA in different media composition incubated at 25 °C

MeCN in PBS (vol) (%)	Ee values/%		
	7c	**8c**	**9c**
1	28	33	27
5	50	40	72
10	76	44	92
15	**80** (94)[a]	**48** (56)[a]	**94** (>99)[a]
20	74	40	91

Reproduced from Ref. [29] with permission from the Royal Society of Chemistry

[a]Ee values in parentheses are those when incubated at −4 °C

formed in HSA to be *(S,S)*-**5c** by X-ray crystallographic analysis of a derivative. Since the basic molecular skeleton of **8c** and **9c** is common to **5c**, we undertook the transformation experiments between **5c**, **8c** and **9c**.

Although it seemed easy, the correlation between **5c** and **8c** was very difficult to establish. Oxidation of optically resolved **5c** with various oxidizing agents failed. The reduction of **9c** was also unsuccessful. Only one successful transformation was the borane reduction of **8c** to **5c**. Esterification of **8c** or the hydrolysis of **9c** by acid- or base-catalyzed reactions also failed. Esterification of **8c** by trimethylsilyldiazomethane was the only way to connect **8c** with **9c**.

As a result, it was concluded that the main enantiomer of **8c** is (*R*,*R*)-**8c** and the main enantiomer of **9c** is (*S*,*S*)-**9c**. The transformation pathways, the absolute stereochemistry of the major and minor enantiomers obtained in HSA and the their properties related to their behavior in the HPLC and CD spectra are all shown in Chart 15.6 [29].

15.3.5 Section Conclusion

The enantioselective ring closure of diarylethenes incorporated in HSA and the determination of the absolute stereochemistry of the major enantiomers were carried out [29]. Since HSA is a peptide made of chiral amino acids, the transfer of chirality from HSA to the diarylethene molecules during photochemical electrocyclization was efficiently achieved. Since diarylethenes are photochromic, repeated generation and extinction of the chiral molecules on demand are now possible.

It is noteworthy that since serum albumin is a water-soluble peptide which has no particular activities as an enzyme and just transports water-insoluble materials, serum albumin can convey water-insoluble photochromic molecules to any area within an animal through the blood vessels as molecularly isolated forms.

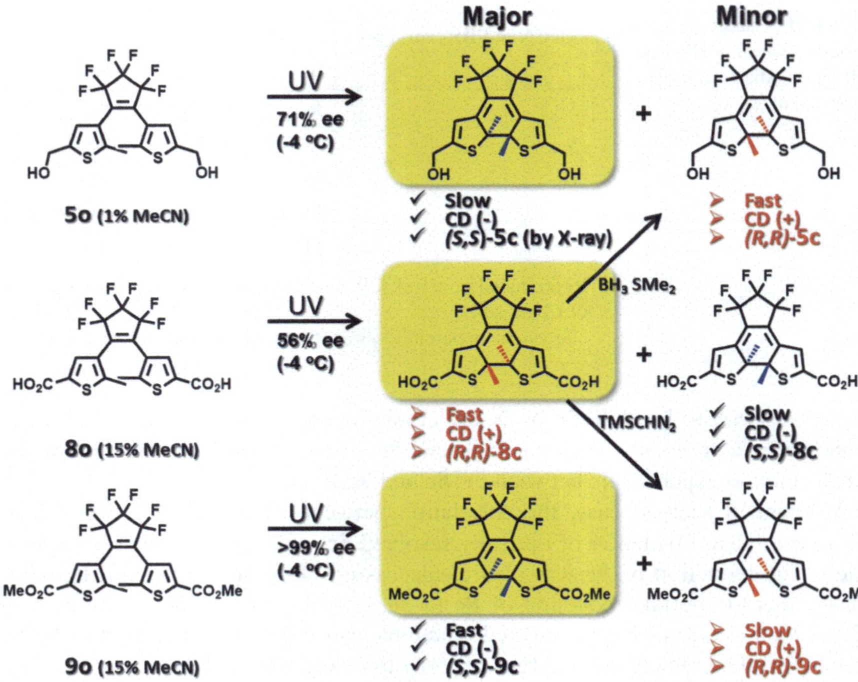

Chart 15.6 Absolute stereochemistry of the major and minor closed forms generated in HSA, their properties and the chemical correlation pathways between the closed forms. Reproduced from Ref. [29] with permission from the Royal Society of Chemistry

15.4 Short-Step Construction of Photochromic Hexatriene Skeletons

In order to realize "integrated multi-responsive photochromic systems," a strong requirement is for the synthesis of photochromic molecules be easy. Here, we show the results of our efforts to develop facile and simple ways for the synthesis of the hexatriene skeleton which may exhibit either thermally irreversible or thermally reversible photochromism depending on their electronic/steric nature.

15.4.1 One-Step Assembly of Arylbutadienes [32]

We have previously reported the synthesis of a thermally irreversible photochromic arylbutadiene **10o** [33]. It exhibited thermally irreversible photochromism, and its absorption maximum wavelength was 445 nm in toluene. If the substituents which may induce electronic perturbation to the π-conjugation system such as the phenyl group or the hexafluoropropano-bridge are carefully removed, the absorption maximum in the visible region may shift toward shorter wavelengths. If the absorption band in the longest wavelength is hidden in the UV region, photochromism will occur by the alternate irradiation of two different UV lights. In such a case, the molecular structure must be exceptionally simple, and the simple structure can considerably facilitate its synthesis. We have thus designed the arylbutadiene **11o** with its possible synthetic pathway depicted in Chart 15.7.

The coupling reaction of two different carbonyl substrates may lead to two homo-coupling products and one cross-coupling product. If the reaction proceeds statistically, the chemical yield of **11o**, the cross-coupling product, does not exceed 50%. However, the McMurry coupling reaction proceeded smoothly to provide **11o** in 65% yield. These results are in good accordance with the fact that the McMurry coupling of an aromatic ketone and an aliphatic ketone predominantly yields the cross-coupling product rather than the homo-coupling products [34].

As expected, the absorption band of **11c** was entirely in the UV region. The photochromic reaction was therefore conducted with two different UV-light irradiations, namely 313 and 366 nm. The spectral changes during photoirradiations are shown in Fig. 15.8. It could thus be confirmed that **11o** shows "invisible photochromism" [35] or "stealth photochromism" [32].

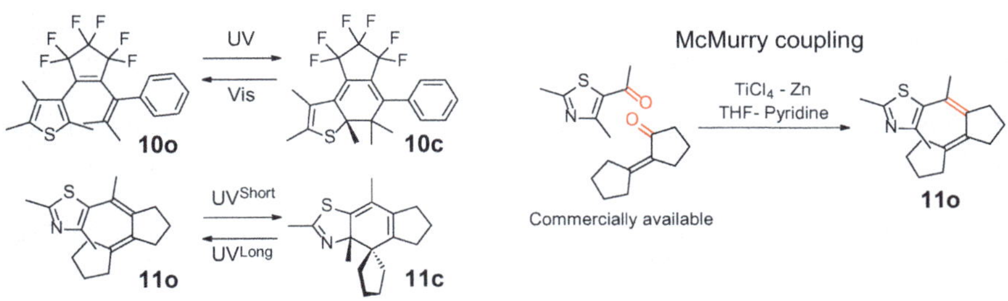

Chart 15.7 Photochromism of arylbutadienes and the one-step synthesis method of **11o**

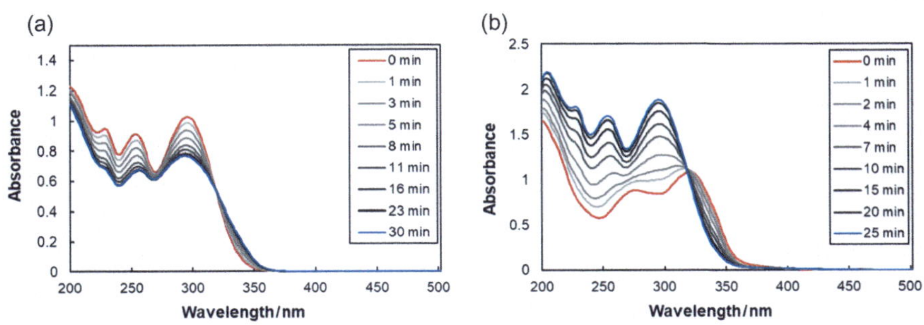

Fig. 15.8 Absorption spectral change of **11** during photoirradiation: **a** irradiation of 313 nm light (0.366 mW cm^{-2}) to hexane solution of **11o** (1.12×10^{-4} mol dm^{-3}), **b** irradiation of 366 nm light (5.23 mW cm^{-2}) to the hexane solution of **11c** (2.38×10^{-4} mol dm^{-3}). Reprinted with permission from [32]. Copyright 2018 American Chemical Society

15.4.2 "Click" Reactions for the Construction of Teraryls [36]

In a previous work, we reported the synthesis and photochromic reactions of **12o** and its acetal **13o** [37], followed by a report on **14o** [27]. These molecules were synthesized by way of the symmetric acetylene **15** as the common intermediate obtained through three steps from 4-iodo-5-methyl-2-phenyl-1,3-thiazole and trimethylsilylacetylene.

The concept of "click chemistry" was proposed by Sharpless and co-workers early in the twenty-first century to express secure, quick, selective, widely applicable and facile reactions [38]. The Cu(I)-catalyzed Huisgen reaction, a representative example of click chemistry, connects two functional groups with the [2 + 3] cycloaddition reaction between an organic azide group and a terminal acetylene group. On the other hand, when Ru(I) is used as the catalyst, the reaction proceeds between an azide group and the internal acetylene. Therefore, when an internal acetylene **15** and an alkyl azide are employed for the Ru(I)-catalyzed Huisgen reaction, the teraryl (4,5-bisthiazolyl-1-substituted-1,2,3-triazole) is formed. We therefore undertook the synthesis of novel bisthiazolyltriazoles from **15** and benzyl azide by the Ru(I)-catalyzed Huisgen reaction. In addition to **15**, **16** and **17** were employed to observe the electronic effects of the substituents attached to the phenyl groups on the photochromic properties of bisthiazolyltriazoles. We thus synthesized bisthiazolyltriazoles **18-20** (Chart 15.8).

Since most of the reported photochromic teraryls are known to show thermally reversible photochromism [39], we expected such properties for these compounds as well. Irradiation of 313-nm light to the MeCN solution of bisthiazolyltriazoles **18o-20o** led to coloration which were all thermally reversible. The changes in the absorption spectra of **18o-20o** in MeCN at 28 °C during 313-nm light irradiation are shown in Fig. 15.9.

The absorption maximum wavelengths of their closed forms are as follows: **18c**: 695 nm, **19c**: 696 nm and **20c**: 739 nm. Although introduction of the electron-donating methoxy groups did not affect the absorption maximum wavelength of the closed form, electron-withdrawing cyano groups induced a large bathochromic shift.

The thermal back reactions (Fig. 15.10) of each compound were examined at three different temperatures, and the first-order reaction rate constants *k* were determined.

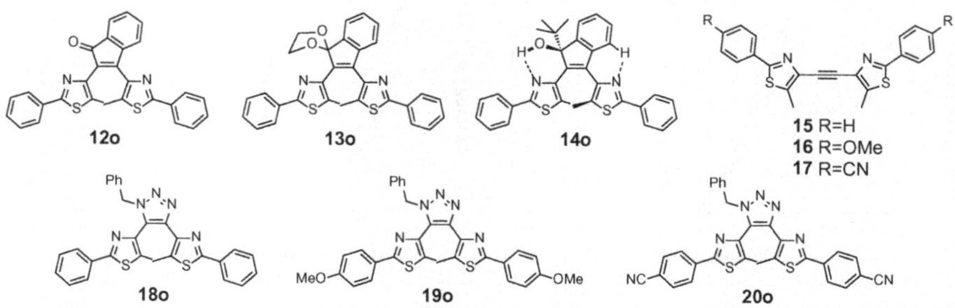

Chart 15.8 Novel teraryls and related compounds

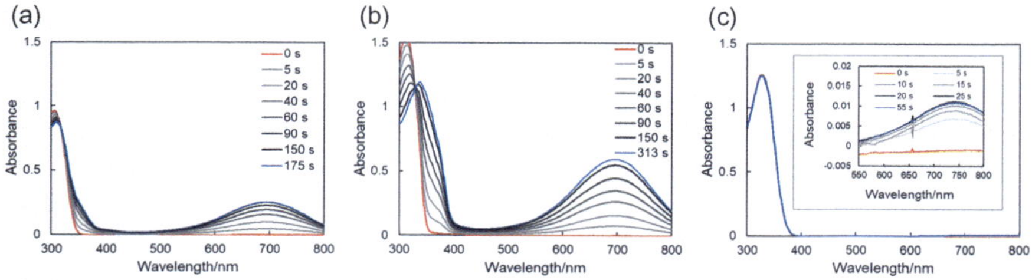

Fig. 15.9 Absorption spectral change of **18o–20o** during 313-nm light irradiation in MeCN at 28 °C: **a 18o** (4.44×10^{-5} mol dm^{-3}), **b 19o** (4.38×10^{-5} mol dm^{-3}), **c 20o** (4.73×10^{-5} mol dm^{-3}). Light intensity: 1.8 mW cm^{-2}. Reprinted from Ref. [36] by the Authors licensed under CC BY 4.0

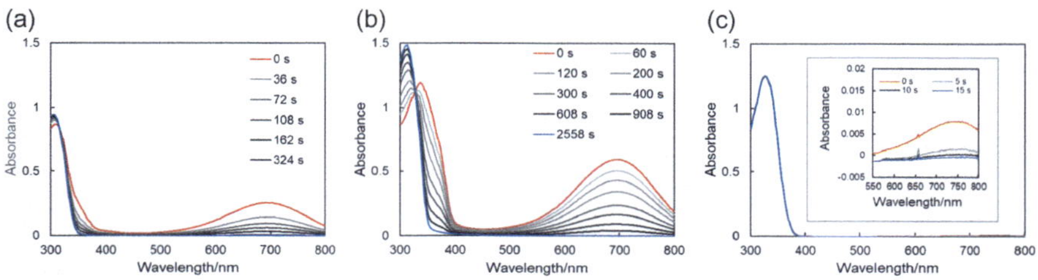

Fig. 15.10 Absorption spectral change of **18c–20c** kept at 28 °C in MeCN. Concentration of compounds is the same as Fig. 15.9. **a 18c**. **b 19c**. **c 20c**. Reprinted from Ref. [36] by The Authors licensed under CC BY 4.0

Table 15.2 Kinetic data of thermal back reactions of **18c-22c** and aromatic stabilization energy of their corresponding aromatic rings

	A/s^{-1}	E_a/kJ mol^{-1}	k (293 K)/s^{-1}	$t_{1/2}$ (293 K)[a]	ASE[b]/kJ mol^{-1}
21c[c]	7.1×10^{11}	112	6.7×10^{-9}[e]	3.3 years	72.9
22c[c]	1.3×10^{9}	85	8.7×10^{-7}[e]	9.2 days	78.6
18c[d]	1.54×10^{13}	87.4	4.18×10^{-3}	166 s	102.0
19c[d]	2.27×10^{12}	86.4	9.54×10^{-4}	727 s	102.0
20c[d]	3.88×10^{13}	82.2	8.99×10^{-2}	7.7 s	102.0

Reprinted from Ref. [36] by The Authors licensed under CC BY 4.0
[a] $t_{1/2}$ (293 K): Half-life at 293 K
[b] ASE: Aromatic stabilization energy of the central aromatic rings when unsubstituted. Data taken from Ref. [42]
[c] In toluene. Data taken from Ref. [40] for **21** and Ref. [41] for **22**
[d] In MeCN
[e] Calculated from $t_{1/2}$ at 293 K

The Arrhenius plots of ln k against $1/T$ gave preexponential factors A and Arrhenius activation energies E_a. The kinetic data of the thermal back reactions of **18c-20c** are shown in Table 15.2, together with literature data on related compounds **21c** [40] and **22c** [41], as shown in Chart 15.9.

The thermal back reaction rate is faster when the aromatic stabilization energy [42] of the central aromatic ring is large. It is natural that the closed forms lose aromaticity for three aromatic rings simultaneously when they cyclize, and the thiazole rings on both sides are common to all compounds.

However, when the thermal back reaction rates of **18c–20c** are compared, that of **20c** is extremely fast though it gradually slows down when the electron-donating character of the substituent on the para positions of the peripheral phenyl groups is larger. This fact is best explained by the push–pull conjugation and resonance structures induced by the substituents. As shown in Chart 15.10, the resonance structures in the brackets are regarded as the intermediates of the thermal back reaction, in which those of **20c** are the most stable because the delocalized electrons on the phenyl groups are stabilized by the cyano groups on their para positions. The thermal

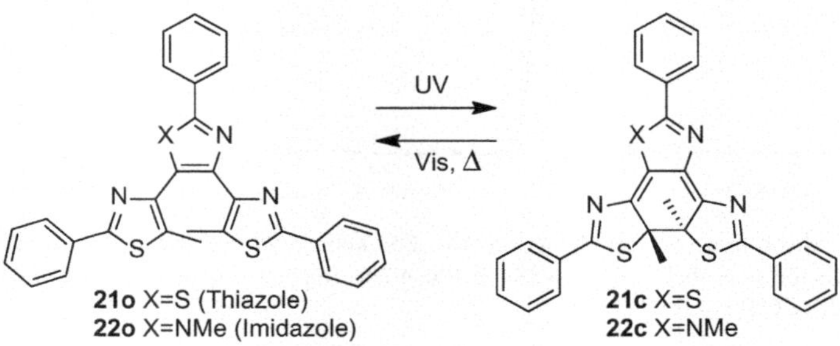

Chart 15.9 Thermally reversible photochromism of closely related teraryls

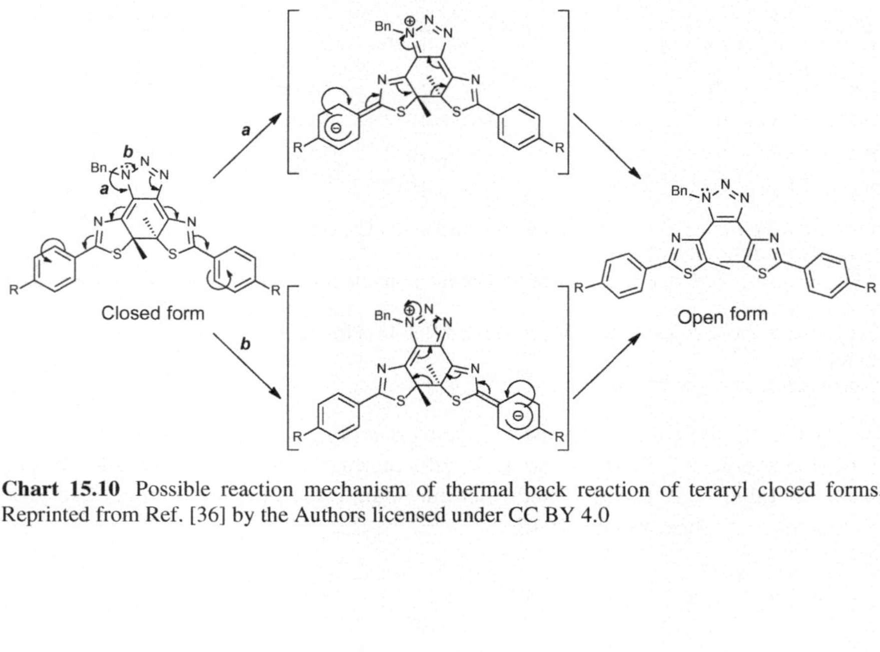

Chart 15.10 Possible reaction mechanism of thermal back reaction of teraryl closed forms. Reprinted from Ref. [36] by the Authors licensed under CC BY 4.0

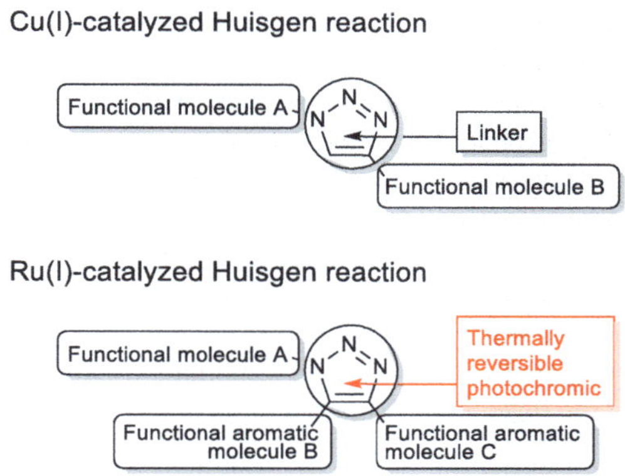

Chart 15.11 Image of conglomerates with highly integrated functions constructed by Ru(I)-catalyzed Huisgen reactions

back reaction rate of teraryls can thus be controlled by the kind of central aromatic rings having different aromatic stabilization energies as well as by the substituents on the peripheral phenyl rings.

Since the Huisgen reactions have so far been used only to connect two functional molecules by an azide group on one molecule and a terminal acetylene unit on the other molecule, the triazole ring thus formed has no particular function, just working as the linker of the two molecules. However, when the Ru(I)-catalyzed Huisgen reaction is employed, three functional groups can gather on the triazole ring. In addition, the 4,5-bisaryltriazole group itself exhibits thermally reversible photochromism. This protocol can, therefore, afford functionally highly integrated molecules (Chart 15.11).

15.4.3 Section Conclusion

In order to create photochromic materials in which functional groups are integrated inter- or intra-molecularly, simple and facile ways to construct the photochromic parts are required. In this section, we have shown two different methods of constructing photochromic hexatriene moieties.

The first method is to use 5-acetyl-2,4-dimethyl-1,3-thiazole and 2-cyclopentylidenecyclopentanone as the commercially available units for the McMurry cross-coupling reaction. The reaction proceeds to give the desired cross-coupling product in 65% yield, which exceeded the statistically expected chemical yield of 50%. The product showed thermally irreversible photochromism in the UV region so that it is referred to as "stealth photochromism" [32].

The second method is to use the Ru(I)-catalyzed Huisgen "click" chemistry between benzyl azide and 1,2-bis(4-methy-2-phenyl-1,3-thiazol-5-yl)ethyne. Different from the Cu(I)-catalyzed Huisgen reaction which works only on the terminal acetylene, the Ru(I)-catalyzed Huisgen reaction works on the internal acetylene. The reaction product is 1-benzyl-4,5-bis(4-methyl-2-phenyl-1,3-thiazol-5-yl)-1,2,3-triazole, which showed thermally reversible photochromism due to the loss of large aromatic stabilization energy upon photocyclization of the hexatriene moiety. It is noteworthy that the thermal ring opening reaction rate can be controlled by the substituents attached to the phenyl rings on the thiazole units [36].

While the triazole unit formed by the Cu(I)-catalyzed Huisgen reaction worked only as a linker between two independent functional groups, the 1,4,5-trisubstituted triazole unit formed by the Ru(I)-catalyzed Huisgen reaction works to build highly integrated functional conglomerates in which the bisaryltriazole array itself works as the thermally reversible photochromic gear.

15.5 Conclusion

The creation of functional photochromic systems in which multiple molecules work cooperatively to produce and transmit information other than color change has been the focus of our most recent research.

At first, we investigated a photochromic spiropyran which generates strong protic acid by visible light irradiation. The proton thus formed was able to continuously change the color of a photochromic dye [9]. The proton generated by visible light also unlocked the photochromic nature of the diarylethene, which otherwise did not show any photochromism with UV light irradiation. This combination of visible and UV light irradiation can thus be regarded as a fail-safe mechanism for this photochromic system [16].

Secondly, we have succeeded in the transfer of chirality with very high selectivity from naturally occurring human serum albumin (HSA) to diarylethenes. It was achieved by the careful selection of the functional groups attached to the diarylethene and by meticulous adjustment of the environmental solution composition [29].

Finally, since the synthesis of photochromic molecules with specific and useful functions requires much effort and preparation, the development of simple methods to construct the hexatriene system is essential. One of the two ways we have described here is to conduct the McMurry cross-coupling reaction of an acetyl-substituted aromatic compound and an enone compound. We used commercially available compounds and succeeded in preparing a stealth photochromic compound in just one step [32]. The other way is to use the Ru(I)-catalyzed Huisgen reaction. It is a type of "click" reaction and is secure, quick, stereochemically selective, facile and can be widely applied. We prepared three 1,2-bis(thiazolyl)-substituted acetylenes which were reacted with benzyl azide. The products showed thermally reversible photochromism, and the thermal back reaction rates were dependent on the nature of the substituents attached to the phenyl groups on the peripheral of the molecules [36].

Thus, "molecularly integrated multi-responsive photochromic systems" could be synthesized through several pathways. We hope these findings can be considered milestones to advance research and lead to useful applications in the field of photochromism.

Acknowledgements The author is deeply grateful to Professors H. Miyasaka (Osaka University), T. Kawai (NAIST), J. Abe (Aoyama Gakuin University) and K. Matsuda (Kyoto University) for the kind assistance and guidance during the research period. Thanks are extended to Drs. T. Ubukata, I. Kawamura and T. Nakagawa of Yokohama National University for their great help to conduct the research, and to the students and postdocs of our laboratory who have carried out the experiments with great efforts. The work shown in this chapter was supported by JSPS KAKENHI Grant Number 26107009 in Scientific Research on Innovative Areas "Photosynergetics," and MEXT KAKENHI Grant Number JP19050004 in Grant-in-Aid for Scientific Research in Priority Areas "New Frontiers in Photochromism (No. 471)."

References

1. Dürr H, Bouas-Laurent H (eds) (2003) Photochromism: Molecules and systems. Elsevier, Amsterdam (Revised Ed.)
2. Crano JC, Guglielmetti RJ (eds) (1999) Organic photochromic and thermochromic compounds, vol 1 and 2. Plenum Press, New York
3. Irie M, Yokoyama Y, Seki T (eds) (2013) New frontiers in photochromism. Springer, Tokyo
4. Tian H, Zhang J (eds) (2016) Photochromic materials: preparation, properties and applications. Wiley-VCH, Weinheim
5. Yokoyama Y, Nakatani K (eds) (2017) Photon-working switches. Springer, Tokyo
6. Irie M (2000) Diarylethenes for memories and switches. Chem Rev 100:1685–1716
7. Tian H, Yang S (2004) Recent progresses on diarylethene based photochromic switches. Chem Soc Rev 33:85–97
8. Irie M, Fukaminato T, Matsuda K, Kobatake S (2014) Photochromism of diarylethene molecules and crystals: memories, switches, and actuators. Chem Rev 114:12174–12277
9. Kusumoto S, Nakagawa T, Yokoyama Y (2016) All-optical fine-tuning of absorption band of diarylethene with photochromic acid-generating spiropyran. Adv Opt Mater 4:1350–1353
10. Shi Z, Peng P, Strohecker D, Liao Y (2011) Long-lived photoacid based upon a photochromic reaction. J Am Chem Soc 133:14699–14703
11. Hoefnage AJ, Hoefnagel MA, Wepster BM (1976) Substituent effects 5. Resonance saturation effects in anilines having a π-donor group in para position. J Am Chem Soc 98:6194–6197
12. Fickling MM, Fischer A, Mann BR, Packer J, Vaughan J (1959) Hammett substituent constants for electron-withdrawing substituents: Dissociation of phenols, anilinium ions and dimethylanilinium ions. J Am Chem Soc 81:4226–4230
13. Uehara S, Hiromoto Y, Minkovska S, Suzuki K, Ubukata T, Yokoyama Y (2012) Photochromic behavior of a bisthienylethene bearing Cu(I)-1,10-phenanthroline complexes. Dyes Pigm 92:861–867
14. Yokoyama Y, Hiromoto Y, Takagi K, Ishii K, Delbaere S, Watanobe Y, Ubukata T (2015) Solubility control of organic acid-base salts by photochromism. Dyes Pigm 114:1–7
15. Dean JA (ed) (1985) Lange's handbook of chemistry, vol 5–53, 13th edn. McGraw-Hill, New York
16. Mahvidi S, Takeuchi S, Kusumoto S, Sato H, Nakagawa T, Yokoyama Y (2016) Gated photochromic system of diarylethene with a photon-working key. Org Lett 18:5042–5045
17. Irie M, Sayo K (1992) Solvent effects on the photochromic reactions of diarylethene derivatives. J Phys Chem 96:7671–7674

18. Grantham PH, Weisburger EK, Weisburger JH (1961) Ionization constants of derivatives of fluorene and other polycyclic compounds. J Org Chem 26:1008–1017
19. Szymański W, Beierle JM, Kistemaker HAV, Velema WA, Feringa BL (2013) Reversible photocontrol of biological systems by the incorporation of molecular photoswitches. Chem Rev 113:6114–6178
20. Fu Y, Han H-H, Zhang J, He X-P, Feringa BL, Tian H (2018) Photocontrolled fluorescence "double-check" bioimaging enabled by a glycoprobe-protein hybrid. J Am Chem Soc 140:8671–8674
21. Putri RM, Zulfikri H, Fredy JW, Juan A, Tananchayakul P, Cornelissen JJLM, Koay MST, Filippi C, Katsonis N (2018) Photoprogramming allostery in human serum albumin. Bioconjugate Chem 29:2215–2224
22. Nakagawa T, Ubukata T, Yokoyama Y (2018) Chirality and stereoselectivity in photochromic reactions. J Photochem Photobiol C: Photochem Rev 34:152–191
23. Yokoyama Y, Uchida S, Yokoyama Y, Sugawara Y, Kurita Y (1996) Diastereoselective photochromism of (R)-binaphthol-condensed indolylfulgide. J Am Chem Soc 118:3100–3107
24. Yokoyama Y, Shiraishi H, Tani Y, Yokoyama Y, Yamaguchi Y (2003) Diastereoselective photochromism of a bisbenzothienylethene governed by steric as well as electronic interactions. J Am Chem Soc 125:7194–7195
25. Yokoyama Y, Shiozawa T, Tani Y, Ubukata T (2009) A unified strategy for exceptionally high diastereoselectivity in the photochemical ring closure of chiral diarylethenes. Angew Chem Int Ed 48:4521–4523
26. Shiozawa T, Hossain MK, Ubukata T, Yokoyama Y (2010) Ultimate diastereoselectivity in the ring closure of photochromic diarylethene possessing facial chirality. Chem Commun 46:4785–4787
27. Ogawa H, Takagi K, Ubukata T, Okamoto A, Yonezawa N, Delbaere S, Yokoyama Y (2012) Bisarylindenols: Fixation of conformation leads to exceptional properties of photochromism based on 6π-electrocyclization. Chem Commun 48:11838–11840
28. Fukagawa M, Kawamura I, Ubukata T, Yokoyama Y (2013) Enantioselective photochromism of diarylethenes in human serum albumin. Chem Eur J 19:9434–9437
29. Kawamura K, Osawa K, Watanobe Y, Saeki Y, Maruyama N, Yokoyama Y (2017) Photocyclization of photoswitches with high enantioselectivity in human serum albumin in an artificial environment. Chem Commun 53:3181–3184
30. Peters T Jr (1996) All about albumin: biochemistry, genetics, and medical applications. Academic Press, San Diego
31. Yokoyama Y, Hosoda N, Osano YT, Sasaki C (1998) Absolute stereochemistry and CD spectra of resolved enantiomers of the colored form of a photochromic dithienylethene. Chem Lett 27:1093–1094
32. Kochi J, Ubukata T, Yokoyama Y (2018) One-step synthesis and stealth photochromism of arylbutadienes. J Org Chem 83:10695–10700
33. Shrestha SM, Nagashima H, Yokoyama Y, Yokoyama Y (2003) Synthesis of novel thermally irreversible photochromic 1-aryl-1,3-butadiene derivatives. Bull Chem Soc Jpn 76:363–367
34. Duan X-F, Zeng J, Lu J-W, Zhang Z-B (2006) Insights into the general and efficient cross McMurry reactions between ketones. J Org Chem 71:9873–9876
35. Fukaminato T, Tanaka M, Kuroki L, Irie, M (2008) Invisible photochromism of diarylethene derivatives. Chem Commun 3924–3926
36. Zhang C, Morinaka K, Kose M, Ubukata T, Yokoyama Y (2019) Click chemistry towards thermally reversible photochromic 4,5-bisthiazolyl-1,2,3-triazoles. Beilstein J Org Chem 15:2161–2169
37. Morinaka K, Ubukata T, Yokoyama Y (2009) Structurally versatile novel photochromic bisarylindenone and its acetal: achievement of large cyclization quantum yield. Org Lett 11:3890–3893
38. Kolb HC, Finn MG, Sharpless KB (2001) Click chemistry: diverse chemical function from a few good reactions. Angew Chem Int Ed 40:2004–2021

39. Kawai S, Nakashima T, Atsumi K, Sakai T, Harigai M, Imamoto Y, Kamikubo H, Kataoka M, Kawai T (2007) Novel photochromic molecules based on 4,5-dithienyl thiazole with fast thermal bleaching rate. Chem Mater 19:3479–3483
40. Nakashima T, Atsumi K, Kawai S, Nakagawa T, Hasegawa Y, Kawai T (2007) Photochromism of thiazole-containing triangle terarylenes. Eur J Org Chem 3212–3218
41. Nakashima T, Goto M, Kawai S, Kawai T (2008) Photomodulation of ionic interaction and reactivity: reversible photoconversion between imidazolium and imidazolinium. J Am Chem Soc 130:14570–14575
42. Cyranski MK, Krygowski TM, Katritzky AR, PvR Schleyer (2002) To what extent can aromaticity be defined uniquely? J Org Chem 67:1333–1338

Chapter 16
Efficient Singlet Fission in Acene-Based Molecular Assemblies

Taku Hasobe

Abstract In this chapter, we introduce singlet fission and related photophysical processes of acene derivatives such as pentacene and tetracene dimers and supramolecular assemblies with gold nanoclusters. Time-resolved spectroscopic measurements revealed the long-lived and high-yield triplet excited state through singlet fission. Moreover, light energy conversion processes such as electron transfer and singlet oxygen generation through singlet fission are discussed.

Keywords Singlet fission · Acene derivatives · Gold nanoclusters

16.1 Introduction

Singlet fission (SF) is a spin-allowed photophysical pathway, in which one singlet exciton (S_1) is converted into two triplet excitons ($2 \times T_1$) in two neighboring molecules [1–3]. It is a promising process to improve light energy conversion properties (i.e., photovoltaics and photocatalysis) [4–6]. To proceed the SF process, the energy-matching condition between the energy of the lowest-lying singlet excited state $E(S_1)$ and energy of two triplet excited states $2E(T_1)$ [i.e., $E(S_1) \geq 2\ E(T_1)$] is definitely required. SF initially enables to generate correlated triplet pairs [(TT)] from a singlet excited state. Then, it subsequently dissociates to the two individual triplet states [$(2 \times T_1)$]. As the reverse process from the correlated triplet state (TT), triplet–triplet annihilation (TTA) also prevents generating $(2 \times T)$ because of the competitive reaction with the dissociation process of (TT) to $(2 \times T_1)$.

Regarding SF molecules, polyacene derivatives satisfy the above condition. Therefore, pentacene derivatives are one of the representative SF compounds considering $E(S_1) = 2.1$ eV and $E(T_1) = 0.8$ eV [7–12]. However, the serious problem of pentacene is relatively low-excitation energy of excited triplet states. This means that the oxidation and reduction properties through triplet excited states become relatively weak. In contrast, tetracene (Tc) is a good candidate for SF considering

T. Hasobe (✉)
Department of Chemistry, Faculty of Science and Technology, Keio University, 3-14-1 Hiyoshi, Kohoku-ku, Yokohama, Kanagawa 223-8522, Japan
e-mail: hasobe@chem.keio.ac.jp

H. Miyasaka et al. (eds.), *Photosynergetic Responses in Molecules and Molecular Aggregates*, https://doi.org/10.1007/978-981-15-5451-3_16

$E(S_1) = 2.6$ eV and $E(T_1) = 1.3$ eV. No attempt has been made to observe the quantitative SF of Tc dimers and molecular assemblies in homogeneous solution because of the competitive reaction between SF and TTA.

Based on the above points, in this chapter, recent developments of singlet fission using tetracene and pentacene derivatives are discussed. In addition to the simple molecular dimers, macroscopic assemblies of pentacene and tetracene using gold nanoparticles were employed. The detailed structural and photophysical properties are discussed here.

16.2 Covalently Linked Pentacene Dimers Bridged by a Phenylene Spacer at *Ortho* and *Meta* Positions

First, simple dimeric forms of pentacene derivatives with different orientations are discussed. We synthesized a series of pentacene dimers linked by a phenylene spacer at *ortho* and *meta* positions to investigate intramolecular orientation-related multi-exciton dynamics [denoted as *o*-$(Pc)_2$ and *m*-$(Pc)_2$] (Fig. 16.1). Especially, in this system, we focused on the singlet fission (SF) and recombination process from correlated triplet pairs (TT) to individual triplet states ($2 \times T_1$). Absorption and electrochemical measurements of *o*-$(Pc)_2$ suggested stronger intramolecular couplings as compared to *m*-$(Pc)_2$. Figure 16.2a demonstrated femtosecond (fs) transient absorption spectra of *o*-$(Pc)_2$ in toluene. The S-S absorption of pentacene unit was initially observed after lase-pulse excitation. After a few picoseconds, the characteristic T–T absorption band of pentacene units increases at 520 nm, whereas the S-S absorption bands at 470 and 555 nm decreased within ca. 15 ps. Therefore, ultrafast intramolecular SF occurs in *o*-$(Pc)_2$. The photophysical trend of *m*-$(Pc)_2$ is largely different from that of *o*-$(Pc)_2$. The corresponding time profile at 520 nm (Fig. 16.2a) revealed the much slower SF process as compared to *o*-$(Pc)_2$. The SF rate constant (k_{SF}) of *m*-$(Pc)_2$ was determined to be 2.1×10^9 s^{-1}, which is much smaller than that of *o*-$(Pc)_2$ (1.2×10^{11} s^{-1}). Namely, the k_{SF} value of *o*-$(Pc)_2$ is two orders of magnitude larger as compared to *m*-$(Pc)_2$. The difference is largely attributable to the chromophore

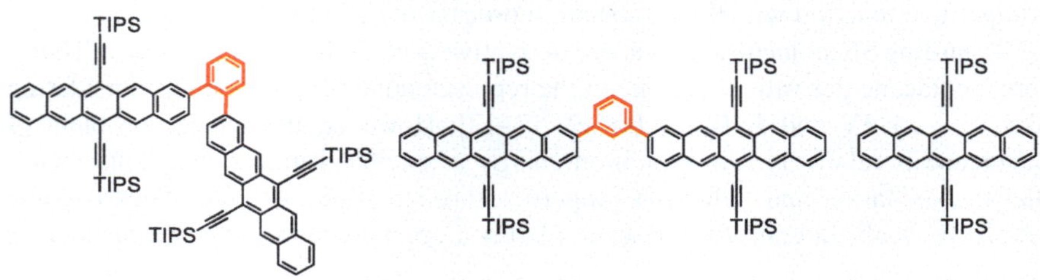

o-$(Pc)_2$ *m*-$(Pc)_2$ Pc-Ref

Fig. 16.1 Structures of pentacene dimers and a reference compound in this study. Reprinted with permission from Ref. [13]. Copyright 2018 American Chemical Society

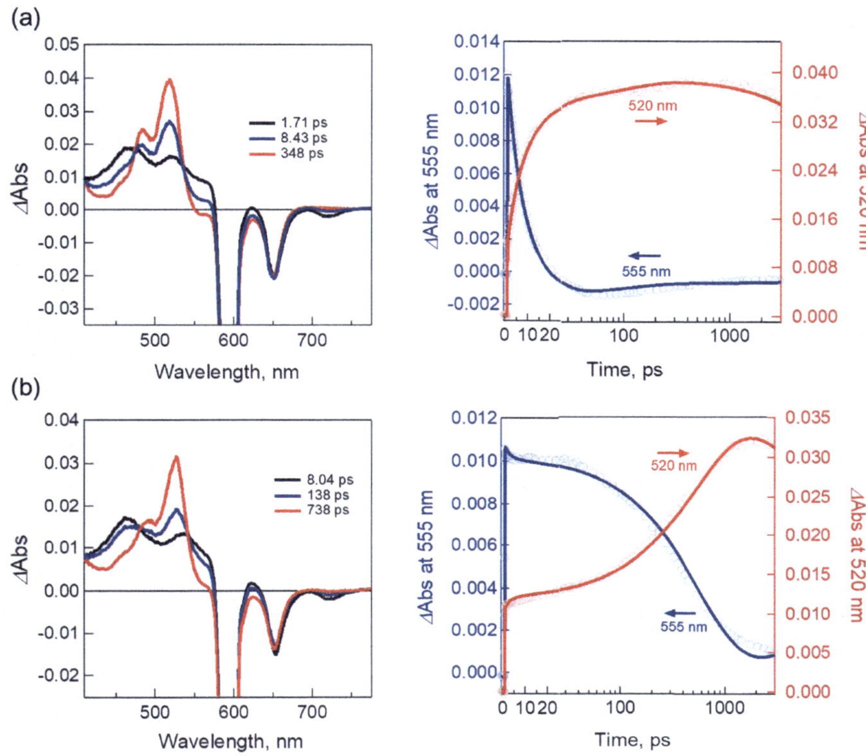

Fig. 16.2 Time-resolved femtosecond transient absorption spectra (left) and corresponding time profiles (right) of **a** *o*-$(Pc)_2$ and **b** *m*-$(Pc)_2$ in toluene (λ_{ex} = 600 nm). Reprinted with permission from Ref. [13]. Copyright 2018 American Chemical Society

coupling between *o*-$(Pc)_2$ and *m*-$(Pc)_2$. The triplet quantum yields of (TT) in these dimers (Φ_{SF}) were accordingly estimated. Noted that theoretical maximum value of Φ_{SF} is 100%. In both systems, we observed efficient Φ_{SF} values such as 99% for *o*-$(Pc)_2$ and 94% for *m*-$(Pc)_2$.

In nanosecond transient absorption measurements, the dissociation process from the (TT) to the individual triplets [(2 × T)] in *m*-$(Pc)_2$ is in sharp contrast to a major recombination process of *o*-$(Pc)_2$. The yields of Φ_T in *o*-$(Pc)_2$ and *m*-$(Pc)_2$ were calculated to be 20% and 105%, respectively. Moreover, time-resolved electron spin resonance measurements strongly suggested that the recombination and the dissociation proceed from quintet state of ^{5}(TT) in *m*-$(Pc)_2$ [14, 15].

Based on the above discussions, we can propose the kinetic models for SF in this system as shown in Scheme 16.1. The summarized parameters were shown in Table 16.1. The emphasis is on the increased k_{DISS} value of *m*-$(Pc)_2$ relative to that

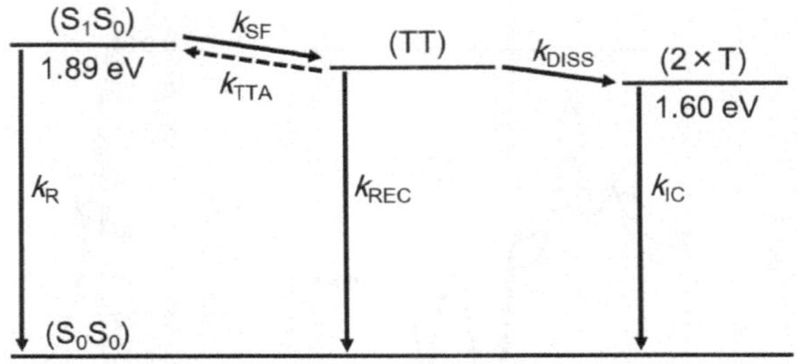

Scheme 16.1 Kinetic process in *o*-$(Pc)_2$ and *m*-$(Pc)_2$. Reprinted with permission from Ref. [13]. Copyright 2018 American Chemical Society

Table 16.1 Summarized Kinetic Parameters and Triplet Quantum Yields in *o*-$(Pc)_2$ and *m*-$(Pc)_2$

	k_R, s^{-1} [a]	k_{SF}, s^{-1} [b]	k_{DISS}, s^{-1} [c]	k_{REC}, s^{-1} [d]	k_{IC}, s^{-1} [e]	Φ_{SF}, % [f]	Φ_T, % [g]
o-$(Pc)_2$	6.5×10^7	1.2×10^{11}	8.9×10^5	7.9×10^6	4.3×10^4	99	20
m-$(Pc)_2$	6.5×10^7	2.1×10^9	3.3×10^6	2.6×10^6	5.0×10^4	94	105

Reprinted with permission from Ref. [13]. Copyright 2018 American Chemical Society
[a]Estimated value by fluorescence lifetime of reference monomer (~15 ns)
[b]Determined by fsTA measurement
[c]Calculated by $k_{DISS} = (\tau_{REC})^{-1} \cdot [\Phi_T/2]$
[d]Calculated by $k_{REC} = (\tau_{REC})^{-1} \cdot [\Phi_{SF} - \Phi_T/2]$
[e]Determined by nanosecond transient absorption measurement
[f]Determined by femtosecond transient absorption utilizing molar absorption coefficients (ε) such as 156,000 M^{-1} cm^{-1} (at 517 nm) for *o*-$(Pc)_2$ and 188,000 M^{-1} cm^{-1} (at 526 nm) for *m*-$(Pc)_2$
[g]Determined by nanosecond transient absorption measurement

o-$(Pc)_2$, whereas the rate constant of k_{SF} in *o*-$(Pc)_2$ is two orders of magnitude greater than that in *m*-$(Pc)_2$. These are largely dependent for the orientations between two pentacenes in *o*-$(Pc)_2$ and *m*-$(Pc)_2$.

16.3 Observation of Quantitative Light Energy Conversion from Singlet Fission (SF) to Intermolecular Two-Electron Transfers by Covalently Linked Tetracene Dimer

Singlet fission (SF) is theoretically able to yield the radical ion pair quantum yield of 200% by sequential photoinduced process from SF to the resulting electron transfer. Especially, the long lifetime of the triplet state is definitely required for intermolecular ET process in bimolecular diffusion-limited reaction. However, no attempt has been made to observe the quantitative two-electron transfer process via SF, so far. Therefore, we newly employed 2,2′-biphenyl-bridged tetracene dimer (denoted as Tet-BP-Tet: SF and electron donor) and chloranil (denoted as Ch: electron acceptor) to observe the quantitative sequential photoinduced process from SF to intermolecular two-electron transfers (Scheme 16.2).

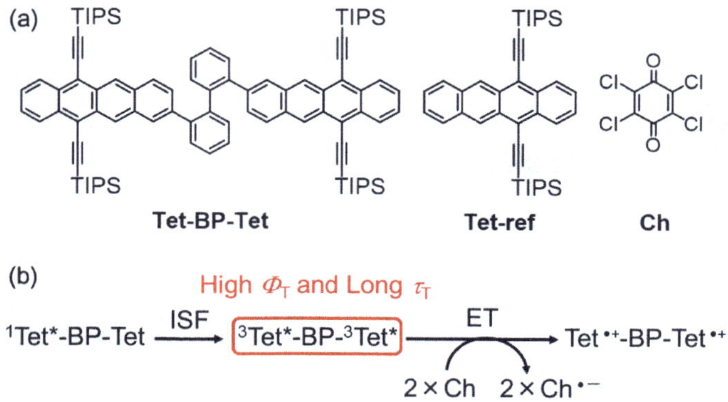

Scheme 16.2 **a** Structures of tetracene dimers, reference monomer and chloranil. **b** A plausible scheme of sequential light energy conversion process from SF to intermolecular two-electron transfers. Reprinted with permission from Ref. [16]. Copyright 2019 American Chemical Society

First, femtosecond transient absorption measurements of Tet-BP-Tet were performed in benzonitrile (PhCN). Although the competitive reactions between SF and TTA were observed, the apparent rate constant of ISF (k_{ISF}) was determined to be 7.1×10^{10} s^{-1} using the decay-associated component analysis. This is 5 orders of magnitude greater than that of the intersystem crossing process (k_{ISC}: 8.5×10^5 s^{-1}). To discuss the transient spectra of Tet-BP-Tet in the longer range, measurements of picosecond transient absorption (psTA) spectra were performed and the individual triplet quantum yield (Φ_T) of Tet-BP-Tet was quantitatively estimated to be 175 ± 5%. Then, the long lifetime species of ^{3}Tet*-BP-^{3}Tet* was evaluated by nanosecond transient absorption spectra. The lifetime of ^{3}Tet*-BP-^{3}Tet* was also calculated to be $\tau_T = 0.29$ ms (i.e., $k_T = 3.4 \times 10^3$ s^{-1}) by monoexponential fitting.

Then, intermolecular electron transfer from ^{3}Tet*-BP-^{3}Tet* (D) to 2Ch (A) was examined (PhCN) by nanosecond transient absorption measurements (Fig. 16.3). The T–T absorption in Tet-BP-Tet was immediately observed (520 nm) after photoexcitation. The decay of T–T absorption of Tet-BP-Tet agrees with the rises of Ch radical anion ($Ch^{•-}$) at 450 nm and Tet radical cation ($Tet^{•+}$) at 1000 nm (Fig. 16.3a). This strongly indicates the occurrence of intermolecular electron transfer from ^{3}Tet*-BP-^{3}Tet* to 2Ch. We obtained the second-order plot for the decay of $Ch^{•-}$ from the absorbance at 450 nm. By the slope of the linear plot, the second-order rate constant of the back electron transfer (from $Ch^{•-}$ to $Tet^{•+}$) was calculated to be 5.0×10^9 M^{-1} s^{-1} [17]. This is very close to the diffusion-limit in PhCN [18–20].

The decay rate constant of the T–T absorption (at 520 nm) significantly increased with an increase of the Ch concentrations (0–0.72 mM) (Fig. 16.3c). Figure 16.3c also shows a linear plot of the observed decay rate constant (k_{obs}) at 520 nm based on the Ch concentrations. The second-order rate constant of intermolecular electron transfer from ^{3}Tet*-BP-^{3}Tet* to Ch was estimated to be 4.8×10^8 M^{-1} s^{-1} from the slope of the linear plot [17], which is slightly smaller than that of diffusion-limited

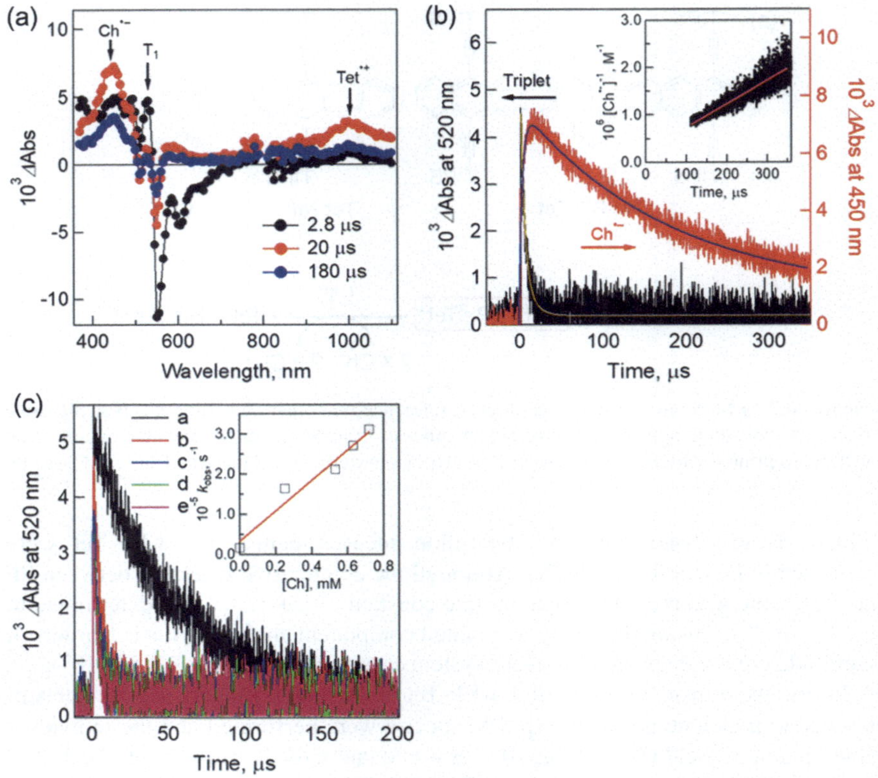

Fig. 16.3 **a** Nanosecond transient absorption spectra of Tet-BP-Tet (50 μM) in the presence of 250 μM Ch in Ar-saturated PhCN ($\lambda_{ex} = 532$ nm). **b** Time profiles at 450 and 520 nm. Inset: second-order plots. **c** Time profiles of absorption at 520 nm in the presence of different concentrations of Ch such as (a) 0 mM, (b) 0.25 mM, (c) 0.53 mM, (d) 0.63 mM, (e) 0.72 mM. Insert: plots of the pseudo-first-order rate constant (k_{obs}) versus the Ch concentrations. Reprinted with permission from Ref. [16]. Copyright 2019 American Chemical Society

value according to the reported value of PhCN (5.6×10^9 M^{-1} s^{-1}) [18–20]. Based on the above results, the intermolecular electron transfer from Tet-BP-Tet to 2Ch was successfully summarized in Fig. 16.4 and Table 16.2.

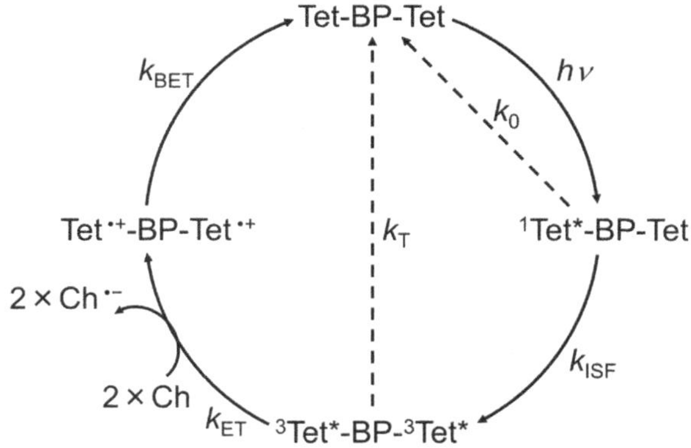

Fig. 16.4 Light energy conversion process in this study. Reprinted with permission from Ref. [16]. Copyright 2019 American Chemical Society

Table 16.2 Summarized parameters of rate constants in this study

k_{ISF}, s^{-1}	k_0, s^{-1}	k_T, s^{-1}	k_{ET}, M^{-1} s^{-1}	k_{BET}, M^{-1} s^{-1}
5.4×10^9	8.5×10^7	3.4×10^3	4.8×10^8	5.0×10^9

16.4 Efficient Generation of Excited Triplet States in Self-assembled Monolayers on Gold Nanoclusters Utilizing Singlet Fission of Pentacene and Tetracene Units

One of the major issues of dye-modified self-assembled monolayers (SAMs) on metal surface and nanoclusters for efficient utilization of solar energy includes the significant deactivation pathway of excited singlet states of dye molecules by surface energy transfer to the metal surface, because the surface-dipole energy transfer process is highly related with Förster theory [22, 23]. Actually, the fluorescence emission significantly decreases (more than 90%) and generation of triplet excited states through intersystem crossing (ISC) is basically negligible. Our concept to obtain long-lived excited states on metal surfaces is accordingly to utilize singlet fission. Therefore, in this study, 6,13-bis(triisopropylsilylethynyl) pentacene-alkanethiolate monolayer-protected gold nanoclusters having different alkyl chain lengths and cluster sizes were successfully synthesized to efficiently generate excited triplet states of pentacenes by singlet fission as shown in Fig. 16.5.

We measured the TEM images and corresponding size-distributions of TP-C11-*X*-MPC. The diameters in TP-C11-*S*-MPC and TP-C11-*L*-MPC are 1.65 ± 0.30 nm and 2.13 ± 0.33 nm, respectively. Additionally, based on the values of elemental analysis, the adsorbed pentacene molecules on one gold nanocluster determined to

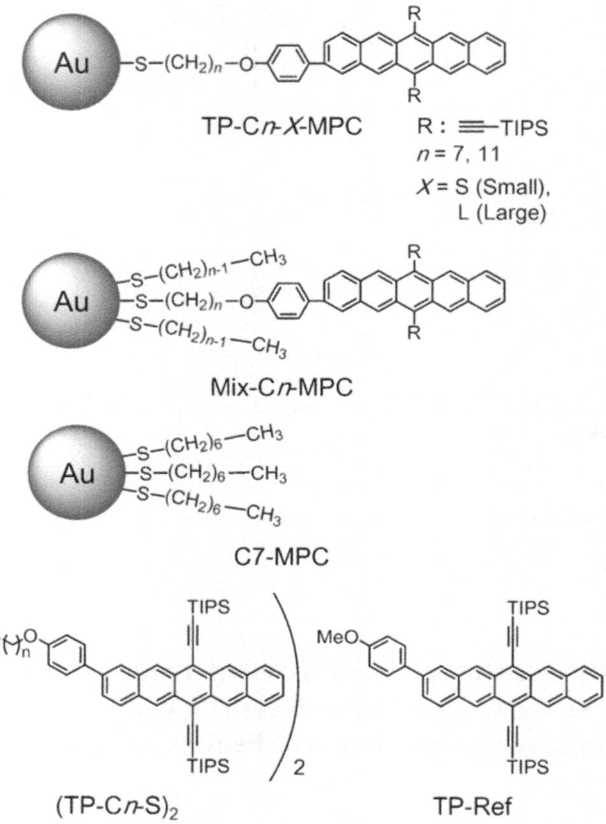

Fig. 16.5 Structures of TP-C*n*-*X*-MPC ($X = S$ and L), Mix-C*n*-MPC, C*n*-MPC, (TP-C*n*-S)$_2$, and TP-Ref used in this study. Reprinted with permission from Ref. [21]. Copyright 2016 Wiley-VCH

be 51 (TP-C11-*S*-MPC) and 75 (TP-C11-*L*-MPC), respectively. These values are very similar to those of the TP-C7-*X*-MPC systems.

Measurement of time-resolved transient absorption spectra of these materials in toluene was performed by femtosecond laser flash photolysis to investigate the ultrafast photodynamics. Transient spectra were measured utilizing 600 or 653 nm fs laser pulse, which mainly excited the TP units. The excitation density was reduced to the level at which excitation of more than two TP molecules attached to one gold nanocluster can be neglected. After laser pulse excitation, the T–T absorption of pentacene units at ca. 520 nm increases, whereas the S-S absorption at ca. 630 nm decreases. The detail photophysical parameters such as rate constants and quantum yields were summarized in Table 16.3. The maximum Φ_T values were estimated to be $172 \pm 26\%$ and $157 \pm 17\%$ in TP-C11-*S*-MPC and TP-C7-*S*-MPC, respectively. The Φ_T values decrease with an increase of nanocluster sizes. The reason is probably attributable to the smaller surface coverage of pentacene units on gold nanoclusters.

Table 16.3 Summarized parameters such as singlet lifetimes (τ_S), SF rate constants (k_{SF}) and triplet quantum yields (Φ_T)

Compound	τ_S, ps (k_{S1}, s^{-1})	k_{SF}, s^{-1c}	Φ_T, %^{d}
TP-Ref	1.5×10^{4a} (6.7×10^7)	–	1.5^e
TP-C7-*S*-MPC	24^b (4.1×10^{10})	2.7×10^{10}	157 ± 17^f
TP-C7-*L*-MPC	54^b (1.9×10^{10})	5.0×10^9	77 ± 15^f
TP-C11-*S*-MPC	21^b (4.8×10^{10})	4.1×10^{10}	172 ± 26^f
TP-C11-*L*-MPC	110^b (9.1×10^9)	2.4×10^9	34 ± 4^f
Mix-C7-MPC	71^a (1.4×10^{10})	–	–
Mix-C11-MPC	150^b (6.7×10^9)	–	–

Reprinted with permission from Ref. [21]. Copyright 2016 Wiley-VCH

[a]Estimated by picosecond fluorescence lifetimes

[b]Estimated by decay component lifetimes

[c]$k_{SF} = 1/\tau$ (TP-C*n*-*X*-MPC) – $1/\tau$ (Mix-C*n*-MPC)

[d]The detail calculation processes are described in the original paper

[e]Estimated by nanosecond transient absorption spectra

[f]Estimated by femtosecond transient absorption spectra

Finally, photophysical behaviors of tetracene-alkanethiol-modified MPCs are discussed (Fig. 16.6). As discussed above, the examples of high-yield and long-lived triplet excited states of tetracene derivatives via singlet fission are extremely limited because of the relative acceleration of the reverse TTA considering the above-mentioned energy matching. Then, we also demonstrated a novel technique utilizing the mixed self-assembled monolayers with two different chain lengths. Especially,

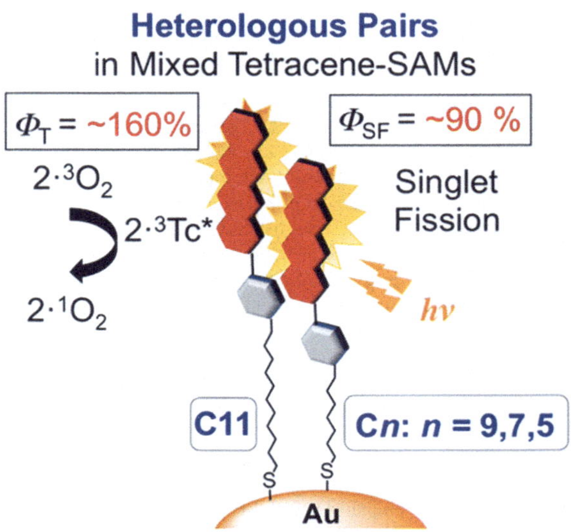

Fig. 16.6 A schematic illustration of tetracene-modified gold nanoclusters. Reprinted with permission from Ref. [24]. Copyright 2019 American Chemical Society

mixed tetracene SAMs on gold nanoclusters, which are synthesized by a tetracene-modified hetero-disulfide with two different chain lengths, demonstrated high-yield SF (Φ_{SF}: ~90%). Individual triplet yields (Φ_T: ~160%) were also estimated by singlet oxygen generation ($^1O_2$).

16.5 Conclusion

In this chapter, we discussed pentacene and tetraene molecular assemblies using covalently linked dimers and monolayered-protected nanoclusters for efficient occurrence of singlet fission. The proper structural control of two acene neighboring molecules enables to demonstrate the generation of high-yield and long-lived triplet excited states, which resulted in the observation of quantitative two-electron transfer process and high-yield singlet oxygen generation. This photophysical information will provide a new perspective for future applications in various areas such as photovoltaics, photocatalysis, electronics and organic synthesis, and phototherapy.

Acknowledgements We deeply appreciate to Professors N. V. Tkachenko, Y. Kobori, Drs. H. Sakai, H. Nagashima, Mrs. D. Kato S. Nakamura, T. Saegusa, R. Inaya for their collaboration. The present work was supported by JSPS KAKENHI Grant Number 17H05270, Grant-in-Aid for Scientific Research on Innovative Areas "Photosynergetics."

References

1. Smith MB, Michl J (2010) Singlet fission. Chem Rev 110(11):6891–6936
2. Miyata K, Conrad-Burton FS, Geyer FL, Zhu XY (2019) Triplet pair states in singlet fission. Chem Rev 119(6):4261–4292
3. Casanova D (2018) Theoretical modeling of singlet fission. Chem Rev 118(15):7164–7207
4. Lee J, Jadhav P, Reusswig PD, Yost SR, Thompson NJ, Congreve DN, Hontz E, Van Voorhis T, Baldo MA (2013) Singlet exciton fission photovoltaics. Acc Chem Res 46(6):1300–1311
5. Krishnapriya KC, Musser AJ, Patil S (2019) Molecular design strategies for efficient intramolecular singlet exciton fission. ACS Energy Lett 4(1):192–202
6. Yang L, Tabachnyk M, Bayliss SL, Böhm ML, Broch K, Greenham NC, Friend RH, Ehrler B (2015) Solution-processable singlet fission photovoltaic devices. Nano Lett 15(1):354–358
7. Chan W-L, Ligges M, Jailaubekov A, Kaake L, Miaja-Avila L, Zhu XY (2011) Observing the multiexciton state in singlet fission and ensuing ultrafast multielectron transfer. Science 334(6062):1541–1545
8. Rao A, Wilson MWB, Hodgkiss JM, Albert-Seifried S, Bässler H, Friend RH (2010) Exciton fission and charge generation via triplet excitons in pentacene/c60 bilayers. J Am Chem Soc 132(36):12698–12703
9. Sanders SN, Kumarasamy E, Pun AB, Trinh MT, Choi B, Xia J, Taffet EJ, Low JZ, Miller JR, Roy X, Zhu XY, Steigerwald ML, Sfeir MY, Campos LM (2015) Quantitative intramolecular singlet fission in bipentacenes. J Am Chem Soc 137(28):8965–8972
10. Zirzlmeier J, Lehnherr D, Coto PB, Chernick ET, Casillas R, Basel BS, Thoss M, Tykwinski RR, Guldi DM (2015) Singlet fission in pentacene dimers. Proc Natl Acad Sci USA 112(17):5325–5330

11. Lukman S, Chen K, Hodgkiss JM, Turban DHP, Hine NDM, Dong S, Wu J, Greenham NC, Musser AJ (2016) Tuning the role of charge-transfer states in intramolecular singlet exciton fission through side-group engineering. Nat Commun 7:13622
12. Sakuma T, Sakai H, Araki Y, Mori T, Wada T, Tkachenko NV, Hasobe T (2016) Long-lived triplet excited states of bent-shaped pentacene dimers by intramolecular singlet fission. J Phys Chem A 120(11):1867–1875
13. Sakai H, Inaya R, Nagashima H, Nakamura S, Kobori Y, Tkachenko NV, Hasobe T (2018) Multiexciton dynamics depending on intramolecular orientations in pentacene dimers: Recombination and dissociation of correlated triplet pairs. J Phys Chem Lett 9(12):3354–3360
14. Tayebjee MJY, Sanders SN, Kumarasamy E, Campos LM, Sfeir MY, McCamey DR (2017) Quintet multiexciton dynamics in singlet fission. Nat Phys 13(2):182–188
15. Weiss LR, Bayliss SL, Kraffert F, Thorley KJ, Anthony JE, Bittl R, Friend RH, Rao A, Greenham NC, Behrends J (2017) Strongly exchange-coupled triplet pairs in an organic semiconductor. Nat Phys 13(2):176–181
16. Nakamura S, Sakai H, Nagashima H, Kobori Y, Tkachenko NV, Hasobe T (2019) Quantitative sequential photoenergy conversion process from singlet fission to intermolecular two-electron transfers utilizing tetracene dimer. ACS Energy Lett 4(1):26–31
17. Kawashima Y, Ohkubo K, Fukuzumi S (2012) Enhanced photoinduced electron-transfer reduction of li + @c60 in comparison with c60. J Phys Chem A 116(36):8942–8948
18. Davis CM, Kawashima Y, Ohkubo K, Lim JM, Kim D, Fukuzumi S, Sessler JL (2014) Photoinduced electron transfer from a tetrathiafulvalene-calix[4]pyrrole to a porphyrin carboxylate within a supramolecular ensemble. J Phys Chem C 118(25):13503–13513
19. D'Souza F, Ito O (2005) Photoinduced electron transfer in supramolecular systems of fullerenes functionalized with ligands capable of binding to zinc porphyrins and zinc phthalocyanines. Coordin Chem Rev 249(13–14):1410–1422
20. Fukuzumi S, Ohkubo K, Imahori H, Guldi DM (2003) Driving force dependence of intermolecular electron-transfer reactions of fullerenes. Chem Eur J 9(7):1585–1593
21. Kato D, Sakai H, Tkachenko NV, Hasobe T (2016) High-yield excited triplet states in pentacene self-assembled monolayers on gold nanoparticles through singlet exciton fission. Angew Chem Int Ed 55(17):5230–5234
22. Thomas KG, Kamat PV (2003) Chromophore-functionalized gold nanoparticles. Acc Chem Res 36(12):888–898
23. Kato D, Sakai H, Saegusa T, Tkachenko NV, Hasobe T (2017) Synthesis, structural and photophysical properties of pentacene alkanethiolate monolayer-protected gold nanoclusters and nanorods: supramolecular intercalation and photoinduced electron transfer with c60. J Phys Chem C 121(16):9043–9052
24. Saegusa T, Sakai H, Nagashima H, Kobori Y, Tkachenko NV, Hasobe T (2019) Controlled orientations of neighboring tetracene units by mixed self-assembled monolayers on gold nanoclusters for high-yield and long-lived triplet excited states through singlet fission. J Am Chem Soc 141(37):14720–14727

Chapter 17
Synergetic Photon Upconversion Realized by a Controlled Toroidal Interaction in Hexaarylbenzene Derivatives

Tadashi Mori

Abstract Triplet-triplet annihilation (TTA) is the most promising photon upconversion (UC) sequence, where a weak light source such as sun light can be used to frequency upconvert the visible to near-infrared light into a photon with a more useful wavelength, applicable in such as the solar cell devices. Although this anti-Stokes shift emission has long been known, and their application has been extensively investigated, less has been explored the effect on the structural modification of the relevant sensitizer and acceptor chromophores. Motivated by a beautiful and functional cylindrical structure of light-harvesting proteins, as well as recent investigation on the cyclic array of nano-wires, we prepared the derivatives of hexaarylbenzene (HAB) containing 9,10-diphenylanthracene and 2,5-diphenyloxazole units as model acceptor systems to investigate the efficiency of TTA-UC and the relevant photophysical processes. The dynamics of propeller-shaped HAB structures as well as toroidal interaction between the peripheral aromatic rings has been well investigated by combined experimental and theoretical studies on their chiroptical properties. Yet preliminary, we found that through the synergetic interaction of the peripheral chromophores, effective energy migration and hopping become possible, to facilitate the efficient TTA-UC, that can be switched on and off by modulation of the strength of toroidal interaction of the relevant HAB emitter. Further possibilities of photosynergetic effects in the TTA-UC process are also discussed.

Keywords Triplet-triplet annihilation · Photon upconversion · Anti-Stokes emission · Toroidal interaction · Exciton delocalization · Hexaarylbenzene derivatives · Photosynergetic effect

---

T. Mori (✉)
Department of Molecular Chemistry, Graduate School of Engineering, Osaka University, Suita, Osaka 565-0871, Japan
e-mail: tmori@chem.eng.osaka-u.ac.jp

H. Miyasaka et al. (eds.), *Photosynergetic Responses in Molecules and Molecular Aggregates*, https://doi.org/10.1007/978-981-15-5451-3_17

17.1 Introduction

Triplet-triplet annihilation (TTA) is one of the most promising upconversion (UC) processes that can frequency convert the lower energy light (with a longer wavelength) into the higher energy photon (with a shorter wavelength) under the relatively weak incident light intensity. Such process is one of the anti-Stokes-shifted emission processes, contrast to the conventional downconverted or Stokes-shifted emission (Fig. 17.1). A comparatively effective UC can be achieved by the TTA process at low photon density (typically <0.1 W cm^{-2}) of non-coherent light that allows to utilize the terrestrial solar irradiation that includes abundant photon in the visible to the near-IR region, potentially applicable for the dye-sensitized solar cell, photovoltaics, artificial photosynthesis, photocatalysis, and so forth. Thus, the TTA-UC has great advantages over the other upconversion processes such as two-photon absorption process that often require intense laser irradiation [1–4]. In addition, the wavelength of light before and after the UC process can be sensibly modulated through the rational modification of triplet donor (sensitizer) and/or triplet acceptor (annihilator/emitter) [5].

The delayed fluorescence from the annihilation of a pair of anthracene and phenanthrene triplet states, that is, eventually established as the TTA process, was first appeared in 1962 [6]. The anti-Stokes fluorescence through the triplet sensitized TTA, that is, the TTA-UC, was also reported in the same year [7]. These phenomena have been readily confirmed by assessing the emission intensity being proportional to the square of the rate of absorption of exciting light. Although the highly efficient TTA-UC in solution have been reported for various systems over the half century, investigations of TTA-UC materials in polymer matrices, film states, solid-states, as well as in molecular self-assemblies are realized only recently, that are unquestionably more stimulating particularly for the application purposes [8–13]. As a latest approach, the TTA-UC nanoparticles made by, for instance, liposome coating, silica-coating, polymer encapsulation, nanodroplet, as well as metal-organic frameworks have been also explored, applicable in bioimaging and photoinduced therapeutic

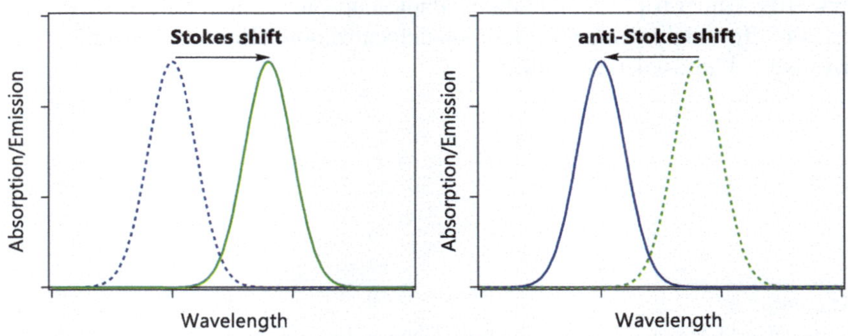

Fig. 17.1 Schematic illustration of the Stokes and anti-Stokes emissions (solid lines), shifted from the corresponding absorption spectra (dotted lines)

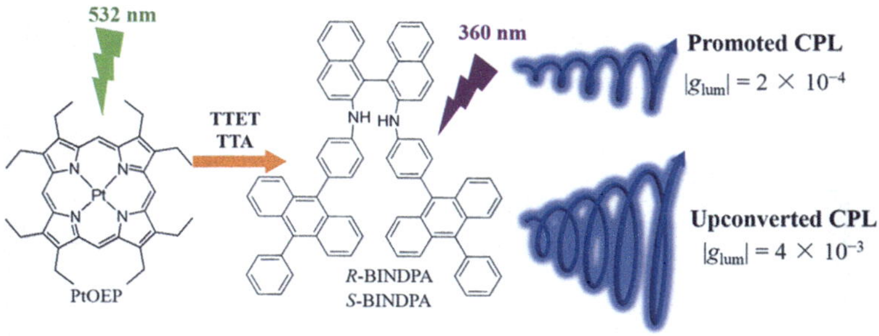

Fig. 17.2 Upconverted circularly polarized luminescence (CPL) shows improved luminescence dissymmetry factor (g_{lum}) than conventional promoted CPL. Reproduced with permission from [23]. Copyright 2017 by the American Chemical Society

agents owing to their biocompatibility [14]. The status quo and limitations of the current TTA-based photon upconverter toward applications in solar cell device have been thoroughly discussed [15, 16], and an outline for improving the UC efficiency for relevant future device has been proposed. As an alternate perspective, circularly polarized luminescent (CPL) materials are of increased attention due to their potential applications in advanced optical technologies such as optical sensing, security paintings, and 3D displays and endoscope devices [17–19]. Recently, the integration of CPL materials with the photon UC via the TTA mechanism has been attracted significant attention for their improved dissymmetry factor (g_{lum}) in the upconverted emission than the conventional promoted emission (Fig. 17.2) [20]. Along with the emission properties of upconverted photon, the excited singlet species formed through the TTA-UC mechanism have been also employed for photochemical transformations such as anthracene dimerization reactions [21, 22].

In order to operate the TTA-UC processes, the triplet sensitizer and the triplet acceptor are usually mixed. The excitation of the sensitizer with a lower energy light generates the excited singlet state of the sensitizer, which is transferred to the triplet state through the intersystem crossing (ISC). The energy of the triplet sensitizer is then transferred to the acceptor, typically by a Dexter-type triplet-triplet energy transfer (TTET). When two excited acceptors in their triplet states meet each other, typically through a diffusion process, they annihilate to form a singlet excited state of acceptor together with a relaxed acceptor in the ground state (TTA), and eventually the former emits fluorescent light with a higher energy. The intensity of this fluorescence emission shows a quadratic dependence against the incident light power, as the TTA occurs by the encounter of two sensitized triplet acceptors. A generalized Jablonski energy diagram summarizing those processes involved in the TTA-UC is shown in Fig. 17.3.

As seen in Fig. 17.3, the process of the TTA-UC relies on a pair of chromophores, i.e., sensitizer and acceptor. The former is sometimes called as donor and the latter is also referred as emitter or annihilator. Crucial parameters of the sensitizer and acceptor properties for efficient TTA-UC processes have been extensively discussed

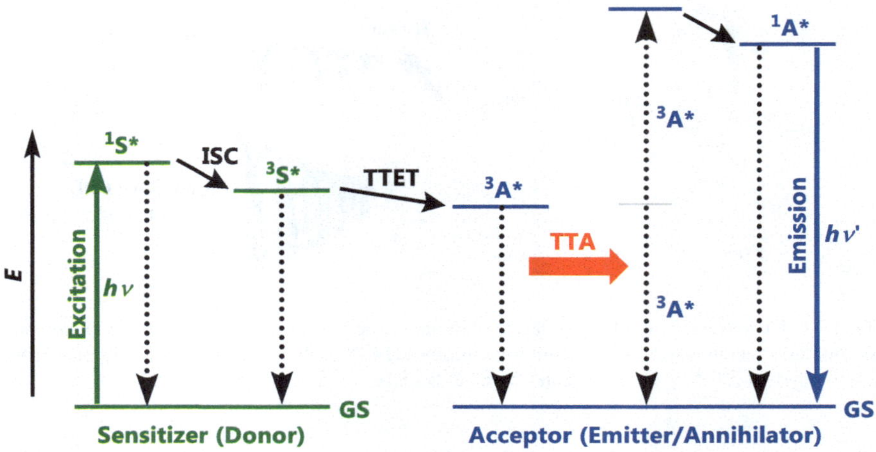

Fig. 17.3 Schematic Jablonski energy level diagram of the TTA-UC process between the sensitizer and the acceptor. Colored solid lines represent radiative processes. $^1S^*$, $^3S^*$, $^1A^*$, and $^3A^*$ are singlet and triplet excited states of sensitizer and acceptor. TTET, ISC, and GS denote triplet-triplet energy transfer, intersystem crossing, and ground state, respectively

and some are briefly described in the following. Firstly, energy levels of the sensitizer and the acceptor need to be appropriate, where the triplet energy level of acceptor lies below that of the sensitizer, and the energy level of singlet excited state of acceptor or emitter is nearly the same or slightly below the combined energy of two triplet excited acceptors [24]. In addition, because the TTET and TTA are bimolecular processes, the higher concentration of triplet sensitizer in its triplet excited state is favored. Thus, the sensitizer requires a large molar absorptivity at the excitation wavelength, a high ISC quantum yield, as well as a long lifetime. Also, a high fluorescence quantum yield of the emissive acceptor is advantageous for the total UC efficiency.

The overall quantum yield of UC process (Φ_{UC}) is a function of all these factors and can be described as follows [25]:

$$\Phi_{UC} = \frac{1}{2} \times f \times \Phi_{ISC} \times \Phi_{TTET} \times \Phi_{TTA} \times \Phi_F$$

where Φ_{ISC}, Φ_{TTET}, and Φ_{TTA} are the quantum yields of ISC, TTET, and TTA processes, respectively, and Φ_F is the quantum yield of fluorescence emission of acceptor. The first 1/2 factor means that two incident photons are combined into one in the overall TTA process and second variable f represents a probability factor that the triplet-triplet annihilation results in a singlet among other possible spin states (triplet and quintet). Inherently, only 1/9 of all annihilated triplets result in the formation of an emissive singlet excited state, leading the spin-statistical Φ_{UC} limit of 11%. However, the triplets formed upon annihilation can be annihilated more than once until they decay via the singlet state, increasing theoretical efficiency up to 40%. Moreover, the quintet state of polyaromatic organic compounds is often inaccessible. Experimentally, the Φ_{UC} value as high as 20% was achieved [3]. Although some factors such as Φ_{ISC} and Φ_F are intrinsic to the chromophore, other factors (Φ_{TTET} and Φ_{TTA}) can be modulated also by external factors such as distance between the chromophores (concentration of a solution) and mobility (viscosity of a solution). To improve the former (i.e., maximizing Φ_{ISC}) by bypassing or minimizing the energy loss through the ISC process, direct singlet-to-triplet absorption materials, the sensitization from chalcogenide and perovskite nanocrystals, and the thermally activated delayed fluorescence behavior have been proposed as alternative triplet sensitization routes [26].

The TTET process between the sensitizer and the acceptor can be treated as a combination of Förster-type resonant energy transfer and Dexter-type mechanisms. However, the former contribution is usually very small since the TTET in the TTA-UC system usually involves forbidden transitions. The Dexter-type electron transfer can be described by the active-sphere or Perrin approximation, which is based on the exponential decay profile of the overlap of the involved orbitals. It has been demonstrated that this assumption is indeed acceptable for dilute solutions of platinum(II) octaethylporphyrin and 9,10-diphenylanthracene at low temperature [27]. Although a relatively large Dexter radius of 2.65 nm was found, the shorter distance of around 1 nm is generally required and definitely preferred for more efficient TTET and TTA processes. Thus, providing translational mobility and/or maximizing the concentrations of the sensitizer and the acceptor, or to make use of (supramolecular) structure or morphology [28], to the point where the average distance between these species within the Perrin limit, are effective general strategies for maximizing the efficiency of the TTET process, despite the possible undesired back electron transfer process.

The triplet pair is also the key functional unit in the TTA-UC. A recent study demonstrated that the solvent polarity and viscosity play a crucial role in promoting emission pathways, consequently the efficient UC, over the excimer formation on the singlet surface which works as energy-loss channels [29].

Experimentally, Φ_{UC} can be determined by using a relative actinometry under precise experimental conditions:

$$\Phi_{UC} = 2\Phi_0 \left(\frac{\eta}{\eta_0}\right)^2 \left(\frac{A}{A_0}\right)^{-1} \left(\frac{I}{I_0}\right)$$

where η, A, and I are refractive index of the solvents, absorbance, and fluorescence intensity of the samples, respectively. The corresponding terms for the subscript 0 are for the reference compounds. Φ_0 refers the fluorescence quantum yield of the reference. A factor of 2 is required as the TTA is two-photon process.

17.2 Dimeric and Multiphoton Annihilators for Efficient TTA

The triplet acceptors typically used for the TTA-UC are anthracene, 9,10-dimethylanthracene, 9,10-diphenylanthracene, perylene, ruburene, pyrene, boron-dipyrromethene, polyfluorenes, and derivatives. Recently, the critical factors for the efficient TTA, such as effect of the conjugation and geometry of acceptor, have been investigated with a series of substituted 9,10-diphenylanthracene derivatives [30]. Accordingly, the intensity of the TTA-UC emission was directly dependent on the ISC rate, which can be modulated by the alignment of the S_1 and T_2 energy levels of acceptor. A relatively less attention has been paid for developing better triplet sensitizers and metal complexes of porphyrin and phthalocyanines are typically used.

Among the various parameters to be adjusted in the TTA-UC processes, the bimolecular (i.e., TTET and TTA) processes become continuously bottleneck, but, in principle, can be circumvented through the synthetic modification of sensitizer and/or acceptor. One of the simplest approaches is to directly connect either sensitizer and acceptor or two or more acceptor molecules. A direct covalent link between the sensitizer and the acceptor allows to improve TTET process that generally requires encounter through the diffusion of the excited triplet state of sensitizer and acceptor [31, 32].

For improving TTA process, dimer and oligomer of anthracene chromophores have been investigated (Fig. 17.4). Recently, anthracene dimers with two different geometries were employed as model emitters in the TTA-UC for fine-tuning of twisted intramolecular charge-transfer intermediate states and thus the emission color by solvent polarity. In addition, the UC quantum yields were found to be 13% with platinum(II) octaethylporphyrin as a sensitizer [33], showing a considerable enhancement of the TTA-UC efficiency. Linear oligomers and two generation dendrimers of covalently linked 9,10-diphenylanthracenes have also been employed as the TTA acceptor, showing substantial positive effect for the UC process, both in solution and in poly(methyl methacrylate) matrix [34, 35]. Further conjugation of 9,10-diphenylanthracene oligomers with ruthenium *tris*-bipyridine sensitizer has been also realized to examine the TTA-UC processes in solution [36]. In this system, however, while an efficient intramolecular TTET occurs, the TTA process only proceeds in an intermolecular manner, leading to a comparatively low quantum yield for the overall UC process.

Theoretical aspects on the diffusion and encounter processes in TTA were recently investigated [37]. Ab initio quantum chemical calculations have been also employed to rationally design the electronic coupling matrix elements between the chromophores as well as the excitation energies in the doubly bridged tetracenes for efficient exciton up- and down-conversion processes [38]. However, to fully understand detailed mechanisms and to develop design principle for the efficient TTA-UC sensitizer and acceptor, more issues remain to be elucidated.

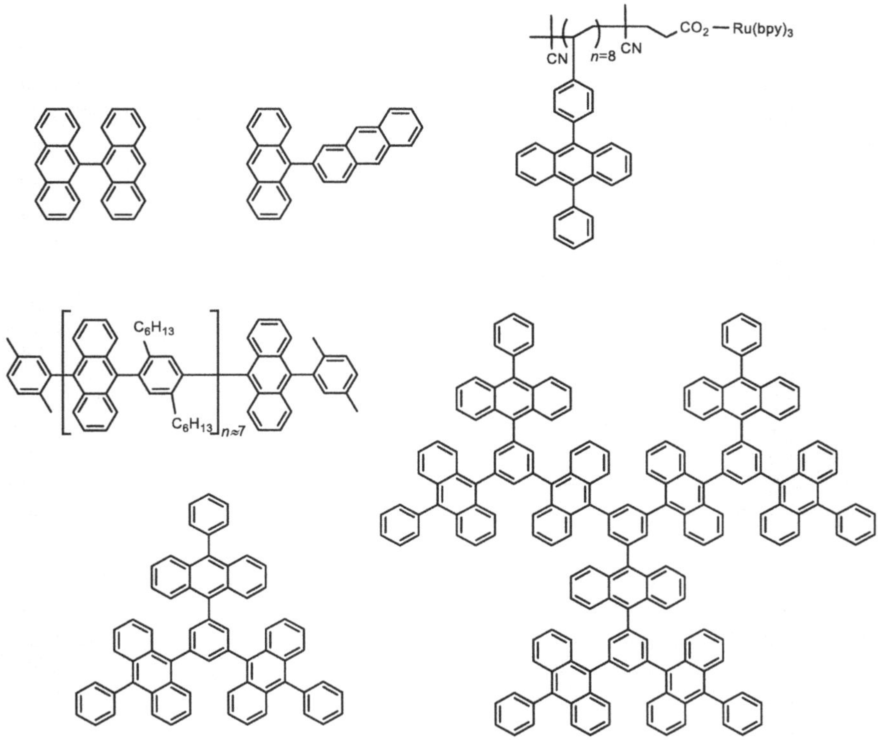

Fig. 17.4 Dimer, oligomer, and dendrimer of anthracene derivatives for better acceptor/emitter in the TTA-UC study employed in the literature

17.3 Structure and Dynamics of HABs

Hexaarylbenzene (HAB) possesses six peripheral aromatic groups against central core benzene molecule in nearly perpendicular orientation. HABs have been much attracted due to their unique electrochemical and photophysical properties for applications in organic light-emitting diodes, photochemical and electrochemical switches, redox materials, molecular receptors, as well as liquid crystalline materials [39, 40]. More specifically, HABs are known to exhibit global electron and/or exciton delocalization over the entire peripheral aryl groups in the three-dimensional donut-shaped scaffold, which is frequently referred to as toroidal interaction [41, 42]. Such interaction cannot be classified as a conventional conjugative π-electronic interaction. In some HABs, the exciton interaction between the aromatic rings in their excited states has been also claimed [43, 44].

The orthogonal array of peripheral aryl groups in HABs, however, is an averaged structure under the certain conditions. For instance, the crystal structures of various HAB derivatives generally bear tilted aromatic rings around the benzene core, providing overall propeller-shaped geometries. More importantly, the dynamics of toroidal interaction in HABs have been less explored and mostly ignored in many situations. We have been independently attracted in the chirality of HABs. By small modification at the periphery of aromatic rings (i.e., attaching chiral auxiliaries), clockwise or counter-clockwise propeller shape is induced in HABs, exhibiting unique chiroptical responses [45–48]. Remarkably, the propeller chirality is not a simple summation of multiple axial chirality, but the peripheral aromatic rings show cooperative or synergetic dynamism in a domino-like movement. Theoretical calculation as well as condensed phase structural study revealed that the rotation of propeller blades is quite feasible, but the overall propeller conformation is easily developed in non-polar solvent at low temperature. Because the propeller chirality is dynamic chirality, the responses were found quite sensible to the environmental variant such as solvent polarity, temperature, as well as hydrostatic pressure. Very recently, propeller chiral HAB composed of sterically congested 3,6-dimethoxycarbazole units has been also documented and the enantiomers were separated by chiral HPLC [49].

A detailed study of environmental variants such as solvent, temperature, as well as pressure, on the propeller chirality of HABs revealed the dynamic nature of conformational equilibria of HABs. Briefly, the clockwise and counter-clockwise propeller conformations are dominant under the low temperature in less polar solvent, but the contribution of whizzing toroids becomes substantial at increased temperature. The solvation to the peripheral aromatic rings becomes more significant in polar solvents. In addition, solvent molecules may intervene the void space between the peripheral aromatic rings that increase the effective tilt angle of the blades against the central benzene ring. All of these factors influence the conformational equilibrium between clockwise and counter-clockwise propeller as well as toroidal structure. Such a unique feature may be explored in advanced functional chiroptical materials and the process of energy and/or exciton transfer and migration. Indeed, we successfully demonstrated that the CPL response from the propeller chiral HABs can be switched on and off, by controlling the dynamism of propeller and toroidal structures (Fig. 17.5).

17.4 Synergetic TTA Through the Toroidal Interaction in HABs

A significant feature shared by photosynthetic organisms is that they capture photons in the form of excitons delocalized over tens of pigment molecules embedded in protein environments of light-harvesting complexes, typically in cylindrical manners. Delocalized excitons created remain well protected and are successfully delivered to the reaction center. It is also known that the conversion of photon into chemical energy is driven by electronic couplings between the pigments that ensure the efficient transport of energy from light-capturing antenna pigments to the destinations (Fig. 17.6). The role of delocalized excitons and tubular arrangement in achieving

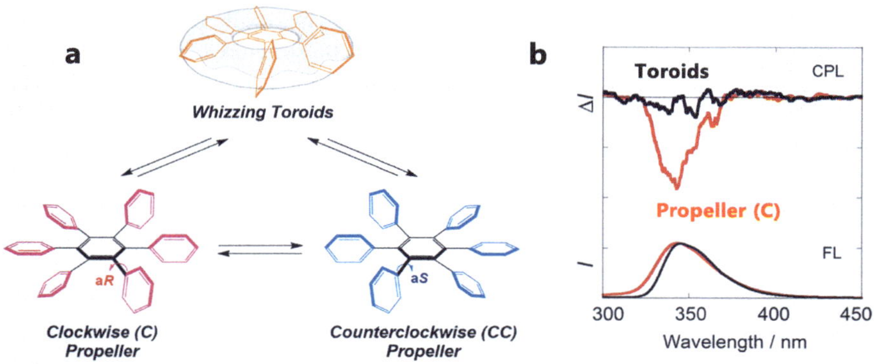

Fig. 17.5 Toroidal interaction and propeller chirality of HABs. **a** Equilibrium between clockwise and counter-clockwise chiral propeller conformations as well as toroidal structure. **b** Switching of CPL of HABs by controlling the preferred conformation, temperature in this case. Reproduced with permission from [46]. Copyright 2017 by the American Chemical Society

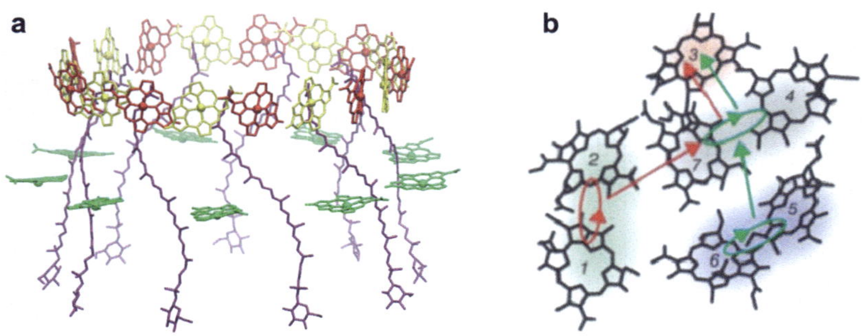

Fig. 17.6 Structures of selected examples of light-harvesting complexes. **a** Cylindrical arrangement of bacteriochlorophyll pigments in the light-harvesting two complex of purple bacteria. Note that upper circular aggregate in B850 ring is formed by α (red) and β (yellow) molecules, whereas γ (green) molecules of B800 constitute the lower circular aggregate. The carotenoids are in purple. Reproduced with permission from [53]. Copyright 2018 by American Physical Society. **b** Arrangement of the seven bacteriochlorophyll *a* molecules in Fenna-Matthews-Olson protein of green sulfur bacteria. Note that two main photoexcitation transfer pathways are indicated by red and green arrows. Reproduced with permission from [54]. Copyright 2005 by Springer Nature

excellent light-harvesting functionality and efficient energy transport mechanisms has been discussed in depth [50–52], but many additional facets are still remained to be elucidated for better understanding. Also, such effect should be more exploited in artificial light-harvesting phenomena, such as the TTA-UC process discussed herein.

As an initial attempt, we focused on the generally used and commercially available photosensitizer, i.e., platinum(II) octaethylporphyrin. This sensitizer has intense absorption in visible green region, long triplet state lifetime of 90 μs and the triplet excited state energy level of 1.92 eV, which is typically used with a combination of 9,10-diphenylanthracene as an efficient emitter owing to the proper energy level alignment. Inspired by beautiful and functional cyclic structures of light-harvesting proteins mentioned above, as well as recent investigation on cyclic array of nanowires [55, 56], we tried to realize the effective energy migration and hopping based on the HAB scaffold with this pair that should eventually be applied for an effective TTA-based photon upconverter. Derivatives of HAB in which two pyrene units are fused in adjacent location were already reported [57], and the fluorescence behavior was found considerably different depending on the *anti/syn* orientation of pyrene units, while the absorption spectra kept comparable.

Consequently, we prepared the HAB derivatives fully substituted by 9,10-diphenylanthracene units to investigate the efficiency of TTA and the relevant photophysical processes (Fig. 17.7). A preliminary investigation revealed that in the ground state the interchromophore interaction was not significant, but the enhanced purple emission of at least 10 times stronger was observed at around 426 nm for this HAB upon photosensitization with green light at 515 nm, as compared with that of parent 9,10-diphenylanthrancene under the comparable conditions. However, more detailed studies were hampered due to the very limited solubility of this HAB. Unfortunately, the situation was not much improved by simply attaching six hexyl groups at the periphery. Nevertheless, this observation clearly demonstrated that the HAB scaffold serves as efficient UC system, which may be switched on and off through a control of propeller dynamics in HAB structure. Currently, we also prepare the HAB derivatives containing 2,5-diphenyloxazole units as an alternate toroidal annihilator.

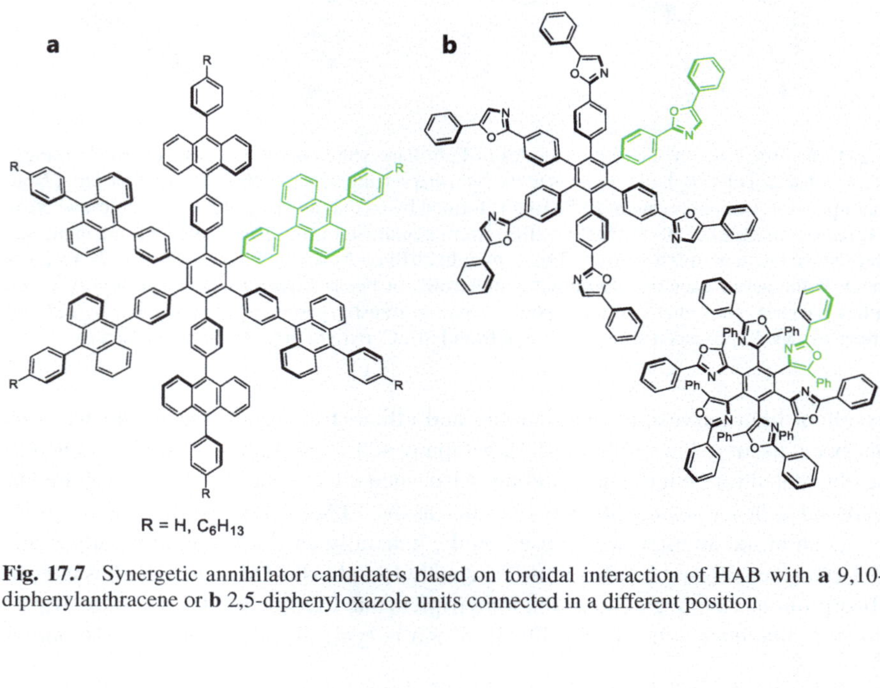

Fig. 17.7 Synergetic annihilator candidates based on toroidal interaction of HAB with **a** 9,10-diphenylanthracene or **b** 2,5-diphenyloxazole units connected in a different position

17.5 Conclusion and Future Perspective

Further studies are certainly needed but a unique cylindrical structure and synergetic dynamism of HAB derivatives were successfully utilized for better TTA-UC processes, where the toroidal interaction between the peripheral aromatic rings plays crucial roles. Thus, aligning multiple chromophores as a cyclic array, like in HAB derivatives exemplified herein, provides a positive synergetic effect for the efficient TTA and such approach was clearly established as a novel design principle for effective annihilator for TTA process.

Additional synergistic effects have been also demonstrated in the literature for the better TTA-UC. For instance, by simply employing multiple (and different) acceptors, the UC quantum yield being significantly improved from the single acceptor system [58]. It has been also reported that the TTA efficiency was improved by using a multi-component acceptor, in which the chromophore of a lower triplet state energy and the chromophore better suited in terms of the corresponding singlet state energy are directly connected [59]. Such an apparently simple modification may provide additional opportunity to further advance the TTA-UC efficiency that should be more exploited in the future design of sensitizer and/or acceptor chromophores.

Acknowledgements Financial supports by Grant-in-Aids for Scientific Research, Challenging Exploratory Research, Promotion of Joint International Research (Fostering Joint International Research), and on Innovative Areas "Photosynergetics" (Grant Numbers JP15H03779, JP15K13642, JP16H06041, JP16KK0111, JP17H05261, JP18K19077, and JP18H01964) from JSPS, by the Asahi Glass Foundation, by the Murata Science Foundation, and by the Research Program of Five-star Alliance in NJRC Mater. & Dev. are greatly acknowledged.

References

1. Zhu X, Su Q, Feng W, Li F (2017) Anti-Stokes shift luminescent materials for bio-applications. Chem Soc Rev 46:1025–1039. https://doi.org/10.1039/C6CS00415F
2. Ye C, Zhou L, Wang X, Liang Z (2016) Photon upconversion: from two-photon absorption (TPA) to triplet-triplet annihilation (TTA). Phys Chem Chem Phys 18:10818–10835. https://doi.org/10.1039/C5CP07296D
3. Zhou J, Liu Q, Feng W, Sun Y, Li F (2015) Upconversion luminescent materials: advances and applications. Chem Rev 115:395–465. https://doi.org/10.1021/cr400478f
4. Singh-Rachford TN, Castellano FN (2010) Photon upconversion based on sensitized triplet-triplet annihilation. Coord Chem Rev 254:2560–2573. https://doi.org/10.1016/j.ccr.2010.01.003
5. Zhao J, Ji S, Guo H (2011) Triplet-triplet annihilation based upconversion: from triplet sensitizers and triplet acceptors to upconversion quantum yields. RSC Adv 1:937–950. https://doi.org/10.1039/C1RA00469G
6. Parker CA, Hatchard CG (1962) Delayed fluorescence from solutions of anthracene and phenanthrene. Proc Chem Soc 147. http://dx.doi.org/10.1039/PS9620000133
7. Parker CA, Hatchard CG (1962) Sensitised anti-Stokes delayed fluorescence. Proc Chem Soc 386–387. https://doi.org/10.1039/PS9620000373

8. Gray V, Moth-Poulsen K, Albinsson B, Abrahamsson M (2018) Towards efficient solid-state triplet-triplet annihilation based photon upconversion: supramolecular, macromolecular and self-assembled systems. Coord Chem Rev 362:54–71. https://doi.org/10.1016/j.ccr.2018.02.011
9. Joarder B, Yanai N, Kimizuka N (2018) Solid-state photon upconversion materials: structural integrity and triplet-singlet dual energy migration. J Phys Chem Lett 9:4613–4624. https://doi.org/10.1021/acs.jpclett.8b02172
10. Li X, Tang ML (2017) Triplet transport in thin films: fundamentals and applications. Chem Commun 53:4429–4440. https://doi.org/10.1039/C7CC00861A
11. McCusker CE, Castellano FN (2016) Materials integrating photochemical upconversion. Top Curr Chem 374:19. https://doi.org/10.1007/s41061-016-0021-7
12. Kimizuka N, Yanai N, Morikawa M (2016) Photon upconversion and molecular solar energy storage by maximizing the potential of molecular self-assembly. Langmuir 32:12304–12322. https://doi.org/10.1021/acs.langmuir.6b03363
13. Simon YC, Weder C (2012) Low-power photon upconversion through triplet-triplet annihilation in polymers. J Mater Chem 22:20817–20830. https://doi.org/10.1039/C2JM33654E
14. Huang L, Kakadiaris E, Vaneckova T, Huang K, Vaculovicova M, Han G (2019) Designing next generation of photon upconversion: recent advances in organic triplet-triplet annihilation upconversion nanoparticles. Biomaterials 201:77–86. https://doi.org/10.1016/j.biomaterials.2019.02.008
15. Frazer L, Gallaher JK, Schmidt TW (2017) Optimizing the efficiency of solar photon upconversion. ACS Energy Lett 2:1346–1354. https://doi.org/10.1021/acsenergylett.7b00237
16. Gray V, Dzebo D, Abrahamsson M, Albinsson B, Moth-Poulsen K (2014) Triplet-triplet annihilation photon-upconversion: towards solar energy applications. Phys Chem Chem Phys 16:10345–10352. https://doi.org/10.1039/C4CP00744A
17. Tanaka H, Inoue Y, Mori T (2018) Circularly polarized luminescence and circular dichroisms in small organic molecules: correlation between excitation and emission dissymmetry factors. ChemPhotoChem 2:386–402. https://doi.org/10.1002/cptc.201800015
18. Tanaka H, Ikenosako M, Kato Y, Fujiki M, Inoue Y, Mori T (2018) Symmetry-based rational design for boosting chiroptical responses. Comms Chem 1:38. https://doi.org/10.1038/s42004-018-0035-x
19. Tanaka H, Kato Y, Fujiki M, Inoue Y, Mori T (2018) Combined experimental and theoretical study on circular dichroism and circularly polarized luminescence of configurationally robust D_3-symmetric triple pentahelicene. J Phys Chem A 122:7378–7384. https://doi.org/10.1021/acs.jpca.8b05247
20. Yang D, Han J, Liu M, Duan P (2018) Photon upconverted circularly polarized luminescence via triplet-triplet annihilation. Adv Mater 1805683. http://dx.doi.org/10.1002/adma.201805683
21. Islangulov RR, Castellano FN (2006) Photochemical upconversion: anthracene dimerization sensitized to visible light by a Ru^{II} chromophore. Angew Chem Int Ed 45:5957–5959. https://doi.org/10.1002/anie.200601615
22. Rao M, Kanagaraj K, Fan C, Ji J, Xiao C, Wei X, Wu W, Yang C (2018) Photocatalytic supramolecular enantiodifferentiating dimerization of 2-anthracenecarboxylic acid through triplet-triplet annihilation. Org Lett 20:1680–1683. https://doi.org/10.1021/acs.orglett.8b00520
23. Han J, Duan P, Li X, Liu M (2017) Amplification of circularly polarized luminescence through triplet-triplet annihilation-based photon upconversion. J Am Chem Soc 139:9783–9786. https://doi.org/10.1021/jacs.7b04611
24. Zhou Q, Zhou M, Wei Y, Zhou X, Liu S, Zhang S, Zhang B (2017) Solvent effects on the triplet-triplet annihilation upconversion of diiodo-bodipy and perylene. Phys Chem Chem Phys 19:1516–1525. https://doi.org/10.1039/C6CP06897A
25. Monguzzi A, Tubino R, Hoseinkhani S, Campione M, Meinardi F (2012) Low power, non-coherent sensitized photon up-conversion: modelling and perspectives. Phys Chem Chem Phys 14:4322–4332. https://doi.org/10.1039/C2CP23900K
26. Yanai N, Kimizuka N (2017) New triplet sensitization routes for photon upconversion: thermally activated delayed fluorescence molecules, inorganic nanocrystals, and singlet-to-triplet absorption. Acc Chem Res 50:2487–2495. https://doi.org/10.1021/acs.accounts.7b00235

27. Monguzzi A, Tubino R, Meinardi F (2008) Upconversion-induced delayed fluorescence in multicomponent organic systems: role of Dexter energy transfer. Phys Rev B 77:155122. https://doi.org/10.1103/PhysRevB.77.155122
28. Fan C, Wu W, Chruma JJ, Zhao J, Yang C (2016) Enhanced triplet-triplet energy transfer and upconversion fluorescence through host-guest complexation. J Am Chem Soc 138:15405–15412. https://doi.org/10.1021/jacs.6b07946
29. Ye C, Gray V, Mårtensson J, Börjesson K (2019) Annihilation versus excimer formation by the triplet pair in triplet-triplet annihilation photon upconversion. J Am Chem Soc 141:9578–9584. https://doi.org/10.1021/jacs.9b02302
30. Serevicius T, Komskis R, Adomenas P, Adomeniene O, Kreiza G, Jankauskas V, Kazlauskas K, Miasojedovas A, Jankus V, Monkman A, Jursenas S (2017) Triplet-triplet annihilation in 9,10-diphenylanthracene derivatives: the role of intersystem crossing and exciton diffusion. J Phys Chem C 121:8515–8524. https://doi.org/10.1021/acs.jpcc.7b01336
31. Kozlov DV, Castellano FN (2004) Anti-Stokes delayed fluorescence from metal-organic bichromophores. Chem Commun 2860–2861. http://dx.doi.org/10.1039/B412681E
32. Yu S, Zeng Y, Chen J, Yu T, Zhang X, Yang G, Li Y (2015) Intramolecular triplet-triplet energy transfer enhanced triplet-triplet annihilation upconversion with a short-lived triplet state platinum(II) terpyridyl acetylide photosensitizer. RSC Adv 5:70640–70648. https://doi.org/10.1039/C5RA12579K
33. Liu H, Yan X, Shen L, Tang Z, Liu S, Li X (2019) Color-tunable upconversion emission from a twisted intramolecular charge-transfer state of anthracene dimers via triplet-triplet annihilation. Mater Horiz 6:990–995. https://doi.org/10.1039/C8MH01586D
34. Dzebo D, Boerjesson K, Gray V, Moth-Poulsen K, Albinsson B (2016) Intramolecular triplet-triplet annihilation upconversion in 9,10-diphenylanthracene oligomers and dendrimers. J Phys Chem C 120:23397–23406. https://doi.org/10.1021/acs.jpcc.6b07920
35. Börjesson K, Gilbert M, Dzebo D, Albinsson B, Moth-Poulsen K (2014) Conjugated anthracene dendrimers with monomer-like fluorescence. RSC Adv 4:19846–19850. https://doi.org/10.1039/C4RA02341B
36. Boutin PC, Ghiggino KP, Kelly TL, Steer RP (2013) Photon upconversion by triplet-triplet annihilation in $Ru(bpy)_3$- and DPA-functionalized polymers. J Phys Chem Lett 4:4113–4118. https://doi.org/10.1021/jz402311j
37. Sato R, Kitoh-Nishioka H, Yanai T, Shigeta Y (2017) Theoretical analyses of triplet-triplet annihilation process of 9,10-diphenylanthracene in solution. Chem Lett 46:873–875. https://doi.org/10.1246/cl.170161
38. Ito S, Nagami T, Nakano M (2017) Rational design of doubly-bridged chromophores for singlet fission and triplet-triplet annihilation. RSC Adv 7:34830–34845. https://doi.org/10.1039/C7RA06032G
39. Vij V, Bhalla V, Kumar M (2016) Hexaarylbenzene: evolution of properties and applications of multitalented scaffold. Chem Rev 116:9565–9627. https://doi.org/10.1021/acs.chemrev.6b00144
40. Lambert C (2005) Hexaarylbenzenes—prospects for toroidal delocalization of charge and energy. Angew Chem Int Ed 44:7337–7339. https://doi.org/10.1002/anie.200502105
41. Sun D, Rosokha SV, Kochi JK (2005) Through-space (cofacial) π-delocalization among multiple aromatic centers: toroidal conjugation in hexaphenylbenzene-like radical cations. Angew Chem Int Ed 44:5133–5136. https://doi.org/10.1002/anie.200501005
42. Rosokha SV, Neretin IS, Sun D, Kochi JK (2006) Very fast electron migrations within p-doped aromatic cofacial arrays leading to three-dimensional (toroidal) π-delocalization. J Am Chem Soc 128:9394–9407. https://doi.org/10.1021/ja060393n
43. Piet JJ, Biemans HAM, Warman JM, Meijer EW (1998) Rapid rotation of energy in the excited state of a circular hexa-carbazole array. Chem Phys Lett 289:13–18. https://doi.org/10.1016/S0009-2614(98)00384-4
44. Khan FA, Wang D, Pemberton B, Talipov MR, Rathore R (2016) Toroidal delocalization of a single electron through circularly-arrayed benzophenone chromophores in hexakis(4-benzoylphenyl)benzene. J Photochem Photobiol, A 331:153–159. https://doi.org/10.1016/j.jphotochem.2016.05.002

45. Kosaka T, Iwai S, Fukuhara G, Imai Y, Mori T (2019) Hydrostatic pressure on toroidal interaction and propeller chirality of hexaarylbenzenes: explicit solvent effects on differential volumes in methylcyclohexane and hexane. Chem Eur J 25:2011–2018. https://doi.org/10.1002/chem.201804688
46. Kosaka T, Iwai S, Inoue Y, Moriuchi T, Mori T (2018) Solvent and temperature effects on dynamics and chiroptical properties of propeller chirality and toroidal interaction of hexaarylbenzenes. J Phys Chem A 122:7455–7463. https://doi.org/10.1021/acs.jpca.8b06535
47. Toyoda M, Imai Y, Mori T (2017) Propeller chirality of boron heptaaryldipyrromethene: unprecedented supramolecular dimerization and chiroptical properties. J Phys Chem Lett 8:42–48. https://doi.org/10.1021/acs.jpclett.6b02492
48. Kosaka T, Inoue Y, Mori T (2016) Toroidal interaction and propeller chirality of hexaarylbenzenes. Dynamic domino inversion revealed by combined experimental and theoretical circular dichroism studies. J Phys Chem Lett 7:783–788. https://doi.org/10.1021/acs.jpclett.6b00179
49. Feng HT, Liu C, Li Q, Zhang H, Lam JWY, Tang BZ (2019) Structure, assembly, and function of (latent)-chiral AIEgens. ACS Mater Lett 1:192–202. https://doi.org/10.1021/acsmaterialslett.9b00116
50. Curutchet C, Mennucci B (2017) Quantum chemical studies of light harvesting. Chem Rev 117:294–343. https://doi.org/10.1021/acs.chemrev.5b00700
51. Mirkovic T, Ostroumov EE, Anna JM, van Grondelle R, Govindjee Scholes GD (2017) Light absorption and energy transfer in the antenna complexes of photosynthetic organisms. Chem Rev 117:249–293. https://doi.org/10.1021/acs.chemrev.6b00002
52. Kilina S, Kilin D, Tretiak S (2015) Light-driven and phonon-assisted dynamics in organic and semiconductor nanostructures. Chem Rev 115:5929–5978. https://doi.org/10.1021/acs.chemrev.5b00012
53. Jang SJ, Mennucci B (2018) Delocalized excitons in natural light-harvesting complexes. Rev Mod Phys 90:035003. https://doi.org/10.1103/RevModPhys.90.035003
54. Brixner T, Stenger J, Vaswani HM, Cho M, Blankenship RE, Fleming GR (2005) Two-dimensional spectroscopy of electronic couplings in photosynthesis. Nature 434:625–628. https://doi.org/10.1038/nature03429
55. Talipov MR, Jasti R, Rathore R (2015) A circle has no end: role of cyclic topology and accompanying structural reorganization on the hole distribution in cyclic and linear poly-p-phenylene molecular wires. J Am Chem Soc 137:14999–15006. https://doi.org/10.1021/jacs.5b09596
56. Talipov MR, Ivanov MV, Rathore R (2016) Inclusion of asymptotic dependence of reorganization energy in the modified Marcus-based multistate model accurately predicts hole distribution in poly-*p*-phenylene wires. J Phys Chem C 120:6402–6408. https://doi.org/10.1021/acs.jpcc.6b00514
57. Wang CZ, Noda Y, Wu C, Feng X, Venkatesan P, Cong H, Elsegood MRJ, Warwick TG, Teat SJ, Redshaw C, Yamato T (2018) Multiple photoluminescence from pyrene-fused hexaarylbenzenes with aggregation-enhanced emission features. Asian J Org Chem 7:444–450. https://doi.org/10.1002/ajoc.201700563
58. Cao X, Hu B, Zhang P (2013) High upconversion efficiency from hetero triplet-triplet annihilation in multiacceptor systems. J Phys Chem Lett 4:2334–2338. https://doi.org/10.1021/jz401213w
59. Turshatov A, Busko D, Avlasevich Y, Miteva T, Landfester K, Baluschev S (2012) Synergetic effect in triplet-triplet annihilation upconversion: highly efficient multi-chromophore emitter. ChemPhysChem 13:3112–3115. https://doi.org/10.1002/cphc.201200306

Chapter 18
π-Electronic Ion-Pairing Assemblies for Photoswitching Materials

Yohei Haketa, Ryohei Yamakado, Kazuki Urakawa, and Hiromitsu Maeda

Abstract Ion-pairing assemblies consisting of appropriately designed oppositely charged π-electronic systems give rise to various functional supramolecular assemblies including crystals and soft materials based on the anisotropic orientation of charged species through electrostatic and other weak noncovalent interactions. The introduction of photo-responsive moiety to the charged species facilitated the formation of photo-responsive ion-pairing assemblies, whose assembling modes were controlled by photoirradiation. Furthermore, pyrrole-based π-system–Pt^{II} complexes were designed and exhibited unique photo-induced excited-state dynamics.

Keywords Ion-pairing assemblies · Dimension-controlled assemblies · Anion-responsive π-electronic molecules · Photo-responsive assemblies · Excited-state dynamics

18.1 Ion-Pairing Assemblies Comprising Charged π-Electronic Systems

18.1.1 Introduction

In recent years, materials comprising ionic species based on the concept of ionic self-assembly (ISA) have attracted much attention in materials science, especially in the fields of functional nanostructured soft materials, surface morphologies, organic-inorganic hybrid materials, etc. [1, 2]. An appropriate combination of designed organic ionic species (anions and cations) affords various ion pairs that form functional ion-pairing assemblies. Among a variety of charged species, π-electronic ions

---

Y. Haketa · K. Urakawa · H. Maeda (✉)
Department of Applied Chemistry, College of Life Sciences, Ritsumeikan University, Kusatsu, Shiga 525–8577, Japan
e-mail: maedahir@ph.ritsumei.ac.jp

R. Yamakado
Department of Organic Materials Science, Graduate School of Organic Materials Science, Yamagata University, Yonezawa, Yamagata 992–8510, Japan

H. Miyasaka et al. (eds.), *Photosynergetic Responses in Molecules and Molecular Aggregates*, https://doi.org/10.1007/978-981-15-5451-3_18

assemble through electrostatic interaction with other noncovalent interactions, fabricating electronically and optically functional dimension-controlled assemblies (e.g., soft crystalline materials, gels, liquid crystals, etc.) [3–7]. Appropriately designed π-electronic ion pairs with suitable geometries and electronic states result in their ordered arrangement and diverse packing modes. As the key concept of ion-pairing assemblies based on π-electronic ion pairs, a charge-by-charge assembly is defined as an assembling mode comprising alternately stacking positively and negatively charged species (Fig. 18.1 left), whereas a charge-segregated assembly results from the stacking of identically charged species by overcoming electrostatic repulsion (Fig. 18.1 right) [6, 7]. The components of ion-pairing assemblies can be categorized into three different types: (i) π-electronic molecules appending ionic units, (ii) planar ion complexes using ion-responsive π-electronic molecules (receptors), and (iii) genuine π-electronic ions. As an example of (i), a simple deprotonation of the acid unit (e.g., carboxy and hydroxy units) of π-electronic molecules by bases with an appropriate countercation would provide ion pairs. Various modifications of electronically neutral receptors facilitated the facile functionalization of the receptor–ion complex ion pairs in the case of (ii). Genuine π-electronic ions, in the case of (iii), are rather rare species due to difficulties in their synthesis. On the basis of the strategies for the preparation of π-electronic ion pairs, modifications of constituent ions result in various functional ion-pairing assemblies. Importantly, the introduction of a stimuli-responsive unit for π-electronic ions is crucial for the formation of ion-pairing assemblies that demonstrate the modulable arrangement of charged components and further, the properties induced by external stimuli. Among various possible stimuli, photoirradiation can be used for the switching of the structures and electronic states of π-electronic systems as well as the resulting assembling modes. A significant point to note is that each assembling mode would show the ordered alignment of the constituent π-electronic ions through the interionic interactions. Furthermore, the photo-excited state of π-electronic species would show high-energy transient states with geometries and properties that differ from those of ground states.

In this chapter, to begin with, the recent progress in the design, synthesis, and assembling behavior of π-electronic ion pairs is briefly summarized. Subsequently,

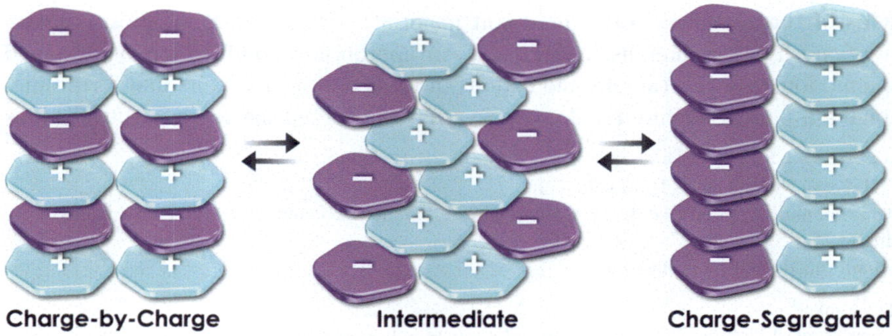

Fig. 18.1 Conceptual diagram of π-electronic ion-pairing assemblies

the assembling behaviors of photo-responsive π-electronic ion pairs are discussed. Finally, the excited-state dynamics of chiral pyrrole-based π-system–Pt^{II} complexes are explained.

18.1.2 Ion-Pairing Assemblies Based on π-Electronic Ions

π-Electronic ions that have planar geometries (genuine π-electronic ions) and exhibit various characteristic electronic properties are suitable for ordered stacking structures owing to the effective electrostatic interaction of charged aromatic cores. Thus, π-electronic ion pairs and their ion-pairing assemblies have been well investigated. In this section, a brief summary of several recent advances is provided followed by the description of the functional π-electronic anions. Pentacyanocyclopentadienide ($PCCp^-$) $\mathbf{1}^-$ [8–11] as a stable and genuine π-electronic anion can be used as the component of ion-pairing assemblies (Fig. 18.2a). In fact, ion pairs $\mathbf{1}^-$-$\mathbf{2}^+$ and $\mathbf{1}^-$-$C_4H_9(CH_3)_3N^+$ with cycloheptatrienyl (Ch^+) and butyltrimethylammonium cation, respectively, formed charge-by-charge and charge-segregated assemblies, respectively, in the crystal state (Fig. 18.2b(i, ii)). On the other hand, X-ray diffraction (XRD) analysis and polarized optical microscopy (POM) observation revealed the formation of the dimension-controlled assembly as a hexagonal columnar (Col_h) mesophase for the $PCCp^-$ ion pair $\mathbf{1}^-$-$(C_{12}H_{25})_3CH_3N^+$ with an aliphatic tridodecylmethylammonium cation (Fig. 18.2b(iii)). The stacking columnar assembly of

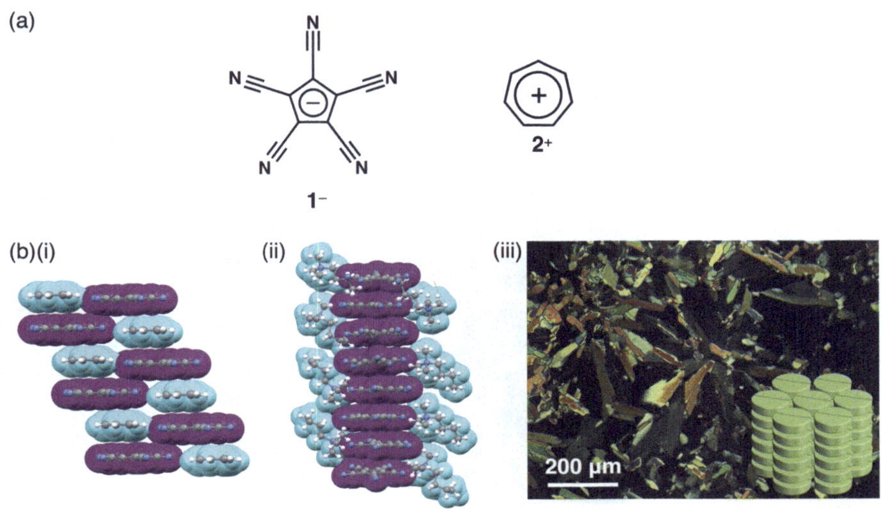

Fig. 18.2 **a** $PCCp^-$ anion $\mathbf{1}^-$ and Ch^+ cation $\mathbf{2}^+$, **b** single-crystal X-ray structures (cyan: cations, magenta: $\mathbf{1}^-$) of (i) $\mathbf{1}^-$-$\mathbf{2}^+$ and (ii) $\mathbf{1}^-$-$C_4H_9(CH_3)_3N^+$, and (iii) POM of $\mathbf{1}^-$-$(C_{12}H_{25})_3CH_3N^+$ at 70 °C upon cooling and packing model

1$^-$-$(C_{12}H_{25})_3CH_3N^+$ showed a hole-transporting property (0.4 cm^2/Vs) in the film prepared by the drop-casting of a $CHCl_3$ solution [11].

π-Electronic anions are less stable than π-electronic cations because the excess electrons in the anions induce oxidation and result in decomposition. On the other hand, the deprotonation of the acid unit introduced in the appropriate π-electronic molecules affords anion-appended π-electronic species. For example, the deprotonation of carboxy and hydroxy units provides corresponding anions with accompanying countercations. The partial negative charges can be delocalized in the π-electronic systems. Dipyrrolylnitrophenols **3a–c** were designed as the stable precursor of the π-electronic anions (Fig. 18.3a). Their deprotonation by tetrabutylammonium hydroxide (TBAOH) formed π-electronic anions **3a**$^-$–**c**$^-$ as TBA^+ ion pairs via the inversion of pyrrole rings and the stabilization of the negative charges by the intramolecular hydrogen bonding of pyrrole NH and anionic oxygen. Density functional theory (DFT) calculation also revealed the delocalization of negative charges in the aryl units with the electron-withdrawing nitro group. Various countercations can be introduced to form π-electronic ion pairs, which gave rise to charge-by-charge assemblies in the crystal state [12]. The deprotonated species of π-extended **3b**$^-$-TBA^+ was more suitable for charge-by-charge assemblies with counter tetraalkylammonium cations (Fig. 18.3b(i)). Meanwhile, deprotonated species bearing aliphatic

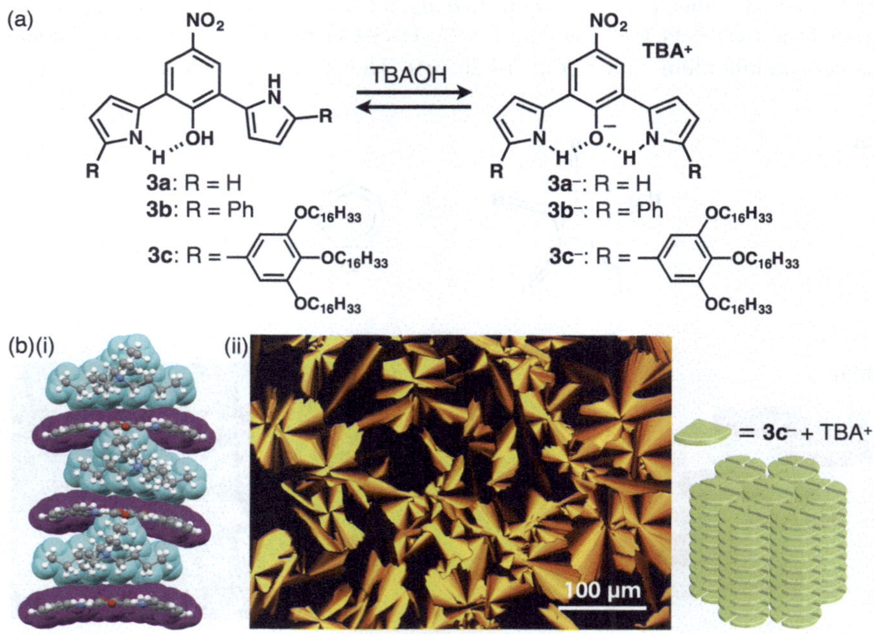

Fig. 18.3 **a** Dipyrrolylnitrophenols **3a–c** and their deprotonated species **3a**$^-$–**c**$^-$ and **b**(i) single-crystal X-ray structure of **3b**$^-$-TBA^+ (cyan: TBA^+, magenta: **3b**$^-$) and (ii) POM image of the mesophase of **3c**$^-$-TBA^+ and the packing model

chains $3c^-$-TBA^+ formed a liquid crystal mesophase of the Col_h structure, as observed by POM (Fig. 18.3b(ii)) [13]. The deprotonation of pyrrole NH unit of appropriately designed pyrrole-based π-electronic systems can also provide corresponding anionic species accompanied by countercations. For example, deprotonation of single NH of dipyrrolylquinoxalines (DPQ) by TBAOH resulted in the formation of ion pairs of anionic DPQ^- and TBA^+. DPQ^--TBA^+ showed a charge-by-charge columnar assembly, as revealed by single-crystal X-ray analysis, whereas the ion pair of aliphatic DPQ^- and TBA^+ formed *n*-octane gel (10 mg/mL) based on the organized fibril morphologies [14].

Larger π-electronic molecules with an acid unit can provide stable anionic species owing to the more effective delocalization of the negative charge in the π-electronic core unit. The deprotonation of *meso*-hydroxy-substituted porphyrin **4** by treatment with TBAOH provided the anionic species $\mathbf{4}^-$ as indicated by the UV/vis absorption and ^{1}H NMR spectral changes (Fig. 18.4a(i)). Single-crystal X-ray analysis of $\mathbf{4}^-$-TBA^+ revealed the exact structure of $\mathbf{4}^-$ with the delocalization of the partial negative charge in the porphyrin core unit. $\mathbf{4}^-$-TBA^+ formed the alternate stacking column of $\mathbf{4}^-$ and TBA^+ (Fig. 18.4a(ii)) [15]. Interestingly, the Pd^{II} complex **4′** coordinated at the *meso*-oxygen and the β-position was obtained from **4** by treatment with $Pd(OAc)_2$,

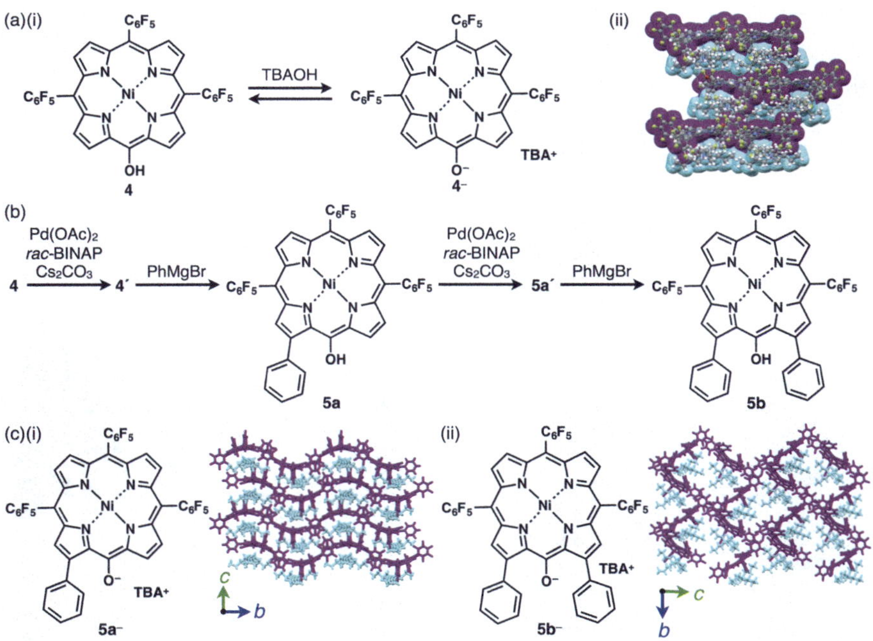

Fig. 18.4 **a** (i) *Meso*-hydroxy-substituted porphyrin **4** and formation of anionic species $\mathbf{4}^-$ and (ii) single-crystal X-ray structure of $\mathbf{4}^-$-TBA^+ (cyan: TBA^+, magenta: $\mathbf{4}^-$), **b** synthesis of β-phenyl-substituted porphyrins **5a,b**, and **c** solid-state packing structures (cyan: TBA^+, magenta: anions) of (i) $\mathbf{5a}^-$-TBA^+ and (ii) $\mathbf{5b}^-$-TBA^+

rac-BINAP, and Cs_2CO_3 (Fig. 18.4b) [16]. The coupling reaction of **5b′** with PhMgBr afforded the β-phenyl-substituted porphyrin **5a**. Subsequently, by similar procedures, **5a** was converted to β-diphenyl-substituted **5b** via the Pd^{II} complex **5b′**. Deprotonation of the OH unit of **5a,b** by TBAOH provided the π-electronic anions **5a**$^-$,**b**$^-$. The ion pairs **5a**$^-$,**b**$^-$-TBA^+ formed charge-by-charge assemblies in the solid-state as elucidated by single-crystal X-ray analysis (Fig. 18.4c) [16]. The introduction of more effectively hydrogen-bonding aryl moieties would stabilize anionic O-appended π-electronic species.

Metal complexes using π-electronic ligands that partially compensate for the positive charges of metal ions can also be used as planar π-electronic charged species. As Au^{III} is a d^8 state, which forms square-planar complexes with suitable ligand species, Au^{III} complexes of porphyrins are π-electronic cationic species that require no axial ligand coordination. In fact, a variety of porphyrin–Au^{III} complexes (e.g., **6a**$^+$,**b**$^+$) were prepared as ion pairs in combination with counteranions such as BF_4^-, PF_6^-, **1**$^-$, and **4**$^-$ (Fig. 18.5a) [17]. *Meso*-tetraarylporphyrin–Au^{III} complexes were found to act as cationic building units of ion-pairing dimension-controlled assemblies as supramolecular gels and thermotropic liquid crystals while also stacking π-electronic ion pairs in the solution state. For example, the ion pair **6a**$^+$-**1**$^-$ displayed a charge-by-charge stacking columnar structure in the crystal state (Fig. 18.5b). Aliphatic ion pair **6b**$^+$-**1**$^-$ showed the exceptionally wide-temperature-range Col_h

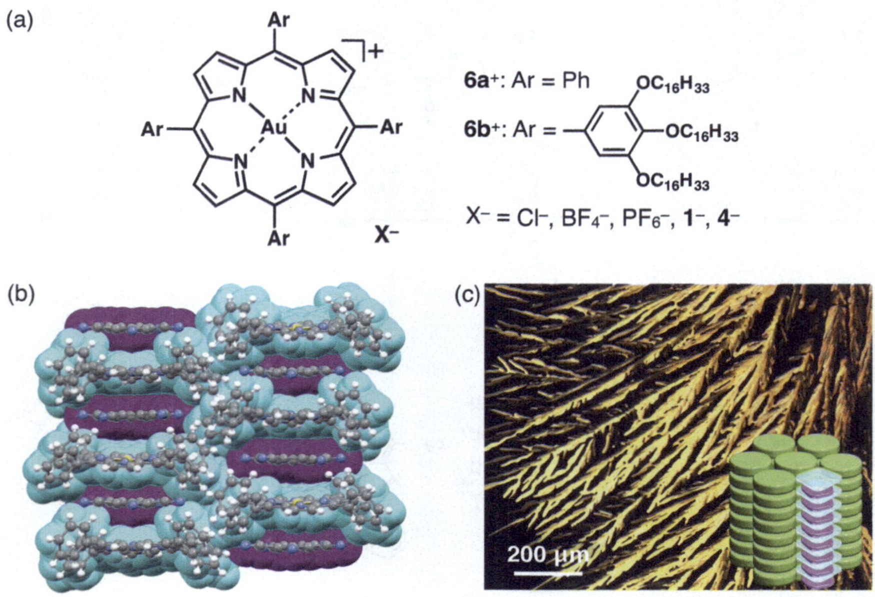

Fig. 18.5 **a** *Meso*-tetraarylporphyrin–Au^{III} complexes **6a**$^+$,**b**$^+$ as ion pairs, **b** single-crystal X-ray structure of **6a**$^+$-**1**$^-$ (cyan: **6a**$^+$, magenta: **1**$^-$), and **c** POM of **6b**$^+$-**1**$^-$ observed at 280 °C upon cooling and the packing model

mesophases from r.t. to nearly 300 °C as observed in POM, which were stabilized by the charge-by-charge stacking of *genuine* π-electronic ions through synergetic π–π and electrostatic interactions (Fig. 18.5c) [17]. Furthermore, the ion pairs based on porphyrin–Au^{III} complexes bearing pentafluorophenyl (C_6F_5) units also formed charge-by-charge and charge-segregated assemblies in the single crystals, depending on the C_6F_5 substitution pattern and the geometries of the anionic species [18].

18.1.3 Ion-Pairing Assemblies Based on Anion-Responsive π-Electronic Molecules

In the course of preparation of various π-electronic anions, the anion complexes of anion-responsive molecules (receptors) were found to be used as *pseudo*-π-electronic anions [6, 7]. The combination of electronically neutral receptors, guest anions, and countercations can provide diverse ion pairs. Among the various anion receptors, a series of dipyrrolyldiketone BF_2 complexes, which show effective anion-binding properties through the inversion of two pyrrole rings, have been investigated (Fig. 18.6a) [19]. Efficient [1 + 1]-type anion-binding behaviors were elucidated by ^{1}H NMR spectral changes, wherein anion-free and anion-complexing signals were independently observed owing to the slow anion-binding processes that were correlated with the pyrrole inversion. So far, various modifications of dipyrrolyldiketone BF_2 complexes were investigated. For example, anion complexes of dipyrrolyldiketone BF_2 complexes **7a–c** (Fig. 18.6a) were used as the components of various ion-pairing assemblies in the form of crystals, supramolecular gels, liquid crystals, etc. [6, 7]. Furthermore, covalently linked oligomers exhibited high anion-binding affinities due to the efficient multiple hydrogen-bonding interactions, forming anion-driven helical structures via pyrrole inversions [20]. The chirality of the helical anion complexes of the dimers was induced by the ion pairing with chiral countercations, exhibiting circularly polarized luminescence [21]. On the basis of their anion-binding behaviors, anion-appended dipyrrolyldiketone BF_2 complexes can provide diverse π-electronic anions as seen in dipyrrolylphenols. In fact, the derivatives substituted with a hydroxy unit at the bridging carbon gave rise to anionic species, whose anionic site underwent interaction with pyrrole NH units through intramolecular hydrogen bonding; however, the deprotonated species in solution were not very stable as they decomposed in 1–2 h [22]. This section demonstrates the tunability of assembling modes based on the modifications of constituent ions in ion-pairing assemblies comprising receptor–anion complexes.

π-Electronic systems with carboxylate units can be used as scaffolds to attach various functional units because the effective carboxylate-binding abilities of dipyrrolyldiketone BF_2 complexes were elucidated in the solution and solid states (Fig. 18.6b) [23]. Single-crystal X-ray analysis of the ion pairs comprising **7a** and tetraalkylammonium salts of benzoate [24] and 3,4,5-trimethoxy-substituted benzoate [25] revealed the formation of [1 + 1]-type carboxylate complexes and

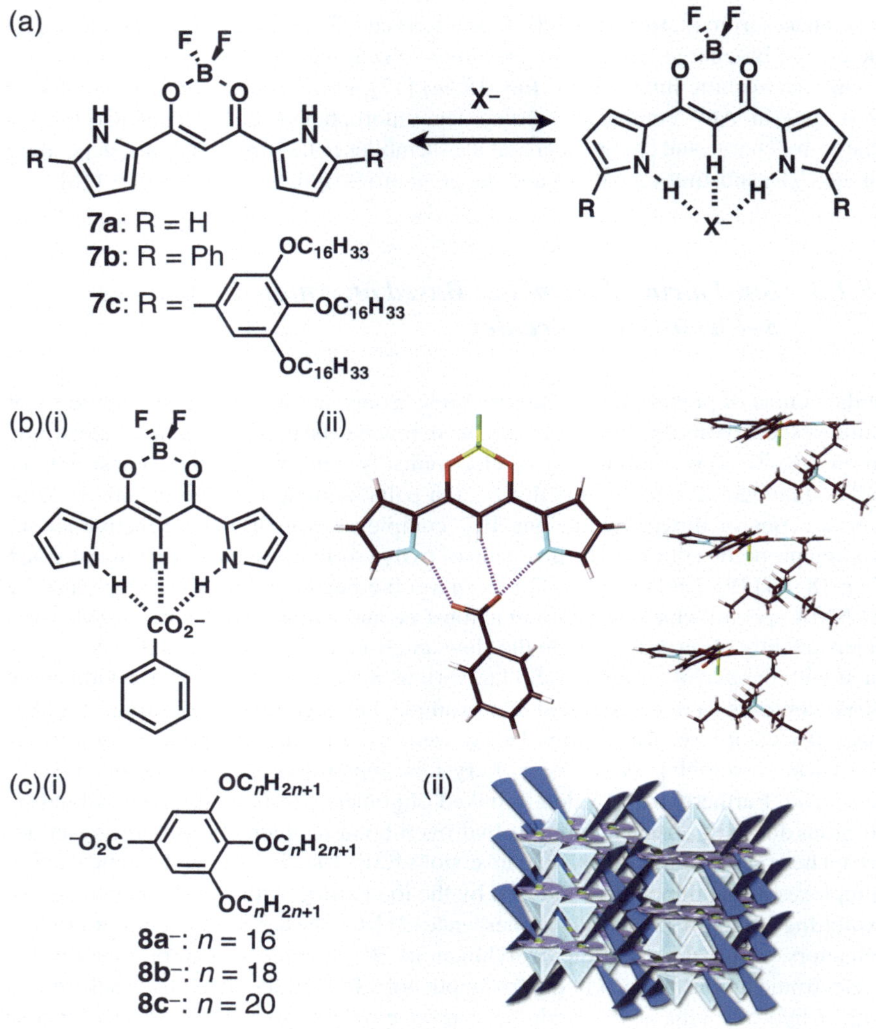

Fig. 18.6 **a** Anion-binding behavior of dipyrrolyldiketone BF_2 complexes **7a–c**, **b**(i) benzoate-binding mode of **7a** and (ii) single-crystal X-ray structure of **7a**·benzoate (left: [1 + 1] complex, right: charge-by-charge assembly) (atom color code: brown, pink, yellow, green, blue, and red indicate carbon, hydrogen, boron, fluorine, nitrogen, and oxygen, respectively), and **c** (i) modified anions **8a**$^-$–**c**$^-$ and (ii) packing model structure of **7a**·**8a**$^-$-TBA$^+$ as the mesophase

charge-by-charge assemblies (Fig. 18.6b(ii)). Such solid-state packing modes are crucial for the further design of dimension-controlled assemblies. For example, the combination of **7a**,**b** with 3,4,5-trialkoxy-substituted benzoates **8a**$^-$–**c**$^-$ as TBA$^+$

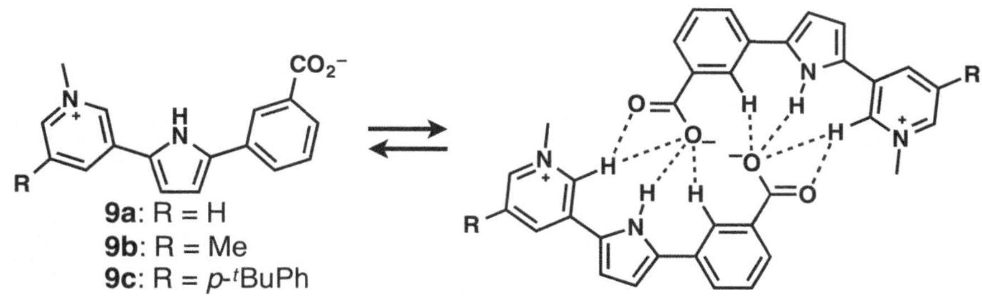

Fig. 18.7 Dimerization behavior of π-electronic zwitterions **9a–c**

salts (Fig. 18.6c(i)) formed lamellar mesophases based on charge-by-charge assemblies (Fig. 18.6c(ii)). These ion-pairing materials showed moderately high charge-carrier transporting properties, as demonstrated in the values of 0.02 and 0.05 cm^2/Vs for **7a·8a**$^-$-TBA$^+$ and **7b·8a**$^-$-TBA$^+$, respectively [25].

Focusing on the components of ion-pairing assemblies, the introduction of a negatively charged site in anion receptors can construct self-assembled dimers with countercations. This can be seen in carboxylate-appended derivatives of dipyrrolyldiketone BF_2 complexes, which exhibit narcissistic self-sorting dimerization behaviors [24]. Such effective self-associating dimeric structures can also be formed by the introduction of a cationic site to anion-appended anion receptors. In fact, zwitterionic π-electronic systems **9a–c** comprising pyridinium and benzoate units on both sides of the pyrrole α-positions were synthesized (Fig. 18.7). Zwitterionic π-electronic systems **9a–c** formed self-associating dimeric structures in DMSO, and the association constant of **9a** that can be considered typically representative was 3300 M^{-1} at 70 °C. Not only hydrogen bonding between the anionic moieties and interaction sites but also charge delocalization stabilized such self-assembled dimerized forms [26]. The further design and synthesis of charged species comprising π-electronic units and their combination with receptor molecules facilitate the formation of dimension-controlled assemblies with fascinating functionalities.

Control of the geometries and electronic states of counter species of π-electronic anions (receptor–anion complexes) is also important to achieve functional ion-pairing assemblies. On the basis of the modifications of diverse π-electronic cationic species, the combination of Cl^- complexes of dipyrrolyldiketone BF_2 complexes and π-ligand–metal complex cations can afford fascinating ion-pairing assemblies. For example, the ion-pairing assembly of **7c** and 4′-hexadecyloxy-substituted $Pt^{II}Cl$ complex of 2,2′:6′,2″-terpyridine (trpy) **10**$^+$ as a Cl^- salt afforded a mesophase based on the Col_h structure of a charge-segregated assembly (Fig. 18.8a) [27]. Furthermore, bis(imidazolium)pyrimidine dication with two Cl^- as counteranions **11a**$^{2+}$·$2Cl^-$ were prepared for the components of ion-pairing assemblies (Fig. 18.8b) [28]. The ion pairs comprising **7c**·Cl^- and bis(imidazolium)pyrimidine dication·Cl^- **11a**$^{2+}$·Cl^- formed a Col_h-based mesophase. Similarly, benzyltrialkylammonium cations **12a**$^+$,**b**$^+$ as Cl^- salts were combined with **7b**, resulting in the formation

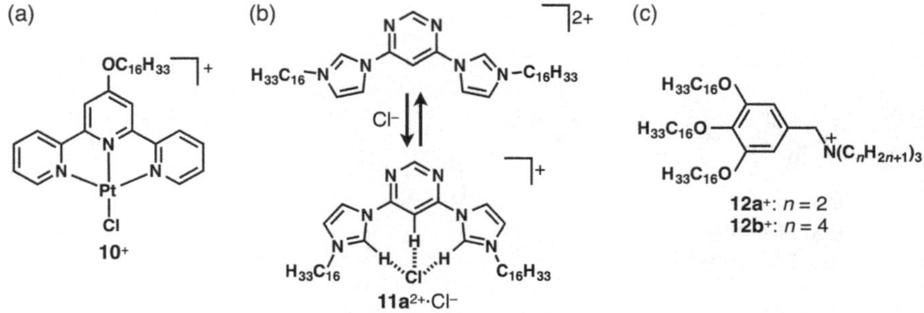

Fig. 18.8 **a** trpy–Pt^{II} complex **10**$^+$, **b** dication **11a**$^{2+}$ and positively charged receptor–Cl$^-$ complex **11a**$^{2+}$·Cl$^-$, and **c** modified cations **12a**$^+$,**b**$^+$

of ion-pairing assemblies as the Col$_h$ mesophases (Fig. 18.8c) [29]. These examples demonstrated the possibility of diverse combinations of anion-responsive π-electronic molecules, modified anions, and coexisting cations, affording highly organized dimension-controlled ion-pairing assemblies. The introduction of functional units to the component(s) would induce the tunable functionalities.

18.2 Photo-Responsive Ion-Pairing Assemblies

18.2.1 How to Prepare Stimuli-Responsive Assemblies

An advantage of ion-pairing materials is the introduction of different functionalities to oppositely charged species by exchanging the constituent charged species. The introduction of various substituents is crucial for the control of the states and assembling modes of ion pairs and their resulting properties. Another strategy to control the states of ion pairs is the introduction of stimuli-responsive moieties whose geometries can be modulated by applying an external stimulus. Stimuli-responsive supramolecular assemblies have attracted much attention because they can yield various nanostructures as potential materials for self-healing, drug delivery, and sensing. Among the various external stimuli, light is a remarkably useful and easy-to-use stimulus [30–32]. Azobenzene, as a photo-responsive unit, shows *trans–cis* isomerization upon photoirradiation; the different geometries of the *trans* and *cis* forms can affect the assembling modes according to their ratios [33]. Photo-induced phase transitions of azobenzene derivatives in bulk materials were often observed in liquid crystals [34–36] including those of ion-pairing systems. However, conditions that allow the photo-isomerization of azobenzenes in the bulk states are restricted because free volumes are required for significant structural changes [37]. In this section, synthesis of photo-responsive ion pairs based on azobenzene and the photo-responsive properties of the resulting dimension-controlled ion-pairing assemblies are introduced (Fig. 18.9).

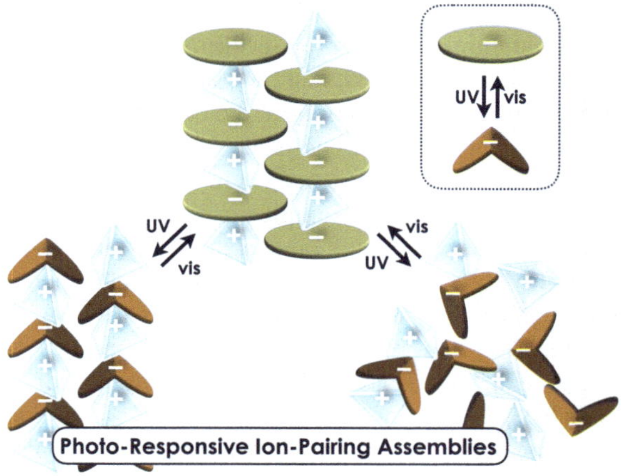

Fig. 18.9 Conceptual diagram of ion-pairing assemblies based on photo-responsive ions and their photo-responsive properties. The combination of various photo-responsive anions and countercations can control the assembled structures and resulting properties

18.2.2 Design and Synthesis of Photo-Responsive Ion Pairs

As shown in Fig. 18.10, ion pairs of azobenzene-appended carboxylates bearing aliphatic chains on a phenyl moiety accompanied with alkylammonium cations were designed to permit the formation of photo-responsive dimension-controlled assemblies [38]. In these ion pairs, flexible alkyl chains and bulky countercation enhance the mobility and generate the free volume around the azobenzene. Assemblies of the ionic species were formed through noncovalent interactions, where the arrangement of the components can be readily modulated by external conditions. In addition, long alkyl chains can provide dimension-controlled assemblies, such as supramolecular gels and liquid crystals, through van der Waals interactions.

Alkyl-substituted azobenzene carboxylate esters $\mathbf{13}^{\mathrm{C}n}$ and $\mathbf{14}^{\mathrm{C}n}$ ($n = 8$, 12, 16) were obtained by the reaction of ethyl 4-nitrosobenzoate and the corresponding aniline derivatives. Subsequent hydrolysis of $\mathbf{13}^{\mathrm{C}n}$ and $\mathbf{14}^{\mathrm{C}n}$ provided azobenzene carboxylic acids $\mathbf{15}^{\mathrm{C}n}$ and $\mathbf{16}^{\mathrm{C}n}$ ($n = 8$, 12, 16). Carboxylate ion pairs $\mathbf{15}^{\mathrm{C}n-}\cdot\mathrm{TBA}^+$ and $\mathbf{16}^{\mathrm{C}n-}\cdot\mathrm{TBA}^+$ ($n = 12$, 16) were obtained by treating $\mathbf{16}^{\mathrm{C}n}$ and $\mathbf{17}^{\mathrm{C}n}$ ($n = 12$, 16) in CH_2Cl_2 with excess TBAOH [25], followed by the removal of the remaining TBAOH by extraction with water (Fig. 18.10). In contrast, the preparation of $\mathbf{15}^{\mathrm{C}8-}\cdot\mathrm{TBA}^+$

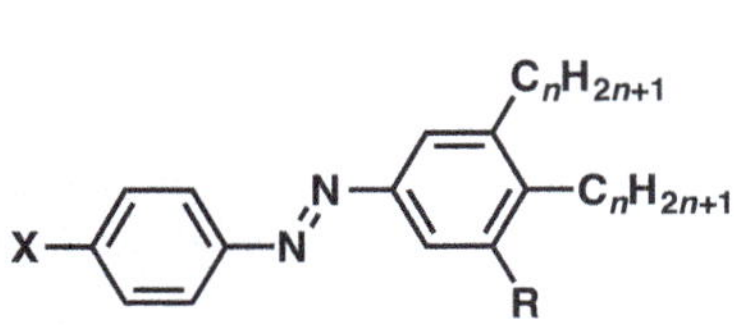

$\mathbf{13}^{\mathrm{C}n}$: R = H, X = CO_2Et
$\mathbf{14}^{\mathrm{C}n}$: R = C_nH_{2n+1} (n = 8, 12, 16), X = CO_2Et
$\mathbf{15}^{\mathrm{C}n}$: R = H, X = CO_2H
$\mathbf{16}^{\mathrm{C}n}$: R = C_nH_{2n+1} (n = 8, 12, 16), X = CO_2H
$\mathbf{15}^{\mathrm{C}n-}$: R = H, X = CO_2^-
$\mathbf{16}^{\mathrm{C}n-}$: R = C_nH_{2n+1} (n = 8, 12, 16), X = CO_2^-

Fig. 18.10 Structures of alkyl-substituted azobenzene carboxylates and their precursors and $\mathbf{16}^{C8-}\cdot TBA^+$ from $\mathbf{15}^{C8}$ and $\mathbf{16}^{C8}$, respectively, failed because the high solubility and low melting points of the resulting materials hindered their purification.

The thermal properties and assembled structures demonstrated the dependence on the degrees of interactions at the carbonyl groups and the number and the length of alkyl chains. Carboxylic acids $\mathbf{15}^{Cn}$ (n = 8, 12, 16), possessing two aliphatic chains, formed lamellar structures in the solid state, and the repeating distances were slightly smaller than the summed length of two molecules, suggesting the formation of a tilted or interdigitated bilayer structure through hydrogen bonding. On the other hand, $\mathbf{16}^{Cn}$ (n = 8, 12, 16), possessing fan-shaped three aliphatic chains, formed discotic columnar structures. Ion pairs $\mathbf{15}^{Cn-}\cdot TBA^+$ and $\mathbf{16}^{Cn-}\cdot TBA^+$ (n = 12, 16) provided lamellar structures through ionic and van der Waals interactions, with the repeating distances corresponding to the summed lengths of azobenzene carboxylates and TBA^+. Interestingly, the ion pair $\mathbf{16}^{C16-}\cdot TBA^+$ provided a stable lamellar phase with repeating distances of 3.8 nm at 22–55 °C only in the heating process, as the metastable crystalline lamellar phase with repeating distances of 4.5 nm was formed by rapid cooling from the isotropic state (>55 °C).

18.2.3 Photo-Responsive Properties of Ion-Pairing Assemblies

The photo-induced phase transitions were examined by POM and XRD analyses under photoirradiation. Although film samples are suitable for these analyses because of the difficulty experienced by light in penetrating thick samples, it is challenging to prepare films of low-molecular-weight species due to the dewetting behavior. This problem was resolved by sandwiching the samples into glass plate cells with 8 μm spacers. The diffraction profiles were obtained by using an X-ray diffractometer equipped with a two-dimensional detector. The X-ray beam (Cu-Kα) was made parallel to the in-plane direction of the cells using pulse controllers. UV light (365 nm) and visible light (436 nm) were irradiated into the cells from an out-of-plane direction (Fig. 18.11a).

In contrast to the ion pairs $\mathbf{15}^{C12-}\cdot TBA^+$, $\mathbf{15}^{C16-}\cdot TBA^+$, and $\mathbf{16}^{C12-}\cdot TBA^+$, which showed the photo-induced phase transitions from ionic crystals to liquids (Fig. 18.11b), the birefringence of $\mathbf{16}^{C16-}\cdot TBA^+$ at 40 °C was completely transformed under the photoirradiation at 365 nm (100 mW/cm^2), suggesting a phase transition. UV (365 nm, 100 mW/cm^2) irradiation onto the sample induced a gradual disappearance of the lamellar phase (Phase A) with repeating distances of 3.8 nm, as well as the appearance of a new crystalline lamellar phase with a repeating distance of 3.4 nm (Phase B) (Fig. 18.12a(i–iii)). The calculated lengths of $\mathbf{16}^{C16-}$ including the countercation are approximately 3.9 and 3.3 nm for the *trans* and *cis* forms, respectively, (Fig. 18.12b). Thus, the difference between Phase A and Phase B can be rationalized by the *trans*-to-*cis* photo-isomerization. In addition, XRD peaks assignable to Phase A and Phase B were independently observed, indicating that the phase transition from

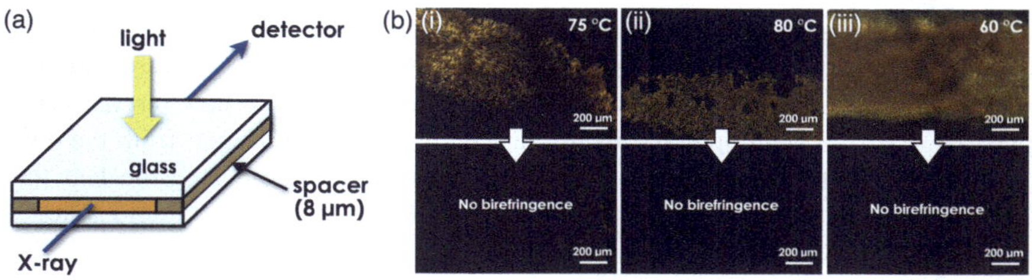

Fig. 18.11 **a** Schematic illustration of a setup for in situ XRD measurement under photoirradiation. UV (365 nm, 100 mW/cm^2) and visible light (436 nm, 60 mW/cm^2) penetration was confirmed in this system. **b** The photo-induced phase transition of (i) $\mathbf{15}^{C12-}$·TBA$^+$, (ii) $\mathbf{15}^{C16-}$·TBA$^+$, and (iii) $\mathbf{16}^{C12-}$·TBA$^+$ in a sandwiched cell (8 μm spacer) observed by POM under UV (365 nm, 100 mW/cm^2) irradiation at 75, 80, and 60 °C, respectively: initial state (top) and after 5 min (bottom)

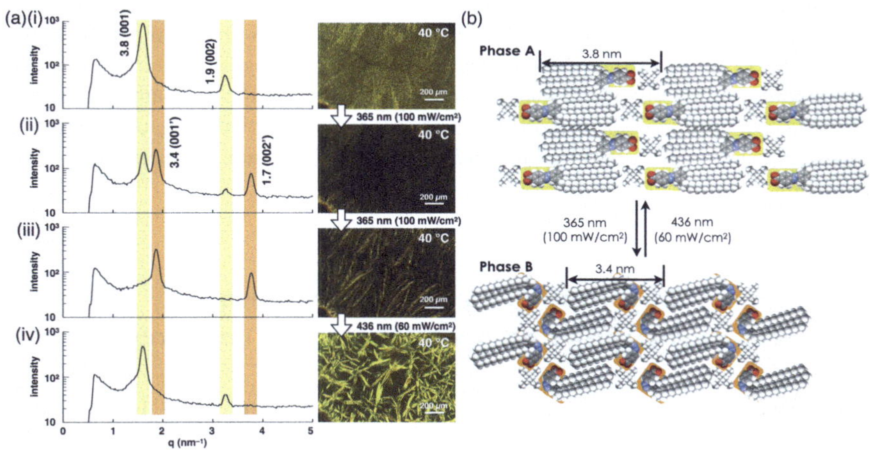

Fig. 18.12 **a** Changes in XRD patterns (small-angle region) and corresponding POM images of $\mathbf{16}^{C16-}$·TBA$^+$ in a sandwiched cell (8 μm spacer) at variable temperatures at (i) 40 °C, (ii, iii) 40 °C under UV (365 nm, 100 mW/cm^2) irradiation with an approximately 0.5 h interval, and (iv) 40 °C under visible light (436 nm, 60 mW/cm^2) irradiation. The diffractions derived from Phase A and Phase B are marked in yellow and orange, respectively. Birefringence became faint or disappeared once during the transition between (iii) and (iv). **b** Schematic illustrations of the assembled arrangements of $\mathbf{16}^{C16-}$·TBA$^+$ and their photo-induced phase transitions

Phase A to Phase B proceeded discontinuously in the respective domains, and Phase A completely disappeared because the remaining *trans* forms at the photo-stationary state (PSS$_{365}$) may be disordered. Thus, two possibilities may arise (i) isomerization of the *trans* isomers induced by exceeding the threshold number of the neighboring *cis* isomers and (ii) transition from Phase A to Phase B through an isotropic phase. Visible light (436 nm, 60 mW/cm^2) irradiation onto the sample that had reached PSS$_{365}$ induced the reverse *cis*-to-*trans* isomerization. As shown in Fig. 18.12a(iv), the XRD pattern assignable to Phase B disappeared, and that assignable to Phase A re-appeared in the small-angle region. In the POM observations, the birefringence disappeared immediately after visible light irradiation and then new birefringence appeared. This transient optically isotropic state is probably different from the optically isotropic ionic liquid state induced by heating, by considering the fact that Phase A was not observed during the cooling process from the ionic liquid state.

As the clearing temperature of $\mathbf{16}^{C16-}$·TBA$^+$ under UV irradiation (52 °C) was slightly lower than that without UV irradiation (55 °C), the photo-induced crystal–crystal phase transition was observed below 52 °C. Thus, the crystalline lamellar phase mainly comprising the bulky *cis* form (Phase B) was fairly stable due to the interactions between the long alkyl chains. In addition, the crystal–crystal phase transition can be repeated by switching between the two irradiation wavelengths (365/436 nm) (Fig. 18.12b), although the liquid–crystal phase transition was not observed at 52–55 °C by visible light (436 nm, 60 mW/cm^2) irradiation. Thus, ion pairs of azobenzene anions possess several advantages for photo-induced phase transitions: (i) the bulky countercations provide free volumes around the azobenzene moieties, (ii) isomerization of the azobenzene unit is not significantly restricted by the packing of alkyl chains on the countercations, because the ion pairs are weakly connected by noncovalent interactions, and (iii) the assembled structures and their properties can be controlled by the modifications of the countercations along with the anionic species [38].

18.2.4 Photo-Responsive Properties of Ion-Pairing Assemblies Based on Receptor–Anion Complexes

On the basis of the high binding constant (K_a) of **7a** for $\mathbf{16}^{C16-}$ (3,100,000 M^{-1}, CH_2Cl_2), which is comparable to that for $CH_3CO_2^-$ (930,000 M^{-1}) [39], the ion pair **7a**·$\mathbf{16}^{C16-}$-TBA$^+$ was formed from a 1:1 mixture of **7a** and $\mathbf{16}^{C16-}$-TBA$^+$ and was purified by recrystallization from diisopropylether [40]. In the ^{1}H NMR spectrum of **7a**, the NH and bridging CH signals were observed at 9.59 and 6.55 ppm, respectively, whereas, in **7a**·$\mathbf{16}^{C16-}$-TBA$^+$, the corresponding signals disappeared with the appearance of new signals at 14.19 and 8.05 ppm. The formation of the receptor–anion complex also caused a shift in the signals of the azobenzene carboxylate (e.g., H^a in Fig. 18.13a shifted from 8.08 to 8.26 ppm). After UV (365 nm) irradiation, new signals from the *cis* form appeared both in $\mathbf{16}^{C16-}$ and **7a**·$\mathbf{16}^{C16-}$-TBA$^+$ (e.g., 7.83 and 8.02 ppm for H^a in $\mathbf{16}^{C16-}$ and **7a**·$\mathbf{16}^{C16-}$-TBA$^+$, respectively). The difference in the ^{1}H NMR chemical shifts between $\mathbf{16}^{C16-}$ and **7a**·$\mathbf{16}^{C16-}$-TBA$^+$ for the *trans* forms, 0.18 ppm (H^a), was slightly smaller than that for the *cis* forms, 0.19 ppm (H^a), suggesting that photo-isomerization from the *trans* to *cis* forms slightly affects the anionic properties of the carboxylate and the interactions in the receptor–carboxylate complexes. In addition, the *trans*/*cis* ratios in the PSS_{365} were estimated as 0:100 for $\mathbf{16}^{C16-}$ and 30:70 for $\mathbf{16}^{C16-}$ in equilibrium with **7a**·$\mathbf{16}^{C16-}$. This difference can be

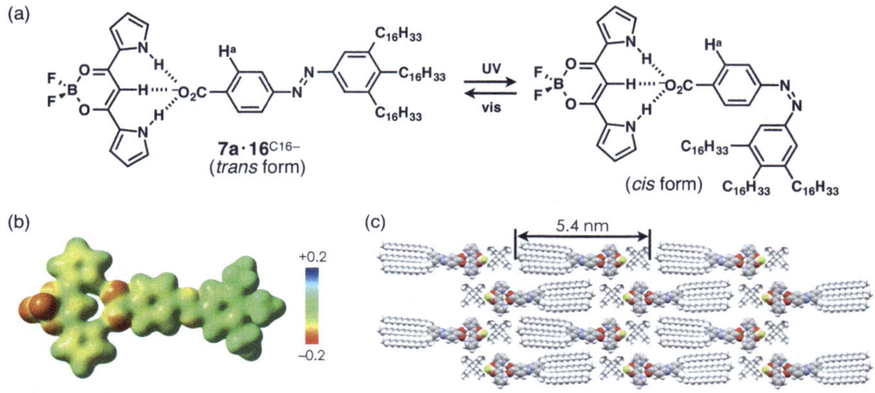

Fig. 18.13 **a** Photo-isomerization of receptor–anion complex **7a**·**16**$^{C16-}$, **b** electrostatic potential map ($\delta = 0.01$) of **7a**·**16**$^{C2-}$, and **c** proposed assembled structure of **7a**·**16**$^{C16-}$-TBA$^+$

explained by the emission of **7a** in the carboxylate complex at around 430–550 nm, which induces the isomerization from *cis* to *trans* forms.

Similar to free anion **16**$^{C2-}$, the *trans* form of **7a**·**16**$^{C2-}$, with ethyl groups replacing the hexadecyl groups for facile calculations, showed a more stable conformation than the *cis* form with relative energy of 12.4 kcal/mol. This value was only slightly smaller than the corresponding value for **16**$^{C2-}$ (14.8 kcal/mol). In addition, the N(–H)···O distances between the neighboring pyrrole NH and the carboxylate oxygen in the *cis* form (2.715 and 2.716 Å) were slightly shorter than that of the *trans* form (2.718 Å). The relative energy difference and binding modes of the carboxylate complexes can be explained by the electronic states of the *trans* and *cis* forms of **16**$^{C2-}$. Compared to the *trans* form of **16**$^{C2-}$, wherein the electronegative charge is partially delocalized on the coplanar azobenzene unit that in the *cis* form is delocalized in a single phenyl ring. Therefore, the hydrogen-bonding interaction between the neighboring pyrrole NH and the carboxylate oxygen in the *cis* form of **7a**·**16**$^{C2-}$ is slightly stronger than that in the *trans* form. This result agrees with the observed shifts in the ^{1}H NMR spectra, indicating that the anionic property of the complex was changed by photo-isomerization. Furthermore, electrostatic potential (ESP) diagrams of **7a**·**16**$^{C2-}$ showed the delocalization of the negative charge of **16**$^{C2-}$ in the π-electronic moiety of **7a** by complexation (Fig. 18.13b). The difference between the free anion and the receptor–anion complex affected the interaction between the charged species and the structures of dimension-controlled assemblies.

POM and XRD profiles of **7a**·**16**$^{C16-}$-TBA$^+$ revealed the formation of a crystalline lamellar structure. Similar to **16**$^{C16-}$-TBA$^+$, the repeating distance of 5.4 nm for **7a**·**16**$^{C16-}$-TBA$^+$ corresponded to the summed lengths of **7a**·**16**$^{C16-}$ and TBA$^+$ (Fig. 18.13c). However, in contrast to **16**$^{C16-}$-TBA$^+$, the photo-induced phase transition of **7a**·**16**$^{C16-}$-TBA$^+$ was not observed in POM and XRD using a sandwiched cell. The difference in the behaviors could be due to the strong packing around the receptor–carboxylate complex in **7a**·**16**$^{C16-}$-TBA$^+$ that inhibits the *trans*-to-*cis* isomerization of the azobenzene moiety and provides a photostable assembly even in the presence of photo-responsive azobenzene carboxylates. In other words, the modification of receptors and azobenzene, which affects the packing structures, could control the photo-responsive properties of the dimension-controlled assemblies [40].

In order to improve the photo-responsive behaviors, azobenzene bearing alkyl chains that can enhance the dynamic nature of the molecules [41] on both sides were designed [42]. Introduction of spacers between the azobenzene and carboxylate units can also interfere with the energy transfer to the receptor–anion complex parts.

Azobenzene **18**, bearing an alkanoic acid and an aliphatic chain, was obtained by the hydrolysis of the corresponding alkanoate ester **17** (Fig. 18.14a). Subsequent treatment of **18** with excess TBAOH in THF provided the alkanoate as a TBA ion pair (**18**$^-$-TBA$^+$) [25]. On the basis of the high K_a value of **18**$^-$ for **7a** (220,000 M^{-1}), ion pairs of the receptor–anion complexes and TBA$^+$ (**7a·18**$^-$-TBA$^+$ and **7c·18**$^-$-TBA$^+$) were prepared from the 1:1 mixture of the anion receptors **7a,c** and the alkanoate salt **18**$^-$-TBA$^+$ (Fig. 18.14b). The photo-isomerization properties of **18**$^-$-TBA$^+$ in solution were investigated by ^{1}H NMR spectra, in which the signals originating from *trans* form of **18**$^-$ at 7.86, 7.84, 6.98, and 6.97 ppm almost disappeared with an appearance of new signals derived from the *cis* form at 6.98, 6.87, 6.77, and 6.75 ppm upon undergoing UV irradiation. In addition, in contrast to **7a·16**$^{C16-}$-TBA$^+$, the changes in ^{1}H NMR spectra when **7a·18**$^-$-TBA$^+$ isomerized from the *trans*-to-*cis* forms were similar to those of **18**$^-$ without significant changes in the signals from the receptor unit. These results indicated that the alkyl spacer between the carboxylate and azobenzene units decreased the effect of electronic and steric changes in the azobenzene moiety of the carboxylate and complex. Similar to **16**$^{C16-}$-TBA$^+$, the *cis/trans* ratio at PSS_{365} for **7a·18**$^-$-TBA$^+$ (70/30) was smaller than that of **18**$^-$-TBA$^+$ (99/1). This result can also be explained by the effect of the emission of the receptor at 430–550 nm that induced *cis*-to-*trans* isomerization [39].

7c·18$^-$-TBA$^+$, with long alkoxy chains on the receptor, afforded a supramolecular gel in *n*-octane (10 mg/mL, 3.9 mM) below ca. 10 °C, and a broad UV/vis absorption

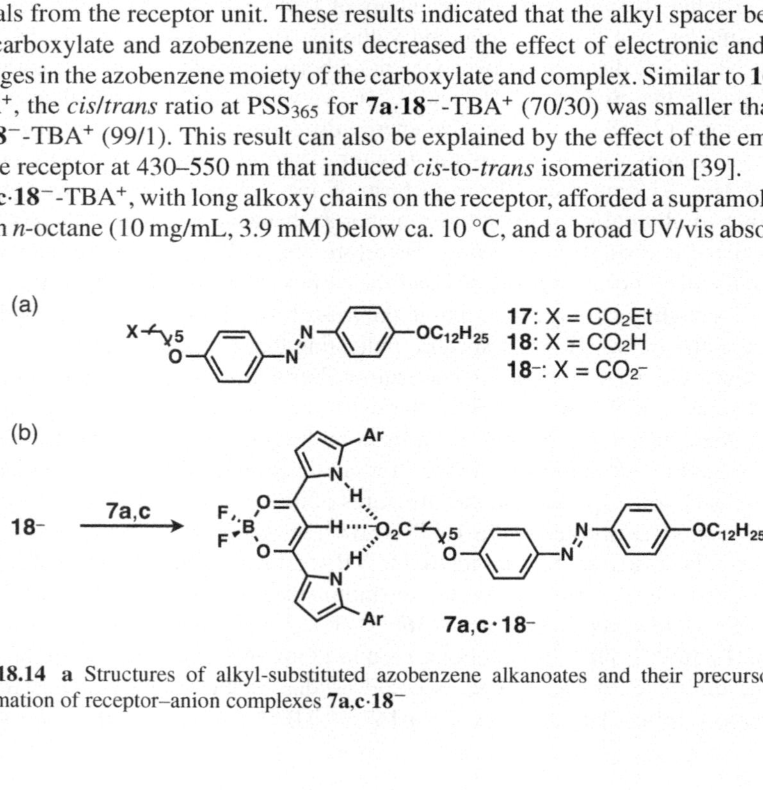

Fig. 18.14 **a** Structures of alkyl-substituted azobenzene alkanoates and their precursors and **b** formation of receptor–anion complexes **7a,c·18**$^-$

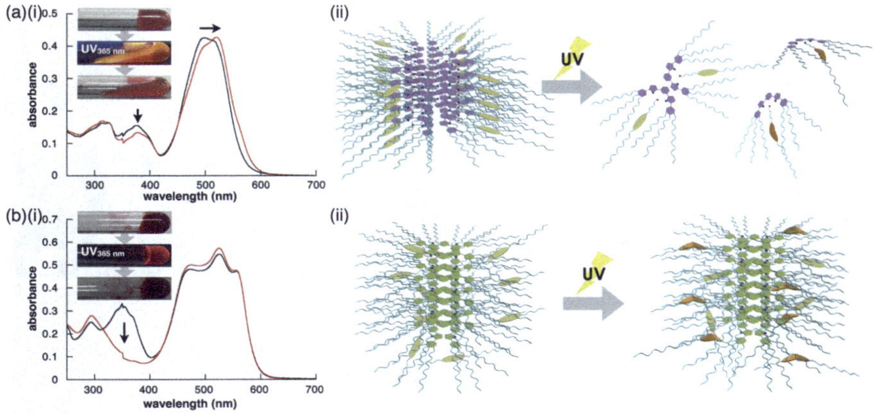

Fig. 18.15 Photo-responsive properties of **a** **7c**·**18**$^-$-TBA$^+$ gel and **b** gel of the 1:1 mixture of **7c** and **17** from *n*-octane (10 mg/mL): (i) UV/vis absorption spectral changes at r.t. (black and red lines indicate before and after UV irradiation (365 nm), respectively) and photographs at <10 °C before, under, and after UV irradiation (top, middle, and bottom, respectively) and (ii) schematic illustration of the assembled arrangements and photo-induced phase transitions (right). TBA$^+$ as a countercation in (**a**) (ii) was omitted for clarity. Adapted with permission from [42]. Copyright 2019 RSC

band at 479 nm derived from the aggregation of the π-electronic system was observed at r.t. (Figure 18.15a(i)). Under photoirradiation (365 nm), the gel transformed into a solution, beginning from the part which was irradiated by the light. In addition, after the reverse isomerization by heating, the supramolecular gel was formed upon cooling. The spectral changes due to the photoirradiation, in which the absorption band at 376 nm decreased and that at 479 nm was red-shifted to 520 nm, indicated that the disaggregation of the π-electronic system occurred upon the conformation change of azobenzene from *trans* to *cis* forms, because the absorption spectrum in the long-wavelength region after photoirradiation was similar to that in diluted conditions without aggregation. Although the exact structure of the assembly in the gel was unclear, the conformation change in the azobenzene moiety bound to the π-electronic system via the carboxylate with an alkyl linker efficiently induced the gel collapse (Fig. 18.15a(ii)). It should be noted that this photo-responsive property was not observed from the 1:1 mixture of **7c** and **17** that bore an ester unit instead of the carboxylate, although the photo-isomerization of **17** was observed (Fig. 18.15b). Thus, the introduction of an azobenzene unit as a part of the anion is an effective method for the preparation of photo-responsive gels based on receptor–anion complexes.

From the XRD profile of **18**$^-$-TBA$^+$ at 50 °C upon heating, a lamellar structure with a repeating distance of 4.18 nm, corresponding to the sum of lengths of **18**$^-$ and TBA$^+$, was observed. Similarly, as observed in the XRD pattern, **7a**·**18**$^-$-TBA$^+$ provided an ordered assembly, derived from the lamellar structure with a repeating distance of 3.63 nm. Despite the formation of a receptor–anion complex, the repeating

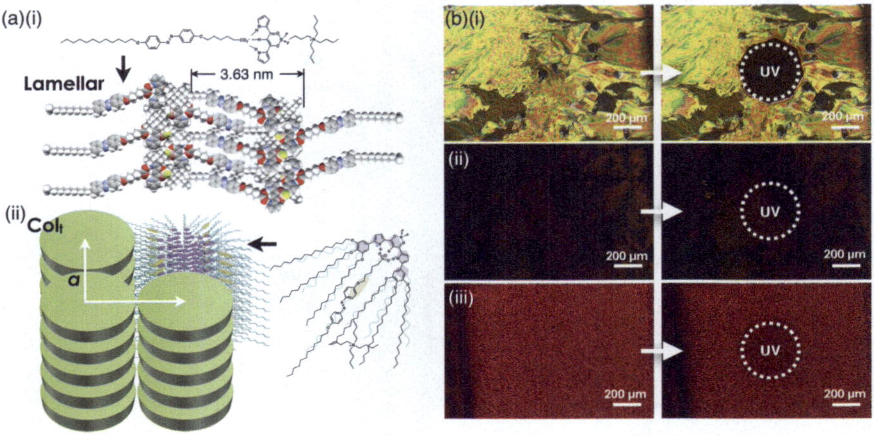

Fig. 18.16 **a** Chemical structures and representation of assemblies of (i) **7a·18**$^-$-TBA$^+$ at 80 °C and (ii) **7c·18**$^-$-TBA$^+$ at 50 °C and **b** photo-responsive properties of (i) **18**$^-$-TBA$^+$, (ii) **7a·18**$^-$-TBA$^+$, and (iii) **7c·18**$^-$-TBA$^+$ investigated by POM at 50, 90, and 70 °C, respectively, in a sandwiched cell under UV irradiation (central spot). Adapted with permission from [42]. Copyright 2019 RSC

distance was smaller than that of the corresponding receptor-free assembly of **18**$^-$-TBA$^+$ probably due to the interdigitation of the alkyl chains (Fig. 18.16a(i)). On the other hand, the XRD profile of **7c·18**$^-$-TBA$^+$ at 50 °C upon heating, suggested the formation of a tetragonal columnar (Col$_t$) structure with $a = 6.6$ nm (Fig. 18.16a(ii)). The fan-like geometry of the pyrrole-inverted receptor–anion complex bearing long alkyl chains in **7c·18**$^-$ provided a discotic columnar structure, whose circular unit contained two to four anion complexes. Thus, assembled structures were controlled by the modifications of the receptors.

In the UV/vis absorption spectra of **18**$^-$-TBA$^+$ upon UV irradiation, the intensity of the absorption band at 343 nm decreased while that at 453 nm increased concurrently. These changes are similar to those seen for photo-isomerization in the solution state. In addition, **7a·18**$^-$-TBA$^+$ and **7c·18**$^-$-TBA$^+$ exhibited only slight photo-isomerization behaviors in contrast to **7a·16**$^{C16-}$-TBA$^+$ [25]. Therefore, the efficiency of energy transfer, which inhibits the photo-isomerization, decreased upon the introductions of an alkyl spacer between the azobenzene unit and carboxylate (receptor–anion complex).

The photo-induced phase transitions of **18**$^-$-TBA$^+$ observed in POM using a sandwiched cell with 8 μm spacers exhibited the disappearance of birefringence under UV irradiation (365 nm) at 25 °C (Fig. 18.16b(i)). On the other hand, the photo-induced phase transition was not observed in **7a·18**$^-$-TBA$^+$, nor was it observed in **7a·16**$^{C16-}$-TBA$^+$ (Fig. 18.16b(ii)), although photo-isomerization was observed in the UV/vis absorption spectra. In addition, unlike the case of the supramolecular gel, birefringence of **7c·18**$^-$-TBA$^+$ showed only a small change under UV irradiation at 70 °C (Fig. 18.16b(iii)). These results indicated that photo-isomerization may occur only on the surface [37], where free volume is available, and may be suppressed by packing of the π-electronic systems and alkyl chains in the solid-state [42].

18.3 Excited-State Dynamics of π-System–Pt^{II} Complexes

18.3.1 Excited-State Dynamics

In general, the structures and reactivity of triplet states are different from those of the corresponding ground states. Thus, it is crucial to investigate the excited-state dynamics, which include radiative and non-radiative processes, as well as the reactivity of excited states because highly efficient photofunctional materials can be fabricated. Applications that depend on the strong absorption of visible light often require transition-metal complexes of π-electronic molecules and their assemblies, which exhibit diverse properties and functions based on their coordination bonds. Among metal complexes, cyclometalated Pt^{II} complexes have gained much attention due to their photophysical properties and show high triplet quantum yields, long-lived triplet excitons, and photocatalytic activities [43]. The photo-induced excitation of the ground (S_0) state to the S_1 state is the trigger for the excited-state dynamics of Pt^{II} complexes (Fig. 18.17). The ground (S_0) state was recovered via the triplet (T_1) state, which was generated by the intersystem crossing from the S_1 state. The efficient photo-absorption in the excitation from ground states and the long lifetime of triplet states are key factors for the development of photo-induced catalysts and building units for use in photo-responsive materials.

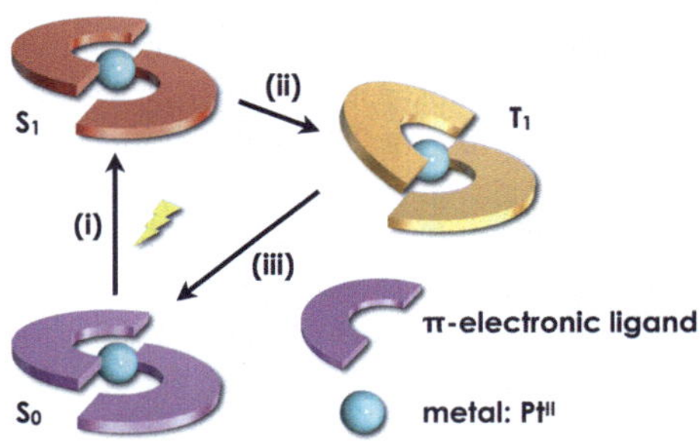

Fig. 18.17 Diagram of the excited-state dynamics of cyclometalated Pt^{II} complexes: (i) excitation from S_0 state to S_1 state by the photoirradiation, (ii) intersystem crossing from S_1 state to T_1 state, and (iii) ground-state recovery

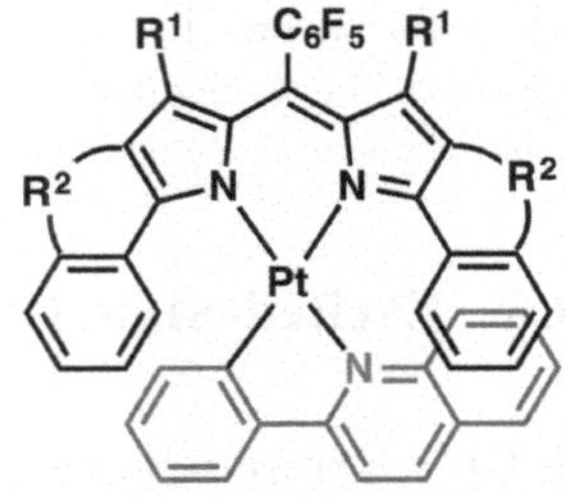

20a: R¹ = H, R² = H, H
20b: R¹ = CH₃, R² = CH₂CH₂

Fig. 18.18 Heteroleptic dipyrrin–Pt^{II} complexes **20a,b**

18.3.2 Pyrrole-Based π-System–Pt^{II} Complexes Exhibiting Excited-State Dynamics with Microsecond Triplet Lifetimes

Heteroleptic Pt^{II} complexes **20a,b**, comprising π-extended dipyrrins and 2-phenylquinoline (pqu), were synthesized (Fig. 18.18) [44]. π-Extended dipyrrins are focused on because they are ligands that efficiently harvest visible light, whereas a C_6F_5 moiety was introduced at the *meso* position as an electron-withdrawing group to enhance the stability of the metal complexes. Furthermore, pqu as a π-extended ligand can induce chirality in Pt^{II} complexes upon cyclometalation.

The sub-picosecond to nanosecond timescale transient absorption measurements with excitation with a femtosecond laser pulse (instrumental response function: ~ 30 fs) revealed the photophysical properties of both the singlet and triplet excited states of **20a,b** (Fig. 18.19a). The ground-state bleach signal at 610 nm and positive absorption signals at 450 and 530 nm for **20a** were observed in the transient absorption at 0.2 ps after the excitation, while for **20b** the bleach signal at 650 nm and a positive absorption signal at 550 nm was observed. The positive absorption bands at 530 nm for **20a** and 550 nm for **20b** decay on a timescale of hundreds of femtoseconds and other positive absorption bands appear at 650 and 870 nm for **20a** and at 700 and 850 nm for **20b**. The evolution-associated spectra obtained through global analyses with singular value decomposition (SVD) of transient absorption spectra showed the excited-state relaxation pathways (Fig. 18.19b). The resulting time constants τ of **20a** were 2 fs, 260 fs, 85 ps, and >10ns, while τ of **20b** were 6 fs, 820 fs, 80 ps, and >10ns. The fastest decay-time components (2 and 6 fs) were excluded from evolution-associated spectra owing to the component derived from coherent artifacts. The τ of 260 and 820 fs were assigned to S_1 state lifetimes because the evolution-associated spectra contain stimulated emission signals most likely from the S_1 states at approximately 700 and 730 nm for **20a,b**, respectively.

In general, third-row transition-metal complexes exhibit very fast intersystem crossing through the strong spin-orbit coupling [45]. Consequently, the S_1 states have very short lifetimes from several tens of femtoseconds to several picoseconds. Therefore, these characteristics suggested that the intersystem crossing from the S_1 to

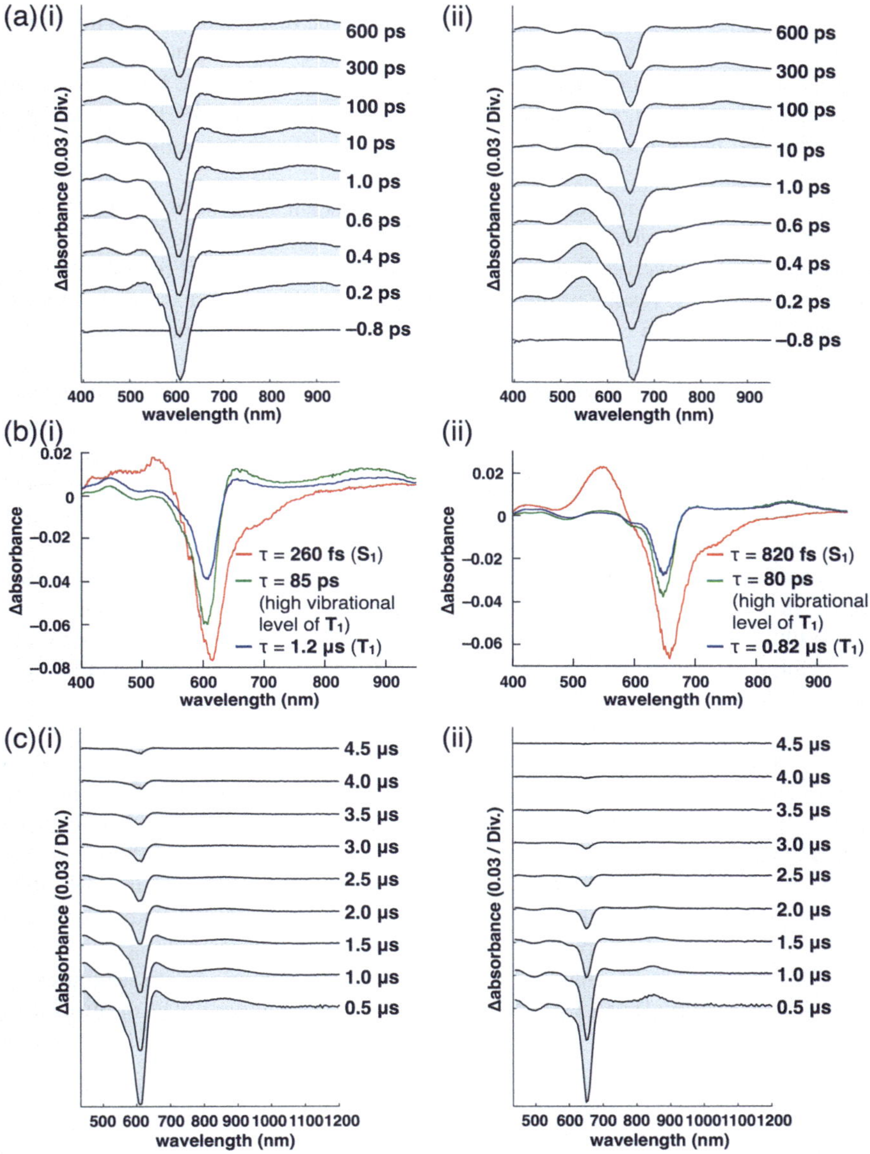

Fig. 18.19 **a** Sub-picosecond to nanosecond timescale transient absorption spectra, **b** evolution-associated spectra obtained through SVD global analysis of (**a**) (T_1 lifetimes were derived from the single exponential decay function of (**c**)), and **c** nanosecond to microsecond timescale transient absorption spectra: (i) **20a** (0.15 mM excited at 600 nm for (**a, b**) and 0.14 mM excited at 610 nm for (**c**) in degassed CH_2Cl_2) and (ii) **20b** (0.13 mM excited at 650 nm for (**a, b**) and 0.12 mM excited at 650 nm for (**c**) in degassed CH_2Cl_2). Adapted with permission from [44]. Copyright 2019 Wiley

T_1 states occurs with time constants of 260 and 820 fs, respectively, and vibrational cooling from the high vibrational level of the T_1 states with lifetimes of 85 and 80 ps, respectively, was observed. With >10 ns time constants, the longest lifetime components can be assigned to the T_1 states, although the exact lifetimes could not be determined using transient absorption measurements on the sub-picosecond to nanosecond timescale.

The nanosecond to microsecond timescale transient absorption measurements with the randomly interleaved pulse train (RIPT) method elucidated the long-lived components, which were 1.2 and 0.82 ms for **20a,b**, respectively (Fig. 18.19c) [46]. On the basis of these results of the transient absorption experiments, the photophysical processes of **20a,b** can be explained by the excited-state diagrams (Fig. 18.20), which consist of three steps in the excited-state intersystem crossing and relaxation processes: (i) fast intersystem crossing from the S_1 states to the high vibrational levels of the T_1 states within 260 and 820 fs for **20a,b**, respectively, (ii) vibrational cooling in the T_1 states (85 and 80 ps, respectively), and (iii) ground-state recoveries (1.2 and 0.82 ms, respectively). The microsecond-order time constants can provide enough time for the diffusion and collision of Pt^{II} complexes in the solution state, which can be used for future applications. Importantly, the original chirality of the ground states of **20a,b** was retained in the excited states. As the properties of excited states, such as relaxation processes, charge transfer, and emission characteristics, can be tuned by appropriate metal centers and ligand molecules, the development of π-extended system–metal complexes and charged metal complexes for functional electronic materials will be investigated.

The excited states can be considered as photo-isomerized structures with geometries and properties that are different from those of the ground states. Thus, the details of the excited-state structures were essential for the discussions on the photo-induced controls of π-electronic molecules. The time-dependent (TD)-DFT calculations at the CAM-B3LYP level of theory with the 6-31 + G(d,p) (LanL2DZ for Pt) of ground- and excited-state structures of **20a,b** revealed the optimized structures and stabilities (Fig. 18.21). The S_1 states of **20a,b** were less stable by 2.43 eV (55.8 kcal/mol) and 2.23 eV (51.3 kcal/mol), respectively, than the corresponding ground (S_0) states. The S_0 states were more stable than the corresponding most stable T_1 state structures by 1.29 eV (29.9 kcal/mol) and 1.14 eV (26.4 kcal/mol), respectively. The optimized S_0

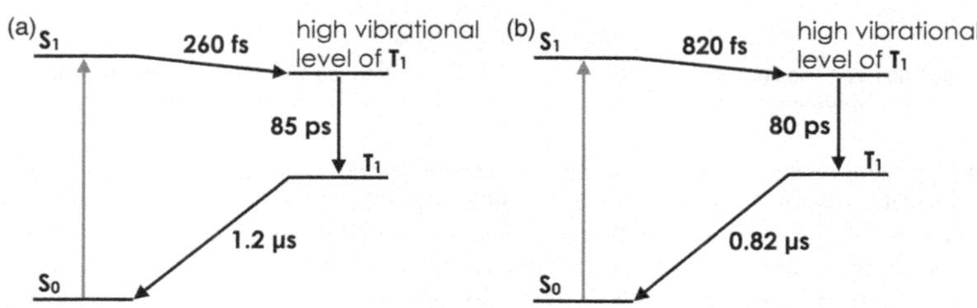

Fig. 18.20 Excited-state diagrams describing the photophysical processes of **a 20a** and **b 20b**. Adapted with permission from [44]. Copyright 2019 Wiley

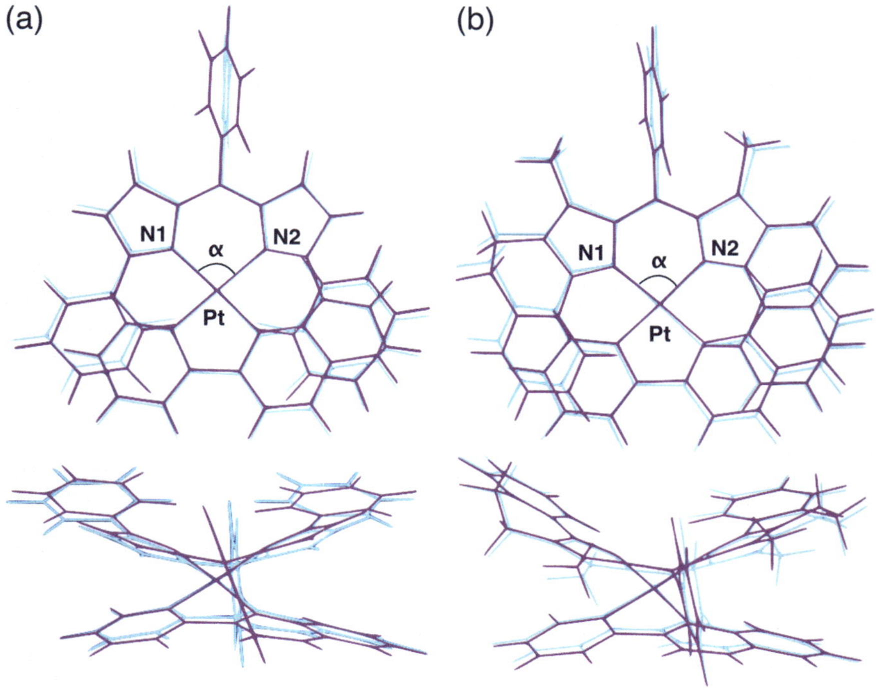

Fig. 18.21 Superposed optimized structures (top and side views) of the S_0 (cyan) and T_1 (magenta) states of **a 20a** and **b 20b**. Adapted with permission from [44]. Copyright 2019 Wiley

and T_1 structures clearly exhibited photo-induced conformation changes. The bite angles with the Pt^{II} center of the dipyrrin ligand (corresponding to α in Fig. 18.21) of the optimized S_0 and T_1 structures in **20a** are 82.9° and 81.5°, whereas the dihedral angles for the dipyrrin moiety (angles between the two pyrrole rings) are almost the same (141.3°). On the other hand, in **20b**, the corresponding bite angles are 81.0° and 79.8° and the dihedral angles are 140.4° and 139.0° for the S_0 and T_1 structures, respectively. The dipyrrin moieties were more distorted in the T_1 states, which retained their chirality [44].

18.4 Conclusion

The key points in the ion-pairing chemistry are the design and synthesis of a variety of ion pairs by the combination of newly synthesized π-electronic ions. The introduction of a photo-responsive unit to π-electronic ions provided ion-pairing assemblies that show a modulable arrangement of ion pairs. The azobenzene carboxylate ion pairs possessing aliphatic chains gave rise to crystalline assemblies that showed a photo-induced crystal–crystal phase transition in the bulk state. Furthermore, negatively charged π-electronic systems prepared by the complexation of anion-responsive molecules with an azobenzene bearing an alkanoate and an aliphatic chain provided unique dimension-controlled assemblies, such as a photo-responsive supramolecular gel. On the other hand, the pyrrole-based π-system–Pt^{II} complexes, which showed excited-state dynamics characterized by fast intersystem crossing and microsecond-order triplet-state lifetimes, are applicable as new chiral photocatalysts through the active use of photo-induced reactive transient states. Appropriately designed π-electronic species, some of which can be charged π-systems, can facilitate the control of the structures, assembling modes, and electronic states by photoirradiation. π-Electronic systems that are further designed and modified will induce tunable functionalities and form fascinating materials possessing electronic and optical functions [47, 48].

Acknowledgements We deeply appreciate the co-authors of the reports related with the contents in this chapter, including Dr. Mitsuo Hara (Nagoya University), Prof. Shusaku Nagano (Rikkyo University), Prof. Takahiro Seki (Nagoya University), Dr. Ryoma Sato (RIKEN), Prof. Yasuteru Shigeta (the University of Tsukuba), Prof. Yoichi Kobayashi (Ritsumeikan University), and Dr. Nobuhiro Yasuda (JASRI), for their collaboration. The present work was supported by JSPS KAKENHI Grant Number 26107007, Grant-in-Aid for Scientific Research on Innovative Areas "Photosynergetics".

References

1. Faul CFJ, Antonietti M (2003) Ionic self-assembly: facile synthesis of supramolecular materials. Adv Mater 15:673–683
2. Faul CFJ (2014) Ionic self-assembly for functional hierachical nanostructured materials. Acc Chem Res 47:3428–3438
3. Binnemans K (2005) Ionic liquid crystals. Chem Rev 105:4148–4204
4. Axenov KV, Laschat S (2011) Thermotropic ionic liquid crystals. Materials 4:206–259
5. Goossens K, Lava K, Bielawski CW, Binnemans K (2016) Ionic liquid crystals: versatile materials. Chem Rev 116:4643–4807
6. Haketa Y, Maeda H (2017) Dimension-controlled ion-pairing assemblies based on π-electronic charged species. Chem Commun 53:2894–2909
7. Haketa Y, Maeda H (2018) Dimension-controlled π-electronic ion-pairing assemblies. Bull Chem Soc Jpn 91:420–436
8. Webster OW (1965) Diazotetracyanocyclopentadiene and its conversion to tetracyanocyclopentadienide and pentacyanocyclopentadienide. J Am Chem Soc 87:1820–1821
9. Less RJ, McPartlin M, Rawson JM, Wood PT, Wright DS (2010) A simple approach to coordination compounds of the pentacyanocyclopentadienide anion. Chem Eur J 16:13723–13728
10. Sakai T, Seo S, Matsuoka J, Mori Y (2013) Synthesis of functionalized tetracyanocyclopentadienides from tetracyanothiophene and sulfones. J Org Chem 78:10978–10985
11. Bando Y, Haketa Y, Sakurai T, Matsuda W, Seki S, Takaya H, Maeda H (2016) Ion-pairing assemblies based on pentacyano-substituted cyclopentadienide as a π-electronic anion. Chem Eur J 22:7843–7850
12. Maeda H, Fukui A, Yamakado R, Yasuda N (2015) Dipyrrolyphenol as a precursor of π-electronic anion that forms ion pairs with cations. Chem Commun 51:17572–17575
13. Maeda H, Takeda Y, Haketa Y, Morimoto Y, Yasuda N (2018) Ion-pairing assemblies of π-electronic anions formed by intramolecular hydrogen bonding. Chem Eur J 24:8910–8916

14. Yamakado R, Ishibashi H, Motoyoshi Y, Yasuda N, Maeda H (2019) Ion-pairing assemblies based on π-extended dipyrrolylquinoxalines. Chem Commun 55:326–329
15. Sasano Y, Yasuda N, Maeda H (2017) Deprotonated meso-hydroxyporphyrin as a stable π-electronic anion: the building unit of an ion-pairing assembly. Dalton Trans 46:8924–8928
16. Sasano Y, Haketa Y, Tanaka H, Yasuda N, Hisaki I, Maeda H (2019) Peripheral modifications of meso-hydroxyporphyrins: formation of π-electronic anions and ion-pairing assemblies. Chem Eur J 25:6712–6717
17. Haketa Y, Bando Y, Sasano Y, Tanaka H, Yasuda N, Hisaki I, Maeda H (2019) Liquid crystals comprising π-electronic ions from porphyrin–Au^{III} complexes. iScience 14:241–256
18. Tanaka H, Haketa Y, Yasuda N, Maeda H (2019) Substitution-pattern- and counteranion-depending ion-pairing assemblies based on electron-deficient porphyrin–Au^{III} complexes. Chem Asian J 14:2129–2137
19. Maeda H, Bando Y (2013) Recent progress in research on anion-responsive pyrrole-based π-conjugated acyclic molecules. Chem Commun 49:4100–4113
20. Haketa Y, Maeda H (2011) From helix to macrocycle: anion-driven conformation control of π-conjugated acyclic oligopyrroles. Chem Eur J 17:1485–1492
21. Haketa Y, Bando Y, Takaishi K, Uchiyama M, Muranaka A, Naito M, Shibaguchi H, Kawai T, Maeda H (2012) Asymmetric induction in the preparation of helical receptor-anion complexes: ion-pair formation with chiral cations. Angew Chem Int Ed 51:7967–7971
22. Sasano Y, Yasuda N, Maeda H (2016) Negatively charged π-electronic systems by deprotonation of hydroxy-substituted dipyrrolyldiketone boron complexes. Chem Asian J 11:3423–3429
23. Maeda H, Ito Y (2006) BF_2 complex of fluorinated dipyrrolyldiketone: a new class of efficient receptor for acetate anions. Inorg Chem 45:8205–8210
24. Maeda H, Kinoshita K, Naritani K, Bando Y (2011) Self-sorting self-complementary assemblies of π-conjugated acyclic anion receptors. Chem Commun 47:8241–8243
25. Maeda H, Naritani K, Honsho Y, Seki S (2011) Anion modules: building blocks of supramolecular assemblies by combination with π-conjugated anion receptors. J Am Chem Soc 133:8896–8899
26. Maeda H, Okubo T, Haketa Y, Yasuda N (2018) Pyrrole-based zwitterionic π-electronic systems that form self-assembled dimers. Chem Eur J 24:16176–16182
27. Sekiya R, Tsutsui Y, Choi W, Sakurai T, Seki S, Bando Y, Maeda H (2014) Ion-based assemblies of planar anion complexes and cationic Pt^{II} complexes. Chem Commun 50:10615–10618
28. Dong B, Sakurai T, Bando Y, Seki S, Takaishi K, Uchiyama M, Muranaka A, Maeda H (2013) Ion-based materials derived from positively and negatively charged chloride complexes of π-conjugated molecules. J Am Chem Soc 135:14797–14805
29. Dong B, Sakurai T, Honsho Y, Seki S, Maeda H (2013) Cation modules as building blocks forming supramolecular assemblies with planar receptor-anion complexes. J Am Chem Soc 135:1284–1287
30. Xu JF, Chen YZ, Wu D, Wu LZ, Tung CH, Yang QZ (2013) Photoresponsive hydrogen-bonded supramolecular polymers based on a stiff stilbene unit. Angew Chem Int Ed 52:9738–9742
31. Xia D, Yu G, Li J, Huang F (2014) Photo-responsive self-assembly based on a water-soluble pillar[6]arene and an azobenzene-containing amphiphile in water. Chem Commun 50:3606–3608
32. Yokoyama S, Hirose T, Matsuda K (2014) Phototriggered formation and disappearance of surface-confined self-assembly composed of photochromic 2-thienyl-type diarylethene: a cooperative model at the liquid/solid interface. Chem Commun 50:5964–5966
33. Bandara HMD, Burdette SC (2012) Photoisomerization in different classes of azobenzene. Chem Soc Rev 41:1809–1825
34. Ichimura K (2000) Photoalignment of liquid-crystal systems. Chem Rev 100:1847–1873
35. Yu H, Ikeda T (2011) Photocontrollable liquid-crystalline actuators. Adv Mater 23:2149–2180
36. Bisoyi HK, Li Q (2016) Light-driven liquid crystalline materials: from photo-induced phase transitions and property modulations to applications. Chem Rev 116:15089–15166
37. Ichimura K (2009) Reversible photoisomerisability and particle size changes of mill-dispersed azobenzene crystals in water. Chem Commun 1496–1498

38. Yamakado R, Hara M, Nagano S, Seki T, Maeda H (2017) Photo-responsive soft ionic crystals: ion-pairing assemblies of azobenzene carboxylates. Chem Eur J 23:9244–9248
39. Maeda H, Terasaki M, Haketa Y, Mihashi Y, Kusunose Y (2008) BF_2 complexes of α-alkyl-substituted dipyrrolyldiketones as acyclic anion receptors. Org Biomol Chem 6:433–436
40. Yamakado R, Hara M, Nagano S, Seki T, Maeda H (2018) Complexation of anion-responsive π-electronic system with alkyl-substituted azobenzene carboxylate providing ion-pairing assemblies. Chem Lett 47:404–407
41. Hoshino M, Uchida E, Norikane Y, Azumi R, Nozawa S, Tomita A, Sato T, Adachi S, Koshihara S (2014) Crystal melting by light: X-ray crystal structure analysis of an azo crystal showing photoinduced crystal-melt transition. J Am Chem Soc 136:9158–9164
42. Yamakado R, Haketa Y, Hara M, Nagano S, Seki T, Maeda H (2019) Photo-responsive dimension-controlled ion-pairing assemblies based on anion complexes of π-electronic systems. Chem Commun 55:10269–10272
43. Williams JAG, Develay S, Rochester DL, Murphy L (2008) Optimising the luminescence of platinum(II) complexes and their application in organic light emitting devices (OLEDs). Coord Chem Rev 252:2596–2611
44. Hirata G, Kobayashi Y, Sato R, Shigeta Y, Yasuda N, Maeda H (2019) Pyrrole-based π-system–Pt^{II} complexes: chiroptical properties and excited-state dynamics with microsecond triplet lifetimes. Chem Eur J 25:8797–8804
45. Penfold TJ, Gindensperger E, Daniel C, Marian CM (2018) Spin-vibronic mechanism for intersystem crossing. Chem Rev 118:6975–7025
46. Nakagawa T, Okamoto K, Hanada H, Katoh R (2016) Probing with randomly interleaved pulse train bridges the gap between ultrafast pump-probe and nanosecond flash photolysis. Opt Lett 41:1498–1501
47. Kita H, Yamakado R, Fukuuchi R, Konishi T, Kamada K, Haketa Y, Maeda H (2020) Switching of two-photon optical properties by anion binding of pyrrole-based boron diketonates through conformation change. Chem Eur J 26:3404–3410
48. Tanaka H, Haketa Y, Bando Y, Yamakado R, Yasuda N, Maeda H (2020) Ion-pairing assemblies of porphyrin–Au^{III} complexes in combination with π-electronic receptor–anion complexes. Chem Asian J 15:494–498

Chapter 19
Photoinduced Morphological Transformation and Photodriven Movement of Objects Using Self-assembly of Amphiphilic Diarylethene in Water

Kenji Higashiguchi and Kenji Matsuda

Abstract Self-assembly of amphiphilic photochromic diarylethene having tri(ethylene glycol) monomethyl ether chains was examined in water from the viewpoint of photoinduced morphological transformation and photodriven movement of objects. Self-assembled supramolecular architectures of the amphiphilic diarylethenes undergo photoinduced macroscopic morphological transformation upon alternate irradiation with UV and visible light. The photoreversible morphological change can be rationalized as a photoinduced phase transition between the high- and low-temperature phases of the lower critical solution temperature (LCST) transition. By using a depletion force in a methylcellulose aqueous solution, an amphiphilic diarylethene hierarchically assembled into bundled fibers, which showed shrinking of more than 100 μm under visible light irradiation. Linearly polarized light induced anisotropic growth of the assembled architecture and the diffusive motion of added polystyrene beads was suppressed in the perpendicular direction to the polarized light. The movement of many objects tracing the movement of a UV-irradiation spot was achieved with the assistance of the photogenerated supramolecular architecture.

Keywords Diarylethene · Supramolecular architecture · Tri(ethylene glycol) monomethyl ether · LCST · Depletion force · Brownian motion

19.1 Introduction

Amphiphilic compounds form self-assembled architectures in water through intermolecular hydrophobic interactions. When the amphiphilic molecules are capable of hydrogen bonding and π–π stacking, the self-assembled structures are affected accordingly. The nanostructures of the supramolecular architecture have been studied

K. Higashiguchi · K. Matsuda (✉)
Department of Synthetic Chemistry and Biological Chemistry, Graduate School of Engineering, Kyoto University, Katsura, Nishikyo-ku, Kyoto 615-8510, Japan
e-mail: kmatsuda@sbchem.kyoto-u.ac.jp

H. Miyasaka et al. (eds.), *Photosynergetic Responses in Molecules and Molecular Aggregates*, https://doi.org/10.1007/978-981-15-5451-3_19

in the case of axially symmetric amphiphiles consisting of an ionic head and alkyl tails [1], and more complex amphiphiles [2]. It is interesting to note that the energy difference between the different kinds of molecular packing is small, so that the packing of the supramolecular architecture in water may change with various external stimuli.

Photoreaction is a well-known technique for changing the packing and morphology of supramolecular architectures in water. Azobenzene, a representative photochromic compound, undergoes reversible photoisomerization between the rod-shaped *trans*-isomer and the bent *cis*-isomer. Such a large change in molecular geometry may result in sizable transformations of the microstructure [3], but the microstructure is often irreversibly destructed [4]. Even if the molecular packing changes reversibly before and after isomerization, the fragmented microstructures do not return to the original position due to diffusion. However, some examples of reversible morphological changes have been reported. Hamada and co-workers showed that a sheet-shaped supramolecular architecture composed of an azobenzene derivative and 1,2-dioleoyl-*sn*-glycero-3-phosphocholine (DOPC) exhibited photoreversible morphological change between sheet and capsule shapes [5]. Kageyama et al. reported that a plate-like assembly composed of another azobenzene derivative and oleic acid show self-oscillations upon irradiation with continuous blue light [6].

Diarylethene, a typical photochromic compound, is also used to form photoresponsive supramolecular architectures [7]. It undergoes reversible photoisomerization between a colorless open-ring isomer and a colored closed-ring isomer. The characteristic photochromic features, e.g., color, quantum yield, and fatigue resistance, can be controlled within a wide range by means of molecular design [8]. We recently studied amphiphilic diarylethenes showing photo- and thermo-responsive morphological changes. Oligo(ethylene glycol) monomethyl ether chain (hereinafter, referred to as "oligo(ethylene glycol) ether"), used here as a hydrophilic part of the amphiphilic diarylethenes, is known to exhibit lower critical solution temperature (LCST) behavior, by which miscibility switches depending on temperature. Below the LCST, the oligo(ethylene glycol) ether is in a hydrated phase by means of hydrogen bonding with the surrounding water molecules and forms an extended chain, and above this temperature, hydrogen bonding ceases to exist because the contribution from the entropy becomes more significant than from enthalpy, estimated from the Gibbs energy of hydrogen bonding. The oligo(ethylene glycol) ether becomes randomly coiled and the chains are tangled with each other. Since miscibility changes drastically near the LCST, this characteristic is often used to bring about reversible morphological changes in the micrometer-sized supramolecular architectures [9].

19.2 Photoinduced LCST Transition and Morphological Changes in Supramolecular Architectures

We recently studied the photoinduced morphological changes in supramolecular architecture and the movement of objects assisted by this architecture. The used amphiphilic diarylethenes had an alkyl chain and two tri(ethylene glycol) ether chains at each end of the diarylethene. As the open- and closed-ring isomers have different LCSTs, LCST behavior was induced by photoirradiation accompanied by the change in the ratio of both isomers [10–15].

The molecule is designed to modify both the photoreactivity and the morphology of the supramolecular architectures. These modifications require undergoing a photoreaction for the cyclization and cycloreversion process under an optical microscope, which means that the chosen diarylethene should have a high conversion ratio to the closed-ring isomer upon irradiation with 365 nm-UV light, generated from the light source of an epi-illumination system. Therefore, 1,2-bis(2-methyl-5-phenyl-3-thienyl)hexafluorocyclopentene [16] was selected because it has an absorption edge near 365 nm as well as a large quantum yield (57%) in a cyclization reaction. In order to give it amphiphilic properties, two tri(ethylene glycol) ether chains were introduced to one of the phenyl groups and an alkyl chain was added to the other as hydrophilic and hydrophobic moieties, respectively. Ether (**1**), ester (**2**), and amide (**3**) are used as linkers between the alkyl chain and the phenyl group (Scheme 19.1). We investigated the photochemical properties and morphological changes for the respective supramolecular architectures.

The synthesis was carried out by the C-H arylation at 5-position of the thiophene rings of the diarylethene, following the procedure shown in Scheme 19.2 [17]. The asymmetric amphiphilic diarylethenes were synthesized using a one pot method with a non-substituted diarylethene, namely 1,2-bis(2-methyl-3-thienyl)hexafluorocyclopentene [18] and an iodobenzene derivative acting as the hydrophobic and hydrophilic moieties, respectively. The synthetic yields were 16,

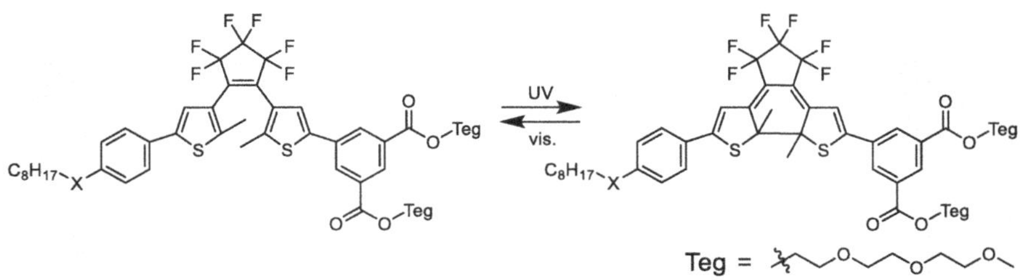

open-ring isomer (colorless)	closed-ring isomer (blue)
1a: X = -CH_2O- (bicontinuous coacervate)	**1b** (long fiber)
2a: X = -OCO- (bicontinuous coacervate)	**2b** (short fiber)
3a: X = -NHCO- (bicontinuous coacervate)	**3b** (bilayer)

Scheme 19.1 Photoisomerization of the amphiphilic diarylethenes

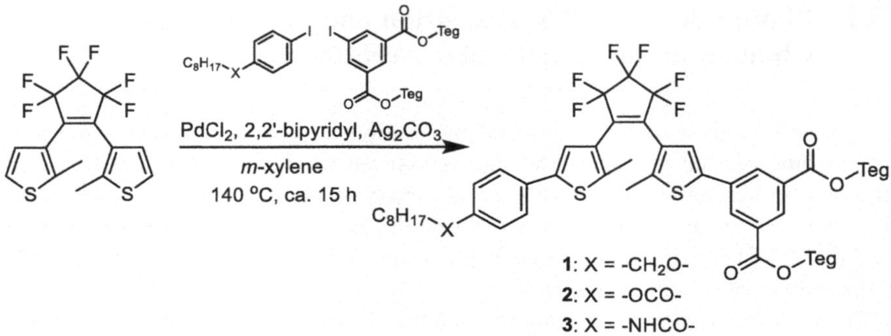

Scheme 19.2 Syntheses of the asymmetric amphiphilic diarylethenes by C–H arylation

25, and 27% for compounds **1**, **2**, and **3**, respectively. Symmetric compounds were also produced as by-products, but the target compounds were easily purified due to significant difference in polarity.

19.2.1 Photoinduced LCST Transition

The change in the absorption spectrum with photoisomerization indicated different behavior for a solution and a suspension. Compound **1** showed ordinary photochromism in an acetonitrile solution, i.e., a colorless solution of **1a** turned blue upon irradiation with UV light (**1b**) and turned back to being colorless upon subsequent irradiation with visible light (Fig. 19.1a). On the other hand, the aqueous suspension displayed a two-step change in the absorption spectrum. Initial white-turbid suspension turned to a blue suspension upon short irradiation with UV light. The absorption band shifted to shorter wavelengths and the color changed to reddish

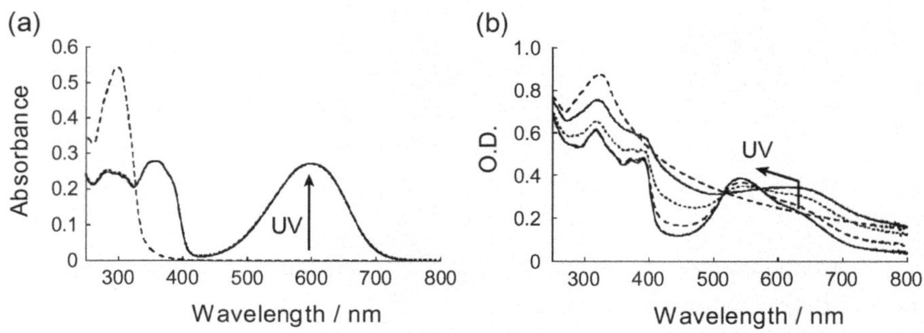

Fig. 19.1 Spectral changes along with photoisomerization from **1a** to **1b** upon irradiation with UV light in **a** acetonitrile solution and **b** aqueous suspension. Adapted with permission from Ref. [10]. Copyright 2015 American Chemical Society

purple upon continuous irradiation with UV light (Fig. 19.1b). The reddish purple suspension returned to the original white-turbid suspension upon irradiation with visible light. Therefore, it can be said that the spectral shift did not originate from the decomposition but the packing change of the closed-ring isomer in the supramolecular architecture, i.e., a change from the random orientation to an H-aggregate (Scheme 19.3).

An LCST phase diagram was obtained from the spectral changes. The transition temperature was estimated from the discontinuous changes in absorption and scattering for the aqueous suspensions with various conversion ratios of the closed-ring isomer (Fig. 19.2a) under controlled temperature conditions. The open-ring isomer formed a dehydrated phase corresponding to the high temperature state even at 0 °C. The transition temperature increased with the increase in the conversion ratio, and finally reached a maximum of 30 °C at 98% of the conversion ratio (Fig. 19.2b). The two-step photochromism, which is shown in Fig. 19.1b, can be explained with reference to the change in the conversion ratio at 20 °C in the phase diagram. At the initial state, the open-ring isomer was in a dehydrated phase of the supramolecular architecture, as described above. It continued to be in that phase until the conversion ratio was up to approximately 40%, which corresponded to the short irradiation. Finally, the LCST transition occurred when the conversion ratio exceeded 40% with the prolonged irradiation.

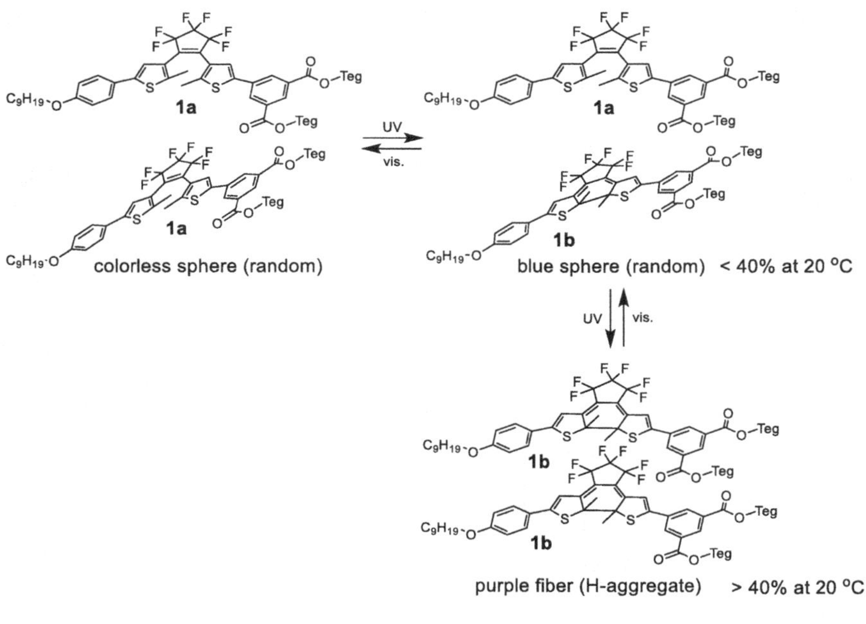

Scheme 19.3 Packing change of the amphiphilic diarylethene in the supramolecular architecture along with photoisomerization

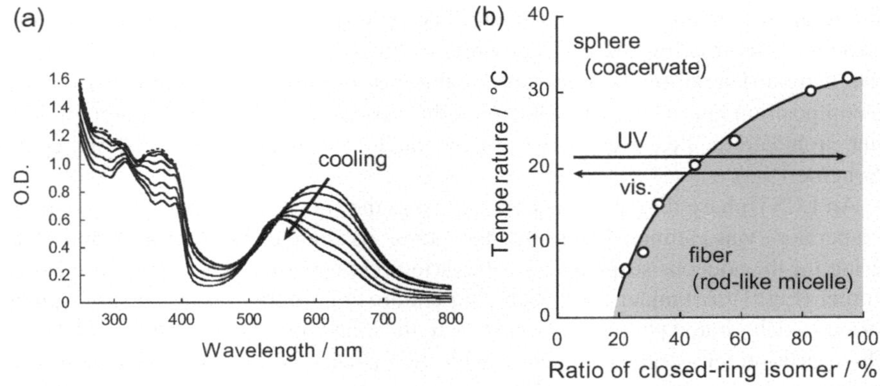

Fig. 19.2 **a** Spectral change of **1b** in aqueous suspension upon cooling from 50 to 10 °C and **b** phase diagram. Adapted with permission from Ref. [10]. Copyright 2015 American Chemical Society

The supramolecular architecture showed a reversible morphological change upon photoirradiation under an optical microscope. The initial white-turbid suspension contained colorless microspheres, which were the supramolecular architectures composed of the open-ring isomer, and the maximum diameter was approximately 10–20 μm. The colorless sphere turned blue upon short irradiation with UV light and disintegrated on being subjected to prolonged irradiation. During the disintegration process, the interface between the sphere and the surrounding water appeared to be hazy, and the color changed to reddish-purple (Fig. 19.3a). Upon subsequent irradiation with visible light, the widely diffused small spheres converged to their original positions, and large microspheres were restored (Fig. 19.3b). The colored spheres showed similar behavior when subjected to temperature control. Interestingly, the

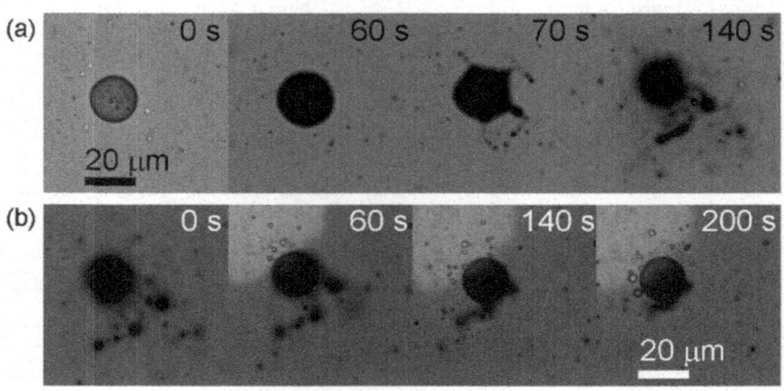

Fig. 19.3 Reversible morphological change of **1** upon irradiation with **a** UV (**1a** to **1b**) and **b** visible (**1b** to **1a**) lights at 15 °C. Adapted with permission from Ref. [10]. Copyright 2015 American Chemical Society

red-purple hazy assembly was not observed under dry conditions and therefore, it can be concluded that the morphological change was caused by an LCST transition as is the same case with a two-step spectral change.

Such micrometer-sized reversible morphological changes were associated with the change in nanostructures of the supramolecular architectures. It was shown by transmission electron microscopy (TEM) measurements that the microsphere, composed of the open-ring isomer (**1a**), was a co-continuous coacervate having water channels rather than a homogeneous droplet (Fig. 19.4a). By contrast, the reddish-purple haze, composed of the closed-ring isomer (**1b**), was a nanofiber without branching, with a diameter of approximately 10 nm and a length of more than 1 μm (Fig. 19.4b). Furthermore, snapshots were obtained halfway during the irradiation process to investigate the morphological change in greater detail. They show that as a result of the irradiation with UV light, nanofibers grew radially from the surface of coacervate (Fig. 19.4c). Upon irradiation with visible light on the red-purple suspension, nanospheres of a granular texture, approximately 10 nm in size, were observed inside the nanofibers (Fig. 19.4d).

The mechanism of reversible morphological change was inferred based on the observations associated with the nanostructure. Disintegration of spheres was caused by the elongation of the nanofibers. Nevertheless the resultant small spheres were

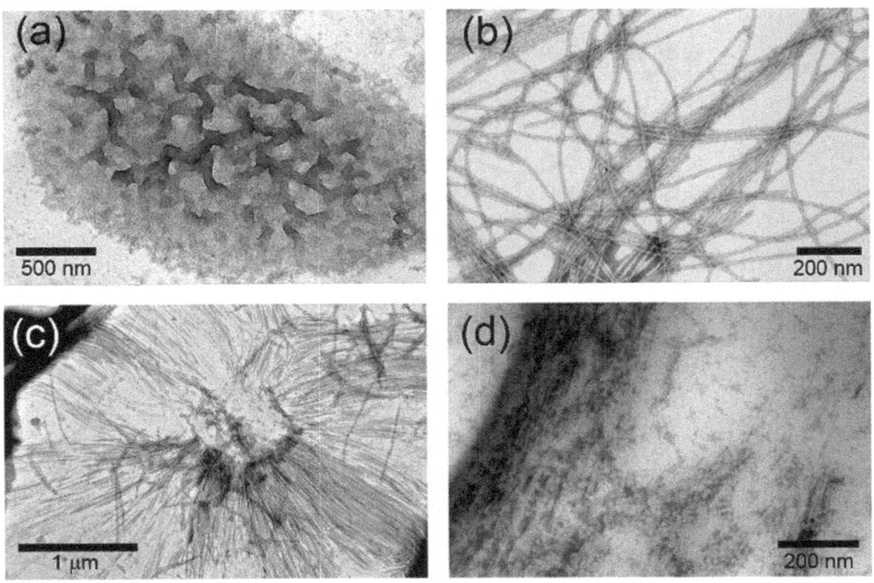

Fig. 19.4 TEM images of the nanostructures of **1**. **a** Open-(**1a**) and **b** closed-ring isomer (**1b**). **c**, **d** show change in nanostructures upon irradiation with UV (**1a** to **1b**) and visible (**1b** to **1a**) lights, respectively. Reprinted with permission from Ref. [10]. Copyright 2015 American Chemical Society

connected by a number of nanofibers, which were invisible under an optical microscope due to the diffraction limit. The transformation back to the nanospheres was initiated at various places on the nanofibers following the reverse mechanism upon irradiation with visible light. Furthermore, the generated nanospheres correspond to the dehydrated phase and aggregated with each other. Therefore, the diffused spheres reverted to their original positions by shrinking of the fibers, and finally the coacervate with water channels was restored (Fig. 19.5).

Thus, LCST transition was caused by a change not only in temperature but also in the conversion ratio accompanied by the photoisomerization. This is due to the fact that the supramolecular architecture composed of amphiphilic diarylethene **1** has different LCST transition temperatures for the two isomers [10]. The micrometer-long nanofibers were considered to be involved with the reversibility of the morphological change. In the next subsection, we examine the differences in characteristics of some of the derivatives and discuss their applications based on these characteristics.

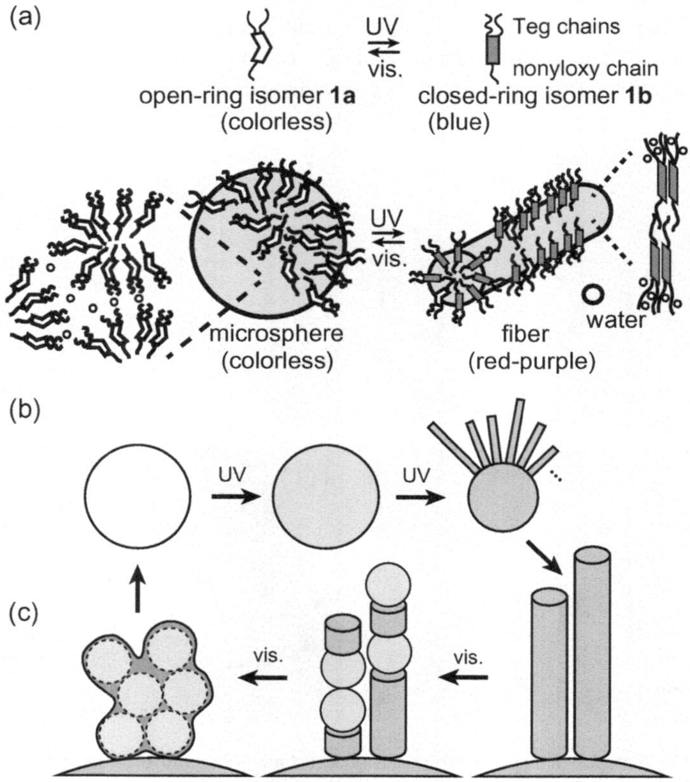

Fig. 19.5 Schematic illustration of the reversible morphological changes. Adapted with permission from Ref. [10]. Copyright 2015 American Chemical Society

19.2.2 Intermolecular Interactions and Supramolecular Architecture

In this subsection, we demonstrate the influence of intermolecular interactions on the morphological changes. The linker moiety was changed from ether (**1**) to ester (**2**) and amide (**3**) in the amphiphilic diarylethene [11]. The molecular shapes of amphiphilic diarylethenes are very similar between all three linkages. However, the existence and nonexistence of hydrogen bonding are expected to affect the photoinduced morphological change because of different intermolecular packing.

The ester and amide took the same π–π stacking structure, but different nanometer-sized structure due to hydrogen bonding. The photochromism of each compound was studied in the aqueous suspension. Hypsochromic shift was observed for **1b**–**3b** upon prolonged irradiation with UV light (Figs. 19.1b, 19.6a, b). A similar behavior of the spectral shifts suggests that the π–π stacking structure composed of the closed-ring isomers, i.e., an H-aggregate, was not affected by hydrogen bonding. However, it was confirmed by TEM measurements that the nanometer-sized structure was different for each supramolecular architecture, composed of the closed-ring isomers **1b**–**3b**. The ester-linked **2b** also formed similar nanofibers, but the length (180 ± 120 nm) was much shorter than that of ether-linked **1b** (>1 μm). Interestingly, the amide-linked **3b** formed flower-like structures made of thin layers (Fig. 19.7). Both the diameter of the nanofiber and the width of the thin layer were approximately 10 nm, which was twice the molecular length. Hence, the nanofibers and thin layers are likely to correspond to a cylindrical micelle and a bimolecular layer, respectively. Since the amphiphilic diarylethene has a fan shape, a cylindrical micelle is the preferred structure when there is no specific interaction other than π–π stacking. In contrast, the amide-linked **3b** formed a bilayer structure, most likely caused by a hydrogen bonding network between the amide linkages.

In addition, the micrometer-sized morphological changes also reflected the difference in the nanometer-sized structures. The initial structures prepared from

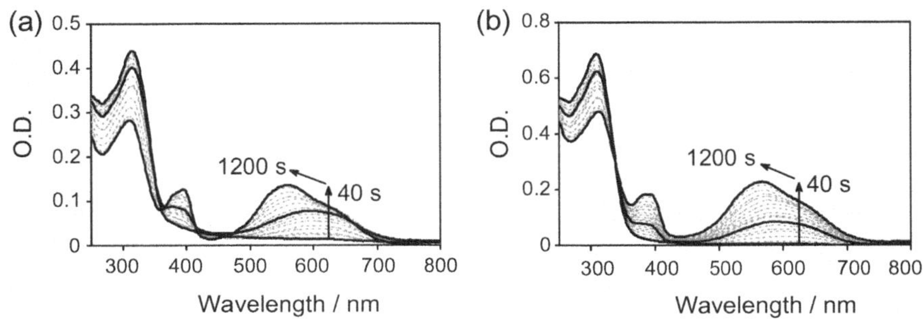

Fig. 19.6 Spectral changes of **2** and **3** along with photoisomerization from the open- to closed-ring isomer upon irradiation with UV light in aqueous suspension. Adapted with permission from Ref. [11]. Copyright 2017 Wiley-VCH

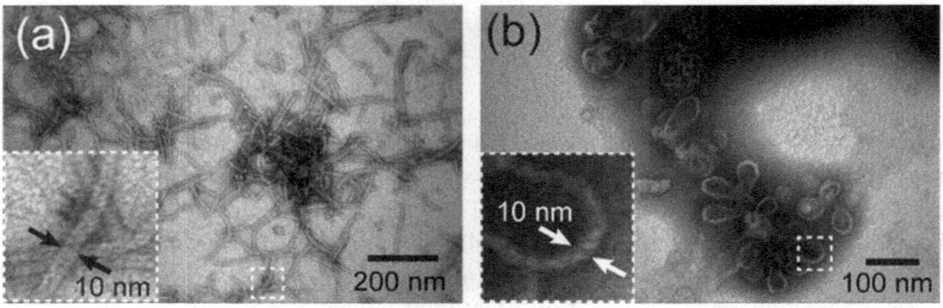

Fig. 19.7 TEM images of the nanostructures of the closed-ring isomer **a 2b** and **b 3b**. Adapted with permission from Ref. [11]. Copyright 2017 Wiley-VCH

the open-ring isomer **1a–3a** were microspheres in all cases. The ether-linked **1b** showed reversible disintegration and convergence, which indicated the formation of micrometer-sized long fibers. The ester-linked **2b**, on the other hand, dissolved and reverted to a few small spheres upon successive photoirradiation with UV and visible lights. The reason for this could be that the short fibers did not connect the spheres, which were dispersed in water (Fig. 19.8a). In the case of the amide-linked **3b**, the supramolecular architecture showed expansion and shrinkage to the original size upon successive irradiation with UV and visible lights, instead of disintegration, as seen in the previous cases. The expanded spheres are considered to have been transformed into spherically aggregated nanosheets (Fig. 19.8b). The suppression of morphological change could have been caused by the hydrogen bonding network. Therefore, to prepare micrometer-sized sheet structures, a different method was employed, i.e., excess amount of water was added into the colored solution of **3b**. This resulted in the formation of large sheet-like structures. A sheet structure quickly shrank to colorless microspheres upon irradiation with visible light (Fig. 19.8c).

19.2.3 Estimation of Molecular Packing by Quantum Yield

The conformation of the open-ring isomer was affected by the tri(ethylene glycol) ether chain, even in the dehydrated state, as suggested by the changes in the quantum yield [12]. In general, the photocyclization quantum yield of the open-ring isomer is influenced by the ratio of the two conformers, namely the photoactive antiparallel and the inactive parallel ones. The conformers mutually convert in solution by free rotation around the single bonds between the aryl groups and the ethene moiety (Scheme 19.4). Therefore, the cyclization quantum yield is close to 50% in solution, but when the antiparallel conformation is made dominant [19], the cyclization quantum yield increases. Additionally, in the crystalline phase, the cyclization quantum yield is often found to be 100% due to the absence of conformational changes [20].

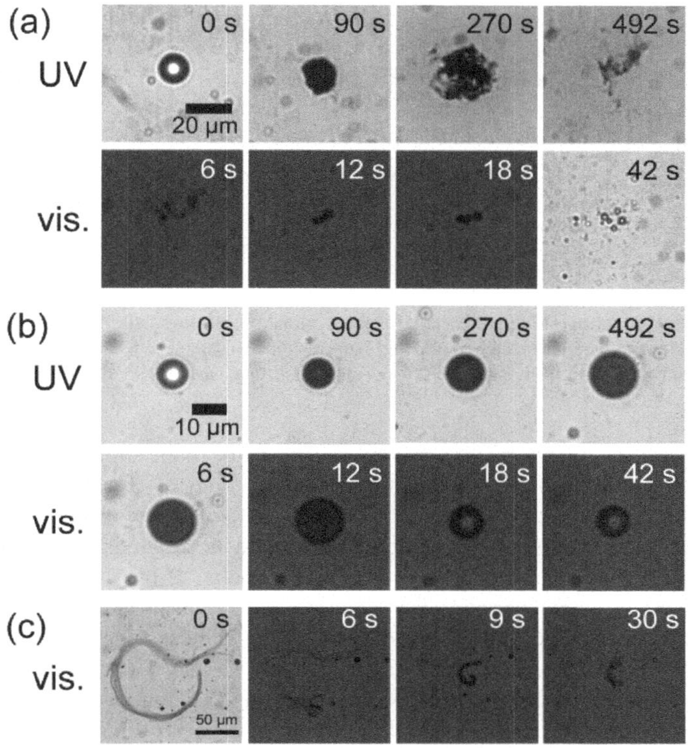

Fig. 19.8 Morphological changes of compound **a 2** and **b 3** upon irradiation with UV and visible lights. **c** The sheet-like structure composed of the closed-ring isomer **3b** and the corresponding morphological change. Adapted with permission from Ref. [11]. Copyright 2017 Wiley-VCH

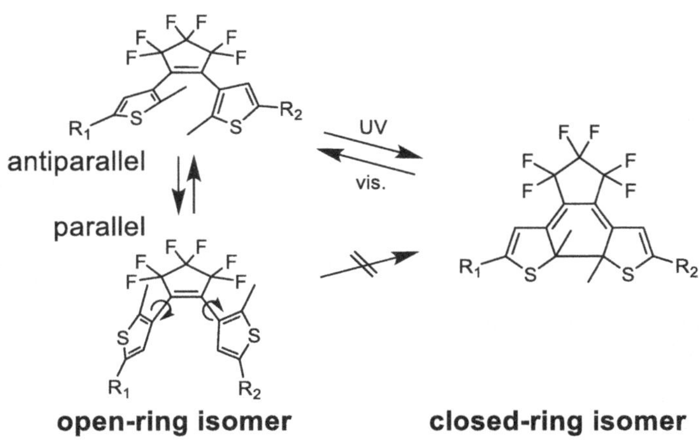

Scheme 19.4 Conformational changes of the open-ring isomer and photocyclization reaction

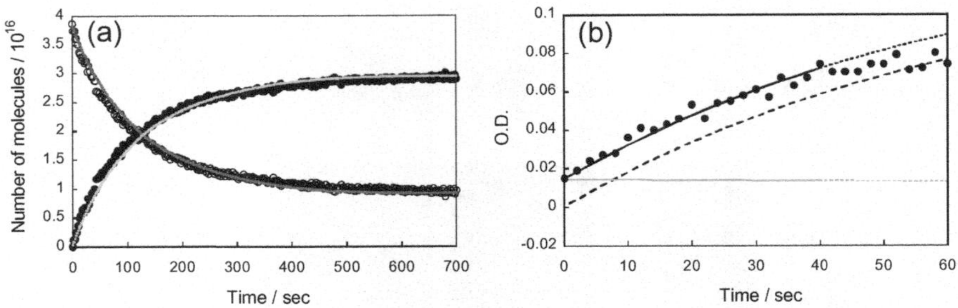

Fig. 19.9 Determination of quantum yield for the cyclization reaction from **3a** to **3b** in **a** solution and **b** aqueous suspension. The curves in **a** and **b** correspond to $\Phi_{3a\text{-}3b} = 59$ and 76%, respectively. Adapted with permission from Ref. [12]. Copyright 2017 the Chemical Society of Japan

The cyclization quantum yield showed a larger value in the supramolecular architecture than in solution. The cyclization quantum yield of **3** ($\Phi_{3a\text{-}3b}$) was found to be 59% upon irradiation with 365 nm-UV light in acetonitrile, in which the amphiphilic diarylethene with an amide linker did not aggregate. The quantum yield of diarylethene in the core structure of **3**, namely 1,2-bis(2-methyl-5-phenyl-3-thienyl)hexafluorocyclopentene was determined as 57% at the same wavelength of irradiation in hexane and is approximately the same as the preceding value. The aqueous suspension contained small spheres was used for the measurement of quantum yield in order to ensure weak scattering. The analysis was completed before the spectral shift triggered by the formation of the H-aggregate because the molar absorption coefficient of each aggregation state was different at the monitored wavelength. The cyclization quantum yield, $\Phi_{3a\text{-}3b}$ was determined as 76% for 365 nm UV light in the supramolecular architecture of dehydrated coacervate state (Fig. 19.9). If the orientation and conformation are random in the dehydrated phase, the cyclization quantum yield should have been the same as that in the solution, and conversely, if the amphiphilic diarylethenes are packed in the bilayer structure, i.e., the hydrophilic and hydrophobic moieties are directed opposite to one another, the cyclization quantum yield should have been close to 100% as the conformation is restricted to the photoactive antiparallel one. The obtained value of the quantum yield suggests that both of the side chains were not completely random, even in the dehydrated phase.

19.3 Motion of Supramolecular Architectures

The supramolecular architecture composed of amphiphilic diarylethene showed reversible changes of nano- and micrometer-sized structures upon irradiation with UV and visible lights, as described above. In order to realize the movement of objects using the supramolecular architecture, the possible approaches are studying the change in morphology itself and that in the physical properties of the suspension. Here, we demonstrate the transduction of the morphological changes of the supramolecular architecture into the movement of objects based on the photoinduced LCST transition using the ether-linked diarylethene **1**, which forms long nanofibers.

19.3.1 Photoinduced Shrinking of Bundled Fibers by Depletion Force

The shrinking of the nanofiber **1b** can be used for unidirectional movement of an external object. However, as described above, since the fibers are generated radially from the microsphere surface (Fig. 19.4c), it is necessary for the nanofibers to align and bundle to control the direction of motion. Therefore, we considered using a depletion force.

Depletion force is the attractive interaction between colloidal particles dispersed in a polymer solution [21]. Since polymer chains form a random coil structure in solution, i.e., a spatially extended structure, an excluded volume, equal to the thickness of the radius of the solute polymers, is generated on the surface of each colloidal particle. This region is known as the depletion layer. The attractive force acts between the colloidal particles in the direction of reduction of the depletion layer exposing to the solvent. The depletion force acts in the order of hundreds of nanometers, whereas hydrophobic interaction acts between molecules. As a result, the depletion force leads to form a coagulated structure of sub-millimeter size. In particular, fibers coagulate and form a bundled structure because the overlap of the depletion layer in parallel alignment is much larger than that in the orthogonal case [22]. Therefore, photoresponsive bundle is expected to be obtained by the use of the supramolecular architecture of amphiphilic diarylethene. Here, we examined the coagulation and photoinduced morphological change of ether-linked **1**, when it was dispersed into an aqueous solution of methylcellulose (MC). The molecular weight of MC is 40,000 and 1–5 wt% aqueous solution was prepared.

The photoinduced morphological change of **1a** showed different behavior depending on the size of assembly and the concentration of the depletant [13]. In the case of pure water, the red-purple hazy assembly is observed, consisting of radially generated nanofibers **1b** as shown in Fig. 19.10a. In contrast, few micrometer-long rod-like structures were observed around the large microsphere in the MC solution (Fig. 19.10b). When the size of the supramolecular architecture is small, the suspension without the depletant shows reddish-purple color upon the photoirradiation due to the homogeneously dispersed nanofibers (Fig. 19.11(a)) whereas in the suspension with MC (4.1 wt%), bundles are generated in the entire field of view (Fig. 19.11b) and the coagulation of the bundles is also observed (Fig. 19.11c). The length of the bundle reached several tens of micrometers, and by stirring the suspension with a pipette, the length of bundle exceeded 1 mm (Fig. 19.12).

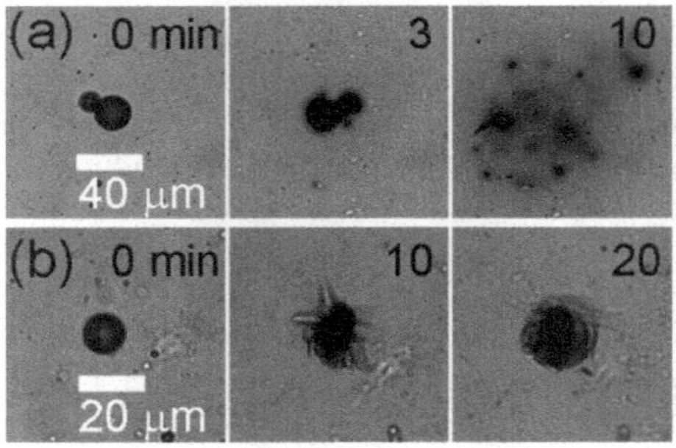

Fig. 19.10 The formation of bundled fibers in **a** 0 and **b** 4.1 wt% MC aqueous solution. Adapted with permission from Ref. [13]. Copyright 2018 the Royal Society of Chemistry

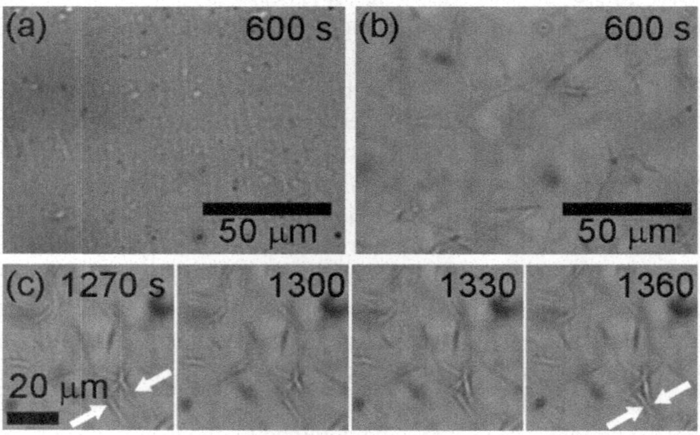

Fig. 19.11 The optical images containing **a** no MC and **b** MC (4.1 wt%). **c** The coagulation of the bundles. Reprinted with permission from Ref. [13]. Copyright 2018 the Royal Society of Chemistry

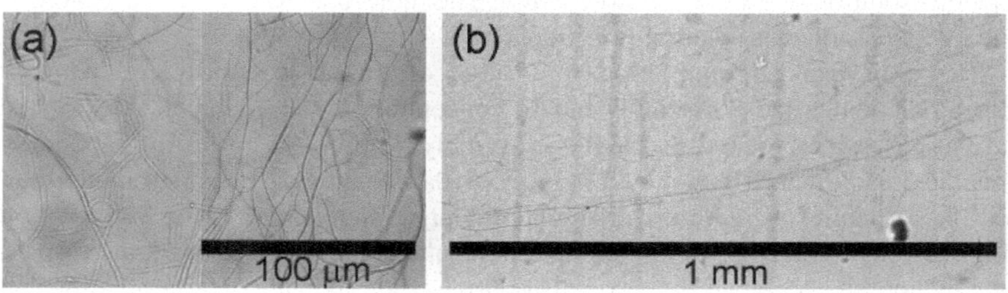

Fig. 19.12 **a** The submillimeter- and **b** millimeter-sized bundles prepared by stirring. Reprinted with permission from Ref. [13]. Copyright 2018 the Royal Society of Chemistry

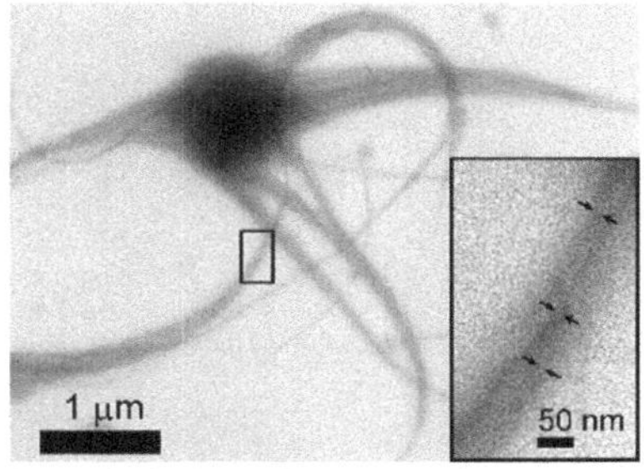

Fig. 19.13 TEM image of the bundle. The arrows indicate the diameter of nanofibers (ca. 10 nm). Adapted with permission from Ref. [13]. Copyright 2018 the Royal Society of Chemistry

The original properties of the nanofiber, such as π–π stacking structure and diameter, were retained even in the bundle coagulated by the depletion force. Intermolecular packing was evaluated from the change in the absorption spectrum upon irradiation with UV light. The spectral shift originating from the H-aggregate was observed in both pure water and 3.4 wt% aqueous solution of MC. The same spectral shift means that the π–π stacking structure between the closed-ring isomers was not affected by coagulation. Furthermore, TEM measurements indicated that the diameter of each nanofiber in the bundle was the same as the original nanofiber, showing that the rod-like micelle structure was retained (Fig. 19.13). Therefore, molecular packing was not affected by the presence of MC and each nanofiber in the bundle is considered to retain the capability of photoinduced morphological change.

The bundled fibers showed photoinduced shrinking by irradiation with visible light (Fig. 19.14). At the initial step of photoirradiation, the further coagulation was observed. The initial length of the bundle was more than 350 μm and it got shrunk to 170 μm under the irradiation conditions of 546 nm and 29 mW cm^{-2}. The speed of the terminal point was 8.8 μm s^{-1}. We achieved submillimeter-sized one-dimensional motion by the bundling of the supramolecular fibers using the depletion force.

19.3.2 *Diffusion of Particle in the Oriented Fibers*

Orientation control of the nanofibers was achieved using linearly polarized UV light [14]. When linearly polarized 365 nm UV light (10–20 mW cm^{-2}) was irradiated on the suspension containing the open-ring isomer **1a**, the photogenerated nanofibers were oriented in the direction of the linearly polarized light, which was confirmed by TEM measurements (Fig. 19.15a). No such orientation was observed in the case of 325 nm UV light (Fig. 19.15b).

The orientation is thought to have been caused by the anisotropic absorption associated with the anisotropic packing of the closed-ring isomer in the nanofiber. The related transition moment was calculated to be 361 nm in the longitudinal direction of the closed-ring isomer (Fig. 19.16). In addition, the packing of the closed-ring isomer in the nanofibers is radial in direction due to its rod-like micelle structure, i.e., the estimated direction of the selective absorption at 365 nm was perpendicular to the

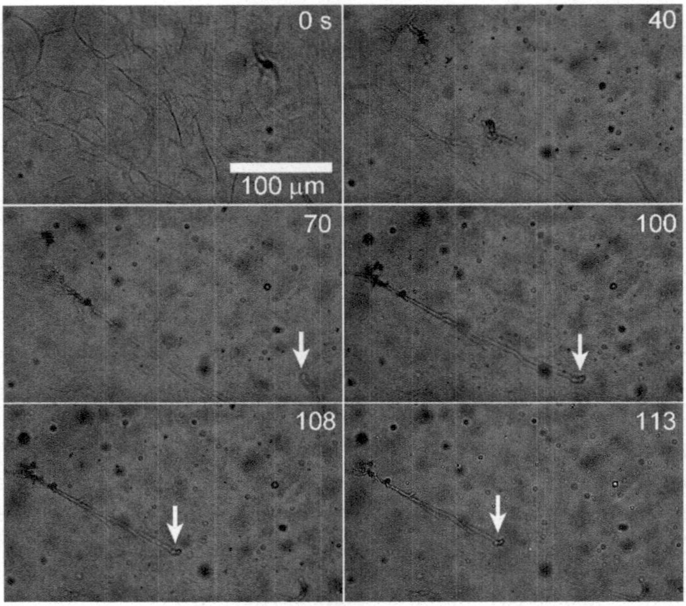

Fig. 19.14 Coagulation and shrinking of a long bundle upon irradiation with visible light. The arrow indicates the terminal point of the bundled fiber. Reprinted with permission from Ref. [13]. Copyright 2018 the Royal Society of Chemistry

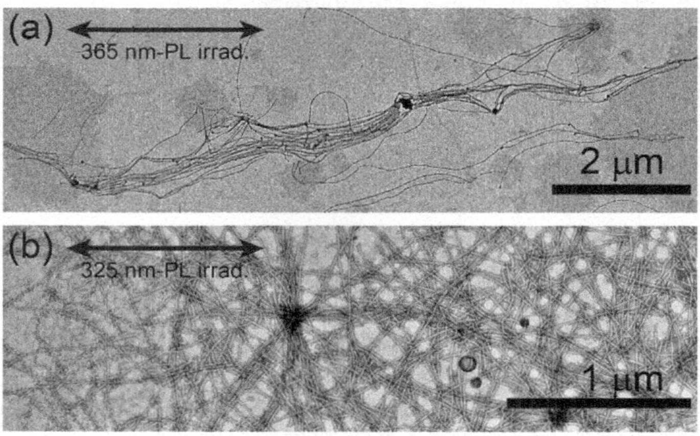

Fig. 19.15 TEM images of fibers generated by irradiation with linearly polarized **a** 365 nm and **b** 325 nm light. Adapted with permission from Ref. [14]. Copyright 2017 Wiley-VCH

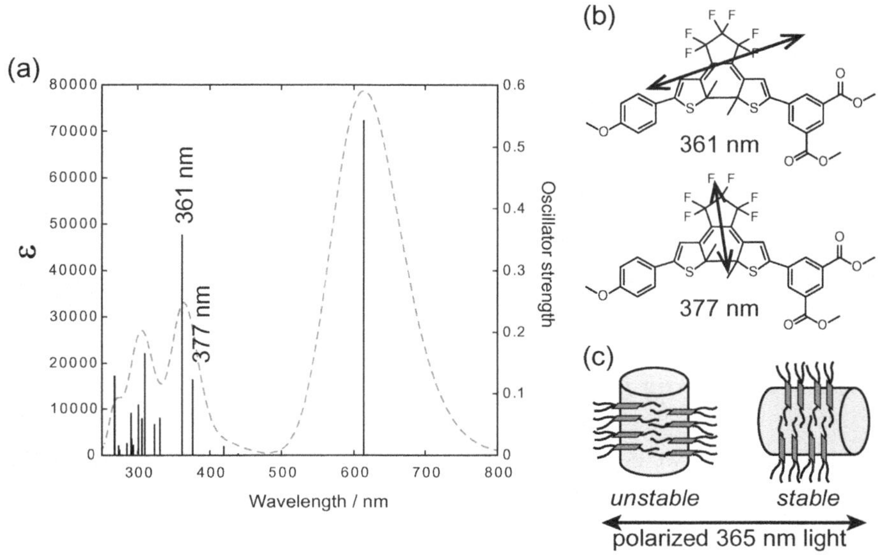

Fig. 19.16 Transition moment of the core moiety of the closed-ring isomer **1b**. **a** Calculated oscillator strengths and spectrum. **b** The direction of transition moment. **c** The direction and stability of the fibers. Adapted with permission from Ref. [14]. Copyright 2017 Wiley-VCH

longitudinal direction of the nanofiber. Therefore, when the directions of the linearly polarized light and the fibers were orthogonal, the transition moment of each molecule became parallel to the linearly polarized light, which is the preferred orientation for light absorption accompanying repetitive isomerization and vibrational relaxation. As a result, the unstable fibers, which were oriented perpendicular to the direction of the polarized light disappeared and only the fibers oriented parallel were allowed to remain in the suspension. On the other hand, when 325-nm light was used instead of 365-nm light, such orientation was not observed. There are many transition moments at around 325 nm, thus the stability of the fiber is considered to be almost the same for all directions of linearly polarized light.

The oriented fiber had an effect on the direction of diffusion of objects. Polystyrene (PS) microbeads were dispersed as tracers in the suspension containing the oriented fibers and the diffusion rate of beads was monitored by studying the Brownian motion. As a result, the mean square displacement (MSD) in the direction perpendicular to the fibers was approximately 30% less than that in the parallel direction (Fig. 19.17). Thus, the oriented nanofibers affected the direction of movement of other objects even when the nanofibers were not bundled.

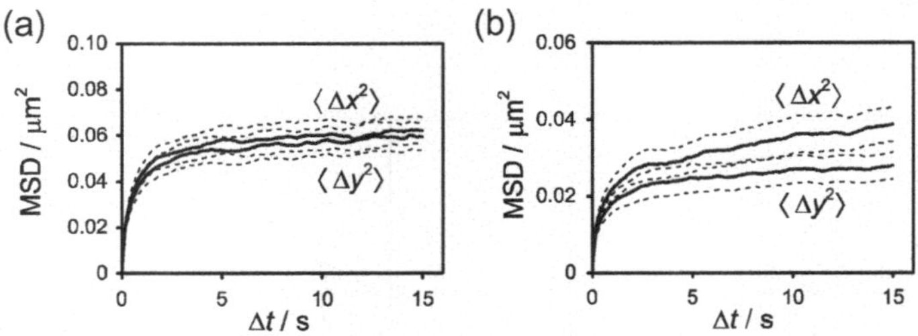

Fig. 19.17 Diffusivity of PS beads in the suspension of fibers upon irradiation with **a** unpolarized and **b** x-polarized light. Adapted with permission from Ref. [14]. Copyright 2017 Wiley-VCH

19.3.3 Clustering in the Focus Point

In this subsection, we describe clustering of objects using viscosity change of the suspension around the UV-focused point [15]. When a viscous liquid moves a static object, the object starts to accelerate quickly due to a transfer of momentum. Based on this theory, we tried to collect the PS microbeads, which were dispersed in the suspension of supramolecular architecture as described in the previous subsection, upon irradiation with UV light under an optical microscope (365 nm, 3.6 W cm^{-2}, spot size 110 μm).

Many PS beads were clustered into the focused point upon irradiation with the UV light (Fig. 19.18). The mechanism of the movement can be explained by the morphology of supramolecular architecture around the PS beads. In the initial state, the viscosity of the suspension was similar to that in pure water despite the presence of the many spherical assemblies around the PS beads (Fig. 19.19a). Subsequently, the viscosity increased on being irradiated with UV light, because the surrounding supramolecular architectures changed the morphology into nanofibers (Fig. 19.19b). Therefore, the movements of PS beads were affected by the change in viscosity accompanying the morphological change. Simultaneously, thermal convection was also generated around the spot because of the heating caused by vibrational relaxation. Since the liquid flowed towards the spot at the bottom of the glass cell, the precipitated PS beads gathered there. However, the force created by the convection

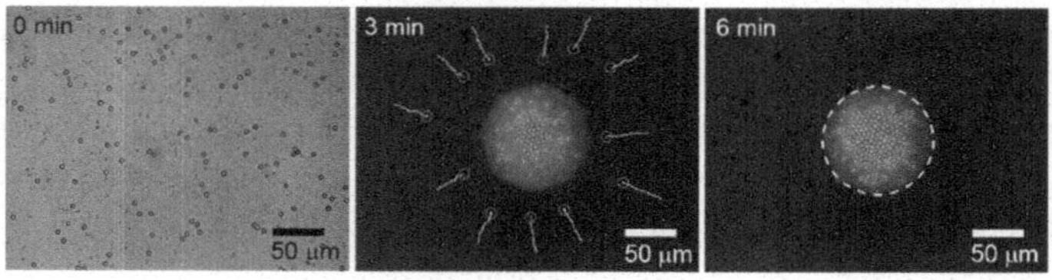

Fig. 19.18 Photodriven clustering of PS beads using a suspension of **1**. Adapted with permission from Ref. [15]. Copyright 2017 American Chemical Society

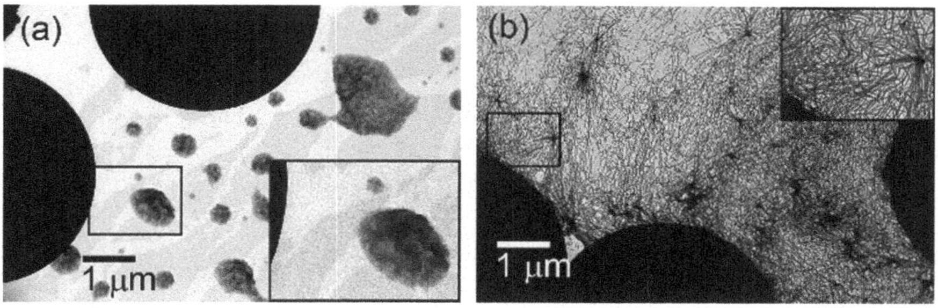

Fig. 19.19 TEM images of the suspension containing PS beads and **a** the open- **1a** and **b** closed-ring isomer **1b**. Reprinted with permission from Ref. [15]. Copyright 2017 American Chemical Society

was not enough to lift up the PS beads. As a result, the PS beads remained and clustered in the spot.

The clustering showed unique characteristics of the surrounding fibers. When the spot was moved, the clustered PS beads followed the spot (Fig. 19.20). However, the beads scarcely diffused from the clustered spot even after the UV irradiation was stopped (Fig. 19.21a, b). This happened because of the existing nanofibers around the clustered beads. Subsequently, the diffusivity was reinstated by the morphological changes to the spherical assembly upon irradiation with visible light (Fig. 19.21c, d). In addition, the analogous compound **4**, in which the closed-ring isomer did not form nanofibers (Scheme 19.5), was used for the same operation as a control experiment. As a result, both the clustering and the suppression of diffusion of PS beads were hardly observed in spite of generation of the thermal convection. The result means that both the factors, namely thermal convection due to UV absorption and the local increase in viscosity due to the formation of nanofibers, were responsible for the clustering process. Thus, the movement of many objects tracing the movement of a UV-irradiation spot was achieved with the assistance of the supramolecular architecture.

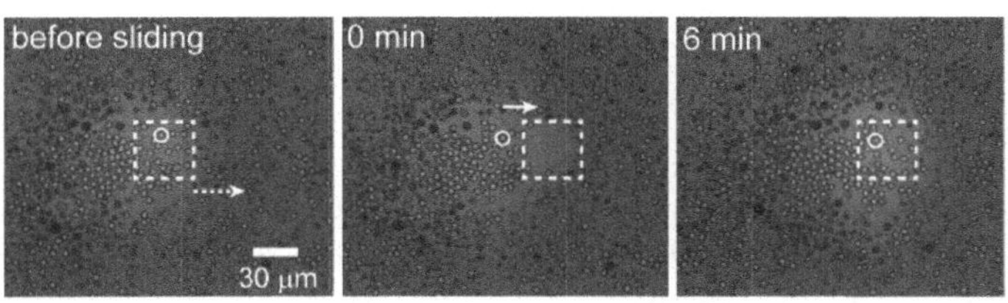

Fig. 19.20 Behavior similar to optical tweezers. **a** Initial state: PS beads in a clustered state, **b** immediately after sliding of the irradiated spot, **c** reclustering. Adapted with permission from Ref. [15]. Copyright 2017 American Chemical Society

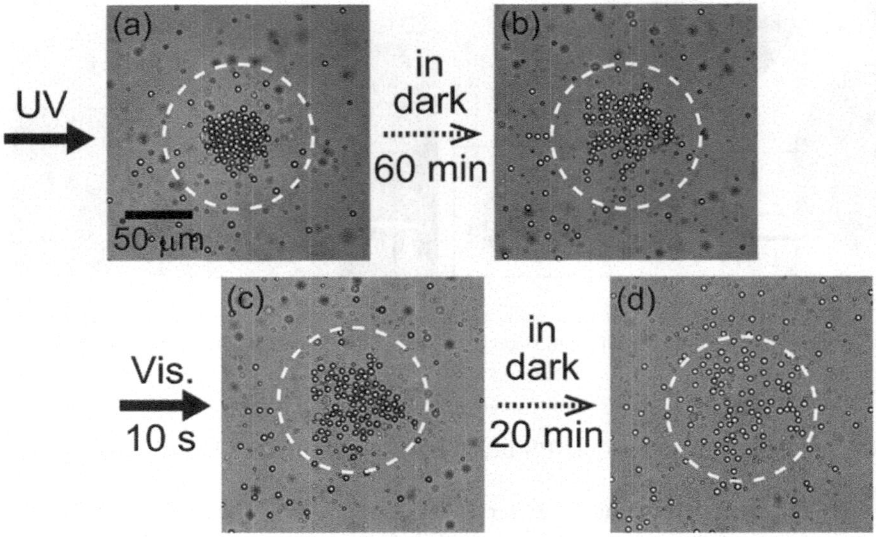

Fig. 19.21 Control of diffusivity by irradiation with UV and visible lights. The PS beads being **a** clustered, **b** trapped, **c** released, and **d** diffused. Reprinted with permission from Ref. [15]. Copyright 2017 American Chemical Society

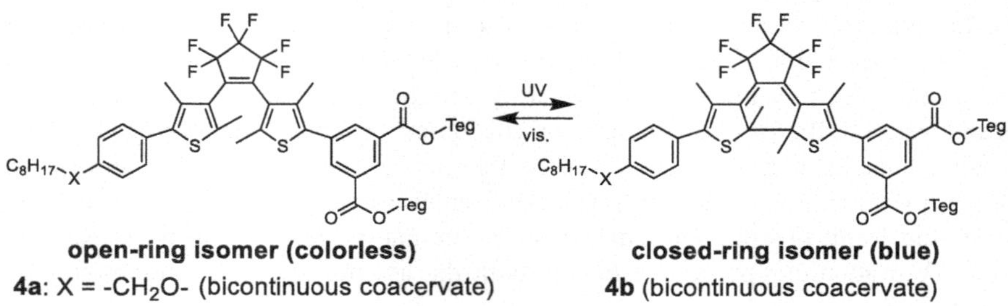

Scheme 19.5 Photoisomerization of the amphiphilic diarylethene **4**

19.4 Conclusion

In this chapter, we discussed photoinduced morphological changes and photodriven movement of objects based on the photoresponsive supramolecular architectures composed of amphiphilic diarylethene. The supramolecular architectures induced LCST transition upon photoirradiation and exhibited micrometer-sized morphological changes under an optical microscope which originated from the differences in nanometer-sized structures. The morphological changes can be controlled by modifying the intermolecular interactions, e.g., by changing the linker moiety. The nanofibers were bundled by the application of a depletion force and the bundle showed submillimeter-sized shrinking. The nanofibers were orientated by linearly polarized

UV light and they were able to affect the diffusion direction of other objects. Moreover, the nanofibers proved to be capable of assisting the clustering of PS beads into a UV-focused point. We expect that the further investigation on various types of morphological changes and external stimuli will lead to the development of more sophisticated functionalities.

Acknowledgements This work was supported by a Grant-in-Aid for Scientific Research on Innovative Areas "Photosynergetics" (JSPS KAKENHI Grant No. JP26107008), from MEXT, Japan and PRESTO, JST (Grant No. JPMJPR12K7). This work was performed under the Research Program of "Dynamic Alliance for Open Innovation Bridging Human, Environment and Materials" in "Network Joint Research Center for Materials and Devices" (No. 2014445, 20162025, and 20172024).

References

1. Israelachvili JN (1992) Intermolecular and surface forces, 2nd edn. Academic Press, London
2. Zhang X, Rehm S, Safont-Sempere MM, Würthner F (2009) Vesicular perylene dye nanocapsules as supramolecular fluorescent pH sensor systems. Nat Chem 1:623–629
3. Kunitake T, Nakashima N, Shimomura M, Okahata Y, Kano K, Ogawa T (1980) Unique properties of chromophore-containing bilayer aggregates: enhanced chirality and photochemically induced morphological change. J Am Chem Soc 102:6642–6644
4. Faure D, Gravier J, Labrot T, Desbat B, Oda R, Bassani DM (2005) Photoinduced morphism of gemini surfactant aggregates. Chem Commun 1167–1169
5. Hamada T, Sugimoto R, Vestergaard MC, Nagasaki T, Takagi M (2010) Membrane disk and sphere: controllable mesoscopic structures for the capture and release of a targeted object. J Am Chem Soc 132:10528–10532
6. Ikegami T, Kageyama Y, Obara K, Takeda S (2016) Dissipative and autonomous square-wave self-oscillation of a macroscopic hybrid self-assembly under continuous light irradiation. Angew Chem Int Ed 55:8239–8243
7. Yagai S, Iwai K, Karatsu T, Kitamura A (2012) Photoswitchable exciton coupling in merocyanine-diarylethene multi-chromophore hydrogen-bonded complexes. Angew Chem Int Ed 51:9679–9683
8. Irie M, Fukaminato T, Matsuda K, Kobatake S (2014) Photochromism of diarylethene molecules and crystals: memories, switches, and actuators. Chem Rev 114:12174–12277
9. Wang L, Zou H, Dong Z, Zhou L, Li J, Luo Q, Zhu J, Xu J, Liu J (2014) Temperature-driven switching of the catalytic activity of artificial glutathione peroxidase by the shape transition between the nanotubes and vesicle-like structures. Langmuir 30:4013–4018
10. Higashiguchi K, Taira G, Ji Kitai, Hirose T, Matsuda K (2015) Photoinduced macroscopic morphological transformation of an amphiphilic diarylethene assembly: reversible dynamic motion. J Am Chem Soc 137:2722–2729
11. Yotsuji H, Higashiguchi K, Sato R, Shigeta Y, Matsuda K (2017) Phototransformative supramolecular assembly of amphiphilic diarylethenes realized by the combination of photochromism and lower critical solution temperature behavior. Chem Eur J 23:15059–15066
12. Higashiguchi K, Yotsuji H, Matsuda K (2017) Determination of quantum yield of photoreaction in solution and in suspension by global fitting of prolonged change of concentration. Chem Lett 46:1564–1566
13. Sakaguchi A, Higashiguchi K, Matsuda K (2018) Bundle formation of supramolecular fibers of amphiphilic diarylethene by depletion force. Chem Commun 54:4298–4301
14. Sakaguchi A, Higashiguchi K, Matsuda K (2017) Anisotropic diffusion of microbeads surrounded by an anisotropically elongated supramolecular diarylethene architecture under linearly polarized light. ChemPhotoChem 1:488–492

15. Sakaguchi A, Higashiguchi K, Yotsuji H, Matsuda K (2017) Photocontrol of clustering, retaining, and releasing of microbeads concomitant with phototransformation of supramolecular architecture of amphiphilic diarylethene. J Phys Chem B 121:4265–4272
16. Irie M, Lifka T, Kobatake S, Kato N (2000) Photochromism of 1,2-bis(2-methyl-5-phenyl-3-thienyl)perfluorocyclopentene in a single-crystalline phase. J Am Chem Soc 122:4871–4876
17. Kamiya H, Yanagisawa S, Hiroto S, Itami K, Shinokubo H (2011) Functionalization of a simple dithienylethene via palladium-catalyzed regioselective direct arylation. Org Lett 13:6394–6397
18. Saika T, Irie M, Shimidzu T (1994) Thiophene oligomers with a photoswitch. J Chem Soc Chem Commun 2123–2124
19. Fukumoto S, Nakashima T, Kawai T (2011) Photon-quantitative reaction of a dithiazolylarylene in solution. Angew Chem Int Ed 50:1565–1568
20. Shibata K, Muto K, Kobatake S, Irie M (2002) Photocyclization/cycloreversion quantum yields of diarylethenes in single crystals. J Phys Chem A 106:209–214
21. Lekkerkerker HNW, Tuinier R (2011) Colloids and the depletion interaction. Springer, Heidelberg
22. Inoue D, Mahmot B, Kabir AMR, Farhana TI, Tokuraku K, Sada K, Konagaya A, Kakugo A (2015) Depletion force induced collective motion of microtubules driven by kinesin. Nanoscale 7:18054–18061

Chapter 20
Functional Photoactive Materials Based on Flexible π Molecules

Shohei Saito

Abstract Flapping molecules (FLAP) have been developed as a new series of photofunctional systems. The molecular design is based on the concept of the rigid/flexible hybridization of π conjugated units. FLAP bearing bulky substituents works as a ratiometric viscosity probe that shows no polarity dependence, while FLAP with long alkyl chains provides a "light-melt adhesive," a highly cohesive columnar liquid crystals that can be melt by ultraviolet light irradiation.

Keywords Flapping molecule · Viscosity probe · Liquid crystal · Adhesive

20.1 Introduction

Applications of π-conjugated molecules cover dyes/pigments and aromatic polymers, and in recent years, they have been used in a wide range of fields as organic electronics materials, fluorescent probes, and building blocks for self-organized and supramolecular structures. However, most of these uses relies on the merit of "rigidity of the skeleton" inherent to the π-conjugated molecular system mainly composed of aromatic rings. For example, porphyrins and fullerenes tend to undergo electron transfer because their molecular shape does not change and they have low reorientation energy [1]. In addition, perylene and fluorescein having a condensed polycyclic structure show slow nonradiative decay when photoexcited because molecular vibration is suppressed, and thereby exhibit high fluorescence quantum yield [2]. Similarly, polycyclic triphenylene has a flat structure and is easy to stack, and it has been put to practical use as a discotic liquid crystal material [3]. However, in a meaning, the fact that the basic molecular skeleton is rigid could be considered that it is difficult to convert the physical properties derived from the flexibility of the structure, and that static physical properties are still expressed. Based on this molecular scientific point of view, we tried to create new molecular technologies by actively giving flexibility to rigid π-conjugated molecules and utilizing the movement of the π-electron skeleton.

S. Saito (✉)
Graduate School of Science, Department of Chemistry, Kyoto University, Sakyo, Kyoto 606-8502, Japan
e-mail: s_saito@kuchem.kyoto-u.ac.jp

H. Miyasaka et al. (eds.), *Photosynergetic Responses in Molecules and Molecular Aggregates*, https://doi.org/10.1007/978-981-15-5451-3_20

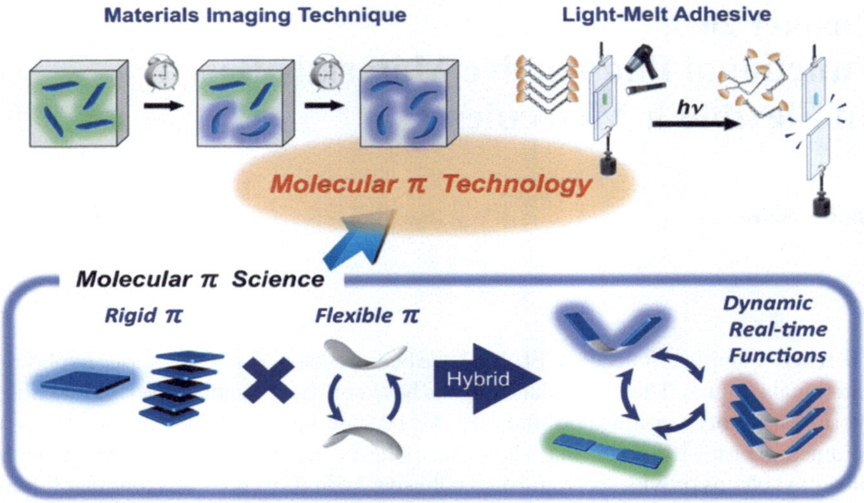

Fig. 20.1 Hybridization of rigidity and flexibility as a new design guideline of functional π conjugated molecules

In other words, by constructing a hybrid π-electron system that combines rigidity and flexibility, in addition to "luminescence properties" and "π stacking ability" derived from a rigid skeleton, "dynamic electronic structure" has been induced with a flexible skeleton (Fig. 20.1) [4, 5].

20.2 Flapping Fluorophore as a Viscosity Probe

Fluorescent probes are widely used as a means of visualizing local environmental changes that occur from time to time in materials that are invisible to the human eye with high spatiotemporal resolution. We developed a flapping luminophore (FLAP) that changes the emission color dramatically in response to the environment, and visualized the material by doping it (Fig. 20.2). To design organic molecules that can be used for such material imaging and emit various types of light, we designed

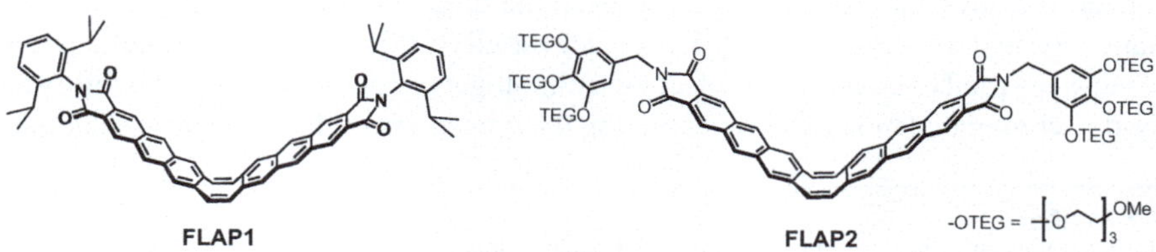

Fig. 20.2 Chemical structures of FLAP1 and FLAP2

FLAP1 which is composed of a rigid anthracene skeleton and a flexible conjugated 8-membered ring (cyclooctatetraene, COT) [6, 7].

FLAP1 shows RGB luminescence with a single component depending on the environment without changing the excitation wavelength. FLAP1 emits green light in solution, blue light in a polymer thin film, and red light in a crystal when excited with ultraviolet light. FLAP1 has a V-shaped structure in the S_0 ground state. When photoexcited, FLAP1 undergoes a quick internal conversion from a higher excited state to the lowest singlet excited state (S_1). In S_1, the molecules trapped by the polymer in the thin film cannot be relaxed freely, and emit blue fluorescence while maintaining almost V-shaped form. On the other hand, after photoexcitation in solution, the molecule undergoes a conformational change flexibly from a V-shaped to a planar forms, and it emits lower-energy green fluorescence. Furthermore, the red fluorescence in the crystal is derived from excimer emission due to intermolecular interaction of rigid anthracene sites.

From the results of transient absorption and time-resolved fluorescence measurements, it was revealed that the time constant of the conformational change in FLAP1 at S_1 was 550 ps in a DMSO solution. When the energy diagram of S_1 was drawn by TD-DFT calculation, it was suggested that there was an energy barrier in the process of structural change from the V-shaped to planar forms (Fig. 20.3).

FLAP has been found to act as a local viscosity probe. That is, the change in conformation from the V-shape to the planar shape due to photoexcitation is suppressed by the increase in the viscosity of the medium, so that the curing of the medium can be visualized by a change in fluorescent color. In recent years, such fluorescent viscosity probe molecules have been actively used for bioimaging, and two methods, the fluorescence ratiometric method and the fluorescence lifetime imaging method,

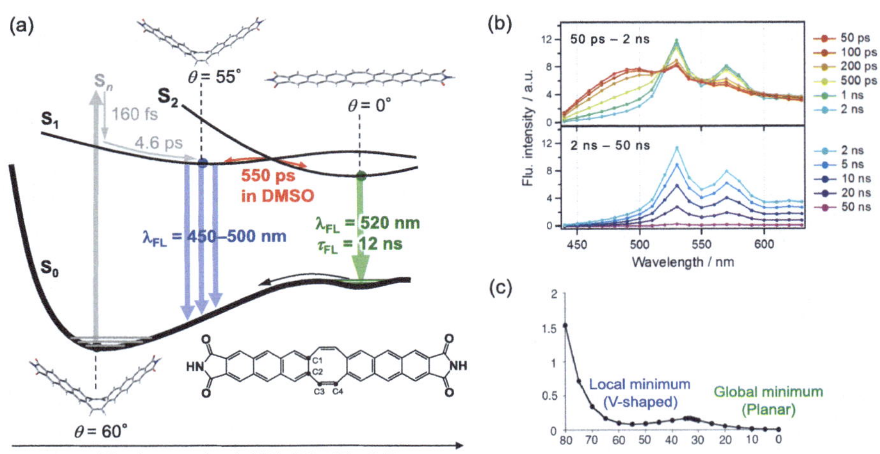

Fig. 20.3 Excited-state dynamics of a flapping molecule. **a** Possible energy profile, **b** time-resolved fluorescence spectroscopy of FLAP1, and **c** calculated energy diagram in S_1. Terminal substituents of FLAP were replaced by H atoms in the calculations

are well known as means for quantitatively analyzing the viscosity distribution. In both methods, however, the so-called "molecular rotor" including BODIPY and cyanine dyes has become widespread as a viscosity probe, and "molecular rotational motion" in the excited state (S_1) is suppressed by a high-viscosity medium [8–14]. On the other hand, FLAP is a local viscosity probe that utilizes the "flapping motion of molecules," and it has dual fluorescence, so that it can be used as a fluorophore to determine viscosity by a ratiometric method. In addition, FLAP2 can selectively sense viscosity rather than polarity, because the fluorescence spectrum of FLAP shows very small polarity dependence (Fig. 20.4). When the fluorescence intensity ratio (I_{461}/I_{525}) of the blue and green emission bands is plotted on the vertical axis and the viscosity is plotted on the horizontal axis, the logarithmic plot provides a linear relationship in the viscosity range of 2.2–100 cP, according to the Förster-Hoffmann

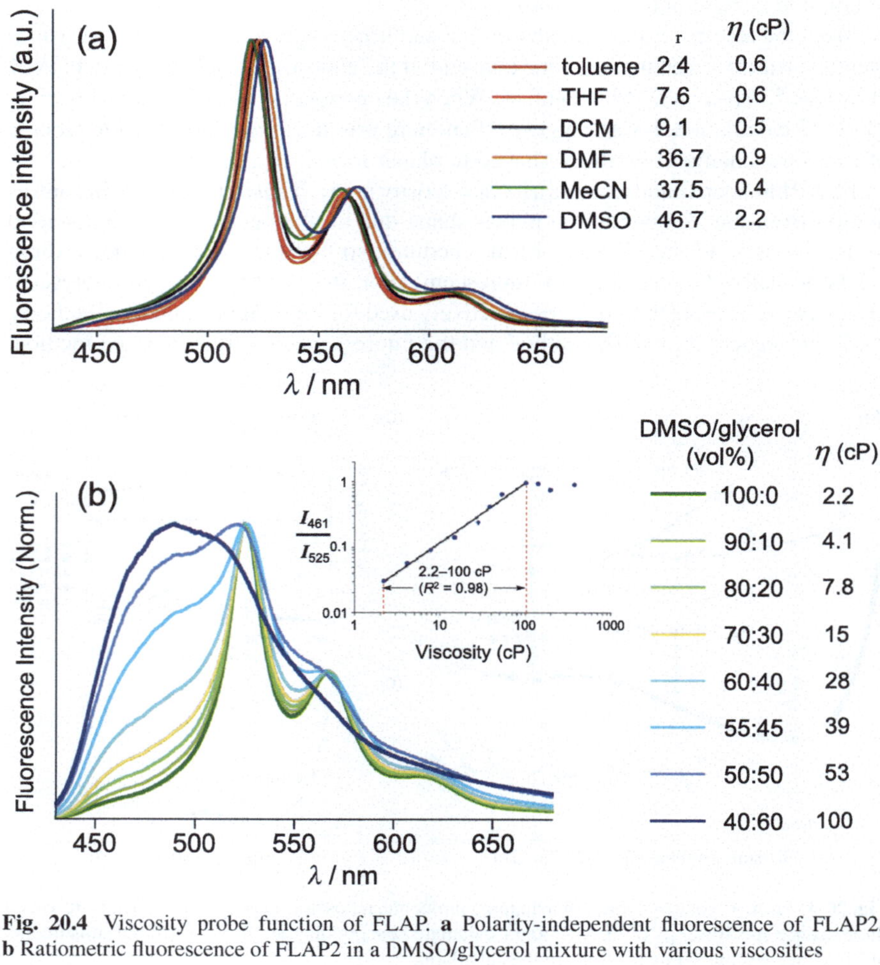

Fig. 20.4 Viscosity probe function of FLAP. **a** Polarity-independent fluorescence of FLAP2. **b** Ratiometric fluorescence of FLAP2 in a DMSO/glycerol mixture with various viscosities

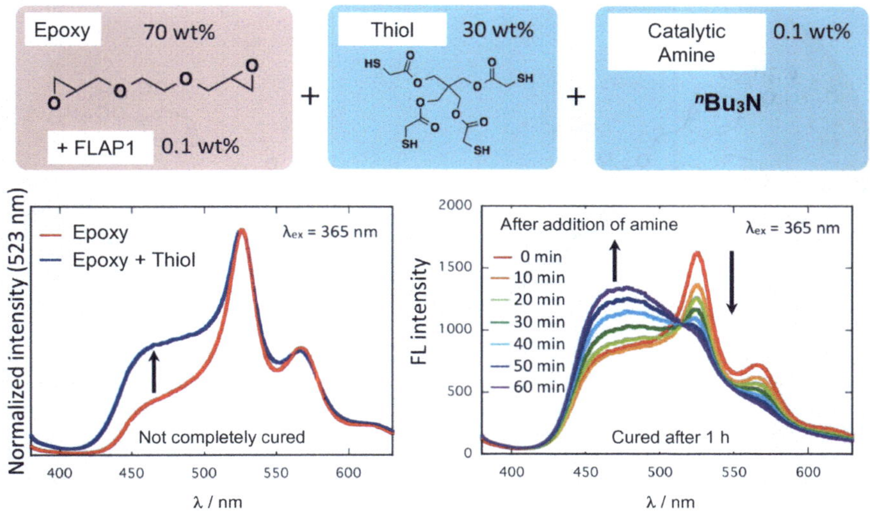

Fig. 20.5 Curing process of an epoxy resin monitored by a FLAP viscosity probe

law [15, 16]. When FLAP is used as a local viscosity probe, the resin curing process can be visualized (Fig. 20.5). Indeed, the process until the adhesive cures at room temperature can be visualized on the spot by adding a small amount of FLAP2 to commercially available transparent adhesives and industrial epoxy resins, in which FLAP2 has bulky substituents to suppress self-aggregation. In addition, it was shown that an area where the curing is not sufficient can be specified without contact.

20.3 Flapping Liquid Crystal as a Light-Melt Adhesive

Controlling the properties of matter with light is an important technology already used in society. For example, photo-curing resins that harden when exposed to light have been used industrially for a wide range of applications such as bonding, sealing, and coating since the 1960s. In contrast, substances that melt when exposed to light have been used in recording materials, and in recent years, they are expected to be used as adhesive materials for temporary fixing that can be separated by light [17, 18].

However, the development of highly functional materials that satisfy the following difficult requirements was necessary to use photoresponsive substances as "adhesive materials that can be melt by light."

(1) Sufficiently high adhesive strength even at high temperature
(2) Adhesive strength greatly reduced by light irradiation
(3) Rapid melting with small energy of light

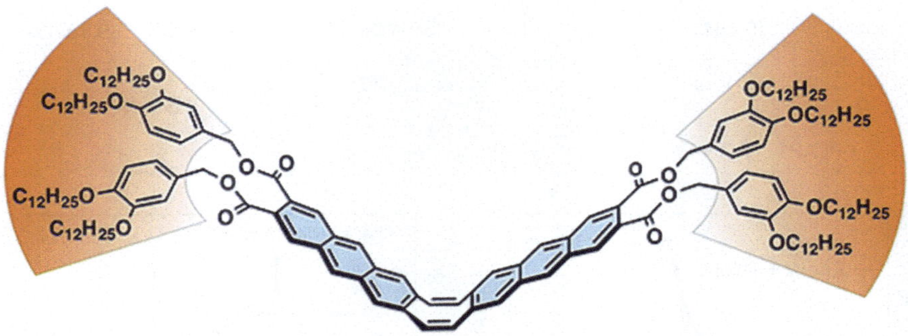

Fig. 20.6 Chemical structure of the light-melt adhesive. The orange moieties bring out the liquid crystal phase of the compound, the blue part causes photoreactions, and the central bent unit allows the molecules to move when exposed to light. The V-shaped molecular structure leads to a strong self-interaction (see Fig. 20.3) and realizes high cohesive force of the material

The development of new adhesive materials that are compatible with both "light-melting function" and "heat-resistant bonding function" is expected, which enables various manufacturing processes where conventional hot-melt adhesive materials cannot be used.

We designed and synthesized a unique molecule that responds to ultraviolet light, and based on this molecule, we developed a columnar liquid crystal material with high cohesive force. As a result, we developed a new functional material that meets all the above-mentioned difficult requirements and named it "Light-Melt Adhesive" (Fig. 20.6) [19].

The light-melt adhesive material was sandwiched between two glass plates to evaluate the adhesive performance. (1) It showed high adhesive strength of 1.6 MPa at room temperature and 1.2 MPa even at a high temperature of 100 °C. When UV light is applied, (2) the adhesive strength decreases by 85% with liquefaction and (3) it can be separated in just a few seconds (small light energy of 320 mJ/cm^2) (Fig. 20.7). In addition, (4) it has a reworkable property that regains its adhesive strength by heat treatment at 160 °C, and (5) it has a fluorescence function that can distinguish between adhesive and non-adhesive states by the difference in fluorescence color. All of these material functions are derived from the molecular design. The following descriptions explain the features of the molecular structure that led to the material functions.

In general, to achieve high adhesive strength, it is necessary to strengthen both the adhesion force to a surface of substrates as well as the cohesive force of the adhesive itself (Fig. 20.8a). If the adhesion force is weak, a specimen of substrates will be separated at the interface, and if the cohesive strength of the adhesive is weak, the specimen will break inside the adhesive material. In the sample specimen using glass substrates and an adhesive material, shear strength did not change regardless of the glass surface condition (hydrophilic or hydrophobic) (Fig. 20.8b). This result showed that the cohesive force of the adhesive itself determines the adhesive strength of the specimen, not the interaction at the glass/adhesive interface. In other words, the

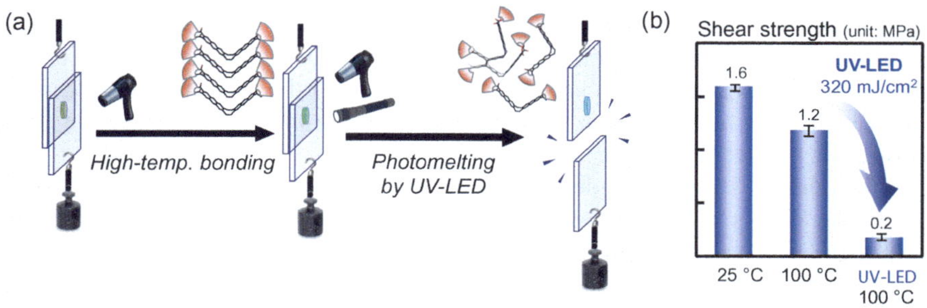

Fig. 20.7 Performance of the light-melt adhesive. The light-melt adhesive can work at 100°C, while hot-melt adhesives cannot be used at high temperature. On the other hand, it can be removed quickly by warming to the temperature range of the liquid crystalline phase and ap-plying ultraviolet light

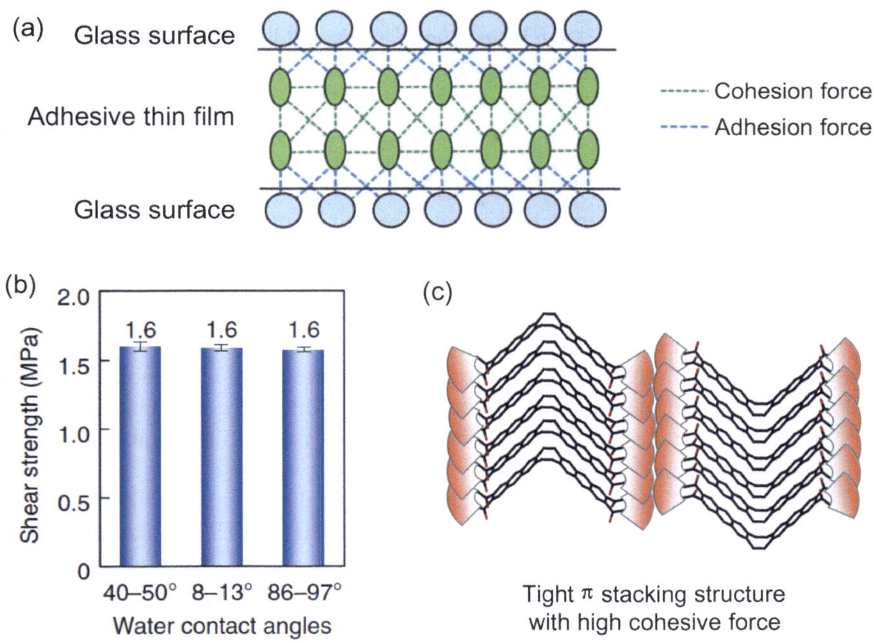

Fig. 20.8 Columnar liquid crystal structure is the key to high shear strength. **a** The adhesive strength is determined by adhesion force (the interaction between the adhesive/substrate interface) and cohesive force (the self-interacting force of the adhesive material). **b** Since the shear strength did not change regardless of the hydrophilicity of the glass substrate surface, it was not the adhesion force but the cohesive force that determined the shear strength of the sample specimen. **c** The V-shaped molecular skeleton strongly interacts with each other to form a columnar liquid crystal structure, thereby generating a high cohesive force

development of materials with high self-interacting property led to the realization of sufficient adhesive strength for temporary fixation, because the adhesion force on the glass substrates was stronger than the cohesive force of the molecular aggregation. The reason of the high adhesion force between the light-melt adhesive and the glass substrate is not yet elucidated.

The high cohesive force is derived from the V-shaped molecular structure that is easy to stack. This V-shaped molecular skeleton interacts strongly with each other to form an ordered structure with columnar π stacking (Fig. 20.8c). By taking advantage of the stacking nature of this molecular framework, we have developed a columnar liquid crystal material with high cohesive force. Generally speaking, liquid crystal is thought of as a material with high fluidity that is used in displays, but the developed columnar liquid crystal material has low fluidity due to strong intermolecular interactions, and the two glass plates are firmly fixed. In addition, heat-resistant adhesion has been achieved by adjusting the temperature indicating the columnar liquid crystal phase to a high temperature range by molecular design.

Some polymer materials have been reported to lose adhesion by various mechanisms when exposed to light. This includes dicing tape that, when exposed to ultraviolet light, molecules are polymerized like a network and the material is cured to induce peeling [20]. On the other hand, liquid crystal materials composed of small molecules are known to show a phenomenon in which the ordered structure of molecules is spontaneously collapsed and liquefied by mixing a few impurities with different molecular shapes. In particular, by using liquid crystal materials based on molecules that change shape when exposed to light such as azobenzene derivatives, molecules that have different shapes (impurities) can be made inside the liquid crystal by light irradiation, and materials that become liquid by light can be created [21, 22]. The functional properties of these functional liquid crystals that change the phase of substances with light have attracted attention for use in memory materials that record information with light. On the other hand, due to the soft nature of liquid crystals, development as an adhesive material has not received much attention until recently [23, 24].

The light-melt adhesive has a columnar liquid crystal structure in which V-shaped molecules are stacked, maintaining high cohesive force (Fig. 20.9a). When ultraviolet light is applied in the temperature range (70–135 °C) with the liquid crystal phase, the V-shaped molecule changes its conformation to a planar form in the lowest singlet excited state, and then those that become close to the next molecule dimerize (Fig. 20.9b). Molecular shape of the dimers generated in this way are unsuitable for ordered stacking, so they act as impurities and destroy the V-shaped molecular packing structure (Fig. 20.9c). As a result, the columnar liquid crystal with strong adhesive strength collapses, and the adhesive strength of the fluid mixture is greatly reduced.

To use a light-melting material as a temporary fixing adhesive in manufacturing processes, it must be removed immediately using a general light irradiation device. For that purpose, it is desirable that photomelting occurs with small energy of light. When light-melt adhesive is sandwiched between two glass plates and irradiated with

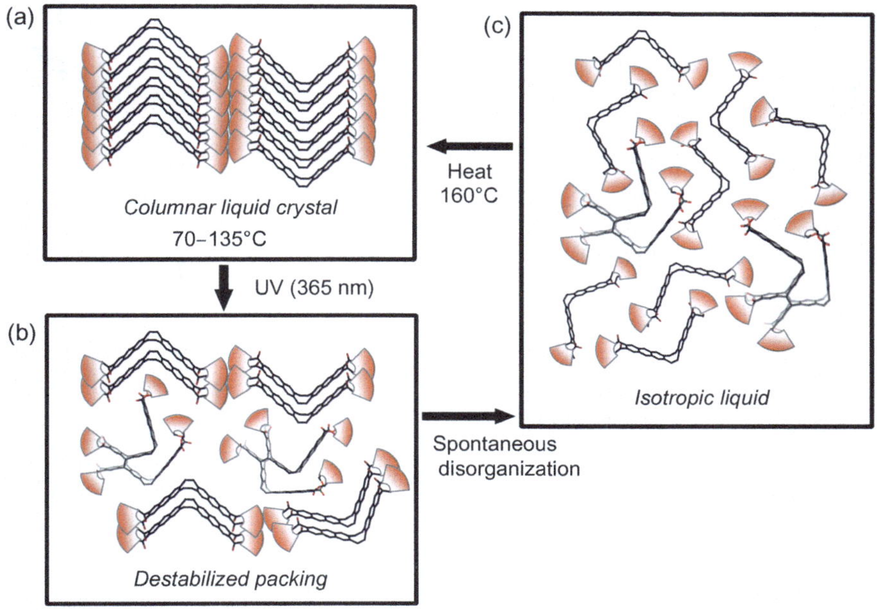

Fig. 20.9 Photomelting mechanism of the light-melt adhesive. Formation of a dimer by photoreaction triggers liquefaction. **a** Before irradiation with light, V-shaped molecules are stacked to form an ordered columnar liquid crystal structure, which maintains a high self-aggregation force. **b** When irradiated with ultraviolet light, some molecules react to form a dimer and work as impurities. **c** The stacked structure of the V-shaped molecules is broken by the formation of impurity dimers, and the adhesive strength is greatly reduced by liquefaction as a mixture

ultraviolet light while warmed by a dryer, it can be melted in just a few seconds with 320 mJ/cm^2 light energy [19].

The following facts are used to achieve this rapid separation. (a) The reaction that the molecule becomes a dimer with light is fast, (b) Even if not all the molecules react, the dimer acts as an impurity and the liquid crystal structure is spontaneously broken and liquefaction proceeds, (c) When a slight interface of the material in contact with the glass melts, separation of the two glass substrates occur. In fact, it was confirmed that the depth at which ultraviolet light reaches the inside of the film of light-melt adhesive is only a few micrometers from the interface of the irradiated glass, and it was separated with the same light energy regardless of the thickness of the film (Fig. 20.10). Through these experiments, it was shown that light-melt adhesive can achieve photo-detachment even with a very small amount of use, and only a small amount of the adhesive remains on the substrate after exposure to light.

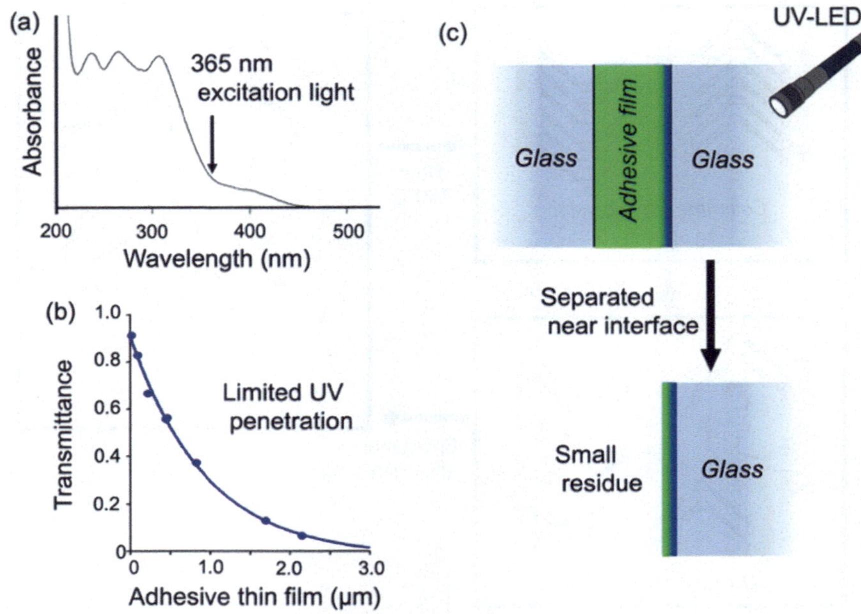

Fig. 20.10 Light absorption and liquefaction near the interface enabling rapid separation. Since the thin film of the light-melt adhesive absorbs ultraviolet light (**a**), the UV light cannot reach deep inside the thin film (**b**), and the thin film melts only in the vicinity of the glass/adhesive interface to which the light is irradiated

20.4 Conclusion

Flapping molecules have emerged as a promising photofunctional systems that can be used as a viscosity probe and a light-melt adhesive. Molecular design guideline based on the rigid/flexible hybridization is universal, and functions of FLAP are still expanding to singlet fission chromophores [25] and flexible mechanophores [26]. FLAP has a potential to change conventional science and technology in materials science.

Acknowledgements we deeply appreciate to Professors H. Sotome, H. Miyasaka, H. Okajima, A. Sakamoto for their collaboration. The present work was supported by JSPS KAKENHI Grant Number JP18H01952, Grant-in-Aid for Scientific Research on Innovative Areas "Photosynergetics" Grant Number JP17H05258, and JST, PRESTO, Grant Number JPMJPR16P6 to S. S.

References

1. Imahori H, Tkachenko NV, Vehmanen V, Tamaki K, Lemmetyinen H, Sakata Y, Fukuzumi S (2001) An extremely small reorganization energy of electron transfer in porphyrin – fullerene dyad. J Phys Chem A 105:1750–1756
2. Kowaka Y, Suganuma Y, Ashizawa N, Nakayama N, Goto H, Ishimoto T, Nagashima U, Baba M (2010) Ultrahigh-resolution laser spectroscopy of the $S_1\ ^1B_{2u} \leftarrow S_0\ ^1A_g$ transition of perylene. J Mol Spectrosc 260:72–76
3. Mori H, Itoh Y, Nishiura Y, Nakamura T, Shinagawa Y (1997) Performance of a novel optical compensation film based on negative birefringence of discotic compound for wide-viewing-angle twisted-nematic liquid-crystal displays. Jpn J Appl Phys 36:143–147
4. Saito S (2019) Flapping molecules for photofunctional materials. In: Yamamoto H, Kato T (eds) Molecular technology: materials innovation. Wiley-VCH, Weinheim
5. Saito S (2019) Rigid-flexible hybrid design for photofunctional molecules and materials. In: Yagai S, Yamada H (eds) Light-active functional organic materials. Jenny Stanford Publishing, Singapore
6. Yuan C, Saito S, Camacho C, Irle S, Hisaki I, Yamaguchi S (2013) A π-conjugated system with flexibility and rigidity that shows environment-dependent rgb luminescence. J Am Chem Soc 135:8842–8845
7. Yuan C, Saito S, Camacho C, Kowalczyk T, Irle S, Yamaguchi S (2014) Hybridization of a flexible cyclooctatetraene core and rigid aceneimide wings for multiluminescent flapping π systems. Chem Eur J 20:2193–2200
8. Haidekker MA, Theodorakis EA (2016) Ratiometric mechanosensitive fluorescent dyes: design and applications. J Mater Chem C 4:2707–2718
9. Kuimova MK (2012) Mapping viscosity in cells using molecular rotors. Phys Chem Chem Phys 14:12671–12686
10. Haidekker MA, Theodorakis EA (2007) Molecular rotors—fluorescent biosensors for viscosity and flow. Org Biomol Chem 5:1669–1678
11. Klymchenko AS (2017) Solvatochromic and fluorogenic dyes as environment-sensitive probes: design and biological applications. Acc Chem Res 50:366–375
12. Klymchenko AS, Kreder R (2014) Fluorescent probes for lipid rafts: from model membranes to living cells. Chem Biol 21:97–113
13. Yang Z, Cao J, He Y, Yang JH, Kim T, Peng X, Kim JS (2014) Macro-/micro-environment-sensitive chemosensing and biological imaging. Chem Soc Rev 43:4563–4601
14. Kowada T, Maeda H, Kikuchi K (2015) BODIPY-based probes for the fluorescence imaging of biomolecules in living cells. Chem Soc Rev 44:4953–4972
15. Kotani R, Sotome H, Okajima H, Yokoyama S, Nakaike Y, Kashiwagi A, Mori C, Nakada Y, Yamaguchi S, Osuka A, Sakamoto A, Miyasaka H, Saito S (2017) Flapping viscosity probe that shows polarity-independent ratiometric fluorescence. J Mater Chem C 5:5248–5256
16. Nakanishi W, Saito S, Sakamoto N, Kashiwagi A, Yamaguchi S, Sakai H, Ariga K (2019) Monitoring fluorescence response of amphiphilic flapping molecules in compressed monolayers at the air-water interface. Chem Asian J 14:2869–2876
17. Hohl DK, Weder C (2019) (De)bonding on demand with optically switchable adhesives. Adv Opt Mater 7:1900230
18. Croll AB, Hosseini N, Bartlett MD (2019) Switchable adhesives for multifunctional interfaces. Adv Mater Technol 4:1900193
19. Saito S, Nobusue S, Tsuzaka E, Yuan C, Mori C, Hara M, Seki T, Camacho C, Irle S, Yamaguchi S (2016) Light-melt adhesive based on dynamic carbon frameworks in a columnar liquid-crystal phase. Nat Commun 7:12094
20. Ebe K, Seno H, Horigome K (2003) UV curable pressure-sensitive adhesives for fabricating semiconductors. I. Development of easily peelable dicing tapes. J Appl Polym Sci 90:436–441
21. Ikeda T (2003) Photomodulation of liquid crystal orientations for photonic applications. J Mater Chem 13:2037–2057

22. Norikane Y, Hirai Y, Yoshida M (2010) Photoinduced isothermal phase transitions of liquid-crystalline macrocyclic azobenzenes. Chem Commun 47:1770–1772
23. Akiyama H, Yoshida M (2012) Photochemically reversible liquefaction and solidification of single compounds based on a sugar alcohol scaffold with multi azo-arms. Adv Mat 24:2353–2356
24. Ito S, Akiyama H, Sekizawa R, Mori M, Yoshida M, Kihara H (2018) Light-Induced reworkable adhesives based on aba-type triblock copolymers with azopolymer termini. ACS Appl Mater Inter 10:32649–32658
25. Yamakado T, Takahashi S, Watanabe K, Matsumoto Y, Osuka A, Saito S (2018) Conformational planarization versus singlet fission: distinct excited-state dynamics of cyclooctatetraene-fused acene dimers. Angew Chemie Int Ed 57:5438–5443
26. Yamakado T, Otsubo K, Osuka A, Saito S (2018) Compression of a flapping mechanophore accompanied by thermal void collapse in a crystalline phase. J Am Chem Soc 140:6245–6248

Chapter 21
Giant Amplification of Fluorescence Quenching in Photochromic Nanoparticles and Crystals

Tuyoshi Fukaminato, Sanae Ishida, Jia Su, Keitaro Nakatani, and Rémi Métivier

Abstract Fluorescence photoswitching properties of novel fluorescent photochromic diarylethene (DAE)-benzothiadiazole (BTD) dyads were studied in a solution, in the nanoparticle state, and in the single-crystalline state. The nanoparticles represent a state-of-the-art system, showing bright red emission, reversible fluorescence photoswitching upon UV-visible irradiation, complete ON-OFF contrast, excellent photostability and fatigue resistance. Most interestingly, upon UV irradiation, the nanoparticles exhibit a complete fluorescence quenching even at very low conversion (<5%) of the photochromic unit. This "giant amplification of fluorescence quenching" originates from a long-range intermolecular Förster resonance energy transfer (FRET) within each nanoparticle, leading to the quenching of ~400 fluorescent molecules for only one converted photochromic molecule. Furthermore, similar efficient fluorescence photoswitching was observed even in the single crystal of a DAE-BTD dyad.

Keywords Fluorescence · Diarylethene · Nanoparticles · Single crystals · Energy transfer · Giant amplification

21.1 Introduction

Molecules which can reversibly change in response to external stimuli are of central importance to the development of molecular devices and memory materials, but also pharmacology and imaging technologies [1–5]. Organic photochromic compounds [6, 7], which reversibly change color under excitation with an appropriate wavelength of light, are good examples of such molecular systems. They present two stable, distinct and addressable states. The difference in geometry and electronic structure

T. Fukaminato (✉) · S. Ishida
Graduate School of Science and Technology, Department of Applied Chemistry and Biochemistry, Kumamoto University, Kurokami, Kumamoto 860-8555, Japan
e-mail: tuyoshi@kumamoto-u.ac.jp

J. Su · K. Nakatani · R. Métivier
PPSM, ENS Paris-Saclay, CNRS, Université Paris-Saclay, Gif-sur-Yvette 91190, France

H. Miyasaka et al. (eds.), *Photosynergetic Responses in Molecules and Molecular Aggregates*, https://doi.org/10.1007/978-981-15-5451-3_21

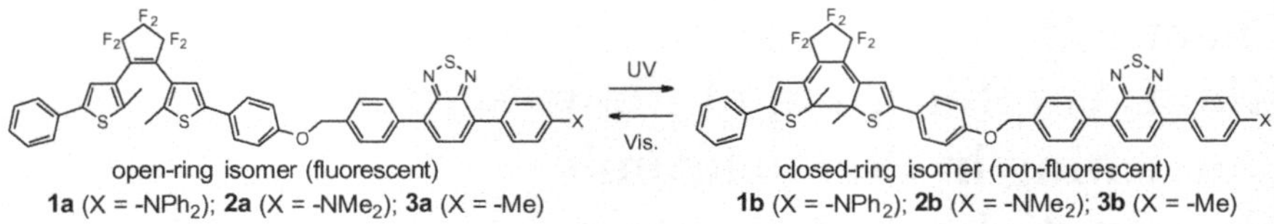

Fig. 21.1 Molecular structures and photochromism of DAE-BTD dyad **1**, **2** and **3**

between the two isomers can be exploited to reversibly control a wide variety of properties. In particular, photoswitching of fluorescence signals has attracted much attention because of its potential in various optoelectronic applications including optical memories, bio-imaging, and photoswitches with high-sensitivity [8–10].

A typical molecular design for fluorescence photoswitchable molecules is based on the combination of fluorescent and photochromic moieties, the latter operating as a switch to control the emission of the former via an energy or electron transfer [11, 12]. In this research field, triggering property changes of a large number of molecules by only a few photons is among the wildest dreams of photochemists and photophysicists, which would represent a significant energy and time saving added value in the field of molecular fluorescent photoswitches [1, 11]. An approach to achieve the above-mentioned dream is to take advantage of intermolecular Förster Resonant Energy Transfer (FRET) [13], which enables a single acceptor to quench the fluorescence of multiple donor fluorophores [14–17]. A wide range of architectures is introduced for this purpose, ranging from natural proteins to artificial molecular systems [18–20]. Especially, organic nanoparticle systems have attracted increased attention due to excellent optical properties, high brightness, easiness of preparation, and biocompatibility [21–24].

In this chapter, the fluorescence photoswitching properties of a diarylethene-benzothiadiazole (DAE-BTD) dyads **1–3** (Fig. 21.1) were explored in a solution, in a suspension of nanoparticles, and in a single-crystalline state [25–27].

21.2 Molecular Design

In order to achieve the efficient fluorescence photoswitching in a nanoparticle state, we designed and synthesized DAE-BTD dyad **1**. BTD derivatives show high brightness, long fluorescence lifetime, and large Stokes shift, providing an interesting red-shifted emission [28, 29]. In particular, they show desirable fluorescence efficiency in aggregated states with restricting solvation-induced quenching by self-assembling [30]. The choice of DAE as the switching component was motivated by its exceptional thermal stability and outstanding fatigue resistance, as well as its photoreactivity in many media, from solution to solid state [31, 32]. In DAE-BTD dyads, photochromic DAE moiety plays the role of trigger to control the emission of fluorophores via energy transfer. Figure 21.2 exhibits the spectral overlaps between absorption and

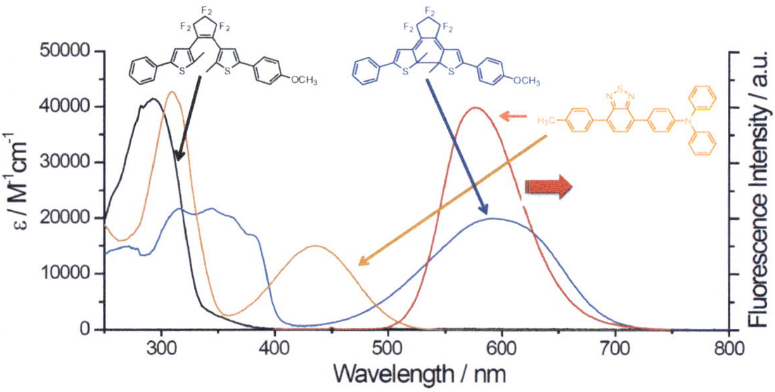

Fig. 21.2 Spectral overlap between each unit of DAE-BTD dyad **1**

fluorescence spectra of each component in DAE-BTD dyad **1**. The absorption band of DAE overlaps with the fluorescence spectrum of BTD in the closed-ring form, while the absorption band of DAE has no overlap with the fluorescence spectrum of BTD in the open-ring form. From these spectra, in DAE-BTD dyad **1**, it is expected that only closed-ring form of DAE works as fluorescence quencher for BTD and therefore the fluorescence intensity can be modulated along with the photochromic reactions of DAE moiety. Based on the spectral properties of DAE and BTD units, Förster distance (R_0) can be calculated to be 6.8 nm.

The introduction of a spacer unit was aimed to break the conjugation and to keep electronic independence of DAE and BTD moieties. In the followings, we will focus on the description of nanoparticles preparations and some examples of organic systems exhibiting fluorescence in the nanoparticle state.

21.3 Fluorescence Photoswitching in Solution

The initial solution of dyad **1a** displayed a pale yellow color originated from the absorption band of BTD moiety. The absorption spectrum of **1a** exhibits two bands at 306 and 437 nm (Fig. 21.3a). Upon irradiation with UV (313 nm) light the solution color changes to blue and a characteristic absorption band located in the visible region appeared, which is results of generation of **1b**. The absorption band at 594 nm grows along with UV irradiation until photostationary state (PSS) and the conversion yield from **1a** to **1b** under irradiation with 313 nm light was estimated to be 93% (determined by UV–Vis absorption and HPLC measurements). Upon subsequence irradiation with visible light at >520 nm, the solution color completely returned to the initial state.

The photochromic reaction induces not only the color change of the solution but also dramatical modulation of the fluorescence intensity (Fig. 21.3b). Dyad **1a**

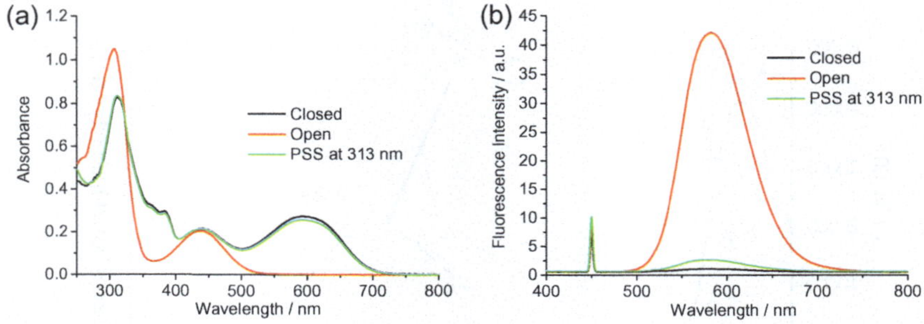

Fig. 21.3 **a** Absorption and **b** fluorescence spectra of DAE-BTD dyad **1a** (red), **1b** (black), and the photostationary state (green) under irradiation with 313 nm light in THF solution

exhibits bright fluorescence in the region between 550 nm and 800 nm (λ_{max} = 620 nm, $\Phi_F = 0.67$) with a large Stokes shift 6754.3 cm^{-1}. The fluorescence decay profile of **1a** at 620 nm shows a single-exponential decay, giving a τ_f value of 7.8 ns, which is slight shorter than BTD in THF solution (8.3 ns). Upon irradiation with 313 nm light, the fluorescence signal dropped down to 8% of initial intensity at PSS. The contrast in the signal is consistent with the population of **1b** at PSS. Indeed, the pure **1b** separated by HPLC exhibits no fluorescence.

The apparent photochromic quantum yield of the cyclization reaction (from **1a** to **1b**) in THF solution was determined to be $\Phi_{app,\mathbf{1a}\rightarrow\mathbf{1b}} = 0.18$. Since only parts of the photons are absorbed by the photochromic DAE subunit in the UV region, this value leads to a corrected photochromic quantum yield ($\Phi_{\mathbf{1a}\rightarrow\mathbf{1b}}$) of 0.46. This value is comparable to the DAE subunit alone ($\Phi_{\mathbf{a}\rightarrow\mathbf{b}} = 0.55$ in CH_3CN [33]). The measured cycloreversion quantum yield $\Phi_{\mathbf{1b}\rightarrow\mathbf{1a}}$ is 5×10^{-3}.

DAE-BTD dyad **1a** exhibits the desirable fluorescence photoswitching properties, such as thermal and photochemical stability, good fatigue resistance and highly emissive property. The mutual effect between DAE and BTD results in an efficient fluorescence modulation of the dyad, which can be reversibly achieved by alternating UV and visible light irradiation. Florescence contrast between ON and OFF states can be maintained after many cycles. However, there are still some limitations of the performance of DAE-BTD dyad in solution. Since the conversion yield at PSS is not 100%, with a residue of fluorescence signal at the PSS which limits fluorescence ON/OFF contrast. Relatively long photoswitching time is also undesirable for further application. In order to overcome those problems, a new strategy of designing nano-switches based on DAE-BTD molecular photoswitching taking advantage of intermolecular FRET is discussed in the following part.

21.4 Fluorescence Photoswitching in Nanoparticle

The nanoparticles of DAE-BTD dyad **1a** was prepared by conventional reprecipitation method [34] and providing a suspension of nanoparticles with size tunability. In such nanoparticles, a high chromophore density is reachable, compared to other types of molecular assemblies or materials, providing a way to maximize the intermolecular FRET efficiency.

Dyad **1a** was well dissolved into THF to obtain a 5.0×10^{-5} M mother solution. Then 0.6 mL of this solution was quickly added into 2.4 mL of distilled water under vigorous stirring. After stirring for 5 min., final transparent suspension of nanoparticles was obtained with a concentration of 1×10^{-5} M. Characterizations of nanoparticles were performed by field-effect scanning electron microscopy (FESEM) and atomic force microscopy (AFM) (Fig. 21.4). Round spherical nanoparticles were observed in SEM and AFM images with diameters ranging from 10 to 60 nm. Averaged diameter of nanoparticles are 25 nm, and 80% of the nanoparticles have a diameter between 15 and 35 nm. Furthermore, based on a molecular weight (1004.14 g/mol) and density (1.4 g/cm^3), one can estimate that a 25 nm nanoparticle contains about 7000 DAE-BTD molecules.

Spectral properties of dyad **1** nanoparticles in a suspension of THF/H_2O mixtures (Fig. 21.5) were almost similar to those in THF solution (Fig. 21.3). Compared to the absorption spectra in THF solution, a red-shift of 4–10 nm was observed for a suspension of nanoparticles. Upon UV light (313 nm) irradiation to a suspension of **1a** nanoparticles, typical absorption spectral change attribute to the photocyclization reaction of DAE moiety and 73% of **1a** molecules converted to **1b** at PSS, which proportion is lower than in THF solution (97%). The photogenerated **1b** nanoparticles can be fully converted back to the initial state upon irradiation with visible (> 520 nm) light.

1a nanoparticles show strong orange fluorescence when they are excited with 450 nm light ($\lambda_{f,max} = 620$ nm, $\Phi_F = 0.65$). The brightness of **1a** NPs is estimated

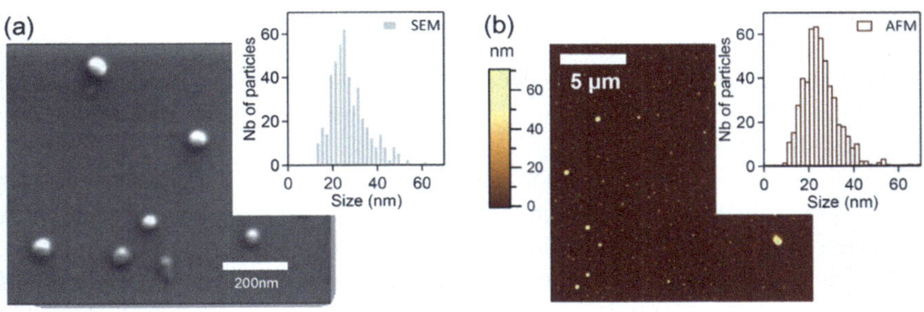

Fig. 21.4 **a** SEM image of a sample of **1a** nanoparticles in a suspension of THF/H_2O (1:4) mixture; Inset: size histogram of **1a** nanoparticles by SEM measurements; **b** AFM image of a sample of **1a** nanoparticles in a suspension of THF/H_2O (1:4) mixture; Inset: size histogram of **1a** nanoparticles by AFM measurements

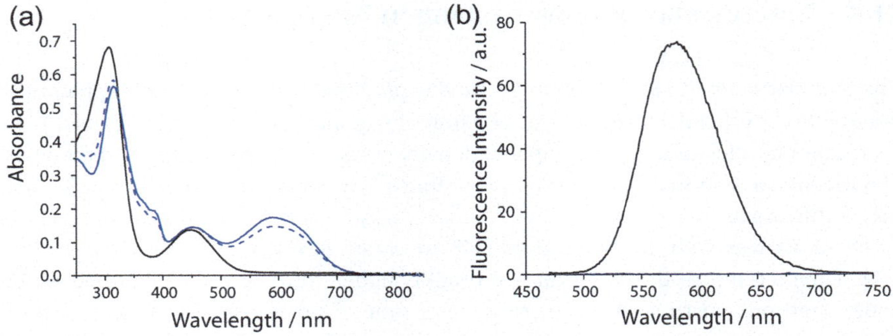

Fig. 21.5 **a** Absorption and **b** fluorescence spectra of DAE-BTD dyad **1a** (black solid-line), **1b** (blue solid-line), and the photostationary state (blue broken-line) under irradiation with 313 nm light in a suspension of nanoparticles

to be 7×10^7 L/(mol•cm) at 450 nm that is more than 100 times brighter than most common quantum dots [35]. After 313 nm light irradiation, fluorescence is dramatically quenched. Dyad **1** nanoparticles at the PSS are non-fluorescent with a 100% fluorescence quenching ratio. A contrast in the fluorescence signal of about 4000-fold was found between pure **1a** nanoparticles and PSS, which was much larger than that in THF solution, although conversion yields of nanoparticles (73%) under irradiation with 313 nm light was lower than in THF solution (97%). The corresponding Förster radius R_0 was calculated to be 4.9 nm in nanoparticles, taking $\kappa^2 = 0.476$ for fixed molecules with random orientations [13]. The fluorescence decay curve of **1a** nanoparticles in a suspension shows a clear deviation from mono-exponential behavior. Proper fitting of the data provided two decay times: a long time-constant (τ_1) 10.1 ns associated with a predominant pre-exponential factor (a_1) 0.79, which represents the major part of the fraction of intensity (f_1) 0.94, and a minor proportion of a shorter time-constant (τ_2) 2.5 ns ($a_2 = 0.21, f_2 = 0.06$). This bi-exponential decay can be tentatively interpreted as the existence of two populations of fluorescent molecules. **1a** molecules included in the core of nanoparticles represent the major part of the fluorescence, and as they are isolated from the solvent medium, contribute to the long fluorescence decay. The population of **1a** molecules located at the surface of nanoparticles can interact with the water molecules, leading to rare fraction of shorter decay.

Fatigue resistance of fluorescence photoswitching in the nanoparticle state was measured to evaluate the reversibility of their color and fluorescence modulation. A suspension containing **1a** nanoparticles was irradiated with 313 nm UV light (2.8×10^{-4} W/cm^2, 10 s) turning to blue color. Fluorescence was quenched in a few seconds of irradiation, although it did not reach a high photochromic conversion. Then the sample was completely bleached by irradiation with 575 nm visible light (3.2×10^{-3} W/cm^2, 1000 s). The steps were alternated many times with good reversibility. No decrease in the maximum fluorescence intensity of **1a** was observed during these cycles.

Comparing to THF solution, the nanoparticle exhibits many advantages. Most interestingly, the fluorescence quenching ratio was 100% with excellent reversibility. When fluorescence intensity at the maximum wavelength of the dyad is plotted as a function of the conversion yield between **1a** to PSS, a linear dependence with a one-to-one ratio is observed for dyad **1** in THF solution (Fig. 21.6a). In solution, this phenomenon is well-expressed by a simple intramolecular FRET process: at any irradiation time, the fluorescence signal is proportional to the residual amount of fluorescent **1a** molecules (vide supra). The minimum level of fluorescence depends on the PSS composition and cannot reach zero.

The same experiment was performed in the nanoparticle state. The correlation plot of the normalized fluorescence intensity versus (*vs.*) the conversion yield exhibits a strong deviation from linearity (Fig. 21.6c). The initial fluorescence of **1a** nanoparticles decreases dramatically at very low conversion yield. More than 90% of the whole fluorescence is quenched for only 1% of **1b**, and the fluorescence can be considered to reach almost zero for only 5% of **1b**. This behavior is reversible under 575 nm visible light irradiation.

The fluorescence quenching of **1a** molecules in THF solution and in a suspension of nanoparticles induced by UV light irradiation can be visualized on a series of cuvettes pictures (Fig. 21.6b and d). The first series of cuvettes (Fig. 21.6b) contain dyad **1** molecules in THF solution whereas the second series contain nanoparticles

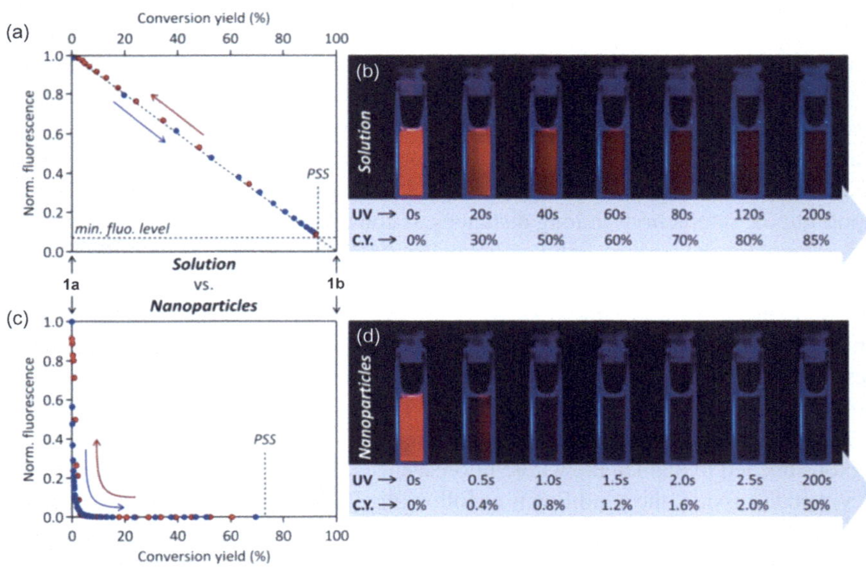

Fig. 21.6 Fluorescence intensity versus conversion yield (C.Y.) correlation plots under increasing UV (blue dots) and visible (red dots) exposure times (**a** and **c**) and photographs of sample cuvettes (**b** and **d**) for DAE-BTD dyad **1**, (**a** and **b**) in solution (2×10^{-6} M in THF) and (**c** and **d**) in a suspension of nanoparticles (1×10^{-5} M in THF/H_2O (1:4)). (From [25] Copyright © 2016 by John Wiley Sons, Inc. Reprinted by permission of John Wiley & Sons, Inc.)

in a suspension of THF/H_2O mixture (Fig. 21.6d) with the same molar concentration of **1a** (1.0×10^{-5} mol/L), respectively. Fluorescence was imaged in the dark under continuous excitation with 436 nm light. Owing to similar fluorescence quantum yield and emission color of **1a** in THF solution and in the nanoparticle state, the initial picture of both series of cuvettes shows almost same bright orange color. Fluorescence in THF solution was quenched gradually and reached a dim state at PSS. In the contrary, the nanoparticles suspension rapidly switched off to a dark state by UV light within less than 2.5 s. This observation suggests significantly different mechanisms of energy transfer processes of dyad **1** molecules in THF solution and in the nanoparticle state.

Such a fascinating "giant amplification effect" can be ascribed by a very efficient intermolecular FRET process within the nanoparticles. The packing of dyad **1** molecules being quite dense in the solid matrix of nanoparticles, a single **1b** molecule can actually play the role of energy acceptor for many neighboring **1a** molecules located within the Förster radius. The intermolecular distance of dyad **1** in diluted THF solution is too large to allow energy transfer between two dyad **1** molecules. For example, in a dyad **1** solution (10^{-6} M), the average distance between two molecules is around 100 nm, which is much larger than its R_0 (6.8 nm). However, in nanoparticles, the molecular distance of dyad **1** molecules in the same nanoparticles is short enough to allow efficient intermolecular energy transfer.

To quantify this giant quenching effect in nanoparticles, simplified estimations have been made in two different approaches. The fluorescence intensity *versus* conversion yield profile in nanoparticles was calculated using straightforward assumptions, considering spherical nanoparticles composed of punctual and evenly space-distributed dyad **1** molecules. This simplified model assumes that a given **1a** molecule is not quenched when the neighboring dyad **1** molecules located at a distance shorter than R_0 are all in the **1a** state, and fully quenched in all other cases. The principle of "FRET-induced giant amplification effect," based on intermolecular dipole-dipole interactions at long distances, is considered to be the main photophysical process for the strong nonlinear behavior of the fluorescence quenching in dyad **1** nanoparticles. The number of quenched **1a** molecules per the population of **1b** unit can be deduced from the initial slope of the curve plotted in Fig. 21.6d, providing a huge amplification factor: around 420 **1a** molecules are estimated experimentally (resp. 310 from the numerical simulation) to be quenched by intermolecular energy transfer to a single **1b** acceptor molecule.

Another way to figure out this amplification effects is to express it in terms of photons needed to quench the whole fluorescence of the nanoparticles. Based on the previous discussion, the condition is as following: (i) the typical size of nanoparticles is around 25 nm containing about 7000 dyad **1** molecules each; (ii) based on the experimental result, only 1% **1b** is enough to quench more than 90% of the whole fluorescence; (iii) each photon absorbed by DAE leads to the conversion to **1b** with an intrinsic quantum yield $\Phi_{\mathbf{1a}\rightarrow\mathbf{1b}} = 0.46$. Consequently, we can deduce that only 150 photons absorbed by the photochromic DAE species induce the quenching of 6300 fluorophores BTD. Then, we can deduce that 42 dyad **1** molecules are quenched per absorbed photon.

21.5 Efficient Fluorescence Photoswitching in Single Crystal

The amplification of fluorescence quenching is not limited to only in the nanoparticle system. Especially, the organic crystals in well-defined relative orientations and distances may provide an ideal scaffold for efficient intermolecular energy transfer compared to the randomly oriented amorphous nanoparticle systems. [36–38] In order to realize further efficient fluorescence photoswitching systems, we tried to make a crystal of dyad **1** under several recrystallization conditions. However, only amorphous solids was obtained for dyad **1**. It is well known that the triphenylamine group is bulky-substituent and it can prevent strong π-π interaction between chromophores. Therefore, the derivatives having this substituent show highly fluorescence property even in the condensed solid state. However, such bulkiness of this substituent is not suitable to prepare the regular packed molecular rearrangement in the crystalline state.

To overcome this issue, we tried to reduce the bulkiness by replacing the triphenylamine group to the dimethylamine group (DAE-BTD dyad **2**) and the methyl group (DAE-BTD dyad **3**) (Fig. 21.7). Consequently, a yellow plate-like crystal was obtained by recrystallization from dichloromethane/acetone mixture solution of dyad **3**.

Single crystal X-ray crystallographic analysis was successfully performed dyad **3a**. Crystal **3a** belongs to the triclinic *P* space group. The crystal structure of crystal **3a** in an asymmetric unit and the molecular packing diagram are shown in Fig. 21.8. Crystal **3a** has two molecules in a unit cell and one molecule in the asymmetric unit.

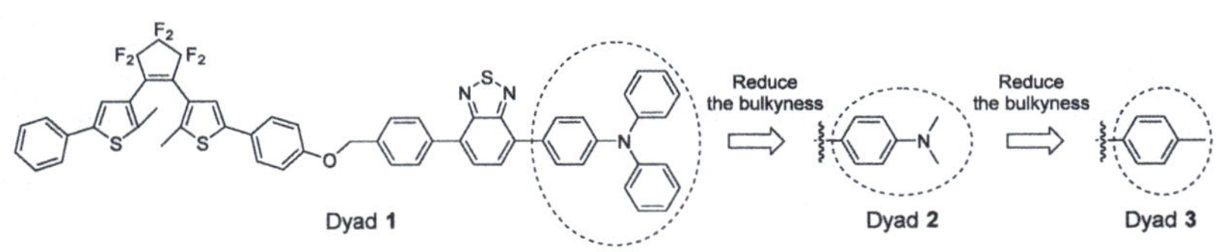

Fig. 21.7 Molecular design strategy to obtain the single crystal of DAE-BTD dyads

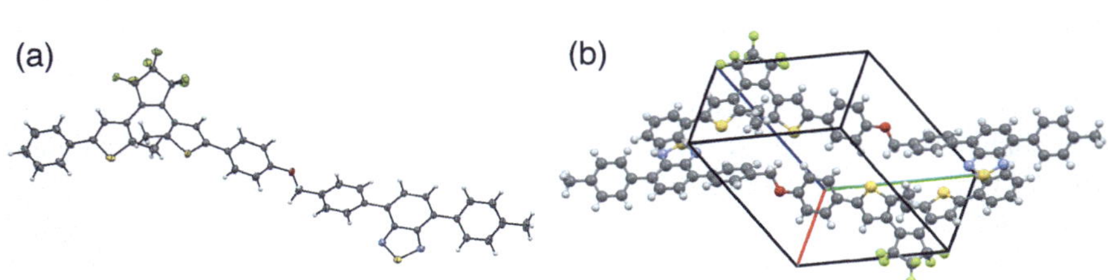

Fig. 21.8 **a** Molecular structure of dyad **3a** and **b** molecular packing. (Reproduced from Ref. [27] with permission from The Royal Society of Chemistry.)

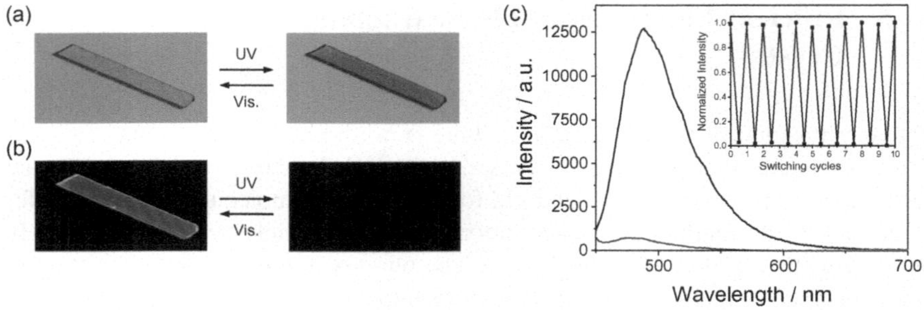

Fig. 21.9 Photographs of the single crystal of **3a**: **a** photo-induced color changes under irradiation with UV (λ = 365 nm) and visible (λ > 450 nm) light and **b** photo-induced fluorescence changes under dark-field observation. The excitation wavelength for the fluorescence detection was 438 nm light. **c** Fluorescence spectral changes of dyad **3** in single crystal state; **3a** (solid-black line) and PSS under irradiation with 365 nm light (solid-blue line). Inset: Peak fluorescence intensity of dyad **3** in the single crystal state over multiple cycles of UV (λ = 365 nm) and visible (λ > 450 nm) lights irradiation. (Reproduced from Ref.[27] with permission from The Royal Society of Chemistry.)

In the crystals, **3a** molecules are fixed with the photoreactive antiparallel conformation and the distance between the reactive carbon atoms is 3.54 Å. The conformation and the distance between the reactive carbon atoms fulfill the requirement for diarylethene molecules to undergo the photocyclization reaction in the single-crystalline phase [39]. The molecular packing arrangement in the crystal shows **3a** molecules alternately packed each other and the linear distances of the adjacent fluorescence units indicate negligible π–π interactions. This arrangement is especially suitable for the fluorescence property in the crystalline phase and efficient fluorescence quenching based on the intermolecular energy transfer from the fluorescent open-ring isomer to the photogenerated closed-ring isomer because the photogenerated quencher (closed-ring isomer) exists close by the fluorescence unit.

Figure 21.9a shows photographs of photo-induced color changes of the single crystal upon alternate irradiation with UV and visible light. Upon irradiation with UV (365 nm) light, the pale-yellow-colored crystal of **3a** turns green, indicating the formation of the blue-colored closed-ring isomer **3b**. This color was thermally stable in the dark and bleached upon irradiation with visible (λ >450 nm) light. Figure 21.9b shows photographs of fluorescence changes of the single crystal under dark-field observation. Initially, the single crystal of **3a** emits strong greenish fluorescence with maxima at around 488 nm under irradiation with 438 nm light. The fluorescence quantum yield of single crystal **3a** was measured to be 0.21. Upon irradiation with UV (365 nm) light, the fluorescence of the crystal suddenly disappeared. Upon irradiation with visible (>450 nm) light, the greenish fluorescence appeared again. Figure 21.9c shows the fluorescence spectral changes of the single crystal **3** along with photocyclization and photocycloreversion reactions. Typically, the photoconversion yield from the open- to the closed-ring isomer in the single-crystalline state is not high (less than 10%) due to the inner-filter effect and therefore the fluorescence on/off contrast is not good as reported previously [40]. However, the fluorescence

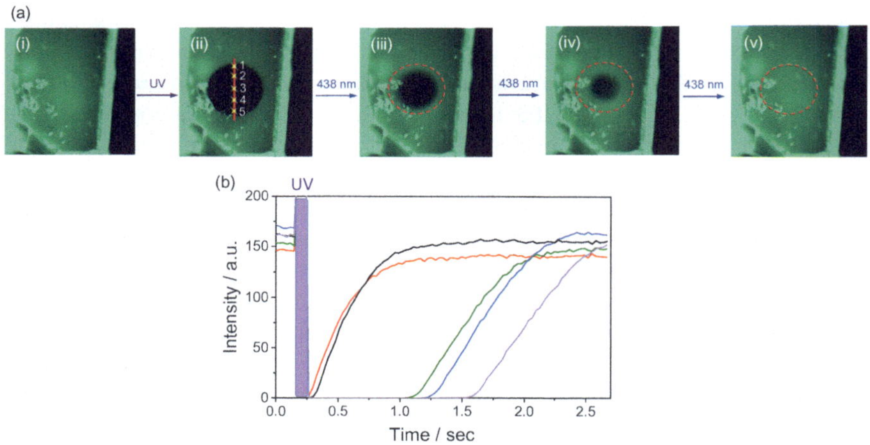

Fig. 21.10 **a** Fluorescence images under excitation with 438 nm light of single crystal **3a**; (i) before UV (390 nm) irradiation, (ii)–(v) fluorescence recovery under 438 nm excitation light after stopping UV irradiation. **b** Fluorescence intensity trajectory corresponding to points 1–5 marked in (a, ii); Point 1; black, Point 2; blue, Point 3; purple, Point 4; green and Point 5; red. (Reproduced from Ref. [27] with permission from The Royal Society of Chemistry.)

signal from single crystal **3a** significantly decreased to almost back ground level, as shown in Fig. 21.9c. Therefore, these results suggested that the efficient intermolecular energy transfer took place in the crystal. The crystals also exhibited fatigue resistance against repeated switching operations (inset of Fig. 21.9c).

Fluorescence photoswitching behavior in the single-crystalline state was further studied under the fluorescence microscope. Figure 21.10a shows fluorescence images of the single crystal of **3a** with area-selective UV light irradiation. The excitation wavelength for the fluorescence imaging was 438 nm. Upon irradiation with very weak UV (390 nm) light through an objective lens to a surface of the single crystal **3a** for 100 ms, the fluorescence signal of the irradiated area instantly decreased to the background level. Such high-contrast fluorescence photoswitching was suggested to the efficient fluorescence quenching of **3a** attributed to the intermolecular FRET process in the regularly oriented single crystals. In such oriented materials, it is anticipated that the long-range intermolecular FRET process takes place efficiently, and therefore a small number of photogenerated closed-ring isomers can quench the fluorescence of a large number of neighboring open-ring isomers. Furthermore, the fluorescence recovery behavior of fluorescence quenching area under irradiation with 438 nm excitation light also suggested that the contribution of nonlinear fluorescence quenching in the crystalline state. Under continuous irradiation with 438 nm excitation light, the size of dark area gradually decreased toward center with recovering the fluorescence signal (Fig. 21.10a, (ii)–(v)). Figure 21.10b shows the intensity trajectories of points 1–5 marked in the figure. The fluorescence intensity of outside region (points 1 and 5) in the irradiation area quickly recovered to original level after stopping UV light irradiation. On the other hand, the fluorescence intensity of the center (point 3) and the middle regions (points 2 and 4) maintained the dark level for a while even after stopping UV light irradiation and then abruptly started to recover the fluorescence intensity. In the typical linear-response fluorescence photoswitching systems, the fluorescence signal of dark area originated from the generation of non-fluorescent closed-ring isomers recovers uniformly under irradiation with visible light. Therefore, this result indicated that the other fluorescence quenching processes should be contributed in the single-crystalline state.

Such unexpected fluorescence recovery behavior in single crystal **3a** might be explained by the following mechanism. The light intensity through an objective lens has Gaussian distribution and such distribution generates small distribution of conversion yield in the irradiation area, where the conversion yield becomes high as the position closes to the center. Such small difference of conversion yield is almost negligible in typical linear fluorescence photoswitching system such as a dye-doped polymer film. However, in the nonlinear system, the fluorescence intensity suddenly and largely changes based on a certain conversion yield. The conversion yield sequentially reaches to the critical point from outside to center in the irradiation area because the conversion yield becomes high with closing to the center part. As a result, the fluorescence recovery behavior with decreasing the size of dark area was observed. Although further quantitative analysis will be necessary to discuss the mechanism more detail, from these fluorescence photoswitching behavior, there is no doubt that the efficient nonlinear fluorescence quenching takes place in the single crystal **3a**.

21.6 Conclusion

We described the giant amplification of fluorescence quenching in photochromic nanoparticles and a single crystal based on the efficient intermolecular energy transfer process in densely packed molecular system. These molecular systems show outstanding state-of-the-art properties in the field of photochromic-fluorescent materials: bright emission, high photostability, and excellent performances in terms of photoswitching, with very high contrast (10,000:1) and fatigue resistance. Such efficient fluorescent photoswitchable nanoparticles and crystals find potential applications, such as in memory devices, sensors, multicolor displays, and bio-imaging work with the minimum number of photons [41, 42].

Acknowledgements We deeply appreciate to Prof. S. Kobatake, Dr. D. Kitagawa, Prof. T. Asahi, Dr. T. Onodera and Dr. Y. Ishibashi for their collaboration. The present work was supported by JSPS KAKENHI Grant Numbers JP15H01076, JP17H05269 in Scientific Research on Innovative Areas "Photosynergetics", JP19H02692 in Scientific Research (B). The author (Sanae Ishida) also acknowledges to Grant-in-Aid for JSPS Research Fellow Number JP18J23127.

References

1. Feringa BL (2007) The ART of building small: from molecular switches to molecular motors. J Org Chem 72:6635–6652
2. Irie M (2008) Photochromism and molecular mechanical devices. Bull Chem Soc Jpn 81:917–926
3. Heilemann M, Dedecker P, Hofkens J, Sauer M (2009) Photoswitches: key molecules for subdiffraction-resolution fluorescence imaging and molecular quantification. Laser Photon Rev 3:180–202
4. Bianco A, Perissinotto S, Garbugli M, Lanzan G, Bertarelli C (2011) Control of optical properties through photochromism: a promising approach to photons. Laser Photon Rev 5:711–736
5. Velema WA, Szymanski W, Feringa BL (2014) Photopharmacology: beyond proof of principle. J Am Chem Soc 136:2178–2191
6. Bouas-Laurent H, Dürr H (2001) Organic photochromism. Pure Appl Chem 73:639–665
7. Zhang J, Zou Q, Tian H (2013) Photochromic materials: more than meets the eye. Adv Mater 25:378–399
8. Yildiz I, Deniz E, Raymo FM (2009) Fluorescence modulation with photochromic switches in nanostructured constructs. Chem Soc Rev 38:1859–1867
9. Yun C, You J, Kim J, Huh J, Kim E (2009) Photochromic fluorescence switching from diarylethenes and its applications. J Photochem Photobiol C: Photochem Rev 10:111–129
10. Fukaminato T (2011) Single-molecule fluorescence photoswitching: design and synthesis of photoswitchable fluorescent molecules. J Photochem Photobiol C: Photochem Rev 12:177–208
11. Irie M, Fukaminato T, Sasaki T, Tamai N, Kawai T (2002) A digital fluorescent molecular photoswitch. Nature 420:759–760
12. Fukaminato T, Doi T, Tamaoki N, Okuno K, Ishibashi Y, Miyasaka H, M Irie (2011) Single-molecule fluorescence photoswitching of a diarylethene-perylenebisimide dyad: non-destructive fluorescence readout. J Am Chem Soc 133:4984–4990
13. Jares-Erijman EA, Jovin TM (2003) FRET imaging. Nat. Biotech 21:1387–1395
14. Fölling J, Polyakova S, Belov V, van Blaaderen A, Bossi ML, Hell SW (2008) Synthesis and characterization of photoswitchable fluorescent silica nanoparticles. Small 4:134–142
15. Trofymchuk K, Prodi L, Reisch A, Mély T (2015) Exploiting fast excitation diffusion in dye-doped polymer nanoparticles to engineer efficient photoswitching. J Phys Chem Lett 6:2259–2264
16. Métivier R, Badré S, Méallet-Renault R, Yu P, Pansu RB, Nakatani K (2009) Fluorescence photoswitching in polymer matrix: Mutual influence between photochromic and fluorescence molecules by energy transfer processes. J Phys Chem C 113:11916–11926
17. Nakahama T, Kitagawa D, Sotome H, Ito S, Miyasaka H, Kobatake S (2017) Fluorescence on/off switching in polymer bearing diarylethene and fluorene in their side chains. J Phys Chem C 121:6272–6281
18. Scholes GD, Fleming GR, Olaya-Castro A, van Grondelle R (2011) Lessons from nature about solar light harvesting. Nat Chem 3:763–774
19. Kundu S, Patra A (2016) Nano strategies for light harvesting. Chem Rev 117:712–757
20. Guo S, Song Y, He Y, Hu X-Y, Wang L (2018) Highly efficient artificial light-harvesting systems constructed in aqueous solution based on supramolecular self-assembly. Angew Chem Int Ed 57:1–6
21. Wu C, Zheng Y, Szymanski C, McNeill J (2008) Energy transfer in a nanoscale multichromophoric system: fluorescent dye-doped conjugated polymer nanoparticles. J Phys Chem C 112:1772–1781
22. Zhang Y, Zhang K, Wang J, Tian Z, Li ADQ (2015) Photoswitchable fluorescent nanoparticles and their emerging applications. Nanoscale 7:19342–19357
23. Trofymchuk K, Reisch A, Didier P, Fras F, Gilliot P, Mely Y, Klymchenko AS (2017) Giant light-harvesting nanoantenna for single-molecule detection in ambient light. Nat Photon 11:657–663

24. Fukaminato T, Ishida S, Métivier R (2018) Photochromic fluorophores at the molecular and nanoparticle levels: fundamentals and applications of diarylethene. NPG Asia Mater 10:859–881
25. Su J, Fukaminato T, Placial J-P, Onodera T, Suzuki R, Oikawa H, Brosseau A, Brisset F, Pansu R, Oikawa H, Métivier R (2016) Giant amplification of photoswitching by a few photons in fluorescent photochromic organic nanoparticles. Angew Chem Int Ed 55:3662–3666
26. Sasaki S, Watanabe T, Ishibasi Y, Fukaminato T, Asahi T (2018) Giant fluorescence modulation induced by UV/Vis excitation of benzothiadiazole nanoparticles doped with diarylethene derivatives. Chem Lett 47:163–166
27. Ishida S, Kitagawa D, Kobatake S, Kim S, Kurihara S, Fukaminato T (2019) Efficient "turn-off" fluorescence photoswitching in a highly fluorescent diarylethene single crystal. Chem Commun 55:5681–5684
28. Suk J, Cheng J-Z, Wong K-T, Bard AJ (2011) Synthesis, electrochemistry, and electrogenerated chemiluminescence of azide-BTA, a D-A-π-A-D species with benzothiadiazole and *N*, *N*-diphenylaniline, and its nanoparticles. J Phys Chem C 115:14960–14968
29. Kato S, Matsumoto T, Shigeiwa M, Gorohmaru H, Maeda S, Ishii T, Mataka S (2006) Novel 2, 1, 3-benzothiadiazole-based red-fluorescent dyes with enhanced two-photon absorption cross-sections. Chem Eur J 12:2303–2317
30. Ishii T, Kei I, Kichise Y, Ogawa M (2012) Red-light-emitting system based on aggregation of donor-acceptor derivatives in polar aqueous media. Chem Asian J 7:1553–1557
31. Irie M (2000) Diarylethenes for memories and switches. Chem Rev 100:1685–1716
32. Irie M, Fukaminato T, Matsuda K, Kobatake S (2014) Photochromism of diarylethene molecules and crystals: memories, switches, and actuators. Chem Rev 114:12174–12277
33. Kobatake S, Terakawa Y, Imagawa (2009) Solvent effect on photochromism of a dithienylperfluorocyclopentene having diethylamino group. Tetrahedron 65:6104–6108
34. Kasai H, Nalwa HS, Oikawa H, Okada S, Matsuda H, Minami N, Kakuta A, Ono K, Mukoh A, Nakanishi H (1992) A novel preparation method of organic microcrystals. Jpn J Appl Phys 31:L1132
35. Resch-Genger U, Grabolle M, Cavaliere-Jaricot S, Nitschke R, Nann T (2008) Quantum dots versus organic dyes as fluorescent labels. Nat Methods 5:763–775
36. Chen P-Z, Weng Y-X, Niu L-Y, Chen Y-Z, Wu L-Z, Tung C-H, Yang Q-Z (2016) Light-harvesting systems based on organic nanocrystals to mimic chlorosomes. Angew Chem Int Ed 55:1–6
37. Sun M-J, Liu Y, Yan Y, Li R, Shi Q, Zhao YS, Zhong Y-W, Yao J (2018) In situ visualization of assembly and photonic signal processing in a triplet light-harvesting nanosystem. J Am Chem Soc 140:4269–4278
38. Sun M-J, Liu Y, Zeng W, Zhao YS, Zhong Y-W, Yao J (2019) Photoluminescent anisotropy amplification in polymorphic organic nanocrystals by light-harvesting energy transfer. J Am Chem Soc 141:6157–6161
39. Kobatake S, Uchida K, Tsuchida E, Irie M (2002) Single-crystalline photochromism of diarylethenes: reactivity-structure relationship. Chem Commun 23:2804–2805
40. Fukaminato T, Kobatake S, Kawai T, Irie M (2001) Three-dimensional erasable optical memory using a photochromic diarylethene single crystal as the recording medium. Proc Jap Acad Ser B 77:30–35
41. Ishida S, Fukaminato T, Kim S, Ogata T, Kurihara S (2017) Sequential red-green-blue (RGB) fluorescence color photoswitching in multicomponent photochromic fluorescent nanoparticles. Chem Lett 46:1182–1185
42. Ishida S, Fukaminato T, Kitagawa D, Kobatake S, Kim S, Ogata T, Kurihara S (2017) Wavelength-selective and high-contrast multicolor fluorescence photoswitching in a mixture of photochromic nanoparticles. Chem Commun 53:8268–8271

Chapter 22
Cooperative Molecular Alignment Process Enabled by Scanning Wave Photopolymerization

Atsushi Shishido, Yoshiaki Kobayashi, Norihisa Akamatsu, Kyohei Hisano, and Miho Aizawa

Abstract Arbitrary and precise control of two-dimensional (2D) molecular alignment patterns over large areas play an important role for developing highly functionalized soft materials and devices. Here we demonstrate a dye-free system for 2D alignment patterning, termed "scanning wave photopolymerization (SWaP)". SWaP utilizes a spatial light-triggered mass flow induced by scanning light to propagate the wavefront to direct molecular order. Macroscopic, arbitrary 2D alignment patterns are generated in a wide variety of optically transparent polymer films from various polymerizable mesogens with sufficiently high birefringence (>0.1) by single-step photopolymerization, without alignment layers or polarized light sources. SWaP successfully inscribed a set of 500 × 300 arrays of a radial alignment pattern with a size of 27.4 μm × 27.4 μm, in which each individual pattern is smaller by a factor of 10^4 than that achievable by conventional photoalignment methods.

Keywords Molecular alignment · Scanning wave photopolymerization · Liquid crystal · Unpolarized light · Photoalignment

22.1 Introduction

Macroscopic alignment control of liquid crystal (LC) films is key to developing next-generation highly functionalized photonic, electronic, mechanical and biomedical organic devices [1–5]. Current methods achieve the large-area alignment of LCs which compose self-assembled hierarchical parallel or perpendicular ordering over various length scales from nanometer to micrometer [6–9]. Such methods use uniform external fields such as elongation, surface rubbing treatment, and electromagnetic or light fields [10–12]. Among these methods, elongation is the most simple and

A. Shishido (✉) · Y. Kobayashi · N. Akamatsu · K. Hisano · M. Aizawa
Laboratory for Chemistry and Life Science, Institute of Innovative Research, Tokyo Institute of Technology, Midori-ku, Yokohama 226-8503, Japan
e-mail: ashishid@res.titech.ac.jp

A. Shishido · Y. Kobayashi · N. Akamatsu
Department of Chemical Science and Engineering, School of Materials and Chemical Technology, Tokyo Institute of Technology, Midori-ku, Yokohama 226-8503, Japan

H. Miyasaka et al. (eds.), *Photosynergetic Responses in Molecules and Molecular Aggregates*, https://doi.org/10.1007/978-981-15-5451-3_22

powerful method for only one-dimensional (1D) molecular alignment. However, the other methods enable to control hierarchical alignment over two dimensions. Of these more advanced 2D techniques, light-driven alignment control (photoalignment) offers the greatest potential to finely control molecular orientation because of its remote and precise influence, and suitability for micro- to nano-fabrication. These advantages lead to many applications that require more complex alignment patterns [13–33].

Here we demonstrate a new method termed scanning wave photopolymerization (SWaP) based on a concept of light-triggered mass flow caused by photopolymerization reaction [34–39]. In SWaP, spatiotemporal scanning of focused guided light is employed to direct LC alignment coincident with the incident light patterns (Fig. 22.1). A desired target pattern of molecular alignment is achieved in a single-step by light scanning since spatiotemporal patterns of the light stimulus propagate as a controlled flow wavefront. This technique thus generates complex alignment patterns with fine control over large areas, in a wide variety of photopolymerizable LC materials, with no need for any added dyes or pre- or subsequent processing steps. As a further advantage over any other techniques, involving mechanical rubbing, electric fields, or heat, the size of the final alignment patterns would be restricted only by light diffraction limits in principle, and thus complexity of any patterns in 2D is then effectively unlimited. Moreover, in an industrial production setting, SWaP

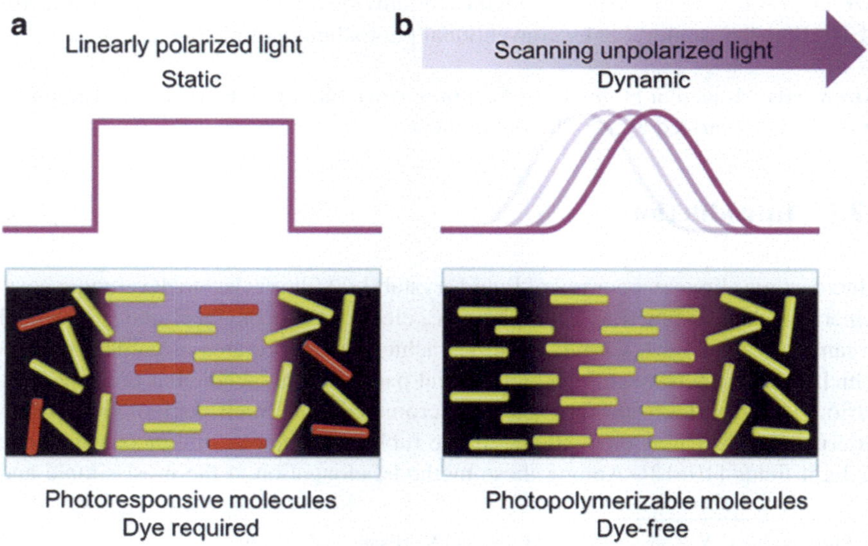

Fig. 22.1 Schematic illustration of photoirradiation for controlling molecular alignment. (**a**) Conventional molecular alignment method. (**b**) New concept proposed for generating molecular alignment by photopolymerization with spatiotemporal light scanning with pattern, termed “SWaP”. Pink and black regions represent irradiated and unirradiated regions, respectively. Reprinted from [67] by The Authors licensed under CC BY 4.0

has advantages such as electronics-free, single-step, noncontact, inexpensive, and, in general, applicable to incorporation directly into existing fabrication production lines.

22.2 One-Dimensional Molecular Alignment in LC Polymer Films by Scanning Wave Photopolymerization

First, we induced one-dimensional molecular alignment by SWaP with unidirectional light scanning with 250-μm-slit ultraviolet (UV) light at the wavelength of 365 nm to photopolymerize the mixture composed by sample 1 including M1, C1, and P1 with 97:3 molar ratio of M1:C1 in a handmade glass cell (Fig. 22.2a). The cell thickness was 3 μm, which shows the maximum degree of molecular alignment induced by SWaP. The photopolymerization temperature was 100 °C at which the resultant polymer exhibits a nematic phase, and then the sample was cooled to room temperature. The obtained LC film was transparent and had no color. Polarized optical microscope (POM) observation of the film under crossed polarizers showed brightness at 45° with the light scanning direction. On the other hand, it became completely dark when the angle between polarizers and the light scanning direction was 0° or 90° (Fig. 22.2b). Detailed POM observations using a tint plate with a retardation of 137 nm elucidate that the film is aligned parallel to the light scanning direction. Furthermore, the resultant film possessed high birefringence of 0.12. This value is comparable with typical birefringence levels required for current LC devices [40]. This result means that the mesogenic units of the monomer M1 were unidirectionally aligned along the light scanning direction. For further investigation of the molecular alignment behavior, we measured the polarized UV-visible (UV-vis) absorption spectra of the resultant film (Fig. 22.2c). Polar plots of the absorption band of cyanobiphenyl moieties exhibited the maximum value along the scanning direction (Fig. 22.2d). We calculated the planar alignment order parameter (S) from the equation $S = (A_{\parallel} - A_{\perp})/(A_{\parallel} + 2A_{\perp})$, where $A_{\parallel}$ and $A_{\perp}$ denote the parallel or perpendicular absorbance which is the direction of polarized incident light. The order parameter value was found to be 0.52. This value demonstrates that the degree of molecular alignment is achieved in similar chemical systems with current photoalignment methods [13, 28–30]. Furthermore, polarized infrared (IR) absorption spectroscopy allowed one to identify functional groups in the films. As a result, the IR spectra revealed that carbonyl moieties were aligned anisotropically besides cyanobiphenyl moieties throughout the film (Fig. 22.2e). On the other hand, the alignment direction of carbonyl moieties was orthogonally oriented to that of the cyanobiphenyl moieties. This result is rationalized by the behavior of mesogens in side-chain LC polymers. The side chains moieties are aligned alongside the polymer main chain under a shear-flow field [41–43], and thus the carbonyl groups generally

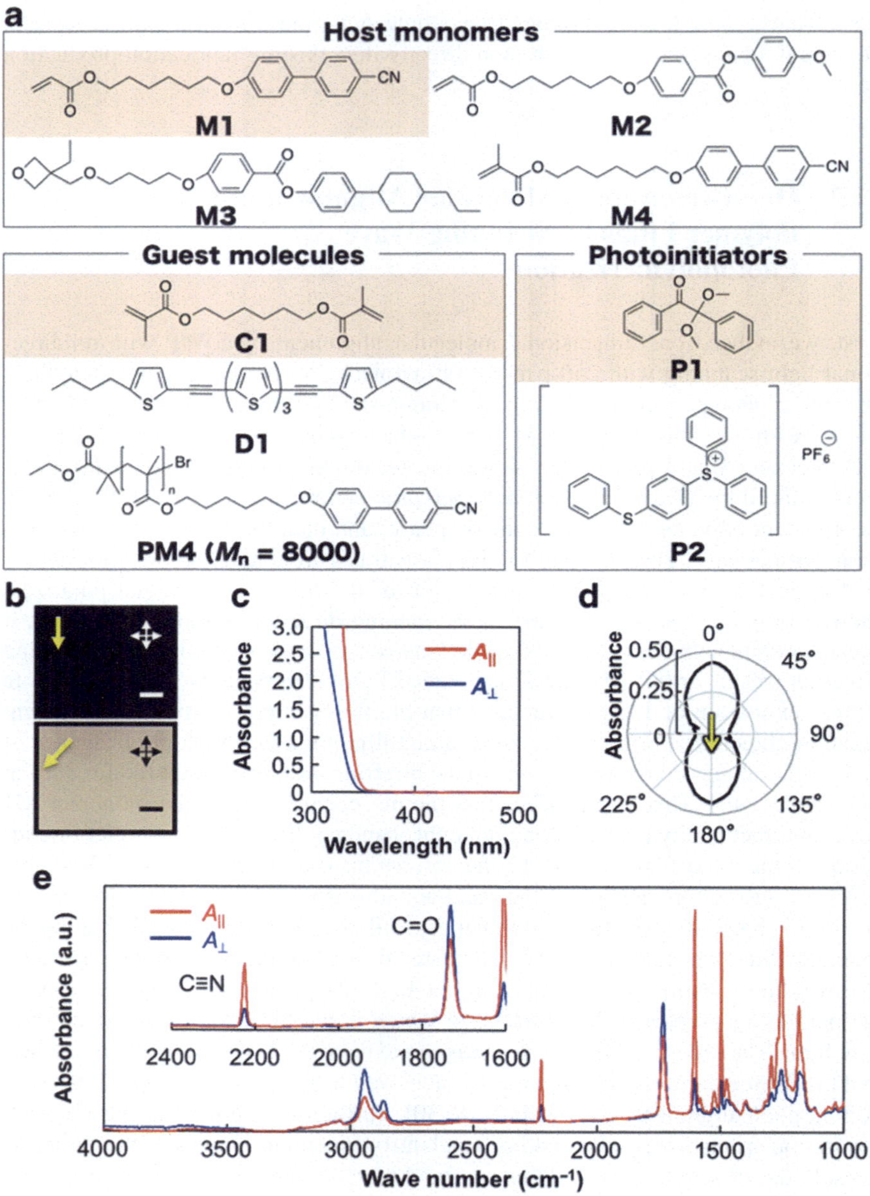

Fig. 22.2 Characterization of molecularly aligned polymer film induced by SWaP. (**a**) Chemical structures used in this study. The structures colored by orange rectangles are compounds for sample **1**. (**b**) POM images of resultant film. Scale bars, 100 μm. Yellow arrows indicate the light scanning direction; black and white arrows depict the direction of the polarizers. (**c**) Polarized UV-vis absorption spectra, (**d**) a polar plot of UV-vis absorbance, and (**e**) polarized IR absorption spectra of a resultant film photopolymerized with 250-μm-slitted UV light scanned in 1D at 20 μm/s. Reprinted from [67] by The Authors licensed under CC BY 4.0

lie perpendicular to the polymer main chain [30, 44]. This suggests that SWaP aligns the polymer main chains along the light scanning direction.

SWaP can also generate unidirectional molecular alignment even above the isotropic temperature of monomer and resultant polymers, which has not been previously achieved in LCs using other photoalignment systems. Furthermore, the polymer films recover their alignment after disappearing under heating above the isotropic temperature of polymer, displaying a thermal memory effect of molecular alignment. The results achieved by SWaP clearly confirm the hypothesis that the polymer main chain aligns along the light scanning direction and might affect neighboring mesogens to align in the same direction. On the other hand, a film photopolymerized by irradiation with uniform light showed a polydomain structure, which is typical texture of nematic LC polymers. These observations elucidate that the scanning light is powerful and important method to control the molecular alignment and to emerge uniform optical anisotropy.

To demonstrate the generality of SWaP, we also tried to induce molecular alignment in various other chemical materials such as monomers, crosslinkers, photoinitiators, and/or nonreactive dye molecules [67, 68]. The photopolymerization conditions were optimized regarding light intensity, temperature, and scanning rate to obtain the highest birefringence and order parameter. POM images and polarized UV-vis absorption spectra of the obtained films revealed that SWaP provided 1D molecularly aligned polymer films which is composed of abovementioned materials. This results clearly indicate that SWaP has simple processing and general materials designing generation capabilities. Although the radical photopolymerization shown above can be commonly used because of its generality of materials design, oxygen needs to be removed to prevent photopolymerization from its disturbance. To avoid this critical issue, we explored a sample of a cationic system, and performed two SWaPs onto a glass substrate and a flexible substrate running under ambient conditions. We successfully generated 1D molecularly aligned polymer films by SWaP without a glass cell or a nitrogen atmosphere. In addition to generality, the processing time for induction of 1D molecular alignment was much decreased by using higher light intensity and increased scanning rate.

22.3 Arbitrary 2D Molecular Alignment Patterns by Scanning Wave Photopolymerization

In SWaP, complex and spatiotemporal scanning of incident light offer arbitrary 2D alignment patterns which is directed by light-triggered diffusion. We defined SWaP as photopolymerization that spatial or temporal scanning light generate the molecular alignment utilizing mass flow caused by molecular diffusion in the non-equilibrium state. Such characteristics indicates that even patterned photoirradiation (e.g., a grating, a lattice, a honeycomb) could provide various complex molecular alignment patterning induced by diffusion at the edge of the pattern.

For precise control of macroscopic or further complex 2D molecular alignment patterns, we conducted SWaP of sample 1 by a digital micromirror device which enables to irradiate spatiotemporal arbitrary UV light patterns. As representatives, we designed three trial alignment patterns over large areas as shown in Fig. 22.3a: (I) a macroscopic radial alignment, (II) an planar cycloidal alignment where the LC director was helically twisted perpendicular to the light scanning direction, and (III) asymmetry alignment such as the word "Tokyo Tech." Fig. 22.3b exhibits the schematic illustrations as how to guide the molecular diffusion with spatiotemporal light scanning: (I) 2D expanding doughnut-shape, (II) 1D scanning of periodic dots, and (III) a 2D scanning along the words. As was abovementioned, to generate a uniform molecular alignment, light intensity and the scanning rate were required for SWaP during the process. We thus optimized the parameters from the resultant retardation in films. POM observation of the resultant films elucidated that they possessed hierarchical alignment patterns over large areas as expected to be induced (Fig. 22.3c). All films showed 2D molecular alignment after light scanning. Detailed POM observation of films indicated that unidirectional molecular alignment was induced along the light scanning direction within the irradiated regions, but at the edge of the irradiated regions, the alignment direction became perpendicular to the scanning direction. Designing the spatially guided photoirradiation, we successfully obtained the molecular alignment with a symmetric uniform radial pattern, low-symmetric cycloidal patterns and asymmetry pattern as shown in Fig. 22.3c.

In addition to large-area alignment patterns, we demonstrated flexible designing for microscopic alignment patterns by SWaP. We irradiated UV light pattern, which

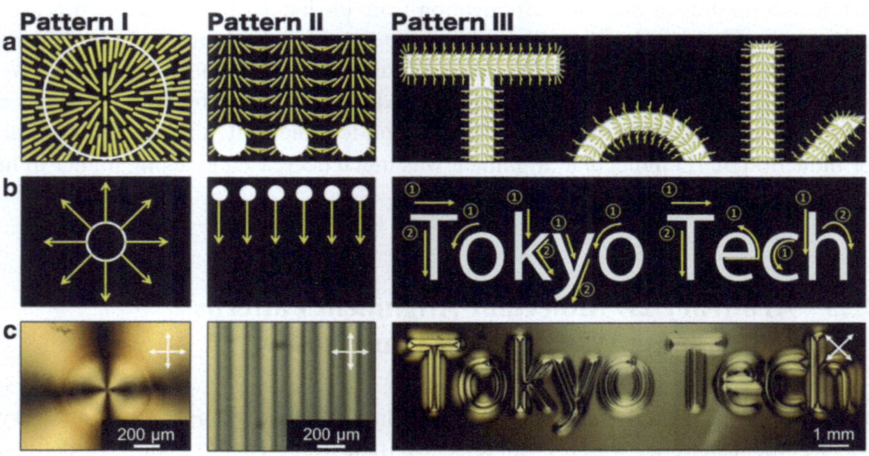

Fig. 22.3 Complex 2D alignment patterns induced by spatiotemporal light scanning (**a**) Schematic representation of desired patterns of mesogenic side-chain unit aligned (yellow rod). (**b**) Irradiated light patterns: (I) two-dimensional expanding doughnut-shape; (II) periodic dots scanned in 1D; (III) scanned along the words "Tokyo Tech". (**c**) Polarized optical micrographs. White arrows depict the direction of the polarizers. Reprinted from [67] by The Authors licensed under CC BY 4.0

has a square lattice with a lattice point distance of 27.4 μm and an irradiated region width of 13.7 μm. POM observation of the resultant film showed an array pattern of radial alignment. The size of a radial alignment was ~27.4 μm, reflecting the same periodicity as a square lattice of irradiated pattern. It must be noted that the size of each radial alignment in the film is much smaller (by a factor of 10^4) and high-resolution than that obtained by means of conventional photoalignment methods [2, 33, 45]. Furthermore, the number of radial alignment patterns reached 500 × 300 throughout the film.

22.4 Single-Step Creation of Polarization Gratings and Diffractive *Q*-Plates by Scanning Wave Photopolymerization with Unpolarized Light

Diffractive waveplates (DWs) show the properties of a polarization diffraction grating and have a molecular orientation pattern based on LCs that shows extremely high diffraction efficiencies. Rational design of DWs is key for the development of some of the greatest emerging technologies in the field of next-generation optics and photonics [17, 45–61]. One of the greatest benefits is that massive optical systems for manipulation of light, which are typically composed of large and complex arrays of optical elements such as lenses, prisms, and polarization converters, can be replaced with several stacked DW films of a few micrometers in thickness. The orientation patterns of LCs govern such optical properties of DWs because LCs with patterned birefringence (Δn) can spatially modulate the phase of light. As the simplest example of such modulation, 1D aligned LCs act as a half-wave plate, which has the ability to convert the phase of incident polarized light into 90° rotated. To exhibit such property, the retardation ($R = \Delta n\, d$; d is a thickness of medium) must be matched with $\lambda/2$ (λ: wavelength of incident light). In the same manner, more complexed 2D orientation patterns of LCs enables one to realize versatile geometrical phase modulations, providing various optical functionalities (e.g., lenses, optical vortex generators, beam steering, etc.) [17, 55, 58]. In particular, LC films with a cycloidal molecular orientation pattern (Fig. 22.3., Pattern II) are known as "cycloidal diffractive waveplates (CDWs)", having attracted a great deal of attention. Such films with orientation patterns meeting the half-wave plate condition diffract the incident light only into the +1st and −1st orders with the polarization state of left- and right-circular polarization, respectively (Fig. 22.4) [46–56]. As a result, CDWs have the potential to open windows for the application of next-generation projection displays.

The recent development of methods to control molecular orientation has readily achieved abovementioned complex 2D alignment patterns of LCs. The most advanced procedure is a "surface photoalignment control" technology [11–15, 29]. With this technology, LC orientation can be controlled through photoreactions of a thin film coated over a substrate surface. Because the films contain photoresponsive

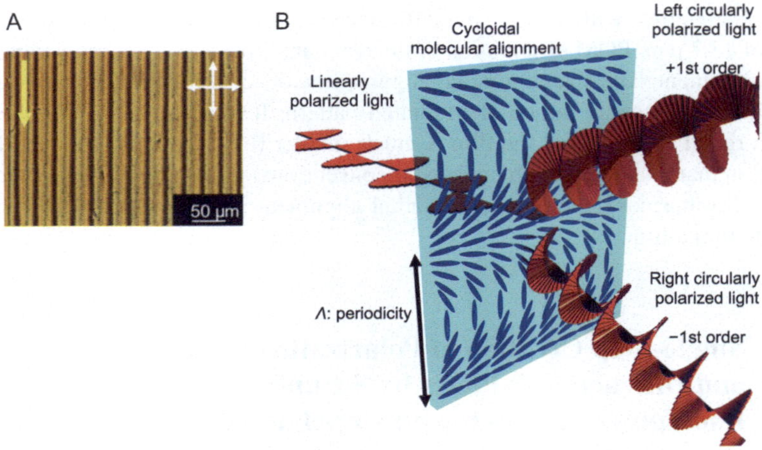

Fig. 22.4 Characterization of CDW fabricated by SWaP. (**a**) POM images of cycloidal molecular alignment under crossed polarizers. White arrows show the direction of polarizers. Yellow arrows depict the light scanning direction. (**b**) Schematic representation of the light diffraction behavior of a film with a cycloidal molecular orientation pattern. A linearly polarized light beam passing through the medium is diffracted into the +1st and −1st orders with left- and right-circularly polarized light, respectively. A left-circularly polarized light is diffracted into the −1st order light with right handedness, and vice versa. Λ denotes the physical periodicity of the cycloidal orientation. Blue ellipses represent LCs. Reprinted with permission from [69] c The Optical Society

dye molecules, the irradiation into a film with uniform polarized light leads to reorientation of dyes perpendicular to the polarization direction of incident light. The LCs over the film surface spontaneously align along the same direction with dyes by cooperative effect. Consequently, 2D molecular orientation patterns were provided by the spatial modulation of the polarization state of incident light. To control the spatial polarization states, state-of-the-art technologies have been proposed by using specially designed optical setups with numerous optical elements: computer-generated holograms, plasmonic photomasks, and polarization holograms [46, 62–65]. The polarization holograms are widely used for fabricating CWDs where one merely exposes a photoresponsive film to two orthogonal circularly polarized beams at the selected position. This method has a great advantage of high spatial resolution (less than tens of micrometers) with highly precise cycloidal orientation patterns. CDWs were recently reported, which has a high diffraction efficiency over 95% and/or with achromatic features ($\Delta\lambda > 200$ nm) [58, 66]. However, challenges remain to be improved such as higher processability, generating CDW over large areas, and avoiding the complex optical setup (e.g., interference of two circularly polarized beams). Although photoalignment technologies can create large-area CDWs using photoalignment of a dye-containing layer and subsequent photopolymerization of aligned reactive mesogens on a photoalignment layer [56, 60], they require spatial and precise modulation of the polarization direction in each pixel. Thus, such complexity leads to increased process time to obtain large-area CDWs.

Here, we propose the single-step fabrication of CDWs and q-plates over large areas by means of SWaP [34, 67–70]. As described above, SWaP has a principal advantage unlike conventional photoalignment methods. Thus, SWaP allows one to generate arbitrary 2D orientation patterns (various DWs) over large areas in a single-step only by scanning spatiotemporal patterned light during photopolymerization. As a first step, cycloidal molecular alignment was fabricated by SWaP using 1D scanning with periodically arranged rod-shaped pattern for less than 3 min (Fig. 22.4a). The resultant polymer film showed a cycloid pattern of $\sim$30 μm pitch over an area. Detailed investigation with a probe laser beam at 633 nm elucidated that the resultant film successfully acts as a CDW with a diffractive efficiency of 50% (Fig. 22.4b). SWaP could precisely and briefly control the complex molecular orientation pattern by scanning the rod-shaped pattern light in one direction. This process can be easily applied to large-area production, and the fabrication of films with precise molecular orientation can be greatly expected.

Next, molecular pattern I acts a waveplate with a specific optical function to convert a phase polarization state of a laser beam into optical axis symmetry, which is now one of the most desired and useful devices [71, 72]. Q-Plates function as optical devices with a passive optical element that is capable to manipulate the spatial distribution of the polarization of a homogenous polarized beam (e.g. a simple Gaussian laser beam). A q-plate can be fabricated simply as a birefringent waveplate with a 2D patterned distribution of the optical axis in the cross section. "Q" indicates a semi-integer topological charge defined as the degree of optical axis distribution. For example, it is known that a film having a π rotated optical axis of the retardation rotated on a laser optical axis works as a $q = 1$ q-plate. As a result, a radial vector beam or an optical vortex were obtained when linearly polarized light or circularly polarized light is introduced to the film, respectively. We confirmed that radial molecular alignment pattern achieved by SWaP successfully acts as a q-plate that converts a linearly polarized beams into a vector beam [70].

22.5 Conclusion

We have demonstrated that arbitrary 2D alignment patterns over large areas is readily achieved by SWaP without any surface treatment, electrodes, or polarized light sources. SWaP offers great potentials for the fabrication of a variety of high-performance LC devices because 2D patterns with complex and high resolution are inscribed, which is challenging or even impossible to achieve with conventional photoalignment techniques. Moreover, SWaP has great economic advantages for enabling to introduce it in existing photoproduction facilities. We believe that SWaP provides a new and powerful pathway for the direct design of highly functional organic materials with arbitrary, fine molecular alignment patterns over large areas.

Acknowledgements This work was supported by the Japan Society for the Promotion of Science (JSPS) KAKENHI grant no. JP17H05250 in Scientific Research on Innovative Areas "Photosynergetics." This work was supported by JSPS KAKENHI grant no. JP17J09899. This work was performed under the Cooperative Research Program of "Network Joint Research Center for Materials and Devices." This work was performed under the Research Program for CORE Lab of "Dynamic Alliance for Open Innovation Bridging Human, Environment and Materials" in "Network Joint Research Center for Materials and Devices." This work was supported by JSPS KAKENHI in Scientific Research on Innovative Areas, "Molecular Engine: Design of Autonomous Functions through Energy Conversion." This work was supported by the Core Research for Evolutional Science and Technology (CREST) program, "Creation of Innovative Core Technologies for Nano-enabled Thermal Management" of JST.

References

1. Lakes R (1993) Materials with structural hierarchy. Nature 361:511–515
2. White TJ, Broer DJ (2015) Programmable and adaptive mechanics with liquid crystal polymer networks and elastomers. Nat Mater 14:1087–1098
3. Hirst AR, Escuder B, Miravet JF, Smith DK (2008) High-tech applications of self-assembling supra-molecular nanostructured gel-phase materials: From regenerative medicine to electronics devices. Angew Chem Int Ed 47:8002–8018
4. Woltman SJ, Jay GD, Crawford GP (2007) Liquid-crystal materials find a new order in biomedical applications. Nat Mater 6:929–938
5. O'Neill M, Kelly SM (2003) Liquid crystals for charge transport, luminescence, and photonics. Adv Mater 15:1135–1146
6. Lehn J-M (2002) Toward self-organization and complex matter. Science 295:2400–2403
7. Kato T (2002) Self-assembly of phase-segregated liquid crystal structures. Science 295:2414–2418
8. Whitesides GM, Grzybowski B (2002) Self-assembly at all scales. Science 295:2418–2421
9. Kato T, Mizoshita N, Kishimoto K (2006) Functional liquid-crystalline assemblies: Self-organized soft materials. Angew Chem Int Ed 45:38–68
10. Schadt M (2009) Milestone in the history of field-effect liquid crystal displays and materials. Jpn J Appl Phys 48:03B001
11. Yaroshchuk O, Reznikov Y (2012) Photoalignment of liquid crystals: basics and current trends. J Mater Chem 22:286–300
12. Seki T, Nagano S, Hara M (2013) Versatility of photoalignment techniques: from nematics to a wide range of functional materials. Polymer 54:6053–6072
13. Chigrinov VG, Kozenkov VM, Kwok H-S (2008) Photoalignment of liquid crystalline materials: physics and applications. Wiley, Hoboken, New Jersey, USA
14. Priimagi A, Barrett CJ, Shishido A (2014) Recent twists in photoactuation and photoalignment control. J Mater Chem C 2:7155–7162
15. Seki T (2016) Light-directed alignment, surface morphing and related processes: recent trends. J Mater Chem C 4:7895–7910
16. Lv J-A, Liu Y, Wei J, Chen E, Qin L, Yu Y (2016) Photocontrol of fluid slugs in liquid crystal polymer microactuators. Nature 537:179–184
17. Kobashi J, Yoshida H, Ozaki M (2016) Planar optics with patterned chiral liquid crystals. Nat Photon 10:389–392
18. Palagi S, Mark AG, Reigh SY, Melde K, Qiu T, Zeng H, Parmeggiani C, Martella D, Sanchez-Castillo A, Kapernaum N, Giesselmann F, Wiersma DS, Lauga E, Fischer P (2016) Structured light enables biomimetic swimming and versatile locomotion of photoresponsive soft microrobots. Nat Mater 15:647–653

19. Gelebart AH, Mulder DJ, Varga M, Konya A, Vantomme G, Meijer EW, Selinger RLB, Broer DJ (2017) Making waves in a photoactive polymer film. Nature 546:632–636
20. Todorov T, Nikolova L, Tomova N (1984) Polarization holography. 1: A new high-efficiency organic material with reversible photoinduced birefringence. Appl Opt 23:4309–4312
21. Ichimura K, Suzuki Y, Seki T, Hosoki A, Aoki K (1988) Reversible change in alignment mode of nematic liquid crystals regulated photochemically by command surfaces modified with an azobenzene monolayer. Langmuir 4:1214–1216
22. Gibbons WM, Shannon PJ, Sun S-T, Swetlin BJ (1991) Surface-mediated alignment of nematic liquid crystals with polarized laser light. Nature 351:49–50
23. Schadt M, Schmitt K, Kozinkov V, Chigrinov V (1992) Surface-induced parallel alignment of liquid crystals by linearly polymerized photopolymers. Jpn J Appl Phys 31:2155–2164
24. Schadt M, Seiberle H, Schuster A, Kelly SM (1995) Photo-induced alignment and patterning of hybrid liquid crystalline polymer films on single substrates. Jpn J Appl Phys 34:L764–L767
25. Medvedev AV, Barmatov EB, Medvedev AS, Shibaev VP, Ivanov SA, Kozlovsky M, Stumpe J (2005) Phase behavior and photooptical properties of liquid crystalline functionalized copolymers with low-molecular-mass dopants stabilized by hydrogen bonds. Macromolecules 38:2223–2229
26. Nishikawa M, Taheri B, West JL (1998) Mechanism of unidirectional liquid-crystal alignment on polyimides with linearly polarized ultraviolet light exposure. Appl Phys Lett 72:2403–2405
27. Kumagai K, Sakamoto K, Usami K, Arafune R, Nakabayashi Y, Ushioda S (1999) Molecular orientation of liquid crystal monolayers on polyimide films exposed to linearly polarized UV light. Jpn J Appl Phys 38:3615–3618
28. Kawatsuki N, Washio T, Kozuki J, Kondo M, Sasaki T, Ono H (2015) Photoinduced orientation of photoresponsive copolymers with N-benzylideneaniline and nonphotoreactive mesogenic side groups. Polymer 56:318–326
29. Seki T (2014) New strategies and implications for the photoalignment of liquid crystalline polymers. Polym J 46:751–768
30. Wu Y, Demachi Y, Tsutsumi O, Kanazawa A, Shiono T, Ikeda T (1998) Photoinduced alignment of polymer liquid crystals containing azobenzene moieties in the side chain. 2. Effect of spacer length of the azobenzene unit on alignment behavior. Macromolecules 31:1104–1108
31. Schadt M, Seiberle H, Schuster A (1996) Optical patterning of multi-domain liquid-crystal displays with wide viewing angles. Nature 381:212–215
32. Fukuhara K, Nagano S, Hara M, Seki T (2014) Free-surface molecular command systems for photoalignment of liquid crystalline materials. Nat Commun 5:3320
33. Ware TH, McConney ME, Wie JJ, Tondiglia VP, White TJ (2015) Voxelated liquid crystal elastomers. Science 347:982–984
34. Hisano K, Kurata Y, Aizawa M, Ishizu M, Sasaki T, Shishido A (2016) Alignment layer-free molecular ordering induced by masked photopolymerization with non-polarized light. Appl Phys Express 9:072601
35. Krongauz VV, Schmelzer ER, Yohannan RM (1991) Kinetics of anisotropic photopolymerization in polymer matrix. Polymer 32:1654–1662
36. Broer DJ, Lub J, Mol GN (1995) Wide-band reflective polarizers from cholesteric polymer networks with a pitch gradient. Nature 378:467–469
37. Hikmet RAM, Kemperman H (1998) Electrically switchable mirrors and optical components made from liquid-crystal gels. Nature 392:476–479
38. Sutherland RL, Tondiglia VP, Natarajan LV, Bunning TJ (2004) Phenomenological model of anisotropic volume hologram formation in liquid-crystal-photopolymer mixtures. J Appl Phys 96:951–965
39. Leewis CM, de Jong AM, van IJzendoorn LJ, Broer DJ (2004) Reaction–diffusion model for the preparation of polymer gratings by patterned ultraviolet illumination. J Appl Phys 95:4125–4139
40. Demus D, Goodby JW, Gray GW, Spiess H-W, Vill V (1998) Handbook of liquid crystals. Wiley-VCH

41. Fridrikh SV, Terentjev EM (1999) Polydomain-monodomain transition in nematic elastomers. Phys Rev E 60:1847–1857
42. Pujolle-Robic C, Noirez L (2001) Observation of shear-induced nematic-isotropic transition in side-chain liquid crystal polymers. Nature 409:167–171
43. Agrawal A, Chipara AC, Shamoo Y, Patra PK, Carey BJ, Ajayan PM, Chapman WG, Verduzco R (2013) Dynamic self-stiffening in liquid crystal elastomers. Nat Commun 4:1739
44. Nair BR, Gregoriou VG, Hammond PT (2000) FT-IR studies of side chain liquid crystalline thermoplastic elastomers. Polymer 41:2961–2970
45. Nersisyan S, Tabiryan N, Steeves DM, Kimball BR (2009) Fabrication of liquid crystal polymer axial waveplates for UV-IR wavelengths. Opt Express 17:11926–11934
46. Berg RH, Hvilsted S, Ramanujam PS (1996) Peptide oligomers for holographic data storage. Nature 383:505–508
47. Bomzon Z, Biener G, Kleiner V, Hasman E (2002) Space-variant Pancharatnam-Berry phase optical elements with computer-generated subwavelength gratings. Opt Lett 27:1141–1143
48. Sarkissian H, Serak SV, Tabiryan NV, Glebov LB, Rotar V, Zeldovich BY (2006) Polarization-controlled switching between diffraction orders in transverse-periodically aligned nematic liquid crystals. Opt Lett 31:2248–2250
49. Marrucci L, Manzo C, Paparo D (2006) Pancharatnam-Berry phase optical elements for wave front shaping in the visible domain: switchable helical mode generation. Appl Phys Lett 88:221102
50. Provenzano C, Pagliusi P, Cipparrone G (2006) Highly efficient liquid crystal based diffraction grating induced by polarization holograms at the aligning surfaces. Appl Phys Lett 89:121105
51. Ishiguro M, Sato D, Shishido A, Ikeda T (2007) Bragg-type polarization gratings formed in thick polymer films containing azobenzene and tolane moieties. Langmuir 23:332–338
52. Shishido A, Ishiguro M, Ikeda T (2007) Circular arrangement of mesogens induced in Bragg-type polarization holograms of thick azobenzene copolymer films with a tolane moiety. Chem Lett 36:1146–1147
53. Nikolova L, Ramanujam PS (2009) Polarization holography. Cambridge University
54. Nersisyan SR, Tabiryan NV, Steeves DM, Kimball BR (2009) Optical axis gratings in liquid crystals and their use for polarization insensitive optical switching. J Nonlinear Opt Phys 18:1–47
55. Nersisyan SR, Tabiryan NV, Steeves DM, Kimball BR (2010) The promise of diffractive waveplates. Opt Photon News 21(3):40–45
56. Shishido A (2010) Rewritable holograms based on azobenzene-containing liquid-crystalline polymers. Polym J 42:525–533
57. Nersisyan SR, Tabiryan NV, Steeves DM, Serabyn E (2013) Improving vector vortex waveplates for high-contrast coronagraphy. Opt Express 21:8205–8213
58. Kim J, Li Y, Miskiewicz MN, Oh C, Kudenov MW, Escuti MJ (2015) Fabrication of ideal geometric-phase holograms with arbitrary wavefronts. Optica 2:958–964
59. Tabiryan NV, Serak SV, Nersisyan SR, Roberts DE, Zeldovich BY, Steeves DM, Kimball BR (2016) Broadband waveplate lenses. Opt Express 24:7091–7102
60. Serak SV, Roberts DE, Hwang J-Y, Nersisyan SR, Tabiryan NV, Bunning TJ, Steeves DM, Kimball BR (2017) Diffractive waveplate arrays. J Opt Soc Am B 34:B56–B63
61. Lee YH, Tan G, Zhan T, Weng Y, Liu G, Gou F, Peng F, Tabiryan NV, Gauza S, Wu ST (2017) Recent progress in Pancharatnam-Berry phase optical elements and the applications for virtual/augmented realities. Optical Data Process Storage 3:79–88
62. De Sio L, Roberts D, Liao Z, Nersisyan S, Uskova O, Wickboldt L, Tabiryan N, Steeves D, Kimball B (2016) Digital polarization holography advancing geometrical phase optics. Opt Express 24:18297–18306
63. Guo Y, Jiang M, Peng C, Sun K, Yaroshchuk O, Lavrentovich O, Wei Q-H (2016) High-resolution and high-throughput plasmonic photopatterning of complex molecular orientations in liquid crystals. Adv Mater 28:2353–2358
64. Ogiwara A, Hirokari T (2008) Formation of anisotropic diffraction gratings in a polymer-dispersed liquid crystal by polarization modulation using a spatial light modulator. Appl Opt 47:3015–3022

65. Nikolova L, Todorov T (1984) Diffraction efficiency and selectivity of polarization holographic recording. Opt Acta 31:579–588
66. Oh C, Escuti MJ (2008) Achromatic diffraction from polarization gratings with high efficiency. Opt Lett 33:2287–2289
67. Hisano K, Aizawa M, Ishizu M, Kurata Y, Nakano W, Akamatsu N, Barrett CJ, Shishido A (2017) Scanning wave photopolymerization enables dye-free alignment patterning of liquid crystals. Sci Adv 3:e1701610
68. Aizawa M, Hisano K, Ishizu M, Akamatsu N, Barrett CJ, Shishido A (2018) Unpolarized light-induced alignment of azobenzene by scanning wave photopolymerization. Polym J 50:753–759
69. Hisano K, Ota M, Aizawa M, Akamatsu N, Barrett CJ, Shishido A (2019) Single-step creation of polarization gratings by scanning wave photopolymerization with unpolarized light. J Opt Soc Am B 36:D112–D118
70. Aizawa M, Ota M, Hisano K, Akamatsu N, Sasaki T, Barrett CJ, Shishido A (2019) Direct fabrication of a q-plate array by scanning wave photopolymerization. J Opt Soc Am B 36:D47–D51
71. Marrucci L, Manzo C, Paparo D (2006) Optical spin-to-orbital angular momentum conversion in inhomogeneous anisotropic media. Phys Rev Lett 96:163905
72. Slussarenko S, Murauski A, Du T, Chigrinov V, Marrucci L, Santamato E (2011) Tunable liquid crystal q-plates with arbitrary topological charge. Opt Express 19:4085–4090

Chapter 23
Ultrafast Energy Transfer of Biohybrid Photosynthetic Antenna Complexes in Molecular Assembly Systems

Takehisa Dewa, Yusuke Yoneda, Masaharu Kondo, Hiroshi Miyasaka, and Yutaka Nagasawa

Abstract Photosynthesis in natural systems is one of the typical examples where the interaction between multiple photons and multiple chromophores leads to the synergetic responses in three-dimensionally arranged molecular aggregates. Accordingly, the investigation of primary events in the excited state could provide important information on the progress in the artificial photosynergetic systems. Along with this line, we have prepared biohybrid light-harvesting complexes to study ultrafast dynamics of excitation energy transfer in the present project. Light-harvesting 2 complex (LH2) from photosynthetic purple bacteria is a molecular assembly, in which multiple photosynthetic chromophores are well organized. To expand high-harvesting activity of LH2, we attached a fluorophore, Alexa647, to lysine residues on the LH2 polypeptide so as to cover broad solar spectrum. Ultrafast energy transfer from Alexa 647 to the intrinsic chromophores B800 and B850 in LH2 was observed using femtosecond transient absorption spectroscopy. Detailed analysis revealed two possible energy transfer pathways. In addition, we made a LH2 mutant bearing reactive Cys residue on the N-terminal region of LH2β polypeptide. The engineered LH2 allows us to provide the defined position of Alexa647. Together with the system in which LH2 was dissolved in micellar solution, a lipid bilayer system in which the bioengineered LH2 was incorporated was applied to control the location of the external fluorophore.

Keywords Light-harvesting complex 2 (LH2) · Photosynthesis · Biohybrid · Ultrafast energy transfer · Fluorophore

---

T. Dewa (✉) · M. Kondo
Department of Life Science and Applied Chemistry, Graduate School of Engineering, Nagoya Institute of Technology, Gokiso-Cho, Showa-Ku, Nagoya 466-8555, Japan
e-mail: takedewa@nitech.ac.jp

Y. Yoneda · H. Miyasaka
Graduate School of Engineering Science, Division of Frontier Materials Science, Osaka University, Toyonaka, Osaka 530-8531, Japan

Y. Nagasawa
Department of Applied Chemistry, College of Life Sciences, Ritsumeikan University, 1-1-1 Noji-Higashi, Kusatsu, Shiga 525-8577, Japan

H. Miyasaka et al. (eds.), *Photosynergetic Responses in Molecules and Molecular Aggregates*, https://doi.org/10.1007/978-981-15-5451-3_23

23.1 Introduction

Enhancing the light-harvesting potential of antenna components in a system of solar energy conversion is an important topic in the field of artificial photosynthesis [1]. For the purpose of energy conversion, the essence of light-harvesting strategy is (1) use of multiple chromophores to absorb photons in the broad solar spectrum and (2) to transfer excitation energy between the chromophores within subpico–few picoseconds to a photocatalytic component [2]. Photosynthetic organisms have diverse light-harvesting antenna systems that have been developed to adapt to their habitats with different or variable intensities and colors of light [3–5]. Such photosynthetic antennae are based on the strategy (1) and (2).

Figure 23.1a shows X-ray crystallographic structure of light-harvesting 2 complex (LH2) from *Rhotopseudomonas* (*Rps.*) *acidophila* strain 10050 [6]. LH2 consists of nine-membered pair of α, β-polypeptides and photosynthetic pigments bacteriochlorophyll *a* (BChl) and rhodopin glucoside (Car). BChl molecules form two types of characteristic molecular array, B800 and B850, named according to their absorption maxima. Light energy absorbed by B800 is immediately transferred to B850 with a time constant of ~700 fs [7, 8]. Carotenoids also function as light-harvesting pigments; light energy absorbed by the carotenoid is transferred to B800 and B850 with the time constant of 200 fs [3]. In this way, light energy absorbed by the photosynthetic pigments is funneled into B850.

LH2 possesses absorption bands in the visible and near infrared regions because of the intrinsic chromophores (BChl and Car) and exhibits low absorbance in the wavelength range of 620–750 nm (Fig. 23.1b). In this study, we constructed LH2 with a covalently attached artificial fluorescent dye (Alexa Fluor 647: A647) as an external light-harvesting pigment to expand the light-harvesting ability without affecting inherent functions. We call the LH2 conjugate LH2-A647 or biohybrid LH2 [9, 10]. Through this approach, we constructed the semi-artificial light-harvesting system that drives ultrafast energy transfer.

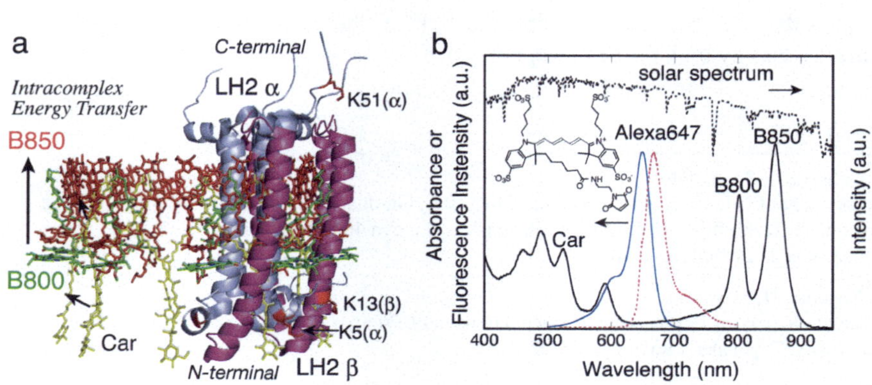

Fig. 23.1 **a** X-ray crystallographic structure of LH2 from *Rhodopseudomonas acidophila* 10050 (PDB: 1NKZ). **b** Absorption spectrum of LH2 (solid black line), absorption and fluorescence spectra of Alexa647 (blue and dotted red lines), and solar spectrum of the geosphere (dotted black line)

23.2 Strategy of Conjugation of Fluorescent Dye to LH2

We chose two types of LH2; one is LH2 from *Rps. acidophila* strain 10050 and the other is the engineered LH2 bearing Cys as reactive site on the N-terminal region of LH2-β polypeptide. Amino acid sequences are shown in Fig. 23.2.

To conjugate the dye to LH2 (from *Rps. acidophila* 10050), A647 bearing maleimide group (Maleimide-C2-Alexa647) was covalently attached to Lys residues, either K5 or K51, on the LH2α-polypeptides via the bifunctional crosslinker reagent sulfo-AC_5-SPDP (Fig. 23.3). Tandem MS spectroscopy confirmed the attachment.

To obtain biohybrid LH2 with A647 attached to defined position, Cys-bearing LH2 was prepared from *Rhodobacter* (*Rba.*) *sphaeroides* IL106-DL2 (pJRD-SpucB-CGG-L2). The Cys was introduced to the N-terminal region of LH2β-polypeptide

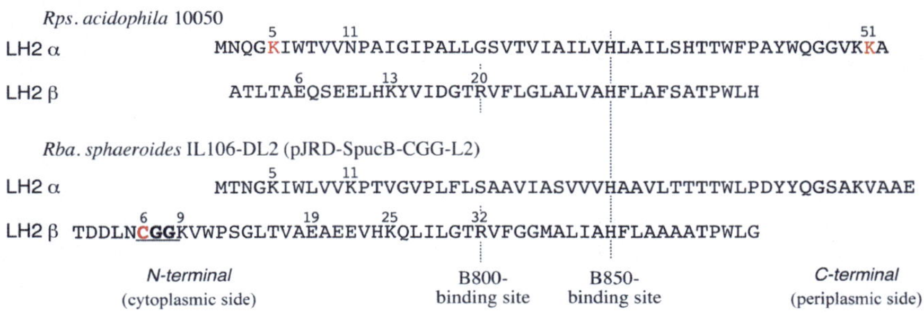

Fig. 23.2 Amino acid sequences of LH2α- and LH2β-polypeptides from *Rps. acidophila* 10050 and engineered *Rba. sphaeroides* IL106-DL2 (pJRD-SpucB-CGG-L2). Reactive amino acid side chains were colored in red. The figure modified with permission from Springer Nature Ref. [10] Copyright 2019

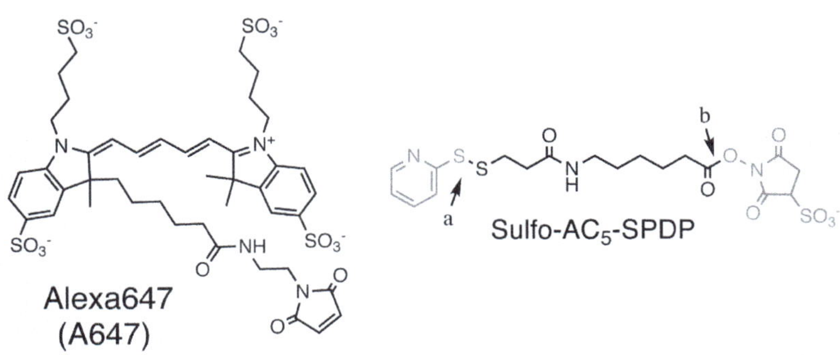

Fig. 23.3 Maleimide-C2-Alexa647 (A647, left) and the bifunctional crosslinker (Sulfo-AC_5-SPDP). The reactive sites with maleimide group (in A647) and Lys residues in the LH2α polypeptide are indicated by the arrows a and b, respectively

(Fig. 23.2). Maleimide-C2-Alexa647 was directly attached to the Cys residue. The numbers of Alexa647 attached to LH2 were ~9 and ~5 for the former and the latter biohybrid LH2s, respectively. Hereafter, we call the two type of LH2 biohybrids as LH2-A647 and LH2-NC-A647.

23.3 Energy Transfer Dynamics of LH2-A647 in the Micellar Solution System [9]

Figure 23.4a shows absorption spectrum of LH2-A647. The spectral profile can be reproduced as a sum of A647 and the intrinsic chromophores (BChl and Car). The absorption bands and circular dichroism (CD) signals of the intrinsic chromophores in LH2 were well preserved, indicating that modification with A647 does not perturb the environment of the chromophores. In addition, CD and linear dichroism spectra suggested no excitonic coupling and particular orientation of A647 attached to LH2.

Upon excitation at 650 nm, significant quenching of A647 and prominent emission from B850 were observed for LH2-A647. This result clearly indicates energy transfer from A647 to B800/B850. Figure 23.4b shows comparison of fluorescence excitation and absorptance (1 – T) spectra of LH2-A647 indicating that the efficient energy transfer of 83% takes place from A647 to B800/B850. When free A647 was simply mixed with LH2 in solution, no energy transfer was observed. This result indicates that the efficient energy transfer is established by conjugation of LH2 and A647 external chromophore.

Energy transfer dynamics was observed with femtosecond transient absorption (TA) spectroscopy. Figure 23.5 shows TA spectra upon excitation at 650 nm. Immediately, after the photoexcitation, the TA spectrum exhibits a negative band with a minimum at 657 nm and positive bands with maxima at 517 and 454 nm within the response time of the apparatus. The negative band is ascribed to the superposition of ground state bleach (GSB) and stimulated emission (SE) of A647, and the positive

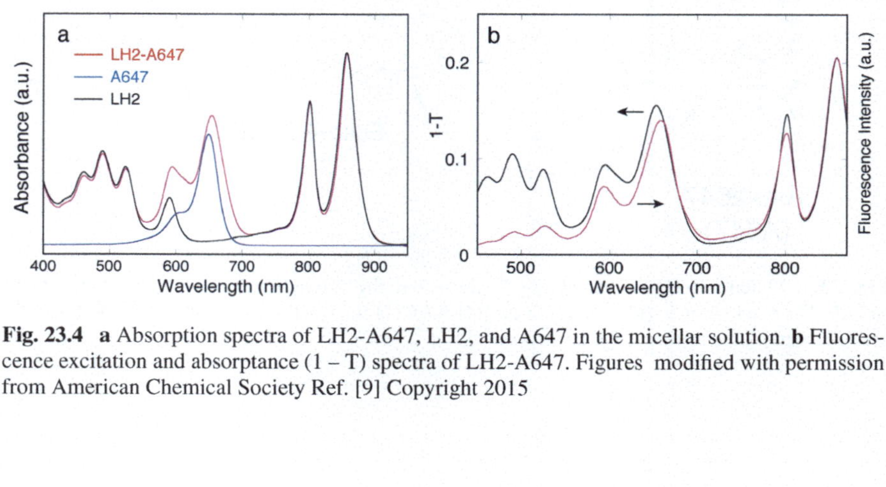

Fig. 23.4 **a** Absorption spectra of LH2-A647, LH2, and A647 in the micellar solution. **b** Fluorescence excitation and absorptance (1 – T) spectra of LH2-A647. Figures modified with permission from American Chemical Society Ref. [9] Copyright 2015

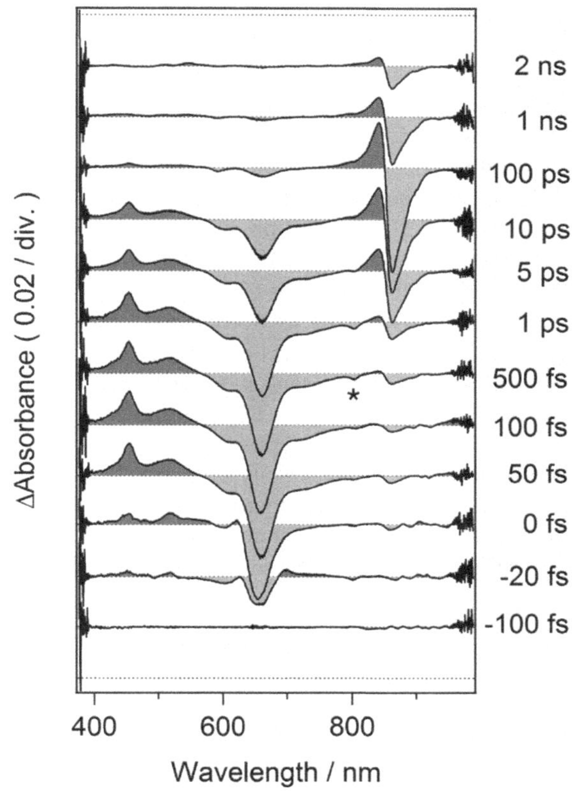

Fig. 23.5 Transient absorption spectra of LH2-A647 excited at 650 nm with a pulse duration of ~17 fs. The time delays are indicated in the figure. The figure modified with permission from American Chemical Society Ref. [9] Copyright 2015

band to the excited state absorption (ESA) of A647. The lifetime of the excited state of free A647 was 1.2 ns. With the increase in the delay time after the excitation, the signal of A647 decays in the picosecond time domain with simultaneous appearance of new positive and negative bands at 842 and 863 nm. These are ascribed to the exciton absorption and the superposition of GSB and SE of B850. This result clearly indicates energy transfer from A647 to B850. In the time domain of 100 fs–1 ps, a small negative band at 803 nm (marked by *) was observed, showing that B800 is also an energy acceptor from A647. The ultrafast energy transfer channel B800 → B850 with the time constant of ~1 ps was confirmed to be maintained in the LH2-A647 through the experiment upon 800 nm excitation. These results suggest that energy transfer pathways consist of A647 → B850 and A647 → B800 → B850.

Applying global analysis, time constants were evaluated to be 260 fs, 4.3 ps, 23 ps, and 1.2 ns. The femtosecond TA measurement indicates that the rate of energy transfer via B800 is faster than that of direct transfer from A647 to B850. Thus, the component with the fastest time constant, 260 fs, likely involves not only structural relaxation and/or ultrafast solvation of A647 in the excited state but also energy transfer by way of B800 to B850. The component with the time constant of 4.3 ps is also assignable to this pathway. The time constant of 23 ps is ascribed to the direct energy transfer of A647 → B850. Considering the position of A647s attached to LH2α-polypeptide, those attached to N- and C-terminal regions, where B800 and B850 are, respectively, in the vicinity, should be responsible for the faster and slower transfer pathways. The longest time constant of 1.2 ns corresponds to the lifetime of A647 and B850.

The model energy transfer diagram is illustrated in Fig. 23.6, and the kinetic equations are derived based on this model. The excited state of A647, A647* was divided into four species (**1**, **1′**, **2**, and **3**) with the time constants of 0.44 ps (6% for **1**), 4.1 ps (34% for **1′**), 23 ps (50% for **2**), and 1.2 ns (10% for **3**). The faster time constants, 0.44 ps and 4.1 ps, correspond to the energy transfer to B800. The 23 ps time constant represents direct energy transfer to B850.

To rationalize the time constants, donor–acceptor distances (R_{calcd}) were evaluated from the Förster mechanism. The faster rate constants, 440 fs and 4.1 ps, predicted 11 and 16 Å of donor–acceptor distances as depicted in Fig. 23.7a. Considering possible areas for A647 moiety with the linkage from K5 and K51 residues, the A647 linked to K5 can be regarded as the donor for B800 but not that of K51. The time constants could arise from transfer to the closest and the second closest B800 BChls (Fig. 23.7a). For B850 as the primary acceptor, on the other hand, the evaluated

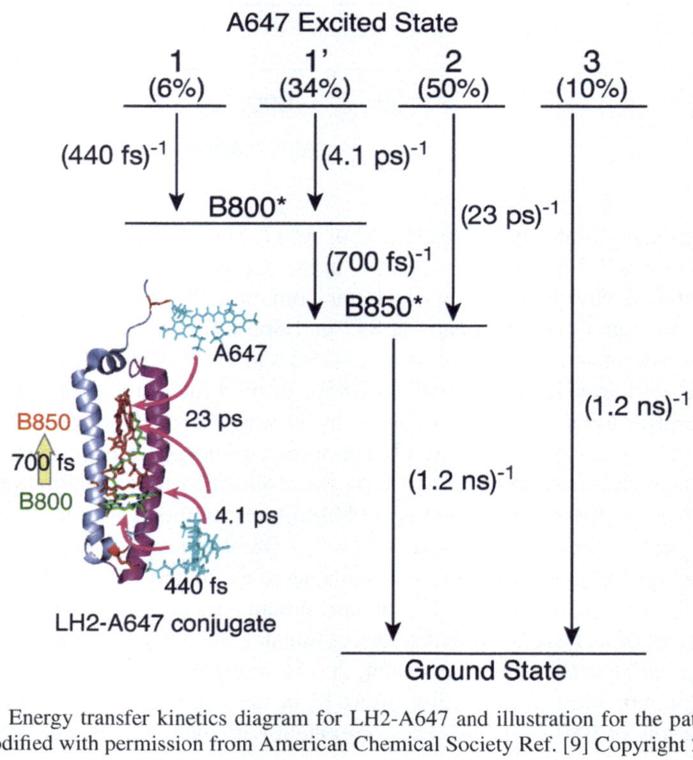

Fig. 23.6 Energy transfer kinetics diagram for LH2-A647 and illustration for the pathway. The figure modified with permission from American Chemical Society Ref. [9] Copyright 2015

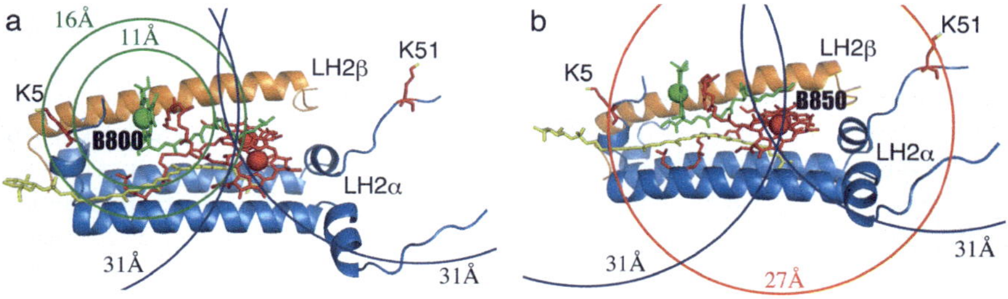

Fig. 23.7 Graphical description of LH2-A647 conjugate consisting of B800, B850, LH2α-, and β-polypeptides. Evaluated donor–acceptor distances R_{calcd} from B800 (**a**) and B850 (**b**) to A647 are depicted by green (**a**) and red circles (**b**). Possible range of A647 attached to K5 and K51 with linkage was estimated to be 31 Å as fully extended conformation, which is depicted by blue circles. The figure modified with permission from American Chemical Society Ref. [9] Copyright 2015

R_{calcd} was 27 Å. The A647 attached to either K5 or K51 is possible to be the direct donor to B850 as shown in Fig. 23.7b.

This work using the photosynthetic component LH2 indicates that by conjugation with appropriate donor chromophores close to B800, the ultrafast energy transfer cascade can be accomplished, where the natural ultrafast channel B800 → B850 is involved.

23.4 Energy Transfer Dynamics of Cys-Bearing LH2 Conjugate (LH2-NC-A647) in Micellar and Lipid Bilayer Systems [10]

As described above, fluorescent dyes attached to a protein scaffold exhibit multi-exponential energy transfer dynamics in many cases [11]. This is probably due to inhomogeneity of the attached chromophores, in terms of their surrounding environments, diverse orientation and donor–acceptor distance, multiple binding sites, and so on. Here, we designed a cysteine-bearing LH2 for the site-selective modification of A647. In this section, we described use of engineered LH2 bearing cysteine (LH2-NC) from *Rba. sphaeroides* IL106 strain. The amino acid sequences were shown in Fig. 23.2. The reactive cysteine was introduced into the N-terminal region on the LH2β-polypeptide, which enables the attachment of A647 maleimide in a site-selective manner proximate to the B800-BChl. Using A647-attached LH2-NC conjugate (LH2-NC-A647), we revealed the dynamics of sequential energy transfer A647 → B800 → B850 in micellar and reconstituted lipid bilayer systems. We have briefly discussed the potential of LH2 complex as a remarkable platform for construction of biohybrid antennae.

Absorption spectrum of LH2-NC-A647 was similar to that of LH2-A647. The intrinsic ultrafast energy transfer was confirmed to be maintained by femtosecond

TA spectroscopy. Therefore, introducing Cys and chemical modification with A647 did not perturb the environment of the intrinsic chromophores. The absence of CD signal from A647 implied no excitonic coupling between A647 moieties.

Energy transfer dynamics was observed using femtosecond TA spectroscopy in detail. Figure 23.8a shows TA spectra of LH2-NC-A647 upon excitation at 650 nm in micellar solution system. After the photoexcitation, positive and negative signals immediately appeared in the range of 400–550 nm and 550–800 nm, respectively. These signals, which result from A647, were assignable to ESA and the superposition of GSB and SE, respectively, as described previously. As time progressed, the signals from the excited state of A647 disappeared within tens of picosecond, correlating with the appearance of positive and negative bands at 840 and 855 nm indicative of B850 excited state formation (ESA and a mixture of GSB and SE, respectively). Such excitation dynamics clearly indicates the energy transfer from A647 to B850 in

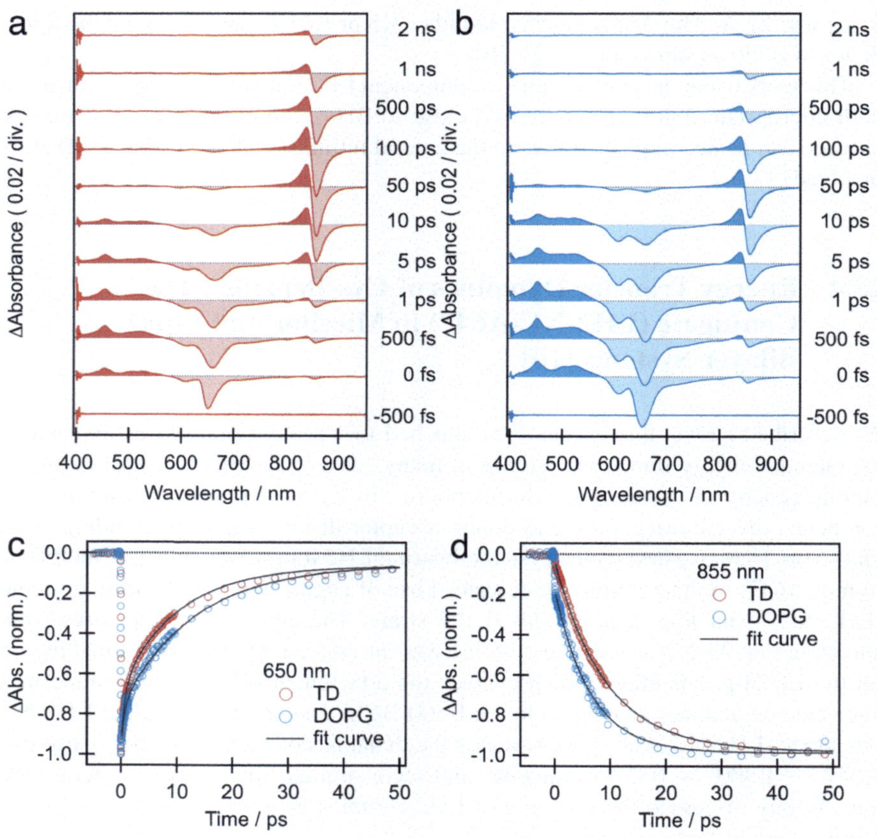

Fig. 23.8 Femtosecond TA spectra of LH2-NC-A647 in a micellar (TD) solution **a** and lipid bilayer (DOPG) systems **b**. Time traces at 650 nm **c** and 855 nm **d** in micellar and lipid bilayer systems. The figure modified with permission from Springer Nature Ref. [10] Copyright 2019

the LH2-NC-A647. For comparison the lifetime of the excited state of free A647 was 1.2 ns [9]. With careful inspection at < 2 ps, a small negative band at 805 nm ascribed to GSB of B800 was observed with the appearance (<1 ps) and disappearance (>1 ps) of GSB of B800. This is a clear evidence demonstrating the mediation of B800 in the energy transfer from A647 to B850. Figure 23.8c and d shows the time traces of ΔA at 650 (c) and 855 nm (d). The former exhibited bi-exponential dynamics (τ_{av} = 7.3 ps), on the other hand, the dynamics of the latter could be appropriately described by monoexponential energy transfer, providing the time constant of 9.1 ± 0.1 ps.

Detailed dynamics were inspected by decay-associated spectra (DAS) obtained through global analysis. The DAS consists of three components: The time constants (amplitude ratio of GSB and SE of A647) were 770 fs (0.24), 10 ps (0.70), and 1.0 ns (0.06). The first component (770 fs) can be assigned to the A647 relaxation in the excited state, such as solvation or structural relaxation [9]. In addition, a small positive signal at 805 nm was observed, indicating energy transfer to B800 in LH2. The time constant of 770 fs was in the same time scale to that for the energy transfer from B800 to B850. The second DAS indicates the energy transfer component with the time constant of 10 ps, which was in good agreement with that evaluated from the time trace at 855 nm, 9.1 ps. The third DAS with 1.0 ns represents the decay in the excited state of B850 [9]. The energy transfer dynamics was also analyzed based on a sequential kinetic model, suggesting that energy transfer from A647 to B800 takes place through single-phase dynamics.

We attempted to modulate environments surrounding A647 moieties through reconstitution of LH2-NC-A647 into a lipid bilayer composed of an anionic phospholipid DOPG, because A647, which is a highly hydrophilic fluorophore, has been reported to be preferentially partitioned in aqueous phase and less soluble in lipid bilayer phase [12]. Therefore, we hypothesized that the A647 moieties should be positioned proximate to or in the exterior region of the interface of the anionic lipid bilayer, away from B800 through the reconstitution. As a result, energy transfer efficiency significantly decreased to 42% (evaluated from an excitation spectrum). This result suggested that part of A647 moieties in the LH2-NC-A647 disengaged from the energy transfer system. The energy transfer dynamics in the lipid bilayer system was observed by femtosecond absorption spectroscopy (Fig. 23.8b). Basic spectral features were same as those for the micellar solution system. Although the appearance of GSB of B800 was not clearly observed in lipid bilayer system, the signal at 805 nm was negative until ~4.5 ps. For comparison, a B800-removed conjugate exhibited very small ΔA intensity, which became positive at ~3.5 ps. These results supported the fact that B800 is also involved in the energy transfer pathway in the lipid bilayer system. The time traces of transient absorbance at 650 and 855 nm for the membrane systems are shown in Fig. 23.8c and d. The rates of energy transfer evaluated from the formation of excited state of B850 were slightly faster than those in the micellar system. Interestingly, the B850 dynamics were again described by monoexponential kinetics, giving a time constant of 6.4 ± 0.1 ps, which is unexpectedly shorter than 9.1 ± 0.1 ps for the micellar system. Note that the rise of the negative signal of B850 (Fig. 23.8d) is faster for the lipid bilayer system, while the decay of the negative A647 signal at 650 nm (Fig. 23.8c) is slower for the lipid bilayer system. The DAS analysis exhibited four components: time constants (amplitude) were 670 fs (0.15), 9.2 ps (0.47), 27 ps (0.35), and 490 ps (0.03). The second component with 9.2 ps lifetime corresponds to the energy transfer, which is longer than 6.4 ± 0.1 ps, due to averaging of slower decay and faster rise of the negative signals at 650 and 855 nm, respectively, for the lipid bilayer system. The time constant in the lipid bilayer system was equivalent to that in the micellar system (10 ps), suggesting that energy transfer pathway are similar, A647 → B800 → B850. Note that the relative amplitude of the 9.2 ps component (0.47) is smaller than the 10 ps component (0.70) for the micellar system, which is responsible for the smaller energy transfer efficiency of 42%. The third DAS at 27 ps likely represents non-radiative deactivation of A647. A647 moieties located near the interface region of the lipid bilayer might be enhanced to be deactivated in a non-radiative way. A LH2 complex tends to cluster in DOPG lipid bilayer [13] due to which intercomplex energy transfer between neighboring B850 rings is facilitated [14]. This results in faster deactivation kinetics of B850 in the lipid bilayer (490 ps) than in the micellar system (1.0 ns). Similar deactivation mechanism can be also taking place for A647 and such enhancement of deactivation will lead to reduction of energy transfer efficiency. Nevertheless, energy transfer dynamics of LH2-NC-A647 chiefly displayed monoexponential kinetics also in the lipid bilayer system.

In order to rationalize the location of A647, the donor–acceptor distances, R_{calcd}s, were calculated based on the Förster mechanism. The R_{calcd}s are depicted in the structure of the LH2 complex as the colored circles (green and red) centered at B800 and B850 (Fig. 23.9; Here, the strucutre of *Rps. acidophila* 10050 (PDB 1NKZ) was adopted because of the lack of crystallographic structure of LH2 from *Rba. sphaeroides*. R_{calcd}s were 19 and 22 Å from B800 to B850 (Fig. 23.9a), respectively, in the micellar system. The latter distance was obviously insufficient for the A647 attached to the N-terminal domain, making direct energy transfer impossible. The R_{calcd} 19 Å from B800 suggested that the energy donor A647 (with the linkage length ~16 Å) was placed near the end of LH2β-polypeptide α-helix (N-terminal side). In the micellar solution, the hydrophobic region of LH2 was covered with detergent molecules as depicted in a and c (the top view from the N-terminal side) [15]. Even if the N-terminal domain were flexible enough to approach B800, detergents bound to LH2 would prevent A647 moieties from being localized at 19 Å external to the hydrophobic region. Concerning the N-terminal domain, the β-polypeptide of LH2-NC was longer than that of LH2 from *Rps. acidophila* 10050; the A647-attached C6 was away from the N-terminal of LH2β-polypeptide of *Rps. acidophila* 10050 by seven amino acids toward the N-terminal (Fig. 23.2). The C6 position could be presumed to be near the cytoplasmic surface of LH2-NC due to the P12 that would induce a bend in the conformation of the β-polypeptide chain. Considering the crystallographic structure of LH2 from *Rps. acidophila* 10050 (PDB: 1NKZ), the N-terminal domain shows a bend at A5 heading toward the LH2α-polypeptide (the direction of the center of circular structure, Fig. 23.9). Taken together, the energy transfer distance of 19 Å from B800 can be explained by the location of A647 (with ~16 Å of the linkage length) positioned near the end of the α-helix of the N-terminal

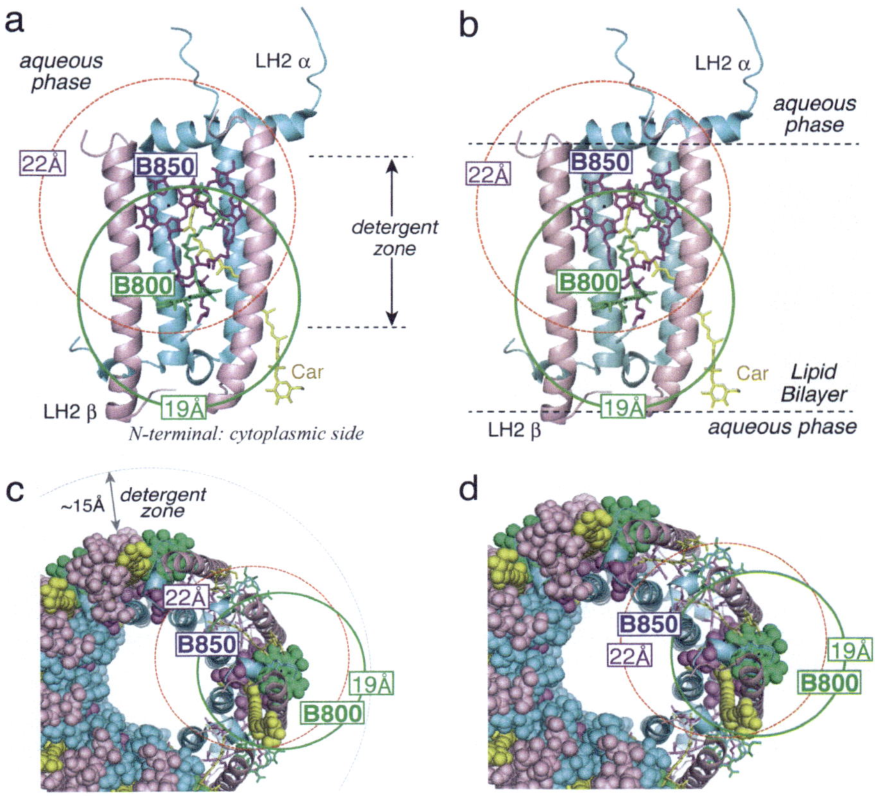

Fig. 23.9 Graphical representation of the donor–acceptor distances, R_{calcd}, estimated by the Förster mechanism for LH2-NC-A647 in the micellar (**a**, **c**) and reconstituted lipid bilayer systems (**b**, **d**). The figure modified with permission from Springer Nature Ref. [10] Copyright 2019

region. Such limits imposed by location and orientation of A647 moieties likely produce the single-phase dynamics of energy transfer to B800.

In the case of the reconstituted lipid bilayer system, R_{calcd} values evaluated were the same, 19 Å from B800 to 22 Å from B850. A possibility of the direct transfer to B850 is clearly ruled out due to the reason mentioned above and an impossibility of partitioning hydrophilic A647 into a lipid bilayer interior. The R_{calcd} value of 19 Å from B800 is realistic, as described above. The A647 moiety is possibly located near the membrane interface. Such an A647 moiety could function as a donor to B800 exhibiting single-phase dynamics. Some portions of A647 moieties in the conjugate were positioned such that they were unfavorable to energy transfer, resulting in deactivation in that region of the membrane interface to give a lower energy transfer efficiency (42%).

In many cases, dynamics of energy transfer from a covalently attached fluorophore to an intrinsic photosynthetic chromophore exhibits complicated multiexponential kinetics. This comes from heterogeneous conditions of fluorophore in or around protein environments. The LH2-NC-A647 demonstrated the single-phase dynamics of sequential energy transfer in the micellar and lipid bilayer systems. Such simple kinetics is quite rare as per our knowledge.

23.5 Conclusion

We described ultrafast excitation energy transfer in the biohybrid LH2s to which artificial chromophores A647 were covalently attached. We found that A647 successfully functions as the external light-harvesting chromophore. The rates of energy transfer were comparable to that takes place in natural systems. For the attachment of A647 to B800- and B850-sides, energy transfer exhibited multiexponential dynamics due to heterogeneity of the environment around A647. In sharp contrast, engineered LH2 bearing the Cys residue on the B800 side exhibited simple monoexponential kinetics of the sequential energy transfer A647 → B800 → B850.

LH2 conjugated with artificial pigments, therefore, will produce a promising motif in the methodology to construct multichromophore devices. This will provide deeper understanding of highly efficient energy transfer processes in natural photosynthetic systems, and clues for construction of highly efficient artificial photosynthetic systems.

Acknowledgements T. D. deeply appreciates to Dr. Tomoyasu Noji and graduate students (Mr. Daisuke Komori, Naoto Mizutani, Daiki Mori, and Ms. Akari Goto) for their contribution to this project. The present work was supported by JSPS KAKENHI Grant Numbers JP26107002, JP16J00627, and 17H05255, Grant-in-Aid for Scientific Research on Innovative Areas "Photosynergetics" and also by JP18H05160 and JP18H05180 in "Innovations for Light-Energy Conversion."

References

1. Inoue H, Shimada T, Kou Y, Nabetani Y, Masui D, Takagi S, Tachibana H (2011) The water oxidation bottleneck in artificial photosynthesis: how can we get through it? An alternative route involving a two-electron process. ChemSusChem 4:173–179
2. Scholes GD, Fleming GR, Olaya-Castro A, Van Grondelle R (2011) Lessons from nature about solar light harvesting. Nat Chem 3:763–774
3. Green BR, Parson WW eds (2003) In: Light-harvesting antennas in photosynthesis. Kluwer Academic publishers
4. Cogdell RJ, Lindsay JG (2000) The structure of photosynthetic complexes in bacteria and plants: an illustration of the importance of protein structure to the future development of plant science. New Phytol 145:167–196
5. Blankenship RE (2014) In: Molecular mechanisms of photosynthesis. Wiley Blackwell, UK
6. McDermott G, Prince SM, Freer AA, Hawthornthwaite-Lawless AM, Papiz MZ, Cogdell RJ, Isaacs NW (1995) Crystal structure of an integral membrane light-harvesting complex from photosynthetic bacteria. Nature 374:517–521

7. Hess S, Feldchtein F, Babin A, Nurgaleev I, Pullerits T, Sergeev A, Sundström V (1993) Femtosecond energy transfer within the LH2 peripheral antenna of the photosynthetic purple bacteria Rhodobacter sphaeroides and Rhodopseudomonas palustris LL. Chem Phys Lett 216:247–257
8. Joo T, Jia Y, Yu JY, Jonas DM, Fleming GR (1996) Dynamics in isolated bacterial light harvesting antenna (LH2) of *Rhodobacter sphaeroides* at room temperature. J Phys Chem 100:2399–2409
9. Yoneda Y, Noji T, Katayama T, Mizutani N, Komori D, Nango M, Miyasaka H, Itoh S, Nagasawa Y, Dewa T (2015) Extension of light-harvesting ability of photosynthetic light-harvesting complex 2 (LH2) through ultrafast energy transfer from covalently attached artificial chromophores. J Am Chem Soc 137:13121–13129
10. Yoneda Y, Kato D, Kondo M, Nagashima KVP, Miyasaka H, Nagasawa Y, Dewa T (2020) Sequential energy transfer driven by monoexponential dynamics in a biohybrid light-harvesting complex 2 (LH2). Photosynth Res 143:115-128
11. Dutta PK, Lin S, Loskutov A, Levenberg S, Jun D, Saer R, Beatty JT, Liu Y, Yan H, Woodbury NW (2014) Reengineering the optical absorption cross-section of photosynthetic reaction centers. J Am Chem Soc 136:4599–4604
12. Hughes LD, Rawle RJ, Boxer SG (2014) Choose your label wisely: water-soluble fluorophores often interact with lipid bilayers. PLoS ONE 9:e87649
13. Sumino A, Dewa T, Noji T, Nakano Y, Watanabe N, Hildner R, Bösch N, Köhler J, Nango M (2013) Influence of phospholipid composition on self-assembly and energy-transfer efficiency in networks of light-harvesting 2 complexes. J Phys Chem B 117:10395–10404
14. Pflock TJ, Oellerich S, Krapf L, Southall J, Cogdell RJ, Ullmann GM, Köhler J (2011) The electronically excited states of LH2 complexes from Rhodopseudomonas acidophila strain 10050 studied by time-resolved spectroscopy and dynamic monte carlo simulations. II. homo-arrays of LH2 complexes reconstituted into phospholipid model membranes. J Phys Chem B 115:8821–8831
15. Prince SM, Howard TD, Myles DAA, Wilkinson C, Papiz MZ, Freer AA, Cogdell RJ, Isaacs NW (2003) Detergent structure in crystals of the integral membrane light-harvesting complex LH2 from *Rhodopseudomonas acidophila* strain 10050. J Mol Biol 326:307–315

Part III
Advanced Construction of Meso- and Macroscopic Cooperative Photoresponsive Molecular Assemblies and Elucidation of Their Mechanisms

Chapter 24
Biomimetic Functions by Microscopic Molecular Reactions in Macroscopic Photoresponsive Crystalline System

Kingo Uchida, Ryo Nishimura, Hiroyuki Mayama, Tsuyoshi Tsujioka, Satoshi Yokojima, and Shinichiro Nakamura

Abstract Photoresponsive crystalline systems by microscopic molecular photoreactions to induce biomimetic functions are prepared using photochromic diarylethenes. Upon UV irradiation to crystals of open-ring isomer of diarylethene, needle-shaped crystals are formed on the surface by self-aggregation of the photogenerated closed-ring isomers. By using the photoinduced crystal growth technique, we prepared both superhydrophilic and superhydrophobic surfaces. Our technique is then successfully applied to prepare the double roughness surfaces by mimicking the structure of natural lotus leaf to show the bouncing ability of water droplet suggesting the importance of this double roughness structure. By a fine control of surface structures, we also prepared an object transportation system exploiting the bending behavior of surface-assembled diarylethene crystals. This system can transport polystyrene beads on a photo-actuated smart surface to a desired area by changing the light incidence irradiation directions. As further biomimetics, we prepared a hollow crystal showing photosalient phenomena. After packing small beads in the hollow, the crystal broke with scattering the included beads resembling the *Impatiens*.

Keywords Crystal growth technique · Diarylethene · Double roughness surface · Object transport system · Photosalient effect · Impatiens

---

K. Uchida (✉) · R. Nishimura
Department of Materials Chemistry, Ryukoku University, Seta, Otsu, Shiga 520-2194, Japan
e-mail: uchida@rins.ryukoku.ac.jp

H. Mayama
Department of Chemistry, Asahikawa Medical University, 2-1-1-1 Midorigaoka-Higashi, Asahikawa, Hokkaido 078-8510, Japan

T. Tsujioka
Department of Arts and Sciences, Osaka Kyoiku University, Asahigaoka 4-698-1, Kashiwara, Osaka 582-8582, Japan

S. Yokojima
School of Pharmacy, Tokyo University of Pharmacy and Life Sciences, 1432-1 Horinouchi, Hachioji, Tokyo 192-0392, Japan

S. Nakamura
RIKEN Cluster for Science, Technology and Innovation Hub, Nakamura Laboratory, 2-1 Hirosawa, Wako, Saitama 351-0198, Japan

H. Miyasaka et al. (eds.), *Photosynergetic Responses in Molecules and Molecular Aggregates*, https://doi.org/10.1007/978-981-15-5451-3_24

24.1 Introduction

Diarylethenes are well-known photochromic compounds which undergo cyclization reaction to form colored closed-ring isomer upon UV irradiation, while they revert to the original colorless open-ring isomer upon visible-light irradiation. The most striking features of the photochromic reactions are thermal stability of both isomers, high quantum yields of the reaction, high fatigue resistance, and photoreactivity in the crystalline state [1, 2]. We have found the photoinduced crystal growth of needle-shaped crystals of a closed-ring isomer upon UV irradiation to single crystal and microcrystalline surface of a diarylethene followed by storage at 30 °C for 24 h in the dark [3]. Such photoinduced topographical changes on the surface are due to self-aggregation of photogenerated molecules of the closed-ring isomer. The photoinduced crystal growth technique is considered to be one of the molecular assembling techniques.

On the other hand, photoinduced bending of a diarylethene crystals reported by Irie et al. is also the system in which microscopic molecular photoreaction induces the macroscopic changes of the system [4]. We aimed to build up the system where bending crystals are aligned just like crystalline brush and they bend by photoirradiation to transport an object on the system. In the present chapter, we describe the molecular assembled crystalline system in which photoinduced molecular change introduced the macroscopic functional changes of the system including surface topographical changes above mentioned.

24.2 Photoinduced Topographical Changes of Diarylethene Derivatives Providing Control of the Superhydrophobicity

In 2006, we have reported photoinduced topographical changes of a photochromic diarylethene crystalline thin film of 1,2-bis(2-methoxy-5-trimethylsilylthien-3-yl)perfluorocyclopentene (**1o**) via eutectic mixture state as shown in Fig. 24.1. Upon UV irradiation (254 nm) to the surface of a single crystal of **1o** (Fig. 24.1b), the color became deep blue within 5 min. After 10 min irradiation with the light followed by storage in the dark, the surface of the crystal was covered with needle-shaped crystals of **1c** (Fig. 24.1c) [3, 5]. Upon irradiation of the surface with visible light (λ > 500 nm), the color disappeared, and the needle-shaped crystals melted. Changes in surface topography of microcrystalline film prepared by coating the chloroform solution of **1o** on a glass substrate were also found [3, 5, 6]. The explanation of the mechanism of crystal growth and melting of the needle-shaped crystals by alternate irradiation with UV and visible light is given by a phase diagram (Fig. 24.1d). The shape of the crystals of **1o** and **1c** is cubic and needles, respectively, in the case of recrystallization from the solutions of these isomers. The needle-shaped crystals of **1c** tend to grow in one direction along the A-axis of the crystal [3, 5, 6]. Upon

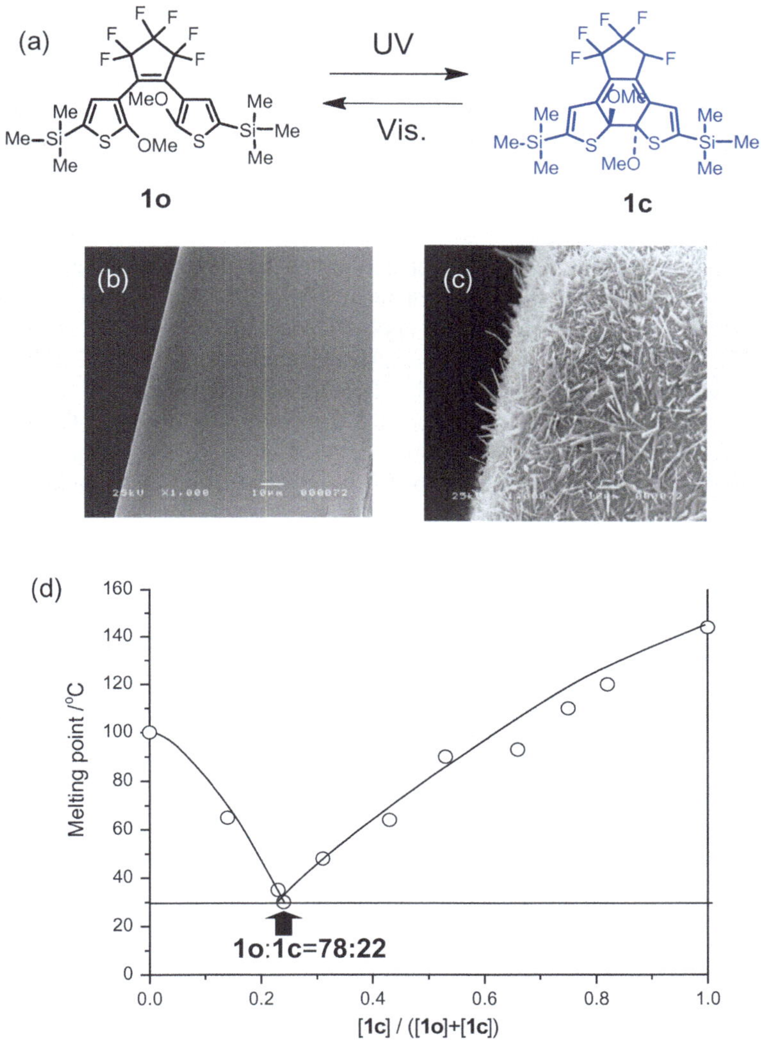

Fig. 24.1 Photoinduced topographical changes on the surface of a single crystal of **1o**. **a** Molecular structures of **1o** and **1c**. Scanning electron microscope (SEM) image of a crystal surface of **1o** **b** before UV irradiation, and **c** after UV (254 nm, 12 W) irradiation for 10 min followed by storing for 24 h at 30 °C in the dark. (SEM images: ×1000). After visible-light (λ > 500 nm, 500 W, 20 min) irradiation to the surface of (**b**), the surface reverts to the flat one like (**a**). **d** Phase diagram of mixtures of open- (**1o**) and closed-ring (**1c**) isomers. The melting point changes upon irradiation with UV light (313 nm) on the crystal surface of **1o** due to the change in molar ratio by the photoisomerization

UV irradiation, **1o** is converted to **1c**. When the molar ratio of **1c** exceeds 22%, crystal of **1c** with fibril structure grows on the film surface. The surface covered with needle-shaped crystals of **1c** showed superhydrophobicity on which contact angle (CA) and sliding angle (SA) of a water droplet are 163° and 2°, respectively. They are almost the same as those on natural lotus leaf. Therefore, photocontrol of the superhydrophobicity (CA is larger than 150°) of the surface was achieved [3, 5, 6].

Such photoinduced topographical changes are observed for the other photochromic diarylethene derivatives **2o–4o**. Figure 24.2 shows the molecular structures. For the single crystal of **2o**, Irie et al. reported the photoinduced topographical changes by step-valley formation mechanism (SVM) [7]. We examined the difference between SVM and the melt- and crystal-growth mechanism (CGM) observed for **1o**. The key parameter is the glass transition temperature (T_g). At 25 °C, the small step was formed upon irradiation with UV light to the crystal surface of **2o**. At 30 °C which is around the T_g of **2o**, the steps on the surface were blurred [8]. These results indicate that SVM changed to CGM for the photoinduced topographical changes of diarylethenes around T_g. The step and valley formations were not observed

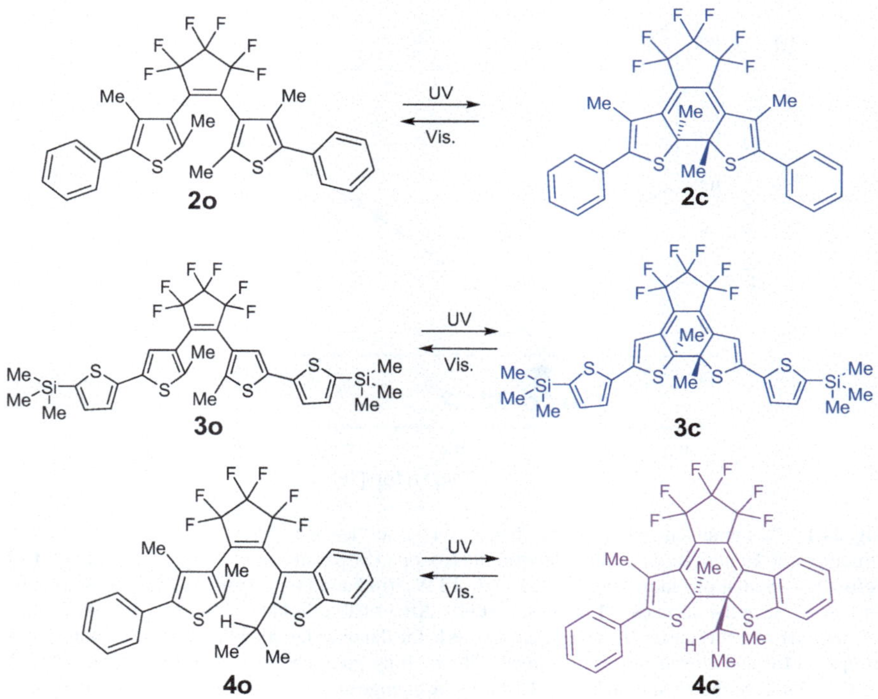

Fig. 24.2 Molecular structures of open- (o) and closed-ring (c) isomers of the diarylethenes **2, 3**, and **4**

above T_g. Then, we monitored the crystal growth **1c**, **2c**, **3c**, and **4c** on the microcrystalline surfaces of the open-ring isomers above T_g, where the T_g of **1o**, **2o**, **3o**, and **4o** are −13.0, 27.9, 31.3, and 24.7 °C, respectively. The shapes of photogenerated crystals of their derivatives are needles for **1c** and **3c**, cubic for **2c**, and plates for **4c** as shown in Fig. 24.3 [8]. They are attributed the crystal structure of the closed-ring

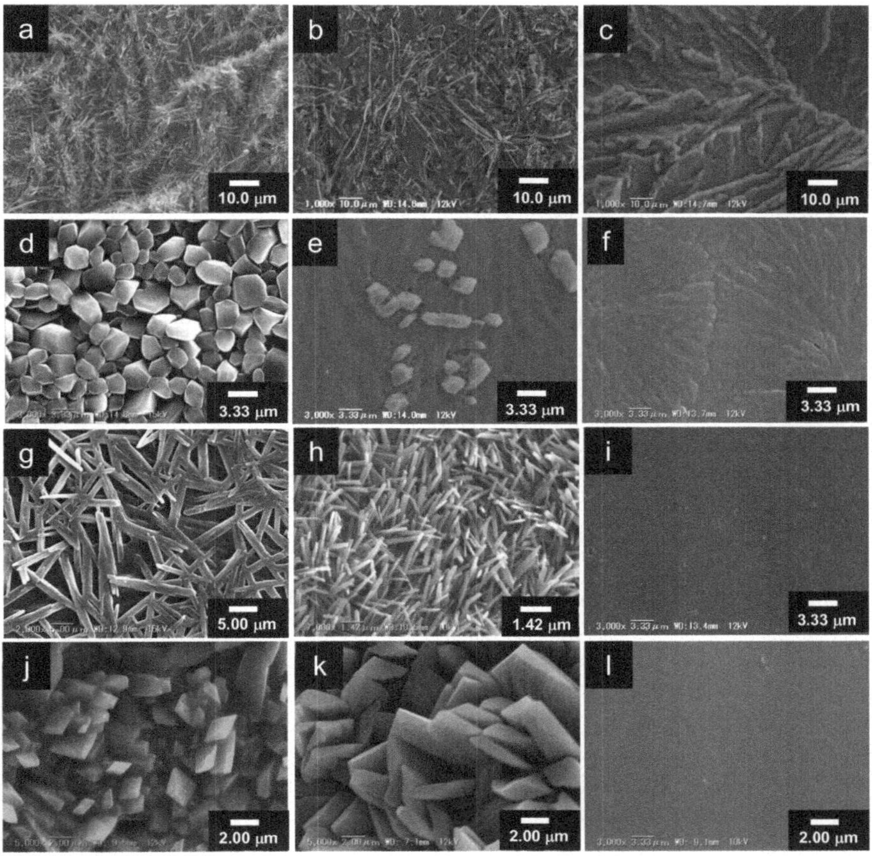

Fig. 24.3 SEM images of the diarylethene microcrystalline surfaces after UV irradiation followed by storage in the dark. **a**–**c**: surface of **1o**, **d**–**f**: surface of **2o**, **g**–**i** surface of **3o**, and **j**–**l** surface of **4o**. **a**, **d**, **g**, and **j** Stored at the eutectic temperature of the mixture of the open- and closed-ring isomers. **b**, **e**, **h**, and **k** Stored at the T_g or at the temperature above the T_g and below the eutectic temperatures, which are indicated below for each case. **c**, **f**, **i**, and **l** Stored at the temperature below the T_g. **a** Stored at 30 °C for 15 h, **b** stored at 8 °C which is above the T_g and below the eutectic temperature for 48 h, **c** stored at–30 °C for 1 week, **d** Stored at 120 °C for 24 h, **e** stored at 30 °C of the T_g for 72 **h**, **f** stored at 8 °C for 1 week, **g** stored at 141 °C for 20 min, **h** stored at 30 °C of the T_g for 12 h, **i** stored at 8 °C for 7 days, **j** stored at 100 °C for 3 h, **k** stored at 80 °C which is above the T_g for 48 **h**, **l** stored at 8 °C for 7 days. Magnitude: **a**, **b**, **c**, **i**: ×3000, **d**, **e**, **f**: ×1000; **g** ×2000; **h** ×7000; **j**, **k** ×1000, **l**: ×3000

isomers. The size of the crystals is inversely proportional to the molecular size. For example, **1c** and **3c** form needle-shaped crystals. Since the molecular size of **1c** is smaller than **3c**, the size of needle-shaped crystals of **1c** is much larger than that of **3c** (Fig. 24.3).

24.3 Photoinduced Topographical Changes of a Diarylethene Derivative Providing Control of the Superhydrophilicity

Photoresponsive superhydrophobic surfaces with the lotus effect were prepared in the previous section. In contrast, it was difficult to prepare photoresponsive superhydrophilic surfaces; in order to construct a superhydrophilic surface, it is required to prepare a derivative of diarylethene that is hydrophilic enough to reduce a contact angle of less than 65° on its flat surface, but not dissolve in water. Here, the critical contact angle of 65° is based on the report that a contact angle less (more) than 65° on flat polymer surfaces decreased (increased) when the surface roughened [9].

We prepared the photoresponsive superhydrophilic surface of microcrystalline films of diarylethene derivatives with ionic liquid structures [10]. A series of diarylethene derivatives that we have synthesized has pyridinium salt moieties. We found that a derivative **5o** (upper panel in Fig. 24.4) shows photoinduced reversible topographical changes upon alternate irradiation with UV and visible light and that a superhydrophilic surface was generated (Fig. 24.4) [10].

The molecular structures, the estimated crystal shapes from the crystal units, and the phase diagram of **5o** and **5c** are shown in Fig. 24.4. The melting points of **5o** and **5c** were 203.2 and 190.2 °C, respectively. The eutectic temperature was 161.1 °C, where the ratio of **5o** and **5c** was 66: 34. The T_g of **5o** was measured to be 64.4 °C by DSC when heating the glassy state of **5o**, which was formed by cooling down the melted **5o** after reaching the melting temperature. As we have discussed for the photoswitching of hydrophobic surfaces, photoinduced crystal growth took place at temperatures above T_g, and in fact, the current photoinduced topographical changes were observed at 70 °C, which is slightly higher temperature than the T_g.

The microcrystalline film of **5o** was prepared by coating with an acetonitrile solution on a glass substrate and by evaporating the solvent. The surface was irradiated with UV light for 5 min, and the surface was melted and became flat (Fig. 24.5 right panel) where the ratio of **5o** and **5c** was nearly 4: 6. It is the hydrophilic surface, whose contact angle was less than 65°. Thus, a superhydrophilic surface can be realized by roughening the surface [9]. After visible-light ($\lambda > 430$ nm) irradiation for 5 min onto the surface followed by maintaining the temperature of the film at 70 °C for 24 h, bumpy crystals of **5o** grew on the surface, and the contact angle became 0° (Fig. 24.5 left panel) [10].

The reversible formations of the flat and the superhydrophilic surfaces were observed by alternate irradiation with UV and visible light at least three times.

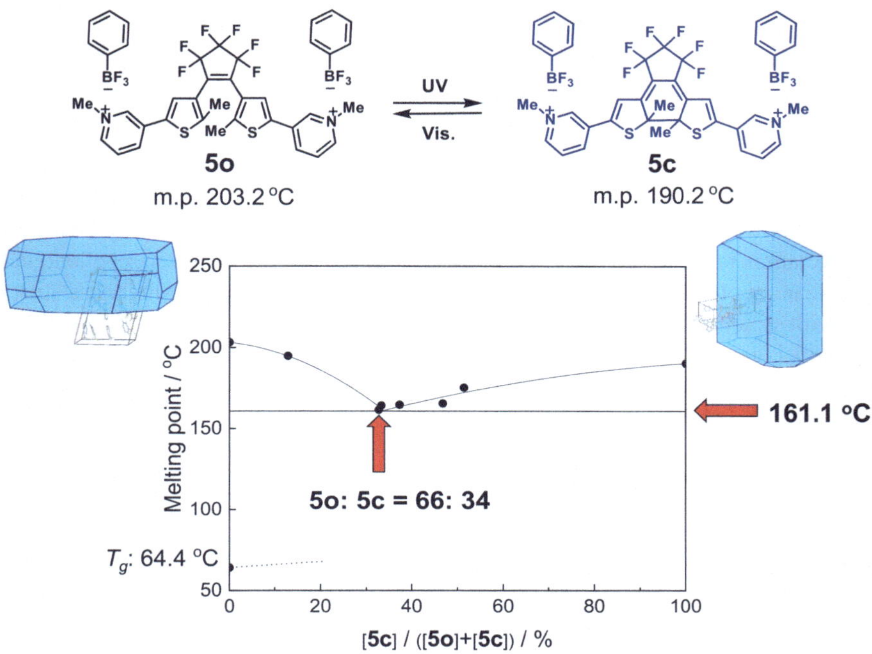

Fig. 24.4 Molecular structures of open- (**5o**) and closed-ring (**5c**) isomers of a diarylethene having ionic liquid structures, the estimated crystal shapes, and the phase diagram of **5o** and **5c**

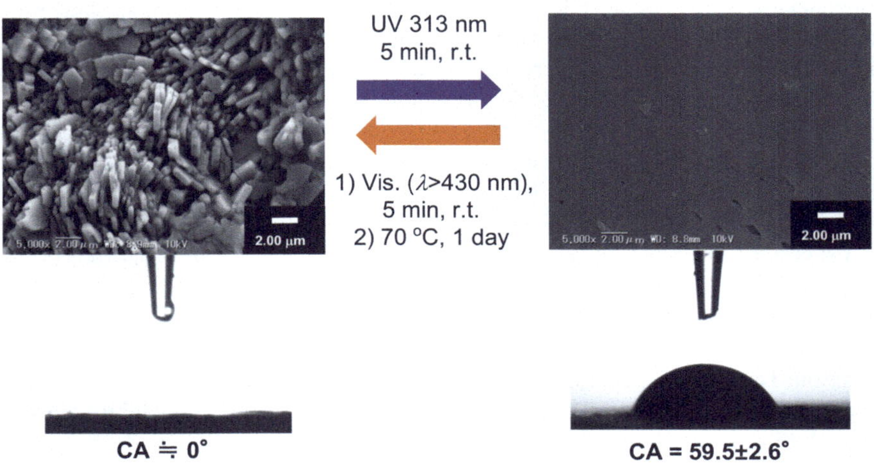

Fig. 24.5 Photoswitching of the superhydrophilicity. (Upper panel) Photoinduced topographical changes on microcrystalline surface of open-(**5o**) and closed-ring (**5c**) isomers of a diarylethene. (Lower panel) CA changes of a water droplet on the surface

24.4 Photoinduced Fabrication of Double Roughness Surface Mimicking Lotus Leaf and the Superhydrophobicity

In nature, many plants and insects have superhydrophobic surfaces with double roughness structure. For example, the lotus leaf and legs of a water strider have the double roughness structure. Historically, the "lotus effect" found by Barthlott and Neinhuis is the milestone [11]. They investigated the super water-repellent and self-cleaning effect of the lotus leaf and attributed it to the double roughness structure with structures and nanostructures on its surface, together with the hydrophobic properties of epicuticular wax [12]. As for the self-cleaning, the bouncing of a water droplet plays an important role, and artificial double roughness structures have been prepared [13, 14]. Here, we report a guiding principle to prepare a superhydrophobic surface with the double roughness structure of a diarylethene microcrystal by using crystal growth technique (CGT). The prepared surface bounced a water droplet on it [15, 16].

The procedure to prepare the double roughness structured microcrystalline surface of **1o** is shown in Fig. 24.6. Initially, the microcrystalline surface of **1o** (Fig. 24.6-1) was irradiated with UV light. Then, the colorless **1o** near the surface is photoisomerized to be **1c** with blue color (Fig. 24.6-2). By storage at 50 °C, the needle-shaped crystals of **1c** grew on the surface (Fig. 24.6-3). Due to Ostwald ripening, the crystals covering the surface changed from the needle-shaped to the rod-shaped crystals of **1c** after storage for 24 h at 50 °C in the dark (Fig. 24.6-4) [5]. To make the double roughness structure, we need to grow needle-shaped crystals on rod-shaped crystals as in Fig. 24.6-9a, i.e., first make **1o** on rod-shaped crystals by irradiation with visible light, then grow needle-shaped crystals of **1c** by irradiation with UV light. However, the rod-shaped crystals of Fig. 24.6-4 were not thick enough; the crystals melt upon visible-light irradiation.

To make thick rod-shaped crystals of **1c**, we further applied the Ostwald ripening. The surface of Fig. 24.6-4 was irradiated with UV light again and stored it at 70 °C for 3 h in the dark (Fig. 24.6-5). Then, the small rod-shaped crystals of **1c** grew on the surface (Fig. 24.6-6), and were incorporated in the larger rod-shaped crystals (Fig. 24.6-7a). The average width and length of these rod-shaped crystals at the state shown in Fig. 24.6-7a were 10 and 21 μm, respectively.

Upon visible-light irradiation to the surface of Fig. 24.6-7a, cycloreversion reaction from **1c** to **1o** proceeded, and the surface was covered with cubic-shaped crystals of **1o** (Fig. 24.6-8a) on the 10-μm rough structures. The size of each bumpy rod-shaped crystal was almost the same as that of a projection on a lotus leaf covered with plant wax tubes. The enlarged crystals were not melted [5, 6].

Upon UV irradiation to the surface, each rod-shaped crystal was covered with needle-shaped crystals of **1c** of 0.2–0.5 μm width and 3–5 μm length. Thus, double roughness structures were obtained (Fig. 24.6-9a). The needle-shaped crystals were much smaller than those grown on the flat microcrystalline surface of **1o** due to the small supply of **1c** crystals to form needle-shaped crystals on each projection. The

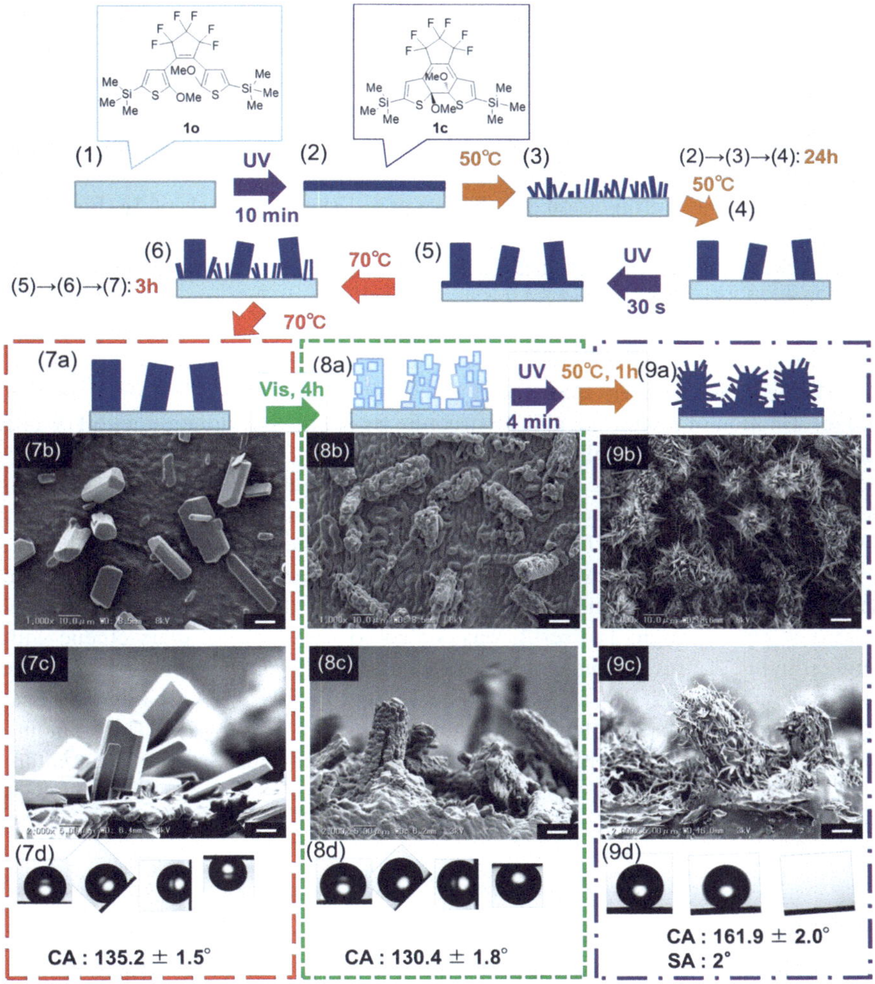

Fig. 24.6 Process of the preparation of double roughness structures. (1–6,7a,8a,9a): the surface illustrated by pale blue and dark blue colors correspond to crystals of **1o** and **1c**, respectively. SEM images of each surface: (7b) a surface with rod-shaped crystals of **1c**, (8b) the surface after irradiation with visible light to that of 7b, (9b) the surface after second UV irradiation to that of 8b followed by storing it at 50 °C for 1 h in dark; (7c, 8c, 9c) the side view of the surfaces of 7b, 8b, and 9b, respectively; (7d, 8d, 9d) water droplet (1.5 μL) on surfaces of 7b, 8b, and 9b, respectively. Scale bars and magnifications, 10 μm and 1000 times for 7b–9b; 5 μm and 2000 times for 7c–9c

SEM images of the surfaces in Fig. 24.6-7a,8a,9a are shown in Fig. 24.6-7b,8b,9b, respectively. Corresponding SEM images of cross sections of samples in Fig. 24.6-7b,8b,9b are shown in Fig. 24.6-7c,8c,9c, respectively. The size distribution of the microstructures on natural lotus leaf was reported by box-counting fractal analysis [17]. The size distribution of the double roughness structure was very similar to that of natural lotus leaf.

The CAs of water droplets on the surfaces of rod-shaped crystals of **1c** (Fig. 24.6-7a,7b,7c) and cubic crystals of **1o** (Fig. 24.6-8a,8b,8c) were around 130°. The water droplets were pinned on the surface, even when the surfaces were upside down (Fig. 24.6-7d,8d). The CA and SA on the surface with double roughness structures were 161.9 ± 2.0° and 2°, respectively. The advanced (CA_{ad}) and receding (CA_{rec}) CAs of a water droplet were 162.0 ± 1.6° and 158.3 ± 1.4°, respectively. Thus, a remarkable lotus effect was observed (Fig. 24.6-9d). The reversibility of the melting and reformation of double roughness structure by alternate irradiation with UV and visible light were successfully achieved.

The static CAs and SAs of water droplets on the microcrystalline surface with double roughness structure resembled those obtained for single roughness surfaces of the same diarylethene **1**. To further examine the hydrophobic property of the surface, a water-droplet bouncing experiment on the diarylethene microcrystalline surface with double roughness structures was carried out by dropping the water droplet from $h = 1.8$ mm high. The result was compared with those of diarylethene with single roughness structure and natural lotus leaf (Fig. 24.7). On the lotus leaf, water droplets bounced (Fig. 24.7a), while bouncing was not observed on a diarylethene microcrystalline surface with single roughness structure (Fig. 24.7b). The bouncing was reproduced only on the microcrystalline surface with double roughness structure (Fig. 24.7c). The importance of double roughness structure over single roughness structure is thus confirmed.

24.5 Dual Wettability on Diarylethene Microcrystalline Surface Composited with Two Types of Needle-Shaped Crystals of Two Diarylethene Derivatives

In nature, many plants and insects have specific interesting properties with specific rough structures. The termite wing has a specific property of wetting in contact with a water droplet: it adsorbs water mist, whereas larger water droplets are bounced on the surface [18]. It is called "dual wettability." The surface of termite wing is covered with two types of projections of different sizes. One is a hair-like projection with about 50 μm long and 1 μm wide, whereas the other is a star-shaped projection of 5–6 μm in height and width. These hairs protect the wing from the contact with large water droplets, whereas the small projections work to keep the small droplets on the surface of the wing, followed by gathering to build up large ones to remove them via the hair arrays. Here, we reproduce the dual wettability of termite wing by

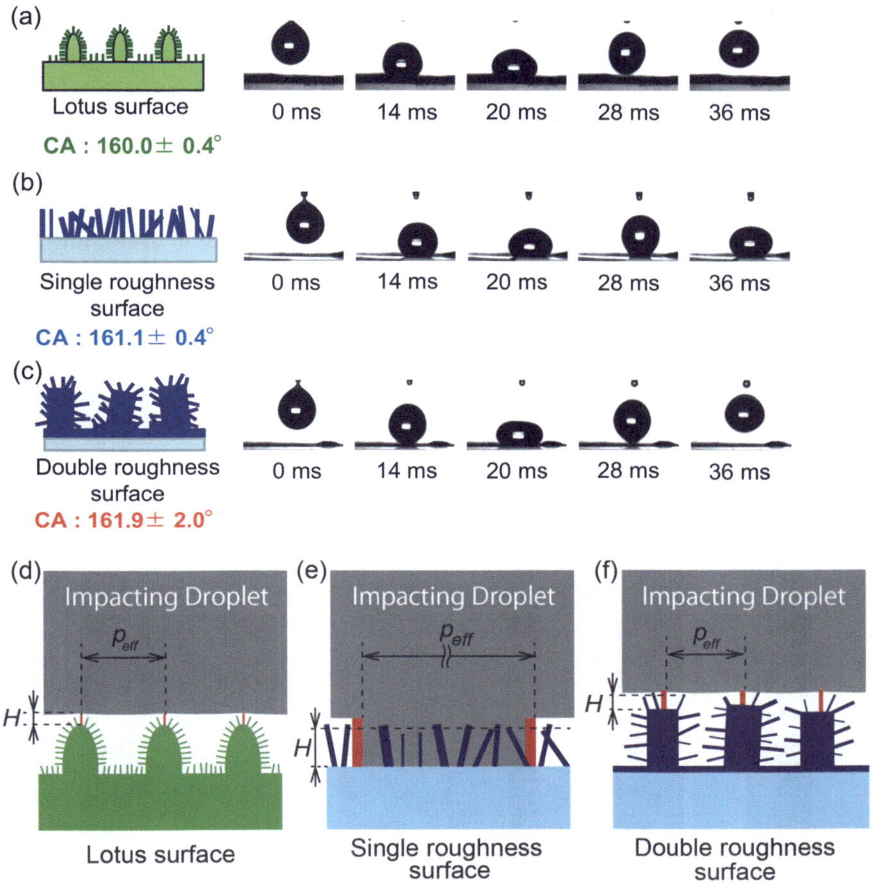

Fig. 24.7 Comparison of bouncing phenomena of a water droplet (7.6 μL) on different surfaces. Water-droplet-bouncing phenomena on **a** lotus leaf, and on diarylethene microcrystalline surfaces with **b** single and **c** double roughness structure ($h = 1.8$ mm). Illustration of the parameters determining Laplace pressure (P_L) on **d** lotus leaf, and **e** single and **f** double roughness structures. P_L is determined by height H of surface structures and effective pitches p_{eff} between almost perpendicularly standing needles indicated by red color

a photoinduced crystal growth technique [19] by mixing two derivatives **1o** and **3o** with expecting that two types of crystals grow on the surface with different sizes just like the structure on termite wing. A solution containing equimolar amount of **1o** and **3o** was coated onto a glass plate. After the evaporation of the solvent in vacuo, the surface of the microcrystalline was irradiated for 5 min with 313-nm light followed by storage at 30 °C. This temperature is higher than both of the T_g of **1o** and **3o** crystals. Therefore, crystals of **1c** and **3c** grew on a softened surface above T_g [8]. After storage of 9 days at this temperature in the dark, the surface was covered with two types of needle-shaped crystals of **1c** and **3c** (Surf1c + 3c). SEM images at different angles with different mixture of **1c** and **3c** are shown in Fig. 24.8(a–f). The larger needle-shaped crystals grown on the mixed microcrystalline surface are crystals of **1c**, and their lengths and widths are about 16 and 1.5 μm, respectively, whereas the smaller ones are crystals of **3c** with lengths and widths around 1.9 and 0.2 μm, respectively. These sizes are quite consistent with those observed for homo microcrystalline surfaces.

The photoinduced CGT of **3o** was already studied in detail [20]; however, the mixture of two diarylethenes may produce the surface with the sum of two different characteristics or completely different characteristics.

The obtained photogenerated rough surface with the two different-sized needle-shaped crystals of **1c** and **3c** shows specific bouncing ability mimicking a termite wing as we have intended. Then, we analyzed the surface structures to clarify the relation between the wettability by monitoring the different sizes of water droplet on the surface [21]. Water droplets with the volume of 7.6 ± 0.6 μL were poured from

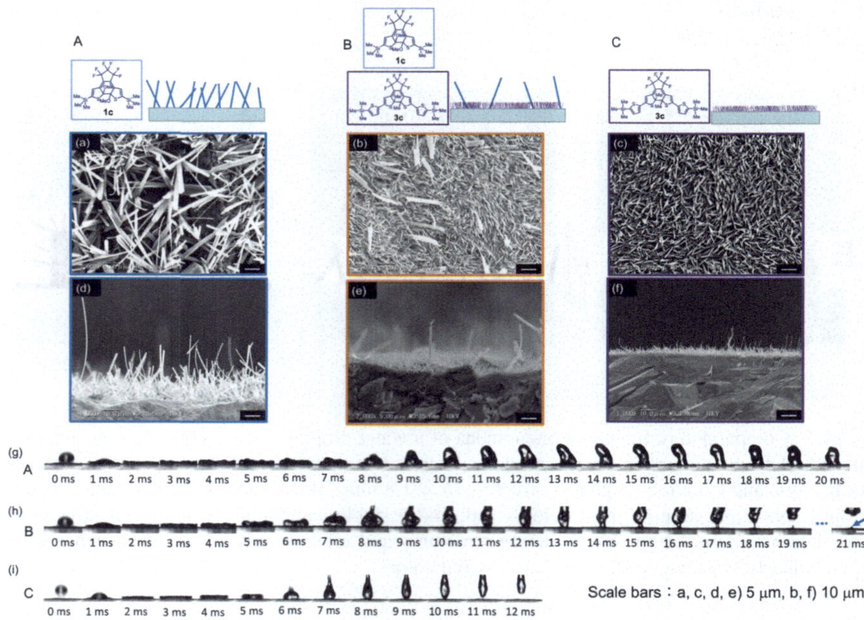

Fig. 24.8 Comparison of the photoinduced surface topologies. (Upper panels) **A**: Surface of **1c**, **B**: Surface of **1c** + **3c**, and **C**: Surface of **3c**. Illustrations of pale blue substrate and dark blue crystals correspond to crystals of open- and closed-ring isomers, respectively. (Middle panels) SEM images of each surface; **a** and **d**: surface with needle-shaped crystals of **1c**, **b,** and **e**: surface with needle-shaped crystals of **1c** and **3c**, **c** and **f**: surface with needle-shaped crystals of **3c**. Scale bars: 5 μm for images of **a**–**c** and **e**, which are magnified 2000 times; 10 μm for images of **d** and **f**, which are magnified 1000 times. (Lower panels) **g** Optical images of a non-bouncing droplet on surface A. **h** Optical images of a bouncing droplet on surface B. A small droplet remained on the surface (indicated by a blue arrow). **i** Optical images of a bouncing droplet on surface C. Scale bar of the last image in h is 1 mm

a height of 10 cm onto these surfaces for monitoring the bouncing phenomena. The microcrystalline surface consists of only **1o** forming a rough surface of larger needle-shaped crystals shows water-adhesive properties as shown in Fig. 24.8g, whereas the microcrystalline surface consists of only **3o** forming a rough surface of smaller needle-shaped crystals shows water-bouncing properties (Fig. 24.8i). We released a water droplet to the surface B with two types of needle-shaped-crystals of **1c** and **3c** in the same way and found a small portion of water remained on the surface (Fig. 24.8h). On this surface in which needle-shaped crystals worked to provide either water adhesion or repellency is a question. These results indicated that on surface B, the longer needle-shaped crystal of **1c** achieved water-capturing capability, whereas the smaller needle-shaped crystals of **3c** achieved water-repelling capability.

Then, we measured the sizes of all droplets contacting the surfaces. By comparing the size distributions of the droplets, the bouncing and non-bouncing ranges are separated by mixing the two diarylethenes.

On the surfaces of A and C, the difference between the average of the size of a droplet showing bouncing and that showing non-bouncing behavior is relatively small, at 76.2 and 66.8 μm, respectively. On the other hand, this difference for B was rather larger (155.1 μm). These results indicate that by mixing the two diarylethene derivatives, the difference between bouncing and non-bouncing water droplets became dramatically separated, i.e., the dependence on the size of water droplets was clearly distinguished for both types of wettability (Fig. 24.9).

24.6 Object Transportation System Consists of Bending Crystals

The design of an object transportation system exploiting the bending behavior of surface-assembled diarylethene crystals is reported. Since Irie and his co-workers reported the photoinduced bending behavior [4], photoinduced bending, rolling and twisting [22], and elegant demonstration of an iron ball lifting and gear rotating on a bending crystals [23, 24]. Herein, we design the array of bending crystals with cooperative photoinduced bending to carry the object far from the UV incidence, just like photo-actuated cilia of inspired *paramecium aurelia* (Fig. 24.10) [25]. Molecular packing of **6o** in the crystal is shown in Fig. 24.11. There are two conformers of *M*- and *P*-helix molecules, and they are alternatively combined with intermolecular hydrogen bonding along long axis (b-axis) of the crystal.

The crystal array of the bending crystals of **6o** was prepared by sublimation; however, direct sublimation of **6o** to obtain an amorphous film of **6o** with a nearly flat surface with dotted microstructures whose average diameter and height were 0.30 μm and 0.43 μm, respectively (Fig. 24.12a). Then the dotted surface was coated with Au–Pd alloy by magnetron sputtering, and then the surface with the Au–Pd alloy was used as a target substrate resulted in the accelerated growth of thin platelet crystals of **6o** on each of the dots on the glass substrate (Fig. 24.12b). The average length and

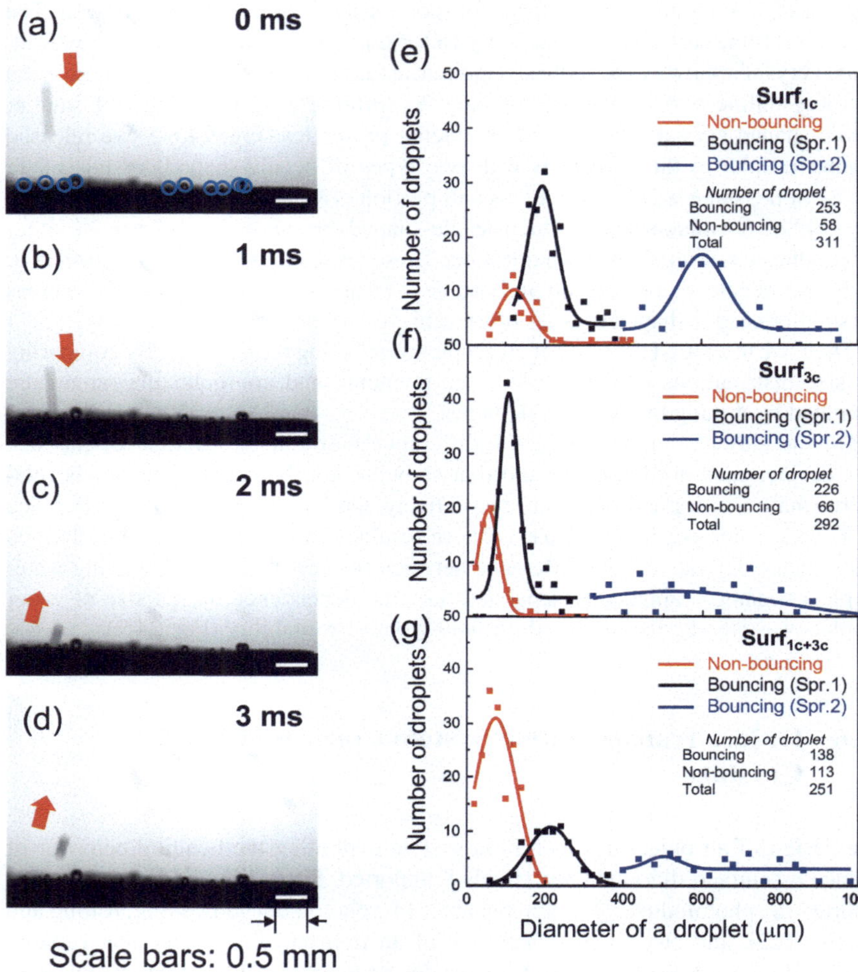

Fig. 24.9 Optical images of a microdroplet and size distributions. **a–d** Optical afterimages during bouncing of a water droplet on Surf1c + 2c in 3 ms. Red arrows indicate direction of motion of the droplet. Blue circles in **a** are droplets adsorbing on the surface. Scale bars for the optical images in **a–d** are 0.5 mm. **e–g** Distributions of diameters of water droplets contacted with Surf1c, Surf3c, and Surf1c + 3c. **e** Surf1c, **f** Surf3c, and **g** Surf1c + 3c. In **e–g**, red dots: diameter distribution of non-bouncing droplets, black dots: diameter distribution of bouncing droplets generated by sprayer 1 (Spr. 1: Generating the microdroplet of 40–400 μm in diameter), and blue dots: diameter distribution of bouncing droplets generated by sprayer 2 (Spr. 2: generating the microdroplets of 400–1000 μm in diameter). Number of droplets used for preparation of distribution is shown inside the graphs (e)–(g)

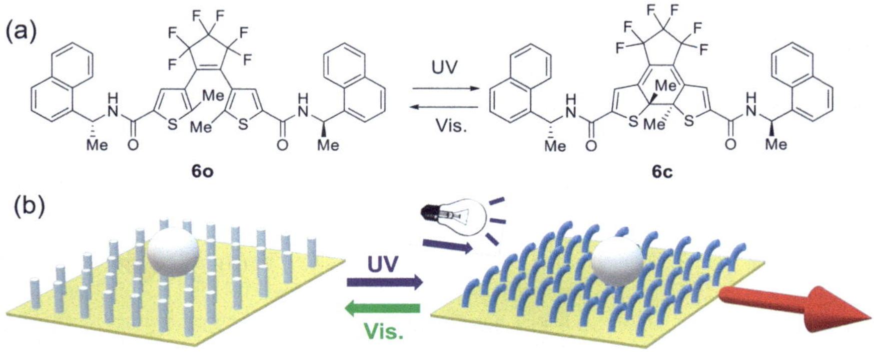

Fig. 24.10 **a** Molecular structure of the open- (left) and closed-ring (right) isomers of diarylethene derivative **6** and **b** the concept of the photoreactive object transportation system

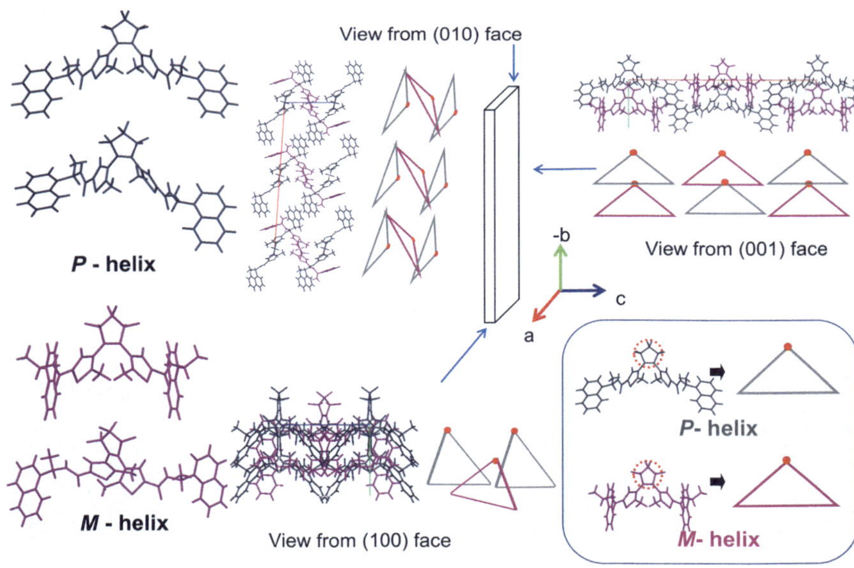

Fig. 24.11 Packing of molecules of **6o** in the single crystal

width of the platelet crystals of the **6o** on the glass substrate was 384.5 and 10.5 μm, respectively (Fig. 24.12c,d).

The long axis of the platelet crystals in Fig. 24.12 is assigned to the b-axis in Fig. 24.11. Upon UV irradiation from both (100) and (001) surfaces, the crystal bends far from incidence because short axis of **6o** molecules expands at photo-irradiated surface.

To the surface with array of bending crystals of **6o**, UV light was irradiated from left side. Then, all crystals bent to right side cooperatively. On the surface,

Fig. 24.12 **a** Packing of molecules of **6o** in the single crystal. (scale bar for **c**: 3.33 μm)

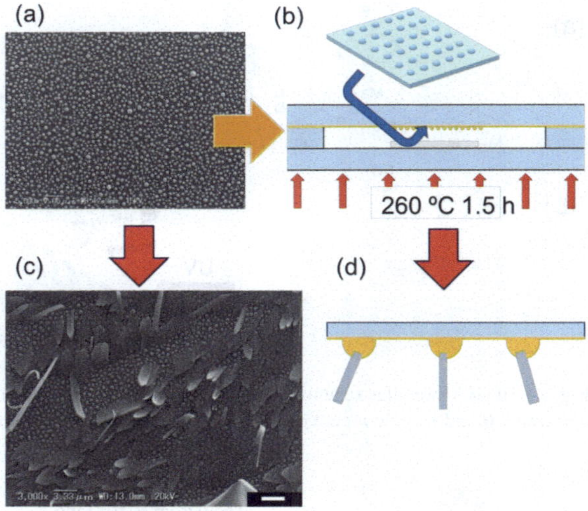

a polystyrene bead (PB) with 1 mm diameter was placed. Then, the UV light was irradiated. Initially, (a) UV-light was irradiated from the bottom of the screen, the PB moved to top side and then (b) visible light was irradiated orthogonal to the surface followed by UV-light irradiation from the top of the screen. The PB moved to the bottom direction as shown in Fig. 24.13. These results indicate the cooperative bending of the DAE crystals works well to transport a PB. This smart surface will be applicable for remote-controlled object transportation in various environments and holds potential for the development of soft robots.

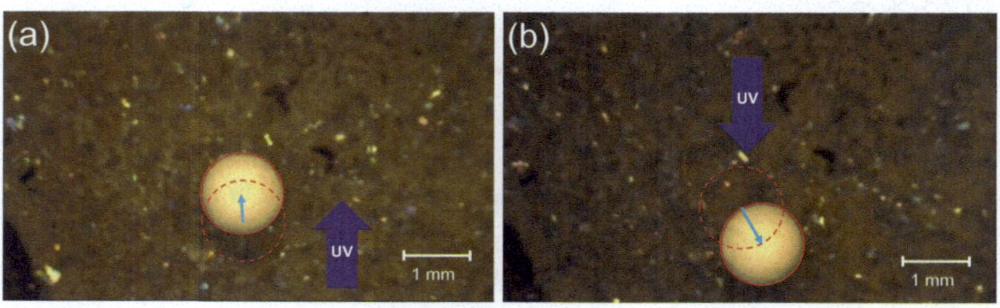

Fig. 24.13 Photographs showing the transportation of the PB upon UV and visible-light irradiation. **a** After UV-light irradiation from the bottom of the screen, **b** after visible-light irradiation orthogonal to the surface followed by UV-light irradiation from the top of the screen. Scale bars: 1 mm

24.7 Photosalient Phenomena of Hollow Crystals of a Diarylethene

In diarylethene crystals, the most commonly observed phenomena have been bending and twisting [4, 22]. On the other hand, light induced scattering of organic crystals are well known as photosalient effect [26]. A photosalient effect was also observed in non-hollow crystals of some diarylethenes [27, 28]. Recently, we synthesized a diarylethene **7o** having a perfluorocyclohexene ring with two thiazole rings and reported photochromism in solution and a crystalline state as well as the photosalient effect [29]. The thin crystals of **7o** fabricated by sublimation at normal pressure showed bending, jumping, and scattering upon UV irradiation.

A diarylethene **7o** with a perfluorocyclohexene ring formed hollow crystals by sublimation under ambient pressure. Upon UV irradiation of the crystals, they showed remarkable photosalient phenomena and scattered into small pieces. The speed of the flying debris released from the crystal by UV irradiation exceeded several meters per second. It is due to the larger molecular size changes from **7o** to **7c** (Table 24.1, Fig. 24.14), which is much larger than those of corresponding five-membered ring derivative [29].

By sublimation of **7o** under ambient pressure, hollow crystals were appearing during sublimation in 5% yield. They have photoresponsive properties, which have never been reported previously. Molecular packing of the hollow crystal was analyzed at SPring-8 (synchrotron radiation facility), and it was found to be the same crystal structure as that observed in the bulk crystal. The hollow structure of **7o** can effectively achieve photoinduced scattering. The formation of such a hollow structure is attributed to the crystal growth from surrounding microcrystals by Ostwald ripening. During this growth, it is difficult to supply **7o** for the central part of the standing crystals (Iwanaga model) [30]. Optical and SEM images of the hollow crystal are shown in Fig. 24.15a, b. Upon UV irradiation in the direction of the white arrow, small cracks were generated on the UV-irradiated surface (Fig. 24.15d). This occurred because the standing crystal fixed on the substrate did not scatter into small pieces.

Next, small plastic beads (1.0-μm diameter) containing fluorescent dye were packed into the hollow crystal, and the crystal was irradiated with UV light

Table 24.1 Crystal data for **7o** and **7o-UV**

	7o	**7o-UV** (after 400 nm light irradiation for 5 min)
a/Å	13.3855(4)	13.4391(6) + 0.40%
b/Å	25.1206(9)	25.2408(14) + 0.48%
c/Å	14.7089(5)	14.6414(8) − 0.46%
α/°	90	90
β/°	97.3991(10)	97.8337(16) + 0.45%
γ/°	90	90
V/Å^3	4904.7(3)	4920.2(4) + 0.32%

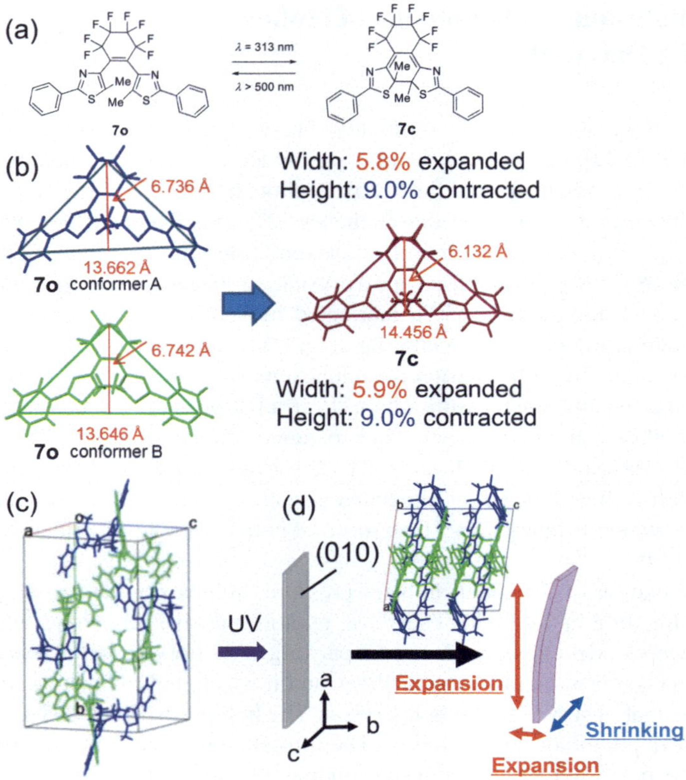

Fig. 24.14 **a** Molecular structures of the open- (left) and closed-ring (right) isomers of diarylethene derivative 7, **b** Changes of molecular sizes from **7o** to **7c** in the crystal, **c** Molecular packing of **7o** in the unit cell, **d** photoinduced bending of thin crystal of **7o** and the relation to the unit cell

(Fig. 24.15e). The beads immediately scattered with the deformation and breaking of the walls, in the manner of seeds bursting from *impatiens* (Fig. 24.15f). This phenomenon implies that the photoresponsive hollow crystal works as a kind of soft robot which scatters inclusions by action of light [31].

24.8 Conclusion

We described the fabrication of biomimetic crystalline surfaces. They show specific wetting properties owing to self-aggregation of photogenerated molecules. These systems realize that the molecular-level aggregation, which is microscopic, induces the macroscopic surface functions. At first, we prepared the switching system in which the aggregation of photochromic molecules is reversible. Secondly, we

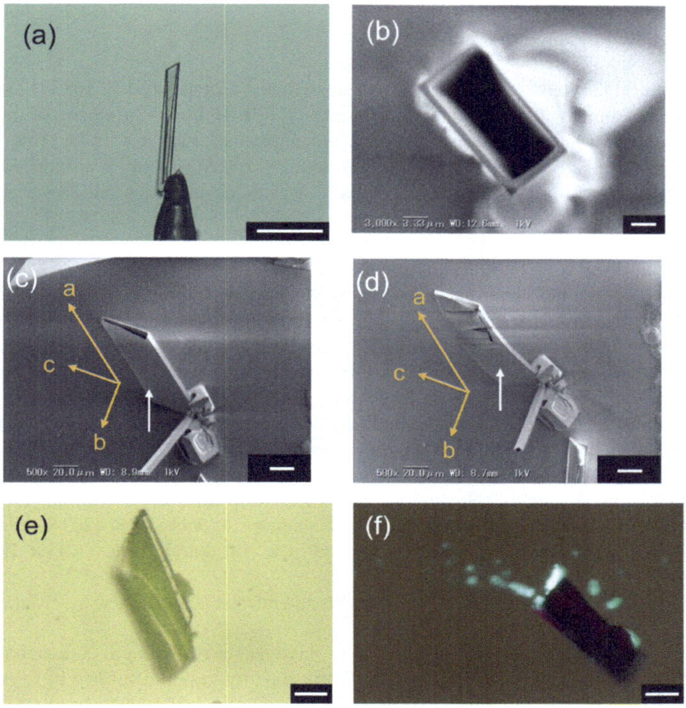

Fig. 24.15 **a** A hollow crystal of **7o** prepared by sublimation under normal pressure, **b** A SEM image of a hollow crystal from top view, **c** A hollow crystal in SEM chamber before UV irradiation, **d** A hollow crystal in SEM chamber after UV irradiation, **e** A hollow crystal filled with fluorescent beads, **f** Scattering of beads upon UV irradiation. Scale bars: 100 μm for **a** 3.3 μm for **b**, 20 μm for (**c**–**f**)

prepared a crystalline brush consisting of photochromic bending crystals; the systems could transfer the object on the surface, following the direction of incident light. In this system, molecular movement upon photoirradiation cooperatively works to realize the function of object transport. Finally, we presented the photoresponsive broken system (i.e., photosalient effect) in which hollow-shaped crystal included small fluorescent beads scattered the beads upon UV irradiation with breaking the crystals. Photosalient phenomena were already well known, however, in this study, we gave a real function to the phenomena.

Acknowledgements We deeply appreciate to Professors M. Irie, H. Miyasaka, M. Morimoto, K. Tawa, and B. L. Feringa for their collaboration. The present work was supported by JSPS KAKENHI Grant Number 26107012, Grant-in-Aid for Scientific Research on Innovative Areas "Photosynergetics."

References

1. Irie M (2000) Diarylethenes for memories and switches. Chem Rev 100:1685–1716
2. Irie M, Fukaminato T, Matsuda K, Kobatake S (2014) Photochromism of diarylethene molecules and crystals: memories, switches, and actuators. Chem Rev 114:12174–12277
3. Uchida K, Izumi N, Sukata S, Kojima Y, Nakamura S, Irie M (2006) Photoinduced reversible formation of microfibrils on a photochromic diarylethene microcrystalline surface. Angew Chem Int Ed 45:6470–6473
4. Kobatake S, Takami S, Muto H, Ishikawa T, Irie M (2007) Rapid and reversible shape changes of molecular crystals on photoirradiation. Nature 446:778–781
5. Uchida K, Nishimura N, Izumi N, Yamazoe S, Mayama H, Kojima Y, Yokojima S, Nakamura S, Tsujii K, Irie M (2010) Phototunable diarylethene microcrystalline surfaces: lotus and petal effects upon wetting. Angew Chem Int Ed 49:5942–5944
6. Nishikawa N, Uyama A, Kamitanaka T, Mayama H, Kojima Y, Yokojima S, Nakamura S, Tsujii K, Uchida K (2011) Photoinduced reversible topographical changes on diarylethene microcrystalline surfaces with bio-mimetic wetting properties. Chem Asian J 6:2400–2406
7. Irie M, Kobatake S, Horichi M (2001) Reversible surface morphology changes of a photochromic diarylethene single crystal by photoirradiation. Science 291:1769–1772
8. Fujinaga N, Nishikawa N, Nishimura R, Hyodo K, Yamazoe S, Kojima Y, Yamamoto K, Tsujioka T, Morimoto M, Yokojima S, Nakamura S, Uchida K (2016) Photoinduced topographical changes on microcrystalline surfaces of diarylethenes. Cryst Eng Comm 18:7229–7235
9. Xia F, Jiang L (2008) Bio-inspired, smart, multiscale interfacial materials. Adv Mater 20:2842–2858
10. Takase K, Hyodo K, Morimoto M, Kojima Y, Mayama H, Yokojima S, Nakamura S, Uchida K (2016) Photoinduced reversible formation of a superhydrophilic surface by crystal growth of diarylethene. Chem Commun 52:6885–6887
11. Barthlott W, Neinhuis C (1997) Purity of the sacred lotus, or escape from contamination in biological surfaces. Planta 202:1–8
12. Tian Y, Su B, Jiang L (2014) Interfacial material system exhibiting superwettability. Adv Mater 26:6872–6897
13. Lu Y, Sathasivam S, Song J, Crick CR, Carmalt CJ, Parkin IP (2015) Robust self-cleaning surfaces that function when exposed to either air or oil. Science 347:1132–1135
14. Bird JC, Dhiman R, Kwon H-M, Varanasi KK (2013) Reducing the contact time of a bouncing drop. Nature 503:385–389
15. Nishimura R, Hyodo K, Sawaguchi H, Yamamoto Y, Nonomura Y, Mayama H, Yokojima S, Nakamura S, Uchida K (2016) Fractal surfaces of molecular crystals mimicking lotus leaf with phototunable double roughness structures. J Am Chem Soc 138:10299–10303
16. Nishimura R, Mayama H, Nonomura Y, Yokojima S, Nakamura S, Uchida K (2019) Crystal growth technique for formation of double roughness structures mimicking lotus leaf. Langmuir 35:14124–14132
17. Nishikawa N, Mayama H, Nonomura Y, Fujinaga N, Yokojima S, Nakamura S, Uchida K (2014) Theoretical explanation of the photoswitchable superhydrophobicity of diarylethene microcrystalline surfaces. Langmuir 30:10643–10650
18. Watson GS, Cribb BW, Watson JA (2010) How micro/nanoarchitecture facilitates anti-wetting: an elegant hierarchical design on the termite wing. ACS Nano 4:129–136
19. Uchida K, Nishimura R, Hatano E, Mayama H, Yokojima S (2018) Photochromic crystalline systems mimicking bio-functions. Chem Eur J 24:8491–8506
20. Nishikawa N, Kiyohara H, Sakiyama S, Yamazoe S, Mayama H, Tsujioka T, Kojima Y, Yokojima S, Nakamura S, Uchida K (2012) Photoinduced formation of superhydrophobic surface on which contact angle of a water droplet exceeds 170° by reversible topographical changes on a diarylethene microcrystalline surface. Langmuir 28:17817–17824

21. Nishimura R, Hyodo K, Mayama H, Yokojima S, Nakamura S, Uchida K (2019) Dual wettability on diarylethene microcrystalline surface mimicking a termite wing. Comms Chem 2: article number 90
22. Uchida K, Sukata S, Matsuzawa Y, Akazawa M, de Jong JJD, Katsonis N, Kojima Y, Nakamura S, Areephong J, Meetsma A, Feringa BL (2008) Photoresponsive rolling and bending of thin crystals of chiral diarylethenes. Chem Commun 326–328
23. Morimoto M, Irie M (2010) A diarylethene cocrystal that converts light into mechanical work. J Am Chem Soc 132:14172–14178
24. Terao F, Morimoto M, Irie M (2012) Light-driven molecular-crystal actuators: rapid and reversible bending of rodlike mixed crystals of diarylethene derivatives. Angew Chem Int Ed 51:901–904
25. Nishimura R, Fujimoto A, Yasuda N, Morimoto M, Nagasaka T, Sotome H, Ito S, Miyasaka H, Yokojima S, Nakamura S, Feringa BL, Uchida K (2019) Object transportation system mimicking the cilia of *paramecium aurelia* making use of the light-controllable crystal bending behavior of a photochromic diarylethene. Angew Chem Int Ed 58:13308–13312
26. Naumov P, Sahoo SC, Zakharov BA, Boldyreva EV (2013) Dynamic single crystals: kinematic analysis of photoinduced crystal jumping (the photosalient effect). Angew Chem Int Ed 52:9990–9995
27. Colombier I, Spagnoli S, Corval A, Baldeck PL, Giraud M, Leaustic A, Yu P, Irie M (2007) Diarylethene microcrystals make directional jumps upon ultraviolet irradiation. J Chem Phys 126:011101
28. Kitagawa D, Okuyama T, Tanaka R, Kobatake S (2016) Photoinduced rapid and explosive fragmentation of diarylethene crystals having urethane bonding. Chem Mater 28:4889–4892
29. Hatano E, Morimoto M, Hyodo K, Yasuda N, Yokojima S, Nakamura S, Uchida K (2016) Photosalient effect of a diarylethene with a perfluorocyclohexene ring. Chem Eur J 22:12680–12683
30. Iwanaga H, Yoshiie T, Yamaguchi T, Shibata N (1981) Growth mechanism of hollow CdSe and CdS crystals. J Cryst Growth 51:438–442
31. Hatano E, Morimoto M, Imai T, Hyodo K, Fujimoto A, Nishimura R, Sekine A, Yasuda N, Yokojima S, Nakamura S, Uchida K (2017) Photosalient phenomena that mimic *impatiens* are observed in hollow crystals of diarylethene with a perfluorocyclohexene ring. Angew Chem Int Ed 56:12576–12580

Chapter 25
Photoresponsive Molecular Crystals for Light-Driven Photoactuators

Seiya Kobatake and Daichi Kitagawa

Abstract Various types of photomechanical motion have drawn much attention because there is a potential to create photomechanical actuators from molecular-scale to macro-scale. To construct photoactuators, it is necessary to utilize a molecular assembly with a small free volume. Photochromic compounds undergo photoreversible isomerization between the original colorless isomer and the photogenerated colored isomer upon alternating irradiation with UV and visible light. Among many known photochromic compounds, diarylethenes undergo photochromic reactions even in the crystalline phase. The present review introduces recent development in the study of photomechanical crystals including crystal shape changes, bending velocity, dependence of the bending behavior on irradiation wavelength, the behavior in mixed crystal, new types of photomechanical motion, and applications. These photomechanical behaviors are based on geometrical structure changes in the crystalline phase and can be applied to macro-sized light-driven photoactuators.

Keywords Photochromism · Diarylethene · Actuator · Photomechanical · Switching

25.1 Introduction

Molecules that change their molecular shape in response to heat and light have been developed as mechanical materials in a very small region at the molecular level, as known in the Nobel Prize in Chemistry 2016 "for the design and synthesis of molecular machines". Such molecules are expected to be applied not only as individual molecules but also as supramolecular structures that function by combining them. However, the construction of mechanical materials and micromachines in the mesoscopic region (nanometer to micrometer) that is slightly larger than the molecular size is not so easy because the movement of each molecule cannot be directly linked

S. Kobatake (✉) · D. Kitagawa
Graduate School of Engineering, Osaka City University, Sugimoto, Sumiyoshi-ku, Osaka 558-8585, Japan
e-mail: kobatake@a-chem.eng.oasaka-cu.ac.jp

H. Miyasaka et al. (eds.), *Photosynergetic Responses in Molecules and Molecular Aggregates*, https://doi.org/10.1007/978-981-15-5451-3_25

to the deformation of the material due to a large free volume around the molecule. It is necessary to use a molecular assembly with a small free volume.

The phenomenon in which a chemical species reversibly isomerizes between two isomers having different colors upon photoirradiation is called photochromism. Various physicochemical properties such as conductivity, dielectric constant, refractive index, redox potential, and fluorescence are also reversibly changed by the photochromic reaction. Azobenzene, spiropyran, furylfulgide, and diarylethene are representative photochromic compounds. Among them, diarylethene has excellent properties such as thermal stability of both isomers, high durability, high sensitivity, and rapid response [1, 2]. When a diarylethene is irradiated with ultraviolet (UV) light in a colorless state called an open-ring isomer, the molecular structure is changed to a colored state called a closed-ring isomer. Moreover, when the colored closed-ring isomer is irradiated with visible light, it returns to the original colorless open-ring isomer. Many diarylethenes exhibit photochromism not only in solution but also in the crystalline state. Crystals are considered as a molecular assembly with a small free volume in which molecules are regularly arranged and packed, and the photomechanical behavior of diarylethene crystals associated with the photochromic reaction is investigated. This chapter focuses on photomechanical organic crystals, especially diarylethene crystals, and describes future developments of this research field.

25.2 Discovery of Photomechanical Crystals

The first example of a molecular crystal that bends reversibly in response to light (photomechanical crystal) is the rhodium semiquinone complex crystal reported by Abakumov and Nevodchikov [3]. When thin acicular crystals were irradiated with light in the charge-transfer absorption band having an absorption maximum near 1600 nm, it was confirmed that the crystals bend 45°. It has been concluded that this bending behavior is due to the formation of Rh–Rh bonds between the complexes within the crystal. However, since oxygen is required for the cleavage of the Rh–Rh bond in the reverse reaction, it does not return exactly to the original shape. Also, if left in the presence of air all day long, the bending ability is lost.

Since then, little attention has been paid to research on photomechanical crystals, and during that time, research was focused on photochemical reactions in solid and crystalline states. In 2001, a photoreversible crystal surface morphology change in a diarylethene single crystal was reported [4]. In the diarylethene crystal with a size of several millimetres, a step of 1 nm or valley is formed with UV light irradiation, and the original smooth surface is formed by visible light irradiation. This result reflects the reversible molecular structure change of the diarylethene molecule due to UV and visible light irradiation. In 2006, Bardeen et al. found that crystal nanorods of an anthracene derivative were elongated by 15% due to [4 + 4] photodimerization [5]. This is different from an intramolecular reaction such as the cyclization and cycloreversion reaction of diarylethene, and utilizes the volume change in the crystal due to the intermolecular reaction. In 2007, a reversible crystal shape change of diarylethene

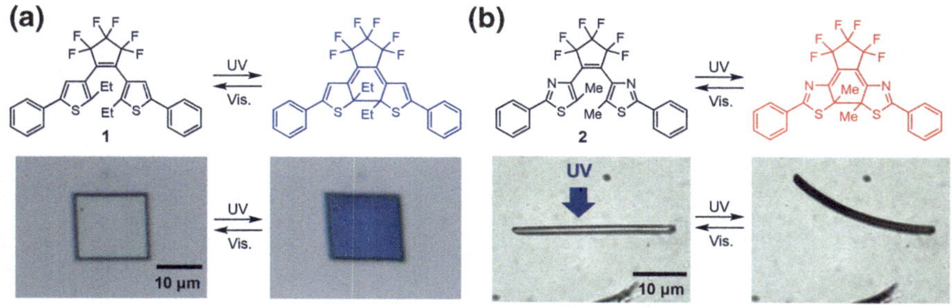

Fig. 25.1 Photoreversible crystal shape changes of diarylethene microcrystals **1** and **2**

single crystals by photoirradiation was reported (Fig. 25.1) [6]. When UV light is irradiated to a small single crystal of diarylethene **1** with a thickness of several hundreds of nanometers, the entire crystal reacts almost uniformly, changing the color of the crystal from colorless to blue, and from a nearly square shape to a rhombus shape. Such crystal shape change could be repeated more than 100 times. Furthermore, when UV light was irradiated to a rod-like crystal of diarylethene **2**, the crystal bent toward the direction of the incident UV light. Moreover, it returned to the original straight shape crystal when irradiated with visible light. Furthermore, using this bending behavior, the crystal can hit silica particle that is 50 times heavier than its own weight. This result suggests the possibility that photoresponsive molecular crystals can be used as light-driven actuators at the micro-scale. Starting from this study, research was conducted on the photomechanical behavior of small single crystals of various photochromic compounds. Examples of compounds exhibiting crystalline photomechanical behavior are shown in Fig. 25.2. It has become clear that photochromic compounds such as diarylethenes [6–24], azobenzenes [25–30], furylfulgides [31], salicylideneanilines [32–34], anthracene derivatives [5, 35–40], 4-chlorocinnamic acid [41], 1,2-bis(4-pyridyl)ethylene salt [42], benzylidenedimethylimidazolinone [43], and others [44, 45] exhibit photomechanical behavior using micro-sized single crystals. The typical photomechanical motions are contraction, elongation, bending, separation, twisting, curling, and so on. In particular, the bending behavior is the most common photomechanical motion reported so far.

25.3 Bending Behavior of Diarylethene Crystal

The bending phenomenon induced by photoirradiation will be described here using diarylethene crystals as an example. When a diarylethene molecule is isomerized from the open-ring form to the closed-ring form, the long axis of the molecule contracts, the short axis of the molecule elongates, and the thickness of the molecule decreases as shown in Fig. 25.3. The slight change in molecular structure accompanying the photochromic reaction causes a photomechanical phenomenon of the

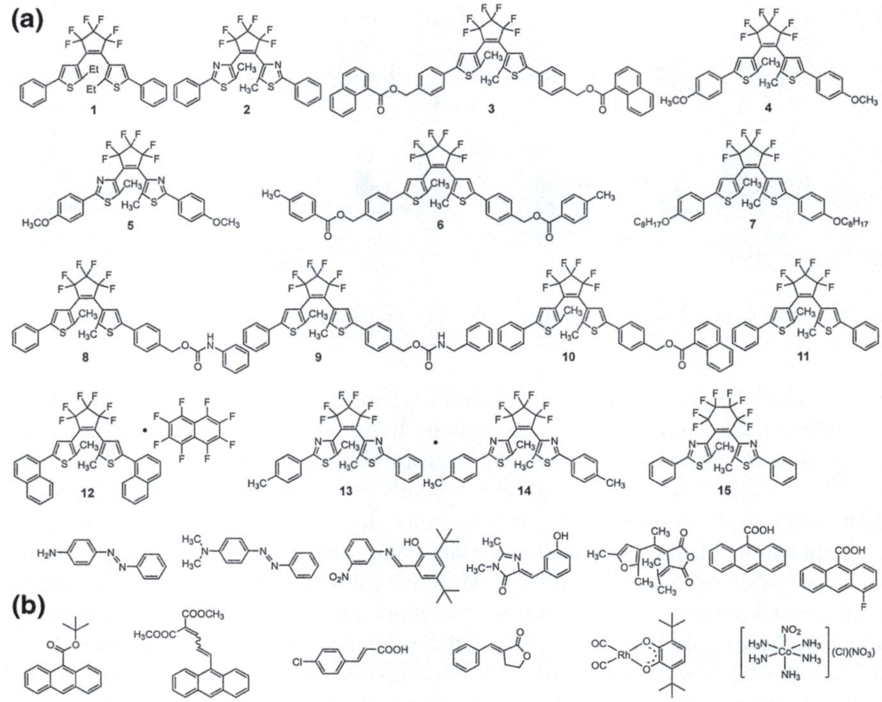

Fig. 25.2 **a** Diarylethene derivatives and **b** others that exhibit crystalline photomechanical response

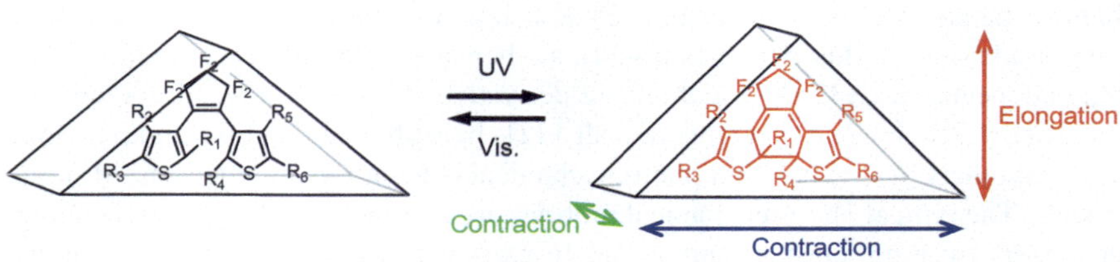

Fig. 25.3 Geometrical structures of diarylethene open- and closed-ring isomers

diarylethene crystal. As described above, when UV light is irradiated onto a thin crystal, the entire crystal reacts almost uniformly, resulting in the crystal shape change. This crystal shape change is observed when the molecular packing is the same even if the molecular structure is different. When light is applied non-uniformly to a slightly thicker crystal, the photochromic reaction proceeds only on the crystal surface, and the crystal bends like a bimetal. The contraction/elongation and bending deformation of the crystal are basically the same phenomena. When the rod-like crystal of diarylethene **2** is irradiated with UV light, the crystal bends in the direction of the incident light and returns to the original rod-like crystal when irradiated with visible light (Fig. 25.1b). On the other hand, crystals that bend away from the incident UV light have been reported. This means that the molecular packing of diarylethene molecules in the crystal plays an important role in the bending direction. Furthermore, such a bending behavior is observed when the thickness of the crystal is about several μm, but the crystal does not bend when the thickness of the crystal is increased. The dependence of the bending velocity on the crystal thickness has been clarified by analysis using a bimetal model. Furthermore, by using Timoshenko's bimetal model analysis, it became possible to compare the bending speed between crystals of different compounds.

Further detailed studies on the bending behavior using diarylethene crystals have been conducted. When the rod crystal of diarylethene **2** is irradiated with 365 nm light, it bends toward the light source, but when irradiated with 380 nm light, it once bends to the opposite side of the light source and then bends toward the light source [17]. Furthermore, it was found that the behavior by visible light irradiation was also different from each other depending on the irradiation direction (Fig. 25.4). It was concluded that this is because the depth of the photochromic reaction from the crystal surface differs by changing the irradiation wavelength of UV light. This is an interesting result that the photomechanical behavior changes depending on the irradiation wavelength even if the same crystal is used.

Polarized light irradiation can be another method to control the depth of the photochromic reaction from the crystal surface based on the electronic transition dipole moment of the aligned molecule. Upon irradiation with polarized UV light, the rod-like crystal **3** and **4** exhibited photoinduced bending behaviors. The initial

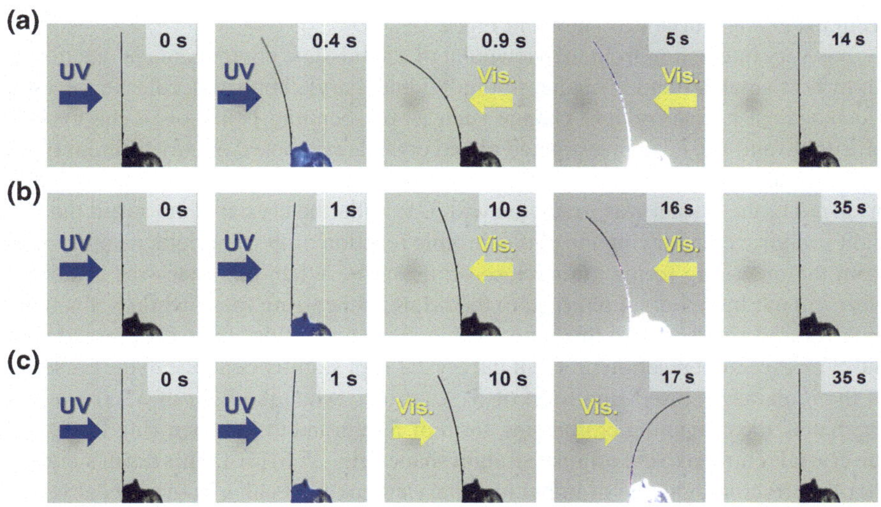

Fig. 25.4 Photoinduced bending behavior of diarylethene crystal **2** upon irradiation with 365 (**a**) and 380 nm light (**b**, **c**) from the left side. The visible light was irradiated from the right side (**a**, **b**) and the left side (**c**). Reproduced from Ref. [17] with permission of the PCCP Owner Societies

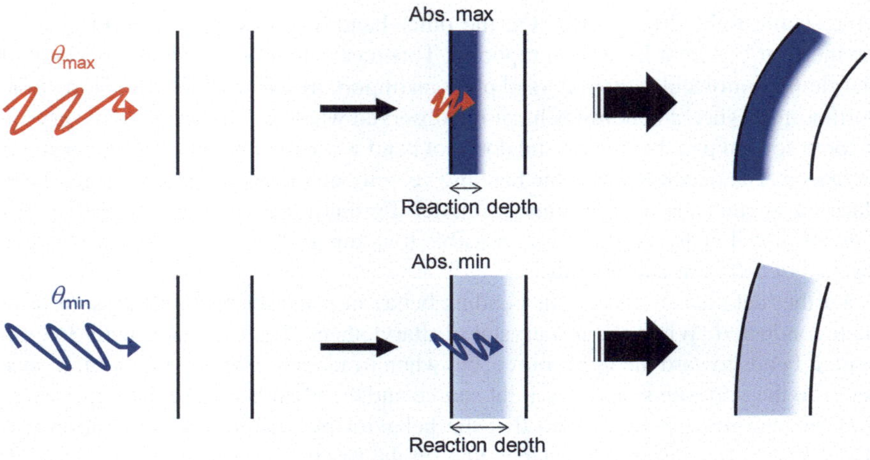

Fig. 25.5 Illustration for difference in the photoinduced bending speed of diarylethene crystals depending on the polarization angle upon irradiation with polarized UV light

velocity of curvature change was well correlated to the absorption anisotropy of the open-ring form in the crystal, which indicates that the reaction depth from the crystal surface plays an important role in the photomechanical crystal bending. Moreover, the unique bending behavior such as "swing-forward" and "swing-backward" motions was observed using polarized UV and visible light irradiation [24]. The mechanism of the unique motions could be well explained by considering the reaction depth from the crystal surface as shown in Fig. 25.5.

It is very interesting from the viewpoint of a solid-state photochemical reaction to know how small structural changes of individual diarylethene molecules in a crystal work against a macroscopic change such as the bending behavior of the crystal. Morimoto and Irie et al. investigated mixed crystals composed of two different types of diarylethenes. When a mixed crystal in which diarylethenes **4** and **5** are randomly arranged in the crystal was irradiated with UV light, the crystal bent toward the UV light source accompanying the photochromic reaction of the diarylethene molecules from the open-ring forms to the closed-ring forms. When this state was irradiated with 750 nm light, only **4** returned to the original open-ring form, while **5** remained in the closed-ring form. At this time, the bending angle of the crystal decreased with the cycloreversion reaction of **4**, but the crystal kept slightly bent due to the presence of the remaining closed-ring form of **5**. After that, when the light of >500 nm was irradiated, the remaining closed-ring form of **5** returned to the open-ring form, and the crystal returned to the original straight shape (Fig. 25.6) [16]. This result suggests that the structural change of individual diarylethene molecules in the crystal works additively to the crystal bending. A study on the effect of the UV light irradiation intensity on the bending velocity of crystals also supports this result. That is, since the bending speed is proportional to the number of reacted molecules of diarylethenes in the crystal, it is proportional to the UV light irradiation intensity (Fig. 25.7) [21].

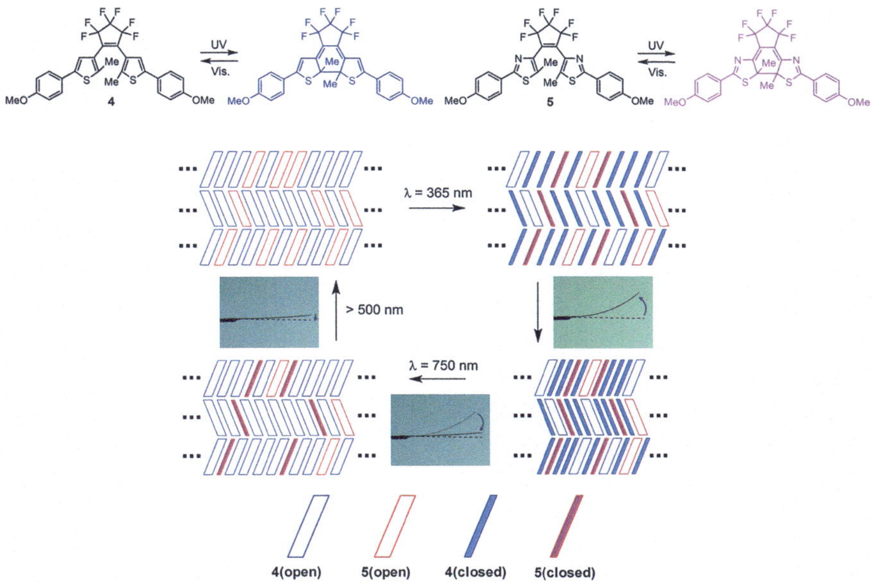

Fig. 25.6 Schematic illustration of the reactions in the mixed crystal of **4** and **5** upon irradiation with 365 nm and 750 nm light and the bending response of the crystal. Adapted from Ref. [16] with permission of The Royal Society of Chemistry

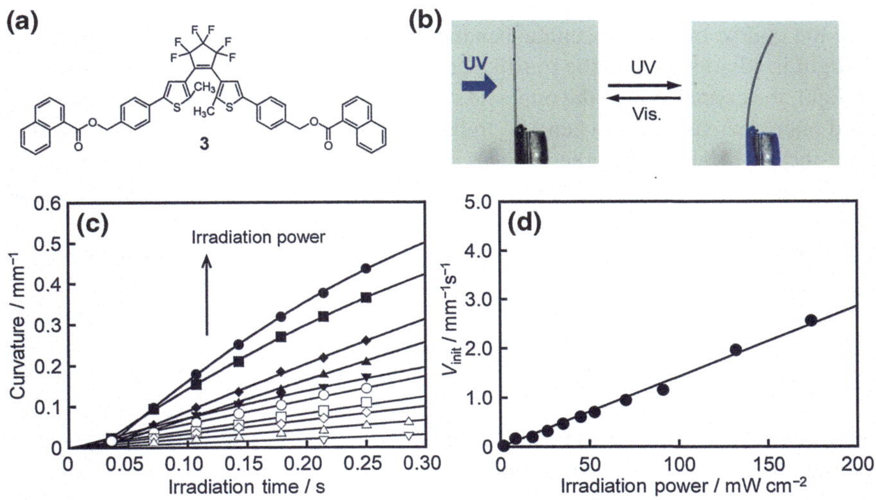

Fig. 25.7 **a** Molecular structure of **3**, **b** photographs of photoinduced bending behavior of crystal **3**, **c** curvature change as a function of UV irradiation time, and **d** the initial velocity of curvature change (V_{init}) as a function of the power of the incident UV light for crystal **3** (thickness: 6.7 μm). The power of the incident UV light was 174 (●), 132 (■), 91.0 (◆), 70.2 (▲), 52.7 (▼), 44.8 (○), 35.1 (□), 26.3 (◇), 17.6 (△), and 8.17 mW cm^{-2} (▽) for (**c**). Adapted with permission from Ref. [21]. Copyright 2017 American Chemical Society

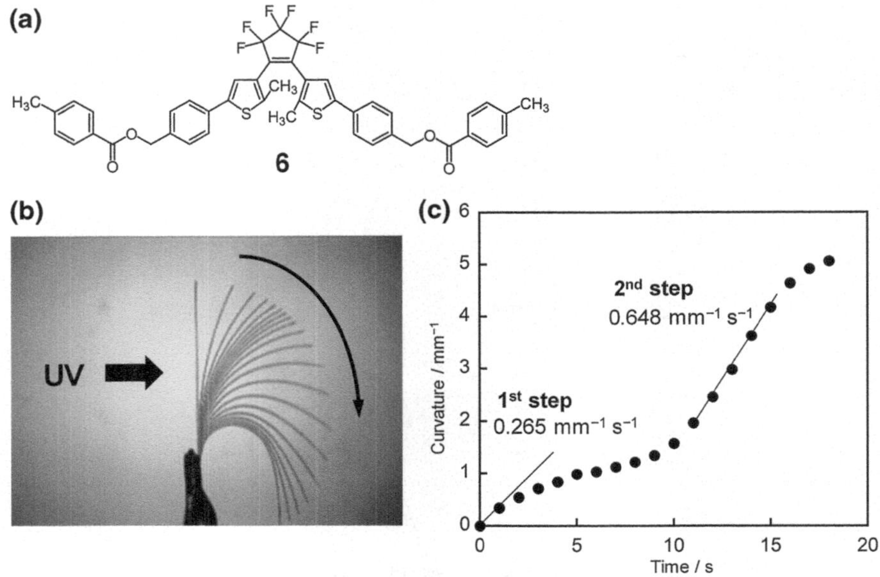

Fig. 25.8 Photoinduced stepwise bending behavior of diarylethene crystal **6** and curvature change relative to irradiation time of UV light. The photograph is a superimposition of 20 frames taken at intervals of 1 s. Adapted from Ref. [19] with permission of The Royal Society of Chemistry

In the course of study, peculiar bending behavior was confirmed by continuing UV light irradiation. When the rod-like crystal of diarylethene **6** was irradiated with UV light, the crystal bent to the opposite side of the UV light source, and the bending speed once decreased, but when UV light was continuously irradiated, the bending speed increased again. In other words, a stepwise bending behavior in which the bending speed changes along the way was observed (Fig. 25.8) [19]. This behavior cannot be explained only by the fact that the structural change of the diarylethene molecule acts additively on the bending of the crystal. Bending only in the first stage could be reversibly repeated more than 100 times by alternative irradiation with UV light and visible light. On the other hand, when UV light was irradiated until the second stage bending occurs, followed by visible light irradiation, the crystal gradually exhibited no stepwise bending behavior. In other words, the bending velocity of the crystal became no to change during the bending motion. Moreover, it was found that when the UV light irradiation was stopped during the second stage of bending, the crystal continued to bend even though no photoirradiation was applied. From melting point and powder X-ray diffraction patterns, it was concluded that the stepwise bending behavior is due to an irreversible phase transition or a decrease in crystallinity caused by photoisomerization of diarylethene molecules from the open-ring to the closed-ring form.

In contrast, crystals of diarylethene **7** were found to undergo a reversible thermodynamic single-crystal-to-single-crystal phase transition accompanying a change in crystal length, which was clarified by differential scanning calorimetry measurement, X-ray crystallographic analysis, and direct microscopic observation of the crystal length [22]. Upon irradiation with UV light, the rod-like crystal exhibited unusual photomechanical behavior. The crystal slowly bent away from the incident light at first and quickly bent away from the light source on the way. After continuous irradiation with UV light, the crystal suddenly returned to the initial straight shape as shown in Fig. 25.9. From the investigation in detail, it was revealed that the unusual bending is based on the combination of the photochromic reaction and the phase transition. Moreover, the photoirradiated crystal exhibited thermal bending behavior by changing temperature as shown in Fig. 25.10. This is a new type of reversible mechanical material. These results suggest the possibility that unique photomechanical behavior appears by combining complex elements other than photochromic reactions.

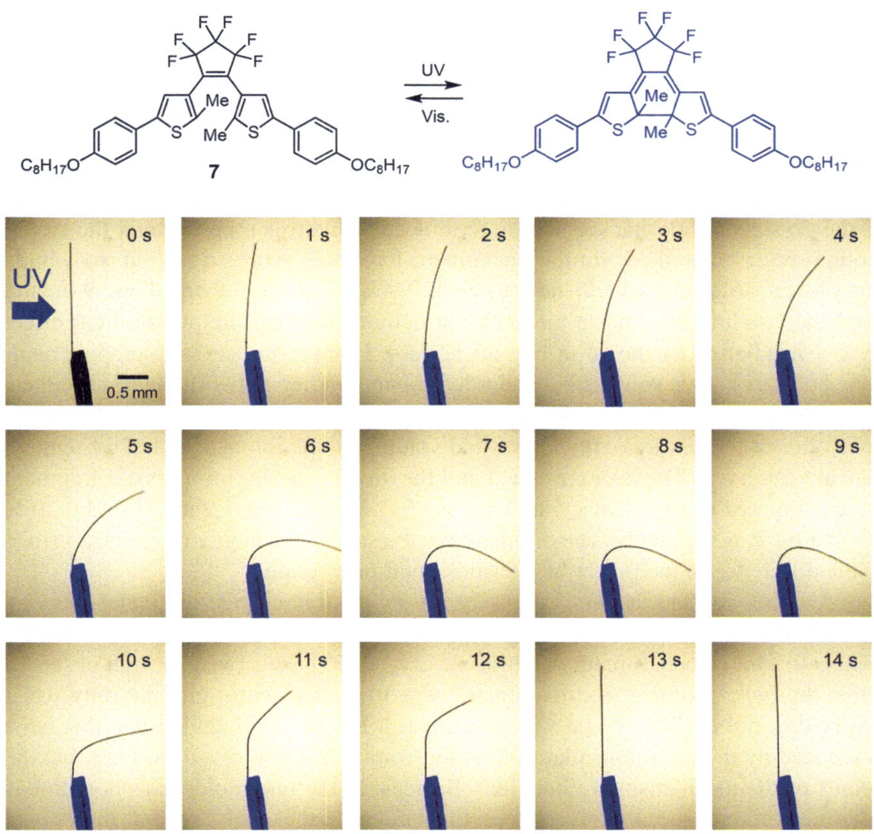

Fig. 25.9 Photomechanical behavior of diarylethene **7** upon irradiation with UV (365 m) light from the left side. Reprinted with permission from Ref. [22]. Copyright 2017 American Chemical Society

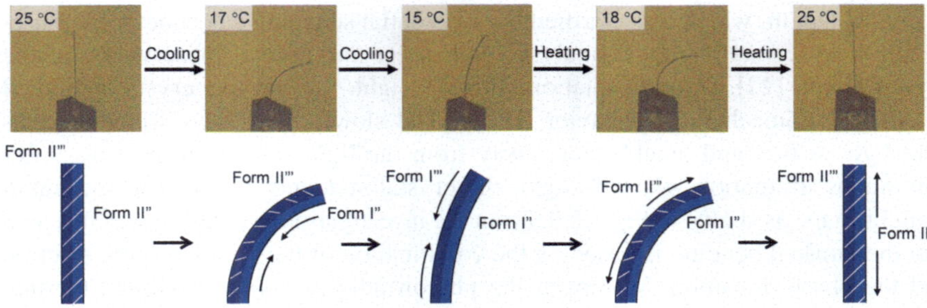

Fig. 25.10 The thermal bending behavior of the photoirradiated crystal **7** by changing temperature. The initial crystal was prepared by irradiation with UV light from the left side. Reprinted with permission from Ref. [22]. Copyright 2017 American Chemical Society

25.4 Novel Photomechanical Phenomena

In recent years, new photomechanical phenomena other than contraction, elongation, and bending have been reported. Irradiation of UV light to crystals of nitropentaamine cobalt complex confirmed the "photosalient effect" in which the crystals jump after the induction period (Fig. 25.11) [45]. It is concluded that this is because the small structural change of the nitro group due to light irradiation generates strain in the crystals tightly packed with intermolecular hydrogen bonds, and the stored strain is released at once. Similar phenomena have been reported in gold isocyanide complexes [46] and diarylethene crystals [20, 47, 48]. Diarylethene **8** and **9** having urethane bonding unit in the molecular structure exhibited the photosalient effect by UV irradiation as shown in Fig. 25.12 [20]. From X-ray crystallographic analysis, both crystals have intermolecular hydrogen bonding network at their urethane bonding units in the crystal, resulting in strong intermolecular interaction. The influence of crystal size (area), thickness, and volume on the initial speed of a fragment was not observed. The average speed and the kinetic energy for a crystal fragment were estimated to be 4.7 m s^{-1} and 3.2×10^{-12} J for **8** and 1.9 m s^{-1} and 0.50×10^{-12} J for **9**, respectively. The difference in the speed and energy might be ascribed to the difference in the power of intermolecular interaction.

On the other hand, as an example of a very flexible change in crystal shape, a photoinduced twisting phenomenon was found, in which a spiral is formed when UV light is irradiated on a 9-methylanthracene crystal [37]. This twist was not observed when the bulk crystal or rod-like crystal was irradiated with light, but only when a very thin ribbon-like crystal was irradiated. It has been concluded that this twist is induced by the intermolecular photodimerization reaction of anthracenes, but the details of the mechanism are not clear. Besides, in the intermolecular photodimerization of anthracene, the reverse reaction proceeds thermally, so that the crystal shape also thermally returns to the original state. Such a photoinduced twisting phenomenon has also been confirmed in diarylethene crystals of **10** (Fig. 25.13) [12]. In this system, since intramolecular photocyclization and photocycloreversion

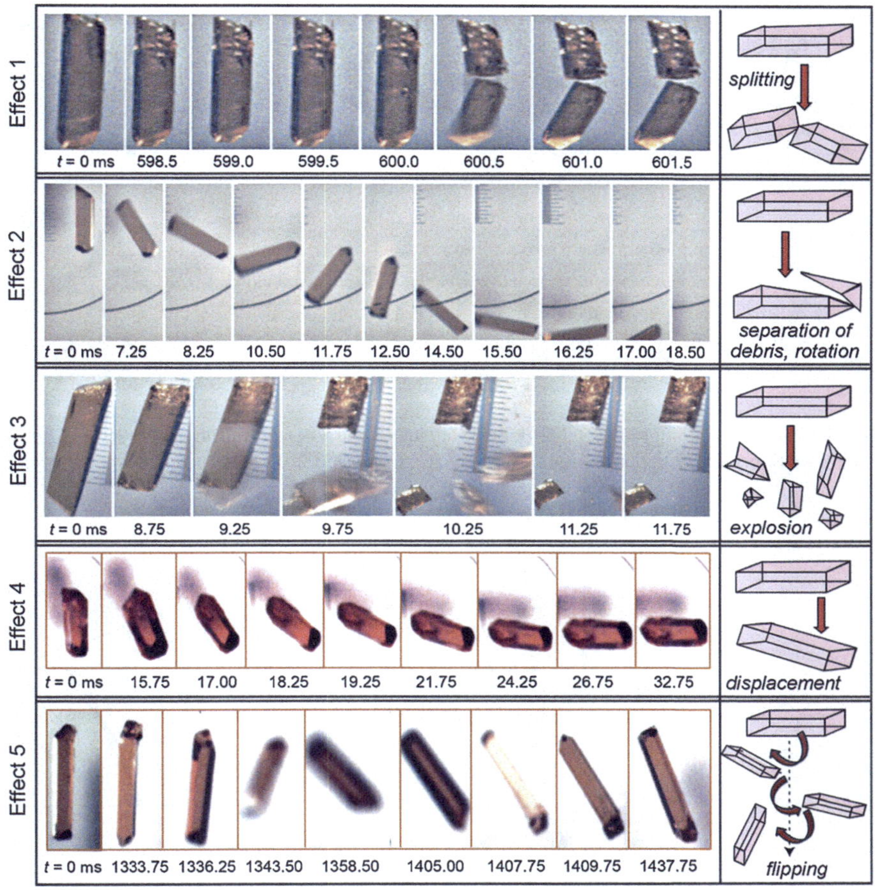

Fig. 25.11 Series of snapshots and sketches of the five kinematic effects of UV-excited crystals of $[Co(NH_3)_5(NO_2)]Cl(NO_3)$. Adapted with permission from Ref. [45]. Copyright 2013 Wiley-VCH Verlag GmbH & Co. KGaA, Weinheim

of diarylethene molecules are used, the crystal shape is not restored thermally and reversibly changes by irradiation with UV light and visible light. Further, the right-handed or left-handed spiral formation can be controlled according to the crystal plane to which UV light is irradiated. It has been clarified that the spiral formation is because the induction period exists before the crystal deformation occurs in addition to the diagonal contraction of the crystal. It was also confirmed that when dimethyl-2-(3-anthracen-9-yl)arylidene) malonate nanowire crystals were irradiated with light, the crystals were irreversibly deformed into a coil shape [39]. In this crystal, not an intermolecular [4 + 4] dimerization reaction of anthracene, but an E $\leftrightarrow$ Z isomerization reaction in the molecule proceeds, whereby the crystal/amorphous mixed-phase generated in the crystal causes internal strain, causing a dramatic shape change.

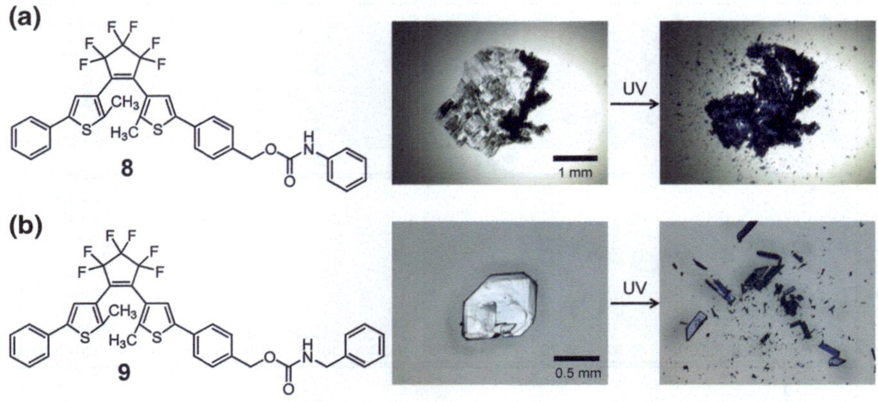

Fig. 25.12 Optical photographs of diarylethene crystals **8** and **9** before and after UV (365 nm light) irradiation. Reprinted with permission from Ref. [20]. Copyright 2016 American Chemical Society

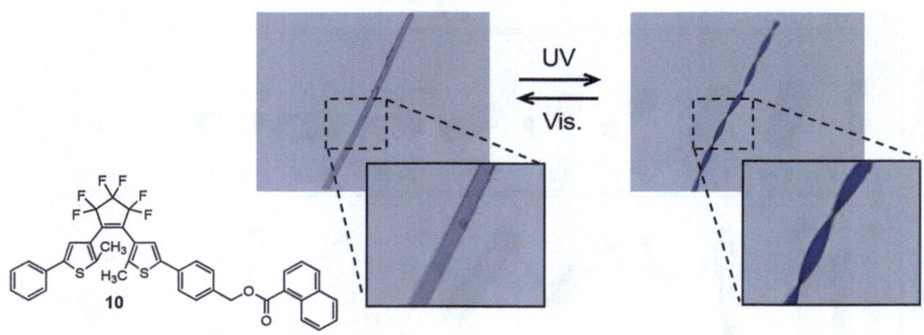

Fig. 25.13 Photoinduced twisting of diarylethene crystal **10**

Although many new and unique photomechanical phenomena have been reported in this way, it is difficult to predict and design photomechanical behavior from the molecular structure, and it is no exaggeration to say that the discovery depends on serendipity. Therefore, it is needed to find something that can be controlled other than the molecular structure. Bardeen et al. found a new photomechanical phenomenon by devising a crystal fabrication method [49]. When 4-fluoroanthracenecarboxylic acid was recrystallized by changing pH, defects were generated during the crystal growth process and branched X-shape crystals were obtained. When the X-shape crystal was irradiated with UV light, it rotated in one direction like a ratchet (Fig. 25.14). It is concluded that the rotation direction depends on the chirality of the branched crystal itself. Furthermore, it was found that crystals with a large number of branches can be obtained by crystallization while carrying out hydrolysis of ester in a solution, and photomechanical phenomena that the silica nanoparticles scattered around are collected in one place have also been reported [50].

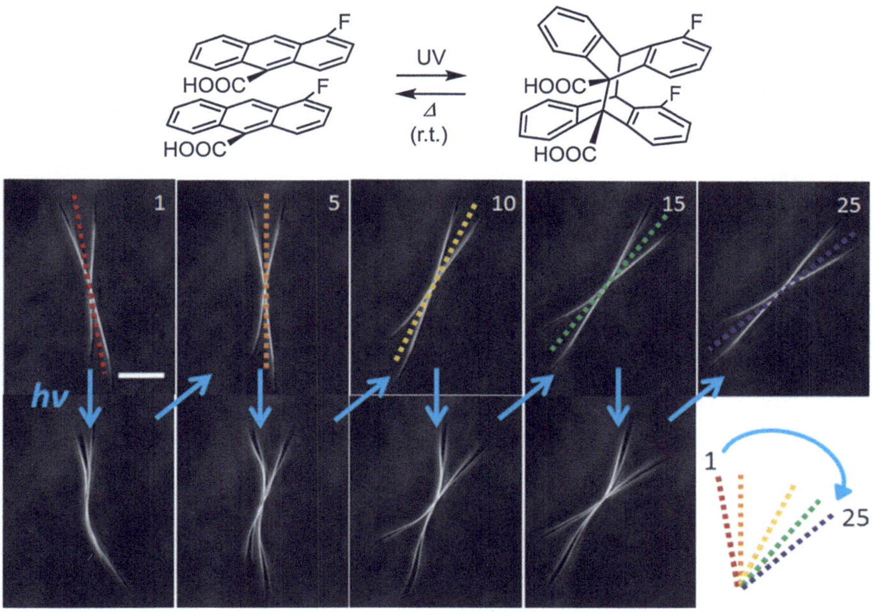

Fig. 25.14 Optical microscopy images of an X-shaped crystal that rotates clockwise after each irradiation period. The cycle was repeated 25 times. 5 cycles of the X-shaped crystal before and after irradiation are illustrated. Scale bar: 20 μm. Adapted with permission from Ref. [49]. Copyright 2016 Wiley-VCH Verlag GmbH & Co. KGaA, Weinheim

As described above, the irradiation condition is also one of the factors to control photomechanical behavior. Here, the effect of the irradiation direction on the photomechanical behavior was investigated [23]. Diarylethene **11** can be crystallized into the ribbon-like crystals by a sublimation method, while the recrystallization from organic solvents yields the block crystals. Regarding the ribbon crystal, the photomechanical behavior depending on the irradiation direction was investigated in detail. The definition of the illumination direction is depicted in the upper right side of the image in Fig. 25.15. Upon irradiation with UV light from the tip of the crystal (incident light angle = 0°), the crystal twisted into a helicoid shape. On the other hand, when UV irradiation was incident on the ribbon at larger angles, it gradually transformed into a cylindrical helix shape. Eventually, the crystal exhibit bending motion when the ribbon crystal was irradiated with an angle of 90°. Thus, the mode of photomechanical motion depends on the incident light angle. This result means that the photoinduced strain tensor in the crystal (and thus the mode of photomechanical deformation) can be controlled by the direction of UV light irradiation.

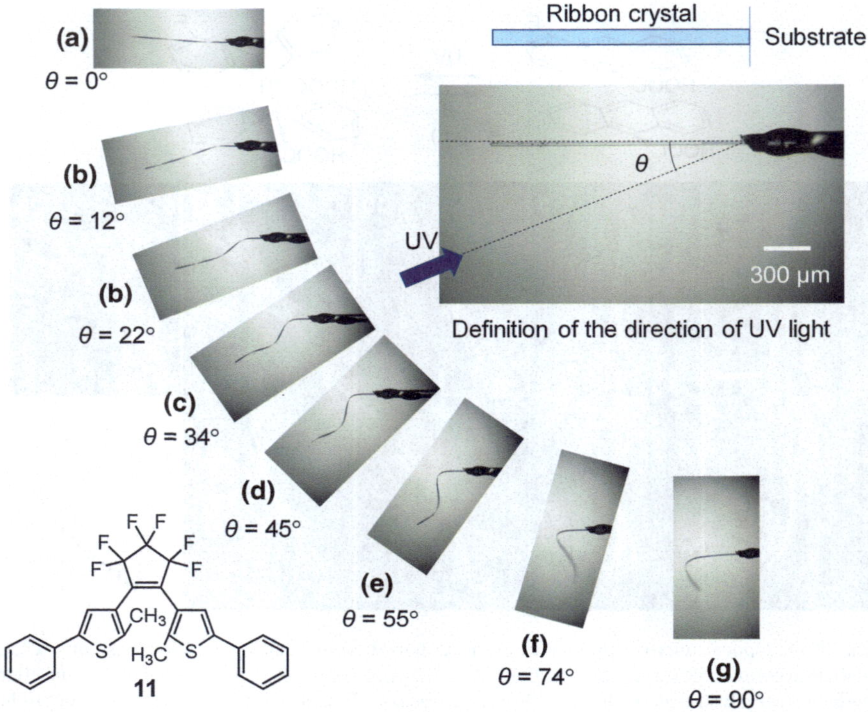

Fig. 25.15 Different twisting motions, ranging from a helicoid to a cylindrical helix, of the ribbon crystal **11** depending on the angle of the incident light. Adapted with permission from Ref. [23]. Copyright 2018 American Chemical Society

25.5 Application of Photomechanical Behavior

The behavior and mechanism of photomechanical crystals have been described so far, but there have also been reports of examples of using these photomechanical crystals as actual photoactuators. The cocrystal of diarylethene **12** and perfluoronaphthalene bends to the direction opposite to the incident light when irradiated with UV light, and returns to the original state when irradiated with visible light. This bending behavior could be repeated 250 times or more without crystal deterioration. Since the photochromic reaction of diarylethene takes place within 10 picoseconds, its bending speed is expected to be high. In fact, it was confirmed that the crystal had already been deformed 5 microseconds after the pulse laser irradiation. The bending behavior was also observed at an extremely low temperature of 4.7 K. In addition, it has succeeded in lifting a lead ball of about 50 mg, which is 200 times its own weight, by 1 mm or more (Fig. 25.16) [9]. This mechanical stress is as much as 44 MPa, which is more than 100 times that of muscle.

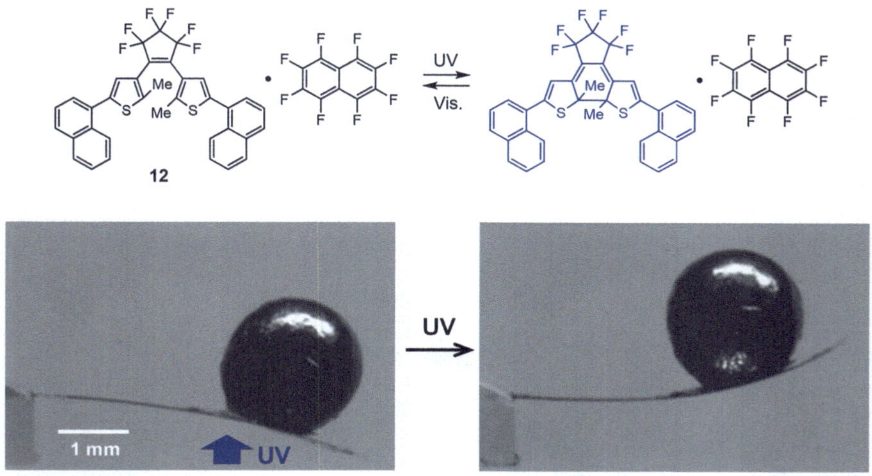

Fig. 25.16 Photomechanical work of molecular crystal cantilevers made of diarylethene **12**/perfluoronaphthalene cocrystal. Adapted with permission from Ref. [9]. Copyright 2010 American Chemical Society

The mixed crystal of diarylethenes **13** and **14** bends toward the light source side by irradiation with UV light and returns to the original state by irradiation with visible light. This bending behavior could be repeated 1000 times or more without crystal deterioration. This crystal was confirmed to be bendable in a very wide temperature range of 4.7 to 370 K and in water. The crystal can be moved freely by adjusting the irradiation intensity of UV light and visible light, and a gearwheel has been successfully turned by utilizing the photomechanical motion (Fig. 25.17) [11].

The bending behavior when gold is vapor-deposited on one side of a rod-like crystal of diarylethene **2** has been investigated [15]. It was confirmed that the crystal bends in a reversible manner even after gold deposition, and is less likely to bend than before gold deposition. This is because Young's modulus of the diarylethene crystal is about 3 GPa, whereas that of gold is as large as 83 GPa. As shown in Fig. 25.18, an electric circuit was prepared by contacting a gold-deposited diarylethene crystal with metal wire, and a current flowed when a voltage was applied. By alternately irradiating UV light and visible light here and bending the crystal, it was possible to switch the current on and off in a reversible manner, and it could be repeated more than 10 times. Current switching using such gold-deposited crystals has also been reported in rotaxane molecular crystals [51].

Diarylethene **15** can be crystallized into the rod-like crystals with a hollow structure by a sublimation method under normal pressure. Upon UV irradiation, the hollow crystal exhibited the photosalient effect, which the crystal broke into pieces explosively [52]. The speed of crystals flying away was measured as 2.67 m s^{-1}. To mimic the behavior of "Impatiens", small fluorescent beads were inserted into the hollow

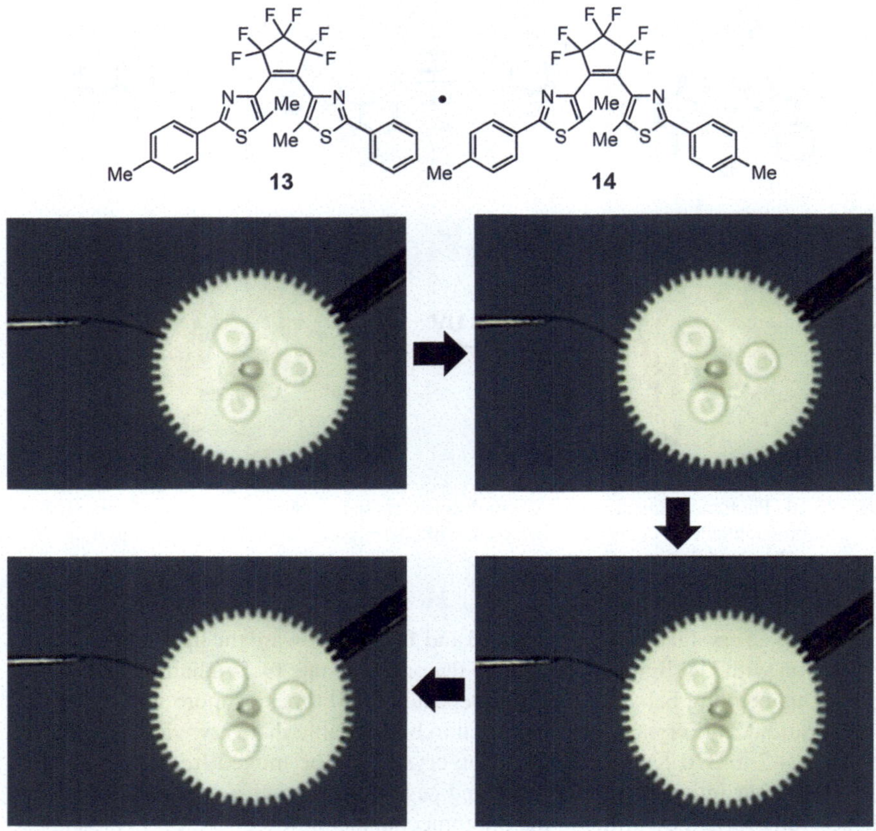

Fig. 25.17 Gearwheel rotation operated by a mixed crystal of diarylethene **13** and **14**. Adapted with permission from Ref. [11]. Copyright 2012 Wiley-VCH Verlag GmbH & Co. KGaA, Weinheim

crystal and UV light was exposed, resulting in the scattering of the beads by deformation and bursting of the hollow structure as shown in Fig. 25.19. This is the first example of the utilization of photosalient phenomena.

25.6 Summary

This chapter mainly described the photomechanical behavior of photochromic diarylethene molecular crystals. The photomechanical motions are contraction, elongation, bending, separation, twisting, curling, and so on. Among them, the bending is the most common photomechanical motion. The bending velocity largely depends on

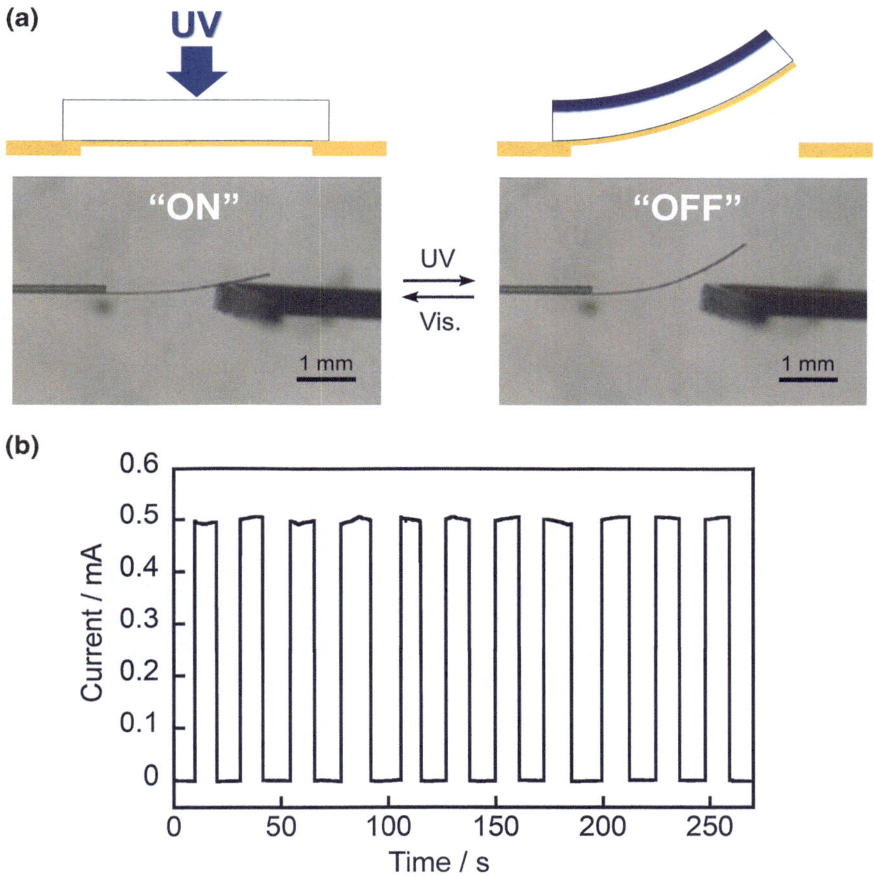

Fig. 25.18 Photoreversible current switching operated by a gold-coated diarylethene crystal of **2**. The crystal thickness is 6.2 μm. The gold thickness is 21 nm. The applied voltage is 1.0 V. The external resistance is 2 kΩ. Adapted from Ref. [15] with permission of The Royal Society of Chemistry

the irradiation wavelength, polarization direction, irradiation surface, and so on. The difference in the velocity is due to the difference in reaction depth caused by UV light irradiation. In some cases, not only simple bending behavior but also cylindrical helix and twisting were observed by changing the irradiation direction. These achievements play an important role in the development of photoactuators of molecular crystals.

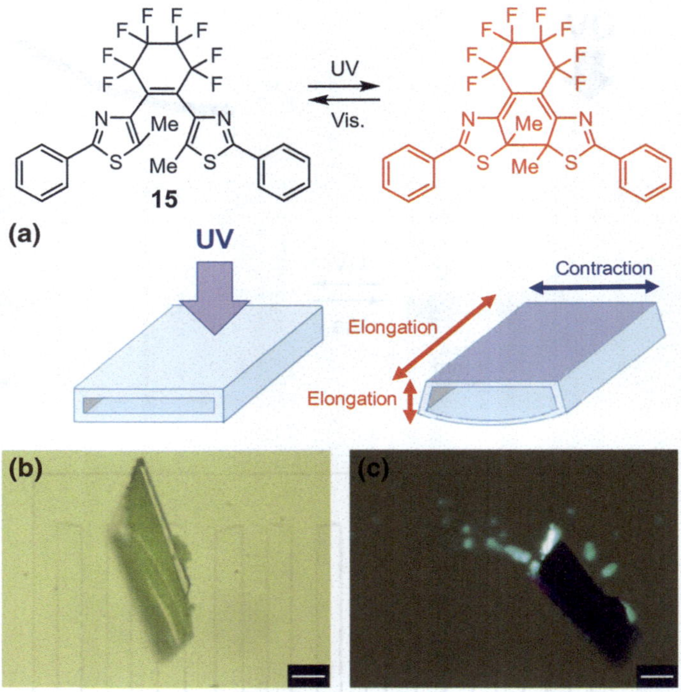

Fig. 25.19 UV light-induced photosalient effect of a hollow crystal of **15**, mimicking *Impatiens*: **a** deformation of the hollow crystal upon UV irradiation, **b** a hollow crystal packed with fluorescent polystyrene beads with 1-mm diameter, **c** UV-induced scattering of fluorescent beads. Scale bars = 20 μm. Adapted with permission from Ref. [52]. Copyright 2017 Wiley-VCH Verlag GmbH & Co. KGaA, Weinheim

Acknowledgements We deeply appreciate Professors M. Irie, C. J. Bardeen, Drs. H. Nishi, R. Tanaka, Mrs. T. Okuyama, H. Tsujioka, K. Kawasaki, A. Hirano for their collaboration. The present work was supported by JSPS KAKENHI Grant Number JP26107013, Grant-in-Aid for Scientific Research on Innovative Areas "Photosynergetics".

References

1. Irie M (2000) Diarylethenes for memories and switches. Chem Rev 100:1685–1716
2. Irie M, Fukaminato T, Matsuda K, Kobatake S (2014) Photochromism of diarylethene molecules and crystals: memories, switches, and actuators. Chem Rev 114:12174–12277
3. Abakumov GA, Nevodchikov VI (1982) Thermomechanical and photomechanical effects observed on crystals of a free-radical complex. Dokl Akad Nauk SSSR 266:1407–1410
4. Irie M, Kobatake S, Horichi M (2001) Reversible surface morphology changes of a photochromic diarylethene single crystal by photoirradiation. Science 291:1769–1772
5. Al-Kaysi RO, Mueller AM, Bardeen CJ (2006) Photochemically driven shape changes of crystalline organic nanorods. J Am Chem Soc 128:15938–15939

6. Kobatake S, Takami S, Muto H, Ishikawa T, Irie M (2007) Rapid and reversible shape changes of molecular crystals on photoirradiation. Nature 446:778–781
7. Uchida K, Sukata S, Matsuzawa Y, Akazawa M, de Jong JJD, Katsonis N, Kojima Y, Nakamura S, Areephong J, Meetsma A, Feringa BL (2008) Photoresponsive rolling and bending of thin crystals of chiral diarylethenes. Chem Commun 326–328
8. Kuroki L, Takami S, Yoza K, Morimoto M, Irie M (2010) Photoinduced shape changes of diarylethene single crystals: correlation between shape changes and molecular packing. Photochem Photobiol Sci 9:221–225
9. Morimoto M, Irie M (2010) A diarylethene cocrystal that converts light into mechanical work. J Am Chem Soc 132:14172–14178
10. Kobatake S, Hasegawa H, Miyamura K (2011) High-convertible photochromism of a diarylethene single crystal accompanying the crystal shape deformation. Cryst Growth Des 11:1223–1229
11. Terao F, Morimoto M, Irie M (2012) Light-driven molecular-crystal actuators: rapid and reversible bending of rodlike mixed crystals of diarylethene derivatives. Angew Chem Int Ed 51:901–904
12. Kitagawa D, Nishi H, Kobatake S (2013) Photoinduced twisting of a photochromic diarylethene crystal. Angew Chem Int Ed 52:9320–9322
13. Kitagawa D, Kobatake S (2013) Crystal thickness dependence of photoinduced crystal bending of 1,2-bis(2-methyl-5-(4-(1-naphthoyloxymethyl)phenyl)-3-thienyl)perfluorocyclopentene. J Phys Chem C 117:20887–20892
14. Kitagawa D, Kobatake S (2014) Crystal thickness dependence of photoinduced crystal bending of 1-(5-methyl-2-(4-(p-vinylbenzoyloxymethyl)phenyl)-4-thiazolyl)-2-(5-methyl-2-phenyl-4-thiazolyl)perfluorocyclopentene. Photochem Photobiol Sci 13:764–769
15. Kitagawa D, Kobatake S (2015) Photoreversible current ON/OFF switching by photoinduced bending of gold-coated diarylethene crystals. Chem Commun 51:4421–4424
16. Ohshima S, Morimoto M, Irie M (2015) Light-driven bending of diarylethene mixed crystals. Chem Sci 6:5746–5752
17. Kitagawa D, Tanaka R, Kobatake S (2015) Dependence of photoinduced bending behavior of diarylethene crystals on irradiation wavelength of ultraviolet light. Phys Chem Chem Phys 17:27300–27305
18. Kitagawa D, Iwaihara C, Nishi H, Kobatake S (2015) Quantitative evaluation of photoinduced bending speed of diarylethene crystals. Crystals 5:551–561
19. Kitagawa D, Tanaka R, Kobatake S (2016) Photoinduced stepwise bending behavior of photochromic diarylethene crystals. CrystEngComm 18:7236–7240
20. Kitagawa D, Okuyama T, Tanaka R, Kobatake S (2016) Photoinduced rapid and explosive fragmentation of diarylethene crystals having urethane bonding. Chem Mater 28:4889–4892
21. Hirano A, Hashimoto T, Kitagawa D, Kono K, Kobatake S (2017) Dependence of photoinduced bending behavior of diarylethene crystals on ultraviolet irradiation power. Cryst Growth Des 17:4819–4825
22. Kitagawa D, Kawasaki K, Tanaka R, Kobatake S (2017) Mechanical behavior of molecular crystals induced by combination of photochromic reaction and reversible single-crystal-to-single-crystal phase transition. Chem Mater 29:7524–7532
23. Kitagawa D, Tsujioka H, Tong F, Dong X, Bardeen CJ, Kobatake S (2018) Control of photomechanical crystal twisting by illumination direction. J Am Chem Soc 140:4208–4212
24. Hirano A, Kitagawa D, Kobatake S (2019) Photomechanical bending behavior of photochromic diarylethene crystals induced under polarized light. CrystEngComm 21:2495–2501
25. Koshima H, Ojima N, Uchimoto H (2009) Mechanical motion of azobenzene crystals upon photoirradiation. J Am Chem Soc 131:6890–6891
26. Koshima H, Ojima N (2012) Photomechanical bending of 4-aminoazobenzene crystals. Dyes Pigm 92:798–801
27. Bushuyev OS, Singleton TA, Barrett CJ (2013) Fast, reversible, and general photomechanical motion in single crystals of various azo compounds using visible light. Adv Mater 25:1796–1800

28. Bushuyev OS, Tomberg A, Friscic T, Barrett CJ (2013) Shaping crystals with light: crystal-to-crystal isomerization and photomechanical effect in fluorinated azobenzenes. J Am Chem Soc 135:12556–12559
29. Taniguchi T, Fujisawa J, Shiro M, Koshima H, Asahi T (2016) Mechanical motion of chiral azobenzene crystals with twisting upon photoirradiation. Chem Eur J 22:7950–7958
30. Taniguchi T, Asahi T, Koshima H (2019) Photomechanical azobenzene crystals. Crystals 9:437
31. Koshima H, Nakaya H, Uchimoto H, Ojima N (2012) Photomechanical motion of furylfulgide crystals. Chem Lett 41:107–109
32. Koshima H, Takechi K, Uchimoto H, Shiro M, Hashizume D (2011) Photomechanical bending of salicylideneaniline crystals. Chem Commun 47:11423–11425
33. Koshima H, Matsuo R, Matsudomi M, Uemura Y, Shiro M (2013) Light-driven bending crystals of salicylidenephenylethylamines in enantiomeric and racemate forms. Cryst Growth Des 13:4330–4337
34. Takanabe A, Tanaka M, Johmoto K, Uekusa H, Mori T, Koshima H, Asahi T (2016) Optical activity and optical anisotropy in photomechanical crystals of chiral salicylidenephenylethylamines. J Am Chem Soc 138:15066–15077
35. Al-Kaysi RO, Bardeen CJ (2007) Reversible photoinduced shape changes of crystalline organic nanorods. Adv Mater 19:1276–1280
36. Zhu L, Al-Kaysi RO, Dillon RJ, Tham FS, Bardeen CJ (2011) Crystal structures and photophysical properties of 9-anthracene carboxylic acid derivatives for photomechanical applications. Cryst Growth Des 11:4975–4983
37. Zhu L, Al-Kaysi RO, Bardeen CJ (2011) Reversible photoinduced twisting of molecular crystal microribbons. J Am Chem Soc 133:12569–12575
38. Zhu L, Agarwal A, Lai J, Al-Kaysi RO, Tham FS, Ghaddar T, Mueller L, Bardeen CJ (2011) Solid-state photochemical and photomechanical properties of molecular crystal nanorods composed of anthracene ester derivatives. J Mater Chem 21:6258–6268
39. Kim T, Al-Muhanna MK, Al-Suwaidan SD, Al-Kaysi RO, Bardeen CJ (2013) Photoinduced curling of organic molecular crystal nanowires. Angew Chem Int Ed 52:6889–6893
40. Koshima H, Uchimoto H, Taniguchi T, Nakamura J, Asahi T, Asahi T (2016) Mechanical motion of molecular crystals induced by [4 + 4] photodimerization. CrystEngComm 18:7305–7310
41. Kim T, Zhu L, Mueller LJ, Bardeen CJ (2012) Dependence of the solid-state photomechanical response of 4-chlorocinnamic acid on crystal shape and size. CrystEngComm 14:7792–7799
42. Sun JK, Li W, Chen C, Ren CX, Pan DM, Zhang J (2013) Photoinduced bending of a large single crystal of a 1,2-bis(4-pyridyl)ethylene-based pyridinium salt powered by a [2 + 2] cycloaddition. Angew Chem Int Ed 52:6653–6657
43. Naumov P, Kowalik J, Solntsev KM, Baldridge A, Moon JS, Kranz C, Tolbert LM (2010) Topochemistry and photomechanical effects in crystals of green fluorescent protein-like chromophores: effects of hydrogen bonding and crystal packing. J Am Chem Soc 132:5845–5857
44. Nakai H, Matsuba K, Akimoto M, Nozaki T, Matsumoto T, Isobe K, Irie M, Ogo S (2016) Photoinduced bending of rod-like millimetre-size crystals of a rhodium dithionite complex with n-pentyl moieties. Chem Commun 52:4349–4352
45. Naumov P, Sahoo SC, Zakharov BA, Boldyreva EV (2013) Dynamic single crystals: kinematic analysis of photoinduced crystal jumping (the photosalient effect). Angew Chem Int Ed 52:9990–9995
46. Seki T, Sakurada K, Muromoto M, Ito H (2015) Photoinduced single-crystal-to-single-crystal phase transition and photosalient effect of a gold(I) isocyanide complex with shortening of intermolecular aurophilic bonds. Chem Sci 6:1491–1497
47. Colombier I, Spagnoli S, Corval A, Baldeck PL, Giraud M, Leaustic A, Yu P, Irie M (2007) Diarylethene microcrystals make directional jumps upon ultraviolet irradiation. J Chem Phys 126:011101
48. Hatano E, Morimoto M, Hyodo K, Yasuda N, Yokojima S, Nakamura S, Uchida K (2016) Photosalient effect of a diarylethene with a perfluorocyclohexene ring. Chem Eur J 22:12680–12683

49. Zhu L, Al-Kaysi RO, Bardeen CJ (2016) Photoinduced ratchet-like rotational motion of branched molecular crystals. Angew Chem Int Ed 55:7073–7076
50. Al-Kaysi RO, Tong F, Al-Haidar M, Zhu L, Bardeen CJ (2017) Highly branched photomechanical crystals. Chem Commun 53:2622–2625
51. Chen KJ, Tsai YC, Suzaki Y, Osakada K, Miura A, Horie M (2016) Rapid and reversible photoinduced switching of a rotaxane crystal. Nat Commun 7:13321
52. Hatano E, Morimoto M, Imai T, Hyodo K, Fujimoto A, Nishimura R, Sekine A, Yasuda N, Yokojima S, Nakamura S, Uchida K (2017) Photosalient phenomena that mimic impatiens are observed in hollow crystals of diarylethene with a perfluorocyclohexene ring. Angew Chem Int Ed 56:12576–12580

Chapter 26
Interplay of Photoisomerization and Phase Transition Events Provide a Working Supramolecular Motor

Yoshiyuki Kageyama

Abstract Autonomous molecular-based microrobots have not been created despite progress in our understanding of the chemistry of molecular motors and machines. In the field of chemistry, the design and synthesis of molecular structures remain an ongoing and diverse challenge. The creation of a system in which molecular structures interact represents an additional challenge. In order to functionalize a molecule for a motor, it must be exposed to a specific reaction field. Synchronization of the molecules involved is also required. In this chapter, we discuss our research results on self-oscillatory flipping motions of azobenzene-containing assemblies. Briefly, reversible photoisomerization of an azobenzene derivative occurs under steady light irradiation. In coordination with phase transition events, a cycle modulated by the components involved leads to repetitive structural changes. When this azobenzene-based assembly is placed in water, the assembly exhibits autonomous swimming. This macroscopic light-powered motion is the result of the self-organization of a large number of nanometer-scale molecules and it provides an example of the potential for mechanical work to be performed by molecular assemblies. We subsequently describe the driving concept of far-from-equilibrium dynamics, namely a dissipative structure by which a spatial pattern maintains dynamic behavior to drive the mechanical motion observed.

Keywords Molecular motor · Self-organized motion · Hierarchical structure · Azobenzene photoisomerization · Collective dynamics · Active matter · Limit-cycle dissipative structure

Y. Kageyama (✉)
Faculty of Science, Hokkaido University, Sapporo, Hokkaido 060-0810, Japan
e-mail: y.kageyama@sci.hokudai.ac.jp

H. Miyasaka et al. (eds.), *Photosynergetic Responses in Molecules and Molecular Aggregates*, https://doi.org/10.1007/978-981-15-5451-3_26

26.1 Introduction

26.1.1 Achieving Mechanical Work with Molecular Motors

The creation of active matter and active transporters is currently an exciting area of research. For chemists, the synthesis of molecules which continuously and mechanically perform work is a fundamental task in the preparation of such active molecular devices. From this viewpoint, synthetic molecular motors have been proposed to date [1–3]. Similar to the enzymatic molecular motors in biological systems, molecular motors generally involve a molecular-based catalyst which spontaneously repeats a cyclic structural change while converting a source of energy (e.g., a chemical substrate, electric power, or light) into mechanical energy. The mechanical energy generated by these molecular motors is the sum of kinetic energy and potential energy. In this chapter, the definition of "molecular motor" extends beyond the traditional focus on salient and repetitive structural changes in molecules (Fig. 26.1). For example, if a molecule achieves even a small structural change that converts energy to mechanical potential energy, it is considered a molecular motor in this chapter. Conversely, a molecule that does not attain an energy conversion to mechanical energy will not be considered a molecular motor in the near future, even if chemists traditionally recognize it as a molecular motor. The latter is based on the observation that molecular motors have traditionally been defined based on structural equations of molecules, and not necessarily on actual and realized functions.

In addition to light-powered molecular motors (or rotators) described by Feringa [4–9], several light-powered [10, 11], redox-activated [12–14], and chemical-fueled [15–17] molecular motors have also been proposed. Chemical-fueled molecular motors provide work with a small supply of energy, and thus, the development of

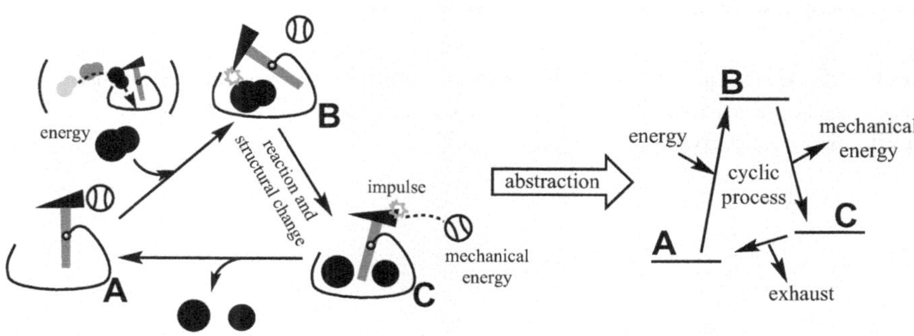

Fig. 26.1 Basic concept of the molecular motor model discussed in this chapter. Left: An illustration of an object which accomplishes a mechanical task by consuming a substrate. A, B, and C indicate the states (structures) of the molecule or super-molecule which constitutes the motor. In this chapter, we also describe molecules which do not produce mechanical work or exhaust as a type of molecular motor

such motors is of great interest. To achieve spatially unidirectional motion of a molecular motor, molecular-level chirality is key. However, it is not sufficient due to the principle of microscopic reversibility. To realize an effective and powerful molecular motor that performs work, a nonlinear process is required in the motor system to control the reverse process [18–20]. Recently, time-delayed chemical processes, such as that employed in a "molecular information ratchet" [16, 21–24], have been considered a good choice for generating molecular motors [25–28]. To date, light-powered motors have been developed more rapidly than chemical-fueled motors. This is partly because microscopic reversibility is not a concern for light-powered molecular motors [25, 29–31].

However, even if excellent molecular motors are identified, the task remains to have these motors act on another object to produce work. It is a straightforward concept that the mechanical energy of a molecular motor in a monodispersed solution is converted immediately to thermal energy, without mentioning the physical principle of Onsager's reciprocal relation [25, 32]. Given the importance of friction for actual work, it is important that the spatial and temporal scales of a working system be considered. A relative size difference between a motor and an object is required in a mesoscale system. The concept represents a challenge in the field of molecular motor research. One promising approach is to place a molecular motor on a large substrate and construct an anisotropic reaction field [33]. A pioneering example is light-powered electron transportation through a vesicular membrane to generate a proton-motive force [34, 35] or electromotive force [36]. In this model, the molecular motor (generally referred to as the "sensitizer") cyclically changes its shape as a result of changes in structure to achieve an excited state, followed by the generation of a radical/ion state. These cyclic changes are represented with triangle, square, and circle symbols, respectively, in Fig. 26.2a. Another well-known example involves the long-term rotation of small rods on cholesteric liquid crystals with light-powered molecular motor dopants [37–40]. To achieve a continuous motion with relative size difference, synchronization of molecular motion is required (Fig. 26.2b). If molecular-level motions are not synchronized, the system falls into a static steady-state and no further mechanical motion is expected (Fig. 26.2c). Thus, it is necessary to keep a system far-from-equilibrium state via a nonlinear process [41].

26.1.2 Motion of Macroscopic Objects

In this chapter, we focus on the synchronization of molecular motors in regard to a "photosynergetic" aspect. As mentioned above, one of the advantages of chemical-fueled molecular motors is their energy efficiency. In cooperation with thermal fluctuations, the energy efficiency of structural transitions can potentially reach up to 100% [42]. However, it is rather difficult to accommodate the injection and elimination of chemical substrates when constructing an artificial motorization system in a solid phase. To date, most studies of self-organized motion using chemical processes have

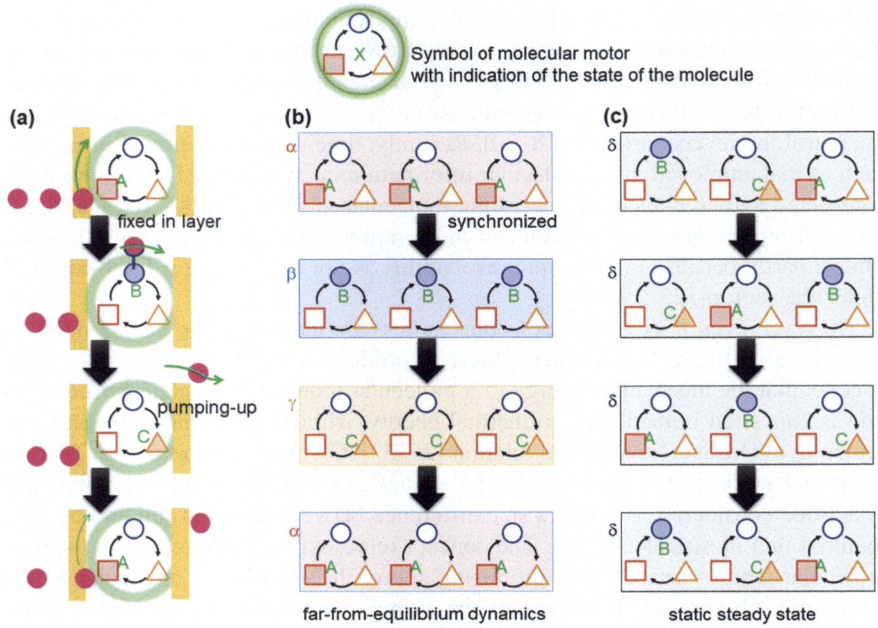

Fig. 26.2 Strategies for achieving mechanical work by a molecular motor. **a** Molecular motors can work individually by being placed in an anisotropic reaction field. This concept is realized in transport systems that exist across biological membranes. **b** Mesoscale materials can assemble into molecular motors. By synchronizing a molecular motor to create a cyclic process, a mesoscopic supramolecular motor is expected to be realized. **c** If a system does not achieve synchronization, the material will not produce a work function

used systems constructed in a liquid phase, or systems which employ diffusion and a surface-based reaction [43]. In contrast, a light-powered molecular motor has its fuel supplied in a bulk phase without contact or discharge being an issue. To achieve work by a molecular motor, synchronizing of motor function is a fundamental issue. Therefore, the use of light-powered molecular motors is preferred given the current chemistry available, despite the low energy efficiency of these motors.

To date, macroscopic, continuous dynamics of synchronized light-powered molecular motors have mainly been realized by using polymer films [44–51]. In these studies, a self-shading mechanism where light triggers a morphological change in a polymer causes auto-switching of light-absorption efficiency. An overview of the periodic cycle for this model is summarized in Fig. 26.3. The relative location of the light source, the polarization state of the light, and/or the anisotropic structure of the polymers involved provoking the structural change which occurs. It is important to note that it is a shape change that affects the light-receiving property to modulate the region for photoreaction and its reaction velocity. As a result, repetitive dynamics are observed. In a system exhibiting periodic motion, a time delay, existing in relation to the change in polymer shape which occurs after isomerization

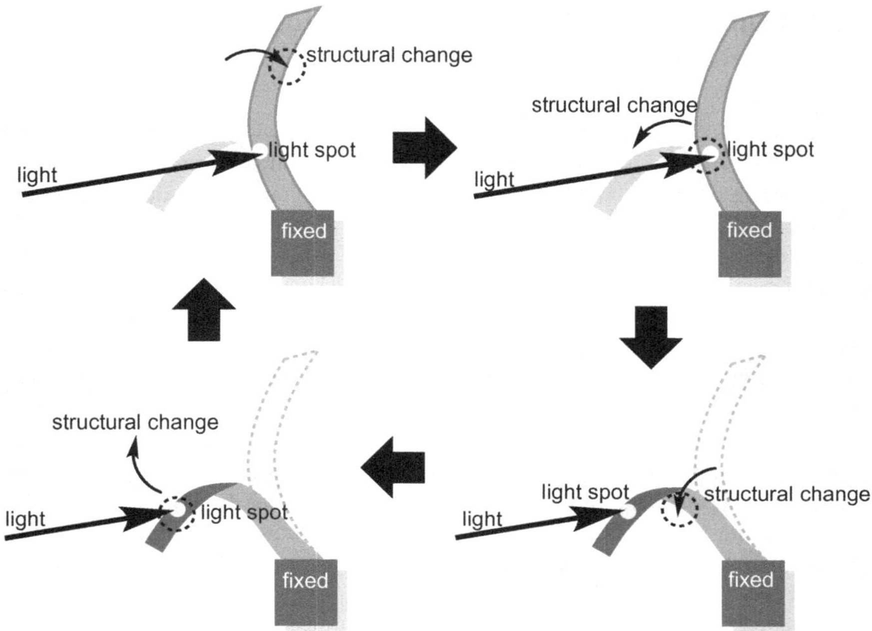

Fig. 26.3 Schematic illustration of a cycle which provides self-continuous flipping motion of light-responsive polymers fixed on a solid stage. The motion is powered by directional light. Structural changes occur after photoisomerization, and this automatically induces a change in the irradiated face (position)

of the molecule, prevents the system from reaching equilibrium [52]. However, if the polymer's structural change occurs within the same time span as the photochemical process involved, the system will not achieve self-continuous dynamics. Indeed, the experimental results described by White and coworkers indicate that a decrease in incident light intensity slows down the photochemical process and the amplitude from the so-called equilibrium position becomes closer [45]. Very recently, when a polymer film was affixed to a solid frame, locomotive objects working on a solid surface [51] or a water surface were realized. In these systems, the direction of motion was determined according to the direction of the incident light.

As shown above, "reversible molecular transformation", "spatial anisotropy", and "time-delayed switching of reaction efficiency" are a set of conditions that are required to achieve a system with self-continuous macroscopic dynamics. By applying this set of requirements to an object, autonomous work independent from direction by external anisotropy is expected. Unfortunately, there have been very few examples of successful systems. However, our group previously reported a self-oscillatory flipping motion by azobenzene assemblies under steady light irradiation [53]. Thus, a discussion of our ongoing research and results for the latter light-powered motion system is subsequently provided.

26.2 Realization of Self-Oscillatory Flipping Motion of Azobenzene Assemblies

The theory of self-organization in continuous dynamics was established based on a series of studies of dissipative structures [54]. This type of structure is dynamic, yet it maintains an ordered configuration, in the presence of an energy source. In regard to photochemical self-oscillation, Nitzan and Ross first proposed that a reversible photochemical process which involves immediate thermal relaxation and a time-delayed reversible thermal process may form a dissipative structure (Fig. 26.4a) [55]. According to their theory, temporal patterns form as follows. A photo-process-induced fluctuation in temperature causes a shift in the equilibrium of a thermal process. The behavior of this system then modulates the velocity of the photo-process. However, under regular experimental conditions, keeping the system from achieving "equilibrium" (namely, achieving an attractive state in the presence of energy supply) is difficult without involving additional processes due to the required steep change in temperature. Experimentally, oscillations in fluorescence have been observed in solutions containing 10^{-3} mol/L solutes [56]. Further studies concluded that convection of light-irradiated solutions in optical cuvettes mediates the fluctuation in fluorescence observed (Fig. 26.4b) [57, 58].

Alternatively, we focused on shifts in reaction rates as a result of phase transitions. Since the reaction rate of a photochemical process depends on the conditions of the molecule involved, a phase shift represents an opportunity to alternate the reaction kinetics of a photochemical process. Correspondingly, a photoisomerization-triggered molecular realignment of an assembly was recognized as a good target for generating a system characterized by self-continuous dynamics.

Azobenzene is a well-known photochromic compound that undergoes a *trans*-to-*cis* and *cis*-to-*trans* photoisomerization event via n-π* excitation. This event generally occurs at a wavelength of 450 nm. We first designed a system composed of an azobenzene derivative and a fatty acid, since the shape of a molecular assembly of fatty acids has been shown to depend on the activity of protons surrounding the assembly [59]. By changing the molecular association externally, a shift in proton

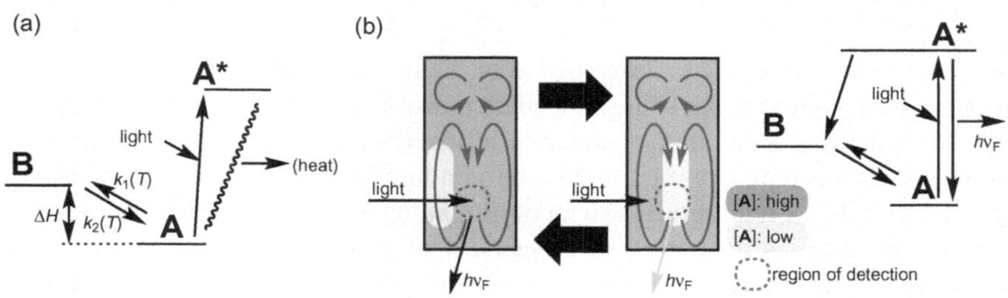

Fig. 26.4 A solution-phase dissipative structure formed by a photochemical process. **a** This model was proposed by Nitzan and Ross in 1973. **b** The model was confirmed experimentally with observations of fluorescence fluctuation

activity and reassembling of molecules are predicted. Correspondingly, our system was designed so that photoisomerization of the azobenzene derivative would trigger realignment of the fatty acid molecules to modulate the kinetics of the photoisomerization events. A helical assembly of oleic acid (**1**) and an amphiphilic azobenzene (**2**) never showed repetitive motion under steady light irradiation [60]. Rather, a multi-step recoiling motion was observed involving acid-dissociation behavior of a carboxyl group [61]. Vesicular, tubular, and block-like assemblies also did not exhibit repetitive dynamics [60]. However, we eventually observed that plate-like assemblies of **1** and **2** exhibited repetitive flips under steady blue-light irradiation (Fig. 26.5) [53].

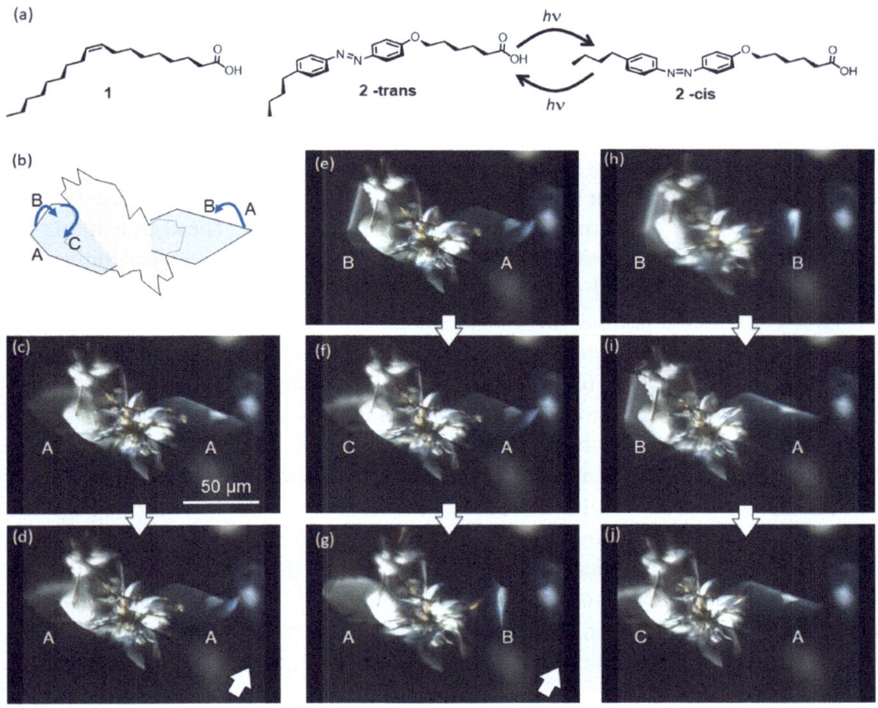

Fig. 26.5 Self-continuous flipping motion observed for a co-assembly of oleic acid (**1**) and an azobenzene derivative (**2**) by Ikegami in 2016 [53]. **a** The chemical structures of **1** and **2** are shown along with the photochromism step of **2**. **b** A generalized illustration of the flipping motion of azobenzene with states A-C and A and B indicated. **c–j** Video imaging of the flipping motion of a **1** crystal detected with a differential interference contrast microscope equipped with a mercury lamp unit ($\lambda_{irr} = 435$ nm). The left side of the crystal exhibited alternating flips (cyclically around A, B, and C), while the right side of the crystal exhibited singular flips between the A and B states. A video is available from the website listed in Ref. [53]

26.3 Mechanism for Self-Continuous Flipping Motion

A co-assembly of molecules **1** and **2** leads to two types of self-oscillatory flipping motion under blue-light irradiation [53]. One type involves a single flipping event, while the second type involves a multi-step flipping event. The mechanistic study described in this section targets the singular flipping motion, while the multi-step motion is described below in Sect. 26.4, and in further detail in [62]. When a dispersion of the co-assembly was analyzed in situ with X-ray diffraction (XRD), the small-angle profile showed one-peak and two-peaks in the dark versus under blue-light irradiation, respectively (Fig. 26.6a). These results indicate that a new phase co-exists with the original phase under light irradiation. Figure 26.6b shows the time-course behavior of the self-oscillatory motion detected under various intensities of steady 435-nm light. Figure 26.6c shows the time-course behavior of another assembly under 435 and 470-nm light, while Fig. 26.6d shows the time-course of thermal behavior of the co-assembly after light irradiation. Taken together, these results indicate that one-cycle of the flipping motion of the co-assembly consists of four steps. Step 1 is a *trans*-to-*cis* photoisomerization event, while step 3 is a *cis*-to-*trans* photoisomerization event. Meanwhile, reversal of the isomerization events occurs due to phase transitions that occur in steps 2 and 4. These transitions are caused by shifts in the kinetic properties of **2**. Slope values for the logarithmic relationship between incident light power and the duration times of steps 1 and 3 are 1 (Fig. 26.6b), indicating that these steps are photon number-determined processes. In other words, to shift from the photo-process events (e.g., steps 1 and 3) to the phase transition-processes (e.g., steps 2 and 4), the isomer ratio of the assembly must achieve a particular threshold. This means that the self-oscillation property of this system is not chaotic, but rather represents a type of limit-cycle (Fig. 26.7). Therefore, the assembly charges internal energy (the sum of the mechanical stress and the enthalpy of molecular formation) up to its threshold by accepting light in step 1. The assembly then discharges part of the energy with morphological changes in steps 2 and 4. Overall, these events demonstrate that this assembly has the ability to apply mechanical work to other objects periodically. For example, when the co-assembly is placed in water, it applies impulses to the water which manifests as a swimming motion [53]. Details regarding this system are reported elsewhere [63, 64].

26.4 Analogues also Exhibiting a Self-Oscillatory Flipping Motion

The simple mechanism described above may also be observed in other assemblies composed of photochromic compounds. However, until now, we have not investigated photochromic compounds other than azobenzene derivatives. To date, we have successfully generated several self-oscillatory motions in co-assemblies of several

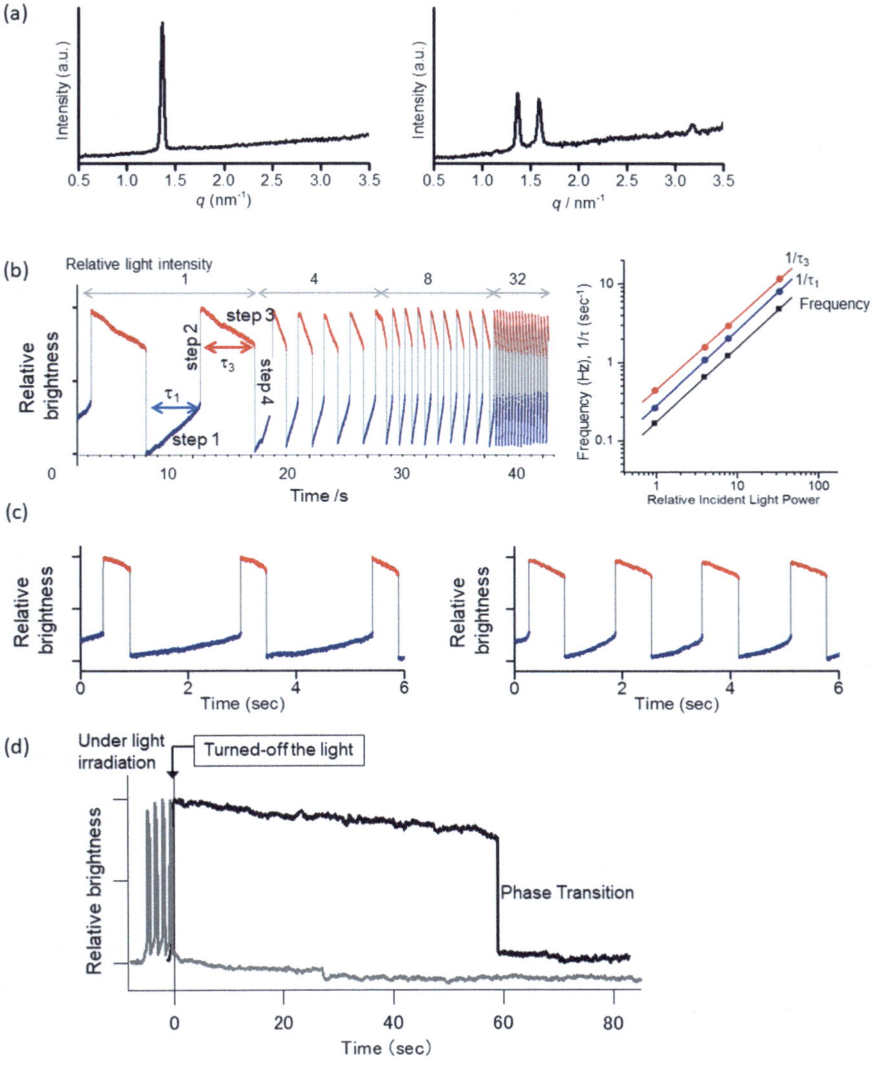

Fig. 26.6 Characteristics of the light-powered periodic flips of an azobenzene-containing co-assembly. **a** The *in situ* small-angle XRD profiles of the co-assembly under dark conditions (left) and under blue-light irradiation (right). **b** Time profile of the periodic motion of the co-assembly under irradiation with various intensities of steady 435-nm light (left). Light power dependencies of duration time (τ_1, and τ_3) and frequency (right) are shown. **c** Time profiles of the periodic motion of the co-assembly under irradiation with 470-nm light (left) versus 435-nm light (right). **d** Behavior of the co-assembly after removing the light source. When the co-assembly was in its bent-shape phase and the light source was removed, thermal relaxation via a phase transition was observed. All figures were produced from the original data [53]

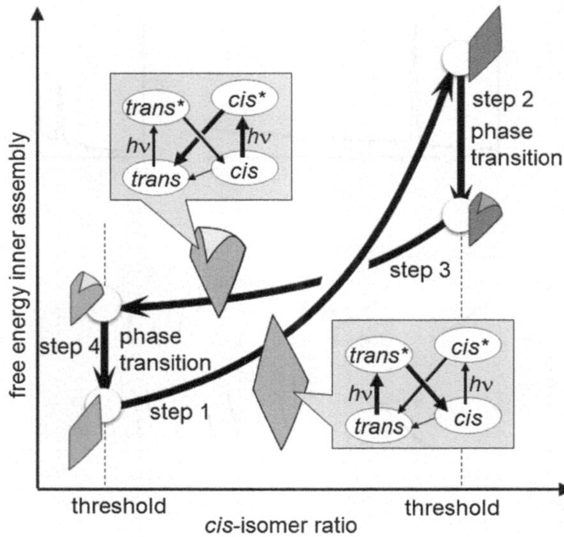

Fig. 26.7 Schematic of the energy cycle employed for the limit-cycle self-oscillation of an azobenzene-based assembly which undergoes a photochemical process under steady light irradiation

fatty acids and azobenzene derivatives, examples of which are shown in Fig. 26.8 [65]. A waved, ribbon-like assembly of stearic acid, **1**, and **2** exhibits a multi-step frog-kick motion [53]. A plate-like assembly of the same compounds was observed when they were maintained at 80-Hz for more than 5 min [66]. Many of the hexagonal or parallelogram plate-like assemblies of fatty acids and azobenzene derivatives exhibit self-oscillatory flipping motions, while ribbon-shape assemblies exhibit a wavy motion [65, 66]. Many varieties of motions have been observed, and there appears to be some trends in the relationship between molecular structure and the spatial pattern of the motion observed. The details of these trends are currently being investigated [65].

We have been particularly interested in one of the polymorphs of a single-component crystal of **2**, which showed repetitive winding dynamics during exposure to blue-light irradiation [62]. The crystal has a symmetry-broken space-group (P_1) with six independent molecules in each unit-cell [67]. Additional characteristics of this crystal include a spare-and-dense layered structure, as well as an anisotropic structure related to a directive flipping motion under non-directional light irradiation. Based on the fact that the average isomer ratio in its steady-state was less than 8%, we have assumed that a particular molecule undergoes isomerization to

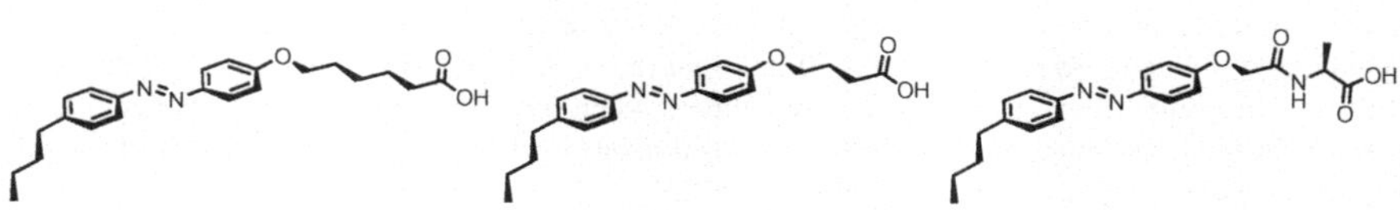

Fig. 26.8 Structures of the azobenzene molecules which generate light-powered self-oscillation motions when they assemble with fatty acids

trigger a phase transition of the crystal, and then it accepts feedback in the manner described above. Meanwhile, the other molecules serve as building blocks to prevent the assembly from breaking down. Our most recent results further indicate that the latter molecules also play a role in selecting the path for the periodic cycle of this system. As a result, periodic changes in shape are varied according to the externally defined light-polarity.

26.5 Conclusion and Remarks Regarding Self-Organized Molecular Motors

The inventions of the water mill and windmill have proved very important for advances in agriculture. Meanwhile, the invention of the steam engine has allowed humans to move objects as small as electrons to objects as large as ships. The key to all of these inventions is that the objects involved are connected to motors. There is no exception. Thus, when we create a self-moving microrobot, we must have a suitable motor.

To date, organic chemists have synthesized molecules that undergo salient changes in their structures. If a cyclic change occurs when steady energy supply is available, this has been considered a molecular motor according to traditional definition. On the other hand, if the change in structure requires alterations in the conditions of the molecule to achieve the change in structure, this has been considered a molecular machine. In parallel, material scientists have worked on generating microrobots which move under the direction of remote control. In these microrobots, molecular machine has been employed rather than molecular motors [68, 69], because an unsynchronized molecular motor is never worked. However, to attain an autonomous function in microrobot, a suitable molecular motor system is needed.

To date, our ability to obtain work from a molecular motor remains insufficient. One reason may be that we need to better understand molecular self-organization. A series of theoretical studies related to dissipative structures have clearly indicated how autonomous energy conversion can occur in a mesoscale molecular system. A dissipative structure involves a dynamic whereby multiple nonlinear processes are coupled in a feedback manner. As a result, they form a spatial and temporal system which exhibits self-continuous dynamics compared with the original process shown in Figs. 26.2 and 26.7. The importance of dissipative structures is their far-from-equilibrium behavior: the distance from so-called equilibrium (not thermal equilibrium) is related to the capability of the mechanical function which a system can accomplish. Moreover, this distance from equilibrium is required to be maintained despite the generation of external disturbances as the system acts on external objects. Therefore, the limit-cycle-type self-oscillation system described in this chapter appears to be the best system for developing a motorization system [70].

Thus, while strategies for creating such dynamics at the molecular level have not been successful to date, recent efforts to construct a multi-molecular motor system appear to be a promising approach for attaining a working molecular system in the near feature.

Acknowledgements The author deeply appreciates to Professor Sadamu Takeda, Dr. Tomonori Ikegami, and Dr. Hiroyasu Sato for their collaboration. The work was supported by JSPS KAKENHI Grant Number JP17H05247 in Scientific Research on Innovative Areas "Photosynergetics", JSPS KAKENHI Grant Number JP18H05423, and JST PRESTO Grant Number JPMJPR13K6.

References

1. Kay ER, Leigh DA, Zerbetto F (2007) Synthetic molecular motors and mechanical machines. Angew Chem Int Ed 46:72–191
2. von Delius M, Leigh DA (2011) Walking molecules. Chem Soc Rev 40:3656–3676
3. Kassem S, van Leeuwen T, Lubbe AS, Wilson MR, Feringa BL, Leigh DA (2017) Artificial molecular motors. Chem Soc Rev 46:2592–2621
4. Koumura N, Zijlstra RWJ, van Delden RA, Harada N, Feringa BL (1999) Light-driven monodirectional molecular rotor. Nature 401:152–155
5. Feringa BL (2001) In control of motion: from molecular switches to molecular motors. Acc Chem Res 34:504–513
6. ter Wiel MKJ, van Delden RA, Meetsma A, Feringa BL (2003) Increased speed of rotation for the smallest light-driven molecular motor. J Am Chem Soc 125:15076–15086
7. van Delden RA, Koumura N, Schoevaars A, Meetsma A, Feringa BL (2003) A donor–acceptor substituted molecular motor: unidirectional rotation driven by visible light. Org Biomol Chem 1:33–35
8. Vicario J, Walko M, Meetsma A, Feringa BL (2006) Fine tuning of the rotary motion by structural modification in light-driven unidirectional molecular motors. J Am Chem Soc 128:5127–5135
9. Wezenberg SJ, Feringa BL (2018) Supramolecularly directed rotary motion in a photoresponsive receptor. Nat Commun 9:1984
10. Ragazzon G, Baroncini M, Silvi S, Venturi M, Credi A (2014) Light-powered autonomous and directional molecular motion of a dissipative self-assembling system. Nat Nanotechnol 10:70–75
11. Yu J-J, Zhao L-Y, Shi Z-T, Zhang Q, London G, Liang W-J, Gao C, Li M-M, Cao X-M, Tian H, Feringa BL, Qu D-H (2019) Pumping a ring-sliding molecular motion by a light-powered molecular motor. J Org Chem 84:5790–5802
12. Pezzato C, Nguyen MT, Kim DJ, Anamimoghadam O, Mosca L, Stoddart JF (2018) Controlling dual molecular pumps electrochemically. Angew Chem Int Ed 57:9325–9329
13. Cheng C, McGonigal PR, Schneebeli ST, Li H, Vermeulen NA, Ke C, Stoddart JF (2015) An artificial molecular pump. Nat Nanotechnol 10:547–553
14. Kudernac T, Ruangsupapichat N, Parschau M, Maciá B, Katsonis N, Harutyunyan SR, Ernst K-H, Feringa BL (2011) Electrically driven directional motion of a four-wheeled molecule on a metal surface. Nature 479:208–211
15. Collins BSL, Kistemaker JCM, Otten E, Feringa BL (2016) A chemically powered unidirectional rotary molecular motor based on a palladium redox cycle. Nat Chem 8:860–866
16. Wilson MR, Solà J, Carlone A, Goldup SM, Lebrasseur N, Leigh DA (2016) An autonomous chemically fuelled small-molecule motor. Nature 534:235–240
17. Fletcher SP, Dumur F, Pollard MM, Feringa BL (2005) A reversible, unidirectional molecular rotary motor driven by chemical energy. Science 310:80–82

18. Astumian RD (2007) Design principles for Brownian molecular machines: how to swim in molasses and walk in a hurricane. Phys Chem Chem Phys 9:5067–5083
19. Pezzato C, Cheng C, Stoddart JF, Astumian RD (2017) Mastering the non-equilibrium assembly and operation of molecular machines. Chem Soc Rev 46:5491–5507
20. Astumian RD (2017) How molecular motors work – insights from the molecular machinist's toolbox: the Nobel prize in chemistry 2016. Chem Sci 8:840–845
21. Serreli V, Lee C-F, Kay ER, Leigh DA (2007) A molecular information ratchet. Nature 445:523–527
22. Alvarez-Pérez M, Goldup SM, Leigh DA, Slawin AMZ (2008) A chemically-driven molecular information ratchet. J Am Chem Soc 130:1836–1838
23. Carlone A, Goldup SM, Lebrasseur N, Leigh DA, Wilson A (2012) A three-compartment chemically-driven molecular information ratchet. J Am Chem Soc 134:8321–8323
24. Martin CJ, Lee ATL, Adams RW, Leigh DA (2017) Enzyme-mediated directional transport of a small-molecule walker with chemically identical feet. J Am Chem Soc 139:11998–12002
25. Astumian RD (2018) Trajectory and cycle-based thermodynamics and kinetics of molecular machines: the importance of microscopic reversibility. Acc Chem Res 51:2653–2661
26. Astumian RD (2019) Kinetic asymmetry allows macromolecular catalysts to drive an information ratchet. Nat Commun 10:3837
27. Rex A (2017) Maxwell's demon—a historical review. Entropy (Basel, Switzerland) 19:240
28. Zhang L, Marcos V, Leigh DA (2018) Molecular machines with bio-inspired mechanisms. Proc Nat'l Acad Sci 115:9397–9404
29. Credi A, Venturi M, Balzani V (2010) Light on molecular machines. ChemPhysChem 11:3398–3403
30. Silvi S, Venturi M, Credi A (2011) Light operated molecular machines. Chem Commun 47:2483–2489
31. Kathan M, Hecht S (2017) Photoswitchable molecules as key ingredients to drive systems away from the global thermodynamic minimum. Chem Soc Rev 46:5536–5550
32. Onsager L (1931) Reciprocal relations in irreversible processes. I Phys Rev 37:405–426
33. Watson MA, Cockroft SL (2016) Man-made molecular machines: membrane bound. Chem Soc Rev 45:6118–6129
34. Steinberg-Yfrach G, Liddell PA, Hung S-C, Moore AL, Gust D, Moore TA (1997) Conversion of light energy to proton potential in liposomes by artificial photosynthetic reaction centres. Nature 385:239–241
35. Steinberg-Yfrach G, Rigaud J-L, Durantini EN, Moore AL, Gust D, Moore TA (1998) Light-driven production of ATP catalysed by F_0F_1-ATP synthase in an artificial photosynthetic membrane. Nature 392:479–482
36. Mizushima T, Yoshida A, Harada A, Yoneda Y, Minatani T, Murata S (2006) Pyrene-sensitized electron transport across vesicle bilayers: dependence of transport efficiency on pyrene substituents. Org Biomol Chem 4:4336–4344
37. Eelkema R, Pollard MM, Vicario J, Katsonis N, Ramon BS, Bastiaansen CWM, Broer DJ, Feringa BL (2006) Nanomotor rotates microscale objects. Nature 440:163
38. Bosco A, Jongejan MGM, Eelkema R, Katsonis N, Lacaze E, Ferrarini A, Feringa BL (2008) Photoinduced reorganization of motor-doped chiral liquid crystals: bridging molecular isomerization and texture rotation. J Am Chem Soc 130:14615–14624
39. Thomas R, Yoshida Y, Akasaka T, Tamaoki N (2012) Influence of a change in helical twisting power of photoresponsive chiral dopants on rotational manipulation of micro-objects on the surface of chiral nematic liquid crystalline films. Chem Eur J 18:12337–12348
40. Kim Y, Tamaoki N (2019) Photoresponsive chiral dopants: light-driven helicity manipulation in cholesteric liquid crystals for optical and mechanical functions. ChemPhotoChem 3:284–303
41. As an example of non-photochemical motion, a catalyst-pendent-macrocycle sliding on a polymer chain has been reported: Hashidzume A, Kuse A, Oshikiri T, Adachi S, Okumura M, Yamaguchi H, Harada A (2018) Toward a translational molecular ratchet: face-selective translation coincident with deuteration in a pseudo-rotaxane. Sci Rep 8:8950

42. Since friction is required for generating work from a molecular motor, a value of 100% does not indicate that a particular molecular motor will be good for an assembled working system
43. Nakata S, Pimienta V, Lagzi I, Kitahata H, Suematsu JN (2019) Self-organized motion. Royal Society of Chemistry, Cambridge
44. Yamada M, Kondo M, Mamiya J, Yu Y, Kinoshita M, Barrett C, Ikeda T (2008) Photomobile polymer materials: towards light-driven plastic motors. Angew Chem Int Ed 47:4986–4988
45. Serak S, Tabiryan N, Vergara R, White TJ, Vaia RA, Bunning TJ (2010) Liquid crystalline polymer cantilever oscillators fueled by light. Soft Matter 6:779–783
46. White TJ, Tabiryan NV, Serak SV, Hrozhyk UA, Tondiglia VP, Koerner H, Vaia RA, Bunning TJ (2008) A high frequency photodriven polymer oscillator. Soft Matter 4:1796–1798
47. Vantomme G, Gelebart AH, Broer DJ, Meijer EW (2017) Preparation of liquid crystal networks for macroscopic oscillatory motion induced by light. J Vis Exp e56266
48. Vantomme G, Gelebart AH, Broer DJ, Meijer EW (2018) Self-sustained actuation from heat dissipation in liquid crystal polymer networks. J Polym Sci A: Polym Chem 56:1331–1336
49. Gelebart AH, Vantomme G, Meijer EW, Broer DJ (2017) Mastering the photothermal effect in liquid crystal networks: a general approach for self-sustained mechanical oscillators. Adv Mat 29:1606712
50. Vantomme G, Gelebart AH, Broer DJ, Meijer EW (2017) A four-blade light-driven plastic mill based on hydrazone liquid-crystal networks. Tetrahedron 73:4963–4967
51. Gelebart AH, Jan Mulder D, Varga M, Konya A, Vantomme G, Meijer EW, Selinger RLB, Broer DJ (2017) Making waves in a photoactive polymer film. Nature 546:632–636
52. Kageyama Y (2019) Light-powered self-sustainable macroscopic motion for the active locomotion of materials. ChemPhotoChem 3:327–336
53. Ikegami T, Kageyama Y, Obara K, Takeda S (2016) Dissipative and autonomous square-wave self-oscillation of a macroscopic hybrid self-assembly under continuous light irradiation. Angew Chem Int Ed 55:8239–8243
54. Nicolis G, Prigogine I (1977) Self-organization in nonequilibrium systems. John Wiley Inc, Canada
55. Nitzan A, Ross J (1973) Oscillations, multiple steady states, and instabilities in illuminated systems. J Chem Phys 59:241–250
56. Yamazaki I, Fujita M, Baba H (1976) Oscillatory behavior in photoreaction of 1,5-naphthyridine in cyclohexane solution. Photochem Photobiol 23:69–70
57. Laplante JP, Pottier RH (1982) Study of the oscillatory behavior in irradiated 9,10-dimethylanthracene/chloroform solutions. J Phys Chem 86:4759–4766
58. Bose RJ, Ross J, Wrighton MS (1977) Aperiodic and periodic oscillations in fluorescence intensity from irradiated chlorocarbon solutions of anthracene and 9,10-dimethylanthracene. J Am Chem Soc 99:6119–6120
59. Kageyama Y, Ikegami T, Hiramatsu N, Takeda S, Sugawara T (2015) Structure and growth behavior of centimeter-sized helical oleate assemblies formed with assistance of medium-length carboxylic acids. Soft Matter 11:3550–3558
60. Kageyama Y, Tanigake N, Kurokome Y, Iwaki S, Takeda S, Suzuki K, Sugawara T (2013) Macroscopic motion of supramolecular assemblies actuated by photoisomerization of azobenzene derivatives. Chem Commun 49:9386–9388
61. Kageyama Y, Ikegami T, Kurokome Y, Takeda S (2016) Mechanism of macroscopic motion of oleate helical assemblies: cooperative deprotonation of carboxyl groups triggered by photoisomerization of azobenzene derivatives. Chem Eur J 22:8669–8675
62. Kageyama Y, Ikegami T, Satonaga S, Obara K, Sato H, Takeda S (2020) Light-driven flipping of azobenzene assemblies — sparse crystal structures and responsive behavior to polarized light. Chem Eur J DOI: https://doi.org/10.1002/chem.202000701
63. Obara K, Kageyama Y, Takeda S (2019) Swimming motion of micro-sized thin crystals with autonomous and rapid flipping under blue light irradiation. In: The 99th annual meeting 2019 of chemical society of Japan, Kobe, Japan, 2I7–50
64. Obara K, Kageyama Y, Takeda S. Unpublished result
65. Kageyama Y, Ikegami T, Satonaga S, Yazaki D, Takeda S. Unpublished result

66. Ikegami T (2018) Realization of mechanical self-oscillatory motion of azobenzene-containing crystal driven by steady light. Doctoral thesis Hokkaido University
67. Crystal data is deposted in CCDC1833362
68. Zeng H, Wasylczyk P, Wiersma DS, Priimagi A (2018) Light robots: bridging the gap between microrobotics and photomechanics in soft materials. Adv Mat 30:1703554
69. Zhang X, Chen L, Lim KH, Gonuguntla S, Lim KW, Pranantyo D, Yong WP, Yam WJT, Low Z, Teo WJ, Nien HP, Loh QW, Soh S (2019) The pathway to intelligence: using stimuli-responsive materials as building blocks for constructing smart and functional systems. Adv Mat 31:1804540
70. Haken H (1978) Synergetics—an introduction (Japanese transl.). Springer, Berlin, Heidelberg

Chapter 27
Crawling and Bending Motions of Azobenzene Derivatives Based on Photoresponsive Solid–Liquid Phase Transition System

Yasuo Norikane, Koichiro Saito, and Youfeng Yue

Abstract In this chapter, the concept of the photoinduced solid–liquid phase transition and its wide area of possible application are introduced. Also, two macroscopic motions induced by the synergetic action of isomerization of azobenzene molecules are described: firstly, the crawling motion of crystals of azobenzene derivatives on a glass surface by light irradiation with two or one light sources. The motion of the crystals is continuous, and it can be generated by simple experimental setups. Secondly, the reversible bending motion of liquid crystalline polymer network consisting azobenzene monomer as a crosslinker. This polymer also changes its glass transition temperature reversibly upon light irradiations.

Keywords Azobenzene · Crystal · Liquid · Glass · Photoisomerization · Crawling · Bending

27.1 Introduction

Nature has developed fascinating molecular systems and organisms that convert light energy into chemical energy and kinetic energy. They also use light as information for communication or signal that triggers for living activities. Scientists have been inspired those natural systems as research subjects to learn the mechanism and even to mimic to fabricate typical functions of living systems. In particular, the motion of objects driven by light stimuli has attracted significant attention, because light is a powerful energy source and an excellent tool for triggering and controlling the direction of the motion. Various artificial systems have been established to move objects in molecular size to centimeter size. For example, optical manipulation or optical trapping and transporting are successful tools in optics, physics, biology, and chemistry [1]. Laser beam produces optical forces, and it is used as optical tweezers and optical torque to manipulate in microscale to nanoscale. The advantage in this

Y. Norikane (✉) · K. Saito · Y. Yue
Research Institute for Advanced Electronics and Photonics, National Institute of Advanced Industrial Science and Technology (AIST), Tsukuba, Ibaraki 305-8565, Japan
e-mail: y-norikane@aist.go.jp

H. Miyasaka et al. (eds.), *Photosynergetic Responses in Molecules and Molecular Aggregates*, https://doi.org/10.1007/978-981-15-5451-3_27

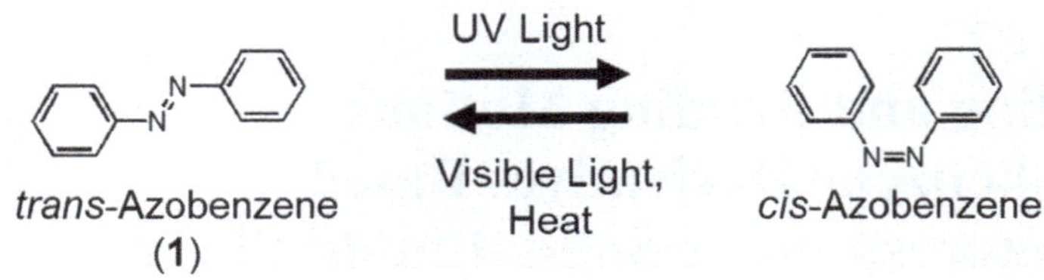

Fig. 27.1 Photoisomerization of azobenzene (**1**)

method is its precision in the position of the target. However, a special set of apparatus is required, and it is difficult to manipulate a large number of target at once. On the other hand, light has been used to make self-propelling motion of artificial materials [2]. In this system, light energy is used to initiate photocatalytic reactions, photolytic decomposition, photothermal effect, and photochromic reactions to produce propulsion force to induce motion of materials and objects. In this kind of self-propelling materials, light energy is first absorbed by the active species, and then, the energy is converted into kinetic energy. Generally, the driving force of this motion is the formation of an asymmetric field of interfacial potential, temperature, or chemical products.

Azobenzene (**1**) is one of the most well-known photochromic compounds that exhibits photoisomerization between the trans and cis isomers (Fig. 27.1) [3]. Due to the difference of the absorption spectra of these two isomers, trans-to-cis and cis-to-trans isomerizations are dominant with UV and visible (blue) light, respectively. The cis isomer is metastable that thermal cis-to-trans isomerization takes place in the dark. Azobenzene shows efficient photoisomerization in solution and liquid crystal media, because the high fluidic property of the media provides a large degree of free volume or sweep volume for the isomerization [4, 5]. In crystal phase, however, the isomerization rarely occurs in simple azobenzene. For example, grounded crystals of azobenzene do not show any change in the IR spectra when the crystals were exposed to the UV light [6]. This is due to the lack of the free volume in the crystal packing of azobenzene. However, the photoresponsivity is drastically improved by the molecular design of azobenzene. The crystal of an azobenzene derivative exhibited photoisomerization in the crystal phase, and it induced the melting of the crystal by light irradiation (the photoinduced solid–liquid phase transition) (Fig. 27.2). This

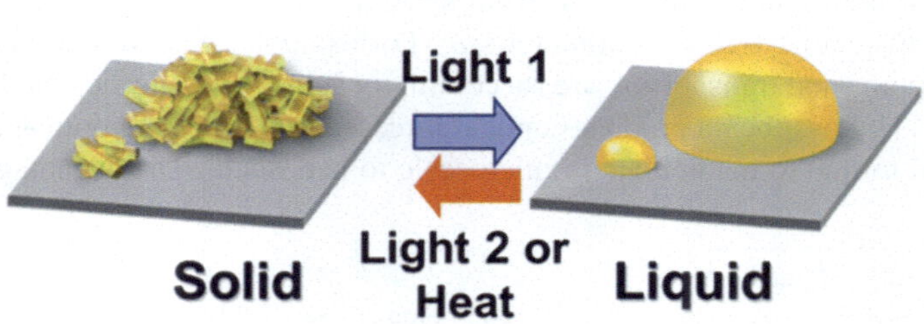

Fig. 27.2 Schematic representation of the photoinduced solid–liquid phase transition

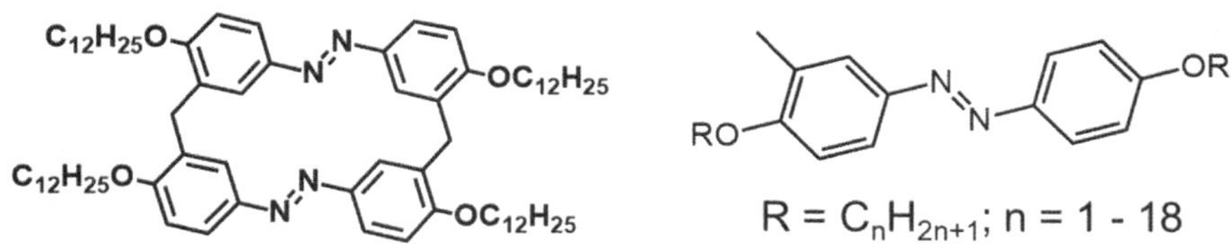

(2) (3-Cn)

Fig. 27.3 Compounds showing the photoinduced solid–liquid phase transition

phenomenon was firstly reported in azobenzene derivatives having a cyclic molecular structure containing two (**2**, Fig. 27.3) or three azobenzene units in a ring with long alkoxy chains [7–9]. The cyclic azobenzene **2** melts upon irradiation with UV light at room temperature, although its melting point is 100 °C. Temperature increase is unavoidable during the photoirradiation experiment, because the photoexcited molecules release energy as a heat through the deactivation processes. However, the increase is only up to 3 K with irradiation of 200 mW cm^{-2} at 365 nm, revealed by the experiment carried out using rod-shaped azobenzene of **3**-C6, that also shows the photoinduced crystal-to-liquid phase transition [10]. In addition, the photoinduced phase transition does not depend on the light intensity, but melting takes place based on the total photon dose. Therefore, the heating effect is not the main reason for the phase transition, but the photoisomerization of azobenzene is the driving force. The photoinduced solid–liquid phase transition has been applied to numbers of smart materials [11].

27.2 Crawling Motion of Azobenzene Crystals on a Solid Surface

27.2.1 Crawling Motion Induced by Two Light Sources

We have investigated numbers of azobenzene derivatives to assess whether the compounds show the photoinduced solid–liquid phase transition or not. We conduct experiments of the phase transition in a manner that crystal powders or crystalline film was irradiated by UV or visible light under the observation of microscope. Usually, the photoirradiation is carried out by alternate exposure of UV and visible light. However, the two lights were left on when we were investigating the crystal powder of 3,3'-dimethylazobenzene (**4**). To our surprise, the crystals seemed to crawling on the glass surface [12]. It should be noted that the motion is directional, and the crystals move away from the UV light source. Despite the crystals change their shape during the motion, the molecular orientation of the crystals is maintained. It is notable that the motion is generated by a simple experimental setup such as light-emitting diodes or Hg lamp in a fixed position. No expensive apparatus such as laser or positioning device is required. As another advantages, no special treatment on the surface, such as chemical modification [13–15], spatial gradients [16], or application of ratchet potential [17], is required.

Crystals of **4** with the size of ca. 20-30 μm were placed on a cover glass, and then, UV (365 nm) and visible (465 nm) light were irradiated as shown in Fig. 27.4. The compound **4** shows trans-to-cis and cis-to-trans isomerizations with UV and visible light irradiation, respectively. Furthermore, **4** exhibits the photoinduced crystal-to-liquid phase transition upon UV irradiation. During the simultaneous irradiation with UV and visible light, the crystals of **4** move away from the UV light and toward the visible light. It should be noted that the motion occurs regardless of whether the crystals are single crystalline, polycrystalline, or a distorted shape. In the sharp contrast, liquid droplet of **4**, generated by supercooling or already melted by photoirradiation, did not move at all. The velocity of the crystal motion, however, varies by crystal. The average velocity of the crystal motion was ca. 1.5 μm min^{-1}. From these observations, as a prerequisite, the phase of the sample should be in the crystal phase to observe this motion.

Optimization was carried out to find the best combinations of light intensity of UV and visible light when two light sources are fixed angles of $\theta_{UV} = \theta_{VIS} = 45°$, $\varphi = 180°$ as shown in Fig. 27.5. The motion of crystals was analyzed statistically, and the average velocity was plotted against the light intensity of each light. When the intensity of UV light was too strong than that of visible light, the crystals melted to droplets and did not move. When the intensity of visible light was too strong than that of UV light, no morphological change or motion of the crystals was observed. The balance of the light intensity is required to obtain the motion. The optimized condition for the crawling motion was found at UV and visible light intensities of 200 and 50–60 mW cm^{-2}, respectively, and it was found that the average velocity was 2.0 μm min^{-1}.

The crystals climbed vertically on a wall of glass as shown in Fig. 27.5, when the entire experimental setup was turned sideway. This result clearly indicates that the crawling motion observed is not caused by gravity or other artefacts, but is a result of photoirradiation.

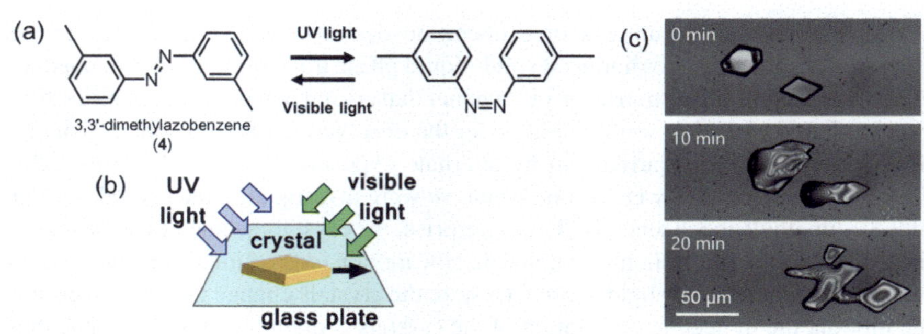

Fig. 27.4 **a** Photoisomerization of 3,3'-dimethylazobenzene (**4**). **b** Schematic diagram of the photoinduced crawling motion of the crystals, **c** microphotographs of the crawling motion

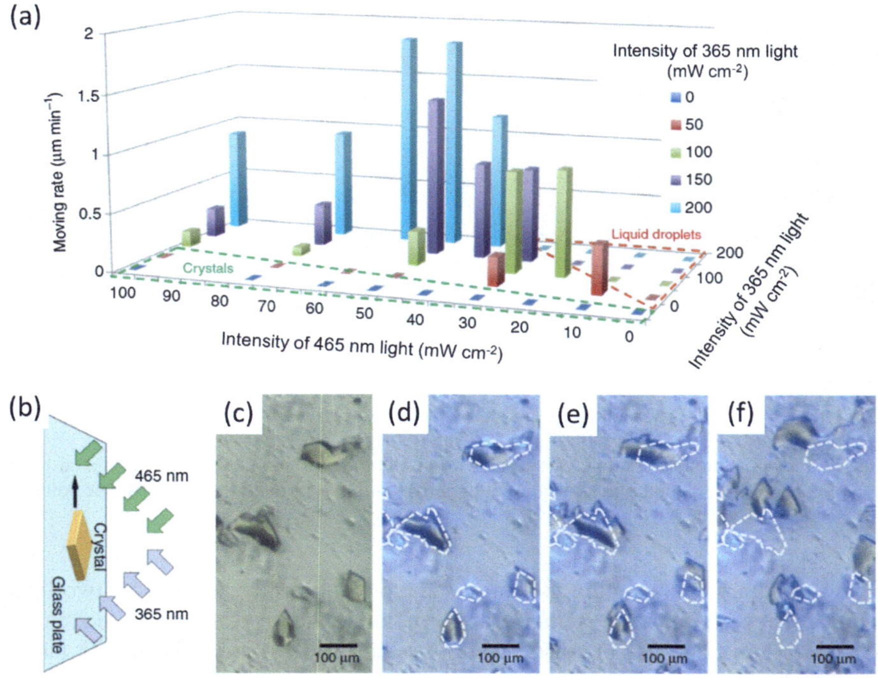

Fig. 27.5 **a** Dependence of light intensity on the average velocity of the crawling motion. **b** Schematic illustration of the vertical motion. **c–f** Microscope images of the vertical motion of crystal of **4** after irradiation for $t = 0$ **c**, 2.5 **d**, 5 **e**, and 10 min **f**. Reprinted from [12] by The Authors licensed under CC BY 4.0

Irradiation with various angles of light sources was carried out to investigate the effect of the irradiation angle. The angles were scanned in the ranges of 25–45° and 20–45° for θ_{UV} and θ_{VIS}, respectively, while fixing the intensities of the UV and the VIS light at 200 and 60 mW cm^{-2}, respectively. It was found that the velocity of crystals relies more on the UV light angle than on the visible light angle. At higher angles on θ_{UV}, gave higher velocity.

It is noteworthy that the direction of molecular orientation of crystal is maintained although the crystal shape is highly deformed during the motion. By using polarizing optical microscope, the direction of optical anisotropy of the crystal was maintained. In the case of a single crystal, shown in Fig. 27.6, the front edge of the crystal grows, whereas the rear edge shrinks when the crystal moves. Two orientations of crystal were investigated, the irradiation from the direction parallel or perpendicular to the longer diagonal axis of the crystal. In both cases, the angles of the corners of the crystal were maintained at ca. 56° and 124° during the motion. By using laser confocal microscopy, three-dimensional measurements during the crystal motion were carried out. It was revealed that the crystals of initial thickness of ca. 1-4 μm became thinner during the motion, especially at the front edge.

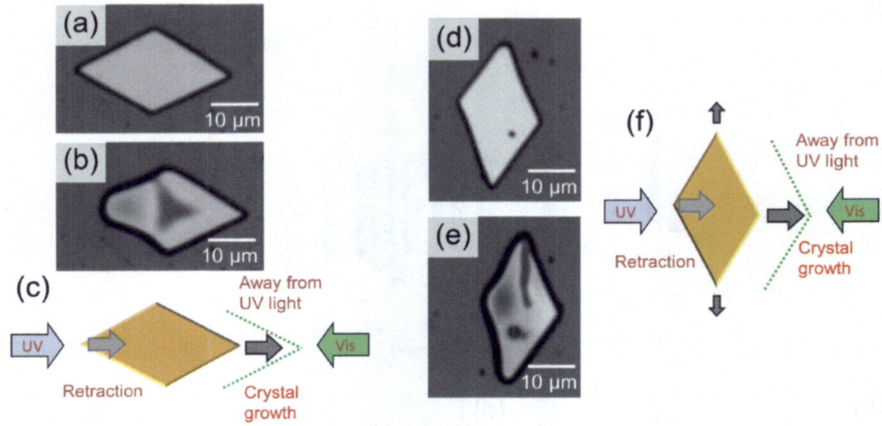

Fig. 27.6 **a–c** Motion of a single crystal of **4** when the longer diagonal axis of the crystal is **a–c** parallel and **d–f** perpendicular to the direction of irradiation. Reprinted from [12] by The Authors licensed under CC BY 4.0

In this experimental setup, the entire sample is uniformly exposed by two light sources. The light intensity is almost uniform within the field of view of the microscope (<1 mm), because the distances to the light sources are substantially larger than the dimension of the view. Exposure of two light sources brings about melting and crystallization on the surface of the crystals, and an equilibrium between the solid and liquid phases is reached locally. However, when the crystal moves, this equilibrium is shifted to the opposite direction at the two edges. The cause of the shift is a small difference in the light intensity and/or difference in the wettability contact angle of melted state and crystal.

Similar crawling motion was observed in azobenzene (**1**). No morphological change or motion was observed when the crystals of **1** were irradiated under room temperature. However, they exhibited crawling motion at 50 °C. This result agrees well with the observation that the crystals of **1** do not show a phase transition by light irradiation, but they do at 50 °C. It is notable that these results show that the crawling motion would occur in other molecular systems if they show reversible crystal-to-liquid phase transitions by light.

27.2.2 Crawling Motion Induced by One Light Source (Negative Phototaxis)

In the case of **4**, the lifetime of the cis isomer is long (several days), and thus, it takes long time to recrystallize the photo-melted state in the dark at r.t. On the other hand, in azobenzenes having electron-donating or -withdrawing substituent, the lifetime of the corresponding cis isomer is known to be shorter. Especially, substituting an electron-donating amino group on the 4-position causes a decrease in the thermal isomerization barrier by increasing the electron density in the π* orbital [3]. While trans-to-cis isomerization occurs during irradiation, fast cis-to-trans thermal isomerization takes place even during the irradiation. As light irradiation is not required to achieve the cis-to-trans isomerization, only a single light source is needed to control the photostationary state of this system. Thus, we attempted to use amino-substituted azobenzenes to induce the crawling motion by using only one light source. We successfully found that the crystals of 4-(methylamino)azobenzene (**5**) exhibited the crawling motion as shown in Fig. 27.7. The crystals of **5** exhibited the crawling motion away from the light source (negative phototactic behavior) [18]. Here, positive phototaxis and negative phototaxis mean the motion toward and away from the light source, respectively. In addition, UV light, which is frequently used in the experiment of azobenzene research, is not required for the motion.

In methanol solution, the lifetime of cis isomer was approximately 0.6 s, determined by the change of the absorption spectra from the photostationary state. The crystal of **5** has its melting temperature of 87 °C and exhibited the photoinduced crystal-to-liquid phase transition when exposed to visible light at 465 nm. Recrystallization started soon after the irradiation was discontinued.

Crystals on a glass plate were exposed to visible light (465 nm) using the experimental setup where the LED light source was positioned at a 30°. The irradiation resulted in the translational motions of the crystal in the direction away from the light source (Fig. 27.7). Notably, this negative phototactic behavior was observed without any surface modification of glass substrate or scanning of light source. In this experiment, an untreated cover glass and an LED with a fixed position were used. As can be seen in Fig. 27.7, the crystal does not need to be a single crystal.

The dependence of the light intensity was studied by using the 465 nm light. The rate of the motion increased with light intensity up to 200 mW cm^{-2}, although the motion was not observed at low intensity (50 mW cm^{-2}), probably due to the

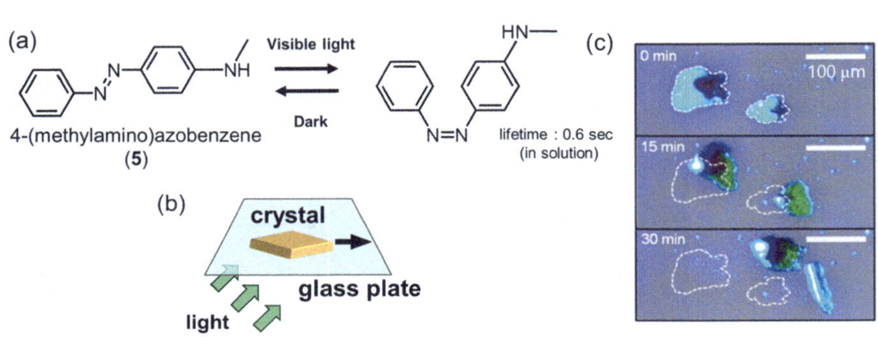

Fig. 27.7 **a** Photoisomerization of 4-(methylamino)azobenzene (**5**). **b** Schematic diagram of the photoinduced crawling motion of the crystals, **c** microphotographs of the crawling motion. The white dashed line represents the initial position of the crystals. Reproduced from [18] by permission of The Royal Society of Chemistry

insufficient melting (Fig. 27.8). Instead of using 465 nm light, a light source of 405 nm was used and the similar crawling motion was observed. In a contrast to the case of 465 nm irradiation, the light intensity dependence seems to be smaller. Although the motion was observed even at an intensity of 50 mW cm^{-2}, the velocity did not increase with the increase of the light intensity to 200 mW cm^{-2}. The plausible reason is that the melting behavior is changed in accordance with different ratios of isomers at different light sources.

White light irradiation was carried out to demonstrate the motion of the crystals. Although the rate was slow, the negative phototactic motion was observed when a white LED was illuminated to the crystals. Furthermore, a solar simulator (xenon lamp) was used and the motion was achieved as shown in Fig. 27.9. These experimental results clearly show that the crystal motion can be induced by sunlight. Small actuators or soft robots driven by sunlight will be possible.

In this molecular crystal system, oblique light incidence causes a slight difference in the light intensity between the front and rear edges of the crystals. It may produce a gradient in light intensity on the surface of a crystal. The rear edge of crystal, which is closer to the light source, is sufficiently melted. At the front edge, on the other hand, the recrystallization occurs faster than at the rear edge. Due to the rapid thermal isomerization of **5**, the motion is induced by the photochemical trans-to-cis photoisomerization and it induces melting at the rear side and the thermal cis-to-trans isomerization to recrystallize the crystal at the front side.

It should be mentioned about the position of the light source. In this experiment, the photoirradiation had to be done from the back face of the glass substrate. We attempted to move the crystals with the position from the front face of the sample, but no evident motion was observed. This is in a sharp contrast to the case of compound **4** and **1**, where two light sources were positioned from the front face of the sample. This may relate to the amount of the melted state in a crystal. Further study has to be done to distinguish the different mechanism of these crawling motions.

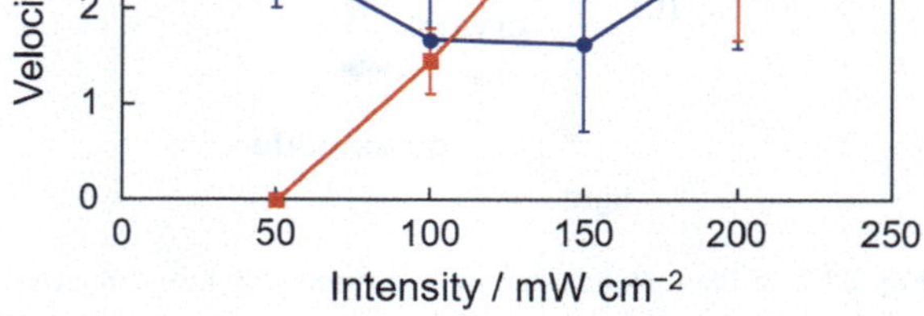

Fig. 27.8 Dependence of light intensity on the average velocities of crawling motion at 465 nm (red line) and 405 nm (blue line) irradiations. Reproduced from [18] by permission of The Royal Society of Chemistry

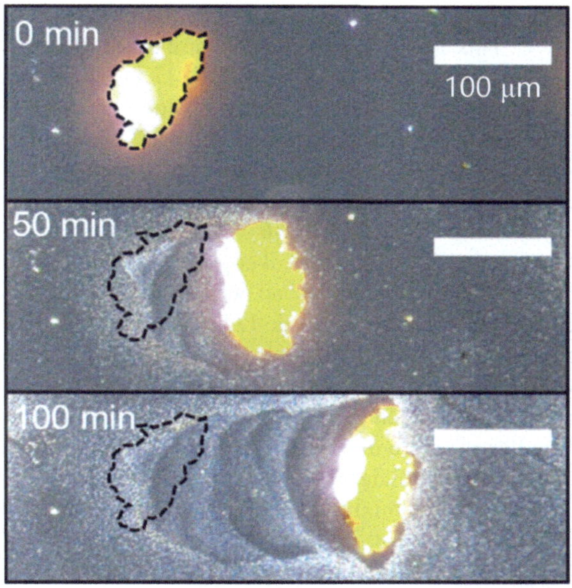

Fig. 27.9 Negative phototactic behavior of crystals under white light irradiation. Optical micrographs of crystals of **5** on a glass during irradiation with solar-simulated light. Images were captured after a lapse of 0, 50, and 100 min. The black dashed line represents the initial position of the crystals. Reproduced from [18] by permission of The Royal Society of Chemistry

27.3 Photomechanical Bending in Polymer Film with Photoswitchable Glass Transition Temperatures

As described earlier, the molecular design of introducing a methyl group at the 3-position of azobenzene drastically enhanced the photo-induced phase transition properties in the compound **3**. According to this molecular design strategy, a monomer (**6**), having two methacrylate groups at the both termini, has been developed (Fig. 27.10), and the copolymer consisting of this monomer exhibits photomechanical bending

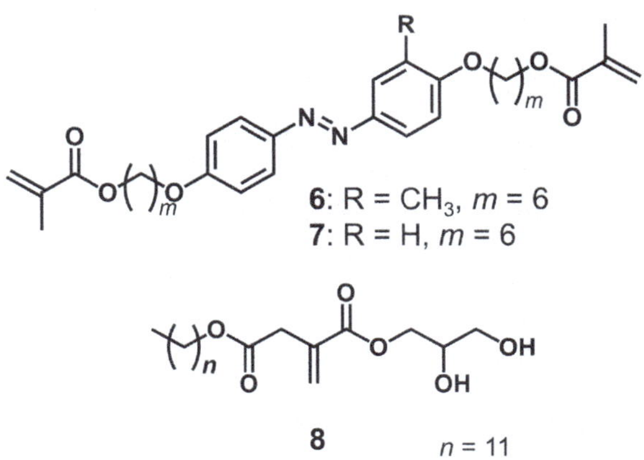

Fig. 27.10 Molecular structures of monomers. Compounds **6** and **8** were used for the synthesis of polymer films having photomechanical property

[19]. The azobenzene **6** exhibits a crystalline phase with its melting temperature of 65 °C. This crystal shows a rapid phase transition from the crystal to the liquid phase upon irradiation with UV light (Fig. 27.11). In contrast, the azobenzene monomer without the methyl group at the 3- position (**7**) did not show the photoinduced phase transition, although its melting point (73 °C) is similar to that of **6**. This fact strongly supports substitution at the 3-position in azobenzene as a robust strategy for molecular design, as it significantly changes the molecular packing in the crystal phase.

By using the azobenzene **6** as a cross-linking monomer, a free-standing polymer film, produced by free-radical copolymerization with **8**, has a liquid crystalline (LC) property. The compound **8** had been originally used in functional hydrogels [20]. The compound **8** exhibits smectic LC as confirmed by polarizing optical microscope. It should be noted that in the literature, there are limited types of nematic LC monomers which had been mainly used to synthesize the photoresponsive LC polymers that possess photomechanical properties [21–23]. The monomer **8** can self-assemble into lamellar structure due to its LC property, and the long alkyl chain in both **6** and **8** may provide sufficient free volume for efficient photoresponsive property.

A pre-heated mixture containing **6**, **8** and an initiator was pulled into a molecular alignment cell (2 × 2 cm, spacing = 5 or 10 μm) by capillary force and heating the cell at 60 °C (1 h) and then at 125 °C (24 h) gave a photoresponsive free-standing film. The characterization of the film using X-ray diffraction (XRD) revealed that the layer spacing of the lamellar structure was about 3.8 nm, which was smaller than the simulated conformation. This result indicates that the molecules of **8** formed the smectic C LC phase. The POM observation revealed that the mesogenic groups are aligned to a unidirectional molecular orientation.

It was found that the *Tg* (glass transition temperature) value of the **6/8** copolymer decreased upon UV irradiation from 29 to 16 °C, recovering in the dark or through visible light irradiation (Fig. 27.12). The polymer film with trans isomer gave higher *Tg* than room temperature, while the same film in the cis isomer shows a lower *Tg* after UV irradiation. The mechanical property of the polymer film was measured by tensile tests before and after UV irradiation. Initially, the polymer film showed a relatively higher Young's modulus (140 MPa). On the other hand, the polymer film

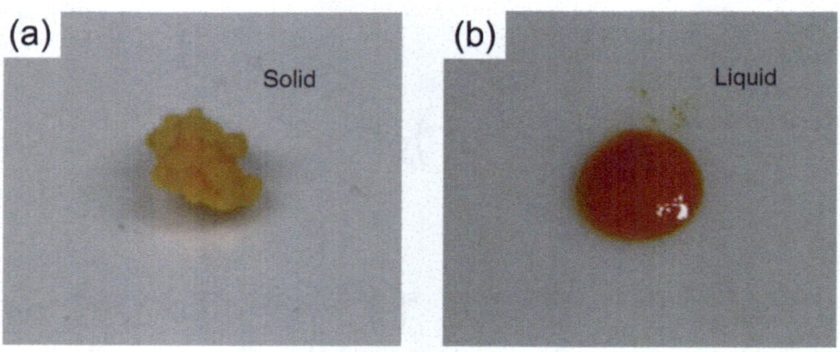

Fig. 27.11 Photographs of **6** before **a** and after **b** UV irradiation (365 nm, 125 mW cm^{-2}, 6 s) at room temperature. Reprinted from [19] by The Authors licensed under CC BY 4.0

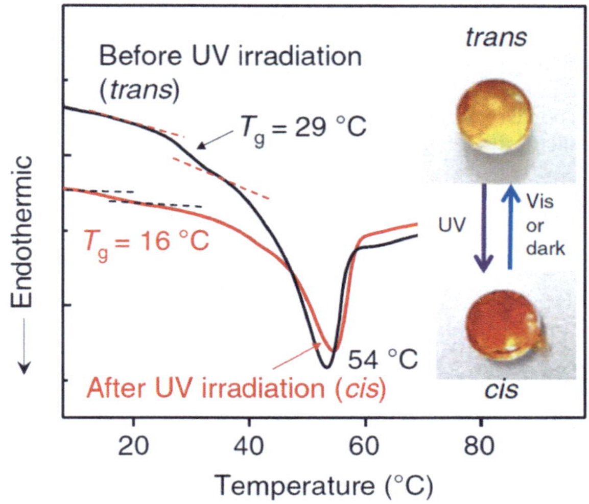

Fig. 27.12 Differential scanning calorimetry (DSC) profiles of the film before (black line) and after (red line) the irradiation of UV light, showing the glass transition temperature is shifted by the irradiation. The photographs show the color change of the DSC sample by the irradiation. Reprinted from [19] by The Authors licensed under CC BY 4.0

after UV irradiation showed a much lower Young's modulus (65.6 MPa) but a higher percentage elongation (strain). These results also indicate that the exposure to UV light decreased the *Tg* and thus decreased the Young's modulus of the polymer films. The photoresponse of the film was confirmed by UV/vis spectroscopy and revealed that the photoresponse was fast. The fast decreasing in the absorbance at 365 nm under UV light (4 s) and even faster for the back isomerization under visible light (2 s).

This polymer film with the photoswitchable *Tg* property is beneficial for generating efficient photomechanical actuations. The bending and recovery of the polymer film were repeated by switching UV and visible light alternatively (Fig. 27.13). When exposed to the UV light, the film bent toward the light source. When the blue light

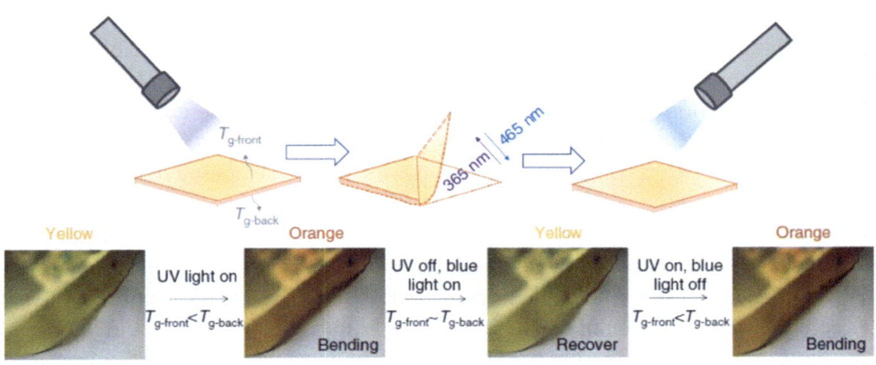

Fig. 27.13 Photomechanical response of the copolymer consisting **6** and **8**. UV light cause bending and visible light straighten the film. Reprinted from [19] by The Authors licensed under CC BY 4.0

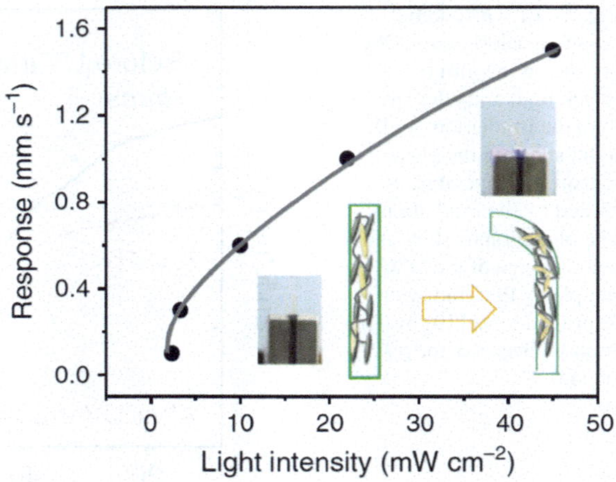

Fig. 27.14 Dependence of the UV light intensity on the photoresponse speed of the polymer consisting **6** and **8**. Reprinted from [19] by The Authors licensed under CC BY 4.0

(465 nm) was irradiated, the film restored to its original flat state. The color of the film changed from yellow to orange when it shows bending and returned to yellow when it becomes flat again.

The bending speed of the polymer strongly depends on the light source. The photobending speed increased with increasing intensity of the UV light. The bending speed (displacement/time) improved from 0.3 to 1.5 mm s^{-1}, when the light intensity increased from 3.28 to 45 W cm^{-2} (Fig. 27.14) at room temperature of ca. 20 °C. The film started to oscillate when the film bent to the direction parallel to the light [24, 25]. Compared to other photomechanical bending films, this polymer has a greater efficiency in causing mechanical motion.

27.4 Conclusion

The photoinduced crystal-to-liquid phase transition found in azobenzene derivatives has attracted much attention from widespread areas of research, because of its potential application in smart soft materials through its reusable characteristics. In addition, the phase transition provides novel mechanical motions, such as crawling, swimming [26], and bending motions that are useful for designing active materials and soft robotics.

The crawling motion of the crystal is an interesting phenomenon because an object (crystal) travels or walks on a solid surface under an ambient atmosphere (at solid/air interface). This is in contrast with numerous examples of stimuli-responsive and self-propelling motion of small objects in solution or at the solid/liquid interface. This motion would be one of candidates for use in a "carriage vehicle" to transport objects/chemicals on a solid surface without a flow channel.

The photomechanical bending motion of polymer films attracts much attention because of its potential application to actuators and sensors. There have been many publications about this field; however, most of research clings on to similar molecular design. In this context, our finding of novel molecular design provides a deeper insight into the photomechanical LC polymers utilizing the photoisomerization of azobenzene.

Throughout our research, the motion of photoresponsive molecules is amplified to the visible phenomena such as phase change, translational motion, and bending motion. Rational molecular design made it possible to induce the molecules to move synergetically and the whole system working.

Acknowledgements We deeply appreciate to Dr. E. Uchida for her collaboration. The present work was supported by JSPS KAKENHI Grant Number 17H05275 in Scientific Research on Innovative Areas "Photosynergetics" and by JSPS KAKENHI Grant Numbers 16K14069, 16H04141, 18H05986, and JP18K14291.

References

1. Gao D, Ding W, Nieto-Vesperinas M, Ding X, Rahman M, Zhang T, Lim CT, Qiu CW (2017) Optical manipulation from the microscale to the nanoscale: Fundamentals, advances and prospects. Light Sci Appl 6:e17039. https://doi.org/10.1038/lsa.2017.39
2. Xu L, Mou F, Gong H, Luo M, Guan J (2017) Light-driven micro/nanomotors: from fundamentals to applications. Chem Soc Rev 46:6905–6926. https://doi.org/10.1039/c7cs00516d
3. Bandara HMD, Burdette SC (2012) Photoisomerization in different classes of azobenzene. Chem Soc Rev 41:1809–1825. https://doi.org/10.1039/c1cs15179g
4. Seki T (2018) A wide array of photoinduced motions in molecular and macromolecular assemblies at interfaces. Bull Chem Soc Jpn 91:1026–1057. https://doi.org/10.1246/bcsj.20180076
5. Seki T (2016) Light-directed alignment, surface morphing and related processes: recent trends. J Mater Chem C 4:7895–7910. https://doi.org/10.1039/c6tc02482c
6. Tsuda M, Kuratani K (1964) Isomerization of cis-azobenzene in the solid phase. Bull Chem Soc Jpn 37:1284–1288
7. Norikane Y, Hirai Y, Yoshida M (2011) Photoinduced isothermal phase transitions of liquid-crystalline macrocyclic azobenzenes. Chem Commun 47:1770–1772. https://doi.org/10.1039/c0cc04052e
8. Uchida E, Sakaki K, Nakamura Y, Azumi R, Hirai Y, Akiyama H, Yoshida M, Norikane Y (2013) Control of the orientation and photoinduced phase transitions of macrocyclic azobenzene. Chem A Eur J 19:17391–17397. https://doi.org/10.1002/chem.201302674
9. Hoshino M, Uchida E, Norikane Y, Azumi R, Nozawa S, Tomita A, Sato T, Adachi S, Koshihara S (2014) Crystal melting by light: X-ray crystal structure analysis of an azo crystal showing photoinduced crystal-melt transition. J Am Chem Soc 136:9158–9164. https://doi.org/10.1021/ja503652c
10. Norikane Y, Uchida E, Tanaka S, Fujiwara K, Koyama E, Azumi R, Akiyama H, Kihara H, Yoshida M (2014) Photoinduced crystal-to-liquid phase transitions of azobenzene derivatives and their application in photolithography processes through a solid–liquid patterning. Org Lett 16:5012–5015. https://doi.org/10.1021/ol502223u

11. Yamamoto T, Norikane Y, Akiyama H (2018) Photochemical liquefaction and softening in molecular materials, polymers, and related compounds. Polym J 50:551–562. https://doi.org/10.1038/s41428-018-0064-4
12. Uchida E, Azumi R, Norikane Y (2015) Light-induced crawling of crystals on a glass surface. Nat Commun 6:7310. https://doi.org/10.1038/ncomms8310
13. Ichimura K, Oh S-K, Nakagawa M (2000) Light-driven motion of liquids on a photoresponsive surface. Science 288:1624–1626
14. Oh S, Nakagawa M, Ichimura K (2002) Photocontrol of liquid motion on an azobenzene monolayer. J Mater Chem 12:2262–2269
15. Berna J, Leigh D, Lubomska M, Mendoza S, Perez E, Rudolf P, Teobaldi G, Zerbetto F (2005) Macroscopic transport by synthetic molecular machines. Nat Mater 4:704–710
16. Chaudhury MK, Whitesides GM (1992) How to make water run uphill. Science 256:1539–1541. https://doi.org/10.1126/science.256.5063.1539
17. Astumian RD (1997) Thermodynamics and kinetics of a Brownian motor. Science 276:917–922. https://doi.org/10.1126/science.276.5314.917
18. Saito K, Ohnuma M, Norikane Y (2019) Negative phototactic behaviour of crystals on a glass surface. Chem Commun 55:9303–9306. https://doi.org/10.1039/c9cc03826d
19. Yue Y, Norikane Y, Azumi R, Koyama E (2018) Light-induced mechanical response in crosslinked liquid-crystalline polymers with photoswitchable glass transition temperatures. Nat Commun 9:3234. https://doi.org/10.1038/s41467-018-05744-x
20. Yue Y, Gong JP (2015) Tunable one-dimensional photonic crystals from soft materials. J Photochem Photobiol C Photochem Rev 23:45–67. https://doi.org/10.1016/j.jphotochemrev.2015.05.001
21. Van Oosten CL, Bastiaansen CWM, Broer DJ (2009) Printed artificial cilia from liquid-crystal network actuators modularly driven by light. Nat Mater 8:677–682. https://doi.org/10.1038/nmat2487
22. Liu D, Broer DJ (2015) New insights into photoactivated volume generation boost surface morphing in liquid crystal coatings. Nat Commun 6:8334. https://doi.org/10.1038/ncomms9334
23. Gelebart AH, Mulder DJ, Varga M, Konya A, Vantomme G, Meijer EW, Selinger RLB, Broer DJ (2017) Making waves in a photoactive polymer film. Nature 546:632–636. https://doi.org/10.1038/nature22987
24. Gelebart AH, Vantomme G, Meijer EW, Broer DJ (2017) Mastering the photothermal effect in liquid crystal networks: a general approach for self-sustained mechanical oscillators. Adv Mater 29:1606712. https://doi.org/10.1002/adma.201606712
25. Gelebart AH, Mulder DJ, Vantomme G, Schenning APHJ, Broer DJ (2017) A rewritable, reprogrammable, dual light-responsive polymer actuator. Angew Chemie - Int Ed 56:13436–13439. https://doi.org/10.1002/anie.201706793
26. Norikane Y, Tanaka S, Uchida E (2016) Azobenzene crystals swim on water surface triggered by light. CrystEngComm 18:7225–7228. https://doi.org/10.1039/C6CE00738D

Chapter 28
Photomechanical Effects in Crosslinked Liquid-Crystalline Polymers with Photosynergetic Processes

Toru Ube and Tomiki Ikeda

Abstract Crosslinked liquid-crystalline polymers (CLCPs) containing photochromic moieties show macroscopic deformation upon irradiation with light. This conversion of light energy into mechanical work is attributed to photosynergetic amplification of changes at multiscale: from nano to macro. This chapter focuses on recent progress of photomobile polymer materials to achieve versatile design as soft actuators. Three approaches are detailed: incorporation of amorphous polymers into CLCPs to control mechanical and photoresponsive properties, reshaping of CLCPs into various three-dimensional architectures through the rearrangement of network structures, and application of two-photon absorption processes to enhance the spatial selectivity of photoactuation.

Keywords Photochromism · Photomechanical effects · Interpenetrating polymer networks · Dynamic covalent bonds · Two-photon absorption

28.1 Introduction

Crosslinked liquid-crystalline polymers (CLCPs) containing photochromic moieties have been extensively studied as photomechanical materials, which show macroscopic deformation upon irradiation with light [1–6]. Feasibility of CLCPs as soft actuators has been demonstrated through achievement of various photoinduced motions, such as bending, twisting, and rotation. The direct conversion of light energy into mechanical work is based on the photosynergetic amplification of changes

---

T. Ube · T. Ikeda (✉)
Research and Development Initiative, Chuo University, 1-13-27, Kasuga, Bunkyo-ku, Tokyo 112-8551, Japan
e-mail: tikeda@tamacc.chuo-u.ac.jp

T. Ube
e-mail: ube@tamacc.chuo-u.ac.jp

T. Ikeda
Technical Institute of Physics and Chemistry, Chinese Academy of Sciences, 29 Zhongguancun East Road, Beijing 100190, China

H. Miyasaka et al. (eds.), *Photosynergetic Responses in Molecules and Molecular Aggregates*, https://doi.org/10.1007/978-981-15-5451-3_28

at multiscale: from nano to macro. Photoisomerization of photochromic moieties leads to order–disorder transition of mesogens and then changes in conformation of polymer chains, resulting in macroscopic deformation.

The development of CLCPs with photomechanical effects is based on the historical establishment of photoswitching methods in liquid-crystal (LC) systems [7]. Azobenzene is a widely used photochromic dye, which shows photoisomerization between a thermally stable *trans*-isomer and a metastable *cis*-isomer (Fig. 28.1a) [8]. Rod-like *trans*-azobenzene stabilizes LC phases, whereas bent *cis*-azobenzene disturbs them. Photoinduced phase transition was reported for LC systems doped with azobenzene in 1987 [9]. Irradiation with UV light in a nematic phase induces LC–isotropic phase transition through *trans*–*cis* isomerization of azobenzene moieties. The LC phase is restored with *cis*–*trans* isomerization, which is induced by visible light or heating. Furthermore, polymeric systems functionalized with mesogens and azobenzene moieties show much stable photoinduced isotropic phases due to the high viscosity, especially below a glass transition temperature. Polymers containing azobenzene moieties in all monomer units show rapid photoinduced phase transition

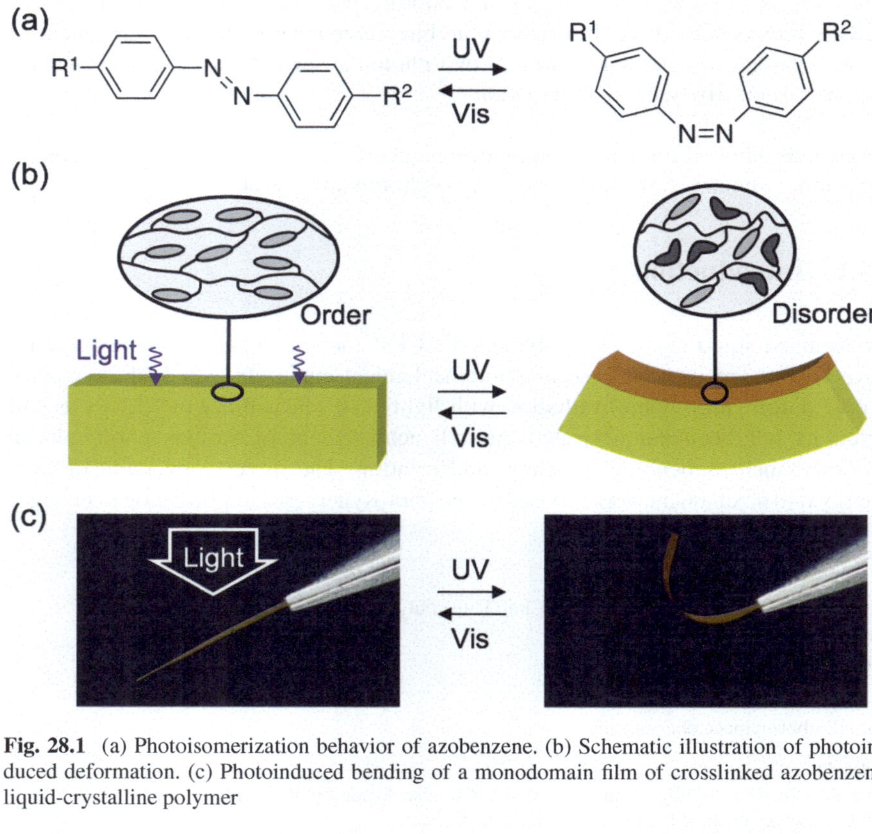

Fig. 28.1 (a) Photoisomerization behavior of azobenzene. (b) Schematic illustration of photoinduced deformation. (c) Photoinduced bending of a monodomain film of crosslinked azobenzene liquid-crystalline polymer

[10]. This reversible photoswitching of molecular order is the basis of the various photosynergetic effects.

In addition, strong coupling between the alignment of mesogens and conformation of polymer chains plays a significant role in CLCPs [11, 12]. The deformation of polymer chains is transformed to macroscopic deformation through crosslinks. Therefore, macroscopic shapes can be altered along with the alignment change of mesogens. Finkelmann et al. prepared CLCPs with uniaxial alignment of mesogens through two-step crosslinking method [13]. Poly(hydorogenmethyl siloxane) is functionalized with mesogens and lightly crosslinked through hydrosilylation reaction. Then, the film is stretched to induce uniaxial alignment of mesogens, which is fixed by further crosslinking. Through this procedure, homogeneous alignment of mesogens is memorized as initial alignment of the CLCP film. These films show contraction and expansion upon heating and cooling along with nematic–isotropic transition. The strain induced in these systems was 20% in the pioneering work, and has been enhanced to as high as 400% [12, 14]. In situ polymerization method has also been developed to prepare CLCPs with polyacrylate and polymethcarylate backbones [15–20]. In this method, LC mono- and di-functional monomers are polymerized in a cell coated with alignment layers (e.g., rubbed polyimide) to form crosslinked network. As the alignment of mesogens is memorized during the formation of crosslinked network, CLCPs with complicated alignment of mesogens can also be prepared with the aid of patterning technique [21–23].

By combining the photoinduced order–disorder transition in LC systems and the strong coupling between nanostructures and macrostructures in CLCPs, photoisormerization of dyes can be amplified to macroscopic deformation. Finkelmann et al. developed monodomain films of CLCPs with polysiloxane backbone functionalized with azobenzene moieties [24]. Contraction of a film was successfully induced upon irradiation with UV light, which induces *trans–cis* isomerization of azobenzene moieties and disturbs the alignment of mesogens. The initial length is recovered along with *cis–trans* isomerization of azobenzene moieties. Furthermore, CLCP films with high concentration of azobenzene moieties, which were prepared through copolymerization of azobenzene LC monomers and crosslinkers, were found to show reversible bending–unbending behavior upon irradiation with UV and visible light (Fig. 28.1b, c) [25, 26]. This bending behavior is attributed to predominant absorption of UV light by chromophores near the surface of the film, which leads to *trans–cis* isomerization, the alignment change of mesogens, and deformation of polymer chains at the surface region. Furthermore, polydomain films have been found to show axis-selective bending upon irradiation with linearly polarized UV light (Fig. 28.2) [25].

As the photosynergetic transformation from nano to macro includes multiple processes, the output can be affected by various factors. To make the full use of photosynergetic effects to achieve macroscopic response, adequate designs of molecules, materials, and devices are essential. However, the molecular design of CLCPs has

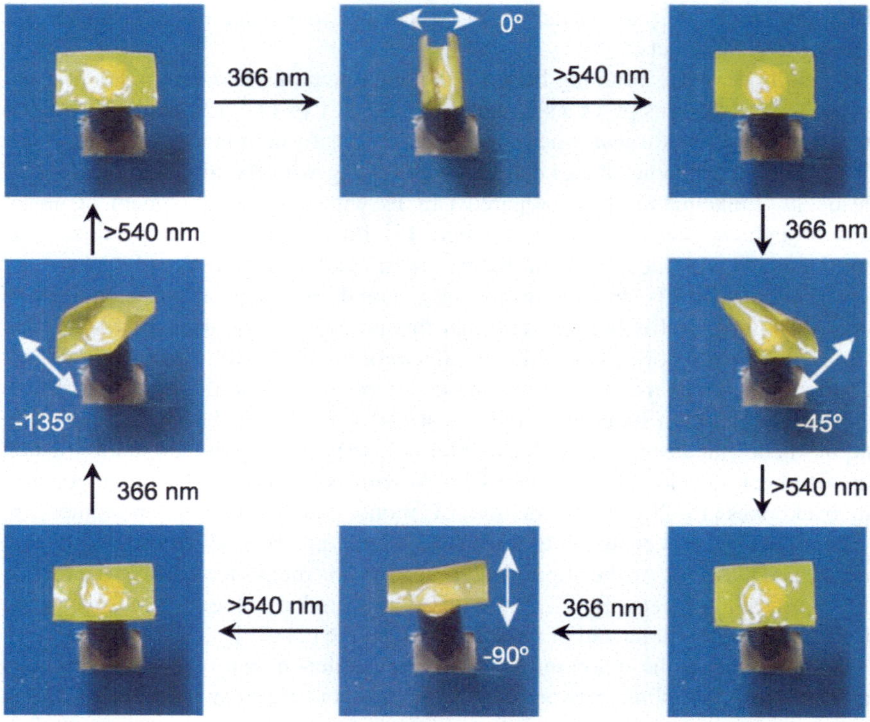

Fig. 28.2 Precise control of the bending direction of a polydomain CLCP film by linearly polarised light. White arrows indicate directions of linearly polarized light. Size of the film: 4.5 mm × 3 mm × 7 μm. (Reproduced from [25])

been limited as they need to show LC properties. Furthermore, macroscopic architectures of CLCPs have also been limited, since two-step crosslinking and in situ polymerization methods yield flat films, which are hardly reshaped due to the shape memory effects of crosslinked networks. Another limitation arises from the penetration depth of light in systems containing chromophores with high concentration: only the region near the surface can contribute to deformation and generation of force. Overcoming these limitations leads to enhancement of photomechanical properties flexible design of architectures, and motions.

This chapter focuses on recent progress of photomobile polymer materials under the concept of photosynergetics. The essence is how to significantly amplify the change at the molecular level to that at macroscopic scale. Three approaches are detailed in the following sections: incorporation of amorphous polymers into CLCPs to control mechanical and photoresponsive properties, reshaping of CLCPs into various three-dimensional architectures through the rearrangement of network structures, and application of two-photon absorption processes to enhance the spatial selectivity of photoactuation.

28.2 Incorporation of Amorphous Polymers into CLCPs

Controlling of mechanical properties is essential for application of CLCPs as soft actuators in various fields. Conventional CLCPs with polyacrylate backbone, which have been prepared by in situ polymerization method, typically suffer from brittleness due to the high crosslinking density. With the purpose of enhancing mechanical toughness of photomobile polymer materials, polymer bilayers (e.g., CLCP and polyethylene [27, 28]) and composites (e.g., CLCP and carbon nanotube [29]) have been developed. In addition, elastic modulus is a key parameter to be adequately controlled for a variety of actual systems. Blending of multiple polymer components is a common strategy to improve mechanical properties of polymer materials. In crosslinked systems, the formation of interpenetrating polymer network (IPN) leads to blending of multiple components at nanometric scales [30]. Structures and functions of constituents can be combined by the formation of IPN, as can be seen in various systems including hydrogels [31, 32] and dielectric elastomers [33]. In combination of LC polymers with various amorphous and semi-crystalline polymers, self-assembled structure of mesogens should be preserved to exert the function of LC polymers. Formation of IPN can be a powerful approach to construct multi-component systems preserving stimuli-responsive properties of CLCPs, since the structure of LC network is memorized by crosslinks. We developed IPN films composed of crosslinked azobenzene LC polymer (PAzo) and various amorphous polymers to improve mechanical and photoresponsive properties of photomobile polymer materials (Fig. 28.3) [34, 35].

We prepared IPN films by sequential formation of CLCP and amorphous networks [34]. First, a PAzo network was formed by copolymerization of azobenzene LC monomers and crosslinkers in the presence of inert LCs. By polymerizing at LC phase in a glass cell coated with rubbed polyimide, we obtained PAzo network with homogeneous alignment. Then, the inert LCs were removed to yield a PAzo network with numerous pores (PAzoTP), which could be used as a template for the formation of IPN structures. After introducing methacrylate monomers—methyl methacrylate (MMA), *n*-butyl methacrylate (BMA) or *n*-dodecyl methacrylate (DDMA)—and crosslinkers into PAzoTP, thermally initiated polymerization was conducted to form the second network. The resultant films (PAzo/PMMA, PAzo/PBMA, and PAzo/PDDMA) showed homogeneous alignment of mesogens, which was confirmed by polarizing optical microscopy and polarized absorption spectroscopy. This result implies that the uniaxial alignment is memorized by crosslinks in the PAzo network, which was preserved even after the incorporation of amorphous components.

Photoinduced bending behaviors of IPN films were observed in comparison with PAzoTP and pristine PAzo films (Fig. 28.4). Deformation of IPN films could be induced in the same way as conventional PAzo films upon irradiation with UV and visible light. The rate of bending increased with the length of alkyl chains of amorphous components. Especially, PAzo/PDDMA films showed much faster bending than the other films.

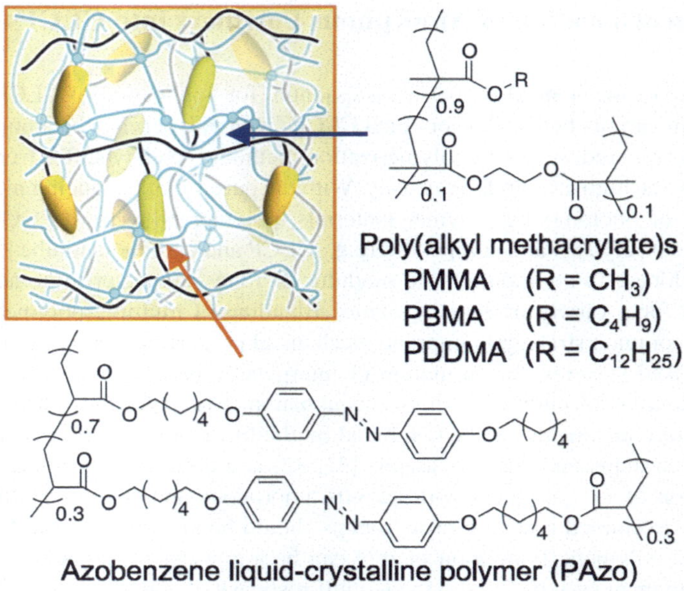

Fig. 28.3 Schematic illustration of IPN. (Reproduced from [34])

Mechanical properties of IPN films were investigated through tensile test. Young's modulus was evaluated from the initial slopes of the stress–strain curves. PAzoTP showed smaller elastic modulus than a pristine PAzo film due to the presence of numerous pores in films. Young's modulus was increased by incorporation of PMMA and PBMA into pores in PAzoTP. On the other hand, PAzo/PDDMA showed smaller Young's modulus than PAzoTP in spite that PAzo/PDDMA contains additional polymer components when compared with PAzoTP. This result implies that incorporation of soft components enhances the motions of PAzo chains. Moreover, the fracture stress and strain could be improved by incorporation of amorphous polymers. Especially, PAzo/PMMA showed significantly higher fracture stress than a pristine PAzo film. The brittleness of CLCP can be compensated by incorporation of amorphous polymers.

The photoresponsive properties of IPN films were directly related with mechanical properties: the films show faster bending with decreasing Young's modulus. The strategy of incorporating soft components into CLCPs was also applied for fabrication of IPN films with PAzo and poly(dimethylsiloxane) (PDMS) [35]. In this case, hydrosilylation reaction was conducted in a PAzoTP film to form crosslinked network of PDMS. Although phase separation was observed in the IPN film, the uniaxial alignment of azobenzene moieties was preserved. The resultant PAzo/PDMS film exhibited about 10 times faster bending behavior upon irradiation with UV light when compared with a pristine PAzo film, thanks to the soft nature of PDMS. Formation of IPN structures was found to be a powerful approach to combine CLCP with various amorphous components, preserving and even enhancing photosynergetic effects.

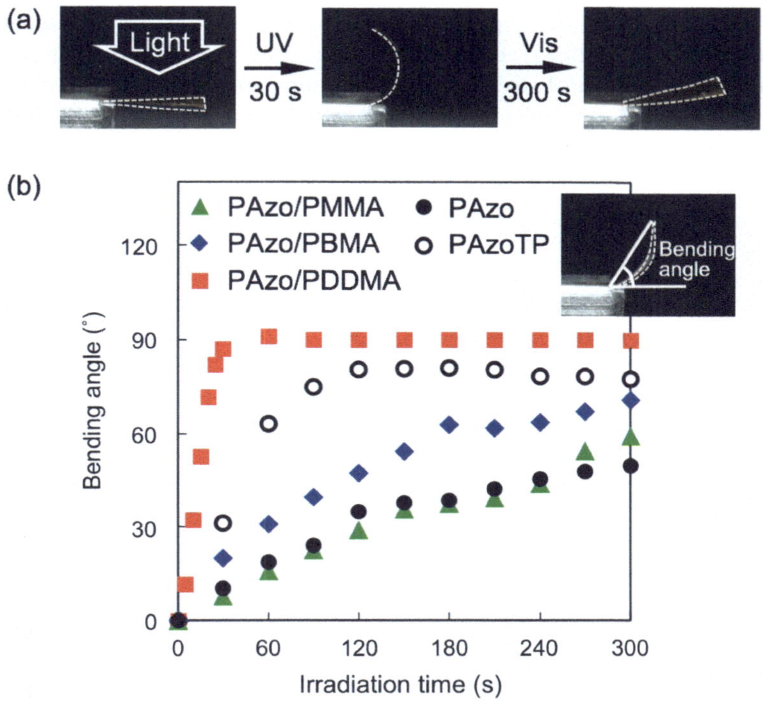

Fig. 28.4 (a) Photoinduced bending behaviour of a PAzo/PDDMA film upon irradition with UV (10 mW cm^{-2}) and visible (40 mW cm^{-2}) light. (b) Bending angles as a function of irradition time of UV light. Size of the films: 3 mm × 1 mm × 16 μm. (Reproduced from [34])

28.3 Reshaping of CLCPs Through the Rearrangement of Network Structures

In application of photomobile polymer materials to actual devices such as soft actuators, CLCPs should be processable into adequate 3D shapes depending on their uses. However, reshaping is a common challenge in crosslinked systems such as elastomers, resins, and gels as they are insoluble and infusible. Especially, in the preparation of CLCPs, conventional processes lead to the formation of flat films. Therefore, construction of CLCPs with 3D structures has been particularly difficult. Leibler et al. developed a new class of polymer materials called vitrimers, which are distinguished from conventional thermoplastics and thermosets [36–38]. Vitrimers are composed of crosslinked networks remoldable under heating through exchange reactions such as transesterification. This strategy has also been applied for thermoresponsive CLCPs based on epoxy networks [39].

We developed LC elastomers containing exchangeable links and photochromic moieties (LCE-**1**, Fig. 28.5a) capable of network rearrangement through transesterification (Fig. 28.5b) [40]. Ester and hydroxy groups were introduced to polysiloxanes through hydrosilylation reaction between poly(hydrogenmethylsiloxane) and vinyl compounds. LCE-**1** was fusible and reshapable under heating at LC phase (Fig. 28.5c), in contrast with CLCPs without dynamic covalent bonds.

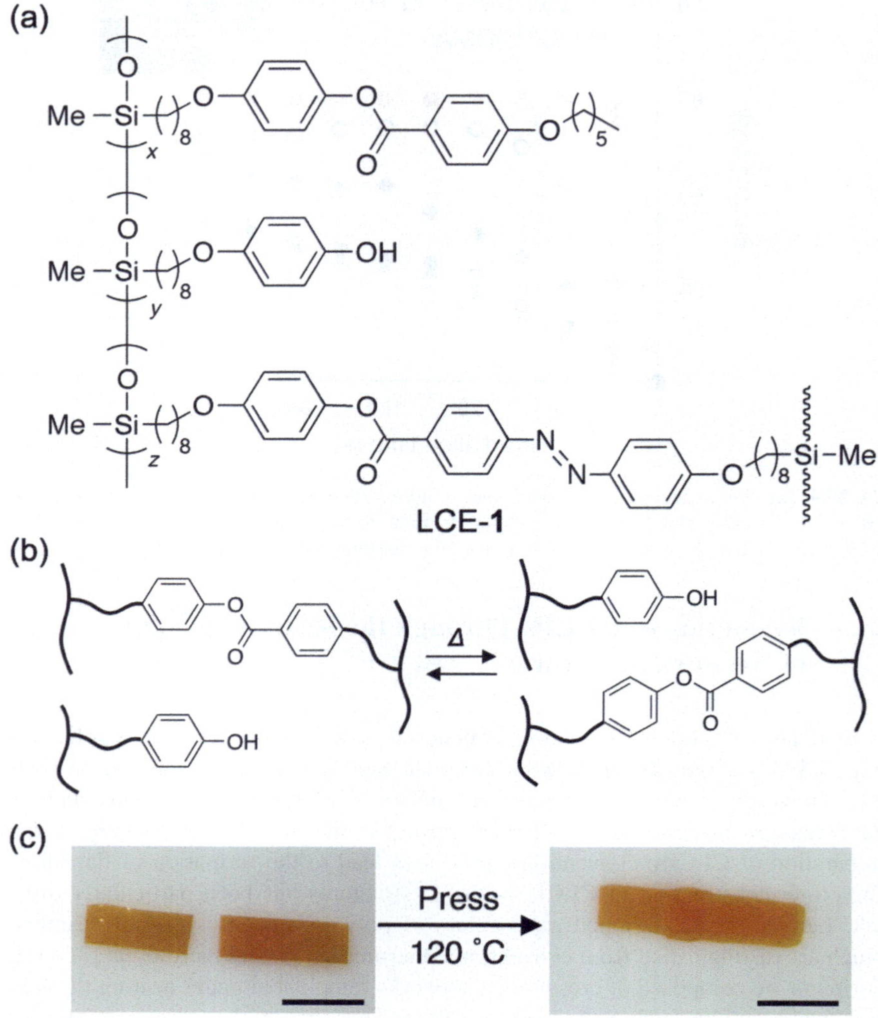

Fig. 28.5 (a) Chemical structure of LCE-1. (b) Schematic illustration of the rearrangement of network structure through transesterification. (c) Fusion test of LCE1. Scale bars: 5 mm. (Reproduced from [40])

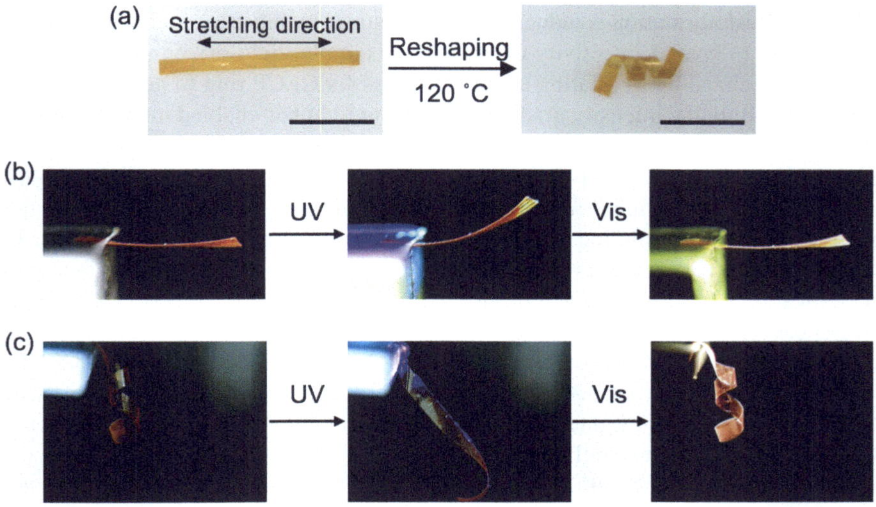

Fig. 28.6 Reshaping of a monodomain film into a spiral ribbon. Scale bars: 10 mm. Size of the film before reshaping: 25 mm × 2 mm × 70 μm. (b) Photoinduced deformation of a monodomain in film upon irradition with UV (365 nm, 35 mW cm^{-2}) and visible light (> 540 nm, 67 mW cm^{-2}). (c) Photoinduced deformation of a spiral ribbon upon irradiation with UV (365 nm, 97 mW cm^{-2}) and visible light (>540 nm, 60 mW cm^{-2}). (Reproduced from [40])

The alignment of mesogens was altered under external force in the same way as conventional CLCPs. With the aid of the rearrangeable network, the alignment induced by external force can be memorized by heating under external stress even after the formation of network structures.

A flat monodomain film of LCE-**1** was reshaped into a spiral ribbon (Fig. 28.6a). Upon irradiation with UV light, the initial flat film showed bending toward the light source (Fig. 28.6b). On the other hand, the spiral ribbon exhibited winding and unwinding motions when irradiated with UV and visible light (Fig. 28.6c). Irradiation with UV light from outside of the sample leads to contraction of the outer surface, resulting in unwinding motions. These examples clearly show that photoinduced deformation of CLCPs strongly depends on the initial shapes of the samples. Therefore, the macroscopic output of photosynergetic effects triggered by isomerization of photochromic moieties can be programmed through initial alignment of mesogens and also through initial macroscopic shapes.

28.4 Photomechanical Effects of CLCPs Induced by Two-Photon Processes

As discussed in the former sections, deformation of CLCPs can be preprogramed through the initial alignment of mesogens and shapes of samples. Another strategy to control 3D deformation is inducing localized strain by inhomogeneous irradiation with the aid of spatial selectivity of light. For example, light-driven plastic motors have been realized by irradiating adequate spots of a CLCP ring to generate torque [27]. Controlling the intensity gradient of incident light has enabled manipulation of fluid slugs in LCP microtubes [41] and wave-like motions of CLCPs [42]. Continuous motions such as oscillation [43, 44] and wave propagation [45] have been demonstrated by self-shadowing effect under controlled incident direction. Therefore, the precise control of irradiation spots is essential to induce sophisticated motions. In conventional CLCP actuators based on one-photon absorption, the surface facing to the light source absorbs actinic light in accordance with Beer-Lambert law, limiting the 3D selection of irradiation spots.

We focused on two-photon absorption (TPA) to induce deformation of CLCPs with high spatial selectivity. As the efficiency of TPA is proportional to the square of the intensity of incident light, chromophores near the focal spot, even those located deeply within a sample, can be selectively excited [46–48]. Application of TPA for photoactuation of CLCPs allows 3D control of actuation spots, which enhances the degree of freedom and preciseness of the photoinduced deformation.

For the design of two-photon actuation systems, CLCPs should contain two-photon chromophores, photochromic moieties, and mesogens. We developed a binary system composed of a stilbene derivative (D4ST) [49] as a two-photon chromophore and azotolane derivatives (A9ABT and DA9ABT) [50, 51] as photochromic mesogens (Fig. 28.7) [52]. The energy transfer between D4ST and A9ABT after two-photon excitation was investigated in solution. Upon irradiation with laser pulses (150 fs FWHM, 600 nm, 1 kHz), D4ST showed fluorescence, which was quenched by A9ABT. Furthermore, A9ABT solution doped with D4ST showed the change

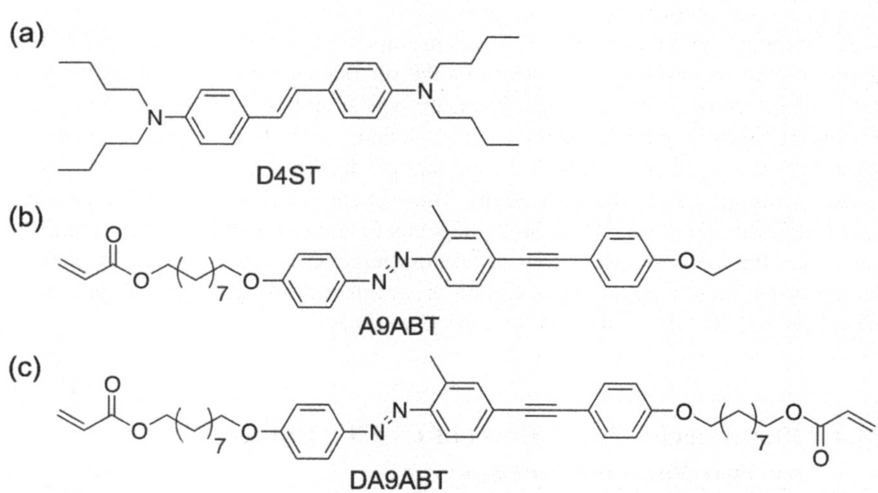

Fig. 28.7 Chemical structure of compounds used for preparation of two-photon driven actuators. (a) Stilbene derivative. (b) Azotolane monomer. (c) Azotolane crosslinker

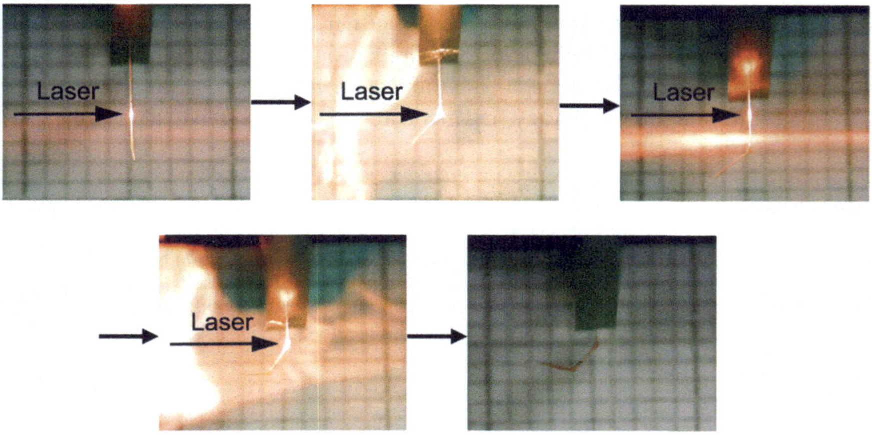

Fig. 28.8 Photoresponsive behavior of the IPN film upon irradiation with femtosecond laser pulses at 600 nm. Size of films: 5 mm × 0.5 mm × 20 μm. (Reproduced from [52])

in absorption spectra by irradiation with laser pulses, showing the similar behavior to A9ABT upon one-photon excitation. These results imply that the energy transfer from D4ST to A9ABT and subsequent *trans–cis* isomerization of A9ABT is brought about by two-photon excitation of D4ST.

We applied IPN for the construction of composite structure of D4ST and azotolane moieties with homogeneous alignment of mesogens [52]. First, A9ABT and DA9ABT were copolymerized in a glass cell coated with polyimide in the presence of inert LCs. After the removal of inert LCs, a mixture of D4ST and DDMA was introduced into the crosslinked azotolane polymer network, and then polymerized to form the IPN structure.

Photoinduced deformation of IPN films was examined under irradiation with fs-laser pulses at 600 nm (Fig. 28.8). Upon focusing the laser pulses, the film deformed at the spot and bent toward the light source. The film showed macroscopic deformation in spite of the fact that the size of the irradiation spot was smaller than 100 μm. This result implies the generation of large strain at the irradiation spot. Furthermore, the direction of the bending could be controlled by tightly focusing laser pulses to control the depth of the irradiation spots. Application of TPA was found to be a powerful approach to precisely control the deformation of CLCPs, which is especially advantageous for microactuators.

28.5 Conclusion

Photoresponsive properties of CLCPs have been enhanced by the formation of IPN structure, controlling the initial shapes of samples, and applying TPA. Light energy is converted to mechanical work through photoisomerization, the alignment change of mesogens, and deformation of polymer chains. This enhancement from nano to macro is a good example of photosynergetic effects. Further improvement of these systems will be made possible by applying novel photochromic molecules, photochemical processes, and assembled systems. In addition, cooperation with various fields, such as robotics, information technology, and medical science, will lead to real application as soft actuators.

Acknowledgements The authors acknowledge the experimental supports provided by Prof. Hiroshi Miyasaka, Prof. Kenji Kamada, Prof. Kenji Katayama, Dr. Hikaru Sotome, Dr. Masayasu Muramatsu, and students at Chuo University, especially Kiyohide Takado, Kyohei Kawasaki, Shota Sasaki, and Ken Minagawa. This work was supported by JSPS KAKENHI Grant Numbers JP15H01095, JP16H04157, JP17H05271, JP17K19161, JP18H04524, JP18K14286, JP26107002, JP26107004, JP26870592.

References

1. Ikeda T, Mamiya J, Yu Y (2007) Photomechanics of liquid-crystalline elastomers and other polymers. Angew Chem Int Ed 46:506–528
2. Ikeda T, Ube T (2011) Photomobile polymer materials: from nano to macro. Mater Today 14:480–487
3. Ube T, Ikeda T (2014) Photomobile polymer materials with crosslinked liquid-crystalline structures: molecular design, fabrication, and functions. Angew Chem Int Ed 53:10290–10299
4. White TJ, Broer DJ (2015) Programmable and adaptive mechanics with liquid crystal polymer networks and elastomers. Nat Mater 14:1087–1098
5. White TJ (2017) Photomechanical materials, composites, and systems: wireless transduction of light into work. Wiley Inc, Hoboken, NJ
6. Ube T, Ikeda T (2019) Photomobile polymer materials with complex 3D deformation, continuous motions, self-regulation, and enhanced processability. Adv Opt Mater 7:1900380
7. Ikeda T (2003) Photomodulation of liquid crystal orientations for photonic applications. J Mater Chem 13:2037–2057
8. Hartley GS (1937) The cis-form of azobenzene. Nature 140:281
9. Tazuke S, Kurihara S, Ikeda T (1987) Amplified image recording in liquid-crystal media by means of photochemically triggered phase-transition. Chem Lett 16:911–914
10. Ikeda T, Tsutsumi O (1995) Optical switching and image storage by means of azobenzene liquid-crystal films. Science 268:1873–1875
11. Warner M, Terentjev EM (2003) Liquid crystal elastomers. Oxford University Press, Oxford, UK
12. Ohm C, Brehmer M, Zentel R (2010) Liquid crystalline elastomers as actuators and sensors. Adv Mater 22:3366–3387
13. Küpfer J, Finkelmann H (1991) Nematic liquid single-crystal elastomers. Makromol Chem Rapid Commun 12:717–726
14. Wermter H, Finkelmann H (2001) Liquid crystalline elastomers as artificial muscles. e-Polymers 1:111–123
15. Broer DJ, Finkelmann H, Kondo K (1988) In-situ photopolymerization of an oriented liquid-crystalline acrylate. Makromol Chem 189:185–194
16. Broer DJ, Mol GN, Challa G (1989) In-situ photopolymerization of an oriented liquid-crystalline acrylate, 2. Makromol Chem 190:19–30
17. Broer DJ, Boven J, Mol GN, Challa G (1989) In-situ photopolymerization of oriented liquid-crystalline acrylates, 3. Oriented polymer networks from a mesogenic diacrylate. Makromol Chem 190:2255–2268

18. Roberts PMS, Mitchell GR, Davis FJ (1997) A single director switching mode for monodomain liquid crystal elastomers. J Phys II 7:1337–1351
19. Thomsen DL, Keller P, Naciri J, Pink R, Jeon H, Shenoy D, Ratna BR (2001) Liquid crystal elastomers with mechanical properties of a muscle. Macromolecules 34:5868–5875
20. Li MH, Keller P, Li B, Wang X, Brunet M (2003) Light-driven side-on nematic elastomer actuators. Adv Mater 15:569–572
21. de Haan LT, Sánchez-Somolinos C, Bastiaansen CMW, Schenning APHJ, Broer DJ (2012) Engineering of complex order and the macroscopic deformation of liquid crystal polymer networks. Angew Chem Int Ed 51:12469–12472
22. Ware TH, McConney ME, Wie JJ, Tondiglia VP, White TJ (2015) Voxelated liquid crystal elastomers. Science 347:982–984
23. de Haan LT, Gimenez-Pinto V, Konya A, Nguyen TS, Verjans JMN, Sánchez-Somolinos C, Selinger JV, Selinger RLB, Broer DJ, Schenning APHJ (2014) Accordion-like actuators of multiple 3D patterned liquid crystal polymer films. Adv Funct Mater 24:1251–1258
24. Finkelmann H, Nishikawa E, Pereira GG, Warner M (2001) A new opto-mechanical effect in solids. Phys Rev Lett 87:015501
25. Yu Y, Nakano M, Ikeda T (2003) Directed bending of a polymer film by light—miniaturizing a simple photomechanical system could expand its range of applications. Nature 425:145
26. Ikeda T, Nakano M, Yu Y, Tsutsumi O, Kanazawa A (2003) Anisotropic bending and unbending behavior of azobenzene liquid-crystalline gels by light exposure. Adv Mater 15:201–205
27. Yamada M, Kondo M, Mamiya J, Yu Y, Kinoshita M, Barrett CJ, Ikeda T (2008) Photomobile polymer materials: towards light-driven plastic motors. Angew Chem Int Ed 47:4986–4988
28. Yamada M, Kondo M, Miyasato R, Naka Y, Mamiya J, Kinoshita M, Shishido A, Yu Y, Barrett CJ, Ikeda T (2009) Photomobile polymer materials—various three-dimensional movements. J Mater Chem 19:60–62
29. Wang W, Sun X, Wu W, Peng H, Yu Y (2012) Photoinduced deformation of crosslinked liquid-crystalline polymer film oriented by a highly aligned carbon nanotube sheet. Angew Chem Int Ed 51:4644–4647
30. Sperling L (1994) Interpenetrating polymer networks—an overview. In: Klempner D, Sperling LH, Utracki LA (eds) Interpenetrating polymer networks, vol 239. Amer Chemical Soc Washington, pp 3–38
31. Gong JP, Katsuyama Y, Kurokawa T, Osada Y (2003) Double-network hydrogels with extremely high mechanical strength. Adv Mater 15:1155–1158
32. Gong JP (2010) Why are double network hydrogels so tough? Soft Matter 6:2583–2590
33. Ha SM, Yuan W, Pei QB, Pelrine R, Stanford S (2006) Interpenetrating polymer networks for high-performance electroelastomer artificial muscles. Adv Mater 18:887–891
34. Ube T, Takado K, Ikeda T (2015) Photomobile materials with interpenetrating polymer networks composed of liquid-crystalline and amorphous polymers. J Mater Chem C 3:8006–8009
35. Ube T, Minagawa K, Ikeda T (2017) Interpenetrating polymer networks of liquid-crystalline azobenzene polymers and poly(dimethylsiloxane) as photomobile materials. Soft Matter 13:5820–5823
36. Montarnal D, Capelot M, Tournilhac F, Leibler L (2011) Silica-like malleable materials from permanent organic networks. Science 334:965–968
37. Capelot M, Montarnal D, Tournilhac F, Leibler L (2012) Metal-catalyzed transesterification for healing and assembling of thermosets. J Am Chem Soc 134:7664–7667
38. Denissen W, Winne JM, Prez FED (2015) Vitrimers: permanent organic networks with glass-like fluidity. Chem Sci 7:30–38
39. Pei Z, Yang Y, Chen Q, Terentjev EM, Wei Y, Ji Y (2014) Mouldable liquid-crystalline elastomer actuators with exchangeable covalent bonds. Nat Mater 13:36–41
40. Ube T, Kawasaki K, Ikeda T (2016) Photomobile liquid-crystalline elastomers with rearrangeable networks. Adv Mater 28:8212–8217
41. Lv J, Liu Y, Wei J, Chen E, Qin L, Yu Y (2016) Photocontrol of fluid slugs in liquid crystal polymer microactuators. Nature 537:179–184

42. Palagi S, Mark AG, Reigh SY, Melde K, Qiu T, Zeng H, Parmeggiani C, Martella D, Sanchez-Castillo A, Kapernaum N, Giesselmann F, Wiersma DS, Lauga E, Fischer P (2016) Structured light enables biomimetic swimming and versatile locomotion of photoresponsive soft microrobots. Nat Mater 15:647–653
43. White TJ, Tabiryan NV, Serak SV, Hrozhyk UA, Tondiglia VP, Koerner H, Vaia RA, Bunning TJ (2008) A high frequency photodriven polymer oscillator. Soft Matter 4:1796–1798
44. Gelebart AH, Vantomme G, Meijer EW, Broer DJ (2017) Mastering the photothermal effect in liquid crystal networks: a general approach for self-sustained mechanical oscillators. Adv Mater 29:1606712
45. Gelebart AH, Mulder DJ, Varga M, Konya A, Vantomme G, Meijer EW, Selinger RLB, Broer DJ (2017) Making waves in a photoactive polymer film. Nature 546:632–636
46. Zipfel WR, Williams RM, Webb WW (2003) Nonlinear magic: multiphoton microscopy in the biosciences. Nat Biotechnol 21:1369–1377
47. He GS, Tan LS, Zheng Q, Prasad PN (2008) Multiphoton absorbing materials: molecular designs, characterizations, and applications. Chem Rev 108:1245–1330
48. Pawlicki M, Collins HA, Denning RG, Anderson HL (2009) Two-photon absorption and the design of two-photon dyes. Angew Chem Int Ed 48:3244–3266
49. Albota M, Beljonne D, Brédas JL, Ehrlich JE, Fu JY, Heikal AA, Hess SE, Kogej T, Levin MD, Marder SR, McCord-Maughon D, Perry JW, Röckel H, Rumi M, Subramaniam G, Webb WW, Wu XL, Xu C (1998) Design of organic molecules with large two-photon absorption cross sections. Science 281:1653–1656
50. Okano K, Shishido A, Ikeda T (2006) Photochemical phase transition behavior of highly birefringent azotolane liquid-crystalline polymer films: effects of the position of the tolane group and the donor–acceptor substituent in the mesogen. Macromolecules 39:145–152
51. Yin R, Xu W, Kondo M, Yen CC, Mamiya J, Ikeda T, Yu Y (2009) Can sunlight drive the photoinduced bending of polymer films? J Mater Chem 19:3141–3143
52. Sasaki S, Ube T, Katayama K, Muramatsu M, Miyasaka H, Ikeda T (2018) Two-photonactuation of crosslinked liquid-crystalline polymers utilizing energy transfer system. Mol Cryst Liquid Cryst 662:53–60

Chapter 29
Femtosecond Pump-Probe Microspectroscopy and Its Application to Single Organic Nanoparticles and Microcrystals

Yukihide Ishibashi and Tsuyoshi Asahi

Abstract A novel femtosecond pump-probe microspectroscope using a femtosecond Ti:Sapphire oscillator as a light source was developed. The fundamental specifications are the temporal resolution of 350 fs, the spatial resolution of 770 nm, and the tunable probe wavelength in the region from 500 to 900 nm. The setup has two modes of the probe light detection; one is back-scattered light detection and another conventional transmitted light detection. The back-scattering mode measurement demonstrated about 20-times higher gain of the transient signal of single nanoparticles compared to the conventional transmittance-mode one. This high-sensitivity enables the single nanoparticle ultrafast spectroscopy of organic nanomaterials which in general are low photo-durability. The femtosecond pump-probe microspectroscopy is fruitful in the researches of solid-state photochemical processes characterized by restricted molecular motions and intermolecular electronic interactions in the confined space. The potential applications of the developed microspectroscopic system are demonstrated for the picosecond excited state dynamics of nm- to μm-sized organic crystals.

Keywords Femtosecond pump-probe spectroscopy · Nanospectroscopy · Single organic nanoparticle · Excited-state dynamics

29.1 Introduction

Over the past two decades, organic nanoparticles have been attracted in a scientific and technological area of materials because they combine molecular properties with nanoparticle characteristics such as optical, electronic, and solubility properties depending on their size and shape. The nanoparticles having the dimension of several

Y. Ishibashi · T. Asahi (✉)
Department of Materials Science and Biotechnology, Graduate School of Science and Engineering, Ehime University, Matsuyama, Ehime 790-8577, Japan
e-mail: asahi.tsuyoshi.mh@ehime-u.ac.jp

Y. Ishibashi
e-mail: ishibashi.yukihide.mk@ehime-u.ac.jp

H. Miyasaka et al. (eds.), *Photosynergetic Responses in Molecules and Molecular Aggregates*, https://doi.org/10.1007/978-981-15-5451-3_29

tens-nm to a few hundred-nm were prepared for various organic compounds, and the nano-aggregation effects on the optical properties such as size-dependent absorption and fluorescence spectra and aggregation-induced emission were examined [1, 2]. The mechanism has been discussed in terms of restricted intramolecular motions and intermolecular electronic interactions in the confined space. Moreover, synergetic interactions between photo-excited molecules and/or the ground-state ones have been attracting increasing interests since novel photochemical reactions hardly realized in solution were induced in solid systems [3–5].

To elucidate the details of photochemical processes in nanoparticles, femtosecond transient absorption spectroscopy is a fruitful technique, which makes it possible to directly observe the ultrafast dynamics of excited states and transient intermediates. It is now recognized that pump-probe experiments for individual nanoparticles are important and indispensable for elucidating the ultrafast photochemical dynamics. Thus, in the recent decade several pump-probe setups combining a far-field microscope and a femtosecond Ti:Sapphire oscillator have been reported and applied to various kinds of nanomaterials such as metal nanoparticles and semiconductor nanowires, and organic nanocrystals [6–13]. In this chapter, we introduce our femtosecond pump-probe microspectroscope using a femtosecond supercontinuum as a probe pulse and compare transient signal sensitivity between the conventional transmittance-mode detection and the back-scattering mode one of the probe light. It was demonstrated that the detection of the back-scattered light from a single nanoparticle permitted 20 times higher gain of the transient signals in comparison with the conventional absorption-based measurement. The high-sensitivity of the pump-probe, back-scattering microspectroscopy allowed us to examine ultrafast excited-state dynamics of single organic nanoparticles which in general are lower photo-durability than inorganic ones. The size-dependent excimer formation dynamics of perylene nanocrystals and the crystalline-phase dependence of the exciton decay for copper phthalocyanine (CuPc) nanorods are demonstrated. Also, some applications of the femtosecond pump-probe microspectroscope to studies on the excited state dynamics of μm-sized single crystals are demonstrated. We will discuss the photoinduced charge separation dynamics in a ps-time scale for a weak charge-transfer complex and the effect of ultrafast transient laser heating on the singlet fission dynamics of rubrene.

29.2 Setup of Femtosecond Pump-Probe Microspectroscopic System

Figure 29.1a shows the schematic illustration of our femtosecond transient absorption microspectroscopic system [13]. A femtosecond Ti:Sapphire laser (Tsunami, Spectra-Physics) pumped by a cw DPSS green laser (Millennia eV, Spectra-Physics) was used as a light source. The output pulse (795 nm, 1 W, 90 fs fwhm, 82 MHz) was

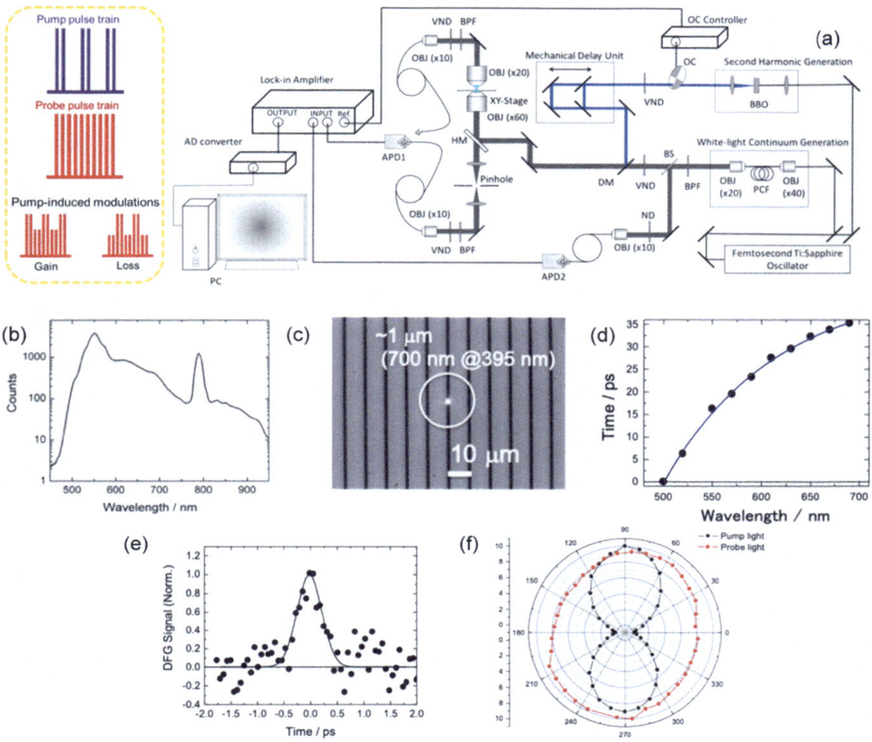

Fig. 29.1 **a** Schematic diagram of the femtosecond pump-probe microspectroscopic system. APD = avalanche photodiode; OBJ = objective lens; BPF = optical bandpass filter; HM = half mirror; PCF = photonic crystal fiber; VND = variable neutral density filter; ND = neutral density filter; BBO = BBO crystal; OC = optical chopper; BS = beam splitter (10%R: 90%T); DM = 400-nm dichroic mirror; AD converter = analog-to-digital converter. **b** Signal intensity distribution of the PCF-generated supercontinuum as a function of wavelength. The spectrum was detected with a modular USB spectrometer (USB2000+, ocean optics). **c** Optical image of probe light at the focusing spot. **d** Measure of the frequency chirp of the PCF-generated supercontinuum. The solid curve is just a guide for the eyes. **e** Cross-correlation function between 397-nm pump and 680-nm probe lights. Both lights were focused on a piece of BBO crystal (0.1 mm thickness) with the same 60× objective lens, and signals of difference frequency generation were detected at different delay times between two lights. Solid line is the calculated curve with a simple gauss function. The FWHM of the correlation is 350 ± 50 fs. **f** Polar plots of the intensity of pump and probe lights versus the polarization angle with respect to a thin film polarizer. Black and red circles are pump and probe light, respectively. The shapes of the plots indicate that the light coming out from the probe is non-linearly polarized, and the pump is linear polarized. Reprinted in part with permission from [13]. Copyright 2016 American Chemical Society

divided into two pulses with the same energy. One of the two pulses was frequency-doubled (397 nm) with a 2-mm thickness BBO crystal, and it was used as a pump pulse, which was chopped at a 20-kHz repetition rate by a high-speed optical chopper to generate transient modulations of target nanocrystal. After passing through an optical delay stage (STM-150, SIGMA Koki) which can control delay time between −50 and 700 ps, the pump pulse was guided into an inverted optical microscope (IX71, Olympus) and focused into the sample with an objective lens (60×, N.A. 0.7) (LUCPLFLN60X, Olympus). The other fundamental pulse was focused into an 80-cm photonic crystal fiber (PCF) (NL-PM-750, Crystal Fibre) to generate a super-continuum in the wavelength range of 500–900 nm as shown in Fig. 29.1b, which was used as a probe pulse. The tuned probe pulse was focused on the sample with the same objective lens co-linearly with the pump pulse. This system has two detection manners; transmitted (absorption-based) light and back-scattering light detections.

The back-scattering light of the probe pulse was collected with a ×60 objective lens and passed through a confocal pinhole, and then the intensity was detected with a fiber-coupled avalanche photodiode module (APD1). In the absorption-based mode experiment, the transmitted light through the sample was collected with a × 20 objective lens, and its intensity was measured with APD1. To extract modulation of weak transient signal by pump light in Fig. 29.1a, the APD1 output was directed into the signal input of a lock-in amplifier (SR830, Stanford Research) referenced to the optical chopper. A part (10%) of the probe light was split before the microscope, and the light intensity detected with the avalanche photodiode (APD2) was used as the reference in a differential amplification of the signals. The output signals of the lock-in amplifier were sent to the analog-to-digital converter (NI USB-6356, National Instruments) and a personal computer to calculate the transient signals induced by the pump pulses. The transient signals were recorded as a function of the position of the translational stage (i.e., delay time between pump and probe pulses). The obtained kinetic data were analyzed by using commercial software and homemade programs.

The fundamental specifications of the developed system are summarized as follows. The diameter of the focusing spot (Fig. 29. 1c) was estimated from images of a reflection at the glass coverslip with a Gaussian function. The diameter of the probe pulse was 590 nm in FWHM, and that of the pump pulse was 730 nm in FWHM. The fluence of the excitation pulse on the microscope stage was varied with a variable neutral density filter (VND) in the range of 0.01–30 mJ cm^{-2} $pulse^{-1}$. To select the appropriate probe wavelength, we inserted an optical bandpass filter (BPF) having 10-nm bandwidth in FWHM (Asahi Spectra) after generating the supercontinuum, Fig. 29.1d shows the frequency chirp of the PCF-generated supercontinuum, which was measured from the time profiles of transient absorbance of an organic laser dye (DCM) solution in the probe wavelength range of 500–800 nm. By taking into accounts the frequency chirp curve, we reconstructed transient absorption spectra from the time profile of the transient signals at each probe wavelength. The FWHM of the cross-correlation between the pump (397 nm) and probe (680 nm) pulses was 350 fs at the sample position (Fig. 29.1c). The polarization of the pump and probe lights were linear and circular, respectively (Fig. 29.1e).

In the case that a sample is larger than the spot size of probe beams (in Fig. 29.2a), the transient signal in transmittance-mode detection will be simply given by $\Delta I_T/I_T$. Here, I_T is the intensity of the transmitted light of the sample without excitation and ΔI_T is the modulation induced by a pump pulse. $\Delta I_T/I_T$ is proportional to the fluence of the pump pulse, and transmittance-mode detection will gain $\Delta I_T/I_T$ enough to be large by setting appropriate values of the pump intensity. On the other hand, in the case of a smaller particle than the spot size of the probe light (Fig. 29.2b), the photodetector (APD1) also detects *stray* light, i.e., the light did not interact with the particle, at the same time, so that the transient signal will be given by $\Delta I_T/(I_T + I_{\text{stray}})$, where I_{stray} is the intensity of *stray* light. It is obvious that as I_{stray} becomes larger than I_T, the gain of transient signals decrease drastically. That is, the transient signal will hardly obtain in the transmittance-mode detection of the probe light. An alternative way to detect the transient signal is to measure the scattering light from the sample nanoparticle. In this case, we will be able to reduce the contribution of *stray* light in the transient signal, and expect to obtain the transient signal of single nanoparticle. In the following, we compare transient signal intensities of the transmittance mode and the back-scattering one in the case of detecting a nanoparticle on a glass slip.

A single nanoparticle on a glass coverslip is placed at the center of a focused Gaussian beam with intensity distribution of $(2P_i/\pi\omega_0^2)\exp[-2x^2/\omega_0^2]$, where x is the radial distance from the beam center, P_i the total power, and ω_0 the beam waist in Fig. 29.3a. The nanoparticle will give the modulation of $-\sigma_{\text{ext}} \times 2P_i/(\pi\omega_0^2)$ in the transmitted light intensity, where σ_{ext} is the particle's extinction cross-section. By taking into account the loss of the incident light by the reflection at the glass surface, the transmission intensity (I_T) is given by $I_T = P_i(1-r) - P_i \times 2\sigma_{\text{ext}}/(\pi\omega_0^2)$. Here, r is the reflectance at the glass surface. On the other hand, in the back-scattering mode, the detected light intensity (I_S) is the sum of the scattering light of the nanoparticle and

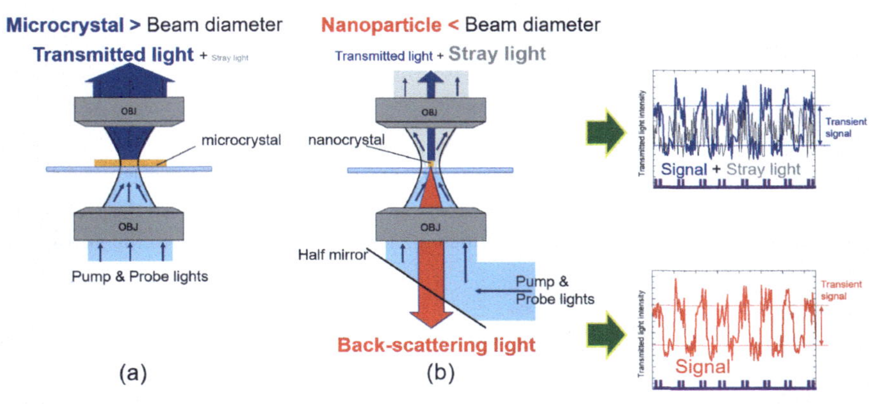

Fig. 29.2 Illustration of comparison between transmittance mode and back-scattering mode detections. **a** Larger microcrystal and **b** smaller nanocrystal than the diffraction limit of pump and probe lights

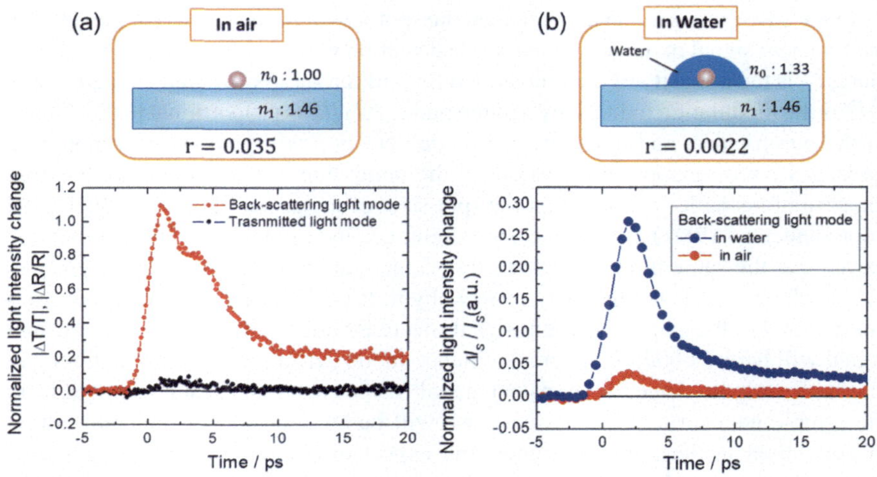

Fig. 29.3 **a** Comparison of transient signal intensities of single gold nanoparticle with a size of 100 nm in air by back-scattering light and transmitted light detections. **b** Comparison of transient signal intensities of single gold nanoparticle with a size of 100 nm in air and in water detected with back-scattering light. Both upper illustrations depict a single gold nanoparticle on glass slip in air (**a**) and covered by water (**b**)

the reflection at the glass surface, and expressed by $I_S = P_i \times r + P_i \times 2\sigma_{\text{scat}}/(\pi\omega_0^2)$, where σ_{scat} is the particle's scattering cross-section. In case of small nanoparticles, whose σ_{ext} and σ_{scat} are much smaller than $\pi\omega_0^2$, the normalized modulation changes in the transmittance-mode ($\Delta I_T/I_T$) and back-scattering ($\Delta I_S/I_S$) one are given by Eqs. (29.1) and (29.2), respectively, by using the change of the extinction cross-section ($\Delta\sigma_{\text{ext}}$) and the scattering cross-section ($\Delta\sigma_{\text{scat}}$) induced by the pump light. Derivations of these equations are in detail described in Ref. [13].

$$\Delta I_T/I_T = -2 \times \Delta\sigma_{\text{ext}}/\{\pi\omega_0^2 \times (1-r)\} \tag{29.1}$$

$$\Delta I_S/I_S = 2 \times \Delta\sigma_{\text{scat}}/\{\pi\omega_0^2 \times r\} \tag{29.2}$$

Also, the reflectance at the glass surface, r is estimated simply from the following equation, where n_0 is the refractive index of the medium, and n_1 is the refractive index of the glass.

$$r = \{(n_0 - n_1)/(n_0 + n_1)\}^2 \tag{29.3}$$

As an example, the transient signal of a single gold nanoparticle with a diameter of 100 nm at the wavelength of its plasmonic resonance peak is discussed. The value of $\pi\omega_0^2$ is calculated to be (2×10^6 nm^2) from the beam waist of 770 nm ($\omega_0 \approx \lambda$/NA) under the experimental condition. Using the value of σ_{ext} and σ_{scat} at 540 nm (2 ×

10^4 nm^2), ($2\sigma_{ext}/\pi\omega_0^2$) and ($2\sigma_{scat}/\pi\omega_0^2$) were calculated to be 0.02. The value of r is 0.035 for a normal incidence of light at the interface between the glass ($n_1 = 1.46$) and air ($n_0 = 1.00$). According to Eqs. (29.1) and (29.2), the ratio of $\Delta I_S/I_S$ to $\Delta I_T/I_T$ is approximate 27 by assuming that the value of $\Delta\sigma_{ext}$ is equal to that of $\Delta\sigma_{scat}$. Consequently, the back-scattering mode will improve signal gain for nanoparticles having a smaller size than the beam west.

To demonstrate a high sensitivity of the back-scattering light detection experimentally, we measured the transient signals of single Au nanoparticle on a cover glass slip. Figure 29.3a shows the comparison of the time profiles of a 100-nm diameter particle between in transmitted light and back-scattering light detections at the probe wavelength of 540 nm. The values of σ_{ext} and σ_{scat} are by chance the same to each other value [14]. In the experiment, the scattering light intensity (I_S) detected with the APD1 was set to be the same to the transmission one (I_T) by tuning the probe pulse intensities before the microscope using the variable neutral density filters. The setups of the lock-in amplifier and the accumulation time (3 min/1 trace) were set to be the same in both experiments. The signal at negative delay times in both experiments showed almost the same noise levels to each other. However, it is obvious that the back-scattering light detection experiment gave a much larger magnitude in the transient signals than that of the transmittance mode. The signal gain in the back-scattering mode was higher by about 20 times than that in the transmittance mode. This experimental value was close to the calculated one (($\Delta I_S/I_S$)/($\Delta I_T/I_T$) = 27) described above. It was demonstrated that the back-scattering light detection could improve the sensitivity of the pump-probe signal. This high-sensitivity is a great advantage in the ultrafast spectroscopic studies of single nanoparticles. We can suppress the intensity of the pump pulse in order to avoid the photodecomposition of the sample in the pump-probe experiments.

According to Eq. (29.2), decreasing the value of r will improve the sensitivity of the back-scattering mode measurement. For example, when the glass coverslip is covered with water, the value of r is calculated to be 0.0022 which is smaller by 15 times than that of the coverslip in air. That is, the transient signal intensity of a nanoparticle is expected to increase 15 times. We have confirmed the idea experimentally as shown in Fig. 29.3b, where the transient signal intensities of the identical single gold nanoparticle (100 nm in diameter) on a glass cover slip covered with and without water. The signal intensity of the Au nanoparticle in water was 10 times larger than that in air.

29.3 Application to Single Organic Nanocrystals

In this section, we describe the applications of the above setup to the investigations of the excited-state dynamics of single organic nanocrystals of perylene and CuPc. First, we explain the results of perylene nanocrystals. Figure 29.4a is the dark-field image of perylene crystals on a glass coverslip prepared by recrystallization from an acetone solution. Target nanocrystal is in the circle and showed a broad fluorescence peak at

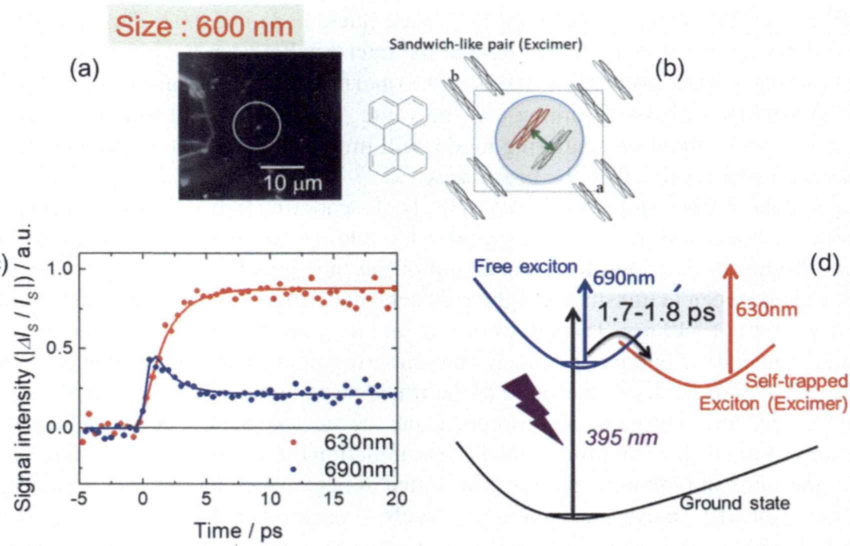

Fig. 29.4 **a** Optical image of target perylene nanocrystal under dark-field illumination. **b** Molecular structure of perylene, and illustration of molecular packing in α-perylene crystal on the basis of Refs. [15–17]. **c** Time profiles of transient signals of single perylene nanocrystal having a size of 600 nm at two probe wavelengths: 630 nm (red) and 690 nm (blue). Pump wavelength was 395 nm. Solid curves are the calculated curves by taking a single-exponential function, constant value, and response function into account. **d** Schematic energy diagram of the excimer formation in α-perylene crystal. Reprinted in part with permission from [13]. Copyright 2016 American Chemical Society

605 nm characteristic to α-form perylene crystal. The crystal size was estimated to be about 600 nm from the spot size (800 nm) in the dark-field image and the diffraction limit of light (500 nm) under the present experimental condition: the objective having NA of 0.70 and the probe light wavelength of 630 nm. It is well-known that in α-form perylene bulk crystal the molecular arrangement is the sandwich-like pair in Fig. 29.4b [15–17] and photoexcitation energy is trapped in this pair as an excimer, and the excimer formation occurred with a time constant of 3 ps [18–20]. Figure 29.4c presents the results of the time profiles at two probe wavelengths of 690 and at 630 nm. A decay component with a time constant of 1.7 ± 0.2 ps at 690 nm a rise one with a time constant of 1.8 ± 0.2 ps at 630 nm were observed. Because the probe wavelengths of 690 and 630 nm correspond to the absorption peak of the monomeric singlet excited state and the excimer state in the crystalline phase, respectively, [19–21] we can safely attribute the time constant of about 1.8 ps to the excimer formation in α-form perylene in Fig. 29.4d, which is comparable to that of the bulk crystal.

We measured excimer formation dynamics smaller nanocrystals having the size distribution of 100 nm to 500 nm in Fig. 29.5a The nanocrystals were deposited on a glass coverslip by spin-coating a drop of the 100-fold diluted aqueous nanocolloid which was prepared with a reprecipitation method [13], and then a target single nanocrystal in the circle was selected under the optical microcsope setup in Fig. 29.5b.

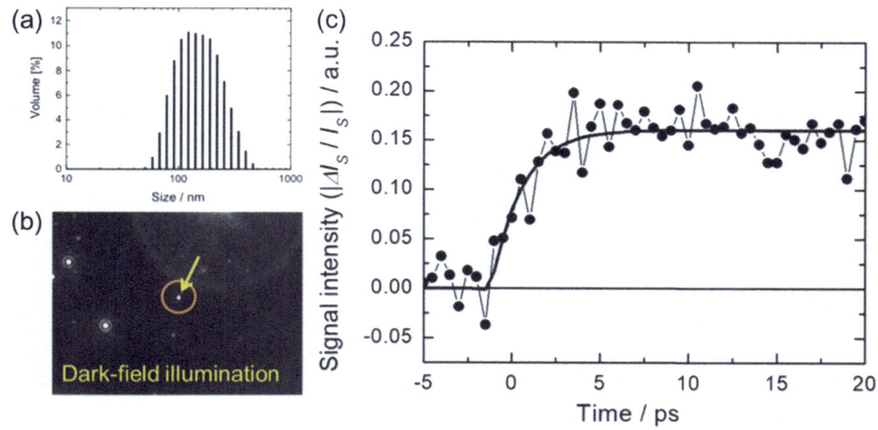

Fig. 29.5 **a** Size distribution of aqueous α-form perylene nanocrystals obtained with the dynamic light scattering measurement. **b** Optical image of the target perylene nanocrystal under dark-field illumination. **c** Time profile of the transient signal of single perylene nanocrystal, monitored at 630 nm. Pump wavelength was 395 nm. Solid curve is the calculated curves by taking a single-exponential function, constant value, and response function into account. Reprinted in part with permission from [13]. Copyright 2016 American Chemical Society

Figure 29.5c shows the time profile of the transient signal at probe wavelength of 630 nm. The time profile clearly showed a rise component and its time constant was calculated to be 1.6 ± 0.2 ps. Consequently, we succeeded in direct observation of the excimer formation dynamics in an individual perylene nanocrystal.

In order to investigate the size effect of the excimer formation, we measured 18 nanocrystals. Probe wavelength was also set to 630 nm. The time constants of 18 single nanocrystals are summarized in Fig. 29.6a. The time constants are scattered from particle to particle in the range of 0 to 3 ps. Interestingly, the relative signal intensity seems to decrease with a decrease in the time constant of the excimer formation, suggesting size dependence. In our previous work on single nanoparticle fluorescence spectroscopy using a far-field fluorescence microspectroscope coupled with an AFM, [22] we demonstrated that the peak wavelength of the excimer emission of nanocrystals shifted from 600 to 585 nm by decreasing of the crystal size from 300 to 100 nm. The mechanism of the size dependence was discussed in terms of a change of intermolecular interactions that arise from the lowering in lattice energy of nanocrystals owing to the large surface-to-volume ratio. That is, a change in the elastic properties of nanocrystal could affect stabilization energy of the excimer formation. From this previous result, it also would be expected that the excimer formation dynamics might depend on the crystal's size. Here, we measured both their pump-probe transient signals and the emission spectra of the identical single nanocrystals. As representative examples, three perylene nanocrystals were chosen, and the results are shown in Fig. 29.6b, c. Each nanocrystal showed the similar time profiles to the sub-micron size crystal as shown in Fig. 29.4b. We estimated the time constant of the

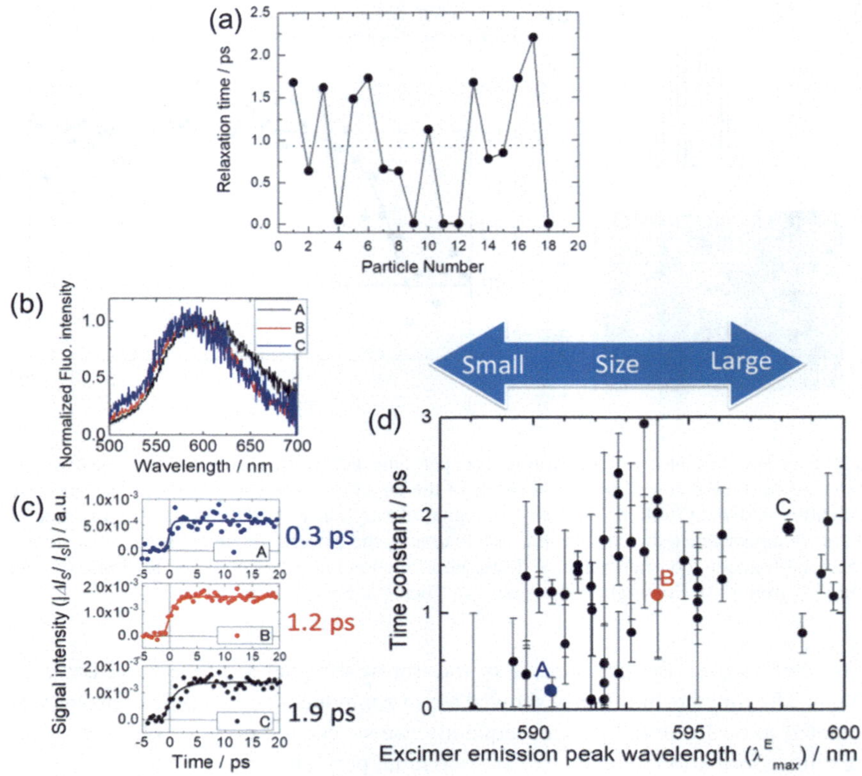

Fig. 29.6 **a** Distribution of the time constant of the excimer formation. **b** Representative emission spectra and **c** time profiles of the transient signals of three single perylene nanocrystals. Solid lines of each panel for (**c**) is the calculated curves by taking a single-exponential function, constant value, and response function into account. **d** Correlation between the obtained time constants of the excimer formation and the excimer emission peak wavelength. Emission peak wavelength was obtained with the peak fitting by a Gaussian function. Reprinted in part with permission from [13]. Copyright 2016 American Chemical Society

excimer formation by analyzing the time profiles at 630 nm using a single-exponential function and taking into account the temporal response function of the pump-probe setup. Figure 29.6d shows the time constant of the excimer formation as a function of the peak wavelength of the excimer emission spectrum. By referring to the size-dependent fluorescence peak in the previous paper, the shift of the excimer emission peak wavelength from 600 to 585 nm corresponds to a decreasing of the crystal size from 300 to 100 nm. Therefore, it can be considered that the plot of the time constant versus the peak wavelength in the figure will represent the size dependence. Some single nanocrystals having the emission peak at the wavelength of shorter than 595 nm showed the time constants of less than 1 ps, which means the faster excimer formation in smaller nanocrystals. These results are the first demonstration of the size-dependent excimer formation dynamics; i.e. energy trapping dynamics, in organic nanocrystals, although the mechanism of the size effect is not cleared yet at the present stage of the investigation.

Next, the exciton dynamics of copper phthnocyanine (CuPc) nanocrystals is discussed. The femtosecond transient absorption spectroscopic studies were reported by several research groups for polycrystalline thin solid films prepared by vapor deposition, and the exciton-exciton annihilation under a dense excitation condition was observed and discussed [23, 24]. Here, we present the results of CuPc nanorods prepared by the laser ablation method [25]. CuPc nanorod colloids having the dimensions of 40 nm in width and 100–1000 nm in length were prepared by ns pulsed laser-induced fragmentation of its microcrystals dispersed in ethyl acetate. We measured femtosecond transient absorption spectra for the nanorod colloid by means of the conventional transient absorption spectroscopy and observed the transient absorption spectra having a strong positive, broad peak in the wavelength range of 400–580 nm. The spectral shape was in good agreement with that in the thin film of β-phase CuPc [24]. Also, the exciton decay dynamics depending on the excitation intensity was similar to that of the thin film.

The CuPc nanorod colloids prepared by the laser fragmentation in liquid include the nanorods of different crystalline phases. The β-phase nanorods are major product and β-phase ones minor [25]. The α- and β-phase nanorods have absorption peaks around 630 nm and 780 nm. The absorption peak ratio of 630 nm to 780 nm was about 0.5 for α-phase nanorod, while the ratio was larger than 1 for β-phase one. So, to examine the exciton dynamics of the nanorods depending on the crystalline phase, we measured exciton lifetimes and the steady-state scattering spectra of individual 20 nanorods. The nanorods dispersed on a glass coverslip were used for single-particle measurements. As representative examples, Fig. 29.7a shows the time profiles of the transient signals and the steady-state scattering spectra of

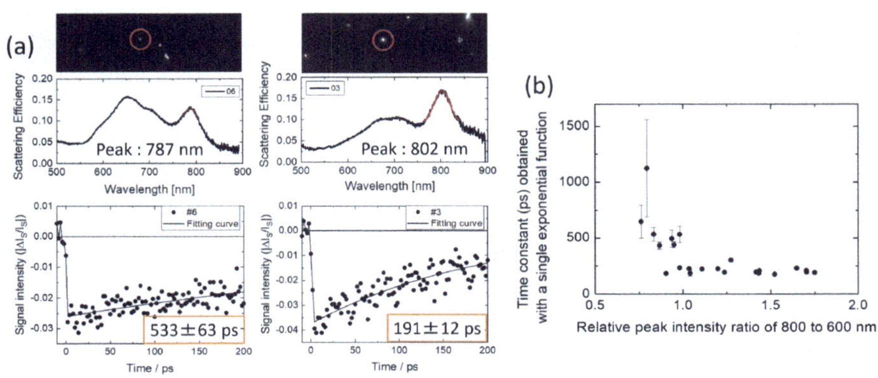

Fig. 29.7 **a** Representative results of CuPc nanonord. Dark-filed images, light scattering spectra, and time profiles of transient signals are depicted from upper to bottom panels. Pump and probe wavelengths are 397 nm and 540 nm, respectively. **b** Correlation between the obtained time constants of S_1 decay and the intensity ratio of the peak wavelength at 800 nm to at 660 nm

two nanorods. The two nanorods showed different exciton lifetimes and spectral shapes. Fig. 29.7b shows the relation between the lifetime and the peak ratio of 800 nm to 660 nm in the scattering spectra. When the peak ratio was larger than 1, the lifetime was 200 ps. The nanorods of the peak ratio larger than 1 were assigned to be β-phase and their lifetime was 200 ps. On the other hand, the lifetimes were scattered from 200 ps to 600 ps for the nanorod having the ratio less than 1. These nanorods could be considered to be α-phase or the mixture of α- and β-phases. We succeeded in demonstrating the exciton relaxation dynamics of CuPc nanorods depending on the crystalline phase.

29.4 Excited-State Relaxation Processes of μm-Sized Single Organic Crystal

In the previous section, we described the results of the pump-probe microspectroscopy in the back-scattering mode and the excited-state dynamics of single organic nanocrystals. Our microspectroscopic system, of course, is fruitfull to examine ultrafast excited state dynamics of μm-sized organic crystals. In the case of μm-sized samples, the transmittance mode (absorption-based) observation permits to get enough transient signals ($>10^{-3}$ OD) because the effect of stray light can be ignored (see Sect. 29.2). Here, a boxcar average (SR205, Stanford Research) was used instead of the lock-in amplifier, and a pocket cell (OG8/10, Avest) to reduce a repetition rate of the oscillator to 10 kHz was inserted after the output of the femtosecond laser oscillator. The repetition rate of the optical chopper for pump pulses was set to 500 Hz. Because the above setup also enables us to obtain time profile of transient absorbance at single probe light wavelength, transient absorption spectra can be reconstructed in the following manner. First, a kinetic trace of transient absorption at single wavelength by changing the delay time between the pump and probe pulses using the optical delay is obtained. Second, time traces at various wavelengths are recorded by considering the frequency chirp of the PCF-generated supercontinuum. Finally, transient absorption spectra are constructed by obtained kinetic traces in arbitrary wavelength range after correcting the frequency chirp at each probe wavelength. The obtained data are analyzed by using commercial software (Igor 6) and homemade programs. All the measurements can be performed under O_2 free condition.In the following, we present the results of two organic solid systems; charge-transfer (CT) complexes and rubrene investigated by using the transmittance mode femtosecond pump-probe microspectroscopic system.

CT interaction in excited states plays fundamental and important roles in a lot of photochemical and photophysical processes in condensed phases such as electron and proton transfer, photoinduced charge transport of organic molecular crystals and polymer films, and so on. Until now, there are enormous reports on synthesis of various kinds of donor (D)-acceptor (A) systems and their excited-state dynamics by time-resolved spectroscopies [26]. Among them, we focus here on the weak ground-state CT complexes, such as methoxy-substituted benzene derivatives as a donor and pyromellitic dianhydride (PMDA) as a acceptor. Picosecond and femtosecond diffuse reflectance spectroscopies (fs-DRSs) has been reported for microcrystalline powders of these complexes [26]. For example, on CT crystals of 1,2,4-trimethoxybenzene (124TMB)-PMDA, a broad PMDA anion-like band around 700 nm appeared in the temporal response of the spectroscopic system (a few picoseconds) with the excitation of the ground-state CT absorption and decayed with a time constant of 6 ps. Because the spectral shape and the lifetime of PMDA anion-like band are obviously proven, measurements of the solid crystal CT interaction dynamics are appropriate to evaluate the developed spectroscopic system.

As a representative example, we present here the results of 1,2,3-trimethoxybenzene (123TMB)-PMDA crystals. The sample CT crystals on a glass plate were grown from the acetonitrile solution containing equimolar amounts of 123TMB and PMDA by slowly evaporating the solvent. A smooth and flat crystal having the size with a few tens of μm was selected under the optical microscope. Figure 29.8b shows time profiles of transient absorbance of 123TMB-PMDA single microcrystal at several probe wavelengths, excited with a femtosecond 395-nm laser pulse. These time profiles are corrected by taking the frequency chirp into account. At all probe wavelengths, an appearance of positive absorbance with the temporal response (350 fs) was followed by the decay to zero in 50 ps. On the basis of these profiles, we constructed transient absorption spectra of 123TMB-PMDA single microcrystal and displayed the results in Fig. 29.8c. Immediately after the excitation, a broadband around 680 nm was observed, and the spectral shape became sharp together with the peak shift to 690 nm with delay time. At 2 ps, the shape was similar to that of the reported PMDA anion radicals [26]. After 2 ps, the transient absorption spectra monotonously decayed without any appreciable spectral change. Precisely to discuss the CT interaction, the time profile at 690 nm was analyzed with a double-exponential function of rise and decay components. The obtained time constants are 1.0 ps for the rise and 8.5 ps for the decay in Fig. 29.8d. The decay component was in good agreement with the reported value by fs-DRS, which was assigned to the lifetime of the relaxed CT excited state. That is, the electronic excitation is localized in one D-A pair at 2 ps after the excitation. However, because the temporal resolution of our spectroscopic system is superior to that of the reported fs-DRS, the rise component of 1.0 ps was clearly observed in our developed system. Considering slight changes in the spectral shape in 2 ps, the rise component was assigned to the relaxation from Franck–Condon state to the relaxed CT excited state.

Figure 29.9 shows the transient absorption spectra and the time profiles of CT complex crystals of 1,2,4-trimethoxybenzene (124TMB)-PMDA, and 1,3,5-trimethoxybenzene (135TMB)-PMDA. In both cases, the PMDA anion radical was observed at 2 ps after the excitation and decayed to the baseline. The time profiles at the peak of the PMDA anion radical were analyzed with a double-exponential function, and obtained results are as follows: 0.8 ps (rise) and 6.0 ps (decay) for 124TMB-PMDA, and 1.2 ps (rise) and >600 ps (decay) for 135TMB-PMDA. We firstly found that the decay of the relaxed CT excited state in each CT crystal was

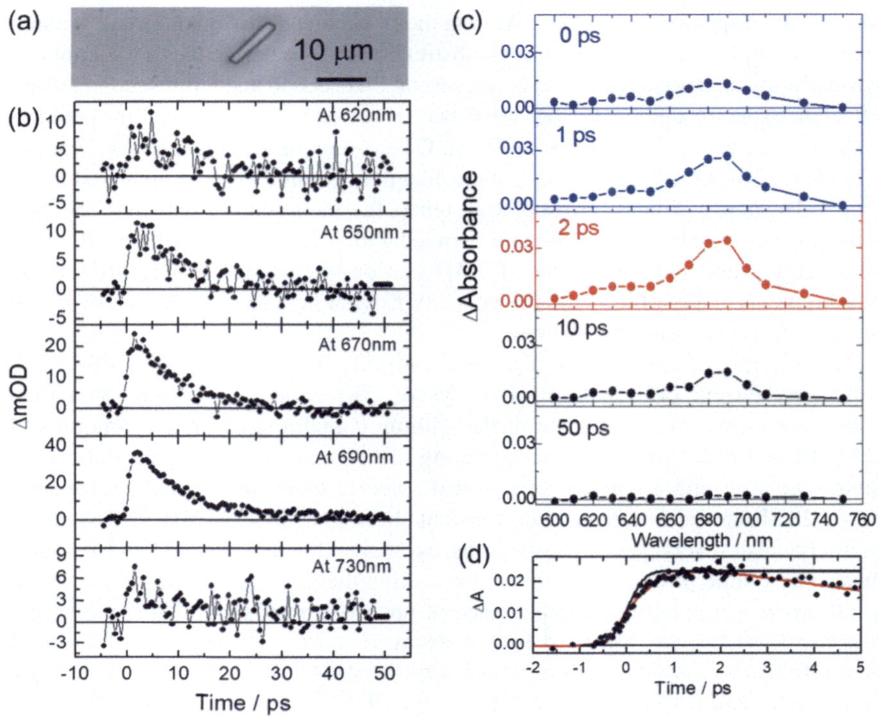

Fig. 29.8 **a** Optical image of the target 123TMB-PMDA CT microcrystal. **b** Corrected time profiles of transient absorbance at five prove wavelengths by taking the frequency chirp into account. **c** Constructed transient absorption spectra on the basis of the time profiles. **d** Time profile of the transient absorbance at 690 nm in the delay time range of −2 to 5 ps. Red solid line is the calculated curve with a double-exponential function with two time constant of 1.0 ps (rise) and 8.5 ps (decay) and pulse duration of 350 fs FWHM. Blackline is the temporal response function

consistent with the reported value, while the rise component was approximate 1 ps independent of donors.

Next, we present the results of rubrene single crystals. Rubrene is well-known to show efficient singlet fission in the crystalline phase. A singlet excited molecule shares its excitation energy with a neighboring molecule in its ground state, resulting in the formation of two triplet excited molecules [27, 28]. The singlet fission requires the energy level of the singlet excited (S_1) state, $E(S_1)$, to be at least the same as twice that of the lowest triplet excited state, $E(T_1)$, i.e., $E(S_1) \approx 2 \times E(T_1)$. In the case of rubrene, $E(S_1)$ is 2.23 eV and $E(T_1)$ is 1.14 eV, and small activation barrier, $E(S_1)–2xE(T_1) \approx –0.05$ eV exists. This indicates singlet fission from the S_1 state is only possible by thermal activation on the tens of picoseconds time scale, so-called thermally activated fission (TAF). Also, direct fission (DF) from the upper excited singlet states, S_N is another fission channel, which competes with the nonradiative relaxation to S_1 on a sub-ps time scale. Ma et al. revealed these two pathways of the

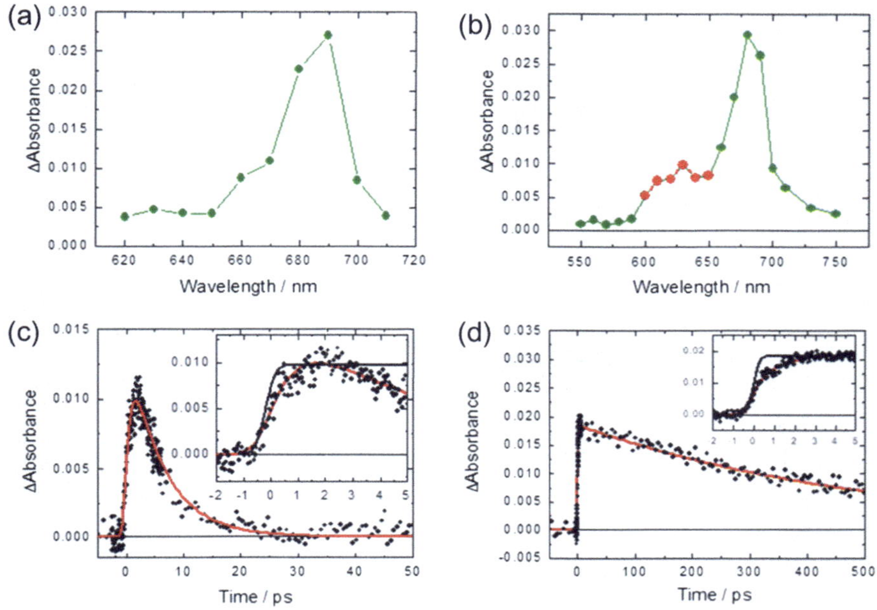

Fig. 29.9 Transient absorption spectra of **a** 124TMB-PMDA and **b** 135TMB-PMDA CT microcrystals, monitored at 2 ps after excitation. Red part of the spectrum in 135TMB-PMDA shows the absorption band of 135TMB cation radical. Time profiles of the transient absorbance of **c** 124TMB-PMDA and **d** 135TMB-PMDA CT microcrystals probed at the peak of transient absorption. Red solid lines are the calculated curve with a double-exponential function and pulse duration (350 fs), and black ones are the temporal response function

singlet fission by ultrafast transient absorption spectroscopy [28]. In relation to TAF, the reported values of the time constant are widespread in the range of 5–50 ps even at room temperature [27–29]. The large scatter could be concerned with the effect of a bimolecular annihilation process of the S_1 state competing with the T_1 state formation because a large number of excited-state molecules are easily formed under femtosecond pulse excitation with high laser fluence. Here, we report the excitation laser fluence dependence of the triplet state formation in a rubrene microcrystal and discuss the influence of dense photoexcitation on TAF.

Figure 29.10a shows the transient absorption spectra of the rubrene microcrystal having a thickness of 2.1 μm, excited with a femtosecond 397-nm laser pulse. The preparation of the sample rubrene crystals was described in ref. 29. The 397-nm excitation (3.12 eV) corresponds to the excitation to higher vibrational energy level of the S_1 state having the excess energy of 0.89 eV against the 0–0 transition of the S_1 state (2.23 eV). A broad absorption which is assigned to the hot S_1 state and/or higher singlet excited (S_N) state was observed immediately after the excitation. A sharp absorption band with a peak of 510 nm due to the absorption of the T_1 state grows up with two time constants of 2.8 ps (62%) and 28 ps (38%) in Fig. 29.10b.

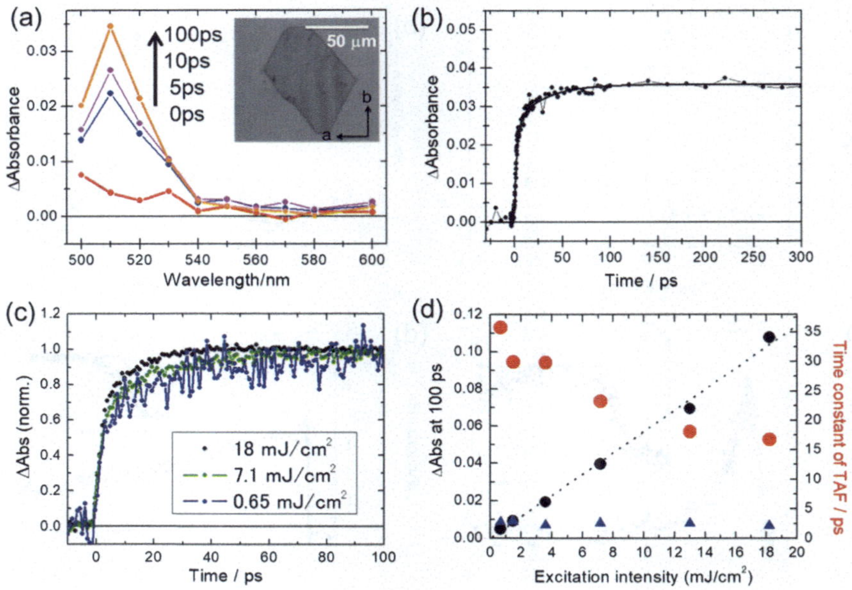

Fig. 29.10 **a** Transient absorption spectra of rubrene single microcrystal at four delay times of 0, 5 10, and 100 ps. Inset depicts the optical image of the target microcrystal with a thickness of 2.1 μm. The polarization of the pump light was set along the b-axis of the crystal. **b** Time profile of the transient absorbance due to the triplet absorption at 510 nm. Laser fluence was 7.1 mJ cm^{-2} $pulse^{-1}$. **c** Excitation laser fluence dependence of time profiles at 510 nm. The results are normalized by the value of the absorbance at 100 ps. **d** Relationship between the two time constants (blue: DF and red: TAF) and excitation intensity, together with excitation intensity-dependent transient absorbance at 100 ps (the amount of the triplet state molecules). Dotted line is the guide for the eyes. Reproduced from Ref. [29] with permission from the European Society for Photobiology, the European Photochemistry Association, and the Royal Society of Chemistry

The shorter time constant corresponds to the triplet state formation via the DF, and the slower one of a few tens of ps time is assigned to the TAF [28].

Next, we measured the time profiles at 510 nm at different laser fluences of 0.65–18 mJ cm^{-2} $pulse^{-1}$ in Fig. 29.10c. From the curve fit analysis with a double-exponential function for each time trace, the fast component of 2.5 ± 0.3 ps due to DF was independent of the laser fluences, while the second component ranged from 35 to 17 ps with increasing in the fluence, and the contribution (36 ± 5%) was independent of the fluence in Fig. 29.10d. These results indicate that the time constant of the triplet formation through the TAF is exclusively influenced by the excitation laser fluence. We considered the reason why S_1 state lifetime was shortened with the laser fluence to be singlet–singlet annihilation process, and then evaluated the concentration of the excited molecules (cm^{-3}), M_{ex} with the following equation.

$$M_{ex} = \varepsilon \times \rho \times N_p \tag{29.4}$$

Here, ε, ρ, and N_p represent the molar absorption coefficient, 1.0×10^3 (M^{-1} cm^{-1}) at the excitation wavelength, the molecular density, 2.21 (M) and the number of photons at the excitation area (cm^{-2}), respectively. At the fluence of 18 mJ cm^{-2} $pulse^{-1}$, the M_{ex} was calculated to be 8.0×10^{19}. Considering the concentration of the ground-state molecules (molecules cm^{-3}) ($M_{gr} = \rho \times N_A \times 10^{-3}$), N_A is Avogadro constant), 6% of the ground-state molecules are excited simultaneously by single shot of the femtosecond excitation laser pulse. This value, which showed three times higher as compared to the reported one by Ma et al. [29], could be enough to consider singlet–singlet annihilation process. When the annihilation process takes place together with the singlet fission, the yield of T_1 state formation should tend to decrease, resulting in the saturation tendency of the transient absorbance of the T_1 state at 510 nm against the laser fluence. The singlet–singlet annihilation process leads to the formation of a higher excited-state molecule and a ground-state one; $S_1 + S_1 \rightarrow S_N + S_0$, and the S_N state molecule could generate two T_1 state molecules through the singlet fission. Consequently, two S_1 state molecules are converted to two T_1 state molecules. Through the annihilation process , the amount of the T_1 state molecules decreases when S_1-S_1 annihilation takes places with the normal singlet fission. However, the experimental result in Fig. 29.10d presents a linear relationship between the excitation intensity and the amount of the T_1 state. Here, because both singlet fissions finished in 100 ps, transient absorbance at 100 ps was used as the amount of the T_1 state. From this result, we concluded that the effect of the singlet–singlet annihilation on the singlet fission processes could be negligible in the present experiments, and it is no important to consider the mechanism of the excitation intensity-dependent time constant of the T_1 state formation.

As another possible mechanism to explain the excitation laser fluence dependence of the TAF, the influence of transient local heating in a ps time scale is considered in Fig. 29.11a, b. As stated above, rubrene molecules are excited to the higher vibrational levels of the S_1 state by the 397-nm femtosecond excitation and relax quickly to lower levels of the S_1 state through the vibrational relaxation process. This indicates that the photothermal conversion will induce temperature elevation at the excitation volume of the rubrene microcrystal in a few ps after excitation, leading to the acceleration of the triplet state formation through the TAF. So, we assume that the excess energy of the photoexcitation (397 nm) from $E(S_1)$ converts into heat in a few ps and that the heat dissipation is ignored, and estimated the local temperature elevation (ΔT) at focal spot of the excitation laser by using Eq. (29.2).

$$\Delta T = \kappa/(\rho \times C) \times F \times (h\nu_{397\,\mathrm{nm}} - E(S_1))/(h\nu_{ex}) \quad (29.5)$$

Here, F, ρ, and C are the laser fluence (J cm^{-2}), the molecular density (M), and the heat capacity, 609 (J K^{-1} mol^{-1}), respectively. κ is the absorption coefficient, which is the inverse of skin depth of the excitation light. The skin depth is determined to be approximately 1.9 μm from the ε and the thickness. $h\nu_{397\,\mathrm{nm}}$ is the photon energy of the excitation laser and set to be 3.12 eV and $E(S_1)$ is the 0–0 transition energy of rubrene crystal (2.23 eV). The value of ΔT was estimated to be 19 K at the laser fluence of 18 mJ cm^{-2} $pulse^{-1}$. The temperature elevation will be enough to

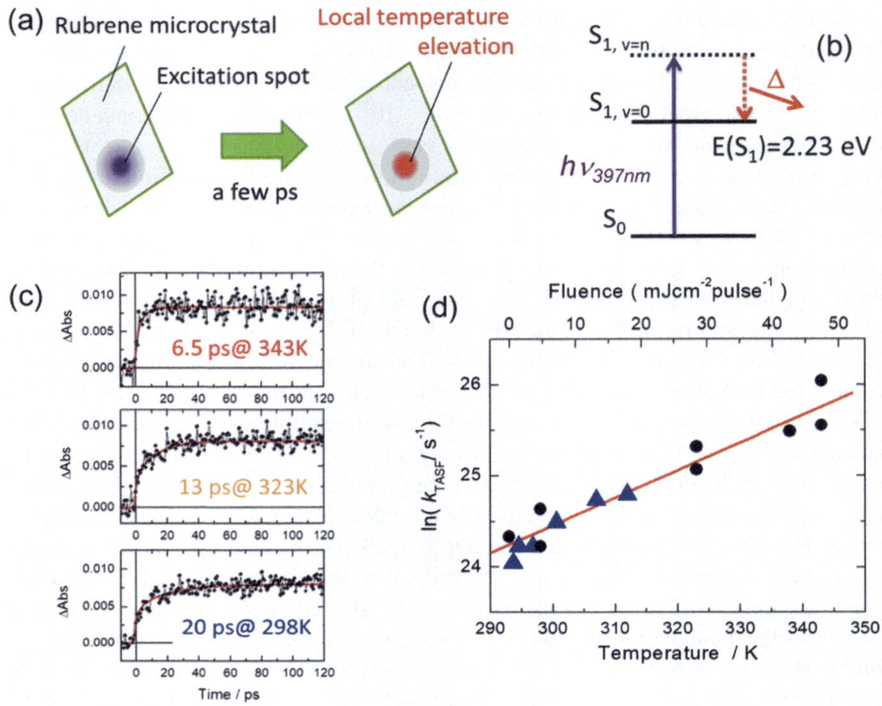

Fig. 29.11 **a** Illustration of local temperature elevation via rapid photothermal conversion in single rubrene microcrystal and **b** its schematic diagram upon a femtosecond 397-nm excitation pulse. **c** Temperature dependence of the time profiles of the transient absorbance at 510 nm. Laser fluence was 0.64 mJ cm^{-2} $pulse^{-1}$. Solid lines are the convolution curves calculated with a double-exponential function and a temporal response function. **d** Relationship between the rate constant of the triplet state formation via TAF and the temperature, together with the relation between 293 K + ΔT calculated by Eq. (29.5) in the text and the rate constant at each excitation fluence on the basis of the excitation fluenece dependence. Solid line is the guide for the eyes. Reproduced from Ref. [29] with permission from the European Society for Photobiology, the European Photochemistry Association, and the Royal Society of Chemistry

accelerate the triplet state formation through the TAF, because the energy difference between $E(S_1)$ and $2 \times E(T_1)$ is very small (–0.05 eV) in rubrene crystal.

In order to clarify the effect of the local temperature elevation, we also examined the temperature dependence of the triplet state formation at a weak excitation intensity of 0.65 mJ cm^{-2} $pulse^{-1}$ to avoid the influence of laser heating. In this experiment, we used a temperature-controlled stage which changes the sample temperature in the range of 293–343 K. Figure 29.10c clearly show that the formation time of the T_1 state became shorter (from 30 to 5 ps) with the increase of the temperature from 293–343 K. Figure 29.10d presents the temperature dependence of the rate constant and the relationship between the transient temperature, 293 K + ΔT, calculated by Eq. 29.5 and the rate constants at each excitation intensity. The excitation intensity dependence was in good agreement with temperature dependence. Therefore, we can safely conclude that transient local heating in a ps time scale owing to the nonradiative vibrational relaxation from higher vibrational level to lower ones in the S_1 state will accelerate the triplet state formation through the TAF.

29.5 Conclusion

We described the novel femtosecond pump-probe confocal microspectroscopic system and its application to single organic crystals having the size of nm to μm. By using normal transmitted light as a probe light, we demonstrated the excited-state dynamics of individual μm-sized crystals such as CT complex and rubrene [29]. On the other hand, back-scattering light detection made it possible to observe transient signals of single organic nanocrystals which size was smaller than the diffraction limit of the visible light. We succeeded in the measurement of the excimer formation of α-phase perylene single nanocrystal and found out the size-dependent excimer formation dynamics in a ps time scale [13]. We expect our microspectroscopic system can open new photophysical processes with nanometer size and picosecond time scale for organic solids.

Acknowledgements This work was supported by JSPS KAKENHI Grant Number JP26107011 to T. A. in Scientific Research on Innovative Areas “Photosynergetics”, and 19K05211 to Y. I. The authors would like to thank Mr. O. Udaka, Mr. Y. Inoue, Mr. R. Kawasaki, Mrs. M. Murakami, and Mr. R. Kihara who helped us directly in obtaining our experimental results in this project.

References

1. Asahi T, Sugiyama T, Masuhara H (2008) Laser fabrication and spectroscopy of organic nanoparticles. Acc Chem Res 41:1790–1798
2. Mei J, Leung NL, Kwok RT, Lam JW, Tang BZ (2015) Aggregation-induced emission: together we shine, united we soar! Chem Rev 115:11718–11940
3. Asahi T, Suzuki M, Masuhara H (2002) Cooperative photochemical reaction in molecular crystal induced by intense femtosecond laser excitation: photochromism of spironaphthooxazine. J Phys Chem A 106:2335–2340
4. Suzuki M, Asahi T, Masuhara H (2005) Cooperative photochemical reaction mechanism of femtosecond laser-induced photocoloration in spirooxazine microcrystals. Chem Phys Chem 6:2396–2403
5. Doan SC, Kuzmanich G, Gard MN, Garcia-Garibay MA, Schwartz BJ (2012) Ultrafast spectroscopic observation of a quantum chain reaction: the photodecarbonylation of nanocrystalline diphenylcyclopropenone. J Phys Chem Lett 3:81–86
6. Itoh T, Asahi T, Masuhara H (2001) Femtosecond light scattering spectroscopy of single gold nanoparticles. Appl Phys Lett 79:1667–1669
7. Cating EEM, Pinion CW, Christesen JD, Christie CA, Grumstrup EM, Cahoon JF, Papanikolas JM (2017) Probing intrawire, interwire, and diameter-dependent variations in silicon nanowire surface trap density with pump-probe microscopy. Nano Lett 17:5956–5961

8. Muskens OL, Del Fatti N, Vallée F (2006) Femtosecond response of a single metal nanoparticle. Nano Lett 6:552–556
9. Hartland GV (2010) Ultrafast studies of single semiconductor and metal nanostructures through transient absorption microscopy. Chem Sci 1:303
10. Grumstrup EM, Gabriel MM, Cating EM, Pinion CW, Christesen JD, Kirschbrown JR, Vallorz EL, Cahoon JF, Papanikolas JM (2014) Ultrafast carrier dynamics in individual silicon nanowires: characterization of diameter-dependent carrier lifetime and surface recombination with pump-probe microscopy. J Phys Chem C 118:8634–8640
11. Fischer MC, Wilson JW, Robles FE, Warren WS (2016) Pump-probe microscopy. Rev Sci Instrum 87:031101
12. Van Goethem EM, Pinion CW, Cating EEM, Cahoon JF, Papanikolas JM (2019) Observation of phonon propagation in germanium nanowires using femtosecond pump-probe microscopy. ACS Photon 6:2213–2222
13. Ishibashi Y, Asahi T (2016) Femtosecond pump-probe microspectroscopy of single perylene nanoparticles. J Phys Chem Lett 7:2916–2951
14. van Dijk MA, Tchebotareva AL, Orrit M, Lippitz M, Berciaud S, Lasne D, Cognet L, Lounis B (2006) Absorption and scattering microscopy of single metal nanoparticles. Phys Chem Chem Phys 8:3486–3495
15. Cohen MD, Haberkorn R, Huler E, Ludmer Z, Michel-Beyerle ME, Rabinovich D, Sharon R, Warshel A, Yakhot V (1978) Pair-states in α-perylene crystal. A theoretical study. Chem Phys 27:211–216
16. Nishimura H, Yamaoka T, Mizuno K-I, Iemura M, Matsui A (1984) Luminescene of free and self-trapped excitons in α- and β-perylene crystals. J Phys Soc Jpn 53:3999–4008
17. Walker B, Port H, Wolf HC (1985) The two-step excimer formation in perylene crystals. Chem Phys 92:177–185
18. Furube A, Murai M, Tamaki Y, Watanabe S, Katoh R (2006) Effect of aggregation on the excited-state electronic structure of perylene studied by transient absorption spectroscopy. J Phys Chem A 110:6465–6471
19. Mizuno K, Furukawa M, Matsui A, Tamai N, Yamazaki I (1988) Double- to triplet-minima change in the adiabatic potential energy curve and exciton relaxation dynamics in α-perylene crystals. In: Ultrafast phenomena VI, Springer series in chemical physics, vol 48. Springer, New York, pp 492–494
20. Ma L, Tan KJ, Jiang H, Kloc C, Michel-Beyerle ME, Gurzadyan GG (2014) Excited-state dynamics in an alpha-perylene single crystal: two-photon- and consecutive two-quantum-induced singlet fission. J Phys Chem A 118:838–843
21. Asahi T, Matsune H, Yamashita K, Masuhara H, Kasai H, Nakanishi H (2008) Size-dependent fluorescence spectra of individual perylene nanocrystals studied by far-field fluorescence microspectroscopy coupled with atomic force microscope observation. Polish J Chem 82:687–699
22. Asahi T, Tamai N, Uchida T, Shimo N, Masuhara H (1995) Nonlinear excited-state dynamics of a thin copper phthalocyanine film by femtosecond transient grating spectroscopy. Chem Phys Lett 234:337–342
23. Gulbinas V, Chachisvilis M, Valkunas L, Sundström V (1996) Excited state dynamics of phthalocyanine films. J Phys Chem 100:2213–2219
24. Ma Y-Z, Xiao K, Shaw RW (2012) Exciton-exciton annihilation in copper-phthalocyanine single-crystal nanowires. J Phys Chem C 116:21588–21593
25. Kihara R, Imada S, Kawai T, Asahi T (2019) Fabrication of nanorods colloids of copper hexafluorophthalocyanine by nanosecond-pulse laser fragmentation in organic solvents. App Surf Sci 478:532–538
26. Asahi T, Matsuo Y, Masuhara H, Koshima H (1997) Electronic structure and dynamics of the excited state in ct microcrystals as revealed by femtosecond diffuse reflectance spectroscopy. J Phys Chem A 101:612–616
27. Smith MB, Michl J (2010) Singlet fission. Chem Rev 110:6891–6936

28. Ma L, Zhang K, Kloc C, Sun H, Michel-Beyerle ME, Gurzadyan GG (2012) Singlet fission in rubrene single crystal: direct observation by femtosecond pump-probe spectroscopy. Phys Chem Chem Phys 14:8307–8312
29. Ishibashi Y, Inoue Y, Asahi T (2016) The excitation intensity dependence of singlet fission dynamics of a rubrene microcrystal studied by femtosecond transient microspectroscopy. Photochem Photobiol Sci 15:1304–1309

Chapter 30
Single-Molecule Level Study and Control of Collective Photoresponse in Molecular Complexes and Related Systems

Martin Vacha and Shuzo Hirata

Abstract We studied nanoscale photosynergetic response of molecular complexes and related inorganic and hybrid materials using single-molecule detection and spectroscopy. As examples, we report simultaneous electroluminescence and photoluminescence study of single $CsPbBr_3$ perovskite nanocrystals which revealed the origin of blinking and of reduced quantum efficiency of electroluminescence devices. In addition, we observed linear and quadratic Stark shift on individual nanocrystals in different matrices. For novel I-III-IV semiconductor quantum dots, defect emission was found to originate from multiple sites in one particle, and the phenomenon of blinking was suppressed by multi-color excitation. In hybrid metal nanoparticle-organic dye systems, selective excitation of a localized plasmon in single gold nanorods by the polarization of light was found to enhance Förster energy transfer efficiency by two orders of magnitude. Localized plasmon was also found to cause previously unknown phenomenon of enhancement of triplet Dexter transfer in photon upconversion materials. Nanoscale study of triplet exciton diffusion in molecular solids by visualization of upconversion emission helped to uncover heterogeneity and the role of molecular orientation in the diffusion process.

Keywords Single-molecule spectroscopy · Electroluminescence · Perovskite nanocrystals · Quantum dots · Plasmon enhancement · Energy transfer

30.1 Introduction

Collective photoresponse of conjugated molecular complexes is an important topic both from the point of basic knowledge of molecular photophysics and from the point of potential applications as photo-driven nanoscale devices. Over the past several

M. Vacha (✉)
Department of Materials Science and Engineering, Tokyo Institute of Technology, Ookayama 2-12-1-S8-44, Meguro-ku, 152-8552 Tokyo, Japan
e-mail: vacha.m.aa@m.titech.ac.jp

S. Hirata
Department of Engineering Science and Engineering, The University of Electro Communications, 1-5-1 Chofugaoka, 182-8585 Chofu, Tokyo, Japan

H. Miyasaka et al. (eds.), *Photosynergetic Responses in Molecules and Molecular Aggregates*, https://doi.org/10.1007/978-981-15-5451-3_30

years, we have used single-molecule spectroscopy [1, 2] to study the response of molecular complexes, such as molecular films or π-conjugated polymers, and their inorganic counterparts, semiconducting quantum dots or halide perovskite nanocrystals, as well as hybrid systems composed of noble metal nanoparticles and organic dyes [3–18]. Using different modes of electronic excitation, i.e., linear absorption, localized plasmon enhancement, energy transfer, or charge recombination, we investigate the effect that each excitation mode has on the resulting excited state and its dynamics. We also attempt to actively manipulate the excited states by external stimuli such as electric field or external nanoscale pressure induced by a tip of a scanning probe microscope. In this chapter, we introduce several examples of this photosynergetic research. In Sect. 30.2, we present our recent work on blinking and electroluminescence of single perovskite nanocrystals and I-III-IV semiconductor quantum dots. Section 30.3 introduces the enhancement of energy transfer by localized plasmons of metal nanostructures in hybrid materials. Section 30.4 is devoted to novel microscopic studies of triplet exciton diffusion in molecular solids.

30.2 Blinking and Electroluminescence of Single Perovskite Nanocrystals and I-III-IV Semiconductor Quantum Dots

30.2.1 Electroluminescence and Photoluminescence of Aggregates of Halide Perovskite Nanocrystals

In recent years, halide perovskites have emerged as a promising alternative material for light absorption and charge separation in solar cells [19, 20]. Their photophysical properties, such as large absorption cross section and narrow tunable photoluminescence (PL), have also lead to extensive research aimed at optoelectronic applications, including light-emitting diodes for displays and lighting, or lasers [21, 22]. Especially, confinement of charges in nanoscale-size structures of different dimensionality results in drastic improvement of PL quantum yield (QY) and overall brightness, and perovskite nanocrystals (NCs) have found use as emitters in light-emitting diodes (LEDs). The advantage of their use is that many aspects of the existing technologies developed for organic LEDs can be applied for dispersions of perovskite NCs. However, electroluminescence devices prepared from such NCs have long suffered from low efficiency and there has been no systematic study on the nanoscale origin of the poor efficiencies.

The light-emitting layer in perovskite LEDs typically consists of several tens of nanometers of densely packed perovskite NCs. Such emitting layer is sandwiched between two electrodes and is under external DC bias of several volts. Under such conditions, the photophysical and electrical processes occurring in the emitting layer can be very complex. The excitonic and electronic interactions depend on NC sizes and their distributions, NC distances and orientations, surface and environment of the

NCs, etc. To examine these phenomena on model systems of controllable size, we use the method of single-molecule/particle spectroscopy to study well-defined aggregates of NCs of the perovskite $CsPbBr_3$. We compare PL and electroluminescence (EL) from the same aggregates in a microscopic EL device to get insight into the processes of charge and energy transport between individual NCs within the aggregate [3].

To compare PL and EL on the same single aggregates of $CsPbBr_3$, we prepared EL devices directly on microscope coverslips for use on top of a high N.A. fluorescence microscope. The device has the structure of ITO/PEDOT:PSS/PVK:$CsPbBr_3$:PBD/TPBi/LiF/Al and uses as the emissive layer a spin-coated film of PVK and PBD into which $CsPbBr_3$ NCs are doped. The $CsPbBr_3$ NCs have cube-like shape and average edge size of 16 nm, and their surface is passivated with oleic acid and oleylamine ligands. The self-assembled aggregates have sizes on the order of 100 nm, and each aggregate contains tens to hundreds of individual perovskite NCs.

Results of the simultaneous PL and EL microscopic imaging are shown in Fig. 30.1. The same aggregates appear in both the PL (Fig. 30.1a) and EL (Fig. 30.1b) images but most of them with different intensity. We have shown recently that the excitation rates in both PL and EL emission processes are comparable. The differences in the emission intensities in Fig. 30.1 thus originate from differences in PL and EL QY, due to different fractions of charged NCs in each aggregate. An important finding is that the individual aggregates show the dependence on their emission intensities on time (time traces) in PL and EL. In PL, the emission intensity is constant, as shown in Fig. 30.1c. In contrast, the EL intensity time trace shows strong fluctuations (blinking) in the form of intensity spikes starting from the background level (Fig. 30.1d). Generally, individual NCs do exhibit blinking in PL under ambient conditions. Therefore, the fact that we do not observe any blinking for the aggregates in PL implies that all NCs within the aggregate behave as isolated units and absorb and emit independently, as shown schematically in Fig. 30.1e. The blinking of the individual NCs is averaged out over the aggregate, and only constant PL intensity is observed.

The strong blinking in EL, on the other hand, indicates that only a small number of NCs are actively emitting EL. This has been confirmed by super-resolution analysis of the blinking images. An example of such analysis is shown in Fig. 30.1f, where three distinct emitting spots are evident. On average 3–7 emission spots are EL active in each aggregate, and these spots emit consequentially, i.e., at each time the emission is coming only from one single NC. This fact helps to explain the strong blinking observed in the EL. Further, there is a size distribution between individual NCs within each aggregate, as illustrated in the TEM image in Fig. 30.1g. Due to the quantum confinement effect, the largest NCs within the aggregate have the smallest bandgap and work as traps for charges captured originally at the surrounding NCs. The conductive environment present between the NCs enables efficient migration of the charges to the traps (the largest NCs) from where the EL takes place, as shown schematically in Fig. 30.1h. This phenomenon of charge funneling is thus an important aspect of EL devices composed of densely packed perovskite NCs, especially in the presence of size distribution.

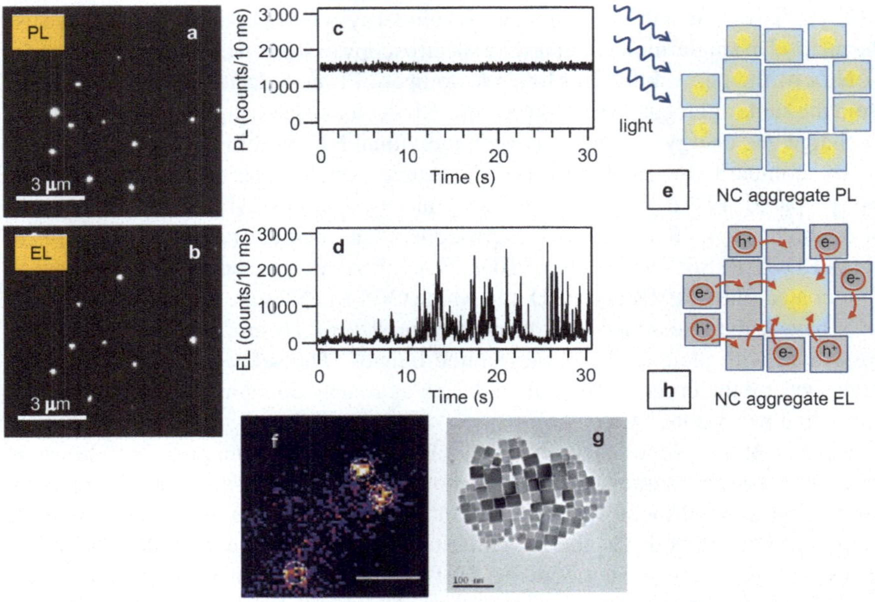

Fig. 30.1 **a** Microscopic PL image of aggregates of $CsPbBr_3$ NCs in EL device; **b** microscopic EL image of aggregates of $CsPbBr_3$ NCs in EL device; **c** intensity time trace of PL of a single aggregate of $CsPbBr_3$ NCs in EL device, excited at 485 nm; **d** intensity time trace of EL of the same single aggregate of $CsPbBr_3$ NCs in EL device, at the bias of 14 V; **e** schematic illustration of PL from an aggregate showing independent emission from individual NCs; **f** super-resolution EL image obtained by analysis of a series of microscopic EL images of aggregates of $CsPbBr_3$ NCs at 14 V over the time of 60 s; scale bar 100 nm; **g** typical TEM image of an aggregate of $CsPbBr_3$ NCs; **h** schematic illustration of EL from an aggregate showing charge migration (red arrows) and emission from the largest NC which works as a trap

It is interesting to note that the distribution of energy bandgaps does not result in the phenomenon of energy (exciton) funneling in PL, and each NC emits individually. Such exciton funneling has been observed extensively in multichromophoric systems, such as, e.g., conjugated polymers [1]. The exciton funneling is driven by dipole–dipole interactions, and prerequisites for efficient exciton transport are spectral overlap between the absorption and emission spectra, and appropriate distance and orientation of the transition dipoles. In the perovskite NC aggregate, some of these prerequisites are not met. While there is a good spectral overlap due to small Stokes shift in $CsPbBr_3$ NCs, the overall large size and the exiting gaps between individual NCs mean that center-to-center distance between two NCs is on average >16 nm. Assuming that the exciton in an NC can be approximated by a single point dipole, the dipole-to-dipole distance of more than 16 nm is far beyond the distance required for efficient energy transfer.

Further information can be obtained from the analysis of the integrated emission intensities in PL and EL over the measurement interval of the time traces, such as the ones in Fig. 30.1c, d. These integrated intensities reflect the total number of photons emitted during the time interval. Since the excitation rates are comparable in PL and EL, the differences in the integrated intensities (number of emitted photons) directly reflect differences between the intrinsic QY in PL and EL. A ratio of integrated EL to integrated PL intensity shows a large distribution which spans a range between 0.12 and 0.66, with the average EL/PL ratio of 0.36. We have previously measured the PLQY of $CsPbBr_3$ NCs dispersed in PVK on ensemble level and found a value of 0.3, which leads to the ELQY value of 10.8%. Assuming further a typical outcoupling efficiency of a LED of 20% gives the external ELQY of about 2%. We can thus conclude that the blinking phenomenon is a crucial factor that contributes to the lower efficiency in electroluminescence of perovskite LED devices.

30.2.2 Stark Effect and Environment Induced Modulation of Emission in Single Halide Perovskite Nanocrystals

In order to get further insight into the blinking phenomenon that is prominently present in the electroluminescence measurements, we systematically examined different external factors that can influence the emission characteristics of $CsPbBr_3$ nanocrystals in working LED devices. In particular, we investigated the impact of an environment and of externally applied electric field on PL characteristics of single $CsPbBr_3$ NCs by performing steady-state and time-resolved single-particle PL imaging and spectroscopy in inert and conductive environments [4]. We used a hole-conducting polymer polyvinyl carbazole (PVK) as the conductive matrix and polymethyl methacrylate (PMMA) as the insulating matrix, respectively, and applied electric field in a capacitance-like device. Also, we studied field-dependent changes in PL peak position and discovered a combination of linear and quadratic Stark effect, which was attributed to polymer matrix-induced polarization.

Single-particle PL imaging and spectroscopy of $CsPbBr_3$ NCs were performed in the polymer matrices of PMMA and PVK. In the time-averaged single-particle images, individual $CsPbBr_3$ appears brighter in PMMA as compared to PVK (Fig. 30.2a, b). Intensity time trajectories of individual particles were monitored to understand this apparent decrease of PL efficiency in the conductive PVK as compared to the inert PMMA. The intensity fluctuations in PMMA mostly show discrete two-state blinking (Fig. 30.2c), i.e., transitions between ON states (high-intensity) and OFF states (at the background level). The intensity traces in PVK, on contrary, show frequent intermediate intensity levels, so-called GRAY states, in addition to the ON and OFF states (Fig. 30.2d). The difference in average intensity (brightness) of individual $CsPbBr_3$ in the two matrices can be attributed to the frequent population of the GRAY states in PVK as compared to PMMA.

The above observations were further complemented by excited-state lifetime measurements on single-particle level. Excited-state lifetime decay curves of

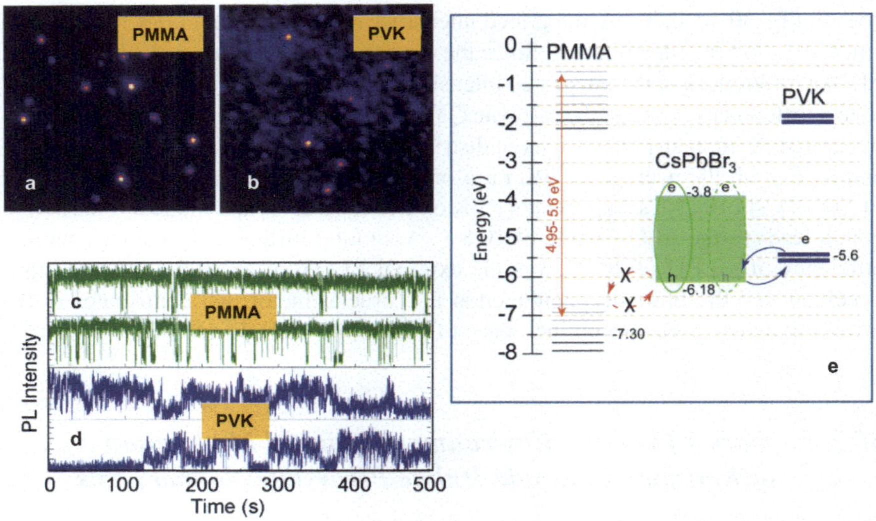

Fig. 30.2 Microscopic photoluminescence (PL) image of single $CsPbBr_3$ nanocrystals in the matrix of **a** PMMA and **b** PVK; PL intensity traces of a single $CsPbBr_3$ nanocrystal in **c** PMMA and **d** PVK; **e** proposed scheme of $CsPbBr_3$ nanocrystal ionization and negative trion formation in the PVK matrix. Reprinted with permission from ACS Nano 13:624–632 (2019). Copyright 2019 American Chemical Society

$CsPbBr_3$ in PMMA obtained as average of several NCs were fitted using two-exponential functions, with the fitting parameters of 4.9 and 25.8 ns. The average lifetime shortened from 13.1 ns in PMMA to 4.8 ns in PVK. In the PMMA matrix, correlation plots between the PL intensity and lifetime for an individual particle were constructed from time-resolved time-tagged imaging (TTTR). The data showed that the shorter lifetime component is correlated with the intermediate GRAY states, and the longer component corresponds to the unperturbed ON emission.

To summarize the observations, we propose a model shown in Fig. 30.2e. Due to the proximity of the energy level of HOMO of PVK (~5.6 eV) and that of the V.B. of $CsPbBr_3$ (~6.2 eV), hole transfer can occur from the excited $CsPbBr_3$ to the PVK. This leads to enhanced photoionization of the $CsPbBr_3$ in the conductive matrix. The hole in the PVK matrix can migrate away from the NC which stabilizes the negative charge on the $CsPbBr_3$ NC. In the PMMA matrix, the larger energy barrier between the $CsPbBr_3$ V.B. and the HOMO of PMMA (>7.2 eV) makes the charge transfer energetically more restricted. We conclude that the observed GRAY states in the PVK matrix originate from the emission of negatively charged excitons (negative trions).

Apart from the effect of the conductive matrix, we examined the effect of applied electric field on the PL intensity and spectra of single $CsPbBr_3$ NCs in the PMMA matrix. The NCs are dispersed in the PMMA film which is sandwiched between ITO and Al electrodes. Under applied electric field, the $CsPbBr_3$ NC blinking pattern as well as the average PL intensity changed significantly. Under a DC field (of +8 V or +0.4 MV/cm) in the PMMA matrix, the intensity of nearly half of the particles examined partially quenched to intermediate (GRAY) levels, with the remaining particles irreversibly quenching to a background OFF level after a few seconds. An example of a PL intensity time trace in Fig. 30.3 shows that the ON intensity level of this NC at zero applied field (Fig. 30.3a) was either suppressed to about 50% to form the GRAY levels or appeared at same ON level but with frequent OFF times under the constant applied DC field of +8 V (Fig. 30.3b).

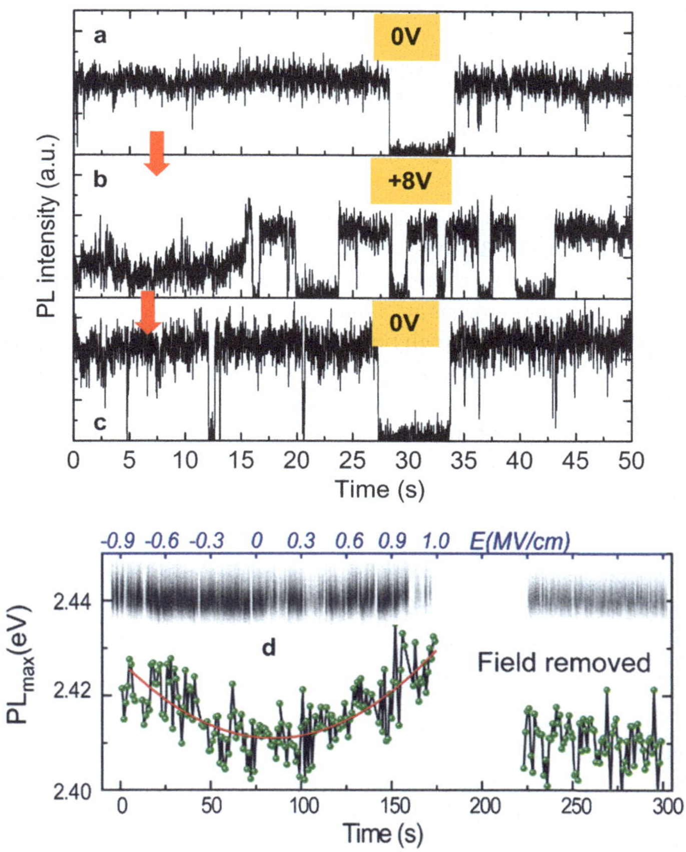

Fig. 30.3 Single-particle PL intensity traces before applying electric field (**a**) with an applied electric field of +8 V (+0.4 MV/cm) (**b**) and after removing the field (**c**) for a single $CsPbBr_3$ NC in PMMA; **d** PL peak maxima (PL_{max}) of a single $CsPbBr_3$ NC in PMMA upon external electric field continuously increasing from – 0.9 to +0.9 MV/cm, and recovery of PL_{max} after the field removal. The data points were analyzed by Gaussian fits of single NC spectra from respective spectral images shown on the top. The red line represents a fit to a combination of linear and quadratic equations. Reprinted with permission from ACS Nano 13:624–632 (2019). Copyright 2019 American Chemical Society

This behavior was reversible and the initial state recovered after removal of the electric field, as shown in Fig. 30.3c. Overall, about 45% of the $CsPbBr_3$ NCs in the PMMA matrix under the influence of the applied electric field show blinking dynamics which is similar to that of $CsPbBr_3$ in PVK under no electric field. This result might reflect electric field-enhanced charging–discharging of $CsPbBr_3$ NCs in PMMA. To summarize, externally applied electric field modifies the PL intensity of single $CsPbBr_3$ NCs in a similar way as the presence of a conductive matrix.

The electric field can potentially modulate also the PL energy, causing the so-called Stark shift. The dependence of a PL spectrum on continuously applied DC electric field within the range of -1.0 and $+1$ MV/cm is plotted on the top of Fig. 30.3d in the form of a spectral image. PL spectra for individual applied voltage values were analyzed by Gaussian fitting, and the peaks are plotted in the bottom of Fig. 30.3d. The observed PL peak energy changes from 2.405 to 2.430 eV within the applied voltage range, and the effect fully reversible. As also seen in Fig. 30.3d, removal of the electric field causes the PL peak to reverse to energy observed at 0 MV/cm during the sweep. The recovery of the PL peak position shows that the observed phenomenon is a reversible effect due to the electric field and is not affected by the degradation of $CsPbBr_3$ NCs during the acquisition of the data. This field-dependent PL emission energy confirms the presence of a DC field-induced, nonlinear Stark effect. Fitting of the experimental data to a combination of a linear and quadratic equations (Fig. 30.3d) indicates both polar and polarizable characters of the electric field-induced phenomenon. Appearance of a linear component of the Stark effect in inorganic halides nanocrystals such as $CsPbX_3$ is generally not expected because of symmetry considerations and the presumed lack of a permanent dipole. However, interactions between $CsPbBr_3$ NCs and the PMMA matrix may cause anisotropic distributions of charges in $CsPbBr_3$ NCs, resulting in a polar character.

30.2.3 Single-Particle Spectroscopy of I-III-VI Semiconductor Nanocrystals: Spectral Diffusion and Suppression of Blinking

Nanocrystals of ternary I-III-VI semiconductors with a direct bandgap are promising materials for various applications due to their excellent optoelectronic properties [23]. Tunable band gap, large extinction coefficient, and low toxicity make them potential alternatives to II–VI semiconductor quantum dots (QD). However, large photoluminescence spectral width and moderate brightness have long restricted their practical use. While it has been recently shown that the large bandwidth can be overcome by capping with an appropriate shell, detailed knowledge of the physical origin of the spectral broadening of the defect-related emission is indispensable for further material development. Here we use single-particle PL spectroscopy on nanocrystals of the composition $(AgIn)_xZn_{2(1-x)}S_2$ to study the origin of the spectral broadening and PL blinking [16].

For the single-particle characterization, we used $(AgIn)_xZn_{2(1-x)}S_2$ NCs with three different compositions that vary in the amount of Zn, i.e., $x = 0.3$, 0.7 and 1.0, respectively. With increasing Zn content, the solution absorption and PL spectra exhibit a blue shift from 1.50 to 2.06 eV, with linewidths between 415 and 440 meV. Single-particle PL spectroscopy reveals that both the transition energies and spectral shapes vary from particle to particle, confirming the inhomogeneous nature of the ensemble PL band. In particular, the single NC spectral linewidths (FWHM) of 230–300 meV are narrower than the ensemble PL for all three compositions but still reach 55–75% of the ensemble PL spectral widths. These large homogeneous linewidths are more than three times broader than those of II-VI semiconductor (CdSe) NCs. To get insight into the homogeneous broadening, we measured time evolution of single NC PL spectra. The spectra of most NCs are dynamic in nature and show spectral diffusion, i.e., random changes of PL positions and linewidths in either spectral direction, such as the example in Fig. 30.4a. The extent of the spectral diffusion can reach up to 46 meV, making the spectral diffusion a significant contribution to the overall PL spectrum lineshape.

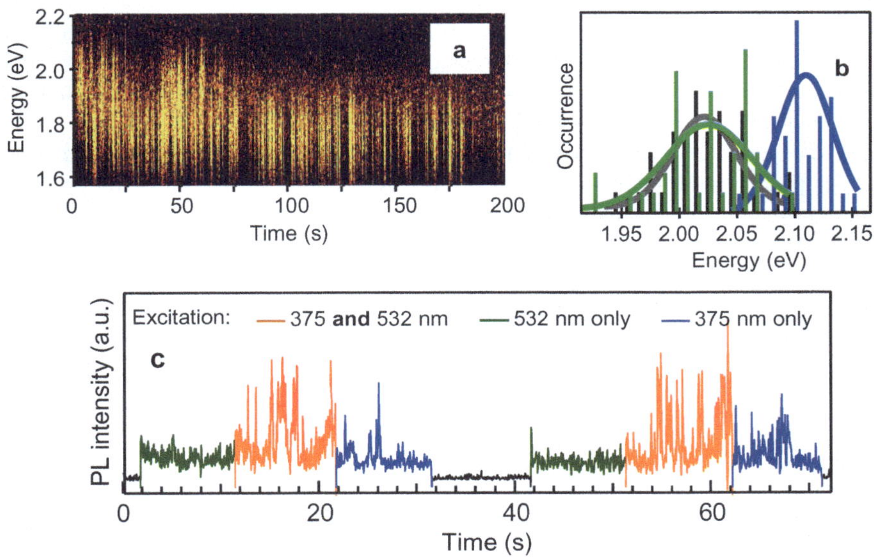

Fig. 30.4 **a** Time evolution of PL spectra (2D spectral plot of PL emission energy vs. time) for one NC of the $x = 0.7$ composition showing spectral diffusion; **b** distributions of single-particle PL positions and their Gaussian fits for spectra appearing upon both 375 and 532 nm excitations (green, black) and for spectra appearing upon 375 nm excitation only (blue). Green represents the 532 nm excitation, black the 375 nm excitation; **c** Samples of PL intensity blinking traces for two single NCs excited sequentially with 532 nm only (green), both 375 and 532 nm (orange) and 375 nm only (blue) lasers, corrected for background differences. Reproduced from Ref. [16] with permission from The Royal Society of Chemistry

Spectral diffusion can originate from multiple defect levels of different energies within one particle. Activation and deactivation of such defect states would lead to spectral changes and blinking. To selectively excite the possible different defects, we used alternately two lasers, one (375 nm) with energy above the bandgap E_g, the other (532 nm) slightly below E_g to measure PL spectra on the same single NC ($x = 0.3$). The spectra that appear upon both excitations have maxima below 2.06 eV, whereas the spectra that appear only upon the 375 nm excitation have maxima around 2.10 eV. Such excitation-dependent differences in PL energy are shown for statistical ensemble of single NCs in distributions NC spectral positions in Fig. 30.4b. The difference in the PL distributions points to the coexistence of two distinct defect states in individual NCs whose recombination energy differs on average by 85 meV. Such multiple defect states within single NCs can be a significant factor contributing to the large homogeneous width in the NCs.

Apart from the differences in PL spectra, excitation at different energies also leads to differences in blinking behavior. To study the blinking phenomenon, we used sequential excitation with 532 nm laser only (10 s), 532 nm and 375 nm lasers simultaneously (10 s), and 375 nm laser only (10 s) and measured PL intensity traces. An example of such traces shown in Fig. 30.4c confirms that the ON times are most frequent during the simultaneous irradiation with both lasers. Statistical analysis of PL intensity obtained with the different excitations on the same NCs shows that the simultaneous irradiation leads on average to 1.85 times more emitted photons compared to each of the individual irradiations.

The difference between the ON and OFF time response to the different irradiation modes points to the involvement of multiple defect states in the emission process. Considering the two coexisting defect states mentioned above, one such state (lower in energy) can be temporarily in the form of a long-lived non-radiative trap that quenches the excitons. Filling of such trap state with the lower-energy laser causes the conduction band exciton to relax into the higher energy defect state from where the emission occurs. This process enables optical modulation of PL blinking and leads to an almost twofold increase in brightness in these semiconductor QDs.

30.3 Enhancement of Energy Transfer by Localized Plasmons of Metal Nanostructures in Hybrid Materials

30.3.1 Selective Turn-on and Modulation of Resonant Energy Transfer in Single Plasmonic Nanostructures by the Polarization of Light

Förster resonant energy transfer (FRET) is a non-radiative process based on dipole–dipole interactions, where excitation energy absorbed by a donor molecule is transferred to an acceptor molecule. The efficiency of FRET strongly depends on spectral overlap between the donor emission and the acceptor absorption, the donor-to-acceptor distance, the relative orientation of the two transition dipoles, and the quantum yield and extinction coefficient of the dyes. The donor–acceptor interaction distance R_{DA} is typically on the order of nm and the energy transfer rate scales as ~R_{DA}^{-6}. This strong distance dependence is utilized in applications of FRET as a nanoruler in biological systems at molecular level. FRET is also at the center of photosynthesis where it ensures transport of the absorbed sunlight toward a reaction center, and novel nature-inspired artificial light-harvesting systems are being explored for potential use as light collectors and energy funnels in solar energy conversion systems. FRET is also used in nanosensors or white light-emitting structures.

In many of these applications, the limited range of R_{DA} may pose restrictions on the design and efficiency of the systems. To increase sensitivity of FRET-based nanosensors or improve the performance of FRET-based light-harvesting systems, it is necessary to enhance FRET efficiencies over larger distances. Introducing plasmonic structures in the vicinity of an energy donor and/or energy acceptor can potentially lead to strong increase of FRET rate, efficiency, and range, especially when the spectral range of energy transfer is overlapping with the plasmon resonance of the plasmonic structure [24].

Here, we fabricated hybrid nanostructures composed of FRET dye pairs (donors and acceptors) covalently attached to gold nanorods (GNR), Fig. 30.5a. Plasmon resonances in GNR are highly sensitive to the polarization of the excitation light due to the GNR anisotropic structure. In addition, the advantages of GNR are their ease of manufacture and spectral tunability. We demonstrate on the level of a single nanorod that the energy transfer process can be switched on and off by changing the polarization of the excitation light relative to the long axis of the GNR [14]. Our study therefore provides an approach for controlling and modulating the FRET process between different fluorescent species.

The hybrid nanoparticles are composed of BODIPY TR NHS dyes as donors and BODIPY 650 NHS dye as acceptors. Both are attached at random positions to a GNR via an alkane chain (Fig. 30.5a). Good overlap of the FRET parameters of the dyes with the GNR plasmon resonance is achieved by using GNRs with a longitudinal plasmon at 636 nm which coincides with the emission of the donor at

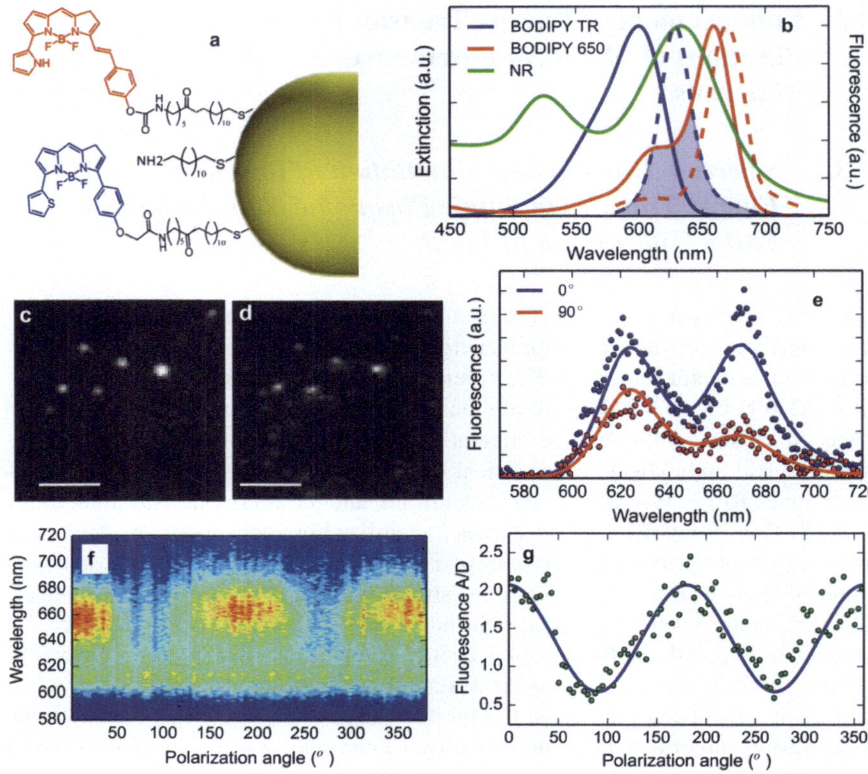

Fig. 30.5 **a** Chemical structures of the donor (blue) and acceptor (red) molecules and attachment to a GNR; **b** solid lines: extinction spectra of donor molecules (BODIPY TR NHS, blue), acceptor molecules (BODIPY 650 NHS, red), and gold nanorods (green). Dashed lines: fluorescence spectra of donor (blue) and acceptor (red) molecules. Overlap between the emission of the donor and absorption of the acceptor is highlighted in blue; **c** dark-field image of single particles of hybrid GNRs; **d** corresponding fluorescence image of the same particles. The scale bars in (**c**, **d**) are 4 μm; **e** symbols—fluorescence spectra of hybrid GNRs excited with 568 nm, for the excitation polarization parallel to the long axis of the nanorod (blue) and perpendicular to it (red). Solid lines—fitting of the experimental data using modified FRET theory; **f** two-dimensional plot of emission spectra from a hybrid GNR as a function of an angle between the polarization of excitation light and the long axis of the nanorod. The spectra are normalized at the donor emission maxima; **g** ratio between the fluorescence intensity of the acceptor and the donor as a function of the angle between the excitation polarization and the GNR long axis. Reproduced from Ref. [14] with permission from The Royal Society of Chemistry

628 nm (Fig. 30.5b). Absorption of the acceptor overlaps both with the longitudinal plasmon and the emission from the donor. There is also at least partial overlap of the plasmon band with the absorption of the donor and fluorescence of the acceptor. The presence of the plasmonic nanoparticle can thus influence not only the FRET process but also the photophysical properties of the individual components of the FRET system.

We studied the effect of localized plasmon on the FRET process by imaging simultaneously dark-field scattering from the GNR and fluorescence from the donor and acceptor molecules attached to the GNR in an optical microscope. From each hybrid GNR, we measured scattering and fluorescence intensity as well as spectra as a function of the polarization of incident light. Dark-field image of hybrid GNRs is shown in Fig. 30.5c, with several bright spots corresponding to scattering from individual particles. The same particles are identifiable in the corresponding fluorescence image in Fig. 30.5d. Orientation of the particles was determined from the polarization dependence of scattering intensity for each GNR. Fluorescence imaging of the GNRs was performed with 568 nm laser which efficiently excites the donor molecules, but also overlaps with the longitudinal plasmon mode. When the excitation light is polarized along the long axis of the GNR, fluorescence spectra from both the donor and acceptor are visible, as shown in Fig. 30.5e (blue dots). The fluorescence intensity of the acceptor is higher than that of the donor in this case. When the excitation polarization is switched to perpendicular to the GNR long axis, the fluorescence intensities of both donor and acceptor drop, but the decrease is more prominent for the acceptor, as seen in the spectra in Fig. 30.5e (red dots). Thus, switching of the polarization of the excitation light by 90° changes the intensity ratio of the donor and acceptor fluorescence, which is a signature feature of changing FRET efficiency. To visualize the selective turn-on of FRET, we plotted the fluorescence spectra as a function of the angle between the polarization of the excitation light and the long axis of the GNR in Fig. 30.5f. Fluorescence intensity from the acceptor molecules increases to a maximum when the polarization is parallel to the long axis of GNR and varies periodically with the polarization angle. Ratio between the fluorescence intensity of the acceptor and the donor (shown in Fig. 30.5g) changes periodically between 2 and 0.5 for the two orthogonal polarizations.

To quantitatively evaluate the results of the fluorescence spectral changes in terms of FRET efficiency due to the GNR localized plasmon, we used a theoretical model which accounts for the enhancement of FRET as well as for quenching of the molecular excitations by the GNR. We approximated the hybrid GNR with a single donor dipole moment and a single acceptor dipole moment. The calculations were carried as fitting of the experimental fluorescence spectra. Results of the fitting are shown by solid lines in Fig. 30.5e. The calculated efficiency of the energy transfer in the presence of the GNR was $E = 0.46$ with a Förster radius of $R_0 = 15.04$ nm. To evaluate the effect of the plasmon enhancement, we kept the geometry of the donor and acceptor dyes and carried out the calculations without the GNR, with resulting FRET efficiency of $E = 0.01$ and Förster radius of $R_0 = 7.17$ nm. From this comparison, we obtained FRET enhancement factor due to the presence of GNR of 46. An average FRET efficiency enhancement obtained from 10 nanoparticles was by a factor of 65. At the same time, the Förster radius on average increased twofold in the presence of the GNR.

In summary, we designed novel hybrid nanoparticles composed of FRET pair dyes covalently attached to GNR. Using single-particle optical microscopy, we have shown that energy transfer from a donor to an acceptor in the presence of a GNR and, consequently, the emission color from the hybrid nanostructure can be reversibly modulated by the state of the polarization of the incident light.

30.3.2 Plasmon Enhancement of Triplet Exciton Diffusion in Photochemical Photon Upconversion

Photon upconversion (UC) based on triplet–triplet annihilation (TTA) between chromophores is a photochemical mechanism that can convert low energy photons to higher energy ones [25]. The TTA-UC process is a sequence of excitation of a singlet state of a donor, intersystem crossing into a triplet state of the donor, triplet–triplet energy transfer from the donor to an acceptor, and TTA between two acceptors leading to upconverted fluorescence from the acceptor singlet state. The TTA-UC mechanism can be optimized by material design to achieve high UC quantum yield, relatively low threshold of the UC emission, and broad and well-tunable excitation and emission characteristics. There is also a considerable potential for utilizing the effect of localized plasmon resonances of metal nanoparticles on the TTA-UC process to further optimize the UC mechanism toward practical applications. Our goal here was to gain insight into the effect of localized plasmons of metal nanostructures on the individual photophysical steps involved in the TTA-UC process. We used optical microscopy to study the properties of plasmonic hybrid nanostructures consisting of silver nanowires combined with a TTA-UC film [11]. We find fundamental differences in the way the localized plasmon influences phosphorescence on one hand and TTA-UC on the other, both in terms of emission intensity and spatial extent of the enhancement phenomena.

The samples, as schematically shown in Fig. 30.6a, are composed of a mesh of silver nanowires (AgNW) randomly dispersed on a microscope cover glass substrate (as seen in the dark-field image in Fig. 30.6b) and coated with a TTA-UC active material. The active material consists of donor molecules of platinum octaethylporphyrin (PtOEP), and acceptor molecules of 9,10-diphenylanthracene (DPA) dispersed in a matrix of poly(methyl methacrylate) (PMMA). Extinction and emission spectra of the donor and acceptor molecules in solution are shown in Fig. 30.6c. The AgNW extinction spectrum also shown in Fig. 30.6c has a maximum around 400 nm and extends beyond 700 nm. This broadband absorption ensures that plasmon resonances of AgNW couple with the absorption and emission of the chromophores dispersed in the film and could potentially mediate and enhance the energy transfer processes.

Microscopic study of the photophysical processes of TTA-UC was carried out by simultaneous imaging of phosphorescence from the donor molecules and UC fluorescence from the acceptor molecules using an image splitter. Microscopic phosphorescence image of the sample excited in the absorption Q-band of the donor PtOEP

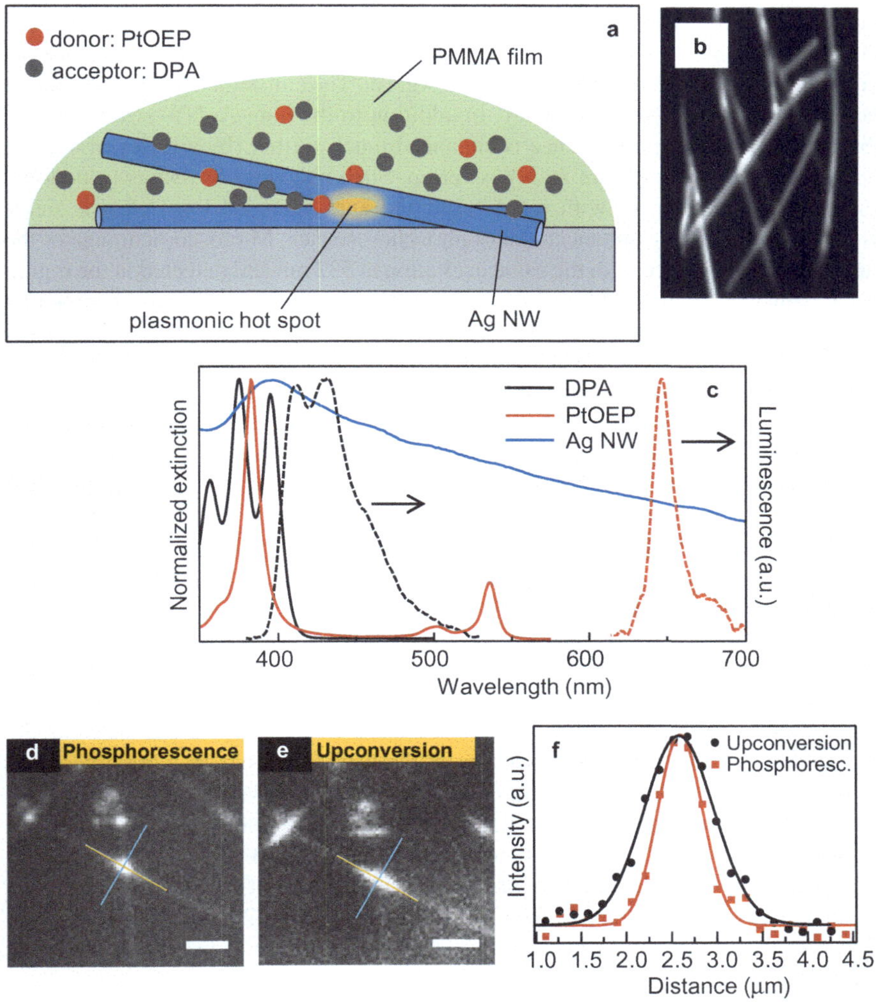

Fig. 30.6 **a** Scheme of the sample structure prepared on a microscope cover glass; **b** dark-field scattering image of the silver nanowires deposited on the cover glass. Scale bar 5 μm; **c** extinction (full lines) and emission (dashed lines) spectra of solutions of donor molecules (PtOEP, red), acceptor molecules (DPA, black) and of suspension of silver nanowires (blue); **d** microscopic image of phosphorescence from donor molecules in the hybrid sample obtained with 532 nm excitation laser; scale bar 2 μm; **e** image of upconversion emission from acceptor molecules in the hybrid sample obtained with 532 nm excitation laser; scale bar 2 μm; **f** 1D intensity profiles along the long axis of the nanowire of upconversion emission (black symbols) and phosphorescence (red symbols); the profiles were taken along a cross section indicated by yellow line in (**d**) and (**e**). Reprinted with permission from J. Phys. Chem. C 121:25479–25486 (2017). Copyright 2017 American Chemical Society

molecules (at 532 nm) is shown in Fig. 30.6d. The phosphorescence is emitted uniformly throughout the film. On top of this uniform background, phosphorescence with higher intensity is emitted close to the AgNW. The emission enhancement along the nanowires is on the order of 1.1. In addition to the long AgNW structures, we observe that the phosphorescence is enhanced much further (by a factor of 4–5) in locations where two or more AgNW overlap. Such nanowire crossings (junctions) form plasmonic hotspots with highly confined and enhanced local electric fields which cause the strong enhancement of phosphorescence. Microscopic image of the same sample area taken with the same excitation at 532 nm and detected in the region of UC emission (below 492 nm) is shown in Fig. 30.6e. The UC and phosphorescence images are very similar. The enhancement of UC intensity on the AgNWs is on the order of 1.3. In the junctions of the AgNWs, the enhancement increased up to a factor of 15. In addition, there is an interesting difference between the UC and phosphorescence patterns near the AgNW junctions. In the UC image, the enhanced emission is clearly elongated along one of the nanowires extending from the junction. This difference is evaluated by plotting 1D cross sections of the emission intensity. In the cross sections along the direction of the AgNW (Fig. 30.6f), the UC (black symbols) is significantly broader than the phosphorescence (red symbols). On the other hand, there is no difference between the UC and phosphorescence in the cross sections taken perpendicular to the AgNWs.

To evaluate and compare the extent of the spatial elongation (dispersion) of phosphorescence and UC emission, we fitted the experimental intensity cross sections with a convolution of a Gaussian function with an exponential function. The Gaussian function represents a point spread function of the microscope, whereas the exponential function describes diffusion of the triplet excitons as a possible cause of the spatial dispersion. An example of such fitting is shown in Fig. 30.5f as solid lines for both phosphorescence (red) and UC emission (black). The fits provide a characteristic parameter of dispersion length. The dispersion lengths were evaluated statistically by fitting the intensity profiles on several tens of AgNW junctions. The average dispersion was 0.5 μm for phosphorescence and 1.2 μm for UC emission. Thus, the UC emission is enhanced by the localized plasmon in AgNW junctions in qualitatively different way than phosphorescence in the same locations, both in terms of intensity enhancement and spatial dispersion.

To summarize the experimental observations, triplet states of the donor PtOEP are generated by the 532 nm excitation within the whole sample and lead to phosphorescence of the donor and UC emission from acceptor. The junctions of individual AgNW work as plasmonic hotspots and further enhance the generation of triplet states of the donor, as evidenced by the increased intensity of phosphorescence from these location (with enhancement factor of 4–5). The locations of the junctions have thus higher density of donor triplets compared to the rest of the sample and work as localized triplet sources. Here, we assume that the hotspots in the AgNW not only generate the higher density of the donor triplets but also enhance the donor–acceptor triplet energy transfer. Further, the plasmons also enhance the acceptor–acceptor triplet–triplet energy transfer (triplet exciton diffusion) within the hotspots and along the AgNW. The result of the enhancement of both types of energy transfer within the hotspot is an increased enhancement of UC emission compared to phosphorescence, whereas enhancement of the triplet diffusion along the AgNW causes a larger spatial dispersion of UC emission along the nanowires. We note that plasmon enhancement of Dexter-type triplet–triplet energy transfer has not been reported, and the concrete mechanism of such process is not completely understood. Further, experimental and theoretical work will be needed to resolve this problem.

30.3.3 Nanoscale Triplet Exciton Diffusion in Molecular Crystals Using Imaging of Single Hybrid Nanoparticles

In the photon upconversion process based on triplet–triplet annihilation, triplet exciton diffusion between the acceptor molecules in a molecular solid is a crucial parameter. Large values of triplet diffusion length L_T result in lowering of the irradiation intensity threshold required for the UC emission and are thus desirable for practical applications using conventional light sources. Consequently, accurate measurement of L_T in UC crystals is very important in the development of the state-of-the-art UC materials working under weak irradiation. However, such accurate measurements are generally difficult in UC crystals because the triplet donor molecules can be dispersed inhomogeneously in the acceptor crystals [26]. Recently, advanced microscopic techniques that use delayed fluorescence imaging or transient absorption imaging have been reported, but these still suffer from the limitation of optical resolution.

For the study of truly localized nanoscale triplet diffusion process, microscopic imaging using localized point-like triplet exciton source with sub-diffraction limited size is one potential way forward. Here, we report direct observation of L_T in UC crystalline systems consisting of donor-functionalized nanoparticles and acceptor molecules. In these systems, individual hybrid donor nanoparticles (HDPs) prepared by chemically attaching donor molecules together with fluorescence dyes to an alumina nanosphere were doped into acceptor molecular crystals to serve as localized point-like triplet exciton generators. In microscopic single-particle fluorescence images under green laser excitation, the individual HDPs in the film generate red emission due to fluorescence of the fluorescence dyes as well as blue UC emission due to fluorescence of the acceptor molecules (Fig. 30.7a). The emission was spectrally separated into blue and red regions and detected simultaneously using a CCD camera. This enabled an analysis of the different color emission patterns from individual HDPs at the same time. Comparative analysis of the two images enabled direct determination of L_T in the acceptor UC crystals.

Pt (II) mesoporphyrin IX (D1) and mesoporphyrin IX-dihydrochloride (FD1) were used as the donor and fluorescent molecules, respectively. Anthracene (AC) or 9,10-diphenylanthracene (DPA) were used as the acceptors. The D1 and FD1 were chemically attached to the hydroxyl-functionalized alumina nanoparticles (size of

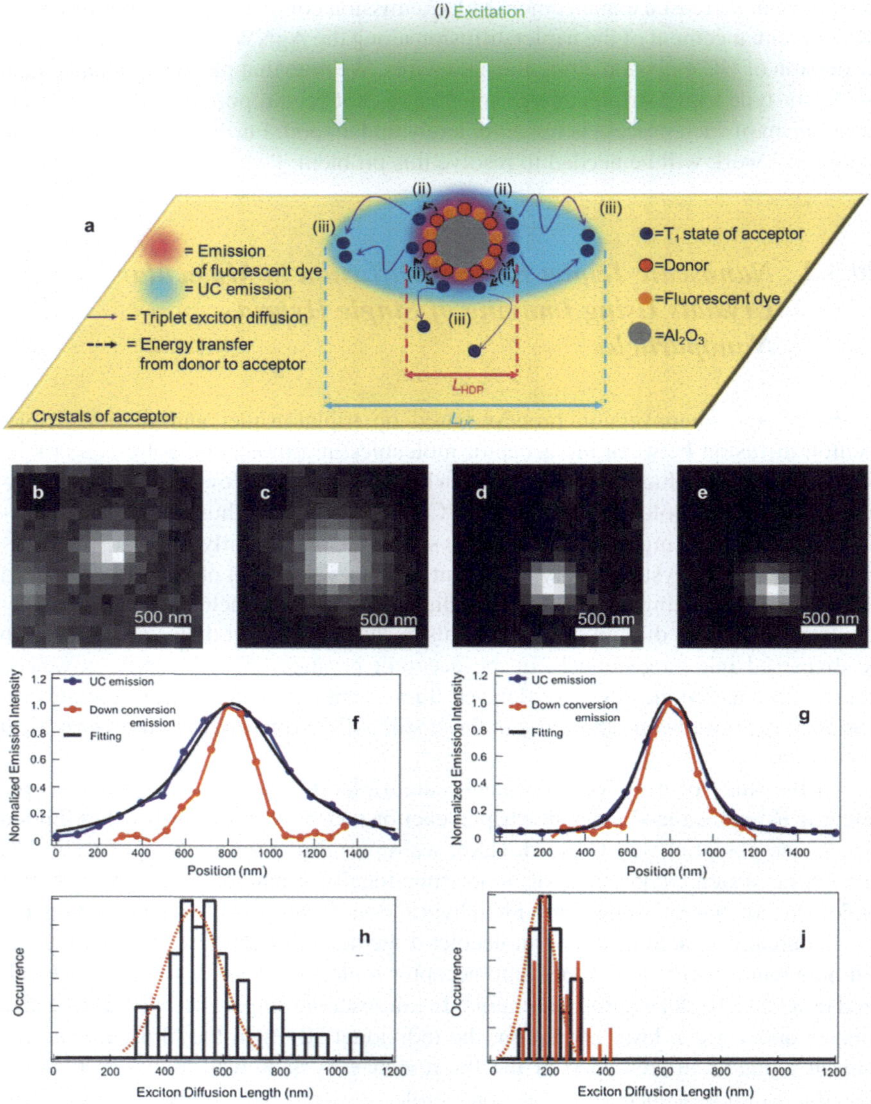

Fig. 30.7 **a** Schematic illustration of the visualization of triplet diffusion length L_T. L_{HDP} and L_{UC} is the size of fluorescence from FD and the size of UC emission from acceptor molecules, respectively. L_{UC} is larger than L_{HDP} due to the triplet exciton diffusion. Analysis of upconversion emission and fluorescence from HDPs. **b**, **c**, **f**, **h** AC acceptor film; **d**, **e**, **g**, **j** DPA acceptor film; **b**, **d** images of fluorescence from a single HDP; **c**, **e** images of UC emission from a single HDP; **f**, **g** one-dimensional intensity cross sections of the spots of UC emission and fluorescence. Black lines represent fits of the UC emission cross section; **h**, **j** histograms of L_T (bars) for AC (**h**) and DPA (**j**). The red solid lines in (**j**) represent a histogram of crystal grain sizes of DPA as analyzed from SEM images of the films. Reproduced from Ref. [12] with permission from The Royal Society of Chemistry

30 nm) to form the HDP. Triplet energy levels of AC and DPA are lower than that of D1, enabling efficient triplet energy transfer from D1 to both types of the acceptors. Diffusion of the acceptor triplet excitons and the consequent TTA result in generation singlet excitons and UC fluorescence. For the sample using AC polycrystalline film as the acceptor, microscopic images of fluorescence from FD1 attached to the HDP and of UC emission from AC in the vicinity of HDP are shown in Fig. 30.7b, c, respectively. Clearly, the size of the light spot in the UC emission is larger than that in fluorescence. The larger size of the UC emission spot results from diffusion of the triplet excitons within the AC films and the subsequent TTA. In the analysis, the CCD images were fitted with two-dimensional convolution function of PSF and singlet exciton density to obtain directly the L_T parameters. Examples of one-dimensional cross sections of the images together with the fits are shown in Fig. 30.7f. The L_T values obtained in this way by measuring UC emission of 46 individual HDPs are summarized in the histogram in Fig. 30.7h. From the histogram, the mean triplet exciton diffusion length L_T in AC polycrystalline film was 491 nm.

For comparison, we used DPA polycrystalline film as the acceptor. Figure 30.7d, e shows the microscopic images of fluorescence from FD1 and of UC emission from DPA, respectively. Figure 30.7g shows horizontal cross sections of the images together with the fits. The profile of UC emission from DPA is just slightly larger than that of fluorescence. By analyzing data from 37 individual HDP spot, we obtained a histogram of L_T as shown in Fig. 30.7j. The mean triplet exciton diffusion length L_T in DPA was 172 nm.

The different values of the mean triplet diffusion lengths L_T obtained for AC and DPA, i.e., 491 nm and 172 nm, respectively, can be ascribed to differences in microscopic and mesoscopic structure of the films. Microscopically, there is a difference in the π orbital overlap in the triplet state between AC and DPA due to the different crystal structure. As a result, the Dexter triplet–triplet energy transfer is more efficient in the AC films. In addition, the AC film is composed of flat crystal flakes with a size of a few micrometers while the crystal grain size in DPA films was on the order of tens–hundreds of nanometers. Size analysis of the DPA crystals measured by SEM is shown as an overlaid histogram in Fig. 30.7j. There is a good correspondence between the two histograms, and the peak of the size histogram at 213 nm is comparable to the observed L_T value of 172 nm. The triplet excitons in DPA are likely quenched after reaching the grain boundary, and the crystal grain size is the limiting factor in the triplet exciton diffusion process.

In summary, direct microscopic imaging of triplet exciton diffusion in polycrystalline upconversion materials was demonstrated. A key feature of the method is that excitation of the donor molecules and generation of high local density of triplet states are confined into spatial location with sub-diffraction limit size. The triplet exciton diffusion in the single-component acceptor medium can be studied individually, and nanoscale location-dependent characteristics of the exciton diffusion process can be obtained.

30.4 Conclusion

In conclusion, we studied collective photoresponse of molecular complexes and related inorganic and hybrid materials on nanometer scales using single-molecule detection and spectroscopy. The single-molecule and single-particle approach have demonstrated its unique abilities and advantages in uncovering mechanisms that lead to decrease of emission quantum efficiency of photoluminescence and electroluminescence due to emission blinking in perovskite nanocrystals and quantum dots, in revealing excitonic and electrical interactions in aggregates of nanocrystals, in demonstrating light polarization selectivity of plasmon enhancement in energy transfer processes and in nanoscale characterization of exciton diffusion in molecular solids. We hope that the current work will stimulate further research using the high potential of this optical nanoscale characterization technique.

Acknowledgements The present work was supported by JSPS KAKENHI Grant Numbers JP26107014, JP26107004, JP15H01082, JP15H01099, JP17H05254, 17H05243, Grant-in-Aid for Scientific Research on Innovative Areas "Photosynergetics" and by the Grant-in-Aid Number 19H02684.

References

1. Vacha M, Habuchi S (2010) Conformation and physics of polymer chains: a single-molecule perspective. NPG Asia Mater 2:134–142
2. Vacha M, Sharma DK, Hirata S (2018) Single-molecule studies beyond optical imaging: multi-parameter single-molecule spectroscopy. J Photochem Photobiol C Photochem Rev 34:121–136
3. Sharma DK, Hirata S, Vacha M (2019) Single-particle electroluminescence of $CsPbBr_3$ perovskite nanocrystals reveals particle-selective recombination and blinking as key efficiency factors. Nat Commun 10:4499
4. Sharma DK, Hirata S, Biju V, Vacha M (2019) Stark effect and environment induced modulation of emission in single halide perovskite nanocrystals. ACS Nano 13:624–632
5. Narushima K, Kiyota Y, Mori T, Hirata S, Vacha M (2019) Suppressed triplet exciton diffusion due to small orbital overlap as a key design factor for ultralong-lived room temperature phosphorescence in molecular crystals. Adv Mater 31:1807268
6. Kameyama T, Kishi M, Miyamae C, Sharma DK, Hirata S, Yamamoto T, Uematsu T, Vacha M, Kuwabata S, Torimoto T (2018) Wavelength-tunable band-edge photoluminescence of non-stoichiometric Ag-In-S nanoparticles via Ga^{3+} doping. ACS Appl Mater Interf 10:42844–42855
7. Hatazaki S, Sharma DK, Hirata S, Nose K, Iyoda T, Kölsch A, Lokstein H, Vacha M (2018) Identification of short- and long-wavelength emitting chlorophylls in cyanobacterial photosystem I by plasmon-enhanced single-particle spectroscopy at room temperature. J Phys Chem Lett 9:6669–6675
8. Uematsu T, Wajima K, Sharma DK, Hirata S, Yamamoto T, Kameyama T, Vacha M, Torimoto T, Kuwabata S (2018) Narrow band-edge photoluminescence from $AgInS_2$ semiconductor nanoparticles by the formation of amorphous III-VI semiconductor shells. NPG Asia Mater 10:713–726

9. Nakamura T, Sharma DK, Hirata S, Vacha M (2018) Intra-Chain aggregates as the origin of green emission in polyfluorene studied on ensemble and single chain level. J Phys Chem C 122:8137–8146
10. Lidster BJ, Hirata S, Matsuda S, Yamamoto T, Komanduri V, Tezuka Y, Vacha M, Turner ML (2018) Macrocyclic poly(p-phenylenevinylene)s by ring expansion metathesis polymerization and their characterization by single-molecule spectroscopy. Chem Sci 9:2934–2941
11. Bujak L, Narushima K, Sharma DK, Hirata S, Vacha M (2017) Plasmon enhancement of triplet exciton diffusion revealed by nanoscale imaging of photochemical fluorescence upconversion. J Phys Chem C 121:25479–25486
12. Narushima K, Hirata S, Vacha M (2017) Nanoscale triplet exciton diffusion via imaging of upconversion emission from single hybrid nanoparticles in molecular crystals. Nanoscale 9:10653–10661
13. Sharma DK, Hirata S, Bujak L, Biju V, Kameyama T, Kishi M, Torimoto T, Vacha M (2017) Influence of Zn on the photoluminescence of colloidal $(AgIn)_xZn_{2(1-x)}S_2$ nanocrystals. Phys Chem Chem Phys 19:3963–3969
14. Bujak L, Ishii T, Sharma DK, Hirata S, Vacha M (2017) Selective turn-on and modulation of resonant energy transfer in single plasmonic hybrid nanostructures. Nanoscale 9:1511–1519
15. Kamada K, Sakagami Y, Mizokuro T, Fujiwara Y, Kobayashi K, Narushima K, Hirata S, Vacha M (2017) Efficient triplet-triplet annihilation upconversion in binary crystalline solids fabricated by solution casting and operated in air. Mater Horizons 4:83–87
16. Sharma DK, Hirata S, Bujak L, Biju V, Kameyama T, Kishi M, Torimoto T, Vacha M (2016) Single-particle spectroscopy of I-III-VI semiconductor nanocrystals: spectral diffusion and suppression of blinking by two-color excitation. Nanoscale 8:13687–13694
17. Lee S, Noda K, Hirata S, Vacha M (2015) Position-dependent three-dimensional diffusion in nematic liquid crystal monitored by single-particle fluorescence localization and tracking. J Phys Chem Lett 6:1403–1407
18. Honmou Y, Hirata S, Komiyama H, Hiyoshi J, Kawauchi S, Iyoda T, Vacha M (2014) Single molecule electroluminescence and photoluminescence of polyfluorene unveils the photophysics behind the green emission band. Nat Commun 5:4666
19. Kojima A, Teshima K, Shirai Y, Miyasaka T. (2009) Organometal halide perovskites as visible-light sensitizers for photovoltaic cells. J Am Chem Soc 131:6050–6051
20. Lee MM, Teuscher J, Miyasaka T, Murakami TN, Snaith HJ (2012) Efficient hybrid solar cells based on meso-superstructured organometal halide perovskites. Science 338:643–647
21. Zhao X, Ng JDA, Friend RH, Tan ZK (2018) Opportunities and challenges in perovskite light-emitting devices. ACS Photon 5:3866–3875
22. Quan LN, García de Arquer FPP, Sabatini RP, Sargent EH (2018) Perovskites for light emission. Adv Mater 30:1801996
23. Allen PM, Bawendi MG (2008) Ternary I-III-VI quantum dots luminescent in the red to near-infrared. J Am Chem Soc 130:9240–9241
24. Hsu LY, Ding W, Schatz GC (2017) Plasmon-coupled resonance energy transfer. J Phys Chem Lett 8:2357–2367
25. Singh-Rachford TN, Castellano FN (2010) Photon upconversion based on sensitized triple-triplet annihilation. Coord Chem Rev 254:2560–2573
26. Mikhnenko OV, Blom PWM, Nguyen TQ (2015) Exciton diffusion in organic semiconductors. Energy Environ. Sci 8:1867–1888

Chapter 31
Cooperative and Hierarchal Photoresponses of Molecular Assembling Processes Probed by Organic Fluorescent Molecules

Fuyuki Ito

Abstract Crystal formation in solution starts from nucleation in the saturated state. The quality of the product and the number, size, and structure of crystals are determined in the initial stage of the nucleation process. Understanding nucleation is the key to control crystal properties including polymorph formation. In this chapter, the fluorescence visualization of the crystal formation process probed by organic fluorescent molecules exhibiting cooperative and hierarchal photoresponses is summarized. Fluorescence observations of evaporative crystallization revealed a two-step nucleation model for both nuclei and polymorph formation. A Dibenzoylmethanatoboron difluoride complex exhibiting mechanofluorochromism and a dipyrrolyldiketone difluoroboron complex derivative displaying polymorphism-dependent fluorescence were studied. The fluorescence from the droplets showed dramatic changes depending on the molecular state, such as monomer, amorphous, and crystal polymorph. This method could be easily used to detect assembly processes by measuring the real-time fluorescence changes under ambient conditions. Furthermore, the formation of polymorphs is most likely affected by the cluster structure prior to nucleation. Therefore, insights into the nuclei precursor clusters in polymorph formation measured by fluorescence changes will enable us to predict the outcome of polymorph formation.

Keywords Evaporative crystallization · Fluorescence color changes · Mechanofluorochromism · Polymorphism-dependent fluorescence · Two-step nucleation model · Liquid-like cluster

---

F. Ito (✉)
Department of Chemistry, Institute of Education, Shinshu University, 6-ro Nishinagano, Nagano 380-8544, Japan
e-mail: fito@shinshu-u.ac.jp

H. Miyasaka et al. (eds.), *Photosynergetic Responses in Molecules and Molecular Aggregates*, https://doi.org/10.1007/978-981-15-5451-3_31

31.1 Introduction

Photochemistry of organic molecular aggregates plays an important role in various systems such as organic light-emitting diodes and organic solar cells. The fluorescence properties of organic molecules in aggregates are different from those of molecules dissolved in solution. Various emission colors have been observed from samples depending on the assembled state, i.e., these samples are cooperative and hierarchal photoresponsive molecular assemblies. To control these emissive properties, it is important to obtain detailed information about the origin and evolution of the emission in the aggregated state, where specific interactions can lead to the appearance of new emission colors.

Crystallization of organic crystals from solution is important for not only separation and purification, but also the formation of single crystals for pharmaceutical production and structure determination and the development of organic solid materials. Because many organic crystals are formed by relatively weak interactions such as van der Waals interactions, the properties of organic molecular solids depend on both their molecular structure and molecular assembly. In other words, molecular crystals are formed under the influence of weak interactions, and thus have different integrated structures depending on slight differences in crystallization conditions such as temperature, solution concentration, and the speed of solvent evaporation. The crystallization process is roughly divided into nucleation and crystal growth. Although crystal formation from solution is a relatively familiar phenomenon, the formation and growth of crystal nuclei in a homogeneous solution is still not well understood. Nucleation in the saturated state is the initial step of crystal formation from the solution [1]. The size, number, structure, and quality of crystals are determined in the initial nucleation process. Therefore, understanding nucleation is important to control crystal properties including polymorph formation. The classical nucleation theory (CNT) is based on the Gibbs energies of surfaces and volume in the growth process without considering molecular identities. The CNT has been widely utilized to simply explain crystallization processes. In this theory, it is assumed that the density and order fluctuations in a supersaturated solution lead to the formation of clusters within the molecular packing, which reflects all possible polymorphs of the solute and crystal nuclei with the same structure as a mature crystal. In some cases, predictions based on the CNT differ from experimental results, suggesting that nucleation from solution must involve a more complex mechanism. Recently, the two-step nucleation model, which includes the formation of metastable precursor clusters with a size of hundreds of nanometers in a dense liquid, was proposed by several groups based on the induction time of crystal formation [2], nuclear magnetic resonance spectroscopy [3, 4], electron microscopy [5], dynamic light scattering [6], and nonphotochemical laser-induced crystallization data [7]. This model assumes the separation of a dense, disordered liquid phase, and fluctuations in its density prior to the formation of crystalline order. In other words, the initial clusters are liquid-like and crystalline order appears in the clusters over time. This model is known to be widely applicable to not only the crystallization of proteins but also that of other organic and inorganic materials and colloids.

Against this background, we conduct real-time measurements of the molecular assembly process using the cooperative and hierarchical fluorescence changes of organic molecules to optimize the fluorescence color modulation properties in the organic solid and understand the relationship between fluorescence color and molecular arrangement including the assembling process. Fluorescence spectroscopy can provide information about the state of molecular aggregates ranging in size from several molecules to the bulk scale because excited molecules show spectral changes that strongly depend on the electronic state, aggregated state or size, and surrounding environment. We have been investigating changes in electronic states associated with cooperative interactions between molecules in organic molecular aggregates. In this chapter, we introduce research based on the concept of monitoring the fluorescence color while crystallization occurs as the solvent evaporates under ultraviolet (UV) light irradiation (evaporative crystallization).

In this chapter, studies of the direct visualization of crystal formation and growth probed by organic fluorescent molecules using fluorescence microscopy and spectroscopy are introduced. We focus on the fluorescence spectral changes of a Dibenzoylmethanatoboron difluoride complex exhibiting mechanofluorochromism [8] and a dipyrrolyldiketone difluoroboron complex derivative displaying polymorphism-dependent fluorescence [9].

31.2 Fluorescence Detection of the Molecular Assembly Process by a Mechanofluorochromic Compound

Dibenzoylmethanatoboron difluoride (BF_2DBM) derivatives possess excellent optical properties such as high fluorescence quantum yield in the solid-state [10], multiple fluorescence colors [11–15], reversible mechanofluorochromic properties [16, 17], and two-photon absorption cross-sections. In particular, BF_2DBM based on the 4-*tert*-butyl-4′-methoxydibenzoylmethane (avobenzone) boron difluoride complex (BF_2AVB) exhibits different emission behavior depending on the crystal phase (polymorph) [18]. BF_2AVB shows not only excellent durability to light irradiation but also high fluorescence quantum yield in the solid-state (~0.5), which is advantageous for photonic applications. In this section, real-time visualization of the two-step nucleation model is investigated by evaluating the fluorescence properties of BF_2DBM solution during solvent evaporation-induced crystallization.

In this work, 4,4′-di-*tert*-butyldibenzoylmethane boron difluoride (BF_2DBMb, Fig. 31.1a) was used as a probe to investigate evaporative crystallization. First, we examined the fluorescence of BF_2DBMb in dilute solution, crystals, and the amorphous state, fluorescence images of which are shown in Fig. 31.1b–d, respectively. These images indicate that the molecular forms of BF_2DBMb such as monomer

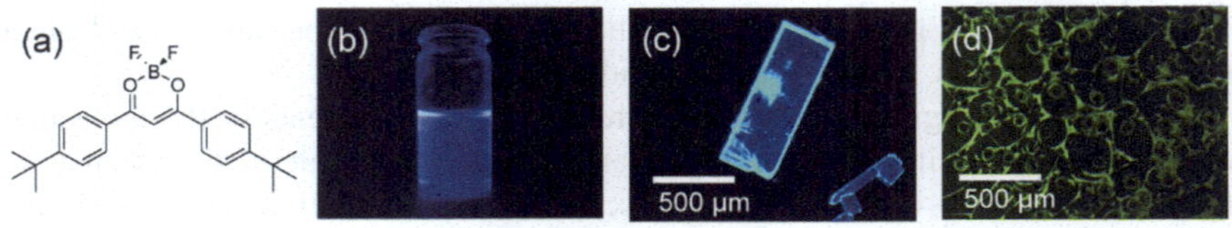

Fig. 31.1 **a** Molecular structure of BF2DBMb. Fluorescence images of BF_2DBMb in **b** 1,2-dichloromethane, **c** crystalline state, and **d** amorphous state under 365-nm UV irradiation

(isolated state) and aggregated states can be distinguished by their different fluorescence colors. Next, we measured the fluorescence changes of BF_2DBMb during evaporative crystallization from solution to detect the molecular assembly process. The fluorescence of the droplet was purple just after droplet formation. The emission color changed from purple to blue via orange. Orange emission from BF_2DBMb was only observed from supersaturated solution. This strongly suggests that BF_2DBMb molecules with orange emission can exist only under non-equilibrium conditions, such as in supersaturated solutions.

To obtain spectroscopic information about the solvent evaporation process, we observed the fluorescence spectra of BF_2DBMb in 1,2-dichloroethane during solvent evaporation. The initial fluorescence spectrum of the droplet exhibited a peak at 433 nm, with shoulders at 415, 460, and 550 nm, corresponding to the emission spectrum of the monomer state. The intensity of the peak around 550 nm originating from the amorphous state increased monotonically over time until 91 s. The intensity of the fluorescence peak at 433 nm decreased from 91 to 95 s. After 95 s, peaks around 445 and 470 nm appeared concomitant with a decrease of the intensity of the 550-nm band. The series of fluorescence spectral changes correspond to the changes in fluorescence images.

The temporal changes in the fluorescence properties of BF_2DBMb described above can be used to explain the molecular assembly process during evaporative crystallization. BF_2DBMb crystals were formed from the solution through the amorphous state. The Raman spectra of each state are slightly different, supporting a change in the molecular interactions in the ground state. The fluorescence spectra were analyzed by nonlinear least-squares fitting with Gaussian peak position and width. As shown in Fig. 31.2a, the relative abundance of the monomer, crystal, and amorphous states was plotted as a function of time. These results imply that crystals of BF_2DBMb are formed from the monomer species through the amorphous state and exhibit hierarchical changes such a consecutive reaction. A transient amorphous state is formed prior to crystal formation. In this case, the orange emission from amorphous species suggests the presence of liquid-like clusters prior to crystallization. A scheme of the crystallization process of BF_2DBMb is shown in Fig. 31.2b.

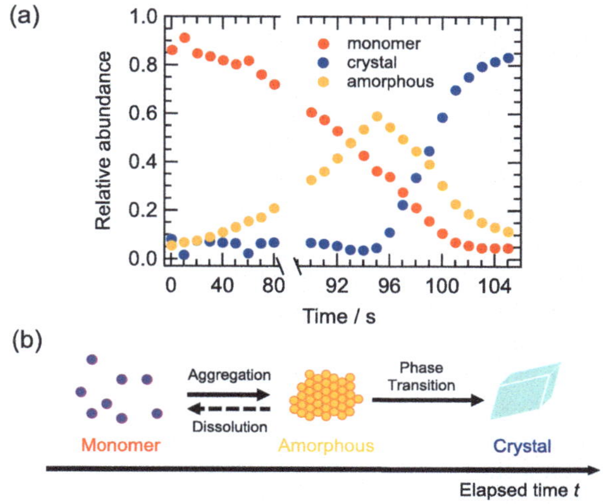

Fig. 31.2 **a** Change in the relative abundance of monomer, amorphous, and crystal states of BF_2DBMb based on time-resolved fluorescence spectral measurements. **b** Schematic representation of the molecular assembly of BF_2DBMb based on the changes in fluorescence spectra

This study clearly confirmed the two-step nucleation model by measuring fluorescence color changes. The present method allowed crystal formation to be observed using a conventional optical detection system under ambient conditions, making it attractive to study the control of organic emissive materials with multiple emissive states or colors depending on their phase.

31.3 Dynamics of the Polymorph Formation Process During Evaporative Crystallization from Solution

The same molecule has different crystal structures, which are called polymorphs [19]. The stability and solubility of each polymorph depend on their crystal packing and structure. In the field of pharmaceuticals, drug solubility affects their behavior in living bodies (bioavailability). Therefore, screening for polymorphic structures is an important step in drug development. The thermal or chemical stability of polymorphs can determine product quality. Based on the above considerations, it is an important industrial concern to selectively obtain polymorphs with desired properties. However, polymorphic control during crystallization from solution has seldom been achieved, which is partly because the underlying mechanism of polymorph formation is poorly understood.

Controlling the liquid-like clusters as nuclei precursors can potentially determine the fate of the nucleation process. For example, glycine has three polymorphs: α, β, and γ. Sugiyama and Masuhara et al. [20] found that the ratio of glycine polymorphs could be regulated by the polarization of the laser light shone into a glycine saturated solution. Garetz et al. [21] investigated the polymorph appearance process in saturated and unsaturated glycine solutions by small-angle X-ray scattering measurements. In nuclei precursor clusters (liquid-like clusters) in glycine solution, it was found that domains formed through the interaction between dimers and solvent molecules. It was suggested that the liquid-like clusters are not perfectly isotropic, but possess an internal packing structure similar to that of the crystal. These experimental results strongly suggest that the liquid-like clusters proposed in the two-step nucleation theory act as important intermediates during the crystallization process including polymorph formation. There are some reports on the polymorph formation process from a theoretical perspective using both CNT and two-step nucleation models [22, 23].

As described in Sect. 31.2, fluorescence visualization was achieved for the evaporation crystallization process of BF_2DBMb in 1,2-dichloroethane [8]. Fluorescence spectroscopy could be used to probe the molecular assembly processes on scales from only a few molecules to the bulk material. The concentration-dependent fluorescence spectral changes of organic molecules during solvent evaporation can provide information about molecular assembly and crystal nucleation and growth [24–26]. This approach has the advantage of being able to detect changes in the surrounding environment in real-time using fluorescence spectral changes as a probe. Moreover, it is expected that the formation of polymorphs will be affected by the cluster structure before nucleation. Therefore, insight about the nuclei precursor clusters involved in polymorph formation obtained by assessing fluorescence changes will enable the prediction of polymorph appearance.

In this section, the 1,3-dipyrrol-2-yl-1,3-propanedione boron difluoride complex **1** (Fig. 31.3a) is reviewed as a model system with three polymorphs showing different luminescence characteristics [27, 28]. It has been reported that the addition of anions to a solution of **1** changes its conformation by inversion (flipping) of two pyrrole rings [15, 29–31]. Aryl substitution at the α- and β-positions of pyrrole can introduce a variety of substituents around the anionic receptor, facilitating the anion-driven modulation of assembly structure, electronic and optical properties, and charge carrier mobility in the bulk state. An innovative design that takes into account the structural and electronic properties of pyrrole molecules has made it possible to obtain attractive molecular systems. From this background, the detection of molecular dynamics during the assembly process is considered to be important for predicting the ordered arrangements of π-electron systems. Despite having the simplest structure (without modification of the pyrrole rings), **1** has three polymorphs abbreviated as **1r**, **1y**, and **1v** [29, 32] that exhibit red, green, and vermilion fluorescence, respectively. The pyrrole rings of the molecules in **1r** and **1y** are oriented toward the BF_2 unit. While **1r** contains two stacked structures with anti-orientation of the BF_2 unit, **1y** forms a stacked structure with syn orientation. The crystal structure of **1v** is different from those of **1r** and **1y**: one of the pyrrole rings in **1v** is inverted and the molecules adopt an anti-oriented stacking structure. In this section, the relationship between the nuclei precursor and polymorph formed is examined by fluorescence detection of the polymorphic expression process based on the spectral changes detected during evaporative crystallization.

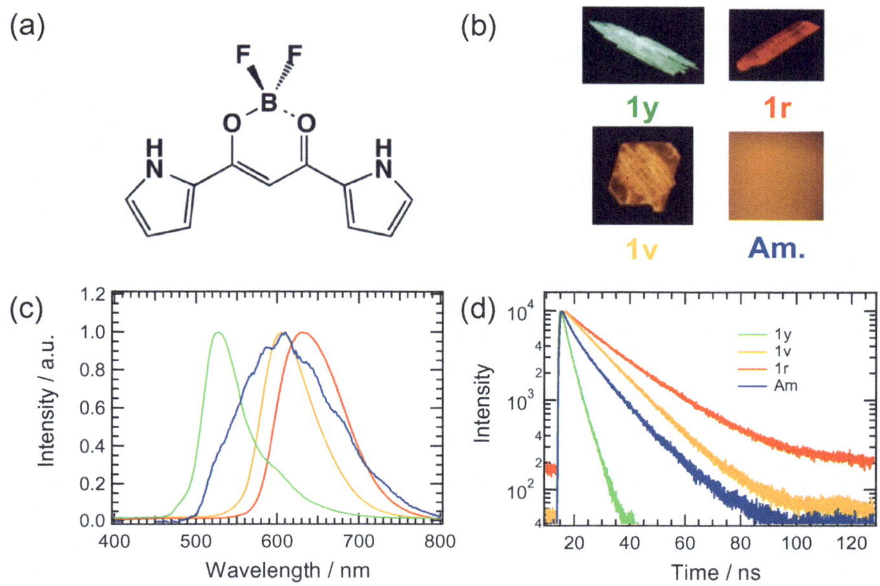

Fig. 31.3 **a** Molecular structure of **1**. **b** Fluorescence images of polymorphs (**1y**, **1v**, and **1r**) and the amorphous state (**Am**) of **1** under 365-nm UV irradiation. **c** Fluorescence spectra of **1y** (green), **1v** (orange), **1r** (red), and **Am** (blue). **d** Fluorescence decay curves of **1y**, **1v**, **1r**, and **Am**

Figure 31.3b shows that the three crystal polymorphs of **1** (**1r**, **1y**, and **1v**) exhibit different fluorescence colors. The fluorescence peaks of **1r**, **1y**, and **1v** were observed at 629, 524, and 604 nm, respectively, as shown in Fig. 31.3c. As mentioned above, the amorphous phase has an important role in the two-step nucleation theory and should, therefore, be considered. The amorphous phase of **1** (**Am**) was fabricated by rapidly freezing a molten droplet over the melting point. A fluorescence image and spectrum of **Am** are shown in Fig. 31.3b, c, respectively. The spectral peak of **Am** observed at 610 nm resembled that of **1v**. The fluorescence decay curves of **1r**, **1y**, **1v**, and **Am** were measured to obtain more information about the fluorescence properties of **1**. The fluorescence decay curves are shown in Fig. 31.3d. The decay profiles of **1r** and **1v** could be approximated by double exponential decays, whereas those of **1y** and **Am** could be described by triple ones. **Am** and **1v** displayed similar fluorescence decay curves. The fluorescence lifetimes of **1r** were 18.80 and 7.05 ns. The lifetimes of **1r** were longer than that of **1y** (0.15 ns). The fluorescence lifetimes of **1v** and **Am** were comparable (3.31 and 3.59 ns, respectively, for the faster component, and 13.14 and 11.98 ns, respectively, for the slower component).

Next, the infrared (IR) spectra of the three polymorphs and **Am** were measured to detect their different intermolecular interactions. The spectra were remarkably different from each other around 3500–3300 cm^{-1}. The IR spectrum of **1r** showed two peaks at 3410 and 3450 cm^{-1} with a shoulder at 3380 cm^{-1}. In contrast, **1v** exhibited a peak at 3370 cm^{-1} with a shoulder at 3390 cm^{-1}, and **1y** displayed a peak at 3420 cm^{-1}. A broad absorption band was observed for **Am** in the range of 3500–3300 cm^{-1} along with a peak located at 3380 cm^{-1}. The IR bands in the region from 3530 to 3480 cm^{-1} can be assigned to the NH stretching mode of pyrrolyl groups. The IR spectral changes of the peaks related to NH bonds reflect the different intermolecular interactions in the polymorphs and amorphous phase of **1**, which depend on the molecular stacking structure. A molecule of **1** forms four hydrogen bonds (N–H···F) in its crystal structure. Although this number of hydrogen bonds is identical in each polymorph, the IR spectral peaks exhibit distinct differences in the region of the NH stretching mode. These findings indicate that the NH stretching vibration mode is affected by intermolecular hydrogen bonding. The broadband observed for **Am** suggests that intermolecular interactions are present at many sites of the molecules in the amorphous phase.

It has been reported that **1y** transforms to **1r** at around 160 °C [32]. Differential scanning calorimetry (DSC) measurements were carried out to investigate the thermal properties of **1y**. The DSC thermograms of **1y** showed two strong peaks at 246 °C ($\Delta H_{m1} = -$ 13.8 J/g) and 253 °C ($\Delta H_{m2} = -$ 103.3 J/g), indicating that the crystals melt in this temperature range. In addition, the DSC thermogram showed a weak endothermic peak at 166 °C ($\Delta H_{trans.} = -4.6$ J/g). The fluorescence color of **1y** also changed from green to red at around 160 °C, as shown in Fig. 31.4. The endothermic DSC peak was identical to the phase transformation from **1y** to **1r**. The strong endothermic peaks at 246 and 253 °C probably reflect phase changes that occur before the melting of **1r**. This phenomenon can be tentatively explained by the following two possibilities. The **1r** crystals melt slightly at 246 °C, then solidify, and finally melt completely at 253 °C. Alternatively, untransformed **1y** in the part of **1r** above 246 °C could melt.

Figure 31.5 shows the fluorescence color changes of a droplet of 3.0 × 10^{-3} mol dm^{-3} of 1 in 1,2-dichloroethane solution during solvent evaporation as a function of time. Blue emission was observed in all regions just after droplet formation. At 25 s, the periphery of the droplet generated radial and dendrite-like solids that exhibited bluish-green emission. The dendrite-like solids extended from the droplet

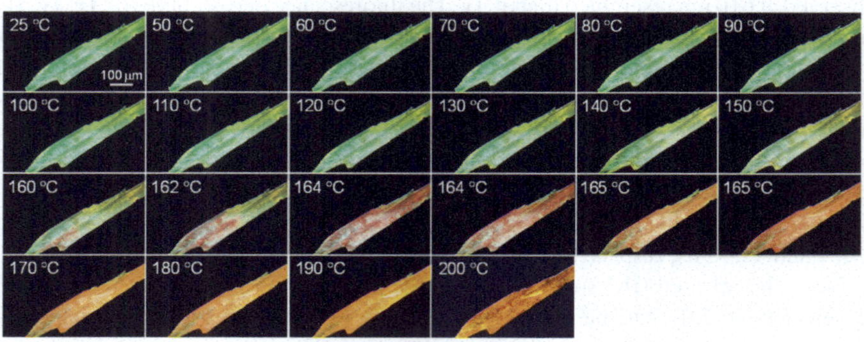

Fig. 31.4 Fluorescence images of **1y** under 365-nm UV light illustrating its fluorescence color as a function of temperature

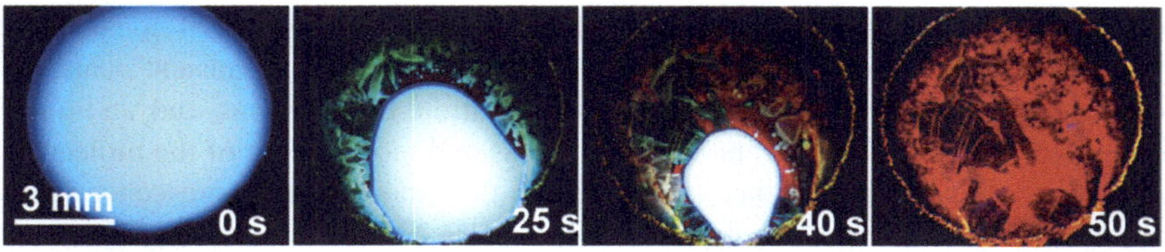

Fig. 31.5 Fluorescence images of a droplet of **1** in 1,2-dichloroethane during solvent evaporation as a function of time under 365-nm UV irradiation

periphery. The bluish-green emission from the radial and dendrite-like solids changed to yellowish-green or yellow after 40 s. As time passed, red fluorescence began to appear at the edge of the droplet. The blue-green emission changed to red and the area of red emission increased as time elapsed. A small region showing bluish-green emission remained in the droplets. When the solvent completely evaporated at 50 s, the emission from the precipitate changed completely to red except in the peripheral region, where red and yellow mixed fluorescence was observed.

The observed emission color changes were quantified by the fluorescence spectral measurements of a droplet of **1** in 1,2-dichloroethane during solvent evaporation. The indicated times of the fluorescence spectra and images are slightly different. Figure 31.6a shows the fluorescence spectra of the droplet. Two emission peaks at 470 and 450 nm were observed from 0 to 75 s. As time elapsed, the peak intensity at 470 nm decreased and the peak gradually shifted to around 500 nm. The peak at around 500 nm disappeared at 100 s, whereas another peak at around 610 nm gradually increased in intensity from 77 s. The final spectrum contained a single peak located at around 612 nm.

To analyze the fluorescence spectral changes semi-quantitatively, it was initially assumed that the fluorescence of the droplet arose from four components—monomer, green, orange, and red species—during solvent evaporation. The fluorescence spectra were approximately composed of five Gaussian peaks, i.e., two peaks for the monomer and three peaks corresponding to green, orange, and red species. However,

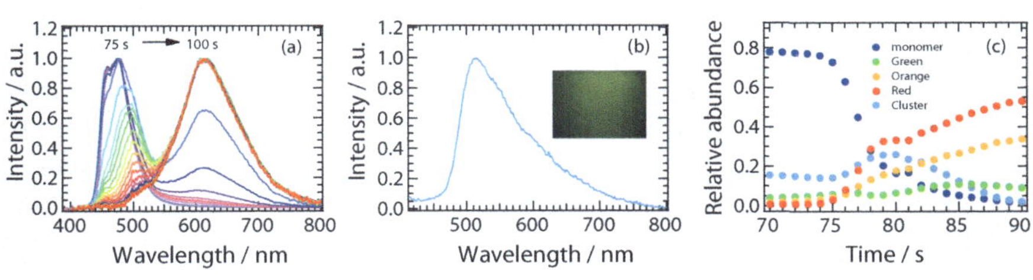

Fig. 31.6 **a** Changes of fluorescence spectra of a droplet of **1** (3.0×10^{-3} mol dm^{-3}) in 1,2-dichloroethane during solvent evaporation. **b** Fluorescence spectrum of molten **1** heated to 235 °C. The inset is a fluorescence image of the molten state of **1** under UV light irradiation. **c** The relative abundance changes of monomer, cluster, and green, red, and orange species as a function of time based on the fluorescence spectra of **1** measured in 1,2-dichloroethane during solvent evaporation

the spectra were unable to be reproduced with these five peaks even considering that the actual fluorescence bands were broader. In particular, a transient peak was observed at around 500 nm during solvent evaporation. In a previous study, a liquid-like cluster that acted as both the crystal nucleus and molten state of the molecules was observed as an intermediate [8]. Based on this finding, the fluorescence spectrum and image of **1** upon heating a sample to over 235 °C were measured, which are depicted in Fig. 31.6b and its inset, respectively. Green emission with a peak at 501 nm was exhibited from the molten state. In the fluorescence spectrum of **1**, the peak at shorter wavelength shifted during solvent evaporation, as shown in Fig. 31.6a. Re-analyzing the fluorescence spectra revealed that the observed spectra at all times during solvent evaporation could be reproduced well by adding a sixth Gaussian peak at 501 nm for the clusters. These findings suggest that the molten state exists as an intermediate during evaporative crystallization.

The relative abundance changes of each component as a function of time are shown in Fig. 31.6c. The monomer fraction steeply decreased after 75 s. The green fraction first increased slightly and then decreased from 76 to 78 s. The orange fraction monotonically increased from 75 s. The red fraction increased after 75 s, became constant between 79 and 81 s, and then increased again. The cluster (molten state) fraction increased from 76 to 80 s when the monomer and green fractions decreased and then decreased with increasing fractions of the orange and red components. The green and red species were assigned to **1y** and **1r**, respectively [32]. Formation of the amorphous state after complete solvent evaporation was confirmed by polarization optical microscopy. Because the fluorescence peaks of **Am** and **1v** appeared at 610 and 604 nm, respectively, it was difficult to distinguish them based on only the fluorescence spectra data. Therefore, the orange fraction was considered to consist of both **Am** and **1v**. The cluster fraction was identified by its similarity to the liquid-like cluster of **1** reported previously [8], which is also consistent with the two-step nucleation model. It is considered that the liquid-like cluster and **1y** are in equilibrium. The generation of **1r** is influenced by that of **1y** because **1r** and **1y** have similar crystal structures. In contrast, because the changes of these three fractions seem to be independent of each other, the orange fraction is unlikely to be influenced by **1y** and **1r**, leading to the conclusion that the orange fraction is not produced by transformation of **1y** or **1r**.

The number density of molecules in the solution increased over time because of the solvent loss during solvent evaporation. A solute dense state was formed from the monomer molecules generated by the fluctuation in the solution; namely, a liquid-like cluster state. These liquid-like clusters formed ordered arrays and acted as crystalline nuclei, from which crystals formed and grew. If the liquid-like clusters cannot form crystal nuclei, they will form disordered aggregates such as an amorphous solid through liquid–liquid separation. Figure 31.7 shows a proposed scheme based on the superposition of fluctuations of the two order parameters density and structure [33]. These two parameters are related to crystal growth and polymorph formation [34]. Liquid-like cluster states can be formed by fluctuation of the monomers in the droplet resulting from solvent evaporation. Then, crystals of the three polymorphs are produced from the liquid-like clusters through the formation of crystal nuclei

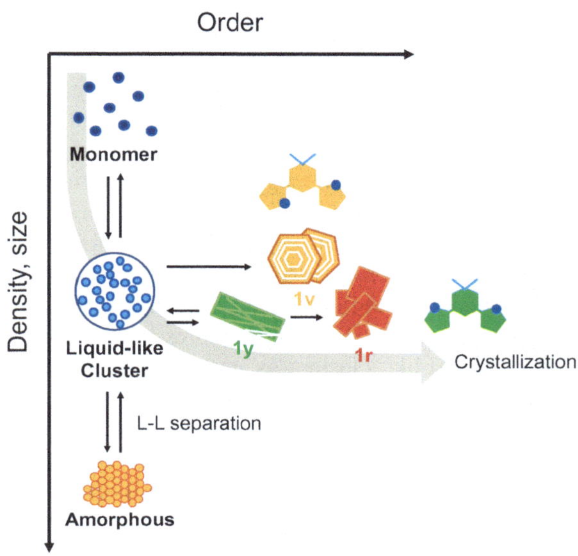

Fig. 31.7 Schematic representation of the assembly process of **1**

driven by the reorganization in the clusters. The most important step in determining the formed polymorph is the intramolecular conformation change of the two pyrrole rings in **1** in the liquid-like cluster. The N atoms of pyrrole rings in **1y** and **1r** are oriented toward the BF_2 unit. In contrast, one of the pyrrole rings in **1v** is inverted, and the molecules stack in an anti-orientation mode. The fluorescence changes of **1** in a droplet from green to red suggest that **1y** transforms to **1r** during solvent evaporation. This transformation can easily occur through intramolecular conformation changes, which is supported by the DSC results. It is envisaged that the fluorescence changes correspond to the visualization of Ostwald's rule of stages. In contrast, the crystallization of **1v** is independent of the formation of **1y** and **1r**, because their intramolecular conformation changes are different. Thus, the crystallization process of **1v** does not depend on those of **1y** and **1r**. The solvent evaporation from the liquid-like cluster without nucleation results in the formation of **Am** through liquid–liquid separation.

The relative stability of each phase is an important factor for the formation of organic polymorphs as well as the density and regularity of aggregates. All four solid states are formed from the intermediate liquid-like cluster state. Thus, it can be proposed that the liquid-like clusters can be considered as a "crucible" that are in a non-equilibrium state at ambient temperature. Crucibles are containers used to make crystals from a molten state at very high temperatures. In this case, the molten state of **1**, which corresponds to a liquid-like cluster state, is regarded as a crucible. The ability to provide detailed information about such crucibles and control the molecular dynamics within crucible will allow the selective generation of desired polymorphs from solution.

31.4 Summary

Research on the crystal formation process of organic fluorescent materials is important to elucidate the photoresponse behavior of structures involving one or more molecules. In addition, the crystal formation process is important for the development of organic solid-state light-emitting materials that exhibit dramatic changes in emission behavior caused by slight differences in molecular aggregation structures and phase changes and is essential for establishing not only molecular design but also crystal engineering molecular assembly design guidelines. Based on the fluorescence visualization of the crystal formation process, the two-step nucleation model was clearly confirmed based on fluorescence color changes. An intermediate state, such as liquid-like clusters, is important in not only the crystal nuclei formation process but also polymorphic expression. The origin of the polymorphic expression is the key to understanding the crystallization conditions that can selectively generate crystals with the desired structure, the molecular dynamics in the initial process of crystal nucleation, and for elucidating the mechanism of crystal polymorphism. The present method has low requirements because crystal formation can be observed with a conventional optical detection system in the ambient atmosphere, especially for control of polymorphism of organic emissive materials with multiple emissive states or colors depending on the phase, such as mechanofluorochromic molecules. This approach can be used to provide a comprehensive picture of crystallization by combination with other observation techniques.

Acknowledgements The authors thank Mr. Jun-ichi Fujimori, Ms. Yukino Suzuki, Ms. Mai Saigusa, and Ms. Narumi Oka (Shinshu University), Dr. Mitsuo Hara and Prof. Takahiro Seki (Nagoya University), Dr. Ryohei Yasukuni and Prof. Marc Lamy de la Chapelle (Paris 13 University), Dr. Yohei Haketa and Prof. Hiromitsu Maeda (Ritsumeikan University), Dr. Tetsuya Miyano, Prof. Norimitsu Tohnai, Dr. Syoji Ito, and Prof. Hiroshi Miyasaka (Osaka University) for their support with the experiments and useful discussions. This work was partly supported by the Nanotechnology Platform Program, JSPS KAKENHI (Grants Numbers JP21750021, JP15H01081, JP17H05253, and JP19H02686), the Iketani Science and Technology Foundation, and the JGC-S Scholarship Foundation.

References

1. Erdemir D, Lee AY, Myerson AS (2009) Nucleation of crystals from solution: classical and two-step models. Acc Chem Res 42:621–629
2. Knezic D, Zaccaro J, Myerson AS (2004) Nucleation induction time in levitated droplets. J Phys Chem B 108:10672–10677
3. Hughes CE, Williams PA, Keast VL, Charalampopoulos VG, Edwards-Gau GR, Harris KDM (2015) New in situ solid-state NMR techniques for probing the evolution of crystallization processes: pre-nucleation, nucleation and growth. Faraday Discuss 179:115–140
4. Hughes CE, Harris KDM (2008) A technique for in situ monitoring of crystallization from solution by solid-state C-13 CPMAS NMR spectroscopy. J Phys Chem A 112:6808–6810

5. Harano K, Homma T, Niimi Y, Koshino M, Suenaga K, Leibler L, Nakamura E (2012) Heterogeneous nucleation of organic crystals mediated by single-molecule templates. Nat Mater 11:877–881
6. Schubertt R, Meyer A, Baitan D, Dierks K, Perbandt M, Betzel C (2017) Real-time observation of protein dense liquid cluster evolution during nucleation in protein crystallization. Cryst Growth Des 17:954–958
7. Garetz BA, Matic J, Myerson AS (2002) Polarization switching of crystal structure in the nonphotochemical light-induced nucleation of supersaturated aqueous glycine solutions. Phys Rev Lett 89:175501
8. Ito F, Suzuki Y, Fujimori J, Sagawa T, Hara M, Seki T, Yasukuni R, de la Chapelle ML (2016) Direct visualization of the two-step nucleation model by fluorescence color changes during evaporative crystallization from solution. Sci Rep 6:22918
9. Oka N, Ito F, Haketa Y, Maeda H, Miyano T, Tohnai N, Ito S, Miyasaka H, Ozeki S (2018) Dynamic polymorph formation during evaporative crystallization from solution: the key role of liquid-like clusters as "crucible" at ambient temperature. Chem Eur J 24:4343–4349
10. Ono K, Yoshikawa K, Tsuji Y, Yamaguchi H, Uozumi R, Tomura M, Taga K, Saito K (2007) Synthesis and photoluminescence properties of BF2 complexes with 1,3-diketone ligands. Tetrahedron 63:9354–9358
11. Zhang GQ, Lu JW, Sabat M, Fraser CL (2010) Polymorphism and reversible mechanochromic luminescence for solid-state difluoroboron avobenzone. J Am Chem Soc 132:2160–2162
12. Zhang GQ, Singer JP, Kooi SE, Evans RE, Thomas EL, Fraser CL (2011) Reversible solid-state mechanochromic fluorescence from a boron lipid dye. J Mater Chem 21:8295–8299
13. Samonina-Kosicka J, DeRosa CA, Morris WA, Fan ZY, Fraser CL (2014) Dual-emissive difluoroboron naphthyl-phenyl beta-diketonate polylactide materials: effects of heavy atom placement and polymer molecular weight. Macromolecules 47:3736–3746
14. Sakai A, Tanaka M, Ohta E, Yoshimoto Y, Mizuno K, Ikeda H (2012) White light emission from a single component system: remarkable concentration effects on the fluorescence of 1,3-diaroylmethanatoboron difluoride. Tetrahedron Lett 53:4138–4141
15. Sakai A, Ohta E, Yoshimoto Y, Tanaka M, Matsui Y, Mizuno K, Ikeda H (2015) Novel fluorescence domain "excited multimer" formed upon photoexcitation of continuously-stacked diaroylmethanatoboron difluoride molecules with fused π-orbital in crystals. Chem Eur J 21:18128–18137
16. Sagawa T, Ito F, Sakai A, Ogata Y, Tanaka K, Ikeda H (2016) Substituent-dependent backward reaction in mechanofluorochromism of dibenzoylmethanatoboron difluoride derivatives. Photochem Photobiol Sci 15:420–430
17. Ito F, Sagawa T (2013) Quantitative evaluation of thermodynamic parameters for thermal back-reaction after mechanically induced fluorescence change. RSC Adv 3:19785–19788
18. Zhang G, Chen J, Payne SJ, Kooi SE, Demas JN, Fraser CL (2007) Multi-emissive difluoroboron dibenzoylmethane polylactide exhibiting intense fluorescence and oxygen-sensitive room-temperature phosphorescence. J Am Chem Soc 129:8942–8943
19. Bernstein J (2008) Polymorphism in molecular crystals. Oxford University Press, New York
20. Rungsimanon T, Yuyama K, Sugiyama T, Masuhara H, Tohnai N, Miyata M (2010) Control of crystal polymorph of glycine by photon pressure of a focused continuous wave near-infrared laser beam. J Phys Chem Lett 1:599–603
21. Erdemir D, Chattopadhyay S, Guo L, Ilavsky J, Amenitsch H, Segre CU, Myerson AS (2007) Relationship between self-association of glycine molecules in supersaturated solutions and solid state outcome. Phys Rev Lett 99:115702
22. Guo C, Wang J, Li J, Wang Z, Tang S (2016) Kinetic pathways and mechanisms of two-step nucleation in crystallization. J Phys Chem Lett 7:5008–5014
23. Karthika S, Radhakrishnan TK, Kalaichelvi P (2016) A review of classical and nonclassical nucleation theories. Cryst Growth Des 16:6663–6681
24. Ye X, Liu Y, Lv Y, Liu G, Zheng X, Han Q, Jackson KA, Tao X (2015) In situ microscopic observation of the crystallization process of molecular microparticles by fluorescence switching. Angew Chem Int Ed 54:7976–7980

25. Ito F, Kakiuchi T, Sakano T, Nagamura T (2010) Fluorescence properties of pyrene derivative aggregates formed in polymer matrix depending on concentration. Phys Chem Chem Phys 12:10923–10927
26. Ito F, Fujimori J, Oka N, Sliwa M, Ruckebusch C, Ito S, Miyasaka H (2017) AIE phenomena of a cyanostilbene derivative as a probe of molecular assembly processes. Faraday Discuss 196:231–243
27. Maeda H (2013) Supramolecular chemistry of pyrrole-based pi-conjugated molecules. Bull Chem Soc Jpn 86:1359–1399
28. Haketa Y, Maeda H (2017) Dimension-controlled ion-pairing assemblies based on pi-electronic charged species. Chem Commun 53:2894–2909
29. Maeda H, Kusunose Y (2005) Dipyrrolyldiketone difluoroboron complexes: Novel anion sensors with C-H center dot center dot center dot X- interactions. Chem Eur J 11:5661–5666
30. Maeda H, Haketa Y, Nakanishi T (2007) Aryl-substituted C-3-bridged oligopyrroles as anion receptors for formation of supramolecular organogels. J Am Chem Soc 129:13661–13674
31. Maeda H, Mihashi Y, Haketa Y (2008) Heteroaryi-substituted C-3-bridged oligopyrroles: potential building subunits of anion-responsive pi-conjugated oligomers. Org Lett 10:3179–3182
32. Maeda H, Bando Y, Haketa Y, Honsho Y, Seki S, Nakajima H, Tohnai N (2010) Electronic and optical properties in the solid-state molecular assemblies of anion-responsive pyrrole-based it-conjugated systems. Chem Eur J 16:10994–11002
33. Talanquer V, Oxtoby DW (1998) Crystal nucleation in the presence of a metastable critical point. J Chem Phys 109:223–227
34. Vekilov PG (2004) Dense liquid precursor for the nucleation of ordered solid phases from solution. Cryst Growth Des 4:671–685

Chapter 32
Fabrication of Charge-Transfer Complex Nanocrystals Toward Electric Field-Induced Resistive Switching

Tsunenobu Onodera and Hidetoshi Oikawa

Abstract Charge-transfer complex exhibits versatile characteristics such as optical properties, conductivity, and magnetism. The physicochemical properties can be controlled through molecular design combining various kinds of electron donors and acceptors. In particular, the degree of charge-transfer γ and the composition ratio of electron donor to acceptor are important factors to determine physicochemical properties. On the other hand, nanocrystallized charge-transfer complex would lead to unique properties, being different from both isolated molecule and bulk crystal. However, the reprecipitation method for common organic nanocrystals cannot be employed due to low solubility of charge-transfer complex in common organic solvents. Therefore, novel nanocrystallization method for charge-transfer complex should be developed. In this chapter, nanocrystallization involving doping process of charge-transfer complex, copper 7,7,8,8-tetracyanoquinodimethane (Cu-TCNQ), will be introduced in detail towards nanoelectronics application. Cu-TCNQ is typical Mott insulator and indicates a unique resistive switching behavior. It is expected that Cu-TCNQ nanocrystals with an excess amount of Cu, namely doped Cu-TCNQ nanocrystals, would show unique resistive switching behavior because of the degree of charge-transfer γ different from bulk crystal.

Keywords Charge-transfer complexes · Nanocrystals · Doping

32.1 Introduction

Charge-transfer complex is one of the interesting research subjects because of versatile characteristics such as optical properties, conductivity, and magnetism [1]. Especially, electric and magnetic properties have been studied intensively because it is possible to control physicochemical properties through molecular design combining various kinds of electron donor and acceptor. The degree of charge-transfer γ and

T. Onodera (✉) · H. Oikawa
Institute of Multidisciplinary Research for Advanced Materials (IMRAM), Tohoku University, Katahira 2-1-1, Aoba-ku, Sendai 980-8577, Japan
e-mail: tsunenobu.onodera.a4@tohoku.ac.jp

H. Miyasaka et al. (eds.), *Photosynergetic Responses in Molecules and Molecular Aggregates*, https://doi.org/10.1007/978-981-15-5451-3_32

the composition ratio of electron donor to acceptor are important factors to determine physicochemical properties of charge-transfer complex [2]. On the other hand, the nanocrystal is occupied in an intermediate state between isolated molecules and corresponding bulk crystals. Therefore, nanocrystallization is utilized to obtain unique properties different from both isolated molecule and bulk crystal [3]. So far, many organic nanocrystals have been fabricated by means of the so-called reprecipitation method; for example, low molecular weight aromatic carbons, ionic dyes, π-conjugated polymers, and so on [4–9]. The crystal size was located in the range of several tens nanometer to sub-micrometers. In the case of crystalline charge-transfer complex on microscale, several fabrication methods, for example, vapor deposition [10, 11], spontaneous electrolysis in organic solvents [12, 13], and chemical and electrochemical reduction methods [14, 15] have been proposed. These fabrication methods, however, were not useful to control the size and shape of crystals on nanoscale. Furthermore, low solubility of charge-transfer complex in common organic solvents disabled us from employing the conventional reprecipitation method for organic nanocrystals. Hence, novel nanocrystallization method of charge-transfer complex should be developed.

32.2 Fabrication of Charge-Transfer Complex Nanocrystals and Their Nanostructure

Cu-TCNQ nanocrystals were fabricated by coprecipitation method with chemical reduction. Namely, methanol solutions of electron donor (Cu) and accepter (TCNQ) were mixed in the presence of a reducing agent as follows [16, 17]. First, TCNQ (0.13–0.21 mmol) was dissolved in methanol (500 mL). The methanol solution was deoxygenated by flow of nitrogen gas and kept at a given temperature. Next, an excess amount of $NaBH_4$ (0.52 mmol) was added into the solution. In this process, the solution turned into bright green in color. TCNQ in the starting methanol solution exists as a TCNQ radical anion ($TCNQ^-$) because of the reduction with $NaBH_4$, which was confirmed by the visible absorption spectrum assigned to $TCNQ^-$ [18]. Next, $CuSO_4{\cdot}5H_2O$ (0.11 mmol) dissolved in methanol (25–150 mL) was added dropwise (0.83 mL min^{-1}) into the vigorously stirred methanol solution containing $TCNQ^-$ and $NaBH_4$. After adding $CuSO_4$, the mixed solution was further stirred at a given temperature for 40 min. In this process, Cu^{2+} in methanol solution of $CuSO_4$ was reduced into Cu^+ with $NaBH_4$. At the same time, insoluble complexes of Cu^+ and $TCNQ^-$ were formed and grew up to be nanocrystals. The resulting methanol dispersion liquid of Cu-TCNQ nanocrystals was bright blue in color. The size of Cu-TCNQ nanocrystals was controlled by the changing amount of TCNQ and $NaBH_4$. On the other hand, it is also possible to prepare Cu-TCNQ nanocrystals by using a copper salt precursor such as $CuSO_4$ and an alkali metal-TCNQ precursor such as Li-TCNQ by cation exchange [16]. A methanol solution (200 μL) was injected into a vigorously stirred solution of Li-TCNQ (0.44–2.2 mM) in methanol (10 mL).

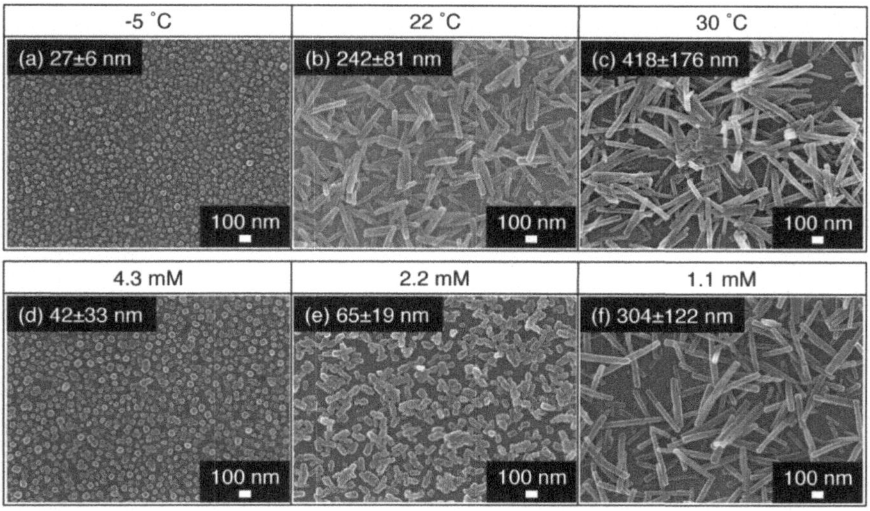

Fig. 32.1 SEM images of Cu-TCNQ nanocrystals fabricated under different conditions of (**a**–**c**) temperature and (**d**–**f**) concentration of $CuSO_4 \cdot 5H_2O$ at 15 °C. Average crystal size was described with standard deviation Ref. [17]

Then, Cu-TCNQ nanocrystals were reprecipitated as Eq. 32.1, and the filtrated with Millipore filter.

$$Cu^{2+} + 2Li^{+}TCNQ^{-} \rightarrow Cu^{+}TCNQ^{-} + TCNQ^{0} + 2Li^{+} \quad (32.1)$$

The formation of Cu-TCNQ nanocrystals was confirmed by scanning electron microscope (SEM: JEOL, JSM-6700F) observation. Figure 32.1 displays SEM images of Cu-TCNQ nanocrystals fabricated under various conditions. As the temperature was raised and the amount of $CuSO_4 \cdot 5H_2O$ decreased, crystal size increased. As a result, we were able to fabricate well-defined Cu-TCNQ nanocrystals, which were rod-like shape within a range from 30 nm to 1000 nm along the longitudinal axis. The composition ratio of Cu to TCNQ was investigated by elemental analysis (J-Science Lab. Co, Micro Corder JM10) in C, H, and N atoms (Found in typical Cu-TCNQ nanocrystals: C, 49.88; H, 1.89; N, 19.34). Interestingly, all of the resulting nanocrystals were found to be Cu:TCNQ=1.3:1 (Calc. for C12H4N4Cu1.3: C, 50.25; H, 1.41; N, 19.54. Calc. for C12H4N4Cu: C, 53.83; H, 1.51; N, 20.93). In order to further investigate the composition of size-controlled Cu-TCNQ nanocrystals in detail, Raman spectroscopy (Tokyo Instruments, Nano finder 30) was performed. In general, charge-transfer complexes of TCNQ exhibit strong Raman bands of TCNQ molecule, whose peak positions are strongly affected by the chemical state of TCNQ [19]. Figure 32.2 shows Raman spectra of Cu-TCNQ bulk crystal and nanocrystals. Three bands at ca. 1200, 1380, and 1620 cm^{-1} in nanocrystals agreed with the C=C–H bending, C–CN stretching, and C=C ring stretching modes of TCNQ in bulk

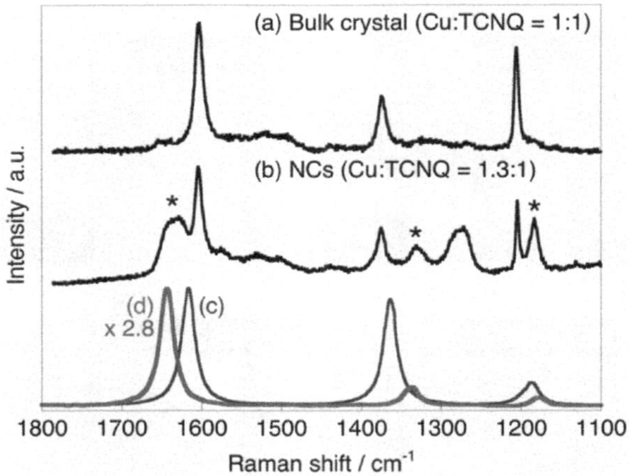

Fig. 32.2 Raman spectra of **a** Cu-TCNQ bulk crystal (Cu:TCNQ=1:1) and **b** nanocrystal (Cu:TCNQ=1.3:1) with 34 ± 12 nm in size, and calculated spectra of **c** TCNQ anion and **d** TCNQ dianion by DFT calculation Ref. [17]

crystal (Cu:TCNQ=1:1), respectively [19]. In addition, new bands were observed in nanocrystals. These new bands were assigned to TCNQ dianion [20], which roughly agreed with DFT (density functional theory, EDF2/6-31G*, Spartan'10 software) calculation as shown in Fig. 32.2d. Furthermore, chemical state of Cu in Cu-TCNQ nanocrystals was also investigated using scanning transmission electron microscopy—electron energy loss spectroscopy (STEM-EELS: FEI, Titan[3] Double Corrector Super-X). Figure 32.3 indicates STEM-EELS spectra of Cu in Cu-TCNQ nanocrystals and bulk crystals. The EELS spectrum of Cu in Cu-TCNQ nanocrystals corresponds to Cu-L_3 edge and agreed with that in bulk crystals, which means

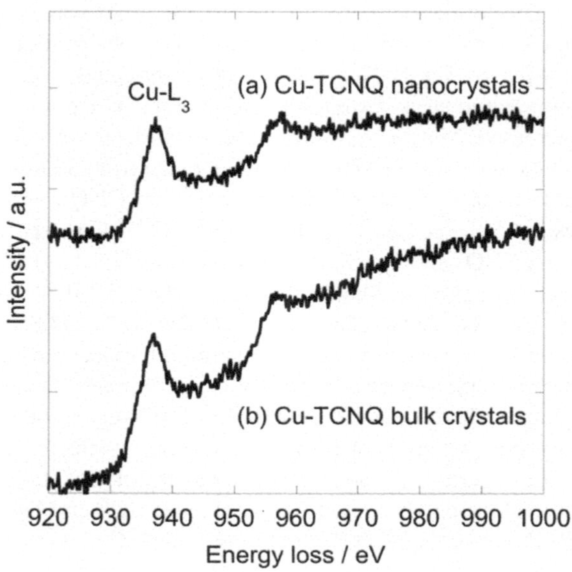

Fig. 32.3 STEM-EELS spectra of **a** Cu-TCNQ nanocrystals (Cu:TCNQ=1.3:1) and **b** bulk crystal (Cu:TCNQ=1:1)

the chemical state of Cu in Cu-TCNQ nanocrystals was mono cation. Actually, it is reported that the main peaks for the Cu-L_3 edge are located at 935.5 eV for Cu, 936.1 eV for Cu_2O, 933.6 eV for CuO, respectively [21]. These facts demonstrate that the excess amount of Cu resulted in the generation of TCNQ dianions in Cu-TCNQ nanocrystals to keep the charge balance between TCNQ and excessive Cu. Moreover, it is noted that the Raman intensity ratio of dianion (C=C stretching mode at 1320 nm) to anion (C=C ring stretching mode at 1380 nm) increased as crystal size increased [17]. This fact suggests that the degree of charge-transfer γ from Cu to TCNQ in nanocrystals would be dependent on crystal size.

Next, crystal structure of Cu-TCNQ nanocrystals was examined by measuring powder X-ray diffraction (XRD) patterns (Bruker, D8 Advance). It is reported that Cu-TCNQ crystal has two polymorphic forms [19]. Phase I has a face-to-face packing structure, in which planar TCNQ molecules are stacked with effective π-π interaction between adjacent TCNQ molecules. The other is phase II with high resistance, which has a tetragonal structure. In a viewpoint of electronic device application, phase I is a very important material because of unique switching behavior between complete and partial charge-transfer states by applying electric field and/or irradiating light [19]. Figure 32.4 shows typical powder XRD patterns of Cu-TCNQ nanocrystals, Cu-TCNQ micro-rods beyond *ca.* 1 μm in size, and bulk crystals (Phases I and II). Interestingly, all of the Cu-TCNQ nanocrystals below *ca.* 1 μm in size were assigned to phase I in spite of composition different from bulk crystal. This fact

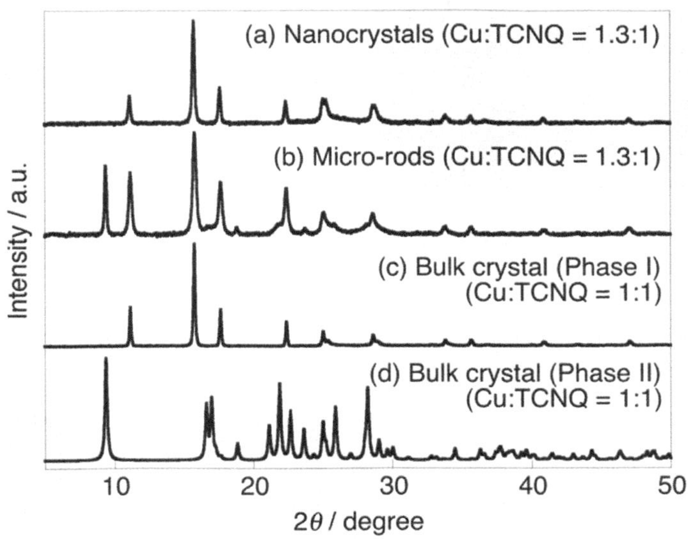

Fig. 32.4 Powder XRD patterns of **a** Cu-TCNQ nanocrystals (Cu:TCNQ=1.3:1) with 34 ± 12 nm in size, **b** micro-rods beyond 1 μm in size, **c** bulk crystal (Phase I) and **d** bulk crystal (Phase II). The XRD pattern **d** was simulated using crystal structure data [19]

might be attributed to the relaxation of lattice strain loaded on Cu-TCNQ nanocrystals with an excess amount of Cu because of the large specific surface area [22]. On the other hand, Cu-TCNQ micro-rods beyond *ca.* 1 μm in size was the mixture of Phases I and II from Fig. 32.4c, d. In large crystals with more ca. 1 μm in size, the strain could be accumulated and not be released smoothly due to a relatively smaller specific surface area. Actually, XRD peak width of nanocrystals was broader than that of bulk crystal (Phase I) in Fig. 32.4 and became broader with increasing nanocrystal size. Figure 32.5 shows scanning transmission electron microscopy—energy dispersive X-ray spectroscopy (STEM-EDS) mapping (FEI, Titan[3] Double Corrector Super-X) of Cu-TCNQ nanocrystals. The spectra were relatively noisy, but it is confirmed that Cu atoms were homogeneously contained in Cu-TCNQ nanocrystals. As a result, the structure of Cu-TCNQ nanocrystals was schematically proposed as shown in Fig. 32.6. The Cu-TCNQ nanocrystal (Cu:TCNQ=1.3:1) is expressed as $Cu_{1.3}(TCNQ^{-})_{0.7}(TCNQ^{2-})_{0.3}$. The resulting Cu-TCNQ nanocrystals, which have the same crystal structure as bulk crystal (Cu:TCNQ=1:1), can be regarded as doped

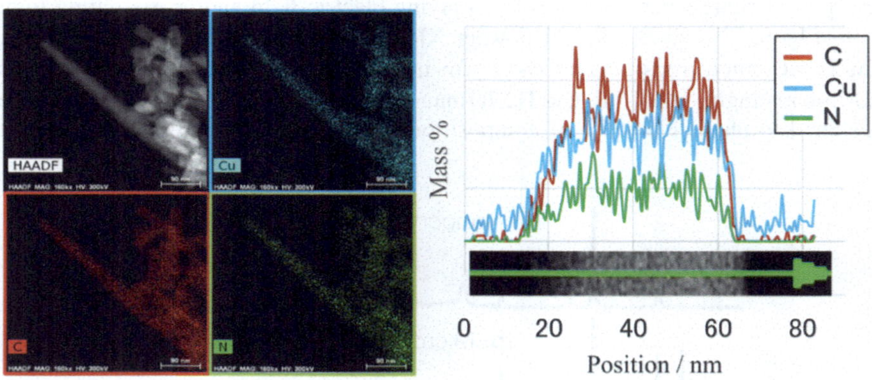

Fig. 32.5 STEM-EDS mappings and line mass profile of Cu-TCNQ nanocrystals (Cu:TCNQ=1.3:1)

Fig. 32.6 Proposed structure of Cu-TCNQ nanocrystals (Cu:TCNQ=1.3:1) and bulk crystal (Cu:TCNQ=1:1). Red and blue discs indicate TCNQ molecule and Cu atom, respectively. Yellow and white circles mean electron and hole, respectively

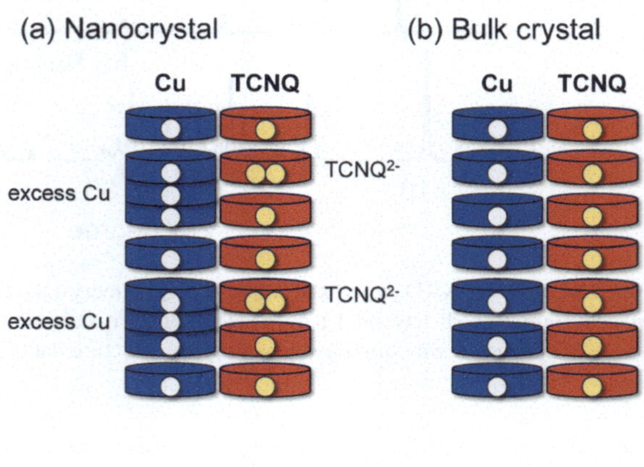

Cu-TCNQ nanocrystals. In other words, an excess amount of Cu was included in nanocrystals, so that excess amount of electron was injected in TCNQ column.

The doped Cu-TCNQ nanocrystals were characterized by UV-Vis-NIR absorption spectroscopy (JASCO, V-570). Figure 32.7 shows UV-Vis-NIR absorption spectra of Cu-TCNQ nanocrystals. The doped Cu-TCNQ nanocrystals were immobilized on a glass substrate as follows. A glass substrate with UV/O_3 treatment was immersed into methanol solution of 3-aminopropyltriethoxysilane under ultrasonic irradiation for 20 min, and then washed with pure methanol under ultrasonic irradiation for 20 min and dried up in air. After the treatment, this substrate was again immersed into the dispersion liquid of doped Cu-TCNQ nanocrystals for several hours, and then carefully washed with pure methanol and dried in a vacuum. Three characteristic peaks were observed at 381 nm, 700 nm, and in the near-infrared (NIR) region. In the absorption spectrum of Cu-TCNQ dissolved in acetonitrile, characteristic peaks for the TCNQ anion radical appear at 409, 744, and 842 nm. Therefore, the peaks at 381 and 700 nm in the nanocrystals are assigned to the TCNQ radical anion [18]. These peaks were shifted and broader than those for Cu-TCNQ dissolved in acetonitrile because the surrounding dielectric environment of the TCNQ anion radical is different in each case. On the other hand, NIR absorption was expected to relate with the existence of TCNQ dianion. Here, it is known that various kinds of metal-TCNQ complexes with partial charge-transfer state ($0 < \gamma < 1$) exhibit an intraband transition from TCNQ radical anion to neutral TCNQ in NIR region beyond 2000 nm [1, 2]. Based on analogous principles, we speculated new NIR absorption peak was attributed to the intraband transition between TCNQ radical anion and dianion. Actually, on-site Coulomb repulsion does not change in the present case. The crystal size dependence of peak energy would be explained by Coulomb repulsion during electron transfer between TCNQ radical anion and dianion in doped nanocrystals.

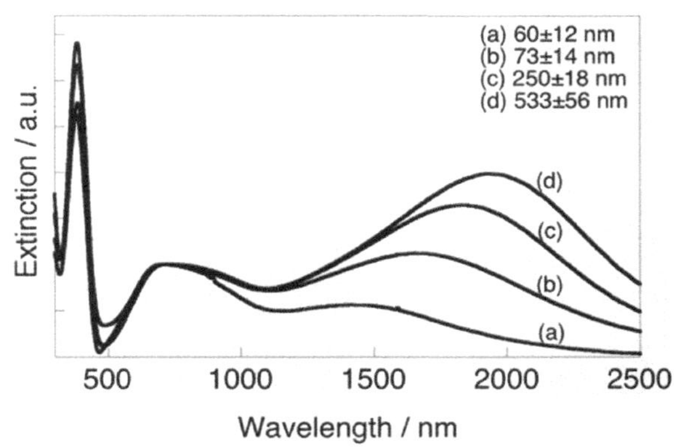

Fig. 32.7 UV-Vis-NIR absorption spectra of Cu-TCNQ nanocrystals (Cu:TCNQ=1.3:1) with different sizes: **a** 60 ± 12 nm, **b** 73 ± 14 nm, **c** 250 ± 15 nm and **d** 533 ± 56 nm Ref. [17]

32.3 Resistive Switching Properties of Doped Cu-TCNQ Nanocrystals

The resistive switching properties of doped Cu-TCNQ nanocrystals (Cu:TCNQ=1.3:1) and bulk crystal (Cu:TCNQ=1:1) were evaluated by measuring the current-voltage curve at room temperature. A sandwich structure was fabricated as a measurement set-up as shown in Fig. 32.8a: the top, middle, and bottom layers were gallium-indium (Ga-In) as a top electrode, Cu-TCNQ nanocrystals, and gold (Au) as a bottom electrode, respectively. As shown in Fig. 32.8b, the Au layer (20 nm in thickness) was vapor-deposited on a membrane filter (Millipore Isopore VMTP), which has a flat surface, and then, a monolayer of Cu-TCNQ nanocrystals was carefully prepared by the filtration of the diluted and purified dispersion liquid of Cu-TCNQ nanocrystals as a measurement sample. A Ga-In spot was placed on the Cu-TCNQ nanocrystals layer. On the other hand, typical Cu-TCNQ bulk crystals were fabricated by spontaneous electrolysis [12, 13] for comparison with doped nanocrystals. A cleaned copper foil was first immersed into a saturated acetonitrile solution of TCNQ. Upon contact with the solution, the black-purple film was observed to appear on the exposed copper surfaces. The resulting Cu-TCNQ crystal was micro-rod with *ca.* 60 μm in size. The measurement of current-voltage curve was performed using a source meter (Keithley 2400 Source Meter) and a dynamic current power supply (Kenwood PAR20-4H).

The electric field-induced resistive switching behavior has been observed successfully in doped Cu-TCNQ nanocrystals. Figure 32.9 shows the typical current-voltage curve of doped Cu-TCNQ nanocrystals (Cu:TCNQ=1.3:1) and bulk crystal

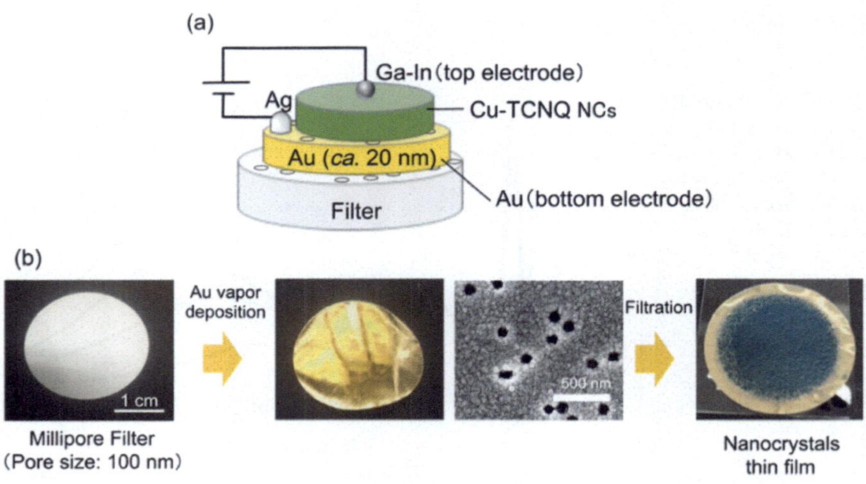

Fig. 32.8 a Experimental set-up for measuring current-voltage curve of doped Cu-TCNQ nanocrystals and **b** fabrication process of doped Cu-TCNQ nanocrystals thin film

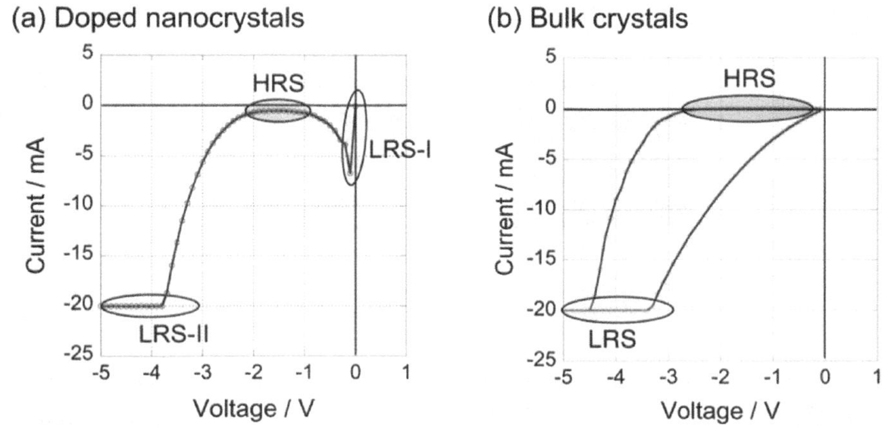

Fig. 32.9 Typical current-voltage curve of **a** doped Cu-TCNQ nanocrystals and **b** bulk crystal. HRS means high resistance state, and LRS-I and LRS-II are low resistance states, respectively

(Cu:TCNQ=1:1). The three-step switching behavior was observed in doped Cu-TCNQ nanocrystal due to the generation of TCNQ dianions as well as TCNQ anion radicals. This behavior was not recorded in Cu-TCNQ bulk crystals (Cu:TCNQ=1:1). In the usual case, as-prepared Cu-TCNQ bulk crystals are first in high resistance state (HRS) because of Coulomb repulsion between electrons as shown in Fig. 32.10. When the electric field is applied, the bulk crystals can switch to low resistance state (LRS) owing to the partial change from the TCNQ anion radical to the neutral TCNQ [19]. When the electric field is reversely swept, the resistance state of bulk crystals returns to the HRS again. Namely, Cu-TCNQ bulk crystal shows reversible stability. On the other hand, as-prepared doped Cu-TCNQ nanocrystals are first in low resistance state (LRS-I) different from bulk crystal, because doped Cu-TCNQ nanocrystals contain TCNQ dianion as well as TCNQ anion radical, and electron in TCNQ dianion can work as a carrier. Actually, on-site Coulomb repulsion does not change by electron transfer from TCNQ dianion to TCNQ anion radical. When the electric field is swept, doped Cu-TCNQ nanocrystals switched to the high resistance state (HRS) because the chemical state of all of TCNQ molecules in columnar structure changes to anion radical. Furthermore, doped Cu-TCNQ nanocrystals were switched again to the other low resistance state (LRS-II) owing to the partial change from the TCNQ anion radical to the neutral TCNQ caused by further application of an electric field. This mechanism is supported by observing new Raman signal at ca. 1450 cm^{-1} assigned to neutral TCNQ in the switched sample as shown in Fig. 32.11 [23]. Cu-TCNQ nanocrystals showed multistep switching induced by electric field as novel properties owing to the generation of TCNQ dianions.

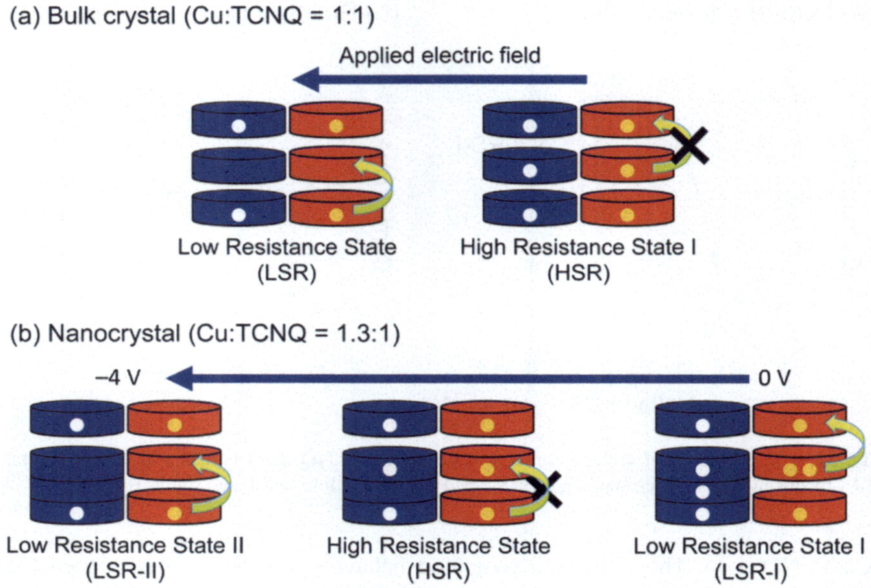

Fig. 32.10 Resistive switching mechanism of Cu-TCNQ bulk crystal (**a**) and doped nanocrystals (**b**). Red and blue discs indicate TCNQ molecule and Cu atom, respectively. Yellow and white circles mean electron and hole, respectively

Fig. 32.11 Raman spectra of doped Cu-TCNQ nanocrystals before and after resistive switching from HRS to LRS-II

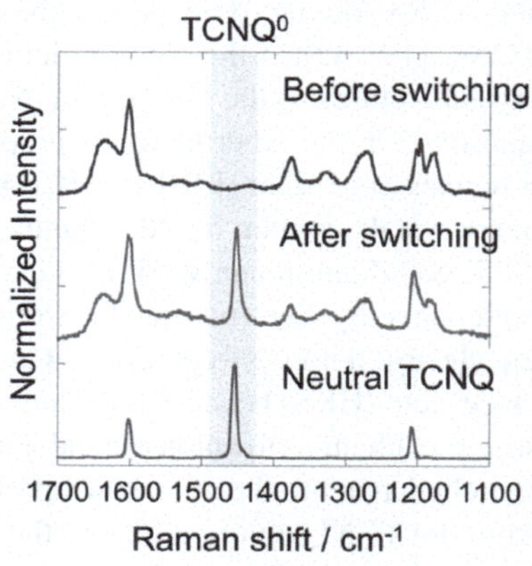

32.4 Conclusion

The doped Cu-TCNQ nanocrystals have been fabricated successfully using coprecipitation method with chemical reduction. The resulting nanocrystals have composition ratio of Cu:TCNQ=1.3:1 and the same crystal structure as bulk crystal (Cu:TCNQ=1:1). The excess amount of Cu resulted in the generation of TCNQ dianions in Cu-TCNQ nanocrystals to keep the charge balance between TCNQ and excessive Cu, which was supported by Raman spectroscopy and STEM-EELS. Their doped nanocrystals also showed a strong absorption peak in NIR region, whose positions were strongly dependent on the content of TCNQ dianion.

The resulting doped Cu-TCNQ nanocrystals exhibited the electric field-induced resistive switching. Especially, multiple switching was observed due to multistep redox reaction of TCNQ dianions. This behavior was characteristic in doped Cu-TCNQ nanocrystals.

Acknowledgements The authors especially thank the late Emeritus Professor Hachiro Nakanishi in Tohoku University (TU, Japan), Prof. Hitoshi Kasai (IMRAM, TU, Japan), Associate Prof. Akito Masuhara in Yamagata University, Mr. Yuichiro Hayasaka (Technical staff, The Electron Microscopy Center, TU, Japan), and D.Sc Kentaro Hiraishi, M.Sc. Satomi Matsuo and M.Sc. Reo Hirose, B.Sc Fumika Matsushita (Dept. of Chemistry, Graduate School of Sci., TU, Japan at that time) for their valuable scientific discussion and great experimental contributions. The present work was supported in part by the Grants-in-Aid for Scientific Research (No. 23655183) from the Ministry of Education, Culture, Sports, Science and Technology, Japan and by JSPS KAKENHI Grant Number 26107002, Grant-in-Aid for Scientific Research on Innovative Areas "Photosynergetics".

References

1. Tanaka J, Tanaka M, Kawai T, Takabe T, Maki O (1976) Electronic spectra and electronic structure of TCNQ complex. Bull Chem Soc Jpn 49:2358–2373
2. Torrance JB (1979) The difference between metallic and insulating salts of tetracyanoquinodimethone (TCNQ): how to design an organic metal. Acc Chem Res 12:79–86
3. Nalwa HS (ed) (2000) Handbook of nanostructured materials and nanotechnology. Academic Press, San Diego
4. Kasai H, Nalwa HS, Oikawa H, Okada S, Matsuda H, Minami N, Kakuta A, Ono K, Mukoh A, Nakanishi H (1992) A novel preparation method of organic microcrystals. Jpn J Appl Phys 31:L1132–L1134
5. Kasai H, Oikawa H, Okada S, Nakanishi H (1998) Crystal growth of perylene microcrystals in the reprecipitation method. Bull Chem Soc Jpn 71:2597–2601
6. Baba K, Kasai H, Okada S, Oikawa H, Nakanishi H (2000) Novel fabrication process of organic microcrystals using microwave-irradiation. Jpn J Appl Phys 39:L1256–L1258
7. Ishii KU, Baba K, Wei Z, Kasai H, Nakanishi H, Okada S, Oikawa H (2006) Mass-production of pigment nanocrystals by the reprecipitation method and their encapsulation. Mol Cryst Liq Cryst 445:177/[467]–183[473]
8. Kaneko Y, Shimada S, Fukuda T, Kimura T, Yokoi H, Matsuda H, Onodera T, Kasai H, Okada S, Oikawa H, Nakanishi H (2005) A novel method for fixing the anisotropic orientation of dispersed organic nanocrystals in a magnetic field. Adv Mater 17:160–163
9. Baba K, Kasai H, Masuhara A, Okada S, Oikawa H, Nakanishi H (2007) Diacetylene nanowire crystals prepared by reprecipitation/microwave-irradiation method. Jpn J Appl Phys 46:7558–7561
10. Oyamada T, Tanaka H, Matsushige K, Sasabe H, Adachi C (2003) Switching effect in Cu:TCNQ charge transfer-complex thin film by vacuum codeposition. Appl Phys Lett 83:1252–1254
11. Xiao K, Ivanov IN, Puretzky AA, Liu Z, Geohegan DB (2006) Directed integration of tetracyanoquinodemethane-Cu organic nanowires into prefabricated device architectures. Adv Mater 18:2184–2188

12. Liu SG, Liu YQ, Wu PJ, Zhu DB (1996) Muitifaceted study of CuTCNQ thin-film materials. fabrication, morphology, and spectral and electrical switching properties. Chem Mater 8:2779–2787
13. Liu Y, Ji Z, Tang Q, Jiang L, Li H, He M, Hu W, Zhang D, Jiang L, Wang X, Wang C, Liu Y, Zhu D (2005) Particle-size control pattering of a charge-transfer complex for nanoelectronics. Adv Mater 17:2953–2957
14. O'Mullane AP, Fay N, Nafady A, Bond AM (2007) Preparation of metal-TCNQ charge-transfer complexes on conducting and insulating surfaces by photocrystallization. J Am Chem Soc 129:2066–2073
15. Neufeld AK, O'Mullane AP, Bond AM (2005) Control of localized nanorod formation and patterns of semiconducting CuTCNQ phase I crystals by scanning electrochemical microscopy. J Am Chem Soc 127:13846–13853
16. Hiraishi K, Masuhara A, Yokoyama T, Kasai H, Nakanishi H, Oikawa H (2009) Fabrication and characterization of size-controlled CuTCNQ charge-transfer complex nanocrystals. J Cryst Growth 311:948–952
17. Onodera T, Matsuo S, Hiraishi K, Masuhara A, Kasai H, Oikawa H (2012) Fabrication of doped Cu-TCNQ nanocrystals and their optoelectronic properties. CrystEngComm 14:7586–7589
18. Jonkman HT, Kommandeur J (1972) The UV spectra and their calculation of TCNQ and its mono- and di-valent anion. Chem Phys Lett 15:496–499
19. Heintz RA, Zhao H, Ouyang X, Grandinetti G, Cowen J, Dunbar KR (1999) New insight into the nature of Cu(TCNQ): solution route to two distinct polymorphs and their relationship to crystalline films that display bistable switching behavior. Inorg Chem 38:144–156
20. Khatkale MS, Devlin JP (1979) The vibrational and electronic spectra of the mono-, di-, and tri-anion salts of TCNQ. J Chem Phys 70:1851–1859
21. Laffont L, Wu MY, Chevallier F, Poizot P, Morcrette M, Tarascon JM (2006) High resolution EELS of Cu-V oxide: application of batteries materials. Micron 37:459–464
22. Takahashi S, Miura H, Kasai H, Okada S, Oikawa H, Nakanishi H (2002) Single-crystal-to-single-crystal transformation of diolefin derivatives in nanocrystals. J Am Chem Soc 124:10944–10945
23. Kamitsos EI, Risen WM (1983) Raman studies in CuTCNQ: Resonance Raman spectral observations and calculations for TCNQ ion radicals. J Chem Phys 79:5808–5819

Chapter 33
Turn-on Mode Photoswitchable Fluorescent Diarylethenes for Super-Resolution Fluorescence Microscopy

Masakazu Morimoto and Masahiro Irie

Abstract Sulfone derivatives of 1,2-bis(2-alkyl-6-aryl-1-benzothiophen-3-yl)perfluorocyclopentene have been developed as a new type of turn-on mode photoswitchable fluorescent molecules for super-resolution fluorescence microscopy. Upon irradiation with ultraviolet (UV) light, the open-ring isomers undergo cyclization reactions to produce the closed-ring isomers, which emit brilliant fluorescence with high quantum yields ($\Phi_f \sim 0.9$). Upon irradiation with visible light, the closed-ring isomers revert back to the initial open-ring isomers and the fluorescence vanishes. The cycloreversion quantum yields of the fluorescent diarylethenes were tuned by appropriate chemical modifications to fulfill the requirements for the different types of super-resolution imaging techniques. A turn-on mode fluorescent diarylethene that shows solvatochromism of fluorescence in the closed-ring form is potentially applicable to multicolor super-resolution imaging of microscopic environmental polarity. Unprecedented reversible photoswitching of fluorescent diarylethene molecules upon irradiation with single-wavelength visible light was found and has been successfully applied to super-resolution fluorescence imaging.

Keywords Diarylethene · Fluorescence · Photochromism · Photoswitch · Super-resolution fluorescence microscopy

33.1 Introduction

Fluorescent molecules with photoswitching ability have been attracting much attention because of their potential applications to optical memory media as well as super-resolution fluorescence microscopy [1–5]. Photoswitchable fluorescent molecules can be constructed by integrating both photochromic and fluorescent chromophores

M. Morimoto (✉) · M. Irie
Department of Chemistry and Research Center for Smart Molecules, Rikkyo University, 3-34-1 Nishi-Ikebukuro, Toshima-ku, Tokyo 171-8501, Japan
e-mail: m-morimoto@rikkyo.ac.jp

M. Irie
e-mail: iriem@rikkyo.ac.jp

H. Miyasaka et al. (eds.), *Photosynergetic Responses in Molecules and Molecular Aggregates*, https://doi.org/10.1007/978-981-15-5451-3_33

in a molecule [6–14]. So far, based on this methodology, a number of photoswitchable fluorescent molecules have been synthesized. These molecules are initially fluorescent, while the fluorescence is turned off by excitation energy transfer or electron transfer when the photochromic unit undergoes photoisomerization.

Super-resolution fluorescence microscopy breaks the diffraction limit of light and provides fluorescence images with nanometer-scale spatial resolution. The imaging techniques are usually classified into two categories, coordinate-targeted and coordinate-stochastic methods. In coordinate-targeted methods, such as reversible saturable (switchable) optical linear fluorescence transition (RESOLFT) microscopy [15, 16], the position of the on- and off-states is determined by a doughnut-shaped pattern of light inducing a turn-on or a turn-off transition in the sample, except at the center with vanishing intensity. Coordinate-stochastic methods, such as photoactivated localization microscopy (PALM) [17] and stochastic optical reconstruction microscopy (STORM) [18], are based on the detection and localization of randomly arising fluorescent spots of on-state single molecules. The reversibly photoswitchable fluorescent molecules can be applied to both RESOLFT microscopy and PALM/STORM. However, the turn-off mode photoswitchable fluorescent molecules are hardly applicable to PALM/STORM because the quenching of fluorescence, in most cases, is imperfect and the imaging methods require a dark background to detect the single-molecule fluorescence. For the application, it is strongly desired to develop turn-on mode photoswitchable fluorescent molecules, which can be efficiently and instantaneously activated to be fluorescent upon photoirradiation.

Although several turn-on mode photoactivatable fluorophores have been reported, the photoactivation quantum yields are relatively low and the reactions are irreversible [19–26]. Recently, a new type of turn-on mode photoswitchable fluorescent molecules has been developed. They are sulfone derivatives of 1,2-bis(2-alkyl-6-aryl-1-benzothiophen-3-yl)perfluorocyclopentene [27, 28]. Although the diarylethene derivatives are initially non-luminous under excitation with visible light, they are activated to emit strong fluorescence upon irradiation with UV light. The photoswitching and fluorescent properties are dependent on the substituents of the benzothiophene rings. In this chapter, we describe photochemical and photophysical properties of the turn-on mode fluorescent diarylethenes **1–8** (Fig. 33.1 and Table 33.1) and how to control the photoswitching performance by chemical modifications for the application to super-resolution fluorescence microscopy. Solvatochromism of photoswitchable fluorescence and unprecedented reversible switching with single-wavelength visible light are also described.

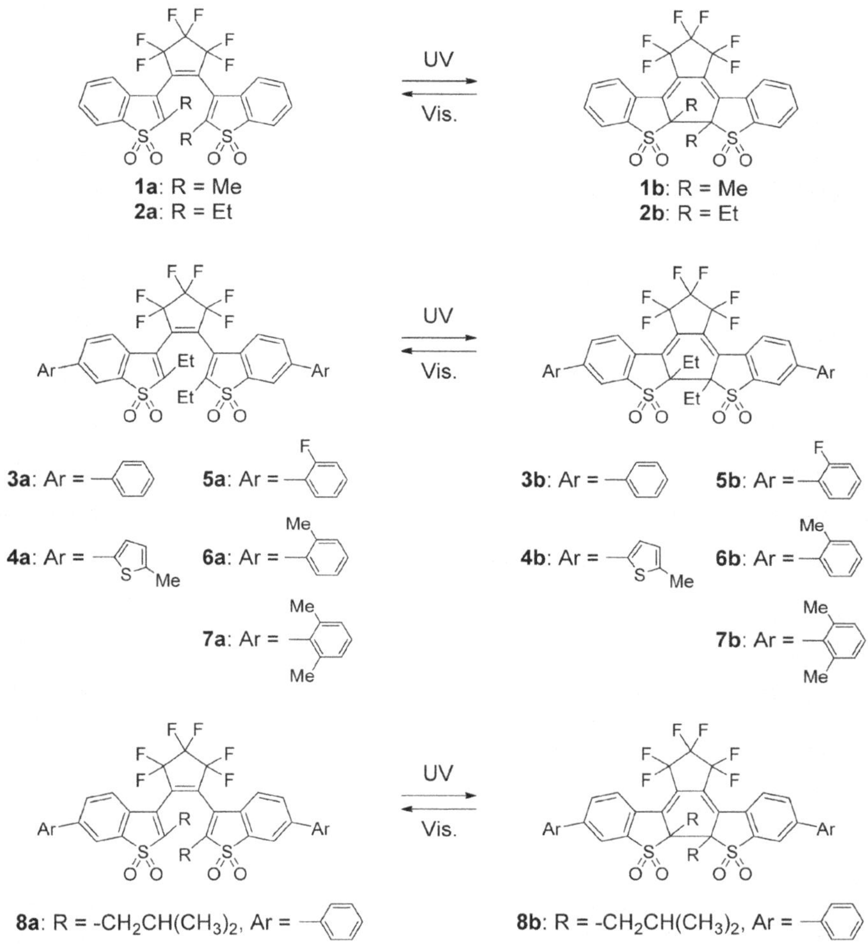

Fig. 33.1 Photoisomerizaiton of fluorescent diarylethenes **1–8**

33.2 Turn-on Mode Photoswitchable Fluorescent Diarylethenes Having Benzothiophene 1,1-Dioxide Groups

A sulfone derivative of 1,2-bis(2-methyl-1-benzothiophen-3-yl)perfluorocyclopentene, **1**, undergoes photochromism and shows fluorescence in both the open- and closed-ring isomers [29–31]. Open-ring isomer **1a** is colorless and has no optical absorption in the visible-wavelength region. Upon irradiation with UV light, **1a** undergoes a cyclization reaction to produce yellow closed-ring isomer **1b**. The absorption maximum of **1b** in ethyl acetate is observed at 398 nm

Table 33.1 Photophysical and photochemical properties of diarylethenes **1–8**

	Open-ring isomer, **a**		Closed-ring isomer, **b**		
	λ_{max}/nm ($\varepsilon/10^4$ $M^{-1}cm^{-1}$)	Φ_{oc}	λ_{max}/nm ($\varepsilon/10^4$ M^{-1} cm^{-1})	Φ_{co}	Φ_f
1	276 (0.37), 308 (0.41)	0.22[a]	398 (2.10)	0.061[d]	0.011
2	275 (0.52), 310 (0.52)	0.28[b]	414 (1.8)	0.18[e]	0.12[i]
3	298 (1.9), 336 (1.5)	0.62[b,c,h]	456 (4.6)	5.9×10^{-4}[f,h]	0.87
4	330 (2.4), 374 (2.5)	0.23[c,h]	506 (5.8)	$<1.0 \times 10^{-5}$[f,g,h]	0.78
5	293 (1.7), 328 (1.4)	0.58[c]	445 (4.2)	2.8×10^{-3}[f]	0.79
6	289 (1.2), 326 (1.0)	0.63[c]	441 (3.7)	3.2×10^{-3}[f]	0.84
7	385 (0.97), 320 (0.82)	0.59[c]	430 (3.5)	1.8×10^{-2}[f]	0.79
8	300 (1.9), 336 (1.5)	0.53[b]	460 (4.2)	1.0×10^{-2}[f]	0.80

λ_{max}: absorption maximum wavelength, ε: molar absorption coefficient, Φ_{oc}: cyclization quantum yield, Φ_{co}: cycloreversion quantum yield, and Φ_f: fluorescence quantum yield. Ethyl acetate and 1,4-dioxane were used as solvents for **1** and **2–8**, respectively

[a]Under irradiation with 312 nm light
[b]Under irradiation with 313 nm light
[c]Under irradiation with 330 nm light
[d]Under irradiation at λ_{max} of closed-ring isomer
[e]Under irradiation with 405 nm light
[f]Under irradiation with 450 nm light
[g]Under irradiation with 488 nm light
[h]Re-examined using a photoreaction quantum yield measuring system (Shimadzu, QYM-01) with a corrected power meter [27, 28]
[i]Measured in ethyl acetate

(Table 33.1), showing a large hypsochromic shift in comparison with the closed-ring isomer of the corresponding unoxidized derivative (515 nm in *n*-hexane) [32]. The fluorescence quantum yields of **1a** and **1b** in ethyl acetate are 0.025 and 0.011, respectively (Table 33.1).

The fluorescent properties are improved by changing the substituents of the benzothiophene 1,1-dioxide rings [33, 34]. When the methyl groups at 2- and 2′-positions (reactive carbon atoms) were replaced with ethyl ones, the fluorescence quantum yield of the closed-ring isomer increased from 0.011 (**1b**) to 0.12 (**2b**) (Table 33.1). Introduction of aryl substituents, such as phenyl or thienyl, at 6- and 6′-positions was found to further improve the fluorescent and photochromic properties. Figure 33.2a shows the absorption and fluorescence spectra of diarylethene **3** having ethyl groups at 2- and 2′-positions and phenyl groups at 6- and 6′-positions in 1,4-dioxane [33]. Open-ring isomer **3a** is colorless and has no absorption in the visible-wavelength region. Upon irradiation with UV light, the colorless solution turns yellow and a new absorption band appears at 456 nm due to the formation of closed-ring isomer **3b**. The absorption coefficient of **3b** is 4.6×10^4 M^{-1} cm^{-1} in

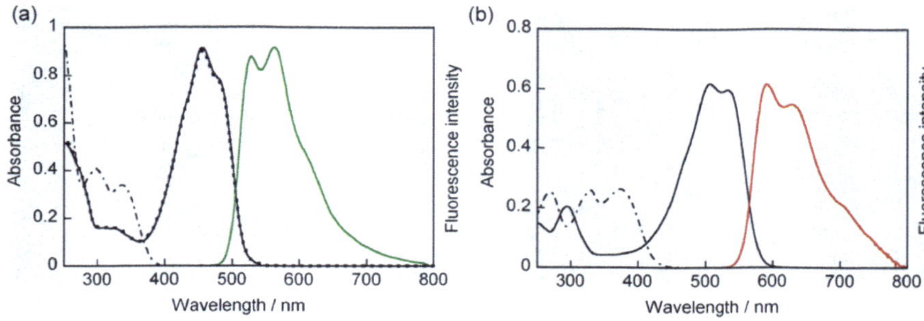

Fig. 33.2 Absorption and fluorescence spectra of diarylethenes **3** (**a** 2.0 × 10^{-5} M) and **4** (**b** 1.0 × 10^{-5} M) in 1,4-dioxane. Black dashed lines: absorption spectra of open-ring isomers **3a** and **4a**, black solid lines: absorption spectra of closed-ring isomers **3b** and **4b**, black dotted lines: absorption spectra of photostationary states under irradiation with 313 nm light for **3** and 365 nm light for **4**, green and red lines: fluorescence spectra of the closed-ring isomers under excitation with 488 nm light for **3b** and 532 nm light for **4b**. The photostationary spectrum of **4** overlaps with the spectrum of the closed-ring isomer due to 100% photoconversion. Reprinted with permission from [33]. Copyright 2011 American Chemical Society

1,4-dioxane, which is twice larger than that of **1b** (2.10 × 10^4 M^{-1} cm^{-1} in ethyl acetate). **3b** exhibits green fluorescence at around 550 nm under excitation with 488 nm light. Both the absorption and fluorescence spectra of **3b** with the extended π-conjugation show significant bathochromic shifts as compared to those of **1b**. The fluorescence quantum yield of **3b** is 0.87 in 1,4-dioxane, which is ~80 times higher than that of **1b**. **3a** has a high cyclization quantum yield of 0.62 in 1,4-dioxane while the cycloreversion quantum yield of **3b** is 5.9 × 10^{-4}, which is much lower than that of **1b** (0.061 in ethyl acetate). The photocycloreversion process is suppressed by the extension of π-conjugation in the closed-ring form [35].

3b exhibits green fluorescence under excitation with 488 nm laser light. Another laser wavelength commonly used is 532 nm. Replacement of the phenyl groups of **3** with thienyl ones shifts the absorption band of the closed-ring isomer to longer than 500 nm. The derivative **4** having 5-methyl-2-thienyl groups at 6- and 6′-positions can be excited with 532 nm laser light. Figure 33.2b shows the absorption and fluorescence spectra of **4** in 1,4-dioxane [33]. Closed-ring isomer **4b** has an absorption maximum at 506 nm ($\varepsilon = 5.8 \times 10^4$ M^{-1} cm^{-1}) and exhibits brilliant red-orange fluorescence at around 620 nm under excitation with 532 nm light. The fluorescence quantum yield in 1,4-dioxane is 0.78. The cycloreversion quantum yield of **4b** is extremely low (<1.0 × 10^{-5}) and **4b** hardly returns to the open-ring isomer by irradiation with visible light.

Figure 33.3 shows photographs of fluorescence switching of 1,4-dioxane solutions containing **3** (left-side solution) and **4** (right-side solution) upon irradiation with 365 nm light under excitation with blue (488 nm) and green (532 nm) light, respectively. Before irradiation with 365 nm light, both of the solutions are dark

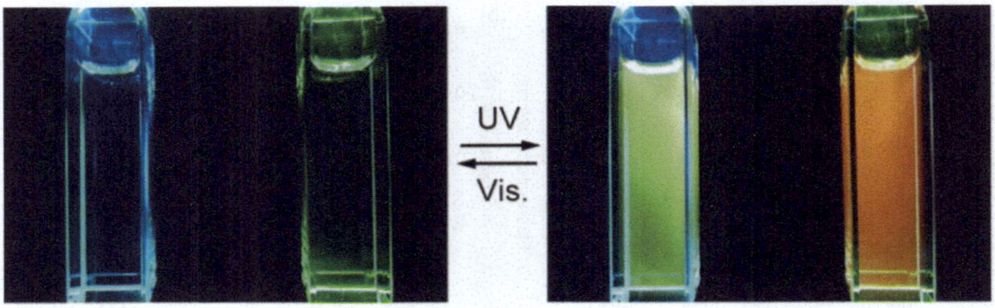

Fig. 33.3 Photographs of 1,4-dioxane solutions of **3** and **4** before and after irradiation with 365 nm light under excitation with 488 nm blue light (left-side solution containing **3**) and 532 nm green light (right-side solution containing **4**). Reprinted with permission from [33]. Copyright 2011 American Chemical Society

and non-luminous. Upon irradiation with 365 nm light, brilliant green and red-orange luminescence instantaneously appeared. These emissions are attributed to the formation of the highly fluorescent closed-ring isomers, **3b** and **4b**.

33.3 Control of the Cycloreversion Quantum Yield by Chemical Modification

Diarylethenes **3** and **4** have high cyclization quantum yields (**3**: 0.62, **4**: 0.23) and readily undergo the cyclization reactions upon UV irradiation, while the cycloreversion (switching-off) quantum yields are quite low (**3**: 5.9×10^{-4}, **4**: $< 1.0 \times 10^{-5}$). Such low cycloreversion quantum yields are useful for PALM/STORM, because in these coordinate-stochastic localization microscopies, sufficient numbers of photons need to be collected for accurate localization of fluorescent spots of single probe molecules in on-states before switching-off reactions take place [17, 18]. On the other hand, coordinate-targeted microscopy, such as RESOLFT microscopy, claims efficient photoswitching of probe molecules from on-states to off-states upon irradiation with a doughnut-shaped low-power laser beam [15, 16]. Therefore, it is required to develop photoswitchable fluorescent molecules with high switching-off quantum yield. According to numerical simulations [36], the spatial resolution of images increases as the switching-off quantum yield increases. To avoid the reduction of the fluorescence intensity during the probing operation, the quantum yield is preferable to be less than 10^{-2}. Taking into consideration of these factors, the optimal switching-off (cycloreversion) quantum yield is concluded to be in the order of 10^{-2}–10^{-3}.

Although the cycloreversion quantum yield of **2b** is 0.18, introduction of phenyl groups at 6- and 6′-positions decreases the yield to 5.9×10^{-4} as observed in **3b**. This suggests that the cycloreversion quantum yield can be controlled by changing the substituents at 6- and 6′-positions, which affect the coplanarity and the π-conjugation length of the closed-ring isomer. In other words, the cycloreversion quantum yield can be controlled by the dihedral angle between the phenyl ring at 6-position (or 6′-position) and the benzothiophene ring. To increase the cycloreversion quantum yield for the application to RESOLFT microscopy, diarylethenes **5–7** having *ortho*-substituted phenyl groups at 6- and 6′-positions of the benzothiophene 1,1-dioxide rings were synthesized [37]. The absorption maxima of the photogenerated closed-ring isomers are located at 456 nm (**3b**), 445 nm (**5b**), 441 nm (**6b**), and 430 nm (**7b**) (Table 33.1). The closed-ring isomers having *ortho*-substituted phenyl groups, **5b–7b**, exhibit a small but significant hypsochromic shift as compared to **3b** having unsubstituted phenyl groups, indicating that **5b–7b** have less extended π-conjugation.

Table 33.1 shows cyclization and cycloreversion quantum yields of **5–7** having *ortho*-substituted phenyl groups. The cyclization quantum yields of **5a–7a** are similar to that of **3a**. On the other hand, the cycloreversion quantum yield varies strongly depending on the *ortho*-substitution of the phenyl groups. The yield of **3b** is as low as 5.9×10^{-4}, while the yield increases to 2.8×10^{-3} (**5b**), 3.2×10^{-3} (**6b**), and 1.8×10^{-2} (**7b**) by introducing fluoro or methyl group(s) at *ortho*-position(s). By replacing *ortho*-hydrogen with fluorine, the yield increased by 5 times. Introduction of the *ortho*-dimethylphenyl groups increased the yield by as large as 30 times. The *ortho*-substitution is an effective way to increase the cycloreversion quantum yield up to 10^{-2}–10^{-3}. The fluorescence quantum yields of **3b**, **5b**, **6b**, and **7b** in 1,4-dioxane were measured to be 0.87, 0.79, 0.84, and 0.79, respectively, and are very similar to each other. The *ortho*-substitution scarcely affects the fluorescent quantum yields.

The *ortho*-substitution of the phenyl groups is considered to decrease the coplanarity of the closed-ring isomers. The dihedral angles between the phenyl and benzothiophene rings were measured by single-crystal X-ray analysis. Replacement of *ortho*-hydrogen with fluorine increases the dihedral angle from 30.2° (**3b**) to 35.9° (**5b**). The angle is further increased up to 61.0° (**6b**) or 65.7° (**7b**) by introducing the *ortho*-methyl group(s). The increase in the dihedral angle lowers the coplanarity of the closed-ring isomer and reduces the π-conjugation length, leading to the hypsochromic shift of the absorption spectrum and the increase in the cycloreversion quantum yield.

Figure 33.4 shows the relationship between the absorption maxima of the closed-ring isomers and the cyclization/cycloreversion quantum yields in less polar 1,4-dioxane and polar ethanol. The fluorescence quantum yields of the closed-ring isomers are also shown. The cyclization quantum yields are higher than 0.58 in 1,4-dioxane and the closed-ring isomers exhibit brilliant green fluorescence, the quantum yield of which is higher than 0.79. The *ortho*-substitution scarcely affects the fluorescence quantum yields but significantly increases the cycloreversion quantum yields by as large as 30 times. The cycloreversion quantum yields of **5b–7b** are in the range of 2.8×10^{-3} to 1.8×10^{-2}. These molecules fulfill the requirements for the application to RESOLFT microscopy.

Another strategy to increase the cycloreversion quantum yield is to introduce isobutyl groups at 2- and 2′-positions (reactive carbon atoms) of the benzothiophene

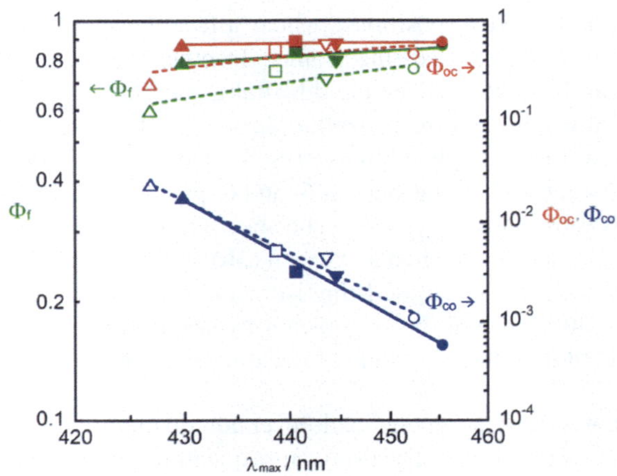

Fig. 33.4 Correlation between absorption maxima of the closed-ring isomers (λ_{max}) and quantum yields of cycloreversion (Φ_{co}, blue) and cyclization (Φ_{oc}, red) of **3** (●,○), **5** (▼, ▽), **6** (■, □), and **7** (▲, △) in 1,4-dioxane (closed marks) and ethanol (open marks). Fluorescence quantum yields (Φ_f, green) of the closed-ring isomers of **3** (●, ○), **5** (▼, ▽), **6** (■, □), and **7** (▲, △) in 1,4-dioxane (closed marks) and ethanol (open marks) were also plotted. Reprinted from [37]. Copyright 2017, with permission from Elsevier

1,1-dioxide rings [27, 38]. As shown in Table 33.1, the cyclization quantum yield of **8b** having isobutyl groups (1.0×10^{-2}) is 17 times larger than that of **3b** having ethyl ones (5.9×10^{-4}) and is also favorable for the application to RESOLFT microscopy.

33.4 Solvatochromism of Photoswitchable Fluorescence

Solvatochromic fluorophores, which change their fluorescent colors in response to solvent polarity, can be used as fluorescent sensors or probes to detect and visualize spatiotemporal heterogeneity of microscopic polarity in biological systems and materials [39, 40]. When the solvatochromic fluorophores acquire additional photoswitching ability, their application is expected to further extend. Solvatochromic fluorophores with photoswitching ability can provide spectrally resolved multicolor super-resolution fluorescence images, in which microscopic polarity in a sample object is mapped with different fluorescent colors [41].

Turn-on mode fluorescent diarylethene **4** having 5-methyl-2-thienyl groups at 6- and 6′-positions of the benzothiophene 1,1-dioxide rings is a photoswitchable solvatochromic fluorophore, which can be potentially applicable to multicolor super-resolution fluorescence microscopy for polarity imaging. The open-ring isomer (**4a**) undergoes a cyclization reaction upon irradiation with UV light and the photogenerated fluorescent closed-ring isomer (**4b**) exhibits solvatochromism of fluorescence [42]. Figure 33.5 shows the absorption and fluorescence spectra of **4b**

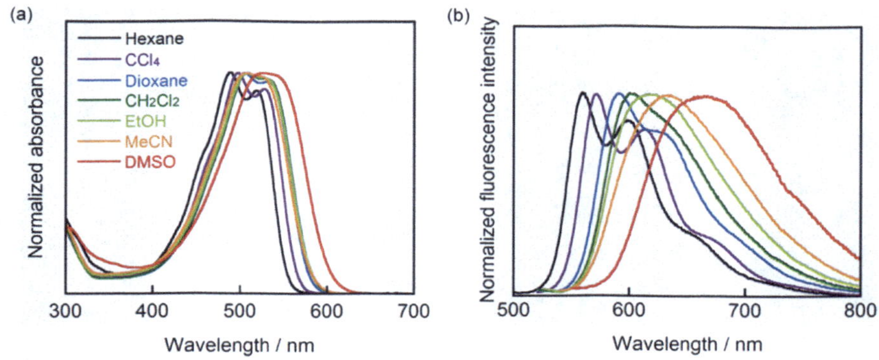

Fig. 33.5 Normalized absorption (**a**) and fluorescence (**b**) spectra of **4b** in *n*-hexane (black lines), CCl_4 (purple lines), 1,4-dioxane (blue lines), CH_2Cl_2 (green lines), ethanol (yellow green lines), acetonitrile (orange lines), and DMSO (red lines). Reprinted from [42]. Copyright 2018, with permission from Elsevier

in *n*-hexane, tetrachloromethane (CCl_4), 1,4-dioxane, dichloromethane (CH_2Cl_2), ethanol, acetonitrile, and dimethyl sulfoxide (DMSO). The absorption spectrum shows a moderate bathochromic shift with increasing the solvent polarity ($\lambda_{max,abs}$: 489 nm in *n*-hexane, 528 nm in DMSO). On the other hand, the fluorescence spectrum exhibits a remarkable bathochromic shift and very large Stokes shifts are observed in polar solvents. Although **4b** shows an emission maximum at 560 nm in *n*-hexane, the maximum shifts to 591 nm in 1,4-dioxane, 618 nm in ethanol, 633 nm in acetonitrile, and 664 nm in DMSO. The fluorescence spectrum shows a significant bathochromic shift as much as 104 nm by changing the solvent from *n*-hexane to DMSO. The solvatochromic property is ascribed to the intramolecular charge-transfer character of **4b** having the electron-donating thienyl groups and the electron-accepting sulfone moieties. In polar solvents, the excited state of **4b** with a large electric dipole moment considerably stabilizes accompanying reorientation of the solvent cage during its lifetime and then undergoes radiative transition with lower-energy (red-shifted) emission. By the analysis of the absorption and fluorescence spectral data using the Mataga–Lippert equation [43–45], the difference in the dipole moments between the excited and ground states of **4b** was estimated to be 12 D. It is worthy to note that **4b** preserves the high-fluorescence quantum yield in a wide range of solvent polarity. **4b** shows high quantum yields in less polar solvents, such as 0.72 in *n*-hexane and 0.78 in 1,4-dioxane. In polar solvents, the yield slightly decreases but remains relatively high, such as 0.62 in ethanol, 0.67 in acetonitrile, and 0.62 in DMSO.

Figure 33.6 shows photographs of fluorescent emissions of **4** in various solvents. Before UV irradiation, the solutions of open-ring isomer **4a** show no emission under excitation with 500 nm light. Upon irradiation with UV light, **4a** undergoes the cyclization reaction to form fluorescent closed-ring isomer **4b** and the solutions emit brilliant fluorescence. **4b** exhibits remarkably different emission colors depending on the solvents. In *n*-hexane, **4b** emits green fluorescence. With increasing the solvent

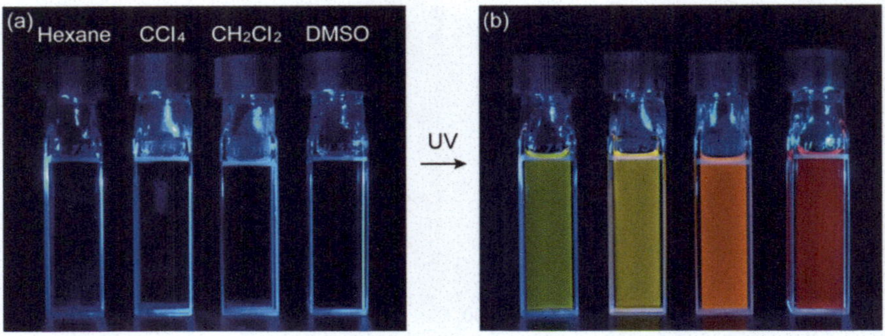

Fig. 33.6 Photographs of fluorescent emissions of **4** in *n*-hexane, CCl_4, CH_2Cl_2, and DMSO under excitation with 500 nm light before (**a**) and after (**b**) irradiation with UV (365 nm) light. Reprinted from [42]. Copyright 2018, with permission from Elsevier

polarity, **4b** shows red-shifted emissions, such as yellow in CCl_4, orange in CH_2Cl_2, and red in DMSO. Such a solvatochromic fluorophore with photoswitching ability can find potential applications to multicolor super-resolution imaging of microscopic polarity in biological cells and materials.

33.5 Reversible Photoswitching by Irradiation with Single-Wavelength Visible Light

Most of diarylethenes, except several specially designed derivatives [46–49], cannot undergo cyclization reactions with visible light (>400 nm), because the open-ring isomers do not have optical absorption in the visible-wavelength region. The visible-light responsive property is advantageous for the application to biological systems because visible light is less phototoxic and can deeply penetrate into the sample. Unprecedented photoreactions of turn-on mode fluorescent diarylethene **3** upon irradiation with single-wavelength visible light were found [50]. Reversible fluorescence photoswitching was induced by irradiation with single-wavelength visible light, the wavelength of which is longer than that of the 0-0 transition of open-ring isomer **3a**.

Figure 33.7 shows photoinduced absorption and fluorescence spectral changes of **3** in 1,4-dioxane. Upon irradiation with UV (365 nm) light, open-ring isomer **3a** readily converts to closed-ring isomer **3b**, which has an absorption band at 456 nm and emits fluorescence at around 550 nm under excitation with 450 nm light. The closed-ring isomer undergoes the cycloreversion reaction upon irradiation with 450 nm light and reverts back to the open-ring isomer. Although both the absorption and fluorescence spectra of **3b** in the visible-wavelength region considerably diminish, noticeable intensity of fluorescence of **3b** remains even after prolonged irradiation with 450 nm light. **3a** has no apparent absorption at 450 nm. Although the closed-ring **3b** isomers are expected to completely revert back to the open-ring **3a** isomers by irradiation

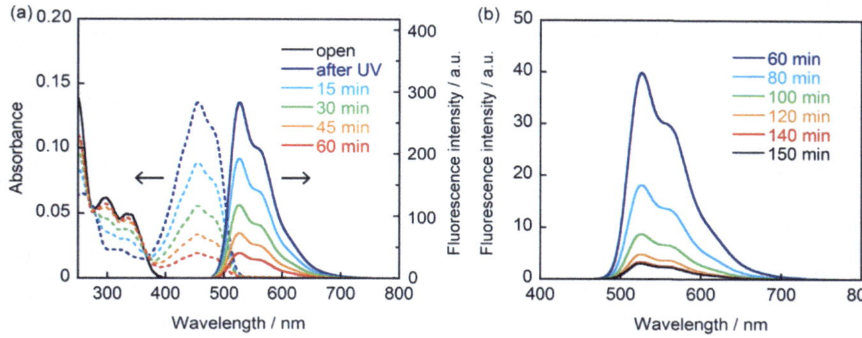

Fig. 33.7 **a** Absorption and fluorescence spectral changes of **3** in 1,4-dioxane (3.2×10^{-6} M) upon irradiation with UV (365 nm) light and subsequent monochromatic 450 nm light. The solid black line showed the absorption spectrum of **3a**. Dashed lines showed the absorption spectra of **1b**. **b** Fluorescence spectral changes of **3b** in 1,4-dioxane upon prolonged irradiation with monochromatic 450 nm light. The fluorescence intensity after irradiation with 450 nm for 150 min was the same as the intensity after irradiation for 140 min. Reprinted with permission from [50]. Copyright 2017 American Chemical Society

with 450 nm light, a tiny amount of **3b** survives. This result indicates that 450 nm light induces the cyclization reaction of **3a** in addition to the cycloreversion reaction.

The reversible photoisomerization reactions are assumed to be induced by irradiation with single-wavelength (λ) light,

$$\text{Open-ring isomer } \mathbf{3a} \underset{\lambda}{\overset{\lambda}{\rightleftarrows}} \text{Closed-ring isomer } \mathbf{3b} \tag{33.1}$$

$$\frac{\mathrm{d}C_\mathrm{B}}{\mathrm{d}t} = (\Phi_\mathrm{AB}\varepsilon_\mathrm{A}C_\mathrm{A} - \Phi_\mathrm{BA}\varepsilon_\mathrm{B}C_\mathrm{B})F \tag{33.2}$$

$$F = \frac{1 - 10^{-(\varepsilon_\mathrm{A}C_\mathrm{A}+\varepsilon_\mathrm{B}C_\mathrm{B})d}}{\varepsilon_\mathrm{A}C_\mathrm{A} + \varepsilon_\mathrm{B}C_\mathrm{B}} \times \frac{10^3 I_0}{d}$$

where Φ_AB, Φ_BA, ε_A, ε_B, C_A, C_B, d, and I_0 are the cyclization quantum yield, the cycloreversion quantum yield, the absorption coefficient of **3a** ($\mathrm{M}^{-1}\ \mathrm{cm}^{-1}$) at λ, the absorption coefficient of **3b** ($\mathrm{M}^{-1}\ \mathrm{cm}^{-1}$) at λ, the concentration of **3a** (M), the concentration of **3b** (M), the cell length (cm), and the intensity of light (einstein $\mathrm{cm}^{-2}\ \mathrm{s}^{-1}$), respectively. When $\varepsilon_\mathrm{A}C_\mathrm{A} + \varepsilon_\mathrm{B}C_\mathrm{B} \ll 1$, the formation of the closed-ring isomers is expressed as follows,

$$C_\mathrm{B} = \frac{\Phi_\mathrm{AB}\varepsilon_\mathrm{A}}{\Phi_\mathrm{AB}\varepsilon_\mathrm{A} + \Phi_\mathrm{BA}\varepsilon_\mathrm{B}} C_\mathrm{o}\left(1 - \mathrm{e}^{-(\Phi_\mathrm{AB}\varepsilon_\mathrm{A}+\Phi_\mathrm{BA}\varepsilon_\mathrm{B})\times 10^3 I_0 t}\right) \tag{33.3}$$

where C_O is the total concentration of **3a** and **3b**. In the photostationary state ($t \to \infty$),

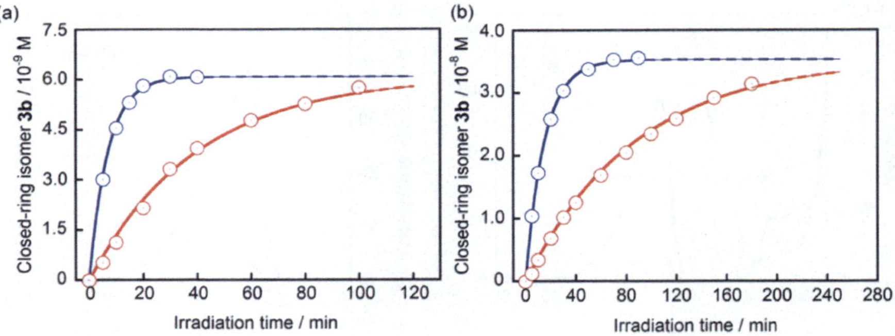

Fig. 33.8 Formations of closed-ring isomer **3b** from pure **3a** (1.0×10^{-5} M) in *n*-hexane (**a**) and CCl_4 (**b**) upon irradiation with monochromatic 450 nm light. The two formation curves were measured by changing the light intensity, 100% (blue) and 25% (red), respectively. Solid lines were simulated by using Eq. (33.3). Reprinted with permission from [50]. Copyright 2017 American Chemical Society

$$\frac{C_B}{C_o} = \frac{\Phi_{AB}\varepsilon_A}{\Phi_{AB}\varepsilon_A + \Phi_{BA}\varepsilon_B} \tag{33.4}$$

Figure 33.8 shows the formation of **3b** from pure **3a** in *n*-hexane and CCl_4 upon irradiation with 450 nm light. Fluorescence was used to detect a very small amount of **3b**, less than 10^{-7} M. The formation profiles were analyzed according to Eq. (33.3). As shown by Eq. (33.4), the very weak absorption coefficients of **3a** (ε_A) at 450 nm can be determined from the concentration ratios of **3a** and **3b** (C_B/C_O) in the photostationary state under irradiation with 450 nm light. Based on numerical simulations of the formation process of **3b** from **3a**, the coefficients at 450 nm in *n*-hexane and CCl_4 were estimated to be 0.084 and 0.19 M^{-1} cm^{-1}, respectively. Although the absorption coefficients were very low, the formation of **3b** was clearly observed upon irradiation with 450 nm light. This is attributed to quite low cycloreversion quantum yields of **3b** (1.3×10^{-3} in *n*-hexane and 6.6×10^{-4} in CCl_4).

According to Wondrazek et al. [51] and Kinoshita et al. [52], the absorption spectra (or the emission excitation spectra) of fluorescent dyes, such as coumarin 7 and rhodamine 6G, show an exponential frequency dependence in the off-resonance region of the absorption tails at room temperature. The tails have been reported to obey the Urbach rule, which is known in the absorption spectra of solid materials, as expressed by the following formula [51–55],

$$\varepsilon(E) \propto \varepsilon(E_{00})\exp\left[-\frac{\sigma(E_{00} - E)}{k_B T}\right] \tag{33.5}$$

where ε, k_B, T, and σ are the absorption coefficient, the Boltzmann constant, the absolute temperature of the sample, and the steepness parameter, respectively.

Figure 33.9 shows the absorption coefficients at 450, 440, 435, and 429 nm, which are estimated from Eq. (33.4), along with the absorption tails of open-ring isomer

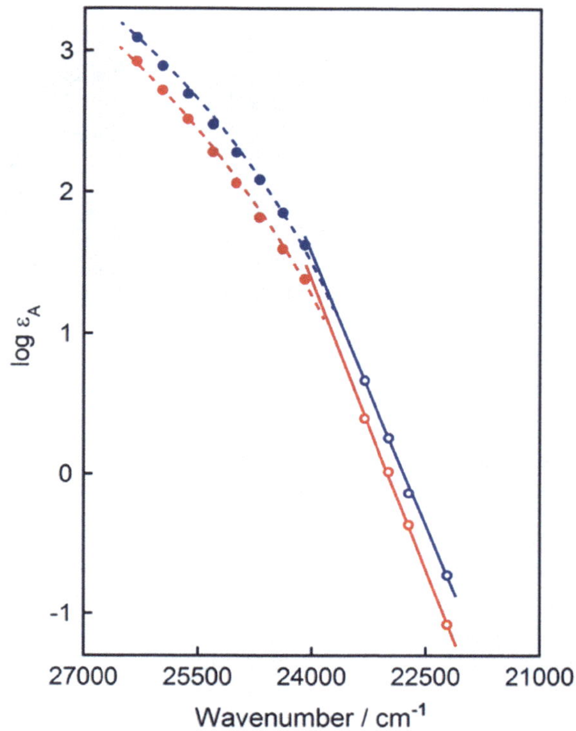

Fig. 33.9 Logarithmic plot of molar absorption coefficients of **3a** (ε_A) in *n*-hexane (red) and CCl_4 (blue) obtained from absorption spectra of **3a** (closed circles) and fluorescence intensities of **3b** in the photostationary states under irradiation with monochromatic visible light (open circles). Reprinted with permission from [50]. Copyright 2017 American Chemical Society

3a. The coefficients obey the exponential frequency dependence at the wavelengths longer than ~415 nm (at the wavenumbers smaller than ~24,000 cm^{-1}). The exponential frequency dependence indicates that the photoreactions with the visible light in the far off-resonance region of the absorption edge are caused by the optical absorption of hot bands or Urbach tails. **3a** in the vibrationally excited states undergoes optical transition into electronically excited states by absorbing the very weak hot bands and the cyclization reaction from **3a** to **3b** takes place. At the same time, the visible light also induces the cycloreversion reaction from **3b** to **3a**. After prolonged irradiation, the system reaches the photostationary state. Both the cyclization and cycloreversion reactions are induced by irradiation with single-wavelength visible light ($420 < \lambda < 470$ nm).

The reversible fluorescence photoswitching with single-wavelength visible light has been successfully applied to super-resolution PALM/STORM imaging with single-wavelength laser light [56]. Figure 33.10 shows a super-resolution fluorescence image of a patterned poly(methyl methacrylate) (PMMA) film fabricated by electron-beam lithography. In the experiment, diarylethene **4** was doped into the film as a photoswitchable fluorescent probe and a continuous 532 nm laser was used as a light source. Under illumination with 532 nm laser light, the wavelength of which is much longer than that of the 0-0 transition of open-ring isomer **4a**,

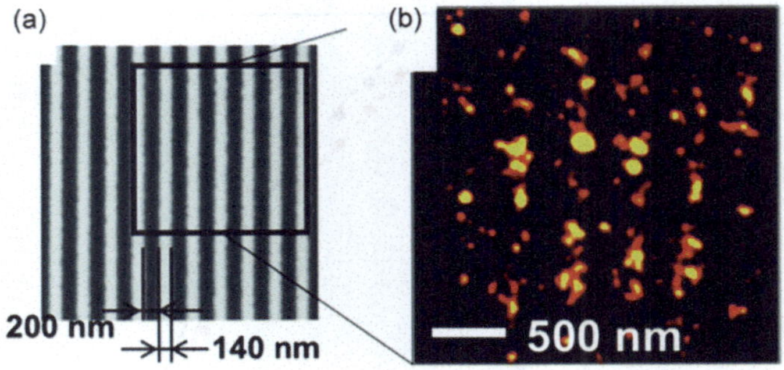

Fig. 33.10 Super-resolution fluorescence imaging with single-wavelength visible laser light. **a** A scanning electron micrograph of the nanostructure of the PMMA on a Si substrate. The dark and bright areas correspond to PMMA and the surface of the Si substrate, respectively. **b** A super-resolution fluorescence image of a part of the PMMA nanostructure. Reproduced from Ref. [56] with permission from the Royal Society of Chemistry

4a molecule undergoes the cyclization reaction by absorbing hot bands or Urbach tails to form closed-ring isomer **4b**. Thus, activation, detection, and deactivation of single-molecule fluorescence are performed by irradiation with single-wavelength visible laser light. Localization analysis on the fluorescent spots of the individual **4b** molecules and subsequent integration of the analyzed images constructed the super-resolution image. The obtained image clearly visualizes the stripe pattern of the PMMA film with ca. 200 nm width and ca. 140 nm separation, as shown in Fig. 33.10. Similar photoswitching of fluorescence with single-wavelength visible laser light was also applied to super-resolution bioimaging using water-soluble fluorescent diarylethene derivatives [57].

33.6 Summary

A new type of turn-on mode fluorescent diarylethenes having benzothiophene 1,1-dioxide groups, which emit strong fluorescence in the closed-ring isomers, has been developed for the application to super-resolution fluorescence microscopies, such as PALM/STORM and RESOLFT microscopy. By chemical modifications, the photoswitching performance of the diarylethenes was tuned to fulfill the requirements for the microcopies. The diarylethene derivative that shows solvatochromism of photoswitchable fluorescence has a potential for the application to multicolor super-resolution imaging of microscopic polarity. The reversible photoswitching upon irradiation with single-wavelength visible light has been successfully applied to super-resolution PALM/STORM imaging. Harnessing the "photosynergy" between fluorescent diarylethene molecules and advanced photoexcitation and imaging principles would lead to the further development of super-resolution fluorescence microscopy, which reveals unexplored structures and functions in biological systems and materials [58–66].

Acknowledgements The present work was supported by JSPS KAKENHI Grant Numbers JP15H01096, JP17H05272, Grant-in-Aid for Scientific Research on Innovative Areas "Photosynergetics."

References

1. Fukaminato T (2011) Single-molecule fluorescence photoswitching: design and synthesis of photoswitchable fluorescent molecules. J Photochem Photobiol, C 12:177–208
2. Zhang J, Zou Q, Tian H (2013) Photochromic materials: more than meets the eye. Adv Mater 25:378–399
3. Fukaminato T, Ishida S, Métivier R (2018) Photochromic fluorophores at the molecular and nanoparticle levels: fundamentals and applications of diarylethenes. NPG Asia Mater 10:859–881
4. Heilemann M, Dedecker P, Hofkens J, Sauer M (2009) Photoswitches: key molecules for subdiffraction-resolution fluorescence imaging and molecular quantification. Laser Photon Rev 3:180–202
5. Fürstenberg A, Heilemann M (2013) Single-molecule localization microscopy–near molecular spatial resolution in light microscopy with photoswitchable fluorophores. Phys Chem Chem Phys 15:14919–14930
6. Yagi K, Soong CF, Irie M (2001) Synthesis of fluorescent diarylethenes having a 2,4,5-triphenylimidazole chromophore. J Org Chem 66:5419–5423
7. Irie M, Fukaminato T, Sasaki T, Tamai N, Kawai T (2002) Organic chemistry: a digital fluorescent molecular photoswitch. Nature 420:759–760
8. Fukaminato T, Sasaki T, Kawai T, Tamai N, Irie M (2004) Digital photoswitching of fluorescence based on the photochromism of diarylethene derivatives at a single-molecule level. J Am Chem Soc 126:14843–14849
9. Fukaminato T, Umemoto T, Iwata Y, Yokojima S, Yoneyama M, Nakamura S, Irie M (2007) Photochromism of diarylethene single molecules in polymer matrices. J Am Chem Soc 129:5932–5938
10. Fukaminato T, Doi T, Tamaoki N, Okuno K, Ishibashi Y, Miyasaka H, Irie M (2011) Single-molecule fluorescence photoswitching of a diarylethene–perylenebisimide dyad: Non-destructive fluorescence readout. J Am Chem Soc 133:4984–4990
11. Golovkova TA, Kozlov DV, Neckers DC (2005) Synthesis and properties of novel fluorescent switches. J Org Chem 70:5545–5549
12. Bossi M, Belov V, Polyakova S, Hell SW (2006) Reversible red fluorescent molecular switches. Angew Chem Int Ed 45:7462–7465
13. Berberich M, Krause AM, Orlandi M, Scandola F, Würthner F (2008) Toward fluorescent memories with nondestructive readout: photoswitching of fluorescence by intramolecular electron transfer in a diaryl ethene-perylene bisimide photochromic system. Angew Chem Int Ed 47:6616–6619
14. Berberich M, Natali M, Spenst P, Chiorboli C, Scandola F, Würthner F (2012) Nondestructive photoluminescence read-out by intramolecular electron transfer in a perylene bisimide-diarylethene dyad. Chem Eur J 18:13651–13664
15. Hell SW, Jakobs S, Kastrup L (2003) Imaging and writing at the nanoscale with focused visible light through saturable optical transitions. Appl Phys A 77:859–860
16. Grotjohann T, Testa I, Leutenegger M, Bock H, Urban NT, Lavoie-Cardinal F, Willig KI, Eggelling C, Jakobs S, Hell SW (2011) Diffraction-unlimited all-optical imaging and writing with a photochromic GFP. Nature 478:204–208

17. Betzig E, Patterson GH, Sougrat R, Lindwasser OW, Olenych S, Bonifacino JS, Davidson MW, Lippincott-Schwartz J, Hess HF (2006) Imaging intracellular fluorescent proteins at nanometer resolution. Science 313:1642–1645
18. Rust MJ, Bates M, Zhuang X (2006) Sub-diffraction-limit imaging by stochastic optical reconstruction microscopy (STORM). Nat Meth 3:793–796
19. Raymo FM (2013) Photoactivatable synthetic fluorophores. Phys Chem Chem Phys 15:14840–14850
20. Lord SJ, Conley NR, Lee HD, Samuel R, Liu N, Twieg RJ, Moerner WE (2008) A photoactivatable push–pull fluorophore for single-molecule imaging in live cells. J Am Chem Soc 130:9204–9205
21. Fölling J, Belov V, Kunetsky R, Medda R, Schönle A, Egner A, Eggeling C, Bossi M, Hell SW (2007) Photochromic rhodamines provide nanoscopy with optical sectioning. Angew Chem Int Ed 46:6266–6270
22. Lee MK, Rai P, Williams J, Twieg RJ, Moerner WE (2014) Small-molecule labeling of live cell surfaces for three-dimensional super-resolution microscopy. J Am Chem Soc 136:14003–14006
23. Kobayashi T, Urano Y, Kamiya M, Ueno T, Kojima H, Nagano T (2007) Highly activatable and rapidly releasable caged fluorescein derivatives. J Am Chem Soc 129:6696–6697
24. Kobayashi T, Komatsu T, Kamiya M, Campos C, González-Gaitán M, Terai T, Hanaoka K, Nagano T, Urano Y (2012) Highly activatable and environment-insensitive optical highlighters for selective spatiotemporal imaging of target proteins. J Am Chem Soc 134:11153–11160
25. Deniz E, Tomasulo M, Cusido J, Yildiz I, Petriella M, Bossi ML, Sortino S, Raymo FM (2012) Photoactivatable fluorophores for super-resolution imaging based on oxazine auxochromes. J Phys Chem C 116:6058–6068
26. Thapaliya ER, Captain B, Raymo FM (2014) Photoactivatable anthracenes. J Org Chem 79:3973–3981
27. Irie M, Morimoto M (2018) Photoswitchable turn-on mode fluorescent diarylethenes: strategies for controlling the switching response. Bull Chem Soc Jpn 91:237–250
28. Morimoto M, Irie M (2017) Turn-on mode fluorescent diarylethenes. In: Yokoyama Y, Nakatani K (eds) Photon-working switches. Springer, Tokyo
29. Jeong Y-C, Yang SI, Ahn K-H, Kim E (2005) Highly fluorescent photochromic diarylethene in the closed-ring form. Chem Commun 2503–2505
30. Jeong Y-C, Yang SI, Kim E, Ahn K-H (2006) Development of highly fluorescent photochromic material with high fatigue resistance. Tetrahedron 62:5855–5861
31. Jeong Y-C, Park DG, Lee IS, Yang SI, Ahn K-H (2009) Highly fluorescent photochromic diarylethene with an excellent fatigue property. J Mater Chem 19:97–103
32. Sumi T, Takagi Y, Yagi A, Morimoto M, Irie M (2014) Photoirradiation wavelength dependence of cycloreversion quantum yields of diarylethenes. Chem Commun 50:3928–3930
33. Uno K, Niikura H, Morimoto M, Ishibashi Y, Miyasaka H, Irie M (2011) In situ preparation of highly fluorescent dyes upon photoirradiation. J Am Chem Soc 133:13558–13564
34. Takagi Y, Kunishi T, Katayama T, Ishibashi Y, Miyasaka H, Morimoto M, Irie M (2012) Photoswitchable fluorescent diarylethene derivatives with short alkyl chain substituents. Photochem Photobiol Sci 11:1661–1665
35. Irie M, Eriguchi T, Takada T, Uchida K (1997) Photochromism of diarylethenes having thiophene oligomers as the aryl groups. Tetrahedron 53:12263–12271
36. Dedecker P, Hotta J, Flors C, Sliwa M, Uji-i H, Roeffaers MBJ, Ando R, Mizuno H, Miyawaki A, Hofkens J (2007) Subdiffraction imaging through the selective donut-mode depletion of thermally stable photoswitchable fluorophores: Numerical analysis and application to the fluorescent protein Dronpa. J Am Chem Soc 129:16132–16141
37. Takagi Y, Morimoto M, Kashihara R, Fujinami S, Ito S, Miayasaka H (2017) Irie M Turn-on mode fluorescent diarylethenes: control of the cycloreversion quantum yield. Tetrahedron 73:4918–4924
38. Morimoto M, Sumi T, Irie M (2017) Photoswitchable fluorescent diarylethene derivatives with thiophene 1,1-dioxide groups: effect of alkyl substituents at the reactive carbons. Materials 10:1021

39. Yang Z, Cao J, He Y, Yang JH, Kim T, Peng X, Kim JS (2014) Macro-/micro-environment-sensitive chemosensing and biological imaging. Chem Soc Rev 43:4563–4601
40. Klymchenko AS (2017) Solvatochromic and fluorogenic dyes as environment-sensitive probes: design and biological applications. Acc Chem Res 50:366–375
41. Moon S, Yan R, Kenny SJ, Shyu Y, Xiang L, Li W, Xu K (2017) Spectrally resolved, functional super-resolution microscopy reveals nanoscale compositional heterogeneity in live-cell membranes. J Am Chem Soc 139:10944–10947
42. Morimoto M, Takagi Y, Hioki K, Nagasaka T, Sotome H, Ito S, Miyasaka H, Irie M (2018) A turn-on mode fluorescent diarylethene: solvatochromism of fluorescence. Dyes Pigm 153:144–149
43. Mataga N, Kaifu Y, Koizumi K (1955) The solvent effect on fluorescence spectrum, change of solute-solvent interaction during the lifetime of excited solute molecule. Bull Chem Soc Jpn 28:690–691
44. Mataga N, Kaifu Y, Koizumi K (1956) Solvent effects upon fluorescence spectra and the dipolemoments of excited molecules. Bull Chem Soc Jpn 29:465–470
45. Lippert E (1957) Spektroskopische Bestimmung des Dipolmomentes aromatischer Verbindungen im ersten angeregten Singulettzustand. Z Electrochem 61:962–975
46. Sumi T, Kaburagi T, Morimoto M, Une K, Sotome H, Ito S, Miyasaka H, Irie M (2015) Fluorescent photochromic diarylethene that turns on with visible light. Org Lett 17:4802–4805
47. Fukaminato T, Hirose T, Doi T, Hazama M, Matsuda K, Irie M (2014) Molecular design strategy toward diarylethenes that photoswitch with visible light. J Am Chem Soc 136:17145–17154
48. Fredrich S, Göstl R, Herder M, Grubert L, Hecht S (2016) Switching diarylethenes reliably in both directions with visible light. Angew Chem Int Ed 55:1208–1212
49. Zhang Z, Wang W, Jin P, Xue J, Sun L, Huang J, Zhang J, Tian H (2019) A building-block design for enhanced visible-light switching of diarylethenes. Nat Commun 10:4232
50. Kashihara R, Morimoto M, Ito S, Miyasaka H, Irie M (2017) Fluorescence photoswitching of a diarylethene by irradiation with single-wavelength visible light. J Am Chem Soc 139:16498–16501
51. Wondrazek F, Seilmeier A, Kaiser W (1984) Ultrafast intramolecular redistribution and intermolecular relaxation of vibrational energy in large molecules. Chem Phys Lett 104:121–128
52. Kinoshita S, Nishi N, Saitoh A, Kushida T (1987) Urbach tail of organic dyes in solution. J Phys Soc Jpn 56:4162–4175
53. Mizuno K, Matsui A, Sloan GJ (1984) Intermediate exciton-phonon coupling in tetracene. J Phys Soc Jpn 53:2799–2806
54. Urbach F (1953) The long-wavelength edge of photographic sensitivity and of the electronic absorption of solids. Phys Rev 92:1324
55. Birks JB (1970) Photophysics of aromatic molecules. Wiley, London
56. Arai Y, Ito S, Fujita H, Yoneda Y, Kaji T, Takei S, Kashihara R, Morimoto M, Irie M, Miyasaka H (2017) One-colour control of activation, excitation and deactivation of a fluorescent diarylethene derivative in super-resolution microscopy. Chem Commun 53:4066–4069
57. Roubinet B, Weber M, Shojaei H, Bates M, Bossi ML, Belov VN, Irie M, Hell SW (2017) Fluorescent photoswitchable diarylethenes for biolabeling and single-molecule localization microscopies with optical superresolution. J Am Chem Soc 139:6611–6620
58. Roubinet B, Bossi ML, Alt P, Leutenegger M, Shojaei H, Schnorrenberg S, Nizamov S, Irie M, Belov VN, Hell SW (2016) Carboxylated photoswitchable diarylethenes for biolabeling and super-resolution RESOLFT microscopy. Angew Chem Int Ed 55:15429–15433
59. Uno K, Bossi ML, Konen T, Belov VN, Irie M, Hell SW (2019) Asymmetric diarylethenes with oxidized 2-alkylbenzothiophen-3-yl units: chemistry, fluorescence, and photoswitching. Adv Optical Mater 7:1801746
60. Uno K, Bossi ML, Irie M, Belov VN, Hell SW (2019) Reversibly photoswitchable fluorescent diarylethenes resistant against photobleaching in aqueous solutions. J Am Chem Soc 141:16471–16478
61. Um SH, Kim HJ, Kim D, Kwon JE, Lee JW, Hwang D, Kim SK, Park SY (2018) Highly fluorescent and water soluble turn-on type diarylethene for super-resolution bioimaging over a broad pH range. Dyes Pigm 158:36–41

62. Nevskyi O, Sysoiev D, Oppermann A, Huhn T, Wöll D (2016) Nanoscopic visualization of soft matter using fluorescent diarylethene photoswitches. Angew Chem Int Ed 55:12698–12702
63. Wöll D, Flors C (2017) Super-resolution fluorescence imaging for materials science. Small Methods 1:1700191
64. Siemes E, Nevskyi O, Sysoiev D, Turnhoff SK, Oppermann A, Huhn T, Richtering W, Wöll D (2018) Nanoscopic visualization of cross-linking density in polymer networks with diarylethene photoswitches. Angew Chem Int Ed 57:12280–12284
65. Nevskyi O, Sysoiev D, Dreier J, Stein SC, Oppermann A, Lemken F, Janke T, Enderlein J, Testa I, Huhn T, Wöll D (2018) Fluorescent diarylethene photoswitches—a universal tool for super-resolution microscopy in nanostructured materials. Small 14:1703333
66. Qiang Z, Shebek KM, Irie M, Wang M (2018) A polymerizable photoswitchable fluorophore for super-resolution imaging of polymer self-assembly and dynamics. ACS Macro Lett 7:1432–1437

Chapter 34
Crystallization Control of the Photoresponsible Diarylethene Film with an Aluminum Plasmonic Chip

Keiko Tawa, Taiga Kadoyama, Ryo Nishimura, and Kingo Uchida

Abstract The photoisomerization between an open-ring isomer (**1o**) and a closed-ring one (**1c**) in diarylethene **1** film and the needle-shaped microcrystal growth subsequent to photoisomerization were in situ measured with an optical microscope. The **1o** film was exposed to UV light on the stage of an upright-inverted microscope and optical and spectroscopic images were in situ observed in order to clarify the mechanism of microcrystal growth in the film. **1o** films were prepared on a glass slide and an aluminum (Al) plasmonic chip, which is the Al-coated substrate with a wavelength-size periodic pattern and can provide the enhanced electric field to the chip surface. A plasmonic chip was expected to promote a photoisomerization and a microcrystal growth. Crystal growth of needle-shaped crystal of **1c** was observed at the center of a film under the UV irradiation from upright side, but not observed under irradiation from inverted side. On the other hand, crystal growth was found at the film edge by the UV exposure even from the inverted side. Therefore, a high mobility of **1c** molecules near the film surface or edge is essential for crystal growth of **1c**. Further, alignment of **1c** molecules also requires the platform of **1o**. The conversion rate from **1o** to **1c** was larger on the Al grating. By the plasmonic enhanced electric field, when the attenuated UV light was exposed to the film edge from inverted side, the needle-shaped crystals were observed only inside the grating at the conversion rate above 60%. Conversion rate to **1c** controlled crystal growth and therefore, crystal growth was promoted inside the grating. In summary, the conversion rate to **1c** above 60%, a mobility of **1c** near the film surface or edge, and the **1o** underlayer platform are required for crystal growth after **1c** alignment.

Keywords Diarylethene · Spectroscopic imaging · In situ microscopic imaging · Cyclization · Crystal growth · Plasmon field

---

K. Tawa (✉) · T. Kadoyama
Department of Applied Chemistry for Environment, School of Science and Technology, Kwansei Gakuin University, Sanda, Hyogo 669-1337, Japan
e-mail: ktawa@kwansei.ac.jp

R. Nishimura · K. Uchida
Department of Materials Chemistry, Faculty of Science and Technology, Ryukoku University, Seta, Shiga, Japan

H. Miyasaka et al. (eds.), *Photosynergetic Responses in Molecules and Molecular Aggregates*, https://doi.org/10.1007/978-981-15-5451-3_34

34.1 Introduction

Diarylethene (DAE) [1–3] is one of the most representative photochromic compounds together with spiropyrans [4, 5], azobenzenes [6–9], and hexaarylbiimidazole (HABI) [10]. Many studies on DAE have been conducted for more than half a century. Physical and chemical properties of various DAE derivatives synthesized have been investigated by spectroscopy [11, 12], electrochemistry [13, 14], and crystallography [15–18], etc. Generally, the open-ring form of DAE converts to the closed-ring form under the ultraviolet (UV) light. The closed-ring form reconverts to the open-ring with visible light. The photoinduced reversible formation of microfibrils on a DAE microcrystalline surface has been reported by Uchida et al. [19–22]. Change in the photoinduced morphology can provide super hydrophobic properties such as in lotus and taro leaves [23]. Micrometer-scale rods or projections indicate a super water-repellent property and the large contact angle of a water droplet is 160° [24–26]. In the previous studies by Uchida et al., the photoinduced tuning of the surface wettability was reported due to the morphological changes in a crystal of photochromic DAE, and in a thin film of 1,2-bis(2-methoxy-5-trimethylsilylthien-3-yl)perfluorocyclopentene (**1**), a biomimetic surface was shown (Chart 34.1). As shown in Fig. 34.1, the absorption maxima assigned to open ring (**1o**) are not found in the visible range but found at $\lambda = 255$ nm ($\varepsilon = 2.8 \times 10^4\ M^{-1}\ cm^{-1}$) and 325 nm ($\varepsilon = 7.7 \times 10^3\ M^{-1}\ cm^{-1}$). The relationship between the surface morphology and the photochromic reaction of **1o** and closed-ring isomer (**1c**) has been experimentally [27, 28] and theoretically [29] studied. Though the morphology of **1c** has been studied by scanning electron microscopy (SEM) up to now [20–22, 27], in SEM measurement, absorbance and morphology have not been precisely measured at the same position as irradiation spot. Therefore, the relationship between the conversion rate and crystallization is not sufficiently understood. In this chapter, the surface morphology was in situ observed with optical microscope and the photoisomerization process was observed with spectroscopic imaging using upright-inverted microscope. In this technique, the growth of microcrystals can be observed at the same position as irradiation spot, and the relationship between macroscopic change (crystallization) and microscopic change (photochromic reaction) can be discussed [30].

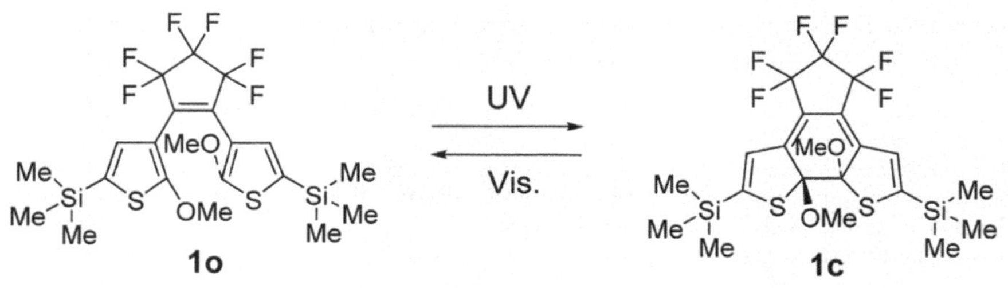

Chart. 34.1 The open-ring (**1o**) and closed-ring (**1c**) isomers of diarylethene **1**. Reprinted with permission from [42]. Copyright 2018 American Chemical Society

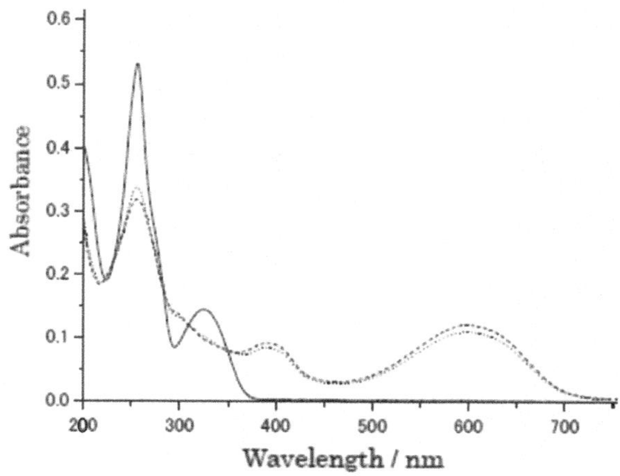

Fig. 34.1 Absorption spectra of diarylethene 1: **1o** form: solid line, **1c** form: broken line. Reprinted with permission from [42]. Copyright 2018 American Chemical Society

Furthermore, as the second topic, the photoisomerization and crystal growths promoted by the enhanced electric field with a plasmonic chip are introduced. A plasmonic chip is the metal-covered substrate with a wavelength-scale periodic structure, and herein, an enhanced electric field based on the grating-coupled surface plasmon resonance (GC-SPR) is excited at the metal surface [31–34]. The GC-SPR condition is described as below [35, 36].

$$k_{\mathrm{spp}} = k_{\mathrm{phx}} + mk_{\mathrm{g}} \quad (m = \pm 1, 2, \ldots) \tag{34.1}$$

$$k_{\mathrm{ph}}\sqrt{\frac{\varepsilon_{\mathrm{d}}\varepsilon_{\mathrm{m}}}{\varepsilon_{\mathrm{d}} + \varepsilon_{\mathrm{m}}}} = k_{\mathrm{ph}} \sin\theta + m \cdot \frac{2\pi}{\Lambda} \tag{34.2}$$

where k_{spp}, k_{phx}, and k_{g} are the wavenumber vectors of the surface plasmon polaritons, incident light of the x component, and grating, and k_{ph}, ε_{d}, ε_{m}, θ and Λ indicate wavenumber of incident light, the complex dielectric constants for the dielectric media and metal, the incident angle, and the pitch size of a grating, respectively. As shown in Eq. (34.2), a resonance angle can be controlled according to the pitch size. The incident light through objective of microscope can be directly coupled with the surface plasmon. The enhanced electric field produced by GC-SPR can be applied to the enhanced fluorescence imaging and promotion of photochemistry reaction. In our laboratory, plasmonic chips coated with silver (Ag) thin films have been applied to sensitive biosensor [37, 38] and bright fluorescence bioimaging [38–40]. Especially, in cell imaging, ten-fold brighter fluorescence images of live breast cancer cells were obtained compared with images observed on a glass substrate [39, 40]. In this chapter, the enhanced electric field was applied to the promotion of photochromic reaction and crystal growth. An aluminum (Al) plasmonic chip can enhance the electric field in the UV wavelength range [41] and can promote a photoisomerization reaction. Therefore, **1o** film prepared on Al plasmonic chip was irradiated with UV light and the reaction process was simultaneously observed with the upright-inverted microscope, i.e., epi-illumination light, transmitted light, and confined light by the plasmonic field were used for local photoisomerization of **1o** and **1c**, and optical images were taken with CCD camera. The needle-shaped crystals studied in the points of direction dependence of UV irradiation from the bottom side or top side, and position dependence at center or edge of a film are introduced. From these results, the growing mechanism of needle-shaped crystals of **1c** is discussed. The possibility of controlling a photoisomerization and a crystallization by the plasmonic chip is studied [42].

34.2 Cyclization by UV Irradiation on the Glass Slide

Photochromic cyclization from **1o** to **1c** and crystal growth of **1c** were in situ observed with an upright-inverted microscope (specially made XI71, Olympus) installing an EM-CCD camera (Luca-r, Andor) and a switchable spectrometer (KNG-CLP-50, Just Solution) on the upright side, and a halogen and a stabilized Xe lamps on the inverted side [41, 43]. The observation and irradiation for cyclization were performed under transmitted light from the inverted side. As an observation light, a yellow light with ~0.2 mW using a halogen lamp and a long pass filter with cut-on wavelength ($\lambda_{\text{cut-on}}$) of 500 nm were used. On the other hand, for irradiation light to cyclize **1o** with 1 mW, a stabilized Xe lamp and an NUA bandpass filter with a center wavelength of 365 nm (bandwidth of 5 nm) were used, and for irradiation to reverse to **1o** with ~200 mW, a stabilized Xe lamp and the long pass filter with $\lambda_{\text{cul-on}}$ of 500 nm were used. The objectives of 40× and 10× were used for the upright side and inverted side, respectively.

Figures 34.2 and 34.3 show the optical images and microscopic spectra, respectively. The mean intensities of transmitted light measured at two positions inside and outside film shown in Fig. 34.2 were divided by the initial values of Fig. 34.2a, individually, and they were converted to absorbance by Lambert–Beer law at the two positions. Figures 34.2a and 34.3a show the optical image and spectra at initial state of the diarylethene **1** films, which were prepared in the dark. As shown in the optical image, a transparent film has no absorption of **1c** (Fig. 34.3a). It suggests that the initial state of a dialyrethene **1** films was **1o** state and the state immediately converted to **1c** just after UV irradiation as found in Fig. 34.2b. The absorption peak shown at ~600 nm was assigned to the **1c** (Fig. 34.3b).

The film thickness of the island-shaped film was evaluated as 7 ± 3 μm by an absorbance of 1.4 (at wavelength of 600 nm) obtained from microscopic measurement. At 1 h after UV exposure, a needle-shaped crystal was found at the film edge (Fig. 34.2c), and the absorbance at peak kept constant until 41 h. In previous studies, randomly growth of needle-shaped crystals from the film surface was shown. Here, the growth of most microcrystals was observed in the inplane radial direction from the outline (edge) of island-shaped film. The high mobility of **1c** is considered to be essential for the molecular alignment and for the growth of crystals, and it can be

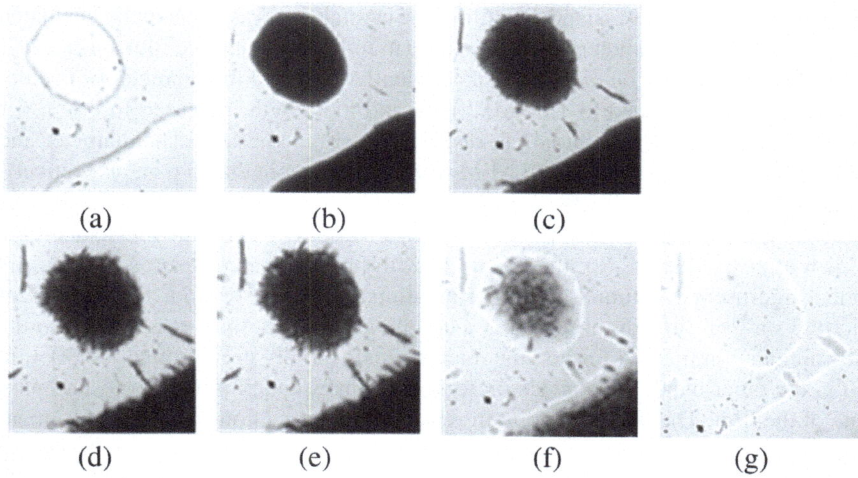

Fig. 34.2 Optical images of diarylethene 1 film prepared on the glass slide: **a** initial state before UV-light irradiation, **b** just after UV irradiation for 5 min, **c** at 1 h after UV irradiation, **d** at 24 h after UV irradiation, **e** at 41 h after UV irradiation, **f** under visible-light irradiation for 1 min, **g** just after visible irradiation for 3 min. Full scale of each image corresponds to 0.8 mm. Reprinted from [30]. Copyright 2018, with permission from Elsevier

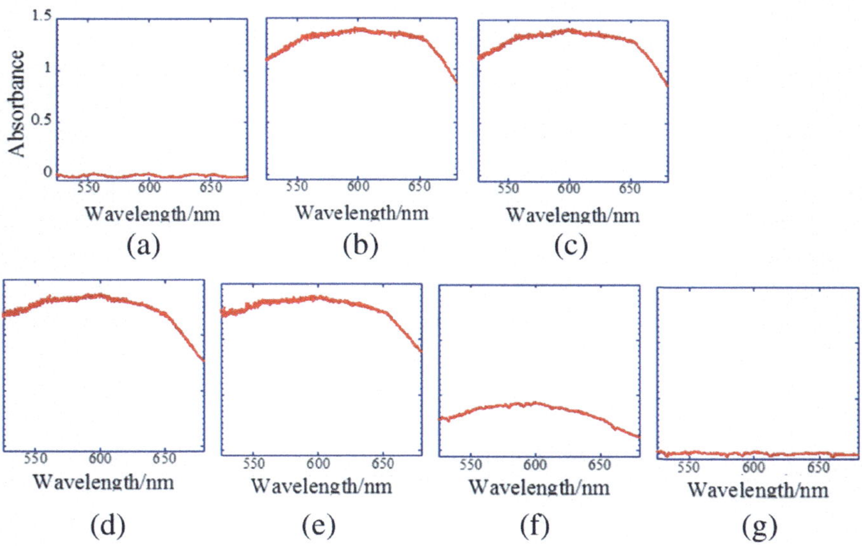

Fig. 34.3 Spectra of diarylethene **1** film (the same film as shown in Fig. 34.2) measured with microscope mounting a spectrometer. The conditions of **a**–**g** are the same as them described in the caption of Fig. 34.2, individually. Reprinted from [30]. Copyright 2018, with permission from Elsevier

allowed for the outline of island-like film. The crystal size was statistically evaluated as 1.2 ± 0.4 μm in diameter and 6.0 ± 5.4 μm in length along the film surface at 1 h after UV exposure. The microcrystals gradually grew and the diameter of needle-shaped crystals increased to 1.6 ± 0.3 μm, and the length increased to 7.6 ± 6.1 and 8.3 ± 6.2 μm after 24 and 41 h, respectively (Fig. 34.2d, e). The diameter and the length of crystals evaluated by AFM were 1-2 and more than 5 μm, respectively, which were consistent with the values evaluated by the optical microscopy. The rate of crystal growth was found to be large within 1 h after UV exposure and to decrease at 24 h after irradiation. So, the crystal growth process was measured in detail by optical microscope within 1 h after irradiation. At initial 10 min, the nucleus of microcrystal was already observed at a certain area. Then, they were spread to the wide area in 30 min, and further, the microcrystals rapidly grew from 30 to 50 min.

On the other hand, the conversion to **1c** was investigated in the photostationary state of the film. The isobestic point in which molar extinction coefficient for open-ring, $\varepsilon_{\text{open}}$, is equivalent to that for closed-ring, $\varepsilon_{\text{closed}}$, was observed at $\lambda = 365$ nm in absorption spectra [21]. The conversion of diarylethene **1** under the photostationary state in a solution is described using quantum yield from **1o** to **1c** (QE_{OC}), QE from **1c** to **1o** (QE_{CO}), and ε for open-ring and closed-ring, as $QE_{\text{OC}}\ \varepsilon_{\text{open}}/(QE_{\text{OC}}\ \varepsilon_{\text{open}} + QE_{\text{OC}}\ \varepsilon_{\text{closed}})$. It is expected to be more than 99%, because QE_{OC} and QE_{CO} were reported to be 0.37 and 0.0015, respectively [21]. However, the values of quantum yield in the film state are known to be around two times larger than their values in a solution [44] and inhomogeneous reaction occurs in the film. Therefore, the conversion in the film is not calculated from this equation. The **1c** area exposed to UV under the microscope was scratched from the film with a small spatula and dissolved in chloroform. Conversion was estimated from the ratio of absorbance (at 600 nm) for the chloroform solution dissolving a film exposed to UV under microscope (1uv) to that for the solution exposed to further UV (2uv) in order to convert entirely up to 99%, i.e., $A_{1\text{uv}}(600\ \text{nm})/A_{2\text{uv}}(600\ \text{nm})$. As a result, the conversion in the film was evaluated as 87 ± 10%, which is smaller value compared with that in DAE solution. An inhomogeneous light intensity may contribute to the lower conversion at the film surface compared with that at the adsorbing face to the glass plate (bottom side), because the irradiation light was absorbed by a film during passing through the film [45].

34.3 Growth of Needle-Shaped Crystals in 1c Film on the Plasmonic Chip

Al plasmonic chip was prepared from a UV-nanoimprint method and Al film coating by rf-sputtering method. The pitch of periodic pattern and the groove depth were 480 nm and 30 nm, respectively. Film thickness of Al was around 40 nm. The relation between crystal growth of **1c** and isomerization to **1c** was investigated at the center part of the film with the Al plasmonic chip. The UV irradiation direction dependence of needle-shaped crystals from either bottom or top of the **1o** film was studied as the first subject. When UV light was irradiated to the **1o** film from the bottom side for 5 min, the irradiation area changed to black. The monochromatic gradation corresponds to the absorption in the visible range assigned to **1c** converted from **1o** [43]. At several hours after UV irradiation, crystals of **1c** did not appear inside and outside the grating, and not even at several days after UV irradiation. On the other hand, needle-shaped crystals of **1c** were observed under the UV irradiation from the top side. Needle-shaped crystal growth of **1c** requires a mobility of **1c** molecules at the center part of a film. Under the UV irradiation from the bottom side, **1o** layer may be still unconverted in the top (film surface) and it can disturb the molecular alignment in **1c** underlayer facing to the substrate.

The irradiation position dependence of crystallization in the film was examined as the second subject. The film edge was exposed to UV from the bottom side for 10 min (1.1 J/cm^2). Exposure area was immediately blacked after UV irradiation. Then, needle-shaped crystals appeared at several hours with UV irradiation from the bottom side and grew after several days. Crystallization was induced at the film edge from the bottom irradiation. This is considered to be attributed to that the mobility of **1c** in the film surface and the film edge is higher than that in the center part of the film. Furthermore, the crystal was never observed at the edge of a too thin-diarylethene **1** film.

In the center part of **1o** film exposed to UV for 5 min from the bottom side with a 100× objective, the area of grating pattern corresponds to upper darker part and area of flat metal corresponds to lower brighter part in Fig. 34.4. The irradiated area was immediately blackened after UV exposure (Fig. 34.4c). As expected for a center part exposed to UV light from the bottom, needle-shaped crystals did not appear even after 20 h incubation (Fig. 34.4d). Then, UV light was irradiated to the film again from the top side for 3 min with 40× objective, in which the irradiation spot area exposed from the top (1.6×10^{-3} cm^2) was larger than that from the bottom (9.3×10^{-4} cm^2) with the 100 × objective. At 7 h after 2nd UV from the top, needle-shaped crystals were observed at the boundary between the irradiation spots from the top and bottom sides (Fig. 34.4g). In the region of the boundary, **1o** only at the film surface is isomerized to **1c** by the irradiation from the top, whereas **1o** at underlayer facing to the chip surface may still remain without conversion, i.e., the **1o** at underlayer in the boundary may make a role of the platform for alignment of **1c**. In contrast, needle-shaped crystals were not found in the center part completely converted to **1c** due to the UV exposure from both top and bottom sides (Fig. 34.4g). According to these results, the platform of **1o** is required for alignment of **1c** and crystals growth.

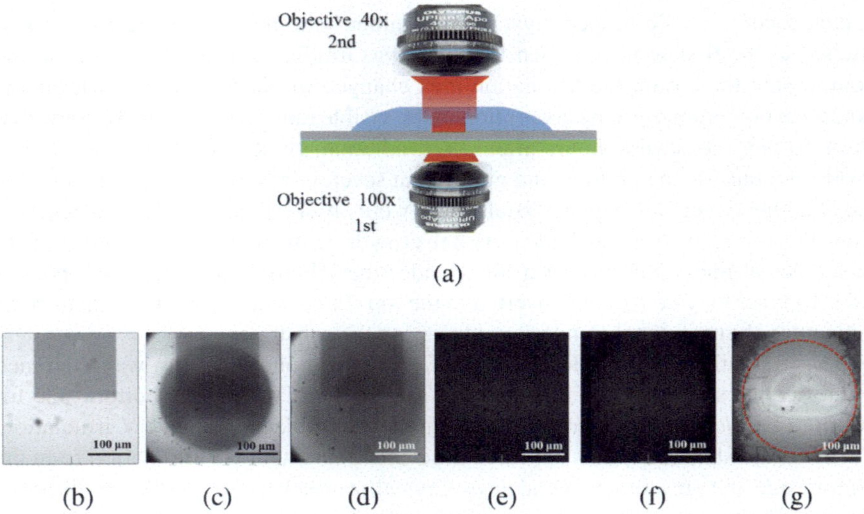

Fig. 34.4 **a** Schematic of the UV-light irradiation with upright-inverted microscope. The optical images of diarylethene 1 film: **b** before UV irradiation, **c** just after UV irradiation for 5 min from the bottom side, **d** at 20 h after UV irradiation from the bottom, **e** under further 3-min UV irradiation from the top side, **f** at 3 h after UV irradiation from the top side, **g** at 7 h after UV irradiation from the top. Reprinted with permission from [42]. Copyright 2018 American Chemical Society

34.4 Conversion Rate from 1o to 1c and Promotion by the Plasmon Field on the al Plasmonic Chip

After the area exposed to UV light with 20× objective from the bottom for 5 min at 1.7 × 10^{-3} W/cm^2 (=0.51 J/cm^2) changed to black, the black part of a film was scratched with a syringe both inside and outside gratings individually and they were separately prepared as chloroform solutions in the dark. The values of conversion from **1o** to **1c** were calculated as described in the Sect. 34.2 and summarized in Fig. 34.5. As the UV irradiation energy increased in both inside and outside of the grating, the conversion became larger. The conversion rate was accelerated inside the grating by a factor of 2 at 0.10 J/cm^2 (the irradiation for 1 min) because of the plasmonic enhanced field. At 0.51 J/cm^2, the SPR-enhanced field promoted conversion rate more than 60% only inside the grating as shown in Fig. 34.5.

On the other hand, the control of crystal growth with the plasmon-enhanced field was shown in Fig. 34.6. At the film edge exposed to UV light for 5 min from the bottom side (0.51 J/cm^2, Fig. 34.6a), the exposed area changed to black (Fig. 34.6c). The needle-shaped crystal was observed only within the grating at 4 h after irradiation (Fig. 34.6d). Further growth of **1c** crystals was shown not outside but inside the grating after 23 h later (Fig. 34.6e). This result was not consistent with the other results.

Creation of crystal nucleus was found to depend on the conversion rate. A glass transition temperature for **1c** is known to be below a room temperature, and the lowest melting point (T_m) is around a room temperature for 3:1-mixture of **1o** and **1c** [21]. **1c** molecules are considered to align cooperatively at the lowest T_m close to a room temperature. Therefore, a creation of microcrystal nucleus may occur over 25% conversion rate (i.e., 3:1-mixture of **1o** and **1c**) [21]. As shown in Fig. 34.5, for the four kinds of UV exposure dose plot (0.10, 0.51, 0.64, and 1.3 J/cm^2), conversion rate was measured inside and outside gratings. Then, the crystal growth was observed for three kinds except for 0.10 J/cm^2 as shown in Fig. 34.7. Figure 34.7a shows no crystal in both sides of grating and flat surface at conversion below 60% a (0.10 J/cm^2 dose). In Fig. 34.7b, the crystals were observed only on the grating area because of the conversion more than 60% (0.51 J/cm^2 dose) though the conversion outside the grating was less than 60%. Finally, the microcrystals were observed in both sides with the conversion more than 60% (Fig. 34.7c: 0.64 J/cm^2 dose).

In summary, the conversion rate to **1c** is important for crystallization and the conversion larger than 60% is required. Crystal growth of **1c** was observed at the film edge or the film surface on the **1o** underlayer platform due to large mobility of **1c**. If the film thickness is too thin, the crystal is not formed. Crystallization can be spatially controlled with the Al plasmonic chip because the plasmonic chip can promote the conversion rate to **1c** by the enhanced electric field.

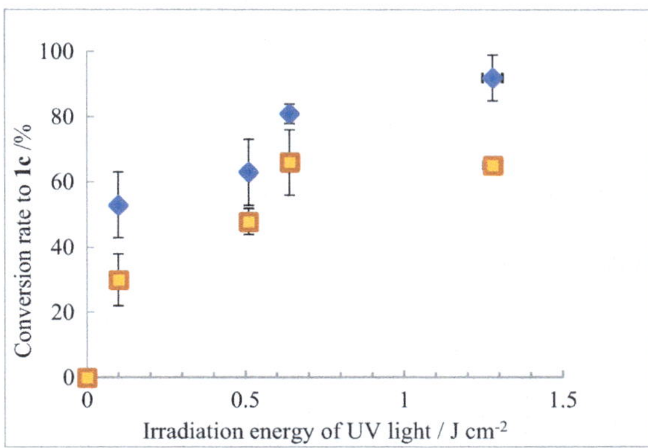

Fig. 34.5 Conversion rate to **1c** plotted against the irradiation energy of UV light. ◆: inside grating, ■: outside grating (flat area). Reprinted with permission from [42]. Copyright 2018 American Chemical Society

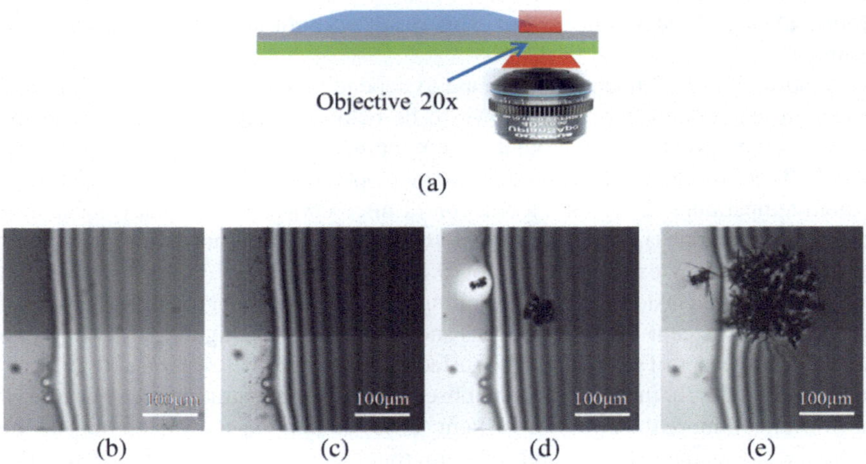

Fig. 34.6 **a** Schematic of microscopic configuration requested for taking optical images. Optical images: **b** before UV irradiation, **c** just after UV irradiation, **d** at 4 h after UV irradiation (micro-crystals detected only inside grating), **e** at 23 h after UV irradiation. Reprinted with permission from [42]. Copyright 2018 American Chemical Society

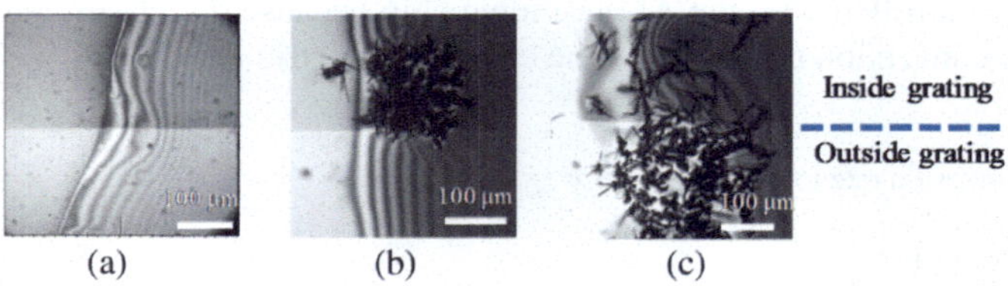

Fig. 34.7 Optical images after UV irradiation of **a** 0.10 J/cm^2 dose, **b** 0.51 J/cm^2 dose, **c** 0.64 J/cm^2 dose

34.5 Conclusions

Plasmonic chip is the grating-patterned substrate with the wavelength-scale pitch covered with a thin metal layer. To provide the enhanced electric field to the chip surface in the wavelength of UV light, Al plasmonic chip is available. Diarylethene **1** films were prepared onto the Al plasmonic chip and glass slide. In the center of a film, needle-shaped crystal of **1c** was formed after exposure to UV light from the top side of a chip, while not formed by UV light from the bottom side of a chip. However, at the film edge, crystal growth was confirmed upon the irradiation from the bottom side. A high mobility of **1c** in the film was required for crystal growth and therefore crystals were observed near the film surface and film edge. Furthermore, the existence of **1o** underlayer platform was found to be essential for crystal growth.

When **1o** converted to **1c** completely, crystals were not formed. **1o** underlayer is necessary for **1c** alignment. In this study, the Al plasmonic chip promoted conversion rate from **1o** to **1c**. By the comparison of the conversion rate between on the grating pattern and on the flat metal, high conversion rate more than 60% is found to be important condition for crystal growth. Therefore, crystal growth can be controlled by the conversion to **1c** which was promoted by the plasmonic field. At conversion more than 60%, **1c** molecules near the film surface and film edge can align on the **1o** underlayer platform and gradually grow to crystals.

Acknowledgements T. K and K. T. thank Toyo Gosei for providing the photocurable resin, and thank Ms. Eri Nakamura for helping our experiment. The present work was supported by JSPS KAKENHI Grant Number JP15H01100 and JP26107012, Grant-in-Aid for Scientific Research on Innovative Areas "Photosynergetics."

References

1. Irie M, Mohri M (1988) Thermally irreversible photochromic systems—reversible photocyclization of diarylethene derivatives. J Org Chem 53:803–808
2. Irie M, Fukaminato T, Matsuda K, Kobatake S (2014) Photochromism of diarylethene molecules and crystals: memories, switches and actuators. Chem Rev 114:12174–12277
3. Tian H, Yang SJ (2004) Recent progresses on diarylethene based photochromic switches. Chem Soc Rev 33:85–97
4. Berkovic G, Krongauz V, Weiss V (2000) Spiropyrans and spirooxazines for memories and switches. Chem Rev 100:1741–1753
5. Minkin VI (2004) Photo-, thermo-, solvato-, and electrochromic spiroheterocyclic compounds. Chem Rev 104:2751–2776
6. Tamai N, Miyasaka H (2000) Ultrafast dynamics of photochromic systems. Chem Rev 100:1875–1890
7. Natansohn A, Rochon P (2002) Photoinduced motions in azo-containing polymers. Chem Rev 102:4139–4175
8. Bandara HMD, Burdette SC (2012) Photoisomerization in different classes of azobenzene. Chem Soc Rev 41:1809–1825
9. Priimagi A, Barrett CJ, Shishido A (2014) Recent twists in photoactuation and photoalignment control. J Mater Chem C 2:7155–7162
10. Hatano S, Abe J (2008) Activation parameters for the recombination reaction of intramolecular radical pairs generated from the radical diffusion-inhibited HABI Derivative. J Phys Chem A 112:6098–6103
11. Fukaminato T, Sasaki T, Kawai T, Tamai N, Irie M (2004) Digital photoswitching of fluorescence based on the photochromism of diarylethene derivatives at a single-molecule level. J Am Chem Soc 126:14843–14849
12. Matsuda K, Irie M (2001) Photochromism of diarylethenes with two nitronyl nitroxides: photoswitching of an intramolecular magnetic interaction. Chem Eur J 7:3466–3473
13. Moriyama Y, Matsuda K, Tanifuji N, Irie S, Irie M (2005) Electrochemical cyclization/cycloreversion reactions of diarylethenes. Org Lett 7:3315–3318
14. Gilat SL, Kawai SH, Lehn JM (1993) Light-triggered electrical and optical switching devices. J Chem Soc Chem Commun 18:1439–1442
15. Irie M, Kobatake S, Horichi M (2001) Reversible surface morphology changes of a photochromic diarylethene single crystal by photoirradiation. Science 291:1769–1772

16. Kobatake S, Takami S, Muto H, Ishikawa T, Irie M (2007) Rapid and reversible shape changes of molecular crystals on photoirradiation. Nature 446:778–781
17. Takami S, Kuroki L, Irie M (2007) Photochromism of mixed crystals containing bisthienyl-, bisthiazolyl-, and bisoxazolylethene derivatives. J Am Chem Soc 129:7319–7326
18. Yamada T, Kobatake S, Muto K, Irie M (2000) X-ray crystallographic study on single-crystalline photochromism of bis(2,5-dimethyl-3-thienyl)perfluorocyclopentene. J Am Chem Soc 122:1589–1592
19. Kobatake S, Shibata K, Uchida K, Irie M (2000) Photochromism of 1,2-bis(2-ethyl-5-phenyl-3-thienyl)perfluorocyclopentene in a single-crystalline phase conrotatory thermal cycloreversion of the closed-ring isomer. J Am Chem Soc 122:12135–12141
20. Nishikawa N, Mayama H, Nonomura Y, Fujinaga N, Yokojima S, Nakamura S, Uchida K (2014) Theoretical explanation of the photoswitchable superhydrophobicity of diarylethene microcrystalline surfaces. Langmuir 30:10643–10650
21. Uchida K, Izumi N, Sukata S, Kojima Y, Nakamura S, Irie M (2006) Photoinduced reversible formation of microfibrils on a photochromic diarylethene microcrystalline surface. Angew Chem Int Ed 45:6470–6473
22. Nishikawa N, Sakiyama S, Yamazoe S, Kojima Y, Nishihara E, Tsujioka T, Mayama H, Yokojima S, Nakamura S, Uchida K (2013) Photoinduced self-epitaxial crystal growth of a diarylethene derivative with antireflection moth-eye and superhydrophobic lotus effects. Langmuir 29:8164–8169
23. Uchida K, Nishikawa N, Izumi N, Yamazoe S, Mayama H, Kojima Y, Yokojima S, Nakamura S, Tsujii K, Irie M (2010) Phototunable diarylethene microcrystalline surfaces: lotus and petal effects upon wetting. Angew Chem Int Ed 49:5942–5944
24. Roach P, Shirtcliffe NJ, Newton MI (2008) Progess in superhydrophobic surface development. Soft Matter 4:224–240
25. Iwaihara C, Kitagawa D, Kobatake S (2015) Polymorphic crystallization and thermodynamic phase transition between the polymorphs of a photochromic diarylethene. Cryst Growth Des 15:2017–2023
26. Ou DP, Yu T, Yang ZY, Luan TG, Mao Z, Zhang Y, Liu SW, Xu JR, Chi ZG, Bryce MR (2016) Combined aggregation induced emission (AIE), photochromism and photoresponsive wettability in simple dichloro-substituted triphenylethylene derivatives. Chemical Science 7:5302–5306
27. Kitagawa D, Yamashita I, Kobatake S (2011) Control of surface wettability and photomicropatterning with a polymorphic diarylethene crystal upon photoirradiation. Chemistry-A European Journal 17:9825–9831
28. Amigoni S, de Givenchy ET, Dufay M, Guittard F (2009) Covalent layer-by-layer assembled superhydrophobic organic-inorganic hybrid films. Langmuir 25:11073–11077
29. Fujinaga N, Nishikawa N, Nishimura R, Hyodo K, Yamazoe S, Kojima Y, Yamamoto K, Tsujioka T, Morimoto M, Yokojima S, Nakamura S, Uchida K (2016) Photoinduced topographical changes on microcrystalline surfaces of diarylethenes. Cryst Eng Comm 18:7229–7235
30. Tawa K, Kadoyama K, Nishimura R, Toma M, Uchida K (2018) In situ optical and spectroscopic imaging of photochromic cyclization and crystallization of a diarylethene film with optical microscopy. J Photochem Photobiolo A Chem 356:397–402
31. Meneghello A, Antognoli A, Sonato A, Zacco G, Ruffato G, Cretaio E, Romanato F (2014) Label-free efficient and accurate detection of cystic fibrosis causing mutations using an azimuthally rotated GC-SPR platform. Anal Chem 86:11773–11781
32. Nootchanat S, Pangdam A, Ishikawa R, Wongravee K, Shinbo K, Kato K, Kaneko F, Ekgasit S, Baba A (2017) Grating-coupled surface plasmon resonance enhanced organic photovoltaic devices induced by Blu-ray disc recordable and Blu-ray disc grating structures nanoscale 9:4963–4971
33. Baba A, Aoki N, Shinbo K, Kato K, Kaneko F (2011) Grating-coupled surface plasmon enhanced short-circuit current in organic thin-film photovoltaic cells. ACS Appl Mater Interfaces 3:2080–2084

34. Wang Y, Dostalek J, Knoll W (2011) Magnetic nanoparticle-enhanced biosensor based on grating-coupled surface plasmon resonance. Anal Chem 83:6202–6207
35. Raether H (1988) Surface plasmons on smooth and rough surfaces and on gratings. Springer, Berlin
36. Knoll W (1998) Interfaces and thin films as seen by bound electromagnetic waves. Annu Rev Phys Chem 49:569–638
37. Tawa K, Umetsu M, Nakazawa H, Hattori T, Nishii J (2013) Application of 300× enhanced fluorescence on a plasmonic chip modified with a bispecific antibody to a sensitive immunosensor. ACS Appl Mater Interfaces 5:8628–8632
38. Tawa K, Kondo F, Sasakawa C, Nagae K, Nakamura Y, Nozaki A, Kaya T (2015) Sensitive detection of a tumor marker α-fetoprotein with a sandwich assay on a plasmonic chip. Anal Chem 87:3871–3876
39. Tawa K, Yamamura S, Sasakawa C, Shibata I, Kataoka M (2016) Sensitive detection of cell surface membrane proteins in living breast cancer cells by using multicolor fluorescence microscopy with a plasmonic chip. ACS Appl Mater Interfaces 8:29893–29898
40. Tawa K, Yasui C, Hosokawa C, Aota H, Nishii J (2014) In situ sensitive fluorescence imaging of neurons cultured on a plasmonic dish using fluorescence microscopy. ACS Appl Mater Interfaces 6:20010–20015
41. Knight MW, King NS, Liu L, Everitt HO, Nordlander P, Halas N (2014) Aluminum for plasmonics. ACS Nano 8:834–840
42. Kadoyama T, Nishimura R, Toma M, Uchida K, Tawa K (2018) Study on the mechanism of diarylethene crystal growth by in situ microscopy and the crystal growth controlled by an aluminum plasmonic chip. Langmuir 34:4217–4223
43. Tawa K, Sasakawa C, Fujita T, Kiyosue K, Hosokawa C, Nishii J, Oike M, Kakinuma N (2016) Fluorescence microscopy imaging of cells with a plasmonic dish integrally molded. Jap J Appl Phys 55:03DF12
44. Kobatake S, Uchida K, Tsuchida E, Irie M (2002) Single-crystalline photochromism of diarylethenes: reactivity–structure relationship. Chem Commun 2804–2805
45. Pariani G, Bianco A, Castagna R, Bertarelli C (2011) Kinetics of photochromic conversion at the solid state: quantum yield of dithienylethene-based films. J Phys Chem A 115:12184–12193

The manufacturer's authorised representative in the EU is Springer Nature Customer Service Centre GmbH, Europaplatz 3, 69115 Heidelberg, Germany. If you have any concerns regarding our products, please contact ProductSafety@springernature.com

Printed and bound by CPI Group (UK) Ltd, Croydon, CR0 4YY
30/09/2026
02225559-0020